YIELDS OF FARMED SPECIES

Constraints and opportunities in the 21st century

Proceedings of Previous Easter Schools in Agricultural Science, published by Butterworths, London

SOIL ZOOLOGY Edited by D.K. McL, Kevan (1955)
THE GROWTH OF LEAVES Edited by F.L. Milthorpe (1956)
CONTROL OF PLANT ENVIRONMENT Edited by J.P. Hudson (1957)
NUTRITION OF THE LEGUMES Edited by E.G. Hallsworth (1958)
THE MEASUREMENT OF GRASSLAND AND PRODUCTIVITY Edited by J.D. Ivins (1959)
DIGESTIVE PHYSIOLOGY AND NUTRITION OF THE RUMINANT Edited by D. Lewis (1960)
NUTRITION OF PIGS AND POULTRY Edited by J.T. Morgan and D. Lewis (1961)
ANTIBIOTICS IN AGRICULTURE Edited by M. Woodbine (1962)
THE GROWTH OF THE POTATO Edited by J.D. Ivins and F.L. Milthorpe (1963)
EXPERIMENTAL PEDOLOGY Edited by E.G. Hallsworth and D.V. Crawford (1964)
THE GROWTH OF CEREALS AND GRASSES Edited by F.L. Milthorpe and J.D. Ivins (1965)
REPRODUCTION IN THE FEMALE ANIMAL Edited by G.E. Lamming and E.C. Amoroso (1967)
GROWTH AND DEVELOPMENT OF MAMMALS Edited by G.A. Lodge and G.E. Lamming (1968)
ROOT GROWTH Edited by W.J. Whittington (1968)
PROTEINS AS HUMAN FOOD Edited by R.A. Lawrie (1970)
LACTATION Edited by I.R. Falconer (1971)
PIG PRODUCTION Edited by D.J.A. Cole (1972)
SEED ECOLOGY Edited by W. Heydecker (1973)
HEAT LOSS FROM ANIMALS AND MAN: ASSESSMENT AND CONTROL Edited by J.L. Montieth and L.E. Mount (1974)
MEAT Edited by D.J.A. Cole and R.A. Lawrie (1975)
PRINCIPLES OF CATTLE PRODUCTION Edited by Henry Swan and W.H. Broster (1976)
LIGHT AND PLANT DEVELOPMENT Edited by H. Smith (1976)
PLANT PROTEINS Edited by G. Norton (1977)
ANTIBIOTICS AND ANTIBIOSIS AGRICULTURE Edited by M. Woodbine (1977)
CONTROL OF OVULATION Edited by D.B. Crighton, N.B. Haynes, G.R. Foxcroft and G.E. Lamming (1978)
POLYSACCHARIDES IN FOOD Edited by J.M.V. Blanshard and J.R. Mitchell (1979)
SEED PRODUCTION Edited by P.D. Hebblethwaite (1980)
PROTEIN DEPOSITION IN ANIMALS Edited by P.J. Buttery and D.B. Lindsay (1981)
PHYSIOLOGICAL PROCESSES LIMITING PLANT PRODUCTIVITY Edited by C. Johnson (1981)
ENVIRONMENTAL ASPECTS OF HOUSING FOR ANIMAL PRODUCTION Edited by J.A. Clark (1981)
EFFECTS OF GASEOUS AIR POLLUTION IN AGRICULTURE AND HORTICULTURE Edited by M.H. Unsworth and D.P. Ormrod (1982)
CHEMICAL MANIPULATION OF CROP GROWTH AND DEVELOPMENT Edited by J.S. McLaren (1982)
CONTROL OF PIG REPRODUCTION Edited by D.J.A. Cole and G.R. Foxcroft (1982)
SHEEP PRODUCTION Edited by W. Haresign (1983)
UPGRADING WASTE FOR FEEDS AND FOOD Edited by D.A. Ledward, A.J. Taylor and R.A. Lawrie (1983)
FATS IN ANIMAL NUTRITION Edited by J. Wiseman (1984)
IMMUNOLOGICAL ASPECTS OF REPRODUCTION IN MAMMALS Edited by D.B. Crighton (1984)
ETHYLENE AND PLANT DEVELOPMENT Edited by J.A. Roberts and G.A. Tucker (1985)
THE PEA CROP Edited by P.D. Hebblethwaite, M.C. Heath and T.C.K. Dawkins (1985)
PLANT TISSUE CULTURE AND ITS AGRICULTURAL APPLICATIONS Edited by Lindsay A. Withers and P.G. Alderson (1986)
CONTROL AND MANIPULATION OF ANIMAL GROWTH Edited by P.J. Buttery, N.B. Haynes and D.B. Lindsay (1986)
COMPUTER APPLICATIONS IN AGRICULTURAL ENVIRONMENTS Edited by J.A. Clark, K. Gregson and R.A. Saffell (1986)
MANIPULATION OF FLOWERING Edited by J.G. Atherton (1987)
NUTRITION AND LACTATION IN THE DAIRY COW Edited by P.C. Garnsworthy (1988)
MANIPULATION OF FRUITING Edited by C.J. Wright (1989)
APPLICATIONS OF REMOTE SENSING IN AGRICULTURE Edited by M.D. Steven and J.A. Clark (1990)
GENETIC ENGINEERING OF CROP PLANTS Edited by G.W. Lycett and D. Grierson (1990)
FEEDSTUFF EVALUATION Edited by J. Wiseman and D.J.A. Cole (1990)
THE CONTROL OF FAT AND LEAN DEPOSITION Edited by K.N. Boorman, P.J. Buttery and D.B. Lindsay (1992)

Proceedings of Previous Easter Schools in Agricultural Science, published by Nottingham University Press

THE GLASSY STATE IN FOODS Edited by J.M.V. Blanshard and P.J. Lillford (1993)
RESOURCE CAPTURE BY CROPS Edited by J.L. Montieth, R.K. Scott and M.H. Unsworth (1994)
PRINCIPLES OF PIG SCIENCE Edited by D.J.A. Cole, J. Wiseman and M.A. Varley (1994)
ISSUES IN AGRICULTURAL BIOETHICS Edited by T.B. Mepham, G.A. Tucker and J. Wiseman (1995)
BIOPOLYMER MIXTURES Edited by S.E. Harding, S.E. Hill and J.R. Mitchell (1995)
MECHANISMS AND APPLICATIONS OF GENE SILENCING Edited by D. Grierson, G.W. Lycett and G.A. Tucker (1996)
PROGRESS IN PIG SCIENCE Edited by J. Wiseman, M.A. Varley and J.P. Chadwick (1998)
PERSPECTIVES IN PIG SCIENCE Edited by J. Wiseman, M.A. Varley and B. Kemp (2003)
CALF AND HEIFER REARING Edited by P.C. Garnsworthy (2004)

Yields of Farmed Species

Constraints and opportunities in the 21st century

R. SYLVESTER-BRADLEY
ADAS

J. WISEMAN
University of Nottingham

NOTTINGHAM
University Press

First published by Nottingham University Press
This reissued original edition published 2023 by 5m Books Ltd www.5mbooks.com

British Library Cataloguing in Publication Data
A catalogue record for this book is available from the British Library

ISBN 9781789183221

Disclaimer

Every reasonable effort has been made to ensure that the material in this book is true, correct, complete and appropriate at the time of writing. Nevertheless the publishers and the author do not accept responsibility for any omission or error, or for any injury, damage, loss or financial consequences arising from the use of the book. Views expressed in the articles are those of the author and not of the Editor or Publisher.

Typeset by Nottingham University Press, Nottingham
Printed and bound by PODWW, UK

PREFACE

These proceedings are from a conference held in June 2004, in the University of Nottingham 'Easter School' series. It was supported by a research grant from the Department for Environment, Food and Rural Affairs. With overseas contributors providing the global context, the meeting considered a wide range of important questions relating to yields of species farmed in the UK. Given the rapid pace of change affecting farming, can we expect the dramatic yield increases of the last century to continue? What are the yield potentials of species that are being farmed now? How will new technologies affect yields achieved in future decades? To what extent will pests, pathogens, competing demands for resources, or scruples about farming processes, curtail our ability to innovate in the production of food?

It is important, for political and economic planning into the future, to recognise the likely extent and rate of further changes in farm yields. A wide range of issues has governed yield improvements through the last 50 years. The challenge in planning and preparing these proceedings has been to anticipate which of these, and which additional issues, may come into play over the next 50 years, and to address these in such a way that credible yield expectations can be derived.

Some yields still appear to be increasing, but rates of increase have not been stable through time and have differed considerably between species. Clearly it is not safe to conclude that increases will continue for the foreseeable future. There may be absolute limits to the yield of a species, and there may be constraints set by the environments (physical, technological, sociological, economic or political) in which a species may be farmed.

The proceedings start by taking an overview of progress in agricultural productivity world-wide. Then the prospects for crops and livestock are assessed in detail. Case studies are presented for the major crop and stock species, looking at recent progress in on-farm performance, physiological potentials, prospects for new technologies, and possible economic and environmental constraints. The concluding papers consider the place of farm productivity in contrasting visions of agriculture in NW Europe over the next 50 years.

In building this long-range view, it has been telling to step back. UK agriculture has become one of the most sophisticated and productive in the world, but the transformation was a distant dream 50 years ago. More than a hectare of land was needed to feed each UK resident at the time of the 1st World War (about three times the available land area). It has largely been due to the collective will of the post-war generation that modern UK agriculture can now feed each of us from less than half a hectare, with food that is both inexpensive and of high quality. Despite an increased population, we have become 80% self sufficient in food, and a significant exporter of agricultural produce.

In the 21st century, as self-sufficiency becomes less important and agriculture assumes broader goals, it seems likely that current rates of increase in productivity of the UK's farmed species must diminish. However, useful new species are still being found, and with our fertile land and above all, our capacity to innovate, we should be optimistic. By foreseeing the important hurdles along the new paths that are being set, particularly towards sustainability, there should be scope to feed others as well as ourselves, and to share the landscape between productive agriculture and activities that fulfil our wider needs.

Roger Sylvester-Bradley
Julian Wiseman

ACKNOWLEDGEMENTS

The organisers would like to thank all invited speakers for their contributions to the meeting and also co-authors of chapters, some who were unable to attend. Sessions were Chaired by the two organisers and, in addition, by Prof Jerry Roberts (University of Nottingham), Prof Ben Miflin (Home-Grown Cereals Authority), Prof Peter Shewry (Rothamsted Research), Prof John Robinson (Scottish Agricultural College) and Sir Ben Gill (past president of the National Farmers Union). The organisers are also most grateful to Dr Peter Berry and Dr Daniel Kindred (ADAS) for editorial assistance and to the Department for Environment, Food and Rural Affairs of England & Wales for funding.

The conference was ably managed by Sue Golds, assisted by Emma Hooley, Hazel Russell, Vicky Onions and Eunice Lee, and well supported by the Catering and Residential Services at Sutton Bonington Campus of the University of Nottingham.

ADAS

CONTENTS

1

DO WE NEED HIGHER FARM YIELDS DURING THE FIRST HALF OF THE 21ST CENTURY?

VACLAV SMIL
Faculty of Environment, University of Manitoba, Winnipeg, Manitoba R3T 2N2, Canada

As I am a generalist fond of looking at interactions of complex systems I am not able to offer any well-founded appraisals of potential yield gains of particular cultivars or any expert evaluation of factors that may hinder or facilitate our quest for higher productivity. That will be your task. Instead, I will make some blunt remarks about the nature of modern agriculture and will provide a variety of essential (and uncomfortably complex) information that is required in order to answer a question that many might think not even worthy of asking: do we need higher yields? I believe that the automatic assumption of the need for substantially higher yields should be strongly challenged and I will address the matter by looking closer at three principal variables that will drive future demand for food: population growth, typical diets and availability of farmland.

Modern food production

Irrational might be the most fitting choice when looking for an adjective that would best describe the existing global food system. How else can one characterize an arrangement whereby all high-income nations lavishly subsidize the degradation of irreplaceable (or only slowly recoverable) natural resources and waste such expensive (and energy intensive) industrial inputs as synthetic fertilizers and herbicides in order to produce obscene surpluses of food whose consumption helps to create an unprecedented incidence of obesity and associated civilizational diseases? At the same time, in too many low-income countries the madness has different methods: ideological obsessions (now at their full tragic display in Zimbabwe or North Korea), systematic neglect of agriculture or endless civil wars (much of the sub-Saharan countries falls into this dismal category) mean that not enough food is produced to cover even basic metabolic needs.

And besides being irrational, all modern farming is also unsustainable. No matter how many times we hear the fashionable mantra about sustainable agriculture, the basic fact is that more than 85% of all food produced worldwide requires considerable to very large fossil fuel subsidies that are used directly as fuel and electricity for field and processing operation, and indirectly as energies used to produce machinery and synthetic chemicals and to breed new cultivars. The dependence is particularly critical as far as nitrogen, the leading plant macronutrient, is concerned: I have demonstrated that without the Haber-Bosch synthesis of ammonia we could not produce food for about 40% of today's world population of just over six billion people. This means that modern agriculture is just an offshoot of a temporary global economy that prospers on the basis of fossil carbon stored in the uppermost layers of the Earth's crust 10^5-10^8 years ago. And modern ocean fishing requires even higher fossil fuel energy subsidies per unit of edible energy than does farming.

And, to eliminate a common misconception, this dependence on fossil energies is actually greater in populous low-income nations of Asia than it is in the US or the UK. Due to their limited areas of farmland and the still growing populations that expect improved diets, countries such as China or Indonesia must rely on higher yields, that is on higher fossil fuel-derived inputs, as the leading mode of rising food production. Some of the already existing (and steadily expanding) gaps are already stunning: average annual applications of nitrogenous fertilizers now amount to less than 60 kg N/ha in the US but they are about 180 kg N/ha in China and they average in excess of 300 kg N/ha in the country's leading rice-producing provinces.

These facts are undeniable: we now rely, worldwide, on agroecosystems whose overall management is determined, at best, by a variety of irrational policies and at worst by astonishing political neglect of the entire sector, and whose functioning is critically, and unsustainably, linked to the use of irreplaceable and obviously finite fossil fuels. But, also undeniably, these inputs and subsidies have resulted in impressive increases of agricultural productivity in general and of crop yields in particular. These accomplishments can be demonstrated on different temporal scales and on all spatial levels — globally, nationally, regionally and locally. Global average of all cereal harvests rose from less than 1 t/ha in 1900 to just over 3 t/ha in 2000. During the last two generations of the 20th century, as short-stalked high-yielding varieties of wheat and rice were adopted worldwide, average cereal harvests in Asia and Latin America had roughly doubled. Dutch and English yields show the rise from less than 2 t/ha of wheat in 1900 to over 8 t/ha by the year 2000.

But, as you well know, even these high yields are only a fraction of the highest recorded harvests on small plots that are near 15 t/ha for both rice and wheat, about 20 t/ha for maize and near 6 t/ha for soybeans. Using these maxima as indicators of potential yield increases is, of course, quite inappropriate and even referring to the best national performances sets up unattainable targets for most of the countries. But I will leave the realistic question — how far can breeding and better agronomic practices push the yields of staple crops during the next one or two generations? — to your expert judgments and will raise instead the two more fundamental considerations: do we really need higher yields and, if so, how much higher do we have to aim? Many factors should go into answering these questions but the state of the current food supply (its adequacy and shortfalls), and three trends that will change the future demand, both by pushing it up or by moderating it -- population changes, typical diets and the availability of arable land -- are obviously the key variables.

Global and national food supply

All one has to do to find out how much food the world, or any particular nation, has at its disposal is to click on the food balance sheet file of the FAOSTAT, the statistical database of the Food and Agriculture Organization of the United Nations in Rome. This electronic repository (http://apps.fao.org) contains, among other statistics, data on land use, irrigation and fertilizer applications, on areas planted to more than a score of crops, their total harvests and average yields, on annual output of all major kinds of meat, eggs and milk as well on fish catches and aquaculture. These figures are the primary inputs for the preparation of FAO's own food balance sheets.

For 2001, the last year for which the sheets are available, we get the average per capita daily availability of about 2,800 kcal/day for the world, nearly 3,300 kcal/day for high-income economies, and 2,800 kcal/day for the low-income countries, with national extremes ranging from more than 3,600 kcal/day in Europe to less than 2,000 kcal/day in parts of the sub-Saharan Africa. These figures are not just widely quoted and often reprinted, they are also used for assessing the extent of national and global malnutrition and the need for future food production increases. But their users should beware: they are dealing with highly heterogeneous mass of figures whose provenance ranges from careful monitoring to mere guesses.

Even an uninitiated observer might wonder who has been monitoring millet harvests in Sudan throughout the decades of the country's relentless civil war, or who counted sheep herds in the Taliban's Afghanistan of the late 1990s -- yet precise annual totals of these variables are listed in FAO's files. A more astute observer will immediately question the reliability of officially estimated food production data (passed by member countries to FAO) in those low-income countries where a high level of traditional subsistence farming, most of whose products do not reach any markets, makes it much more difficult to come up with reliable nationwide harvest figures than in totally commercialized settings. Actually, in too many cases there are no official national figures and the estimates are simply made at FAO's headquarters. My conservative assessment is that between 50-60% of all figures appearing in FAO's data files are estimated in Rome.

While many production totals submitted by individual nations may substantially underestimate actual harvests, many yield averages have been substantially overestimated because of common underestimates of arable land. China provides the best example: until the year 1999 official Chinese claims credited the country with only about 95 million hectares of farmland although it was known for many years, both from new detailed land surveys and from satellite imaging, that the real total is much larger. Finally, in the year 2000 the Chinese government changed the official total to 130 million hectares, FAO uses the total of 137 million hectares and I have shown that the actual total (including orchards and aquacultural ponds) is in excess of 150 million hectares.

Similarly large underestimates, of the order of 15-25%, apply for most of the world's low-income countries and they mean that the actual yields are lower and hence the potential to increase them is higher than indicated by using the official figures. And uncertainties do not end with output data. People and animals do not eat all that is harvested yet, although it is well known that post-harvest losses can be very significant, there has been very little attention paid to these ubiquitous phenomena. In spite of all of these weaknesses, there is no doubt that the FAO figures correctly reflect the obvious split between the average per capita food availabilities in high- and low-income countries. For roughly a fifth of the humanity, food production has reached the levels that range from a very comfortable security margin with per capita rates of 2,800-3,100 kcal/day to indefensible excess of 3,400-3,800 kcal/day.

Moreover, it should be clear even to a casual observer that the principal food production challenge in virtually all affluent countries is not a matter of agricultural practices but rather the management of concentrated masses of domestic animals: the term zooculture would be a much better descriptive term (and this would be true even without the latest dietary madness, the Atkinson diet dominated by animal proteins and fats, that ignores the evolutionary heritage of our species). In contrast, when leaving Brazil and China, the two nations with relatively good food supply aside, the global average per capita food availability for low-income countries is less than 2,500 kcal/day. This would provide adequate nutrition with perfectly equal access to food and with modest labour energy demands.

Neither is, of course, true as inequalities of all kinds are the norm in most low-income countries and as significant portions of their populations are still engaged in moderate to hard physical work. Consequently, there is still a great deal of chronic malnutrition but what is not widely understood is that the frequently cited figures of its global extent are derived by using statistical assumption and not by actually surveying the affected populations. This does not mean that the real figures are substantially different but the estimates may exaggerate the totals by as much as 10-15%. The latest FAO estimate of undernourished people adds up to almost 830 million, or about 14% of the world's population but the real total may be as much as 100 million lower.

The highest shares of undernourished population (about 70% of the total) are, not surprisingly, in Afghanistan and Somalia, but the largest totals are in India and China where there are, respectively, about 20% (or just over 200 million) and just above 10% (or nearly 130 million) people who do not have enough food in order to lead healthy and productive lives. At the same time, China illustrates well the existence of a much less noted phenomenon, namely increasing numbers of people with excessive food intakes that can be found in just about every populous low-income nation. Given China's relatively high average per capita food availability the same procedure that is used to estimate the number of undernourished people indicates that the country may have nearly 200 million people whose daily per capita food supply is in excess of 3,000 kcal. Not surprisingly, when this level has combined with reduced physical activity of relatively affluent urban residents the result has been a dramatic increase of obesity and cardiovascular disease.

And our surprisingly weak understanding extends to actual food intakes and requirements. Even in affluent countries the rates of average food intake are either derived from family expenditure surveys or are based on individual recall: neither of these methods can capture accurately what is actually eaten. We know that Americans (including babies and grannies) do not eat daily the average of 3,700 kcal/capita that are available at retail level: if so the population would be more than 90% obese. On the other hand it is clear that they eat lot more than 2000 kcal/day, the nationwide average calculated from the best available food intake surveys; if so the share of overweight people would not be approaching half of the population. Needless to say, for most low-income countries we have no representative food intake surveys at all.

And, unfortunately, the lack of definite information affects even our understanding of basic dietary requirements needed for healthy and active life. In spite of many indisputable facts, there are still many uncertainties that can be repeatedly exploited by extreme dietary advice that is all too readily absorbed by scientifically illiterate masses of people that will flip from high-carbohydrate diets (the traditional Mediterranean diet loaded with pasta, bread and legumes is perhaps the best example) to high-protein diets (the most recent infatuation justified by pseudo-scientific claims of Robert Atkins) and that will believe that megadoses of some vitamins (particularly C and E) will cure just about any ailment. These litanies could go on –- but the point is already made: our understanding of the existing food supply is much less precise than most people imagine, our knowledge of actual food intakes is highly unsatisfactory and our recommendations for what constitutes the healthiest, optimal diet are a work in never-ending progress. These uncertainties cut both ways.

In the case of enormous Western overproduction they make it difficult to determine if we should reduce our crop harvests by 30% or perhaps (when considering the extent and the implications of our carnivory) by as much as 75%. In the case of clearly insufficient outputs in many African and Asian countries they make it very difficult to quantify the degree of real food deficit, leading easily to exaggerations or underestimates of the real problem. Moreover, even if that could be done satisfactorily one has to consider that in most of those cases it may not be the level of supply but the lack of equity in access to food that is the real problem, and higher availability may do little to change that. All of these realities make it difficult to asses the need for any future productivity increases even before we look at the two key factors that will shape the future demand for food: long-term population trends and dietary improvements and transitions.

Population trends and dietary changes

Global population had already passed two fundamental milestones on its way toward an eventual stabilization, or decline. The relative rate of net increase peaked during the late 1960s (at just over 2.0 in 1967), the absolute rate of increase peaked at 87 million people a year during the late 1980s and it has been declining ever since. Consecutive long-range population forecasts by the UN have been revised downward and the latest one has the 2050 total as low as 7.4 billion, just 21% above the 6.1 billion in the year 2000. The medium forecast for 2050 is for 8.9 billion and the high variant goes up to 10.6 billion: the latter total is highly unlikely and accelerated decline of fertilities strongly indicates that the eventual total will be close to the low forecast.

Whatever the actual increment, all of the net increase will take place in today's low-income countries. Rapid aging, stagnation and population declines of various intensity will be seen in all high- and medium-income countries except those few willing to absorb relatively high immigration (US, Canada, Australia) and the net result will be that that by 2050 the total population of North America, Australia, Japan, South Korea, Taiwan, Singapore, Europe and Russia will be the same as it is now. Growth, albeit at declining rates, will increase populations in East Asia and Latin America. Relatively the most rapid increases will take place in the sub-Saharan Africa and in the Muslim countries.

If the populations totals were the only factor driving the demand for food then the prospects for the next two generations is clear: a mix of stagnation and decline in nations whose populations added up to 1.2 billion people in the year 2000 will produce no net population increase, or even a slight decline, by 2050; continued, but reduced, growth of today's 4.2 billion people in low-income economies will add anywhere between 0.7-1.9 billion people by 2050, and relatively large additions for the poorest countries whose population was about 0.7 billion people at the beginning of this century will at least double their total by 2050. Assuming no change in the availability of farmland and no raise in the overall per capita food supply and composition, this population growth would require no production change (and easily up to 25% decline) for the first group, 15-45% increase for the second, and 100% for the third. Weighted global mean would be then on the order of 30% over a period of 50 years, hardly a challenge for plant breeders.

Constant per capita food supply should be, of course, an unrealistic assumption. Rapidly aging populations will actually require less food per capita due to natural decline in metabolic needs (unless one assumes populations that are nearly 100% obese) and for the sake of better health all Western population should actually decrease their excessive average per capita food intake, particularly their very high consumption of meat whose modern mass production has been also a major source of epizootics and pathogen transfers from animals to people. Populations of the second group of countries will not require any large increments in the overall caloric or protein needs but will expect higher quality of their food supply. Complexity and specificity of dietary transitions makes it very difficult to come up with any reliable quantifications of this demand as historic examples provide arguments for both rapid change and conserved foodways.

Average per capita food supply in affluent nations now has between 55-65% of all protein coming from animal foods, more than 30%, and even more than 40%, of all food energy originating from lipids, and annual meat consumption exceeding one's body weight (70 kg, or even 100 kg/capita). There is a widespread assumption that this diet acts as a universal beacon, that as soon as the rest of the world -- where lipids supply less than 20% of all food energy and meat consumption averages less than 30, and in most countries less than 20 kg/capita — will have the requisite purchasing power it will try to emulate the consumption pattern that now dominates in Europe, North America and Australia.

Dietary attraction of fat is indisputable and its income-dependent nature was perfectly illustrated by recent Brazilian figures: shares calculated for the rural area of the Northeast (Brazil's most impoverished region), for suburban Rio de Janeiro and for urban Sao Paulo (the country's most industrialized region) were, respectively 12, 23 and 28%. Only when people get really rich does their fat intakes saturate and, among the health-conscious segments of such populations, either shifts in favour of plant oils or even declines in absolute terms. Post-1980 China provides perhaps the best example of the often surprisingly rapid transition to eating more meat and lipids as it has moved from a barely adequate diet dominated by staple grains and basic vegetables to a total per capita supply that, according to the FAO, has been actually surpassing the Japanese mean since the mid-1990s (and averages now just over 3000 kcal/day) and that contains increasing shares of meat and fish.

Official Chinese statistics show annual meat consumption rising from only about 11 kg/capita in 1975 to nearly 50 kg by the end of the 1990s. If true this would have been the fastest increase of meat eating in history -- but the same statistical yearbook puts actual per capita purchases of meat during the late 1990s at about 25 kg/capita for urban households (unchanged in a decade!) and the meat consumption of rural families at less than 17 kg, another perfect example of huge information uncertainties. Of course, FAO balance sheets calculate China's average per capita meat supply on the basis of clearly exaggerated production claims, putting it at nearly 49 kg/capita in 1999. When extrapolated "just" to 1-2 billion people that will try to replicate China's economic advance during the next 50 years, such consumption increases would translate into enormous demand for animal feed and, given the limited amount of any new farmland in Asia, to the need for higher maize, cassava, sweet potatoes and soybeans yields.

But these expectations may be vastly exaggerated. In spite of indisputable globalization of tastes, national and regional food preferences are evident around the world and food taboos, now only weakly held in the West, remain strong among nearly two billion Muslims and Hindus. Moreover, further homogenization and globalization of tastes will be accompanied not only by further diffusion of dubious Western dietary patterns but also by diffusion of an alternative dietary attractor. As Seckler and Rock argued, there are two distinct patterns of food consumption representing two alternative attractors in forecasting future composition of per capita food intakes. The Western pattern has the daily mean of more than 3,200 kcal/capita and more than 30% of food energy comes from animal foodstuffs; the Asian-Mediterranean pattern has overall intakes below 3,200 kcal/capita and animal products supply no more than 20-25% of food energy.

There is a clear evidence for this process. Taiwan's high per capita meat intakes have not been matched in Japan where both the average annual meat consumption (around 40 kg/capita) and overall daily food energy supply (about 2800 kcal/capita) have been stable for more than two decades. Consumption of animal foods in the countries with Asian-Mediterranean diet is not moving rapidly toward the Western pattern. For example, in Egypt and Turkey the proportion of meat in average diets has hardly changed in 30 years. And the trend lines of the overall per capita consumption of animal foods in North America and in the European Union have been mostly negative, moving in the direction of the alternative attractor. As for the third group of populations, they will need appreciable increases in per capita food availability if they were to eliminate persistent food deficiencies and malnutrition. Given the latest estimates of malnutrition it would not be unrealistic to aim at increasing the average per capita food supply by at least 25% in order to greatly reduce (but not entirely eliminate) the existing food shortages.

This complex situation makes it possible to posit a number of fairly divergent scenarios and the situation gets even more complicated once we begin to consider the composition of improved diets because higher supply of animal protein can be delivered in ways that will have relatively small impact on the future demand for feed crops or it can be produced in extraordinarily inefficient manner. The two highest efficiency routes are, of course, to supply a significant share of animal proteins from dairy products and from herbivorous aquacultured species. When compared to meat, their production requires only a fraction of feed per unit of protein and any problems with the presence of lactose (lactase deficiency being common in many Asian, African and Latin American population) are limited or disappear entirely with the consumption of partially or fully fermented products. Japan provides an excellent example: per capita consumption of dairy foods, nonexistent in 1945, now amounts to nearly 50 kg/year, about 40% of the high US level!

Should today's low-income countries that aspire to increase their animal protein intakes do so by relying on a combination dominated by dairy products, acquacultured herbivorous fish and poultry, their future needs for feed grain (and hence for yield increases) would be greatly reduced compared to higher intakes of animal foods composed of a mixture of pork, beef and mutton. As Asian populations will grow in coming generations and as the desirability of Mediterranean diets becomes even more appreciated throughout the Western world, there is also a considerable probability that the Asian-Mediterranean attractor will prevail and that the widespread assumption of high income elasticity of demand for meat would not be realized.

Add to this more efficient ways of feeding — transgenic pigs able to produce phytase in their saliva are just one example what future bioengineering approaches can bring — and the overall need for higher harvests by the year 2050 may be then of the order of 50% rather than only 30% needed just to maintain the existing per capita intakes of growing populations. Even when assuming that average crop harvest would have to rise by 60% this means an average annual exponential growth just short of 1% over the period of 50 years, a performance that was greatly surpassed during the past 50 years. At this point we have to add the third key factor that will drive the process, the decreasing availability of arable land.

Agricultural land

Land's importance for food production is obvious: only some 2% of the world's food energy and no more than 7% of all dietary protein come from waters. And so it is particularly unfortunate that we do not know with satisfactory accuracy something seemingly as easy to measure as the total area of farmland. In most cases the available totals underestimate the actually cultivated area and in most low-income countries there is also significant amount of cultivated lands within villages, and even within cities. Nutritional contributions of small, but intensively cultivated, kitchen gardens are important in countries ranging from small Caribbean islands to large economies of Asia. In Java anywhere between 20-35% of village area may be cultivated, and in Sri Lanka 40% of households without any farmland cultivate their own home gardens. Areas of urban home gardens are even more underestimated.

Accounting is complicated by the fact that the extent of cultivated land keeps changing due to the conversion to nonagricultural uses, because of natural or anthropogenic degradation, and also because of outright abandonment of not just marginal but also some very good fields. The latter phenomenon is surprisingly common even in some densely populated areas where new manufacturing opportunities brought the rural labor force into factories and marginalized local farming. This abandonment of good farmland creates isolated patches of land stranded between new factories and roads that can be seen most commonly in rapidly modernizing regions of Asia stretching from India and Korea now including perhaps most prominently China's coastal provinces with their double-digit rates of annual economic growth.

The reality of diverse farmland categories — currently cultivated and officially recorded land, cultivated but unacknowledged fields, formerly cultivated but now fallowed land, recently cultivated but now abandoned land, and farmland abandoned a long time ago but potentially cultivable — presents a difficult classification and accounting challenge in preparing informative national and global accounts. Accuracy of these figures cannot be easily improved by using multispectral satellite imagery: in the absence of a great deal of reliable ground data, spectral confusion between various kinds of land use may result in misclassifying up to a third of the total area.

Cultivated land is being lost through conversions to built-up areas needed to house growing populations and to put in place new industrial and transportation infrastructures. This process will only accelerate during the next two generations as land claims for housing additional billions of people will be much increased by the expansion of economic capacities needed to meet rising developmental aspirations. With greater affluence the growing cities will also require more space for solid waste disposal and for water treatment, and they may also claim more suburban land for recreation. Consequences are naturally most worrying among low-income populous countries whose average per capita farmland availability is already well below 0.2 hectares.

Besides the rapidly industrializing China the highest absolute farmland losses due to growing population and economic modernization will occur in India, Nigeria, Pakistan, Bangladesh, Brazil and Indonesia, the nations where half of the projected increase in the world population is to take place during the next generation. Agriculture itself is also a significant consumer of arable land: irrigation needs about 0.1 ha of land for access, canals, pumping stations and pipes for each ha of effectively watered land. Consolidation of livestock production into larger enterprises will claim more farmland, as will better access roads to remote regions, new machine sheds and fertilizer and grain storages and food and feed processing facilities.

These losses will be partially compensated by bringing new farmland under cultivation. The latest worldwide study by the IIASA used a detailed bottom-up approach as it estimated the extent of land with the potential to grow any of the 21 principal crops ranging from rice to olives through an inventory of agroecological variables determining the productivity of rainfed agriculture, divided the land into very suitable, suitable and marginally suitable categories and subtracted the currently cultivated and built-up land as well as the areas protected in national parks and other reserves. The study concluded that, in addition to roughly 1.5 billion hectares of existing farmland, there are slightly over 1.6 billion hectares of additional land with rainfed cultivation potential, with about 300 million hectares located in areas with sufficient rainfall. As about 60% of this potentially best land is covered by forest and wetlands, converting it into farming would inevitably entail major deforestation and wetland losses.

Undoubtedly, total area of potential farmland is quite large, but its spatial distribution is highly uneven and its initial quality will be generally inferior to the existing cropland. National rates, and hence the eventual global extent, of converting natural ecosystems to future farmland depend on a multitude of constantly changing factors ranging from domestic policies to terms of trade, and from newly tapped national resources to the speed of demographic transition. Most of the potential cropland in low-income countries is roughly split between South America and Africa, with Brazilian cerrado and grasslands of the sub-Saharan Africa representing the greatest arable land reserves. Opening up of this land would mean, above all in the latter case, further extensive tropical deforestation and because of limited transportation links production from much of this new land would be, certainly at least in initial stages, overwhelmingly only for local consumption.

Future land requirements will depend heavily on the composition of their average diets and on the intensity of cultivation. An overwhelmingly vegetarian diet produced by high-intensity cropping would require no more than 700-800 m^2/capita. A fairly balanced Chinese diet of the late 1990s — supplying about 2,800 kcal/day with about 15% of this total coming from animal foodstuffs — is produced from an average of 1,100 m^2/capita by methods ranging from highly intensive cultivation to extensive single cropping (nationwide multicropping index is about 1.5). In contrast, the Western diet, with its high share of meat and dairy products, claims up to 4,000 m^2/capita — but using that total is inappropriate because, as previously noted, very large shares of the available food supply are wasted.

Average daily supply of 2,500 kcal/capita would be quite adequate for all populations where the hardest farming and industrial tasks were mechanized, and if the 30% share of animal products would be composed of about equal parts of dairy products, poultry and pork then even a moderately intensive single cropping would not need more than about 1,500 m^2/capita to provide such a diet. In contrast, high shares of beef could push this rate up to 3,000 m^2/capita. Consequently providing up to 10 billion people, that is more than the UN's medium forecast for the year 2050, with healthy diets would need no more farmland than we are already cultivating, and there would be not even any need for further intensification of cropping.

An encouraging conclusion is that in global terms the availability of farmland is not a limiting factor in the quest for decent nutrition during the next 50 years even should the world's population grow to between 8-10 billion people. If there were no other natural constraints on food production, and if all land-scarce nations would have adequate purchasing power, then land-rich regions should be able to produce enough food even without new major land reclamation efforts; combination of a moderate increase in cultivated area and higher intensity of cropping would assure an appreciable safety margin for the global food supply.

Regional and national assessments for low-income societies tell a different story. About two dozen populous low-income countries already have extreme land scarcity as they cultivate more than 95% of the potentially arable land, supported half of all the population. In regional terms, per capita land availability remains high in Latin America, and more than adequate in sub-Saharan Africa; the greatest concerns exist, and will intensify, in the Middle East, and in South and East Asia. Combination of limited reclamation opportunities and growing space requirements for larger populations means that countries in these regions are facing continuous declines in per capita availability of arable land.

The need for higher yields driven by declines in arable land will be thus highly country-specific. Virtually all countries of East and Southeast Asia and every country in the Middle East will belong to the group with the highest relative declines, while the problem will scarcely arise in Latin America as well as in most of the sub-Saharan Africa. This reality further accentuates the need for sustained research tailored to specific regional conditions. We know that nothing produces such return on investment as the accumulation of innovations but because these research benefits may begin to flow only 10-15 years after launching a particular quest nothing is as important as continuous commitment. Seen from this perspective, the low level of agricultural research throughout the sub-Saharan Africa as well as the declining commitment in China are particularly worrying.

Do we need higher yields?

The only possible, scientifically honest, answer is no and yes. No if these yields are to perpetuate and intensify unsustainable, economically ruinous and nutritionally harmful practices of modern agriculture as it evolved during the latter half of the 20th century in Europe and North America. And no if these yields are to support the extension of these practices to a number of rapidly modernizing countries (or at least to some of their regions) in Asia and Latin America. To dwell solely on factors, conditions and opportunities for increasing crop harvests and for boosting the productivities of domesticated animals is to use a badly truncated view of the world's food system and hence to pursue some very counterproductive solutions and to offer possibly quite misleading conclusions.

Why should we need to boost the productivity and the total output of those agricultural commodities whose consumption has well proven, and quite substantial, health costs? Perhaps only economists can design a system where cost of food as a share of disposable income is constantly falling but the subsidies to produce that food, and the health costs associated with its consumption, are constantly rising, taking away from consumers in taxes (or in bloating government deficits) what they saved on grocery purchases. Western economists have thus perfected a system whereby the ubiquitous obesity, cardiovascular diseases and diabetes can be acquired faster and cheaper by hundreds of millions of people who are directly spending less and less of their income in order to shorten their lifespans but who are indirectly bearing not just the rising cost of health care but also now nearly half a trillion dollars in annual agricultural subsidies.

And resolutely no as long as we ignore profound failings elsewhere in the food system. Nothing illustrates this point better than our inexplicable neglect of food losses. Searching the Web of Science, the world's most comprehensive repository of all scientific publications, shows that among some six million items that database amassed since its inception in 1998 until the end of January 2003 those dealing with post-harvest losses amount exactly to eleven publications — while the publications dealing with the ways to increase crop yields number in tens of thousands! Could this get more irrational? This neglect is particularly astonishing given the relative magnitudes of gains and losses.

On the one hand, there is the endless quest for higher yields so that the annual growth of crop harvests and livestock productivities would at least match the population growth rates of low-income countries. On the other hand, in most of those countries harvest and post-harvest losses of cereals commonly and constantly surpass 10%. A five-year survey of grain losses in China's leading cereal-producing provinces found that about 15% of the grain crops are lost annually during harvesting, threshing, drying, storage, transport and processing; for rice this survey found losses of 7% in harvesting, 2.5% in threshing and drying, 2% in transport, and 5-11% in storage.

Even if I were to assume (unrealistically) that no other low-income country loses more of its staple cereal harvest than China, the rate of 15% translates to about 180 Mt of grain (all low-income countries now harvest nearly 1.2 billion tonnes of cereal grains a year). With an average price of US$ 250 per tonne of cereals this represents an annual loss of about US$ 45 billion. This total is almost exactly equal to total agricultural imports to China, India, Indonesia, Brazil and Mexico. Assuming the total composed of equal amounts of wheat, rice and maize, and diets where these staple grains supply about 70% of daily food energy requirements of 2,200 kcal/capita, the resulting 150 million tonnes of milled cereals would suffice to feed every year about 960 million people, or nearly all of India. And post-harvest losses for tubers, fruits and vegetables, particularly in the tropics, may be higher than 25%, and spoilage can reduce fish catches by an even higher share.

And the already noted differences in feeding efficiency are another perfect example of gains to be made without increased yields. Plant breeders may toil hard to increase crop yields by 1%/year but given the fact that large and increasing shares of crops are fed to animals, incomparably greater gains could be realized by shifts in average diets. Feedlot-fed beef will need as much as

100 kg of concentrate (maize and soybeans) feed in order to produce one kg of protein in lean meat; aquacultured herbivorous fish can produce 1 kg of excellent protein by metabolizing about 12 kg of grain feed, more than an eight-fold difference. There is no way to increase crop yields by a comparable margin but there is a way, over a period of years, to affect a desirable dietary shift from eating concentrate-fed large animals to herbivorous aquacultured fish that would spare enormous amount of natural resources while improving human health. And it turns out that such opportunities are particularly great in modernizing low-income nations that are in the midst of rapid dietary transition.

But we need higher yields in order to produce more food crops, and do so more efficiently, in all but a few low-income countries in Africa and Asia where very low yields result in extensive cultivation of marginal land and in environmental degradation and are unable to supply adequate nutrition. Yes in order to greatly reduce the extent of the existing malnutrition that is attributable to actual shortages of food rather than to highly unequal access to its available supply. And also yes in the case of those modernizing countries where yields have already risen well above the traditional minima but where further intensification can prevent the conversion of forests or wetlands to fields and where better managed use of inputs can actually decrease the loss of nutrients and waste of irrigation water while boosting the productivity.

Bibliography

Further readings to provide detailed analyses of many topics that were merely skimmed in this paper might include:

Bruinsma, J., ed. 2003. *World Agriculture: Towards 2015/2030 An FAO Perspective*. FAO, Rome.

Dyson, T. 1996. *Population and Food: Global Trends and Future Prospects*. Routledge, London.

Fageria, N.K. 1992. *Maximizing Crop Yields*. Marcel Dekker, New York.

Mitchell, D.O. et al. 1997. *The World Food Outlook*. Cambridge University Press, New York.

Runge, C. F., B. Senauer, P. G. Pardey, M. W. Rosegrant. 2003. *Ending Hunger in our Lifetime: Food Security and Globalization*. The Johns Hopkins University Press for the International Food Policy Research Institute, Baltimore, MD.

Smil, V. 2000. *Feeding the World*. The MIT Press, Cambridge, MA.

Smil, V. 2001. *Enriching the Earth*. The MIT Press, Cambridge, MA.

Smil, V. 2002. *The Earth's Biosphere: Evolution, Dynamics, and Change*. The MIT Press, Cambridge, MA.

2

IMPACT OF LIBERALISATION OF FOOD AND LAND MARKETS ON AGRARIAN LAND USE IN THE EU

P. H. VEREIJKEN[1], C.M.L. HERMANS[2] & H.S.D. NAEFF[2]
[1] *Plant Research International, P.O. Box 16, 6700 AA Wageningen, Netherlands. Email pieterh.vereijken@wur.nl*
[2] *Alterra, P.O. Box 47, 6700 AA Wageningen, Netherlands*

Introduction

The EU cannot avoid dismantling its Common Agricultural Policy or CAP (Commission for Agriculture, 2004). For only by abolishing subsidies and protection of agriculture can the EU (and its OECD partners) achieve a comprehensive liberalisation of world trade (OECD, 2002). This will benefit EU consumers (supporting the CAP by €107,000,000.000 in 2002), and the far more important other industries and services (98% of Gross Domestic Production of the EU-25), and developing countries (Kol & Winters 2004, Lips 2004). By abolishing agrarian subsidies (currently half of the EU-budget!), the EU can also acquire the capital needed for other Common Policies such as the development of technology, infrastructure and public safety. Consequently, farmers in the EU must expect increasing competition with each other and farmers outside the EU. This is just a part of the increasing competition within the entire food chain, because the food industry and supermarkets also have to face liberalisation of their markets. Across the entire chain the crucial question is: "who can deliver quality produce for the lowest price?" The EU consumers pose this question to the supermarkets. In order to establish and defend a secure position in EU food markets, the supermarkets then pose the question to their suppliers: the food industry and farmers' collective organisations. In their turn, these pose the question to the individual farmers. If indeed production-limiting measures such as quotas, set-aside and direct income support are abolished, production will increase and so will the surpluses on the EU-food market. The free market will correct this over-production through reduced prices. Only producers who can steadily renew products and increase efficiency will survive the expected price battle. This price battle will inevitably cause a race in technology and a scaling up of production. Thus, the liberalisation of food markets will speed up the transition of agriculture from a small-scale craft to a large-scale industry with ever higher productivity of land and labour. As a result, a lot of agrarian land and labour will become redundant. This paper explores the impact of liberalisation of world food markets on agrarian land use in the EU-15 regions. It also explores the impact of liberalisation of EU-regional land markets. For this is the next logical stage of common development for EU members. If they lift protection against each other's agrarian land users, the fittest holdings of the various agrarian sectors may concentrate in the most suitable regions. Considering the increasing redundancy of agrarian land, it also seems logical that the EU members will eventually lift the protection of agrarian land against non-agrarian land use. This will strongly encourage the transition to non-agrarian land use, notably in the densely populated EU regions with highly competitive non-agrarian sectors. In expectation of this progressive liberalisation of world food markets and EU regional land markets, the regions of the EU-15 are ranked according to their relative rate of transition to non-agrarian land use, based on three simple indicators. By specifying this ranking for any of the main five agrarian sectors, it is possible to identify for each region which agrarian sector is most or least likely to disappear (or persist).

Materials and methods

Three indicators have been used to rank the EU-15 regions according to their relative rate of expected transition to non-agrarian land use:

a. The mean economic size of agrarian holdings in a region indicates the competitiveness of EU-regional agriculture in world food markets. The smaller the holdings, the faster agriculture in a region loses out to agriculture in other regions, leaving the land for non-agrarian use. The economic size of holdings is expressed in European Size Units. ESU is the mean gross margin per unit of production, be it animals or hectares of crops. Its current equivalent is some €1250 (Eurostat, 2004).
b. The mean economic yield per agrarian hectare in a region indicates the competitiveness of regional agriculture in both world food markets and in EU regional land markets. The lower the yield of the land, the faster agriculture in a region loses out to higher yielding agriculture in other regions and/or to non-agrarian users in the region. Transition to non-agrarian land use can happen by buy out of holdings or by holdings changing themselves to non-agrarian land use. The mean economic yield of the land is also expressed in ESU.
c. The number of inhabitants per agrarian hectare in a region indicates the socio-economic pressure of non-agrarian land users to be withstood by agriculture in the regional land market. This pressure is related to the number of inhabitants in the region and growth in their welfare e.g. their needs of more land for living, working and recreation. The higher the number of (ever richer) inhabitants per agrarian hectare, the faster

agriculture in a region loses out to non-agrarian users. The non-agrarian need for land mainly affects agrarian land, since most forests and natural areas are more strongly protected against changes in use.

These three indicators describe the twofold competition agrarian land use is increasingly exposed to, through liberalisation of both world food markets and EU regional land markets. The outcome for regions or for communities will be a slow or fast transition to non-agrarian land-use. The relative rate of this expected transition has been estimated by region (of the EU-15) by ranking the regions from low to high, based on their indicator values. It is assumed that market liberalisation will cause land use transition according to interactions between the three indicators: a. and b. will interact positively with each other and negatively with c. (Figure 2.1).

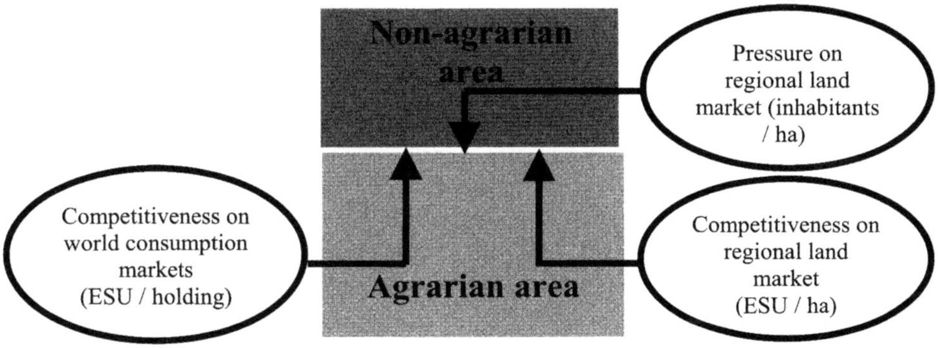

Figure 2.1 Transition to non-agrarian land use as a result of increasing socio-economic pressure and decreasing agrarian competitiveness on world food markets and EU regional land markets.

Therefore, per region, the ranking values based on indicators a. and b. have been added, and the ranking value of indicator c. has been subtracted. The resulting total ranking values represent the relative land use transition rate for each region, from highest to lowest. In the expectation of progressive liberalisation, first of food markets, then of land markets, three transition stages have been explored.

STAGE A: COMPETITIVENESS ON WORLD FOOD MARKETS WILL PREVAIL

Indicator a. was given more weight than the other two indicators together: total ranking values = 3a+b-c. During this stage the CAP is dismantled to achieve liberalisation of world food markets, causing an efficiency race, surplus production and ever lower product prices leading to large holdings out-competing small holdings. So regions will be more likely to face transition to non-agrarian land use the smaller the size of their holdings.

STAGE B: COMPETITIVENESS IN BOTH WORLD FOOD MARKETS AND EU-REGIONAL LAND MARKETS WILL PREVAIL

Indicator b. was given more weight than the other two together: total ranking values = a+3b-c. During this stage the EU-regional land markets are liberalised (as well as world food markets), causing ever higher land prices (and ever lower product prices), leading to holdings with high yielding land and non-agrarian users out-competing holdings with low yields. So regions will be more likely to face transition to non-agrarian land use the less the yield of their agrarian land.

STAGE C: SOCIO-ECONOMIC PRESSURE ON EU-REGIONAL LAND MARKETS WILL PREVAIL

Indicator c. has been given more weight than the other two together: total ranking values = a+b-3c. During this stage EU-regional land markets are maximally liberalised for non-agrarian land use. In urban regions non-agrarian sectors may develop much greater economic strength than agriculture. So regions will be more likely to face transition to non-agrarian land use the greater the inhabitants demand for agrarian land.

The most recent data on these indicators is the so-called 'structural survey' in 2000, which divided the EU-15 into 129 regions (Eurostat 2004, Figure 2.2). Based on these data, regions have been ranked and mapped per stage. Since these data were specified per agrarian sector, the three stages could also be explored for the five main sectors separately. Rather than provide all 15 maps (5 sectors x 3 stages), the results presented show the most and least probable sector(s) per region to disappear in any stage (i.e. three map pairs). This is not just to reduce numbers of maps, but to retain the focus on transition in land use. The maps have been made by comparing the relative transition rates (= total ranking values) of the sectors per region in any stage. Where all five sectors were present in a region (i.e. each >0.1% of its total EU-15 sector area) a sector was deemed *least* likely to disappear if its result exceeded at least three of the other four sectors by 10% of the mean range in transition rates per sector and per stage. A sector has been deemed *most* likely to disappear if its result was exceeded by three of the other four by 10%. Where four sectors were present in a region, sector identification required at least a 10% difference from two of the three other sectors, and where three or two sectors were present, sector identification required at least a 10% difference from one other sector. In this way, a maximum of two sectors per region could be identified as most or least likely to disappear. All the remaining comparisons of sectors in a region have been classified as being of "no significant difference" (including if only one sector was present). Before comparing the relative transition rates of the sectors, they have been corrected for differences in number of regions

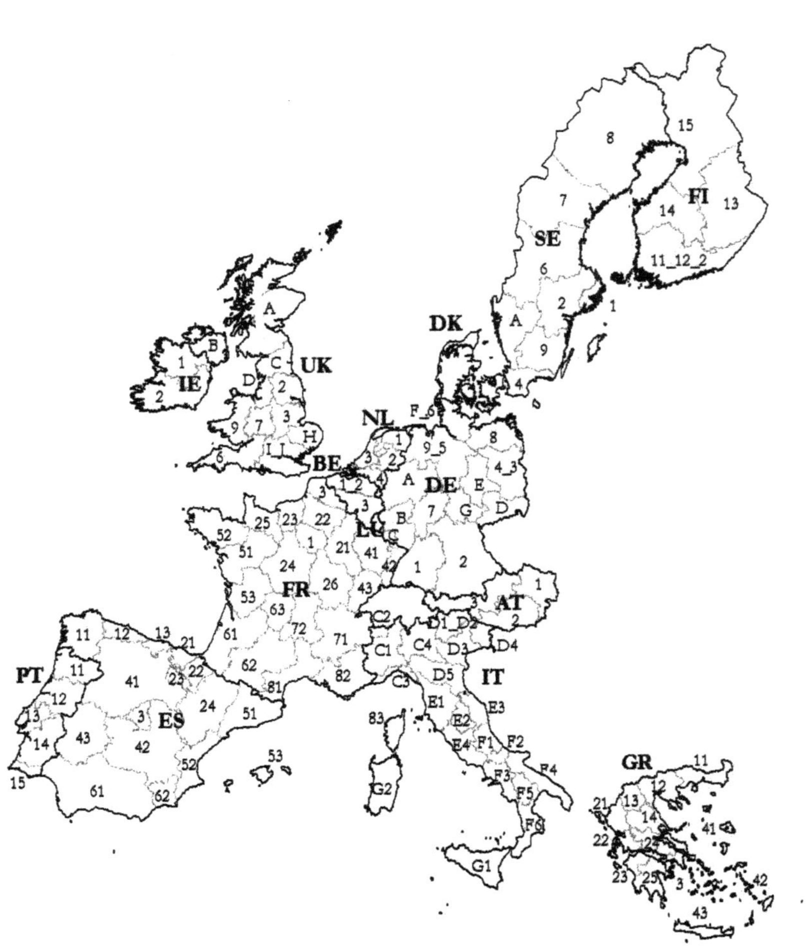

Figure 2.2 Regions of the EU-15 in the Farming Structural Survey (Eurostat, 2004).

Austria (AT): 1 East, 2 South, 3 West.
Belgium (BE):1_2 Brussels and Vlaanderen, 3 Wallony
Germany (DE): 1 Baden-Württemberg, 2 Bayern, 4 Brandenburg with Berlin, 7 Hessen, 8 Mecklenburg-Vorpommern, 9 Niedersachsen, A Nordrhein-Westfalen, B Rheinland-Pfalz, C Saarland, D Sachsen, E Sachsen-Anhalt, F Schleswig-Holstein, G Thüringen.
Denmark, (DK)
Finland (FI):11_12_2 Uusimaa, Etelä, Ahvenanmaa/Åland, 13 Itä, 14 Väli, 15 Pohjois.
France (FR): 1 Île de France, 3 Nord - Pas-de-Calais, 21 Champagne-Ardenne, 22 Picardie, 23 Haute-Normandie, 24 Centre, 25 Basse-Normandie, 26 Bourgogne, 41 Lorraine, 42 Alsace, 43 Franche-Comté, 51 Pays de la Loire, 52 Bretagne, 53 Poitou-Charentes, 61 Aquitaine, 62 Midi-Pyrénées, 63 Limousin, 71 Rhône-Alpes, 72 Auvergne, 81 Languedoc-Roussillon, 82 Provence-Alpes-Côte d'Azur, 83 Corse.
Greece (GR): 3 Attiki , 11 Anatoliki Makedonia, Thraki,12 Kentriki Makedonia, 13 Dytiki Makedonia, 14 Thessalia, 21 Ipeiros, 22 Ionia Nisia, 23 Dytiki Ellada, 24 Sterea Ellada, 25 Peloponnisos, 41 Voreio Aigaio, 42 Notio Aigaio, 43 Kriti
Ireland (IE): 01 Border, Midlands and Western, 02 Southern and Eastern,
Italy (IT): C1 Piemonte, C2 Valle d'Aosta/Vallée d'Aoste, C3 Liguria, C4 Lombardia, D1 Provincia Autonoma Bolzano-Bozen, D2 Provincia Autonoma Trento, D3 Veneto, D4 Friuli-Venezia Giulia, D5 Emilia-Romagna, E1 Toscana, E2 Umbria, E3 Marche, E4 Lazio, F1 Abruzzo, F2 Molise, F3 Campania, F4 Puglia, F5 Basilicata, F6 Calabria, G1 Sicilia, G2 Sardegna.
Luxembourg (LU)
Netherlands (NL): 1 North, 2 East, 3 West, 4 South
Portugal (PT):11 North, 12 Central, 13 Lisboa e Vale do Tejo, 14 Alentejo, 15 Algarve.
Spain (ES): 3 Comunidad de Madrid, 11 Galicia, 12 Principado de Asturias, 13 Cantabria, 21 Pais Vasco, 22 Comunidad Foral de Navarra, 23 La Rioja, 24 Aragón, 41 Castilla y León, 42 Castilla-la Mancha, 43 Extremadura, 51 Cataluña, 52 Comunidad Valenciana, 53 Illes Balears, 61 Andalucia, 62 Región de Murcia,
Sweden (SE): 01 Stockholm, 02 Östra Mellan, 04 Syd, 06 Norra Mellan, 07 Mellersta Norrland, 08 Övre Norrland, 09 Småland med öarna, 0a Väst.
United Kingdom (UK):, 2 Yorkshire and Humberside, 3 East Midlands, 6 South West, 7 West Midlands, 9 Wales, a Scotland, b Northern Ireland, c North East, d North West, h Eastern, i-j London, South East

active in any sector; this varied from 113 for grazers to 76 for granivores (pigs and poultry).

Table 2.1 MEAN REAL VALUES OF PREVAILING INDICATORS PER LIBERALISATION STAGE AND PER CLASS OF RELATIVE TRANSITION RATE (FROM FIGURE 2.4).

Dimensions of prevailing indicator per liberalisation stage	*Highest rate of transition (red)*	*Higher rate of transition (light red)*	*Mean rate of transition (grey)*	*Lower rate of transition (light green)*	*Lowest rate of transition (green)*
A. ESU / holding	5	15	29	51	90
B. ESU / agrarian ha	0.5	0.7	1.0	1.3	2.2
C. Inhabitants / agrarian ha	7	4	2	1	1

Results

Of its total land area of 315 million ha, the EU-15 used 127 million ha (40%) for agriculture in 2000. The regions vary in agrarian area from less than 0.5 million ha in Sweden, Finland and Greece to more than 3 million in Spain (Figure 2.3a). Relative agrarian areas varied from less than 25% of total land area in Sweden, Finland and Greece to more than 60% in Spain, France and United Kingdom (Figure 2.3b). Thus Figure 2.3 should be used to avoid over- or under-estimation of the results.

For any of the stages of progressive market liberalisation (A-C) and for all agrarian sectors, the relative rates of transition to non-agrarian land use (= total ranking values of the three indicators) have been calculated by region. The rates have been mapped as five classes, each class covering 20% of the agrarian area of the EU-15. These classes are based on quite distinct real values of the indicators, over all sectors (Table 2.1).

To summarise the transition maps according to the three expected liberalisation stages (Figure 2.4):

> if competitiveness in world food markets prevails (stage A), transition to non-agrarian land-use will probably achieve high rates in the southern countries, Austria and Sweden and low rates in the north-western countries;
> if competitiveness in both world food markets and EU-regional land markets prevails (stage B), transition to non-agrarian land-use will probably continue as in stage A, except that some regions in Sweden and the UK will show higher rates and most Greek and Italian regions will show lower rates (these regions have relatively small holdings, but high economic yields per ha from permanent crops);
> if socio-economic pressures on EU-regional land markets prevail (stage C), transition to non-agrarian land-use will probably achieve high rates in the Benelux region and most of the UK, Germany, Austria, Italy and

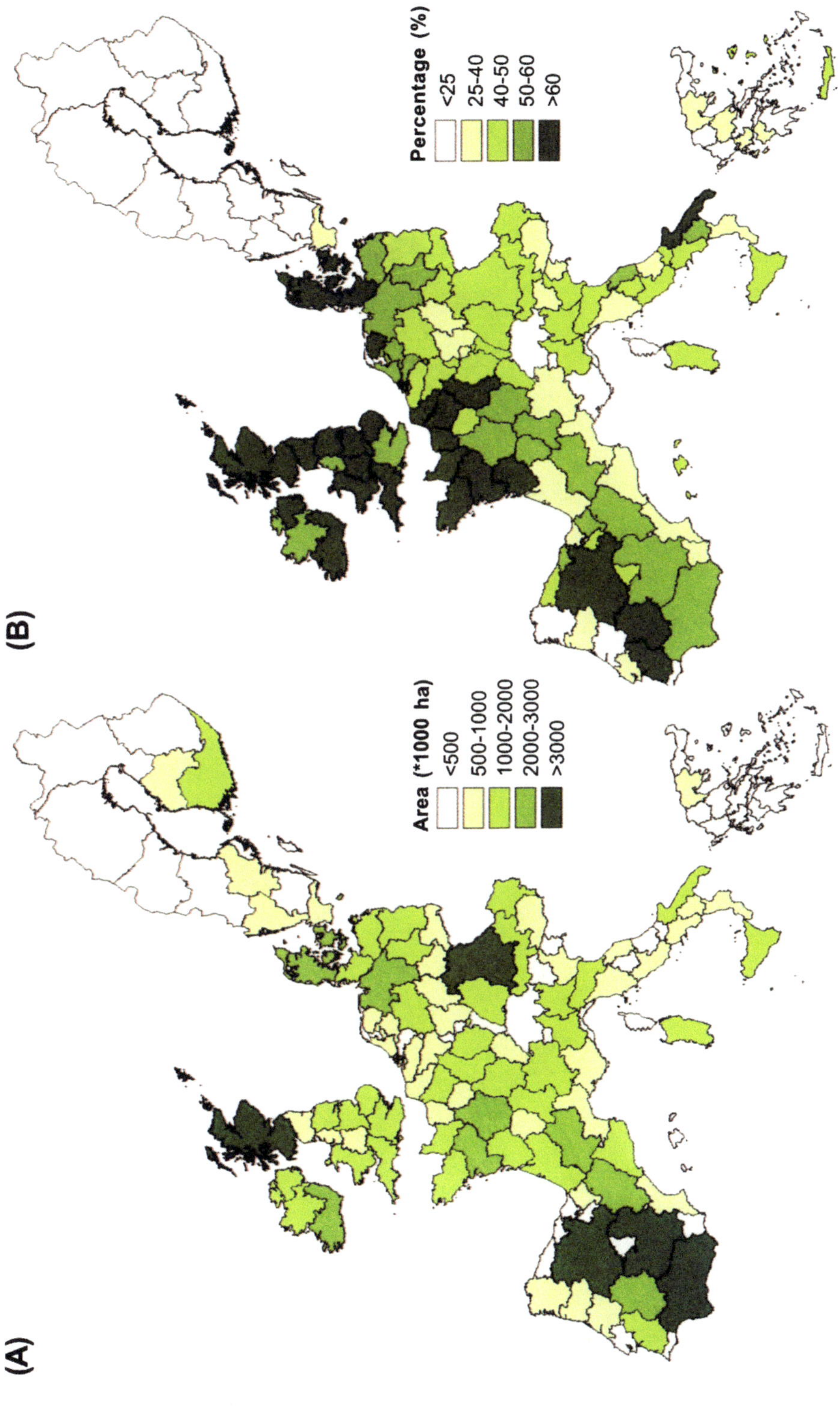

Figure 2.3 Absolute (a) and relative (b) agrarian area of the regions of the EU-15 (Eurostat, 2004).

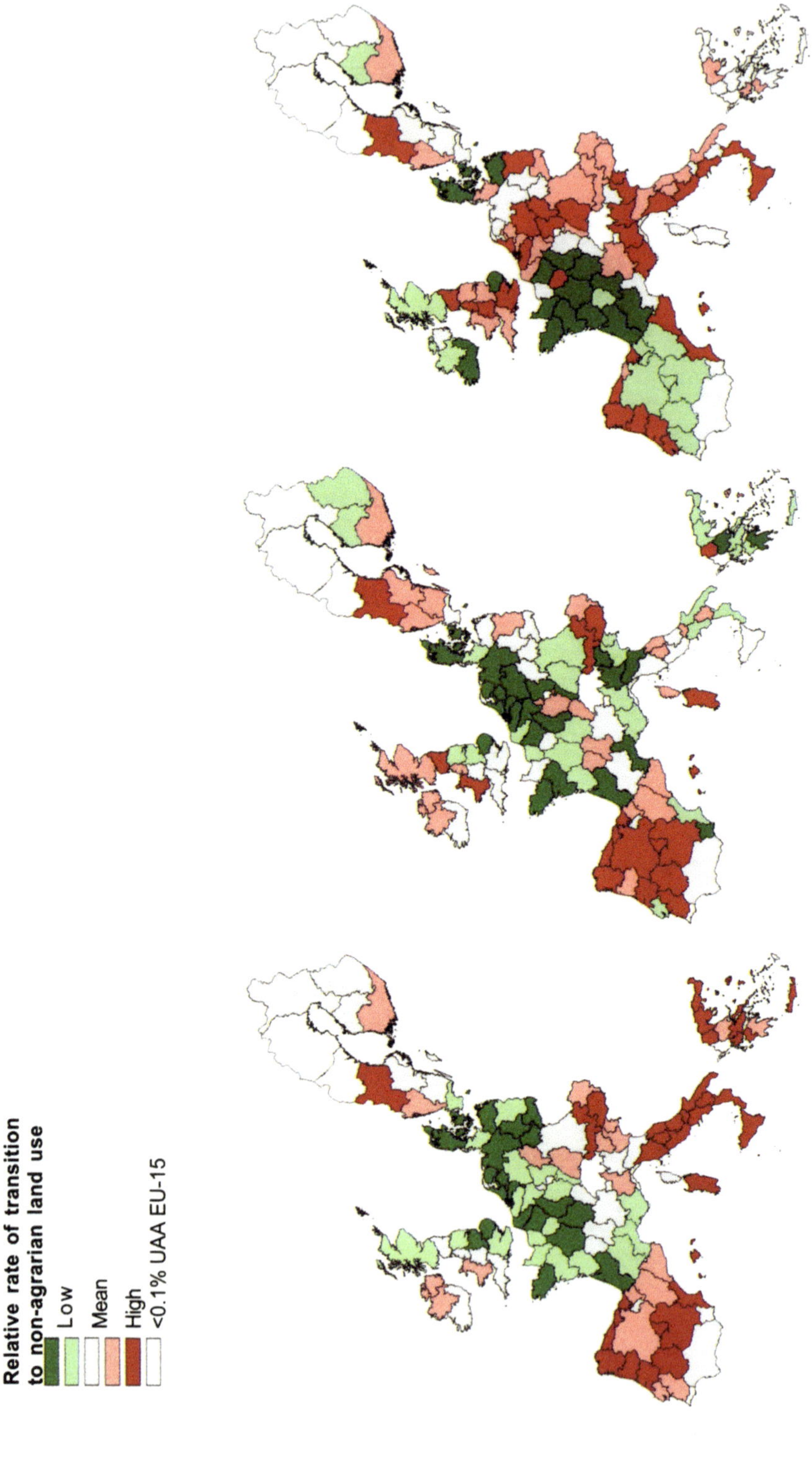

Figure 2.4 Relative rates of transition of the regions of the EU-15 to non-agrarian land use in three stages of progressive liberalisation of world food markets and EU regional land markets (A-C from left to right).

Portugal and low rates in France, Ireland, Scotland, Denmark and central Spain.

Some regions show consistent high or low rates of transition in all three stages. These regions probably will have achieved the highest or lowest degree of transition after the progressive liberalisation of markets. So agriculture may largely disappear (or be least persistent) in Austria, north Portugal and coastal areas of northern Spain and be most persistent in Denmark and the western half of France (Table 2.2). The remaining regions show intermediate patterns of transition (Figure 2.4). For example agriculture in the Benelux and most regions of Germany seems relatively persistent during stages A and B, though could disappear to a relatively high degree in stage C, if the high densities of inhabitants here give rise to use of agrarian land for other purposes.

Table 2. REGIONS WITH HIGHEST OR LOWEST TRANSITION RATES TO NON-AGRARIAN LAND USE, OVERALL LIBERALISATION STAGES (FROM FIGURE 2.4).

Relative transition rates in the 3 liberalisation stages				*Regions of the EU-15*
A	*B*	*C*	*Overall*	*Names (codes according to Figure 2.2)*
High	High	High	Highest	AT; ES11,12,13,21,51,53; FI 11_12_2; ITE2, E3; PT11,12; SE5, A; UK9.
Low	Low	Low	Lowest	DK; FR21, 22, 24, 26, 51, 52, 53, 61;UKH.

In addition to the analysis over all sectors, relative transition rates per sector have been estimated to explore the sector(s) per region most and least likely to disappear at any stage (see Materials and Methods). The results were strikingly straightforward and consistent over all three stages. Therefore, the transition maps can be summarised by sector and country (Figure 2.5a-c) as follows:

- 'Field crops' in Sweden and Finland was the most likely sector to disappear, and was never least likely to disappear.
- 'Grazers' in the UK and south-east France was the most likely to disappear and was the least likely to disappear in Luxemburg, Sweden and Finland.
- 'Permanent crops' in several Italian (non-wine) regions was the most likely to disappear and in the wine regions of France, Italy and Germany was the least likely to disappear.
- Horticulture in northern Greece and Scotland was the most likely to disappear and in southern regions of Germany, Austria, the UK and Italy was the least likely to disappear.
- 'Granivores' in west Germany, Luxemburg, Wallony and many regions of France was the most likely to disappear and in east Germany (stage B), Ireland and many regions of Spain was the least likely to disappear.

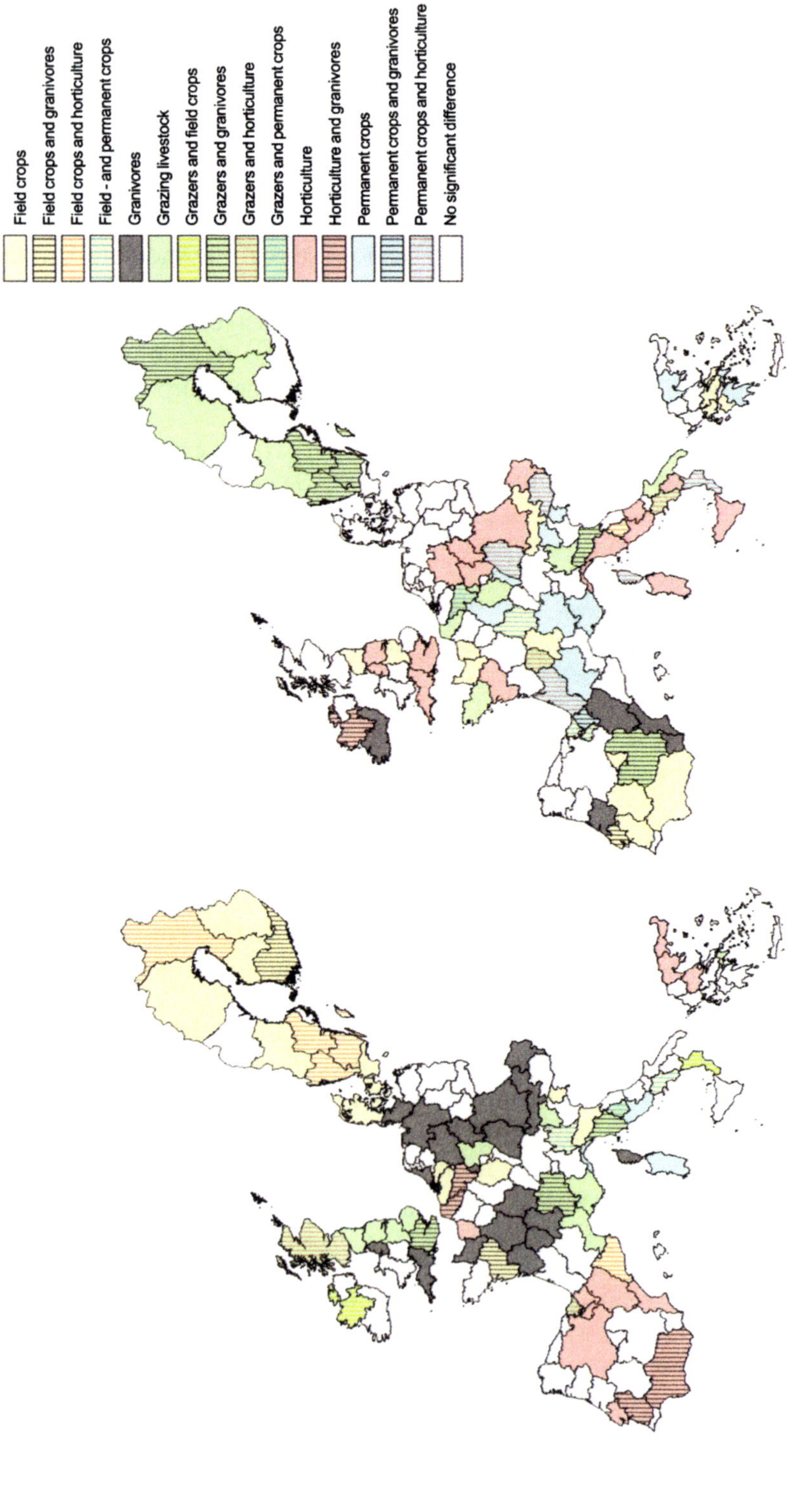

Figure 2.5a Most (left) and least (right) likely sectors to disappear in the EU-15 regions according to liberalisation stage A (competitiveness on international food markets will prevail).

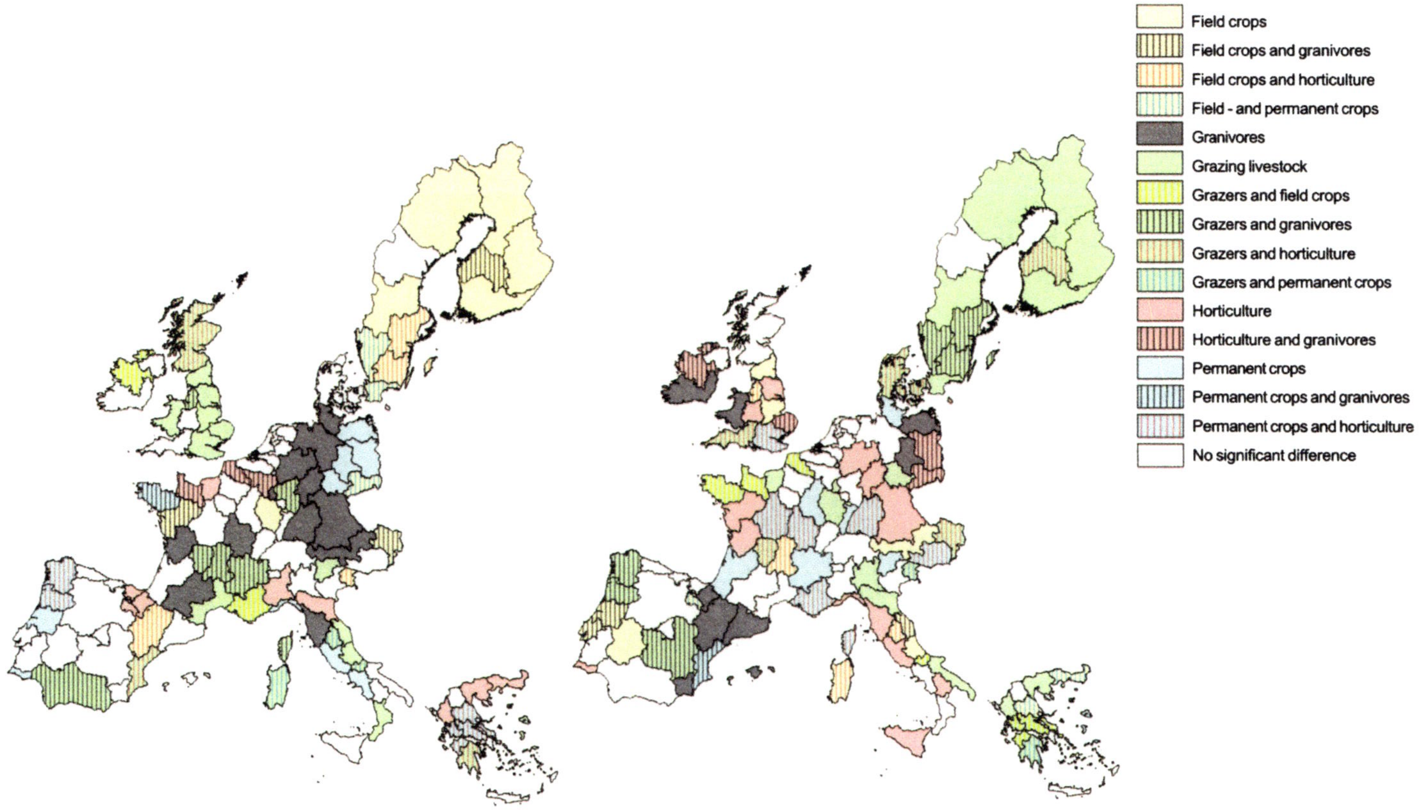

Figure 2.5b Most (left) and least (right) likely sectors to disappear in the EU-15 regions according to liberalisation stage B (competitiveness on regional land markets will prevail).

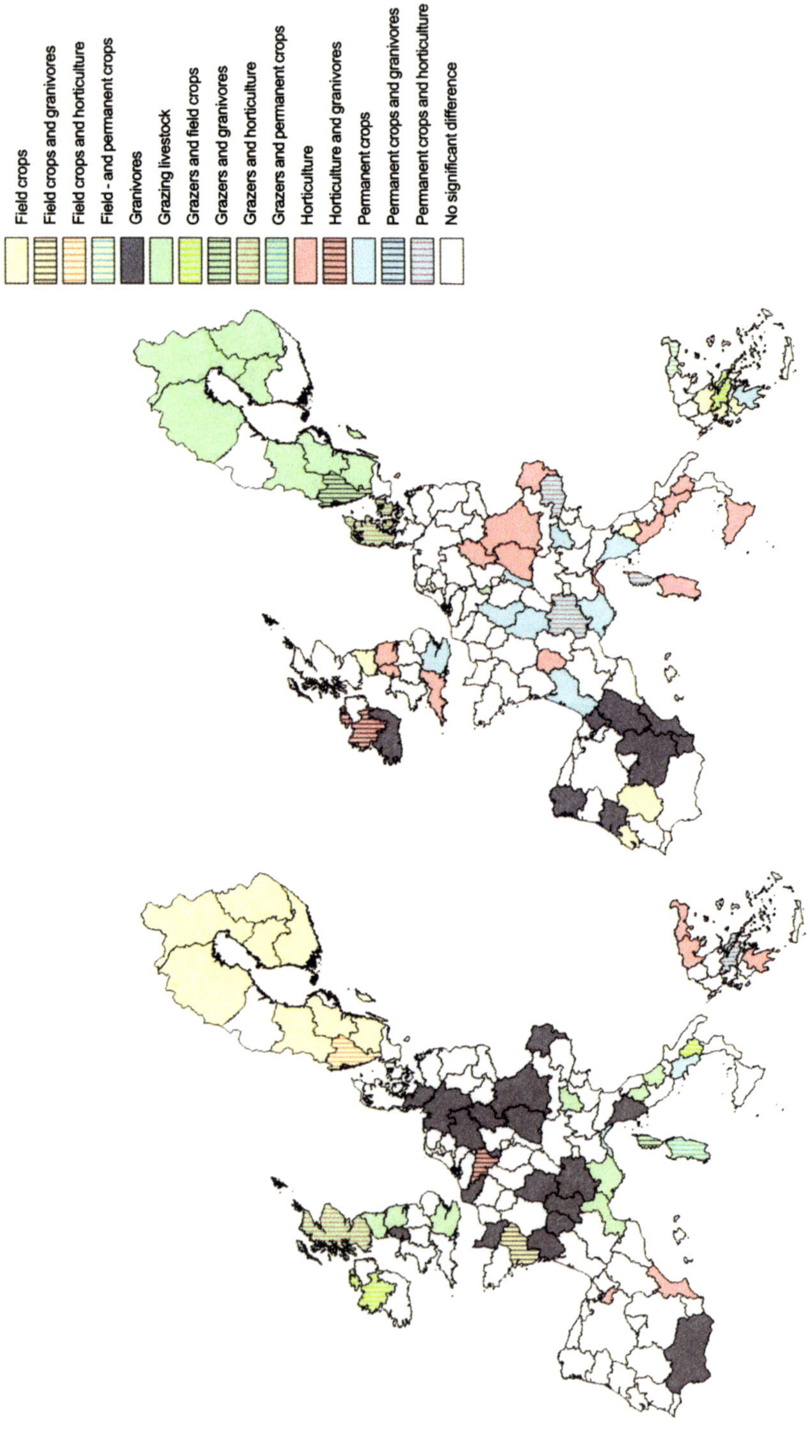

Figure 2.5c Most (left) and least (right) likely sectors to disappear in the EU-15 regions according to liberalisation stage C (socio-economic pressure on regional land markets will prevail).

Discussion

This paper is based on the expectation that the EU will dismantle its CAP to liberalise world food markets for agrarian products, but only as stage A in a progressive liberalisation which will include the EU's regional land markets. At first the land markets will be opened more widely to agrarian land users from other EU members (stage B). As a result of these two stages, agrarian competition and production will be invigorated and a lot of land will become redundant. This will enable a subsequent stage (C) in which regional land markets will also be opened increasingly to non-agrarian land users, insofar as this is compatible with the supply of collective benefits such as nature, landscape and water management. Thus, the progressive liberalisation of food markets and land markets will cause a progressive transition to non-agrarian land use in the EU regions. Absolute rates of transition cannot be predicted, though relative rates can, by ranking the regions according to three quite clear and easy to quantify indicators. Since all three of these proved influential, the ranking values of the regions must be summarised using varying weights, depending on their response at each stage of liberalisation. At stage A agrarian holdings and regions are particularly dependent on competitiveness for consumers, so indicator a. (economic size of the holdings) outweighs the others. At stage B holdings and regions also compete for land, so indicator b. (economic yield per ha) will dominate. At stage C indicator c. will dominate: pressure of inhabitants on regional markets for agrarian land. The use of indicators b. and c. may be challenged; both are reflected in the actual price of the land, since this is affected by socio-economic pressures (or non-agrarian demand), the agrarian (yield related) demand and the regional scarcity (or offer) of land. So in theory, if reliable EU regional land prices were available, these could have been used instead of indicators b. and c. In addition to the lack of a complete set of land prices for each EU region, land prices are volatile and therefore unreliable, especially in urbanised regions. Also, indicators b. and c. are preferred because they enable identification of stages B and C, with first agrarian and then non-agrarian users dominating the land markets. A further point of discussion is the successive weighting of indicators, and particularly the choice of a factor of three. This is arbitrary of course, but greater or smaller weighting would have a limited influence on the ultimate ranking of the regions.

Though it was the only feasible approach, there was uncertainty in estimating the relative transition rates of regions to non-agrarian land use by adding their ranking values for the three underlying indicators. For instance, regions may only differ insignificantly in real values, hence their classification from high to low relative rates could cause exaggeration or could even be deceptive. However, the large differences in real values between the five classes of transition rate confirm their significance (Table 2.1).

The most or least likely sectors to disappear in a region mostly coincide with low or high competitiveness of this regional sector for consumers and land overall EU-regions. Exceptionally, the most likely sector to disappear is the least strong sector of generally strong sectors in its region, or conversely, the least likely to disappear is the best sector of generally weak sectors in its region. Even so, this is significant, since regional sectors, strong or weak compared with similar sectors in other regions, also have to compete with each other for land (and labour). So to survive, a regional sector should withstand three kinds of rivals: sectors in other regions competing for similar consumers and other sectors plus non-agrarian users in its own region competing for land (and labour). The more suitable its land is for other agrarian or non-agrarian land users, the heavier will be the competition for land that regional sectors have to face. If its land is too remote, wet, dry, heavy, stony, steep or high for other sectors, a sector has a comparative advantage in both the land and food markets. 'Grazers' and 'field crops' use most agrarian land in the EU-15 and therefore are each others main rival for land (Table 2.3). In addition, their mean indicator values are almost equal, so overall they are little different in competitiveness for land. Considering the land use by the other three sectors, 'permanent crops' is a minor rival for land in most regions, and 'horticulture' and 'granivores' seem insignificant. However, it should be realised that 'field crops' and 'granivores' are allies, through their exchanges of feeds and manures, and this may give them a comparative advantage in the competition for land with 'grazers'. Given the high mean indicator values of 'horticulture' and 'granivores', these sectors can easily take over land from the others where there is possible expansion in food markets. 'Horticulture' is by far the most difficult sector for any other land user to displace, because of its huge economic yield per ha. This quick overview of strengths and weaknesses of sectors is given to emphasise that the competition for land between sectors is complicated, and that it should be analysed region by region.

Table 2.3 INDICATOR VALUES AND SIZES OF THE MAIN AGRARIAN SECTORS FOR THE WHOLE EU-15 IN 2000 (EUROSTAT, 2004).

EU-15 in 2000	*Grazers*	*Field crops*	*Permanent crops*	*Horticulture*	*Granivores*
A. ESU / holding	24	22	10	55	56
B. ESU / agrarian ha	0.7	0.7	1.7	14.3	3.3
Utilised area (million ha)	63.8	40.8	16.9	0.7	2.6
Holdings (million)	1.41	2.03	2.94	0.19	0.15

The results of this spatial-economic exploration may be useful for various actors. Policymakers and researchers at EU, national or regional levels may use the results in overall transition rates for the three stages (Figure 2.4) and results showing most or least likely sectors to disappear or stay (Figure 2.5) to anticipate the enormous shifts in rural economies and land uses that are expected. Pre-emptive approaches could vary from facilitating innovation and scaling-up of agro-sectors to facilitating alternative employment in non-agrarian activities. Of course, planning should include protection and expansion of collective goods such as landscape, nature and water management. Both agrarian and non-agrarian entrepreneurs may use the results to make basic decisions such as ending, starting, expanding or relocating their activities. Last but not least, the results may help NGO's and democratic institutions to assess and adapt regional plans for the countryside.

It would be interesting to repeat this study as soon as the first data on the 10 new member states are available. Undoubtedly, their entry will augment the twofold competition for consumption and land and thus increase the rate of transition to non-agrarian land use within the EU.

Acknowledgement

The authors thank prof. dr. G. Meester for his critical but most stimulating comments on the draft-paper.

References

Commission for Agriculture (2004). *CAP-reform, a long term perspective for agriculture*. EU, Brussels.

Eurostat (2004). Farm Structure Survey of EU-15 regions in 2000.

Kol, J. and Winters, L.A. (2004). The EU after Cancun: can the leopard change its spots? *European Foreign Affairs Review* **9**: 1-25.

Lips, M. (2004). *The CAP Mid Term Review and the WTO Doha Round; Analyses for the Netherlands, EU and accession countries*. The Hague, Agricultural Economics Research Institute (LEI).

OECD (2002). *Agricultural policies in OECD countries: a positive reform agenda*. OECD, Paris.

3

CROP IMPROVEMENT TECHNOLOGIES FOR THE 21ST CENTURY

BRIAN J LEGG
NIAB, Cambridge CB3 0LE UK

Introduction

Agriculture has been transformed by technology over the last 50 years, with major contributions coming from crop breeding, nutrition, irrigation, mechanisation, the use of herbicides and pesticides, and improvements in management expertise. Over the same period the political agenda has also changed. Fifty years ago, there was a drive for increased production world-wide, but in the last two decades there has been over-production in developed countries, and the emphasis has changed to producing food or non-food products with the quality demanded by the end user, at reduced cost.

There is also concern about the impact that agriculture has on the environment, including air and water pollution and loss of biodiversity. The focus in the developed world has often been on reduced inputs; and research to increase crop yields has been viewed as socially undesirable. In my view this is quite wrong, and given the increasing world population, our aim must be to increase all the beneficial outputs from agriculture, and at the same time minimise harmful effects. The beneficial outputs of food and non-food products are still required but pollution must decrease and agriculture must deliver an increased diversity of plant and animal wildlife. Agriculture must deliver more, not less, and if we can increase crop yields for the same level of environmental impact, that must be a move in the right direction.

In this paper I will give an overview of the technologies that have led to increased yield in the past fifty years, and will also consider how they have made their contribution. Without this understanding it is impossible to predict the further gains that can be made, and the limits of known technology.

When addressing these questions many authors have debated whether there is a limit to crop yield, and have used the term "potential yield". But potential yield is a very academic concept, and it is useful to follow the two definitions of Evans and Fischer (1999). Potential yield is "the maximum yield which could be reached by a crop in given environments, as determined, for example, by simulation models with plausible physiological and agronomic assumptions". This contrasts with yield potential which is "the yield of a cultivar when grown in environments to which it is adapted, with nutrients and water non-limiting and with pests, diseases, weeds, lodging, and other stresses effectively controlled". This paper deals with yield potential, as defined above, though in practice many experiments have not been irrigated and this usually results in yields below potential.

Although yield potential is not a precise concept, it is a useful indicator of what might be achieved with improved technology, and there are many experimental results from around the world to draw on.

Crop yields over the last fifty years

The last 50 years have seen a dramatic rise in the yield of national and world average yields for all crops, and this has been true for most, though not all crops grown under carefully managed experimental conditions. The picture for wheat (Figure 3.1) is typical of results that could be shown for other crop species.

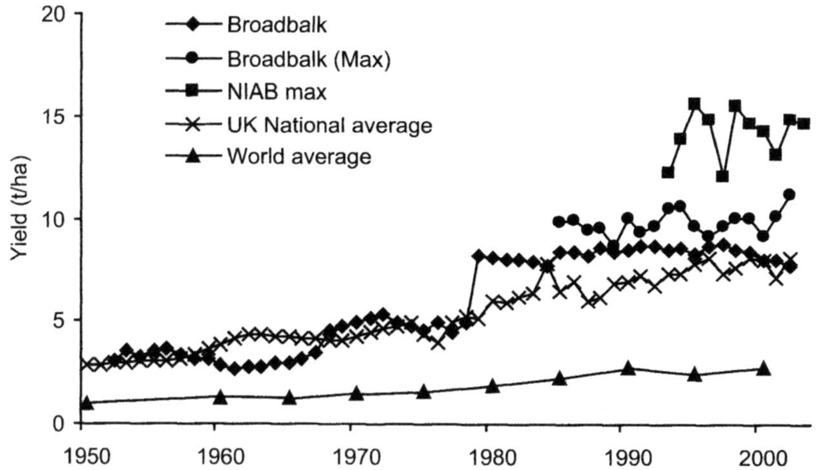

Figure 3.1 Wheat yields over the last 50 years. Data sourced from Rothamsted Research Ltd, Defra, Home-Grown Cereals Authority (HGCA) and the Food and Agricultural Organization of the United Nations (FAO). *(Some data were taken from trials funded by the British Society of Plant Breeders and the Home-Grown Cereals Authority. Full results can be found at www.hgca.com.)*

The experiments on Broadbalk at Rothamsted Research in the UK were started in 1843, to look at the response of wheat to fertiliser applications. The yields in Figure 3.1 are for wheat grown after fallow, and with 147 kg/ha of N plus application of P, K and magnesium. Squarehead Masters had been grown in the experiment from 1905 to 1917, and again from 1943, and there was a marked increase in yield with the introduction of Cappelle Desprez in 1968. There was a further dramatic increase in 1979 when Flanders was first grown, but this also coincided with the introduction of fungicide sprays. Since then there has been little further increase in yield at the same rate of fertiliser application, though the final variety, Hereward, is the highest quality bread wheat and there are now feed wheats that yield about 15% higher. From 1985 onwards additional treatments were introduced with 196 kg/ha N plus farmyard manure, or 294 kg/ha N, and Figure 3.1 also shows the maximum plot yield obtained in each year.

Another valuable source of information comes from the NIAB and HGCA Recommended List trials (funded at various times by Defra, plant breeders and HGCA), which have been running since 1931 for the comparison of new wheat varieties. The trials include replicated plots at 25 sites across the UK, and they receive nationally recommended rates of fertiliser, plus herbicides, growth regulators and a comprehensive fungicide programme. The average yields obtained across the country have been very similar to the Broadbalk yields over recent years, but the highest yielding site each year has ranged from 12 to almost 16 t/ha (Figure 3.1). Over the last 50 years the average UK wheat yield has increased from 3 to 8 t/ha, and over the same period world wheat yields have risen from 1.0 to 2.7 t/ha.

Reilly and Fuglie (1998) showed that the average yields of 11 crop species in the USA from 1939 to 1994 had increased by between 1% and 3% per year, and used a statistical analysis to determine whether the rates of increase in yield are decreasing. This was done by fitting linear, exponential and logarithmic curves through the yield time series, and comparing the quality of fit. For almost all crops the logarithmic model, which would imply that yields were increasing more slowly with time, fitted the data less well than either the linear or exponential, and for half of them this difference was significant at the 5% level.

This result was confirmed for soyabeans by Specht, Hume and Kumudini (1999), who extended the time series from 1923 to 1997 and showed the linear and exponential models to give an equally good fit. They also compared national average yields and the State average for Nebraska with and without irrigation, the winning yields in various yield competitions, and trials to determine maximum yield (reproduced in Figure 3.2).

Competition yields have increased from 4.8 to 6.7 t/ha since 1988, but these are not as high as the yield of 7.3 t/ha obtained in 1968 in a national yield competition. They hypothesise that there is a maximum yield of 8 t/ha for soyabeans in the USA, and that national yields could double in the next 50 years.

Maize yields in the Canada, USA and Europe have increase by approximately 1.5% per year since the 1940s (Tollenaar and Wu, 1999), and the increase has arisen primarily from the ability of new hybrids to tolerate

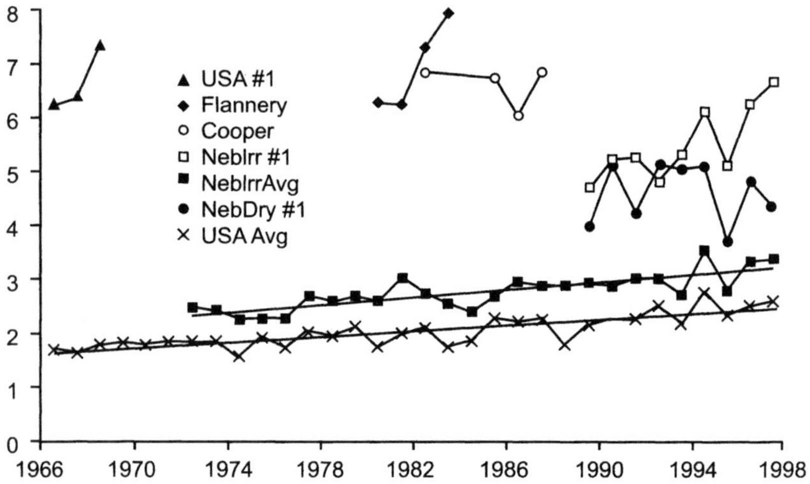

Figure 3.2 Soyabean yields in the USA. Winners of national competitions (USA#1); results of yield maximisation research by R. Flannery and R. Cooper; winners of Nebraska State yield competitions with irrigation (Nebirr#1) and without (NebDry#1); Nebraska State average yield with irrigation (NebirrAvg); and USA average yield (USA Avg). (After Specht *et al.*, 1999)

stress. Through the 1990s, however, the yield of contest winning trials for irrigated maize (Duvick and Cassman, 1999) have not increased. Furthermore, maximum yields for rain-fed maize of 21 to 23 t/ha were reported as far back as the 1970s. They conclude that there is "little compelling evidence that yield *potential* of maize hybrids adapted to the north-central United States has increased during the past 25 years".

The average rice yield for developing countries has doubled from 1.8 t/ha to 3.6 t/ha since the early 1960s (Bruinsma, 2003), a period that has seen the introduction of semi-dwarf varieties including IR8. Peng, Cassman, Virmani, Sheehy and Khush (1999) show that the highest reported yields in the late 1960s and early 1970s was 9 to 10 t/ha from the cultivar IR8, and that the maximum yields of modern varieties under favourable conditions are still the same. Over that period the maximum yields obtained from IR8 have declined from 10 t/ha to 8 t/ha. In other words, 30 years of rice breeding has merely maintained the status quo and not given rise to any further increase in potential.

Technologies for increasing crop yields

PLANT BREEDING AND PHYSIOLOGY

Over the last fifty years most genetic improvements have arisen from conventional phenotypic selection. Typically, the early generations have been screened for disease resistance and for easily visible physiological traits, and later generations have been screened for yield. The parents have frequently been chosen from existing and successful varieties, but crosses with early material and with wild relatives have also been used to acquire new traits including resistance to pests and disease. For some crops, notably maize and many vegetable species, this has been accompanied by hybrid technology. It is widely agreed that of the three-fold increase in yields, approximately 50% can be attributed to the plant breeder, with the rest coming from improved agronomy. Silvey (1994) studied UK national cereal yields over recent decades, and concluded that the proportion that could be attributed directly to plant breeding was 47% for wheat and 50 % for barley. This compares with a figure of 58% for maize in Minnesota (Reilly and Fuglie, 1998) and 50% for the USA as a whole (Duvick and Cassman, 1999).

Within the last decade or so new tools have been added to the plant breeders' armoury, but these would be of no value unless it were physically possible to change the plant phenotype in ways that would give more yield, and unless crop species contained the necessary genetic variation. It is thus important to consider the ways in which breeding has already changed the crop in order to give higher yields.

If a new variety is to have a higher yield than its predecessors, it must either: intercept more solar radiation; convert that radiation into biomass more effectively; or convert a higher proportion of that biomass into harvestable and useful yield. For all but the most intensive agriculture the crop must also be tolerant of a wide range of physical and biological stresses. In practice all of these changes have taken place in one crop species or another, though the relative importance of each varies considerably.

Increased radiation interception

Increased radiation interception can be achieved through either a longer growing season, or a higher percentage interception during the season. In fact both have been achieved in the UK, with the greatest elongation of season being the replacement of spring sown by autumn sown crops. The benefit is magnified in a dry season when spring sown crops are badly affected by drought while deeper rooted autumn crops continue to extract water from below.

Greater interception of radiation has been a major factor for increasing maize yields, and Duvick and Cassman (1999) show that modern cultivars can be grown to advantage at higher plant populations, and thus intercept more radiation, than the earlier cultivars. Authors also refer to the "stay green" factor, which allows radiation to be intercepted and used for longer.

Increased net photosynthesis per unit of light intercepted

At the most fundamental level there have been no increases in photosynthetic efficiency, and McCree (1971), showed that in 22 crop species there is no genetic variation in quantum yield, i.e. units of photosynthesis per unit of light absorbed at low irradiance. The rate of photosynthesis, however, shows diminishing returns at increasing light intensity, and saturates at a value A_{max}. Reynolds, Delgado, Gutiérrez-Rodríguez and Larqué-Saavedra (2000) quoted earlier work that showed a variation in A_{max} in *Triticum* species by up to 36%, though it was associated with other disadvantageous properties, and they also reported significant genetic variation in Emmer wheat. A_{max} is also influenced by environmental factors, and for several crop species there is genetic variation in this response. For a whole canopy it is also possible to increase photosynthesis by arranging the leaves so that most of the light is intercepted at low intensities, where photosynthetic efficiency is highest.

For wheat, Richards (2000) concluded that there has been no genetic increase in net photosynthesis per unit leaf area, and that there is little evidence for an increase in biomass production. Calderini, Dreccer and Slafer (1997) studied seven cultivars of wheat introduced into Argentina from 1920 to 1990, and concluded that there were no systematic differences in canopy light extinction coefficients, pre-anthesis radiation-use efficiency, or pre-anthesis crop growth rates. The poorer performance of the older varieties was caused by lower growth rates after anthesis, and the authors hypothesise that this was due to the grains having a small capacity to accumulate carbohydrates. In other words, the small sink for carbohydrates limited the rate of photosynthesis. Miralles and Slafer (1997) show specifically that the presence of the Rht dwarfing gene in wheat does not affect the timing of occurrence of phenological events, but does cause higher radiation use efficiency from anthesis to maturity. The increase appears to be positively associated with the number of grains set per unit of biomass at anthesis, providing further evidence that radiation conversion to biomass is regulated by sink size.

Increased yield of soyabeans (Specht *et al.*, 1999) has been caused by longer leaf duration through the seed filling period, lower lodging giving increased light interception, and higher stomatal conductance: all leading to an increase in photosynthesis and dry matter accumulation as the seeds fill. There has also been increased N fixation and better tolerance of high plant density. In a high yielding year, higher plant density gave more yield for modern cultivars and less yield for old ones.

Increased leaf angle has increased net photosynthesis in several crops including maize, for which Tollenaar and Wu (1999) quote a 20% yield improvement when leaf angle increases from 30° to 60°. Murchie, Chen, Hubbart, Peng and Horton (1999) showed that modern rice cultivars suffer from light saturation and photo-inhibition for a substantial part of the day, though these effects can be moderated by leaf angle. They also showed that leaf temperatures (at IRRI in the Philippines) were greater than 30°C for all of the photoperiod, reaching a maximum of 37°C, temperatures at which photorespiration would cause large losses of carbon. Although this suggests scope for increasing canopy photosynthesis, Horton (2000) warns that this is unlikely to be achieved in the near future without extreme good fortune.

Increased stomatal conductance causes an increase in leaf internal CO_2 concentration, giving higher rates of photosynthesis. This has been a major factor increasing the yield of irrigated semi-dwarf spring wheat varieties introduced in Mexico between 1962 and 1988 (Fischer, Rees, Sayre, Lu, Condon and Larqué Saavedra, 1998). There is also a beneficial decrease in canopy temperature. For cotton, Lu, Radin, Turcotte, Percy and Zeiger (1998) showed that improvements have come from higher stomatal conductance giving lower leaf temperature during times of heat stress.

Increased harvest index

Austin (1999) concluded that increases in UK wheat yields have come from the use of shorter varieties (with the Rht D1b dwarfing gene or the 1BL/1RS rye translocation) giving higher harvest index; higher N applications combined with greater resistance to lodging; and effective use of herbicides and fungicides. This was demonstrated at Rothamsted Research in 1988 to 1990 when the early variety, Squarehead Masters, was grown with the same treatments as a current high-yielding, breadmaking variety, Brimstone, which has the Rht2 gene (Figure 3.3).

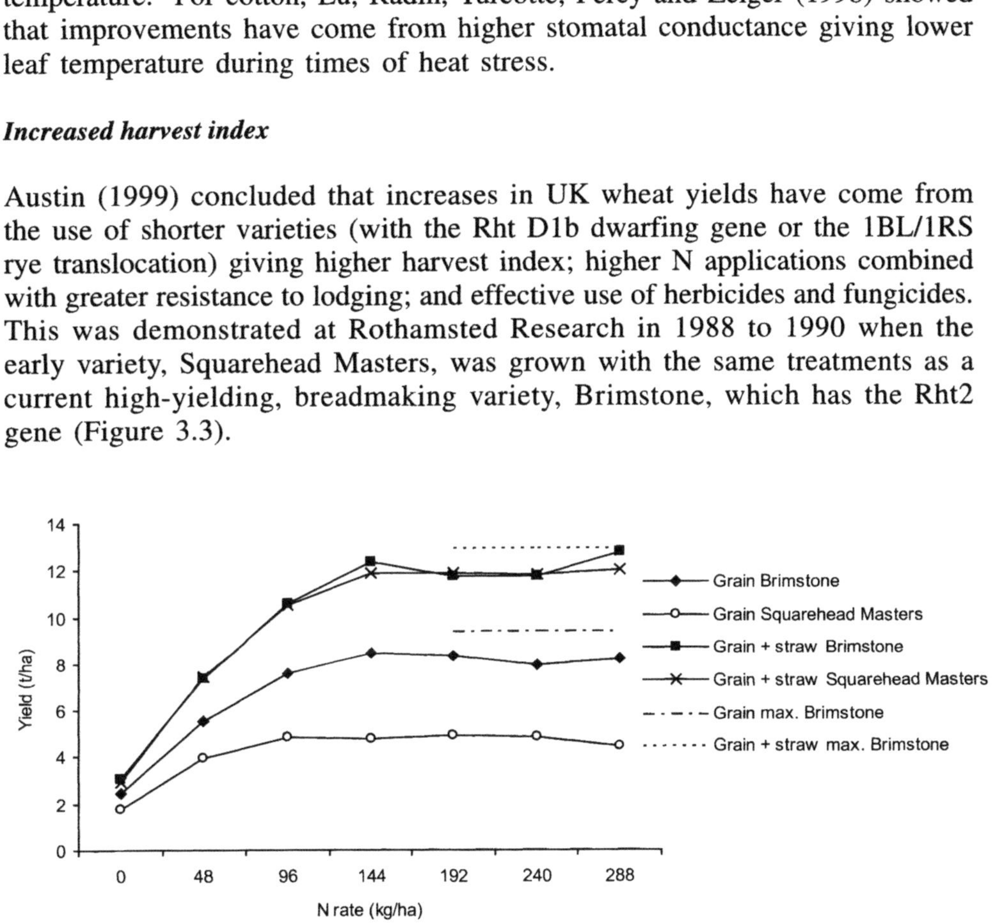

Figure 3.3 Response of wheat grain yield and grain plus straw yield to N fertiliser rates, for an old variety, Squarehead Masters, and a newer variety, Brimstone, with the Rht2 semi-dwarf gene.

The total yield (grain plus straw) was almost identical for the two varieties, and rose to a plateau of 12 t/ha with 147 kg/ha N. The grain yields, however, were very different, reaching a plateau of 4.9 t/ha with 98 kg/ha N for Squarehead Masters, while Brimstone reached a plateau of 8.3 t/ha with 147 t/ha N. The harvest index had increased from 0.43 to 0.70 (based on the straw collected by a plot combine, and not on total biomass). There is no evidence from these results to show that the total biomass production increased with the newer variety, nor that biomass production of the new variety was more responsive to higher N levels.

The yield potential of rice more than doubled to 9 to 10 t/ha with the introduction of dwarf varieties in 1966. This was caused by an increase of about 50% in harvest index, coupled with a 50% increase in biomass.

Stress tolerance

National yields are often determined more by the ability of a cultivar to withstand biotic or abiotic stresses than by the potential that can be achieved in ideal growing conditions.

For maize, increased yields have come almost entirely from increased biomass production, with harvest index remaining static. Tollenaar and Wu (1999) show that yield increases have been associated with increased tolerance to stress including: high plant population densities; drought; low night temperatures and low soil N. Duvick and Cassman (1999) give strong evidence that new maize hybrids respond to high plant population densities, whereas older hybrids did not.

Lee, Staebler and Tollenaar (2002) studied the effects of low temperature (day/night temperatures of 15/3°C cf 25/15°C) on the development of 49 maize inbred lines. They found significant interactions between the cold treatment and genotype, but found that the genotypes that gave the most rapid development at cold temperatures were not the same as those that had the highest rate of carbon dioxide uptake. It follows, therefore, that maximum yields will come from varieties tuned to a specific growing environment, and that the varietal characteristics may need to change throughout the growing season.

For many crops, cultivars vary in their ability to tolerate drought, and for wheat the variation is caused mainly by the ability of some cultivars to maintain harvest index (Foulkes, Scott and Sylvester-Bradley, 2002). The maintenance of harvest index was associated with high levels of stem-soluble carbohydrate reserves.

Hybrids

The highest yielding maize varieties are all hybrids, and heterosis is commonly considered to give a yield increase of 15%. Hybrids have also been prominent among the higher yielding oilseed rape varieties over the last five to 10 years in the UK (Kightley, 1999). But hybrid seed is more expensive to produce, and the earlier hybrids, which were male-sterile, suffered from erratic pollination. There has also been a tendency for the yield of conventional varieties to catch up with hybrids a few years later, and the outcome has been that the area of hybrid oilseed rape varieties has never exceeded 25%. The story for small-grain cereals has been even less encouraging, with hybrid wheat varieties never giving more than a token yield increase, and being sold for niche markets. There is a hybrid six-row winter barley variety, Colossus, on the market in the UK, giving a yield advantage of about 5% over normal in-bred six-row barley varieties, but it is very susceptible to brown rust, and only recommended by the HGCA for use in the wetter north and west regions.

Likely future developments

Over the last 50 years yield increases have arisen from a few major steps forward, plus a wide range of other physiological changes that have been individually quite small but collectively very significant. The most notable step changes have been the introduction of semi-dwarf genes, and dramatic changes in harvest index. Other changes include more erect leaf angles, longer duration of grain filling periods, and greater tolerance to a range of stresses. Most practical plant breeders in the UK believe that continuous incremental improvements will continue for many years yet, as varieties become better adapted to the particular environment in which they are grown. They also believe that this process will become more efficient as we use the tools described below, and gain a greater understanding of the traits that are needed.

But there is also a view among some that major progress will be achieved only by defining the ideal plant structure, or ideotype, and working towards this systematically. This approach has been adopted by Peng *et al.* (1999) who define a new plant type (NPT) for rice and list the characteristics that are essential for higher yields. These are:

i. enhanced leaf growth combined with reduced tillering during early vegetative growth;
ii. reduced leaf growth and greater foliar N concentration during late vegetative and reproductive growth;
iii. a steeper slope of the vertical N concentration gradient in the leaf canopy with a greater proportion of total leaf N in the upper leaves;
iv. increased carbohydrate storage capacity in stems;
v. a greater reproductive sink capacity and an extended grain filling period.

NPT cultivars have been derived from tropical japonica germplasm, and it is expected that hybrids between them and indica cultivars would give a further increase of 10% to 12% in potential yield.

Varieties might also be selected for water use efficiency (WUE). Richards, Rebetzke, Condon and van Herwaarden (2002) propose that greater WUE can be achieved by adopting varieties suitable for earlier sowing, thus reducing soil water evaporation and promoting growth during cooler time of year when rates of transpiration are lower. Another approach is to decrease stomatal conductance, thereby decreasing transpiration relatively more than photosynthesis. Condon *et al.* (2002) showed that while this generally gives higher yields where water supply is limiting, it gives lower yields where it is not. There is an exception to this in peanut where increased WUE has come from increased photosynthetic capacity associated with faster leaf growth rates.

Technologies for the future

Molecular biology has and will provide a large array of new technologies to help the plant breeder, and the role of these technologies has been summarised elsewhere (eg Miflin, 2000; Newbury, 2003; Stuber, Polacco and Senior, 1999; Mayes, Holdsworth, Pellegrineschi and Reynolds, this conference). The most direct way of using molecular biology is through the introduction of genetically modified crop varieties, and despite the reluctance of the European population to accept these technologies there can be little doubt that they will be used elsewhere, and will eventually be accepted in Europe. Dunwell (2000) lists 52 plant species for which field trials have been conducted in the USA, and also lists 21 applications for seven crop species where the intention is to increase yield.

Genetic modification would seem to offer the most likely means by which a major step change in crop yield will be obtained in the medium future. Possibilities that are being explored include the transfer of genes involved in C_4 photosynthesis into C_3 plants, and the modification of key photosynthetic enzymes such as Rubisco; but neither is likely to produce practical benefits in the short term. There are also attempts to transfer the nitrogen fixing capability from legumes to non-legumes: a change that might not increase yield potential, but would increase actual yields in many parts of the world.

In the nearer future, the use of marker assisted selection (MAS) will increase (Koebner, 2003). For maize this technology is already being used by multinational companies, and is claimed to produce a saving, compared to conventional phenotypic selection, of one to two years in the time required to backcross a specific trait into an elite inbred line. For wheat and barley the financial returns are less, and it is Koebner's view that MAS will, for the time being, be confined to the "accelerated selection of a few traits that are difficult to manage using conventional phenotypic selection, for the maintenance of recessive alleles in backcrossing programmes, for pyramiding disease resistance genes and for guiding the choice of parents to be used in crossing programmes".

For the longer term future there is evidence to show that there are many valuable genes to be found in land races and wild relatives of modern crops (Zamir, 2001). As screening techniques improve it is likely that this material will be scanned for useful genes, and that these will be incorporated into modern varieties with or without GM technology. The difficult part is identifying the genes, or groups of genes that contribute to quantitative traits such as yield or bread making quality

FERTILISERS

The average rate of N application to winter wheat in the UK has increased from 20 kg/ha in 1950 to 180 kg/ha in 2000 (Figure 3.4). It can be seen from Figure 3.3 that for Squarehead Masters this would have increased total biomass by a factor of 2.5, but increased its grain yield by a factor of only 1.7. For Brimstone, a more modern variety, the same increase in total biomass was accompanied by an increase in grain yield by a factor of 2.2. Foulkes *et al.* (1998) show that modern wheat varieties are less effective at extracting soil N in the absence of fertiliser N, but that they acquire and use fertiliser nitrogen more efficiently than older varieties. They also show that the economic optimum N level has increased from 150 to 200 kg/ha from 1970 to 1990.

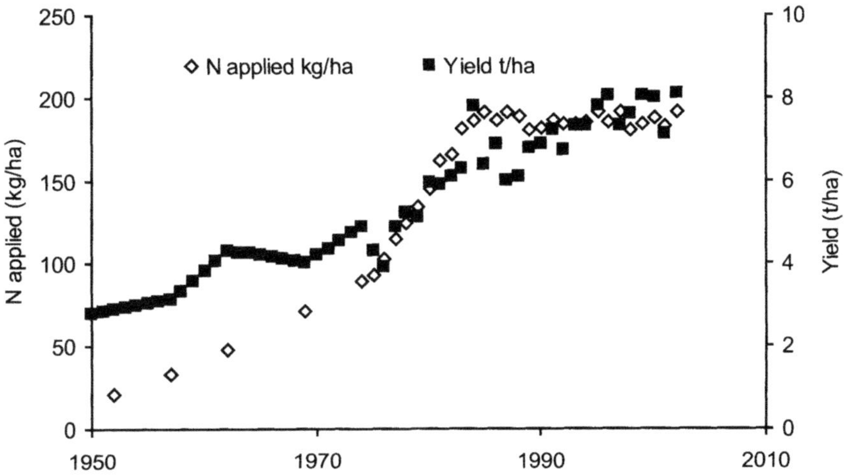

Figure 3.4 UK national N application rates to winter wheat, and average grain yields. Provided by Chris Dawson and Associates, using data from the British Survey of Fertiliser Practice and Defra Statistics.

The total amount of nitrogen applied to wheat in the UK has not increased since the mid 1980s, though crop yields continued to increase until the mid 1990s. Over that period the rates of application of phosphate declined slightly, and the rates of potash were constant (C Dawson, personal communication), so the continuing increase in yields were caused by genetic improvement, and perhaps also by improved management of all inputs.

Since the mid 1990s mean wheat yields in the UK have fluctuated with a maximum of 8.1 t/ha, but without any discernible increase. The reason for the levelling off of yield is not clear, as new varieties have continued to come onto the market with higher yield potential. But over this period application rates of P and K have declined, possibly to below optimum rates. It is also possible that sulphur levels are causing yield reductions, and C Dawson, (personal communication), using results from the British Survey of Fertiliser Practice, shows that only about a quarter of the UK arable area receives sulphur fertiliser, compared to an area of probably a half to two thirds that needs it.

The position on a world scale is shown by an FAO report (Bruinsma, 2003), which gives the average levels of total mineral fertiliser applied in different regions of the world. The table is reproduced as Table 3.1, and shows amounts in 1997/99 that range from 5 kg/ha in Sub-Saharan Africa, to 194 kg/ha in East Asia.

Table 3.1. PAST AND PREDICTED FERTILISER CONSUMPTION TAKEN FROM THE FOOD AND AGRICULTURE ORGANISATION OF THE UNITED NATIONS (BRUINSMA, 2003)

	1961-63	*1979-81*	*1997-99*	*2015*	*2030*	*1961-99*	*1989-1999*	*1997/99-2030*
Per hectare	*kg/ha (arable land)*					*% per annum*		
Sub-Saharan Africa	1	7	5	7	9	4.5	-2.4	1.9
Latin America and the Caribbean	11	50	56	59	67	6.0	0.0	0.6
Near East/ North Africa	6	38	71	84	99	5.7	3.9	1.0
South Asia	6	36	103	115	134	9.5	4.5	0.8
excl India	6	48	113	142	178	8.8	4.3	1.4
East Asia	10	100	194	244	266	8.3	3.6	1.0
excl China	12	50	96	131	92	6.1	3.3	-0.1
All above	6	49	89	102	111	7.7	3.3	0.7
excl China	6	35	60	68	71	6.9	3.2	0.5
excl China & India	7	35	49	58	58	6.0	2.6	0.5
Industrial countries	64	124	117			1.3	0.3	
Transition countries	19	101	29			0.9	-14.4	
World	25	80	92			3.3	0.1	

Note: kg/ha for 1997-99 are for developing countries calculated on the basis of "adjusted" arable land data. For industrial and transition countries no projections of arable land were made.

The use of fertilisers has increased dramatically over the last 50 years, especially in South and East Asia where the average rates are comparable to or higher than those in industrialised countries. None-the-less, there is scope for a further 50% increase and that would give a substantial increase in yield. The situation in Sub-Saharan Africa is particularly worrying: although lower levels of fertiliser use are appropriate for rain-fed agriculture in arid climates, the limitation is often poverty.

PESTICIDES

The impacts of herbicides and insecticides are considered by other authors in this volume (e.g. Clough, and Johnson and Hope). Just fungicides will be considered here.

Although the impact of fungicides was apparent on Broadbalk in 1979 when fungicides were first used (Figure 3.1), the scale of the effect is better shown in the NIAB trials (managed by the HGCA since 2002) where the top yielding varieties have been grown in trials all over the UK with and without fungicides (Figure 3.5).

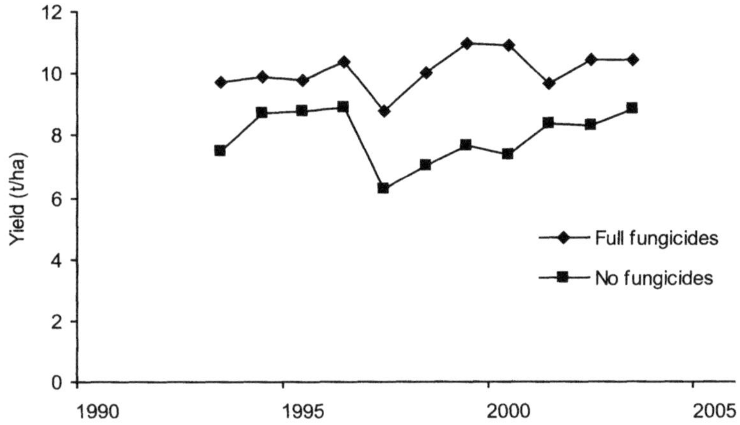

Figure 3.5 Effect of full fungicide treatment on winter wheat in the UK Recommended List trials. *(Some data were taken from trials funded by the British Society of Plant Breeders and the Home-Grown Cereals Authority. Full results can be found at www.hgca.com.)*

These trials have used fungicides at generous rates; as far as possible, all diseases are kept to below 5% so that treated yields show the potential for a healthy crop. On average for the last 11 years the untreated plots have yielded 79% of the treated yields, with loss of yield on untreated varieties varying from 17% to 35%, indicating the genetic variation in overall disease resistance.

Where wheat is grown continuously, yields are generally depressed by pathogens that are not controlled by fungicides. On Broadbalk, from 1979 to 2002, the yield of continuous wheat was only 0.73 of that after a fallow break. (After fallow there will have been more N left in the soil, but this would have been an insignificant effect on plots receiving high levels of N fertiliser.) In NIAB trials the average yield of second wheats has been 86% of first wheats, the depression being mainly caused by take-all. For wheat, therefore, and probably for other crops, there are substantial gains to be made from the judicious use of crop rotations, or perhaps in the future by better control of soil pathogens.

IRRIGATION

Irrigation has been practised for thousands of years, and the scientific principles have been understood for the last 50 years. Basically, maximum crop growth takes place only when water is not restricted. Water needs can be determined either by measuring the water available in the soil, or by using meteorological data to estimate evapotranspiration. As with fertiliser use, there has been a dramatic increase in use of irrigation, with a doubling of the world area irrigated over the last 40 years from 140 to 270 Mha (Bruinsma, 2003). For developing countries the increase has been from 100 to 200 Mha representing an increase from approximately 10% to 20% of the total arable area.

Over this period the technology has improved by evolution rather than revolution, with flood irrigation, overhead sprinklers and drip irrigation all playing their part. FAO (Bruinsma, 2003) estimates that the efficiency with which irrigation water is used (the ratio of consumptive water use to irrigation water extracted) varies from 25% in regions such as Latin America where water is plentiful, to over 40% in the Near East and South Asia where water is scarce and is more highly valued.

For the future, FAO forecasts that in developing countries there will be an additional 40 Mha (20% increase) of land brought under irrigation in the next 30 years, but that the total area harvested will increase by double this as cropping intensity also increases. There is also scope to improve the efficiency of water use through more accurate scheduling, and through better equipment.

The demand for water is likely to be affected by climate change, and K Weatherhead (personal communication) predicts that in 50 years' time the amount of irrigation needed for arable crops in the East Anglian region of the UK will have increased by 30%. At the same time the amount of water available for summer abstraction is likely to fall. Similar effects across the world may have a dramatic effect on irrigation need, and on crop yields where extra water is not available.

ENGINEERING

One could argue that engineering has done nothing to increase crop yields. Nothing is achieved with a tractor, irrigation system or combine harvester that cannot in principle be done with hand tools. But engineering has, of course, had a huge impact on food production by reducing the manpower needed for every operation, and UK arable agriculture now needs only one man for every four or five hundred hectares. The UK is 77% self-sufficient in indigenous food with only 0.9% of the population involved in agricultural production.

Farm mechanisation started several hundred years ago. For example, the first patent on a combine harvester and thresher was taken out in the USA in 1828. Mechanisation has revolutionised agriculture in industrial countries over the last

50 years (reviewed by Finney, 2000). It was observed in 1942 that the cost of human power was 160 times more expensive than electricity (itself an expensive form of power) and that mechanisation was inevitable. The aims of mechanisation, set out in 1944, were to decrease the number of men employed, increase their output, reduce physical energy and drudgery, to improve timeliness of operations, and increase the quality and market value of the produce. This has all been delivered, largely through an inexorable increase in the size and power of farm machines; tractor power in the UK has risen from 15 kW fifty years ago to an average of 85 kW in 2000, and it continues to rise. The outputs of all other farm equipment, including cultivation systems, spraying equipment, harvesters, dryers and stores have risen at similar rates. A world record for ground-based crop spraying was recently claimed in France at 83 ha/hour. Improved quality has come through the increased uniformity of cultivation and fertiliser applications, coupled with more timely application of herbicides and pesticides.

More recently electronics has made an impact on farm equipment;the most advanced machines have GPS guidance and full microprocessor control. Spatially variable field operations (or 'precision agriculture') are now a practical reality (reviewed by Koch and Khosla, 2003), with information being gathered from soil samples, yield meters, or aerial photographs, and subsequent operations being managed variably according to need on a scale of a few tens of metres. Godwin, Richards, Wood, Welsh and Knight (2003a) and Godwin, Wood, Taylor, Knight and Welsh (2003b) based nitrogen applications on previous in-field yields of wheat (two sites) and barley (one site) for three years. They compared a strategy in which the previous high yielding areas received more nitrogen than the low yielding areas, with the opposite strategy, and with uniform application. The results were inconsistent, and they concluded that varying nitrogen according to historical yields was unlikely to give economic benefits. In a more sophisticated analysis Lark (2001) showed that, in a field growing three successive winter barley crops, some areas gave a higher than average yield in one year, but lower than average yield in another. The variation in yield pattern between years was attributed to different soil types, which gave yields that were dependent on rainfall. The areas of field that responded similarly from year to year could be identified using a fuzzy classification of all the data. Lark recognised that this classification is not sufficient on its own to provide the basis for spatially variable field management. It could, however, provide a rational basis for soil sampling, and the identification of areas that should be managed differently.

Godwin *et al.* (2003b) also varied the nitrogen application rate to wheat according to crop density, measured by airborne digital photography immediately before nitrogen application. They found an economic benefit from applying more nitrogen to regions of low crop density and less to regions of high crop density. A knowledge of weed distribution in the field can also inform the application of herbicides, and Stafford and Miller (1996) reported reductions in herbicide use ranging from 7% to 70%.

Although most research and commercial development of precision agriculture has focused on combinable crops, the potential is as great for other crops. For potatoes it is possible to map the distribution of nematodes and reduce pesticide application, and Koch and Khosla (2003) report several trials in which an increase in yield and an improvement in quality were obtained by varying potato seed spacing according to local conditions. Pesticide application to fruit orchards can also be controlled by sensing the foliage density along the row, and Walklate, Cross, Richardson, Baker and Murray (2003) showed that for some orchards this would give a five-fold reduction in chemical applied for the same leaf dose.

Looking further ahead, individual plant care is a realistic possibility. Conventional GPS can already provide location with an absolute accuracy of less than a metre, and an accuracy of one or two centimetres is available, though currently at a high price. Vision systems are also developing rapidly; a system to "see" the rows of sugar beet and guide a steerable hoe is available commercially in the UK. It is conceivable that within the next few decades it will become possible to monitor every plant in the field and treat it according to its individual needs.

At the other end of the spectrum, however, it is salutary to note that there are still large areas of the world where the most appropriate technological development would be to replace human labour with animal power.

Conclusions

The prospects for increasing world food production have been assessed by the FAO (Bruinsma 2003), comparing actual average wheat yields with the agro-ecologically attainable yield under rain-fed conditions in 25 wheat producing countries (Figure 3.6).

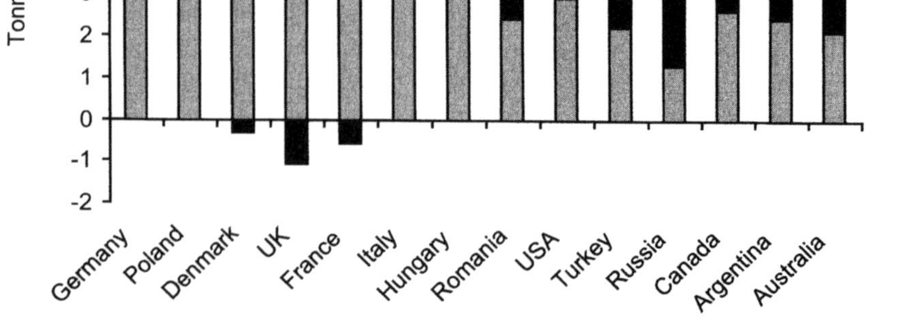

Figure 3.6 Wheat: actual and agro-ecologically attainable yields (rain-fed, high input). (Courtesy of the Food and Agriculture Organization of the United Nations)

They concluded that in all but a few European countries the actual yields were half or less than those that could be attained with known technology and without irrigation. Even in India, where wheat and rice yields appear to be reaching a plateau, the ratio of actual to attainable yield is only 63% for wheat and 53% for paddy rice. It seems reasonable, therefore, to conclude that national and world average crop yields will continue to rise for the next ten or twenty years.

The evidence to support a continuing increase in yield potential, however, is rather weaker, and results from several crops in several countries suggest that the maximum yields attainable in the field have increased little in the last 20 to 30 years. It follows that national yields will eventually reach a plateau unless there is a major breakthrough in plant breeding.

This breakthrough is unlikely to come from conventional crossing and phenotypic selection. It may come by identifying the genes that control major biochemical processes in plants, or possibly by improved understanding of molecular genetics and models that relate the function of many genes to overall crop performance.

Acknowledgements

Results from the Broadbalk Wheat Experiment were provided by Rothamsted Research Limited.

References

Austin, R.B. (1999) Yield of wheat in the United Kingdom. Recent advances and prospects. *Crop Science,* **39**, 1604-1610.

Bruinsma, J. (Ed) (2003) World agriculture: towards 2015-2030 An FAO Perspective *FAO* and *Earthscan Publications Ltd.*

Calderini, D.F., Dreccer, M.F., and Slafer, G.A. (1997) Consequences of breeding on biomass, radiation interception and radiation-use efficiency in wheat. *Field Crops Research,* **52** 271-281. Elsevier Science BV.

Condon, A.G., Richards, R.A., Rebetzke, G.J. and Farquhar, G.D. (2002) Improving intrinsic water-use efficiency and crop yield. *Crop Science,* **42**, 122-131.

Dunwell, J.M. (2000) Transgenic approaches to crop improvement. *Journal of Experimental Botany,* **51,** GMP special issue 487-49.

Duvick, D.N. and Cassman, K.G. (1999) Post-green revolution trends in yield potential of temperate maize in the north-central United States. *Crop Science,* **39,** 1622-1630.

Evans, L.T. and Fischer, R.A. (1999) Yield potential: its definition, measurement and significance. *Crop Science*, **39**, 1544-1551.

Finney, B. (2000) Farm mechanisation in the second half of the 20th century. *RASE Millennium Journal,* 100-109.

Fischer, R.A., Rees, D., Sayre, K.D., Lu, Z.-M., Condon, A.G. and Larque Saavedra, A. (1998) Wheat yield progress associated with higher stomatal conductance and photosynthetic rate, and cooler canopies. *Crop Science,* **38,** 1467-1475.

Foulkes, M.J., Scott, R.K., Sylvester-Bradley, R. (2002) The ability of wheat cultivars to withstand drought in UK conditions: formation of grain yield. *Journal of Agricultural Science*, **138**, 153-169 Part 2.

Foulkes, M.J., Sylvester-Bradley, R. and Scott, R.K. (1998) Evidence for differences between winter wheat cultivars in acquisition of soil mineral nitrogen and uptake and utilization of applied fertilizer nitrogen. *Journal of Agricultural Science,* **130**, 29-44.

Godwin, R.J., Richards, T.E., Wood, G.A., Welsh, J P. and Knight, S.M. (2003a) An economic analysis of the potential for precision farming in UK cereal production. *Biosystems Engineering,* **84,** Issue 4, April 2003, 533-545.

Godwin, R.J., Wood. G.A., Taylor, J.C., Knight, S.M., Welsh J.P. (2003b) Precision farming of cereal crops: a review of a six year experiment to develop management guidelines. *Biosystems Engineering* **84,** Issue 4, 375-391

Horton, P. (2000) Prospects for crop improvement through the genetic manipulation of photosynthesis: morphological and biochemical aspects of light capture. *Journal of Experimental Botany,* **51,** GMP special issue, 475-485.

Kightley, S.P J. (1999) The impact of new technology on winter oilseed rape variety development. *Aspects of Applied Biology,* **55,** Protection and production of combinable break crops.

Kock, B. and Khosla, R. (2003) The role of precision agriculture in cropping systems. *The Haworth Press Inc,* 361-381.

Koebner, R. (2003) MAS in cereals: green for maize, amber for rice, still red for wheat and barley. Marker assisted selection: a fast track to increase genetic gain in plant and animal breeding? *Session 1: MAS in plants, 12-17 Conference organised by Fondazione per la Biotecnologie, Turin, Italy.*

Lark, R.M. (2001) Some tools for parsimonious modelling and interpretation of within-field variation of soil and crop systems. *Soil & Tillage Research,* **58**, 99-111.

Lee, E.A., Staebler, M.A. and Tollenaar, M. (2002) Crop Physiology and Metabolism. Genetic variation in physiological discriminators for cold tolerance - early autotrophic phase of maize development. *Crop Science,* **42**, 1919-1929.

Lu, Z-m., Radin, J.W., Turcotte, E.L., Percy, R. and Zeiger, E. (1994) High yields in advanced lines of Pima cotton are associated with higher stomatal conductance, reduced leaf area and lower leaf temperature. *Physiologia*

Plantarum, **92**, 266-272.
McCree, K.J. (1971) The action spectrum, absorptance and quantum yield of photosynthesis in crop plants. *Agricultural Meteorology,* **9**, 191-216.
Miflin, B. (2000) Crop improvement in the 21st century. *Journal of Experimental Botany,* **51**, No 342 GMP Special Issue 1-8.
Miralles, D.J. and Slafer, G.A. (1997) Radiation interception and radiation use efficiency of near-isogenic wheat lines with different height. *Euphytica,* **97**, 201-208.
Murchie, E.H., Chen, Y.-Z., Hubbart, S., Peng, S. and Horton, P. (1999) Interactions between senescence and leaf orientation determine in situ patterns of phytosynthesis and photoinhibition in field-grown rice. *Plant Physiology,* **119**, 553-564.
Newbury, H.J. (Ed.) (2003) *Plant Molecular Biology*, 265pp, Blackwell Publishing
Peng, S., Cassman, K.G., Virmani, S.S., Sheehy, J. and Khush, G.S. (1999) Yield potential trends of tropical rice since the release of IR8 and the challenge of increasing rice yield potential. *Crop Science,* **39**, 1552-1559.
Reilly, J.M. and Fuglie, K.O. (1998) Future yield growth in field crops: what evidence exists? *Soil & Tillage Research,* **47**, 275-290.
Reynolds, M.P., Delgado, M.I.B., Gutiérrez-Rodríguez, M. and Larqué-Saavedra, A. (2000) Photosynthesis of wheat in a warm, irrigated environment. I: Genetic diversity and crop productivity. *Field Crops Research,* **66,** Issue 1, 37-50.
Richards, R. A. (2000) Selectable traits to increase crop photosynthesis and yield of grain crops. *Journal of Experimental Botany,* **51,** GMP Special Issue, 447-458.
Richards, R.A., Rebetzke, G.J., Condon, A.G. and van Herwaarden, A.F. (2002) Crop physiology and metabolism. Breeding opportunities for increasing the efficiency of water use and crop yield in temperate cereals *Crop Science,* **42,** 111-121.
Silvey, V. (1994) Plant breeding in improving crop yield and quality in recent decades Agribex Symposium *Acta Horticulturae,* No 355, 19-24.
Stafford, J.V. and Miller, P.C.H. (1996) Spatially variable treatment of weed patches. *Precision Agriculture,* Proceedings of the Third International Conference, Minneapolis MN. Ed Robert, P.C., Rust, R.H. and Larson, W.E. American Society of Agronomy, Crop Science Society of America and Soil Science Society of America, 465-474.
Specht, J.E., Hume, D.J. and Kumudini, S.V . (1999) Soybean yield potential - a genetic and physiological perspective *Crop Science,* **39**, 1560-1570.
Stuber, C.W., Polacco, M., and Senior, M.L. (1999) Synergy of empirical breeding, marker-assisted selection and genomics to increase crop yield potential *Crop Science,* **39,** 1571-1583.
Tollenaar, M. and Wu, J. (1999) Yield improvement in temperate maize is

attributable to greater stress tolerance 1998 ASA meeting, Baltimore. *Crop Science,* **39,** 1597-1604.

Walklate, P.J., Cross, J.V., Richardson, G.M., Baker, D.E. and Murray, R.A. (2003) A generic method of pesticide dose expression: application to broadcast spraying of apple trees. *Annals of Applied Biology,* **143**, 11-23.

Zamir, D. (2001) Improving plant breeding with exotic genetic libraries. *Nature Reviews/genetics,* **2**, 983-989.

4

REPRODUCTIVE RATE IN FARM ANIMALS: STRATEGIES TO OVERCOME BIOLOGICAL CONSTRAINTS THROUGH THE USE OF ADVANCED REPRODUCTIVE TECHNOLOGIES

KEVIN D SINCLAIR AND ROBERT WEBB
Division of Agricultural and Environmental Sciences, School of Biosciences, University of Nottingham, Sutton Bonington Campus, Loughborough, Leics., LE12 5RD, UK

Introduction

Low reproductive rate represents a major biological constraint to the efficient and environmentally sustainable production of many farm animal species. It results in, as a proportion, between 0.50 to 0.70 of the total food energy requirement of ruminants to be partitioned towards the maintenance of the parent population (Webster, 1989). This range is in stark contrast to that for modern systems of broiler production, where the estimated proportion of food energy required to maintain the parent population is less than 0.04. Low reproductive rate also hinders production efficiency in large animals by impeding the rate of genetic progress that can be achieved for commercially important traits using conventional methods of selection. This inefficiency is further compounded by the comparatively long generation intervals of these species.

There have been numerous attempts over the years to increase the reproductive rate of farm animals, but with limited success. Attempts have been made to identify and select individuals for increased ovulation rate and litter size. In recent years genetic selection for increased fecundity has benefited greatly from developments in molecular genetics and from improved understanding of the underlying physiological mechanisms that control ovulation rate and uterine receptivity. Selection for increased fecundity results in permanent improvements in reproductive rate but, in some cases, these are not entirely sustainable. Often, all that is required to improve livestock production efficiency is a transitory increase in reproductive rate that leads either to a modest increase in litter size or a significant increase in the number of oocytes or embryos collected for subsequent storage and/or transfer (Telfer *et al.*, 1999). The aim of the present chapter is to consider these issues and examine the potential of recent developments in the reproductive technologies to increase the biological efficiency of farm animals through increases in reproductive rate, and by facilitating the use of advanced methods of genetic selection and modification.

Strategies to improve reproductive rate

GENETIC SELECTION

Cattle

The natural incidence of twin births in cattle is low, typically ranging from 1 to 5% (Guerra-Martinez *et al.*, 1990; Johanson *et al.*, 2001), although specific genotypes and strains within genotypes exist that have much higher (> 10%) natural twin-calving rates (Morris, 1984; Morris and Wheeler, 2002). Environmental factors such as season and nutritional status have only a minor effect on twin-calving rate, whereas twining rates in cattle are known to increase with parity (e.g. from 1 to 9% over the first 5 parities; Ryan and Boland, 1991). Similarly, estimates of the heritability (h^2) of twinning rate in cattle are low at around 0.03 (Morris, 1984) and so the rate of response to selection for twinning is low. However, the estimated h^2 for twinning is greater for cow-families with a high natural twin calving rate (Morris and Wheeler, 2002), suggesting that the practicality of selecting for twinning rate will be greatly enhanced if foundation dams already have an above average rate of twinning. Indeed, although the selection programme for twinning in cattle at the US Meat Animal Research Centre in Nebraska achieved an average annual increase in twinning rate of just 2.2% from 1981 to 1993, the rate of increase for this trait tended to improve with time (Gregory *et al.*, 1997). By 1993, twinning rates at this centre had increased from 3.4% to 33.0%.

Estimates for the h^2 of ovulation rate are comparable to that for twinning rates. Although difficult to record under commercial conditions, the relatively high h^2 of 0.35 for ovulation rates from repeated observations in pubertal heifers, together with the high positive genetic correlation of 0.75 between ovulation rate and twinning rate (Gregory *et al.*, 1997), render this trait a powerful selection criterion to increase reproductive rate in this species. Selection for enhanced ovulation rates with the use of multi-trait threshold models is predicted to lead to more rapid genetic improvements in twinning rates through better identification and selection of genetically superior animals (Van Tassell *et al.*, 1998). furthermore, several studies have now reported putative quantitative trait loci (QTL) for ovulation rate and/or twin-calving rate in cattle on chromosomes 5, 7, 10, 12, 18 and 23 (e.g. Lien *et al.*, 2000; Arias and Kirkpatrick, 2004; Ashwell *et al.*, 2004). As prolificacy is a sex-limited trait expressed relatively late in life, marker assisted selection (MAS) may prove particularly beneficial when selecting for this trait in the future. It is likely, therefore, that significant advances in the ability to select cattle for twining rate will be achieved within the next few years.

Sheep

Considerable variation in ovulation rate exists both between and within breeds of sheep. This is largely attributable to polygenic effects where estimates of the h^2 of this trait range from around 0.05 for less prolific breeds to more than 0.5 in highly prolific breeds (Davis *et al.*, 1998; Webb *et al.*, 1998; Hanrahan, 2003). However, following the discovery that the exceptional prolificacy of Booroola Merinos was attributable to the segregation of a gene with a large effect on litter size (Davis *et al.*, 1982), research focus in recent years has largely been directed towards studies of single gene effects. To date, these studies have identified the inheritance patterns of several naturally occurring genetic mutations that, depending on their pattern of inheritance, can result either in modest increases in ovulation rate or lead to sterility (reviewed by Montgomery *et al.*, 2001; McNatty *et al.*, 2003; Table 4.1). The most recent report by Hanrahan *et al.* (2004) identified mutations in Growth and Diferentiation Factor 9 *(GDF9)* and further mutations in Bovine Morphogenic Protein 15 *(BMP15)* within the Belclare and Cambridge breeds of sheep that are associated with increased ovulation rates. Both of these genes are members of the transforming growth factor beta superfamily, and code for oocyte-specific proteins within the ovarian follicle; it was also shown that animals heterozygous for both a *BMP15* and the *GDF9* mutation have a greater ovulation rate than sheep with either of the mutations alone (ovulation rate of 6.1 vs either 2.5 or 3.3 for *BMP15* and *GDF9* respectively), indicating an additive beneficial effect of these two single mutations. In the immediate future the quest to discover other major single gene effects is likely to continue, although in the longer term the focus is likely to revert again to the understanding of polygenic effects of genes associated with embryo survival.

Pigs

Understanding these polygenic effects of genes associated with embryo survival is certainly the focus of current research work with pigs where, in contrast to mono-ovular species such as cattle and sheep, future selection for increased fecundity in this poly-ovular species will require greater consideration of factors in addition to ovulation rate; factors such as embryo survival and uterine capacity. Mathematically simulated models of litter size in both pigs (Bennett and Leymaster, 1989) and mice (Ribeiro *et al.*, 1997a and b) predict that selection based on an index incorporating ovulation rate, with either uterine capacity or embryo survival, will lead to significantly enhanced responses to selection for increased litter size. These theoretical predictions are supported by empirical data highlighting the biological superiority of the highly prolific Chinese Meishan breed, which can farrow three to five more viable piglets per litter than the European Large White breed. Although the major genes controlling ovulation rate in sheep (i.e. *BMPR1B* and *BMP15*) have been mapped in the

Table 4.1 MUTATIONS IN GENES FOR OOCYTE-DERIVED GROWTH FACTORS ASSOCIATED WITH BOTH INCREASED OVULATION RATE AND STERILITY IN SHEEP

Sheep family (breed)	*Gene*	*Chromosome*	*Gene product*	*Base change*	*Amino acid change*	*Phenotypic ovulation rate*			*Reference*
						+ +	+ *m*	*m m*	
Booroola (Merino)	$FecB^B$	6	BMPR 1B	A - G	Arg - Gln	1.5	2.8	4.6	Wilson *et al.*, 2001
Garole	$FecB^B$	6	BMPR 1B	A - G	Arg - Gln	N/A	N/A	N/A	Davis *et al.*, 2002
Javanese	$FecB^B$	6	BMPR 1B	A - G	Arg - Gln	1.4	2.7	N/A	"
Inverdale (Romney)	$FecX^I$	X	BMP 15	T – A	Asp – Val	1.8	2.9	0	Galloway *et al.*, 2000
Hanna (Romney)	$FecX^H$	X	BMP 15	C –T	Glu – STOP	1.8	2.9	0	"
Belclare	$FecG^H$	5	GDF 9	C –T	Ser – Phe	1.9	2.7	0	Hanrahan *et al.*, 2004
Belclare	$FecX^G$	X	BMP 15	C –T	Glu – STOP	1.9	2.7	0	"
Belclare	$FecX^B$	X	BMP 15	C –T	Ser - Ile	1.9	3.3	0	"
Woodlands (Coopworth)	$FecX2^W$	X	Unidentified maternally Imprinted gene				+0.39†		Davis *et al.*, 2001

+ = wild type allele
m = mutated allele
† = +0.39 refers to an increase in ovulation rate relative to contemporaries

pig (Grapes and Rothschild, 2002; King *et al.*, 2003) their role in influencing litter size through enhanced ovulation rate has not been demonstrated. Indeed the role of a number of other candidate genes including Gonadoptrophin Relseasing Hormone (*GnRH)* (Linville *et al.*, 2001) and FSHß1 (King *et al.*, 2003) that are generally credited with the regulation of ovulation rate across the species has yet to be established. Instead, attention has focused on the role of genes that promote conceptus development (e.g. epidermal growth factor and retinol binding protein), uterine receptivity and luteolysis (e.g. prolactin and cyclooxygenase 2), with the function one gene, the oestrogen receptor (*ER*) (Rothschild *et al.*, 1996), showing the greatest potential as the major single gene influencing litter size in pigs. To date, however, functional studies into the role of the ER in promoting conceptus development and survival have failed to identify clearly its mode of action (e.g. van Rens *et al.*, 2002). The quest to identify QTLs for fecundity in pigs continues, including the identification of other novel candidates for enhanced uterine receptivity (King *et al.*, 2003).

IMMUNOLOGICAL AND HORMONAL INTERVENTIONS

Direct immunological or hormonal approaches to increasing ovulation rate offer several advantages over genetic selection, the most important being the immediate responses that can be induced in a wide range of species and genotypes within species, so obviating the inevitable time lag associated with selection. However, the complexities of ovarian physiology, where the processes of follicular recruitment, growth, selection and ovulation involve both extragonadal (e.g. gonadotrophins) and intragonadal (e.g. steroids and peptides) regulators, inevitably result in highly variable ovulatory responses when such approaches are applied. In mono-ovular species such as the cow it has frequently been difficult to control this variability and to restrict ovulatory responses to just two or three ovulations.

Immunisation

Initial approaches to increasing reproductive rate by immunisation strategies in cattle and sheep during the 1980s focused on the suppression of androstenedione production (reviewed by Webb *et al.*, 1984; Hillard *et al.*, 1995). This lead to the production of a commercially available vaccine (Fecundin®) which, although discontinued in the early 1990s, is once again being manufactured, only by a different company under the trade name of Ovastim. However unlike for sheep, in which immunisation against this androgen results in moderate and reasonably controlled increases in both ovulation rate and lambing rate, responses in cattle are highly variable, but on the whole generally poor. Similarly in pigs, although immunisation against androstenedione can result in a variable increase in ovulation rate, this seldom ever translates into an increase in litter size. However, better success (in terms of both ovulation rate and live pigs born) has recently been achieved in this species by the immunisation of 17α-hydroxyprogesterone, a precursor steroid molecule for androstenedione and one which is produced in large quantities by theca interna cells within the ovarian follicle (Kreider *et al.*, 2001).

Given these limitations with androgens, the purification of inhibin during the late 1980s lead to a flurry of research activity, both to elucidate fully its function in controlling folliculogenesis and to evaluate its role as a means of increasing ovulation rate. Once again, however, although active immunisation strategies were modestly successful in sheep, results in cattle were highly variable. Much of this variability was thought to be due to either the particular peptide sequence, conjugation technique, carrier protein and adjuvant, or immunisation protocol (discussed by Morris *et al.*, 1997). However, recognising that the pattern of follicular dominance in cattle could itself contribute to the variable response to active immunisation, preliminary studies in this species employing an acute programme of passive immunisation against inhibin developed by Campbell *et al.* (1995) for sheep resulted in 67% heifers producing twin ovulations, with the remaining animals producing only single ovulations (Campbell *et al.*, 2004). The possibility clearly exists, therefore, to refine current strategies of immunisation against this ovarian peptide in a way that will enhance ovulation rate and litter size in a controllable manner in cattle.

In keeping with the results of studies investigating the outcome of inactivating mutations in either *GDF9* or *BMP15* (discussed earlier), short-term immunisation against either of these oocyte-derived peptides has recently been shown to increase ovulation rate in sheep with no detrimental effect on embryo development and the ability of immunised ewes to carry pregnancies to term (Juengel *et al.*, 2004). The key to success in this study was due, in part, to the adoption of a short-term immunisation strategy employing a weaker adjuvant (DEAE-dextran) than that (Freunds adjuvant) more commonly used in long-term active immunisation programmes. Consequently, the ovulatory effects observed mimicked that of ewes heterozygous for the Inverdale gene (Table 4.1).

Gonadotrophins

Since the pioneering studies of Ian Gordon and colleagues in the late 1950s, where around 500 cows in the Welsh border counties were treated with pregnant mares serum gonadotrophin (PMSG) in order to study the feasibility of inducing multiple births by pharmacological means (Gordon *et al.*, 1962), there have been few serious and credible attempts to increase the litter size of farm animals by hormonal means. These early studies highlighted the marked individual variability of response to gonadotrophin treatment, with a high incidence of multiple pregnancies and associated complications. More recent studies using follicle stimulating hormone (FSH) (e.g. Davis and Bishop, 1992) merely confirmed these earlier findings in cattle, and further highlighted complications associated with multiple pregnancies (Echternkamp, 1992). Although PMSG can increase ovulation rate in pigs, the increased numbers of embryos that arise following insemination are frequently negated by greater fetal losses within the first 40 days of gestation leading to no improvement in the number of live born (Lubritz *et al.*, 1993). Attempts to increase the reproductive rate of farm animals by hormonal means in recent times have been restricted to just a few studies in developing countries where the timing and level of gonadotrophins administered, were altered in attempt to control ovulation rate (e.g. Galmessa and Prasad, 2003). In developed countries gonadotrophins continue to be administered but in superovulation and embryo recovery programmes, as will be discussed later.

EMBRYO TRANSFER (ET)

As a means of directly increasing reproductive rate, ET offers several advantages over pharmacological approaches. Firstly, litter size can be precisely determined, with little or no risk of cows carrying more than two calves to term. Secondly, the genotype and indeed sex of the offspring produced can be precisely determined. This latter point is particularly important for cattle where the high incidence of placental anastomosis leads to sterility in approximately 90% of female calves that are co-twinned with a male (Kästli, 1978).

Since the pioneering studies of Rowson and colleagues at Cambridge during the early 1970s, a number of other groups have sought to determine the most effective means by which to induce twin pregnancies by ET in cattle. However, one very large study (working with 378 parous crossbred cows; Sinclair *et al.*, 1995a) combined several factors, including the use of *in vitro* produced embryos, into a single experiment that permitted a comprehensive assessment of the relative merits of several approaches that can determine twinning success. The main findings from that study were that 'bred recipients' (i.e. cows that had been artificially inseminated several days prior to ET) receiving a single additional embryo by ET had both the highest pregnancy and twinning rates, and that embryo location (i.e. unilateral vs bilateral) had a significant impact on both pregnancy rate and the proportion of pregnant cows that carried twins (Table 4.2).

During the establishment of pregnancy, embryos can be located either unilaterally (i.e. two embryos in one uterine horn) or bilaterally (one embryo in each of the two horns). The location of embryos does not affect the total number of calves born as a result of embryo transfer, but can influence the ultimate number of calves born within the herd as a whole and the proportion of calves that are twins (Table 4.2). For the first of the two options illustrated in Table 4.2, and under good systems of management, increases of between

20 and 38% in the number of live calves born can be expected following twinning by embryo transfer. For the second option, the predicted production of live calves increases to between 27 and 58%. Pregnancy rates as high as 77% and twinning rates as high as 54% were reported in the study of Sinclair *et al.* (1995a). In a follow up study on commercial premises pregnancy rates of 61% and twinning rates of 28% following the unilateral transfer of two *in vitro* produced embryos were achieved (Sinclair *et al.*, 1995b). These results highlight the fact that, at least in terms of the delivery of live calves, the theoretical predictions of increased output portrayed in Table 4.2 can be realised under good systems of husbandry.

Table 4.2. THEORETICAL OUTPUT OF CALVES FROM A HERD OF 100 SUCKLER COWS FOLLOWING TWINING BY EMBRYO TRANSFER (ET) (BASED ON RESULTS OF SINCLAIR *et al.*, 1995A).

	Unilateral transfer	*Bilateral transfer*
Animals synchronised	100	100
Animals transferred	85	85
Calved of transferred (%)	75	56
Twin calving of transferred (%)	24	45
ET calves	84	86
Option 1:		
Natural service calves	36	52
Option 2:		
Re-synchronisation and ET + natural service calves	43	72
Total calves: Option 1	120	138
Total calves: Option 2	127	158

Option 1 considers naturally mating all those cows that were either not selected for transfer or failed to conceive following transfer. It assumes that all cows ultimately produce calves. Option 2 considers re-synchronisation and a second round of embryo transfers with similar pregnancy outcomes to that achieved during the initial round. Any cows not transferred are naturally mated and ultimately produce calves.

Limitations and biological constraints

The projection of efficiency changes in cattle resulting from genetic selection for improved ovulation rate, or the successful adoption of pharmacological approaches to increase reproductive rate, must consider the distribution of litter size on the economic consequences of modifying this trait. This is because, although most current systems of husbandry in beef-suckler herds can be adapted to accommodate twin pregnancies (Sinclair and Broadbent, 1996), it is uncertain whether many such systems could cope with the increased demands of managing higher (≥ 3 fetuses) pregnancy rates (Echternkamp, 1992). This would most certainly apply to the great majority of dairy herds (Eddy *et al.*, 1991; Beerepoot *et al.*, 1992) where currently the emphasis on selection is for reduced twinning rate (Johanson *et al.*, 2001). It is perhaps reassuring, therefore, that initial attempts to model litter size distributions in cattle indicate that selection for twin-ovulation rates is likely to result in a very small proportion of high-order pregnancies (<5% triplets and 0% quadruplets for 50% multiple births; Bennett *et al.*, 1998; Figure 4.1).

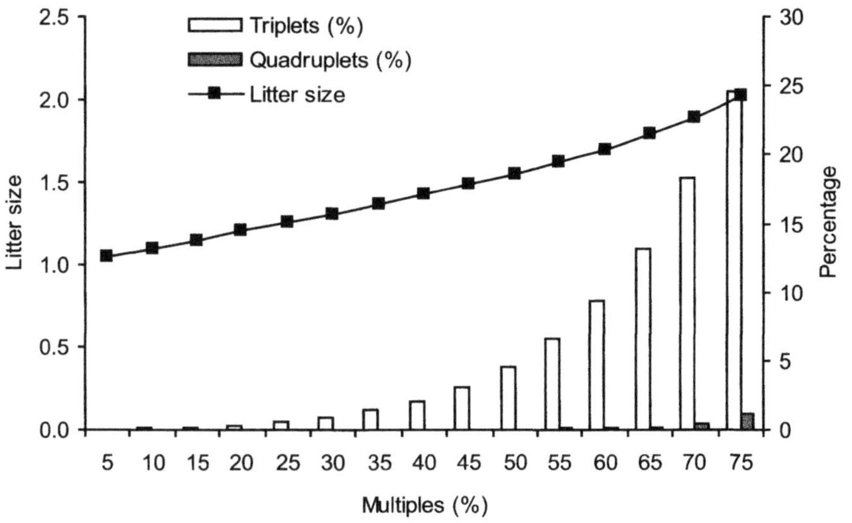

Figure 4.1 Selection for increased ovulation rate in cattle: prediction of litter size distribution from percentage multiple births (adapted from Bennett *et al.*, 1998).

Similarly in sheep, lambs born from large litters are not only lighter, but are also more variable in weight. This increase in variability is determined by events that occur early during pregnancy, as considerable within-litter variation in fetal weight is detectable as early as Day 34 of gestation (Dingwall *et al.*, 1987). Fetal loss at the time of placental attachment reduces the number of placental cotyledonary sites available for the surviving fetuses (Robinson *et al.*, 1980), so decreasing their weight (Rhind *et al.*, 1980). Hence high ovulation rates are associated with greater embryonic wastage and more variable lamb birth weights (Hinch *et al.*, 1985), with important implications for the neonatal health and viability of offspring.

Twinning by ET in cattle can overcome many of these difficulties and our understanding of how to manage twin bearing cows at parturition and during the early post-partum period improved during the 1990s to the point where this method of increasing reproductive rate became commercially viable (Sinclair and Broadbent, 1996). However, the transfer of *in vitro*-produced embryos in ruminant species is associated with an additional range of complications that includes early embryo and late fetal loss, and large size at birth with a variable incidence of congenital abnormalities; a phenomenon referred to as the Large Offspring Syndrome (LOS; Sinclair *et al.*, 2000). The sporadic occurrence of this syndrome has hindered its study and, in countries such as the UK, has also contributed to the low commercial uptake of *in vitro* embryo production and transfer. If the causes of this syndrome can be resolved there remains great potential for the use of this technology to increase the reproductive rate of cattle, not least because of its ability to precisely determine both the gender and genotype of the animals produced.

Application of reproductive technologies in genetic improvement

ARTIFICIAL INSEMINATION

Of the 4 generations of reproductive technology (artificial insemination (AI), *in vivo* embryo collection and transfer, *in vitro* embryo production and transfer, and nuclear transfer and transgenesis) discussed by Thibier and Wagner (2002), AI has had by far the greatest impact on genetic improvement in farm-animal species. This is perhaps best exemplified in the dairy cattle industry where AI has been most effectively used to increase the reproductive rate of bulls (at current levels of technical efficiency, the number of insemination doses from a given sire is in excess of 100,000; Vishwanath, 2003) and in so doing propagates the use of progeny testing schemes. However, the technology has also been very effectively used to promote sire referencing schemes in sheep within the UK (McKelvey, 1999), where the creation of genetic linkages between flocks and the use of best linear unbiased prediction (BLUP) methodologies intensifies selection and thus the rate of genetic improvement (Lewis and Simm, 2000), and in pigs where, although technical difficulties with the cryopreservation of boar semen have hindered its widespread use, AI using fresh semen has also had a major impact on genetic improvement (Singleton, 2001).

The main advantages of AI, compared to the other reproductive technologies, are its ease of use, low cost and consistently good success rates. In small ruminants, however, intrauterine methods for AI are most commonly used and concerns have been raised about the acceptability of this relatively invasive method of insemination, which involves restraining the animal in dorsal recumbancy in a cradle following sedation. Indeed, the Banner Committee (Banner, 1995) recommended that laparoscopic techniques should not be used routinely in sheep breeding and so, in recent years, there has been renewed interest in developing transcervical methods of AI. Such methods of insemination, however, have proven difficult because the funnel-shaped rings of the ovine cervix are not concentrically aligned and the cervical canal is often constricted to less than 3 mm (Halbert *et al.*, 1990). This inhibits the deep penetration of the cervix with an insemination pipette necessary to achieve acceptable conception rates with frozen-thawed semen (Eppleston and Maxwell, 1993). Recent research has highlighted the importance of the inflammatory response in normal cervical dilation and specifically the role of the pro-inflammatory cytokine interleukin-8 (IL-8), which operates within the luminal epithelium and fibroblast-like cells in the connective tissue stroma of the cervix to soften its major structural components through the action of neutrophil derived collagenases (Mitchell *et al.*, 2002). Unfortunately, progesterone-impregnated pessaries inhibit IL-8 gene expression in the ovine cervix and so it is likely that oestrous synchronisation contributes to the low pregnancy rates observed following transcervical AI. Similarly, although pharmacological doses of oxytocin are known to relax the cervix and can facilitate deep penetration with an insemination pipette, the bolus administration of oxytocin is known to have a detrimental effect on pregnancy rates following transcervical AI (King *et al.*, 2004). It may be that simply alternating the dose and/or the frequency of administration of oxytocin may overcome this problem. However, these observations highlight the need for more research effort in this area if alternatives to intrauterine insemination are to be found.

As alluded to earlier, the major barrier to the continued expansion and application of AI in pigs is the relatively poor pregnancy rates achieved with the use of frozen/thawed semen and variability in the survival of spermatozoa following cryopreservation among boars. This latter aspect is not uncommon among the other major farm animal species, where novel techniques are currently being developed to assess objectively sperm quality following the use of modified cryopreservation protocols (Medrano *et al.*, 2002; Kumar *et al.*, 2003). As with other mammalian species, however, only a very small proportion (0.0001-0.001) of fresh porcine spermatozoa have all of the attributes necessary for fertilisation, and conventional cryopreservation techniques employing slow-rate cooling reduces this proportion even further (discussed by Holt and Watson, 2004). This represents a particular challenge for the use of porcine semen where ejaculate volumes are high and sperm numbers are low compared to the other farmed species. However, new techniques are currently being developed which may help overcome some of these limitations. Workers in Israel have developed a new freezing method, 'Multi-Thermal Gradient' (MTG®), designed to increase sperm motility and fertilisation capacity following cryopreservation (Arav *et al.*, 2002). This technology seeks to control the direction as well as the rate of freezing so as to control the velocity and morphology of ice formation both in the surrounding medium and within the spermatozoa, thus minimising cellular damage. The post-thaw motility of semen for a number of species, including boar semen, is reportedly indistinguishable from that of fresh semen.

Although the concept of microencapsulation of spermatozoa for artificial insemination is not new (Nebel *et al.*, 1993), interest in the use of this technique for the enhanced delivery of fresh porcine semen during AI has recently been rekindled. Encapsulation offers several advantages over conventional methods of sperm dilution and storage, the most important of these perhaps being the increase in membrane and acrosome integrity of encapsulated sperm relative to diluted and unencapsulated sperm (Faustini *et al.*, 2004). This is thought to arise through the retention of the various endogenous antioxidants present in seminal fluid that are lost during the normal process of sperm dilution and preparation. The beneficial effects of such treatments are the enhanced life span of spermatozoa during storage, leading to more flexible systems of transport and use, and better pregnancy rates following AI in pigs.

MULTIPLE OVULATION AND EMBRYO TRANSFER (MOET)

The value of MOET within genetic improvement schemes in ruminants is that it facilitates greater selection intensity, and hence rate of genetic gain, by increasing the reproductive rate of females (Lohuis, 1995; Nicholas, 1996). Furthermore, when selection is based on pedigree information and information on full and half-sib families, so the rate of gain can be further increased through reductions in the generation interval (the period between the birth of an animal and the birth of its offspring). Early theoretical predictions of the increase in the rate of genetic improvement achievable following the use of MOET within nucleus breeding schemes (e.g. those of Land and Hill, 1975, for beef cattle) were subsequently revised to account for reduced variance and increased inbreeding, among other factors (Villanueva *et al.*, 1995). Nevertheless, significant improvements in the rate of genetic gain for a number of commercially important traits do arise following the use of conventional MOET methodologies, particularly when information from a variety of sources including progeny is incorporated within the selection index (Nicholas, 1996). The use of more advanced embryo-related technologies such as follicular aspiration and *in vitro* embryo production, embryo splitting (cloning) and embryo sexing present further opportunities for increased genetic gain by facilitating the use of juvenile animals and through increases in the efficiency of embryo production (Lohuis, 1995).

Superovulation and embryo recovery

The use of this technology within nucleus genetic improvement schemes requires that sufficient high quality 'transferable' embryos be collected from comparatively young animals, where the interval between completion of performance test and ovarian stimulation and embryo recovery is short. For example, in order to achieve the target selection intensity of one in three heifers and a generation interval of approximately 2 years in the Simmental MOET project conducted at the Scottish Agricultural College for several years during the early 1990s, each selected heifer was required to produce 15 'Grade 1' embryos during a 12-week period following performance test (Broadbent, 1990). However, superovulatory responses in these young (12-14 months old) animals were generally poor and variable, and the number of 'Grade 1' embryos collected was well below that necessary to achieve the desired number of embryos within the specified time interval following performance test (3 vs 12 embryos; Tregaskes *et al.*, 1996). The timing of embryo recovery relative to the onset of puberty (9.5 months) and the level of body fat at the end of the performance test undoubtedly contributed to the low and variable superovulatory responses in these animals. However, an additional contributing factor will have been the superovulatory protocols adopted at that time when it was customary in cattle breeding programmes to commence ovarian stimulation during the mid cycle, i.e. around Day 8 to 12 following oestrus. No regard was given to follicular status at the time of hormonal treatment. This was primarily due to the fact that follicular monitoring by transrectal ovarian ultrasonography was still in the early stages of development and so not widely practised. It is now known that superovulatory responses are greater and more uniform if ovarian stimulation commences at the time of wave emergence. In terms of optimising superovulatory responses, around 80% of the oestrous cycle is not conducive to commencing ovarian stimulation (reviewed by Bo *et al.*, 2002). Therefore, modern protocols adopt procedures to ensure that ovarian stimulation coincides with wave emergence. These procedures include techniques designed to control follicular dynamics, such as follicular ablation (removal of the dominant follicle synchronises the emergence of a new wave of follicles) and the treatment of animals with pharmacological doses of oestradiol and progesterone (reviewed by Mapletoft *et al.*, 2002). Such protocols have been shown to result in a 40-50% increase in the number of transferable embryos recovered following commercial MOET (Merton *et al.*, 2003).

OPU and IVP

In vivo oocyte recovery by transvaginal ultrasound guided follicular aspiration (Ovum Pick-Up; OPU) and in vitro embryo production (IVP) are now established techniques in cattle (Galli *et al.*, 2001), and facilitate the production of a greater number of viable embryos in a given period of time than the superovulation and *in vivo* embryo recovery procedures described in the previous section. Protocols have been developed and refined that permit the collection of oocytes either once or twice weekly. Twice-weekly aspirations prevent the establishment of a dominant follicle. This, together with the frequency of aspiration, can increase oocyte yield over once-weekly aspirations within a defined period of time (Galli *et al.*, 2001). Once-weekly aspirations with gonadotrophin (FSH) stimulation, however, can increase embryo yields following *in vitro* culture to that achieved for twice-weekly aspirations (Goodhand *et al.,* 1999). Although the number of follicles aspirated is similar in both circumstances (around 25 in heifers), the distribution of follicles between diameter classes differs, with more follicles in the medium-size category when animals are treated with FSH. The number of oocytes recovered may be less in FSH treated heifers aspirated once-weekly compared with non-treated heifers aspirated twice-weekly (larger follicles are technically easier to aspirate than smaller follicles, although they do not always yield an oocyte), but enhanced embryo production rates *in vitro* ensure that embryo yields are similar for both treatments.

In recent years further increases in the yield of transferable embryos following OPU-IVP have arisen as a consequence of our improved understanding of the processes of nuclear and cytoplasmic maturation within the follicle enclosed oocyte (reviewed by Sinclair *et al.*, 2003a). It is beyond the scope of this chapter to consider these processes in detail but, in an attempt to improve the efficiency of *in vitro* embryo production, recent research has focused on events during the final stages of oocyte maturation when the oocyte, arrested at the diplotene stage of prophase I, resumes meiosis. Recognising that meiosis will resume spontaneously on removal from the follicular compartment, attempts have been made to mimic the periovulatory environment prior to follicular aspiration (Blondin *et al.,* 2002; Dieleman *et al.*, 2002), or to delay meiotic resumption during *in vitro* maturation (Sirard, 2001). The former strategy attempts to equip oocytes with the nucleic acids and proteins needed to complete meiosis and fertilisation prior to aspiration, whereas the latter strategy, in recognising that meiosis induces transcriptional arrest, attempts to 'buy' the oocyte time after aspiration and during IVM. Not surprisingly, the benefits are most evident for oocytes aspirated from small antral follicles. These strategies not only improve blastocsyst yields, but the potential of these embryos to develop to term is also enhanced.

Significant advances in our understanding of oocyte and early embryo metabolic requirements in recent years have also led to the formulation and refinement of culture media that obviate the need for somatic support cells (e.g. granulosa and oviductal cells), serum and albumin, whilst ensuring that the changing metabolic requirements of the developing embryo are met (reviewed by Sinclair *et al.*, 2003a). The exclusion of such undefined and potentially pathogen-bearing culture and co-culture constituents is necessary for the continued development of more predictable and biosafe systems for gamete and embryo culture, which can more closely be tailored to the changing metabolic requirements of the pre-implantation embryo during culture whilst reducing the likelihood of disease transmission. However, whilst these systems can support early embryo development, blastocyst yields at present are generally low and cell number and protein content reduced. Furthermore, both energy metabolism and gene expression are altered in embryos cultured in a protein-free environment. Presently the hope is that, as understanding of the changing metabolic requirements of the oocyte and pre-elongation embryo improves

(Sinclair *et al.*, 2003b), so the formulation of sequential embryo culture media that recognise these changing needs (Gardner, 1999; Hasler *et al.*, 2000) will lead to the improvement in post-fertilisation development necessary for both the successful establishment of pregnancy and the birth of normal and healthy offspring.

However, in spite of the many advances in IVP over the last 10 years and the many advantages that OPU-IVP has to offer over conventional superovulation and embryo recovery, at present there are no commercial companies operating in the UK employing such techniques. For most small to medium-sized companies the investment and associated operating costs of running an IVP laboratory alongside an animal practice are prohibitive. Oocyte recovery and in vitro embryo production would have to offer benefits greater than those that have so far been realised commercially for the uptake of this technology to flourish within genetic improvement schemes.

EGG COLLECTION FROM JUVENILES, VELOGENETICS AND EMBRYO SPLITTING (CLONING)

One such benefit might arise following the recovery of oocytes from juveniles which would not only promote the dissemination of embryos from genetically superior animals, but also offer greater potential for accelerated genetic gain by reducing the generation interval even further. This concept was taken to the extreme in the velogenetic schemes of Georges and Massey (1991) who proposed that, in order to reduce the generation interval of cattle, oocytes could be recovered from calves whilst still *in utero*. If such techniques were used in combination with MAS, then the generation interval of cattle could be reduced from between 3 to 6 years to between 3 to 6 months.

Oocyte recovery from juvenile animals can be achieved by the exposure of ovaries using abdominal laparotomy or by employing laparoscopic techniques similar to those used for AI. Early attempts to generate embryos from juvenile ruminants by conventional superovulation and embryo recovery proved unsuccessful due to low and highly variable ovulation rates coupled with low fertilisation rates attributed to the immaturity of the donor reproductive tracts. In contrast, oocytes can be recovered and cultured *in vitro* from lambs and kids as young as 4 weeks of age, coincident with the peak number of vesicular follicles during the pre-pubertal period (Ptak *et al.*, 1999; Baldassarre *et al.*, 2002), and from calves as young as 6 to 8 weeks of age (Armstrong *et al.*, 1997). Such animals generally respond well to gonadotrophin treatment and oocyte recovery and blastocyst production rates *in vitro* are comparable with those of adult animals. Pregnancy rates following embryo transfer, however, tend to be lower and more variable when oocytes are derived from juvenile rather than adult animals. At present the full potential of this technology to increase the rate of genetic gain within livestock improvement programmes has not been realised.

Finally, embryo bisection found limited commercial application in cattle during the 1980s, where the emphasis was on doubling the number of embryos for transfer in an attempt to improve the efficiency of embryo production, rather than producing identical twins (true clones). The technique can be applied from the two-cell stage onwards, but morula- and blastocyst-stage embryos are most frequently used. In one large commercial study, 442 embryos were bisected and, following twin-embryo transfer, produced 441 pregnancies (52.4%) and 551 calves (Leibo and Rall, 1987). As a means of producing clones, however, the technique suffers from a number of limitations. Repeated splitting is not possible and the embryo cannot be dissected into more than four sections. Although the technique requires a high degree of operator skill it does, nevertheless, present the opportunity of increasing the number of transferable embryos within a MOET genetic improvement programme. In future these improvements in efficiency, together with general improvements in methods of IVP, may help persuade those in practice to consider once again this technology.

Further applications of reproductive technologies

GENDER PREDETERMINATION

Gender pre-selection promises to facilitate the generation of offspring in accordance with industry or market requirements and to offer livestock producers the realistic prospect of adopting more judicious breeding policies, with the potential to halve the number of breeding females required to generate replacements. This, in turn, will allow producers to be more selective when identifying dams for rebreeding, therefore greatly enhancing the rate of genetic improvement. Pre-sexed bovine semen from selected dairy sires is already a commercial reality in the UK. As around 90% of dairy cows are mated through AI at present, so the scope for adoption of this technology is great.

The technology also offers the potential to alleviate the ethical dilemma associated with the production of unwanted Holstein male calves whilst increasing the production of more marketable beef-type males (discussed by Sinclair *et al.*, 2003a). Given the low level of use of AI by suckled calf producers it is difficult, at present, to predict the impact that this technology will have on the national beef herd. It can, however, permit the establishment of single-sexed once-bred heifer (SSBH) systems, which have long been recognised to be the most efficient means of producing beef with predicted efficiencies of food utilization approaching that of pig meat production (Figure 4.2). As Taylor *et al.* (1985) explained, this improvement in efficiency arises because the number of older animals (> 1 year of age) in a SSBH system are half that of a traditional suckled-calf production system for the same level of annual output. Similarly for pigs, the once-bred gilt has long been touted to be the most efficient system for producing pig meat with improvements in the marginal efficiency of lean tissue production ranging from 20 to 70% (Brooks and Cole, 1973; MacPherson *et al.*, 1977; Ellis *et al.*, 1996), although some questions regarding the eating quality of meat from the older animal have been raised (Ellis *et al.*, 1996).

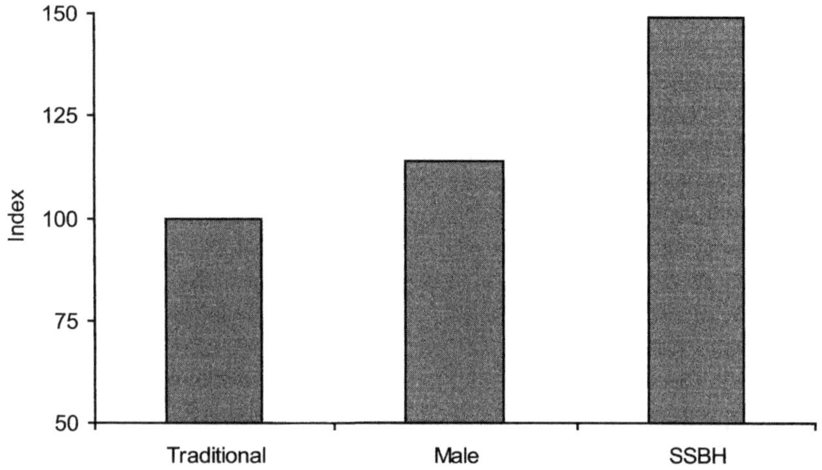

Figure 4.2 Comparison of the relative efficiency of food utilisation (g lean per MJ metabolisable energy) between a conventional suckled calf production system (Traditional), where all the progeny are male (Male), or a single-sexed bred heifer system (SSBH). (Data derived from Talyor *et al.*, 1985).

At present, flow cytometric sorting of sperm offers the most accurate and efficient means for achieving sex pre-selection (Johnson, 2000), although other methods employing immunological and electrophoretic approaches are being investigated (discussed by Seidel, 2003). The flow cytometric technique is based on the principle that female sperm contain about 4% (4.2% ram, 3.8% bull, 3.6% boar) more DNA than male sperm, due to the presence of the larger X chromosome. A fluorescent dye bound to the DNA helix of sperm fluoresces when a stream of semen generated by the sorter is interrogated by a water-cooled argon laser. Fluorescence is quantified and a charge applied to the stream so that, as the droplets pass between charged plates (the polarity of which can be altered), X and Y bearing sperm are differentially deflected into separate container vessels. For bovine semen, modern instruments are capable at the present time of producing 18 million sorted sperm per hour (with a purity of around 90 to 95%) and between one to two thousand straws of semen per week, although the speed of separation keeps increasing. In field studies conducted in North America, pregnancy rates using low dose semen (1.0-1.5 x 10^6 sperm) in heifers were within 90% of those where unsexed, frozen-thawed semen, with 7 to 20 times more sperm per insemination dose, were employed (Seidel *et al.,* 1999). Although pregnancy-rates with the use of sex-sorted semen are noticeably lower in parous, lactating cows, recent advances in deep uterine insemination in cattle (Hunter, 2003) may help overcome this limitation by advancing the site of semen deposition, thereby reducing the volume of inseminate required. Also, reassuringly, although there is some evidence of sperm damage and reduced viability with semen sorting (mainly associated with mechanical stress; Garner and Suh, 2002), there is no evidence to date of any teratogenic effects with the use of this technology (Amann, 1999). Indeed, as Seidel (2003) pointed out, selection of X bearing sperm for use in primiparous heifers may help reduce calving difficulties in these young animals with important beneficial implications for animal welfare. Consequently, the potential of this technology to improve animal welfare as well as increasing production efficiency in the cattle industry is great.

In contrast, given the limitations of low dose semen achieving pregnancy by transcervical AI in sheep, the commercial adoption of gender pre-selection within this sector of the livestock industry in the near future will depend on laparoscopic methods of insemination that deposit sperm close to the utero-tubal junction. The first lambs of predetermined sex using low dose semen deposited in this way were produced in Aberdeen in 1996 (Cran *et al.,* 1997). Similarly in pigs, low sperm numbers preclude conventional methods of AI, which typically deposit around 3 x 10^9 sperm into the cervical canal and require surgical techniques to deposit low doses of sperm (1 x 10^6 to 5 x 10^8 sperm) into the oviducts (Krueger *et al.,* 1999). Until sperm concentrations and dose volumes are increased and/or methods for the transcervical deposition of semen deep into the uterine lumen are developed (similar to that of Martinez *et al.,* 2002), it seems unlikely that this technology will have any significant impact on pig breeding.

CONSERVATION OF FARM ANIMAL GENETIC RESOURCES

Although strictly beyond the remit of the current chapter, brief reference will be made to recent advances in reproductive technologies that could have a direct bearing on the conservation of farm animal genetic resources. This topic was the focus of an international conference held in Edinburgh in November 2002, the proceedings of which have since been published (Simm *et al.*, 2004). Papers at that meeting considered the roles of current and new methods in semen and oocyte technology, and cloning and nuclear transfer technology in genetic conservation and the preservation of genetic diversity. Recent developments in related technologies, such as gamete and embryo cryopreservation, were also considered and recently reviewed by Sinclair *et al.* (2003a). What follows is a brief consideration of some of the main points highlighted from these reviews.

Woolliams (2004) considered the relative merits of the different sources of germplasm (somatic cells, gametes - oocytes and sperm - and embryos) that could be stored and highlighted the technological limitations of working with each of these sources between different species and between different countries. The use of somatic cells may be of greatest value in regions of the world that are not so technologically advanced, but the technology required to generate live born from these cells (i.e. nuclear transfer, NT) requires further refinement (Sinclair *et al.*, 2003a; Wells, 2004; discussed later). As alluded to earlier, semen technology is the most developed of the reproductive technologies and is routinely used in every major farmed animal species (Thibier *et al.*, 2004). The limitation of semen however is that, as a source of germplasm, it exists in the haploid form and therefore a source of oocytes is required in order to complete the genome. Whilst this is possible when living oocyte donors exist, difficulties arise when such donors are not available. Although there have been significant recent developments in the fields of oocyte / ovarian tissue cryopreservation and tissue grafting (Picton, 2004), this is a comparatively new area of biological research with years of development and refinement ahead before it can become commercially available.

However, in spite of these cautionary overtones, there have been a number of high-profile and spectacular developments in recent years which illustrate how scientific endeavour can push back the boundaries of biological science and challenge conventional wisdom and existing dogmas. The birth of Dolly was one such development and this advancement in NT has since paved the way for a number of commercially important applications of both medical and agricultural significance (discussed later). In the context of the conservation of farm animal genetic resources, however, NT has already demonstrated its potential to rescue endangered species and breeds from the verge of extinction, the most prominent demonstration being the cloning of the last surviving Enderby Island cow (Wells *et al.*, 1999). Inter-specific NT between two closely related species (e.g. *Ovis orientalis musimon* and *Ovis aries*; Loi *et al.*, 2001) has been another significant development and presents a means of rescuing endangered species where the number of oocyte donor and embryo-recipient females may be limiting.

Reproductive cloning and transgenics

CLONING

The potential of cloning procedures such as embryo splitting and NT to enhance the rate of genetic improvement in farm animals has been considered by a number of commentators over the last two decades (e.g. Woolliams and Wilmut, 1989; Woolliams, 1989; Nicholas, 1996). They concluded that although the creation of a large number of identical individuals has the potential to increase the accuracy of selection, much of this advantage would be lost in closed nucleus breeding schemes or in situations where resource size (e.g. the number of transfers) is limited, as the reduction in the number of different families reduces selection intensity. Instead, cloning may be more useful in the dissemination of superior genes to commercial herds or flocks. It has the advantage that it can disseminate genes from crossbred or hybrid animals and so exploit the benefits of non-additive genetic effects and heterosis. It is anticipated that the identification and dissemination of genetically superior clones from a nucleus herd would provide a highly significant, albeit a once-only, increase in the average genetic merit of commercial offspring.

Technical advancements in cloning by NT during the 1990s arose with increased understanding of cell-cycle regulation by factors within the cytoplasm and nucleus of the cell (Campbell *et al.,* 1996). Of central importance to the subsequent development of a reconstructed embryo, ensuring normal ploidy and minimising DNA damage, is the stage of the cell cycle of the donor nucleus (karyoplast) in relation to that of the enucleated oocyte (cytoplast) at the point of transfer. Non-activated oocytes arrested at MII are generally preferred as nuclear recipients, particularly when somatic cells are used as nuclear donors, as factors present within the cytoplast at that time help to ensure chromatin remodelling and promote post re-construction embryo development (Heyman *et al.,* 2002). Nuclei that are in the M or G1 phase of the cell cycle are less likely to exhibit chromosomal abnormalities than nuclei in the S or G2 phase of the cycle. However, attempts to further enhance the efficiency of post-reconstruction embryo development and pregnancy outcome in cattle by inducing donor nuclei to enter G0 (quiescence) through the use of cyclin-dependent kinase inhibitors, serum starvation or the establishment of confluence during culture have so far proved unsuccessful (Kasinathan *et al.,* 2001; Gibbons *et al.,* 2002). A further difficulty encountered during porcine NT is the poor developmental potential of reconstructed embryos following artificial activation. This limitation was partially overcome using a successful serial NT system originally developed in mice (Polejaeva *et al.,* 2000). In this system, selected donor cells (preferably in G1 or G0) are transferred into an enucleated MII oocyte as previously described but then, shortly after fusion and artificial activation, the new pronucleus is transferred to an enucleated zygote. This process is thought to facilitate chromatin remodelling in the donor nucleus whilst providing a cytoplasmic environment more suited for subsequent embryo development.

Post-reconstruction embryo development may be further impaired due to a failure on the part of the cytoplast to interact with the donated nucleus and restore its totipotency, and to re-establish an epigenetic state comparable to that acquired during normal fertilisation. Although it is beyond the scope of this chapter to consider these mechanisms in any great detail, it is worth noting that the state of differentiation of the donor nucleus (e.g. somatic vs embryonic), the origin of oocytes used for NT (e.g. nutritional status of the donor animal, follicular background and in vitro maturation), together with the means by which reconstructed embryos are cultured may all contribute to the low and varied pregnancy outcomes observed (Figure 4.3).

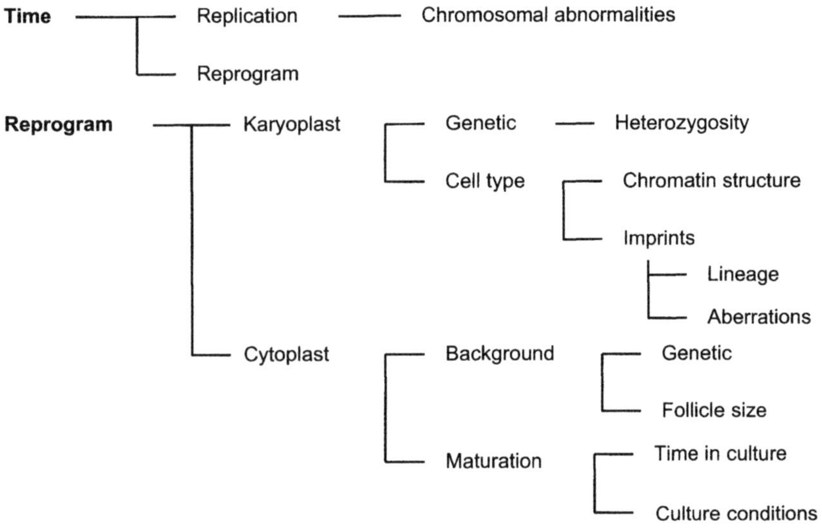

Figure 4.3 Nuclear reprogramming. Factors contributing to low and variable blastocyst rates and pregnancy failures following cloning by nuclear transfer (NT). Factors other than cell cycle synchrony that may account for these losses include time, which following NT is required to both reprogram the donor nucleus and for DNA replication before the first cell division. Mouse studies indicate that nuclei from differentiated cells require more exposure to cytoplasmic factors, so delayed activation may facilitate nuclear reprogramming. Failure to complete DNA replication prior to cytokinesis will result in chromosomal abnormalities, and this is one reason why many reconstructed embryos fail to develop. Recent mouse work also highlights the importance of chromosomal heterozygosity or hybrid vigour in promoting post-reconstruction development. Cell type is important for cytoplasmic factors evolved to receive sperm chromatin that are packed primarily in protamines (adapted for rapid release of DNA on exposure to oocyte cytoplasm), whereas chromosomes in the transferred nucleus will be packed in somatic histones (which may retain a nucleohistone configuration). What's more, sex-specific imprints are normally established in the gametes following erasure of somatic imprints through the germ line, so that the cytoplast may be ill-equipped to re-establish the correct epigenetic state following somatic cell NT, or when donor nuclei are used that already possess imprinting errors, which is often the case with ES cells. Recent reciprocal pronuclear transfer experiments in mice indicate that the maternal genotype can influence the epigenotype of the offspring, and follicle size is important in that oocytes from small antral follicles are less mature. Time and the nature of the culture environment may be critical factors in this respect, for it is unlikely that oocytes from small antral follicles will have adequate time during culture to establish all of the transcripts necessary for nuclear remodelling during in vitro maturation. Finally, nuclear reprogramming is not helped by the fact that a significant amount of cytoplasm usually accompanies the nucleus during transfer. Therefore, factors present within the karyoplast may counter those within the cytoplast to maintain an epigenetic state more akin to that of the donor cell. (Derived from Sinclair and Young, 2002).

At present, although the proportion of reconstructed embryos that develop to the blastocyst stage *in vitro* is increasing, the number of pregnancies that go to term and result in live offspring with no developmental abnormalities is consistently poor across a wide number of species now examined (Sinclair *et al.*, 2003a). These poor results and developmental abnormalities in offspring currently limit the uptake of this technology. These problems are the subject of considerable research effort at present but, even if they were to be resolved, the technical demands and associated costs of cloning are likely to limit its application to situations where the value of the animals concerned is high. The production of transgenic livestock is one such application that fulfils this criterion.

TRANSGENESIS

There have been numerous attempts over the last two decades to modify farm animals genetically in order to alter body growth and composition, intermediary metabolism, protein expression within the mammary gland, disease resistance and the avoidance of tissue rejection following xenotransplantation. These applications were extensively reviewed by Sinclair *et al.* (2003a) and Niemann and Kues, (2003), and some examples are presented in Tables 4.3a and 4.3b. However, these high profile successes belie the fact that current methods of generating transgenic livestock are inefficient and often result in low and variable levels of transgene expression. A number of factors are responsible for these rather poor and variable results including (i) transgene design and (ii) method of transgene integration into the mammalian genome. It is beyond the scope of this chapter to consider these aspects in detail. However, with regard to transgene design, future improvements in the understanding of gene regulation will almost certainly lead to the improved spatial and temporal expression of integrated gene constructs. It is now known, for example, that host sequences surrounding the site of integration can alter transgene expression, an event often referred to as the 'positional effect'. Strategies devised to counter such effects have benefited from recent advancements in our understanding of the many regulatory elements required for gene function, including the presence of boundary elements that serve to insulate the gene from regulatory sequences outside of the expression domain. The transfer of large, chromosomal-sized pieces of DNA (artificial chromosomes), in addition to ensuring position-independent and copy number-dependent effects, also permits the insertion of much longer sequences of DNA (> 1 Mb vs 20-100 kb for conventional gene constructs) required for the synthesis of larger and more complex molecules. Recently, the birth of the first transgenic calves, carrying a human microchromosomal vector encoding human *IgH* and *Ig*l loci was reported (Kuroiwa *et al.,* 2002). In the near future it should be possible to employ more sophisticated molecular techniques, such as the Cre/loxP-specific recombinase system, so successfully used in mice (Nagy, 2000), to conduct conditional-mutagenesis and region-specific mutagenesis in farm animal species, that would allow more subtle modifications to be made to the genome.

Table 4.3a RECENT EXAMPLES OF GENETIC MODIFICATION IN FARM ANIMALS WITH AGRICULTURAL APPLICATIONS

Species	*Gene product*	*Purpose*	*Reference*
(a) Growth and body composition			
Pigs	Growth hormone	Altered growth and body composition in pigs	Pursel *et al.*, 1997
Mice	Myostatin	Mutation: double muscling. Pilot study to establish a transgenic model for muscle growth in pigs	Yang *et al.*, 2001
Sheep	Keratin-IGF-I construct	Modest increase in clean fleece weight with no adverse health effects	Damak *et al.*, 1996
(b) Milk yield and composition			
Pigs	Bovine α-lactalbumin	Increased yields of milk and lactose leading to increased piglet growth	Wheeler *et al.*, 2001
Cows	ß-casein and κ-casein	Increased ß-casein and κ-casein concentrations to improve processing and nutritional value	Brophy *et al.*, 2003
(c) Digestion and intermediary metabolism			
Pigs	Phytase	Hydrolysis of phytate in the alimentary tract. Salivary phytase reduced faecal phosphorous output by 75%	Golovan *et al.*, 2001
Pigs	Δ12 desaturase	Synthesis of Δ12 (n-6) and n-3 fatty acids	Saeki *et al.*, 2004
Sheep	Serine transacetylase O-acetylserine sulphhydrylase	Establishment of a functional cysteine bio-synthetic pathway to promote wool growth	Ward, 2000
Sheep	Isocitrate lyase Malate synthase	Establishment of a functional glyoxylate cycle in ruminants to allow glucose to be synthesised from acetate	Ward, 2000

Table 4.3b RECENT EXAMPLES OF GENETIC MODIFICATION IN FARM ANIMALS WITH BIOMEDICAL APPLICATIONS

Species	*Gene product*	*Purpose*	*Reference*
(a) Mammary gland proteins			
Sheep	Human Factor IX	Haemophilia B: blood clotting protein	Schnieke *et al.*, 1997
Cow	Heavy and light chain hIg	Human polyclonal antibodies: therapeutic purposes	Kuroiwa *et al.*, 2002
Cow	Human lactoferrin	An iron binding glycoprotein involved in innate host defence	van Berkel *et al.*, 2002
(b) Xenotransplantation[a]			
Sheep	α(1,3)-galactosyl transferase	Xenoreactive antigen: ablation to avoid tissue rejection	Denning *et al.*, 2001
	PrP	Scrapie: a transmissible spongiform encephalpathy	
Pigs	α(1,3)-galactosyl transferase	Xenoreactive antigen: as for sheep, but in species of interest Five healthy pigs produced containing one disrupted α1,3-GT allele	Dai *et al.*, 2002

[a] Examples given illustrate targeted gene deletion (knock out) for two genes that represent a barrier to the success of this application.

With respect to methods of transgene integration into the mammalian genome, the inefficient method of microinjecting linear DNA into the pronucleus of one-cell zygotes, as commonly practiced in mice, is currently being superseded by alternative techniques that include retroviral and sperm-mediated methods of gene transfer (Wall, 2002). Whilst the former strategy appears to be efficient for integrating new DNA into the mammalian genome (70% of live pigs born carried the Green Fluorescent Protein-labelled transgene and 94% of these expressed the labelled construct in the study of Hofmann *et al.*, 2003) the latter strategy benefits from its simplicity. Both techniques, however, suffer from many of the limitations of pronuclear microinjection, including the fact that genes can only be added to the genome and the site of integration is random. In contrast, the targeting of foreign DNA to specific sites in the genome of farm animals by homologous recombination offers several advantages and can facilitate both gene deletion (knock out) as well as gene addition (knock in). As such recombination events occur at low frequency, so it is necessary to adopt cell-based strategies of transgenesis that can employ mass methods of transfection such as eletroporation. The major impediment to the use of this approach in farm animals to date has been the failure to identify and isolate embryonic stem (ES) cells from these species. However, the advent of somatic cell NT in the late 1990s permitted the use of alternative somatic cell types (e.g. fetal fibroblasts). The current challenge with this cell type is to overcome their limited number of population doublings and finite lifespan in culture, which can lead to senescence before the necessary genetic modifications and selection of transgenic clonal lines is complete (Kasinathan *et al.*, 2001). A further limitation of this approach lies in the current levels of inefficiency associated with SCNT (discussed earlier). Nevertheless, the study of McCreath *et al.* (2000) demonstrated that the targeting of foreign DNA to specific sites in the genome by homologous recombination can result in the successful creation of transgenic animals. In that study, two gene-targeting vectors (containing a *neo* selectable marker and the human α1-antitrypsin sequence linked to an ovine ß-lactoglobulin expression vector) were used to target the highly expressed and comparatively well characterised *COL1A1* locus. The resulting construct was transfected into fetal fibroblasts which, in turn, were transferred into enucleated oocytes as described previously. As a consequence of this procedure, α1-antitrypsin protein (used in the treatment of cystic fibrosis) was demonstrated to be secreted in the milk of one of the female cloned lambs produced.

Conclusions

STRATEGIES TO INCREASE REPRODUCTIVE RATE

Experience to date indicates that significant improvements in reproductive rate can be achieved through selection, and that the rate of improvement for this trait is likely to be enhanced with the aid of modern MAS techniques. Whilst selection for increased ovulation rate in cattle is unlikely to lead to a significant increases in high-order pregnancies, more consideration of factors such as uterine receptivity and capacity and nutrition will be required to increase the number of live born in species such as the sheep, where currently the focus of attention is on single gene effects for increased ovulation rate. Recent developments in pharmacological strategies to increase ovulation rate have benefited from our improved understanding of the underlying factors that regulate ovarian follicular development. The potential of these techniques to increase the biological efficiency of farm animals will mostly be realised through genetic improvement programmes employing MOET.

APPLICATION OF REPRODUCTIVE TECHNOLOGIES IN GENETIC IMPROVEMENT

Of the four generations of reproductive technology considered by Thibier and Wagner (2002), AI has and will continue to have the greatest impact on animal breeding and livestock improvement. Nevertheless, its full potential has yet to be realised. This technology awaits further refinements in methods of cryopreservation and insemination, particularly in pigs. In contrast, although there have been significant recent advances in methods of superovulation and oocyte/embryo production, it is unlikely that the embryo related technologies will have as great an impact on animal breeding as AI. The recovery of oocytes from juvenile animals, whilst offering great scope for increased rates of genetic improvement, raises a number of welfare and ethical issues which may hinder its future development.

GENDER PRE-DETERMINATION AND THE CONSERVATION OF GENETIC RESOURCES

Modern methods of semen sorting using flow cytometry offer the best means of facilitating gender pre-selection in the foreseeable future. This technology has tremendous potential to revolutionise current systems of animal production, with important benefits for livestock breeding and production efficiency, and also for the environment. Its impact in the short- to medium-term is likely to be greatest in the dairy industry where currently around 90% of cows are AI. Recent developments in gamete and embryo biology, advances in methods of cloning by NT, all considered in this chapter, and progress in methods of tissue cryopreservation and grafting, not considered in detail in this chapter, present numerous opportunities for genetic salvage and for the future conservation of farm animal genetic resources.

CLONING AND GENETIC IMPROVEMENT

Although cloning by NT has long been proposed to increase the rate of genetic improvement and facilitate genetic dissemination, current limitations with this technology mean that, for the foreseeable future, its greatest impact will be to facilitate the use of contemporary techniques for the precise genetic modification of farm animals. Current advances in molecular genetics are providing scientists with the tools that will allow any desired modification to be made to the mammalian genome. Future opportunities to modify animal growth, product composition and quality are limited only by the scientist's imagination and the ethical acceptability of such developments by the general public.

References

Amann, R.P. (1999). Issues affecting commercialization of sexed sperm. *Theriogeneology*, **52:** 1441-1457

Arav, A., Yavin, S., Zeron, Y., Natan, D., Dekel, I. and Gacitua, H. (2002). New trends in gamete's cryopreservation. *Molecular and Cellular Endocrinology*, **187:** 77-81.

Arias, J. and Kirkpatrick, B. (2004). Mapping of bovine ovulation rate QTL; an analytical approach for three generation pedigrees. *Animal Genetics*, **35:** 7-13.

Armstrong, D.T., Kotaras, P.J. and Earl, C.R. (1997). Advances in production of embryos *in vitro* from juvenile and prepubertal oocytes from the calf and lamb. *Reproduction, Fertility and Development,* **9:** 333-339.

Ashwell, M. S., Heyen, D. W., Sonstegard, T. S., Van Tassell, C. P., Da, Y., VanRaden, P. M., Ron, M., Weller, J. L. and Lewin, H. A. (2004). Detection of quantitative trait loci affecting milk production, health, and reproductive traits in Holstein cattle. *Journal of Dairy Science*, **87:** 468-475.

Baldassarre, H., Wang, B., Kafidi, N., Keefer, C., Lazaris, A. and Karatzas, C.N. (2002). Advances in the production and propagation of transgenic goats using laparoscopic ovum pick-up and *in vitro* embryo production techniques. *Theriogenology,* **57:** 275-284.

Banner, M. (1995). Report on Secretary of State's Working Party on the Ethical Implications of Advanced Breeding Techniques for Farm Animals. HMSO, London

Beerepoot, G. M., Dykhuizen, A. A., Nielen, M. and Schukken, Y. H. (1992). The economics of naturally occurring twinning in dairy cattle. *Journal of Dairy Science*, **75:** 1044-1051.

Bennett, G. L. and Leymaster, K. A. (1989). Integration of ovulation rate, potential embryonic viability and uterine capacity into a model of litter size in swine. *Journal of Animal Science,* **67:** 1230-1241.

Bennett, G. L., Echternkamp, S. E. and Gregory, K. E. (1998). A model of litter size distribution in cattle. *Journal of Animal Science*, **76:** 1789-1793.

Blondin, P., Bousquest, D., Twagiramungu, H., Barnes, F. and Sirard, M.A. (2002). Manipulation of follicular development to produce developmentally competent bovine oocytes. *Biology of Reproduction*, **66:** 38-43.

Bo, G. A., Baruselli, P. S., Moreno, D., Cutaia, L., Caccia, M., Tribulo, R., Tribulo, H. and Mapletoft, R. J. (2002). The control of follicular wave development for self-appointed embryo transfer programs in cattle. *Theriogenology*, **57:** 53-72.

Broadbent, P. J. (1990). The NOSCA Simmental MOET project. *British Cattle Breeder's Club Digest,* **45:** 44-48

Brooks, P. H. and Cole, D. J. A. (1973). Meat production from pigs which have farrowed. 1. Reproductive performance and food conversion efficiency. *Animal Production*, **17:** 305-315.

Brophy, B., Smolenski, G., Wheeler, T., Well, D., L'Huillier, P and Laible, G. (2003). Cloned transgenic cattle produce milk with higher levels of b-casein and k-casein. *Nature Biotechnology,* **21:** 157-162.

Campbell, B. K., Gordon, B. M., Tsonis, C. G. and Scaramuzzi, R. J. (1995). The effect of acute immuno-neutralisation of inhibin in ewes during the early luteal phase of the oestrous cycle on ovarian hormone secretion and follicular development. *Journal of Endocrinology*, **145:** 479-490.

Campbell, B. K., William, R. , Gong, J. and Webb, R. (2004). Immunological strategies to boost reproductive efficiency in sheep and cattle without adverse effect on animal welfare. *Proceedings of the British Society of Animal Science*, York, March, 2004: p 259.

Campbell, K.H.S., Loi, P., Otaegui, P.J. and Wilmut, I. (1996). Cell cycle co-ordination in embryo cloning by nuclear transfer. *Reviews of Reproduction*, **1:** 40-46.

Cran, D.G, McKelvey, W.A.C., King, M.E., Dolman, D.F., McEvoy, T.G., Broadbent, P.J. and Robinson, J.J. (1997). Production of lambs by low dose intrauterine insemination with flow cytometric sorted and unsorted semen. *Theriogenology*, **47:** 267 (abstr).

Dai, Y., Vaught, T.D., Boone, J., Chen, S.H., Phelps, C.J., Ball, S., Monahan, J.A. et al. (2002). Targeted disruption of the a1,3-galactosyltranferase gene in cloned pigs. *Nature Biotechnology*, **20:** 251-255.

Damak, S. Su, H., Jay, N. P., Bullock, D. W. (1996). Improved wool production in transgenic sheep expressing insulin-like growth factor I. *Biotechnology,* **14:** 181-184.

Davis, G. H., Montgomery, G. W., Allison, A. J., Kelly, R. W. and Bray, A. R. (1982). Segregation of a major gene influencing fecundity in progeny of Booroola sheep. *New Zealand Journal of Agricultural Research,* **25**: 525-529.

Davis, G. H., Morris, C. A. and Dodds, K. G. (1998). Genetic studies of prolificacy in New Zealand sheep. *Animal Science*, **67:** 289-297.

Davis, G. H., Bruce, G. D. and Dodds, K. G. (2001). Ovulation rate and litter size of prolific Inverdale (Fec XI) and Hanna (Fec XH) sheep. *Proceedings of the Association for Advanced Animal Breeding and Genetics,* **14:** 145-178.

Davis, G. H., Galloway, S. M., Ross, I. K., Gregan, S. M., Ward, J., Nimbkar, B. V., Ghalasasi, P. M., Nimbkar, C., Gray, G. D., Subandriyo, Inouno, I., Tiesnamurti, B., Martyniuk, E., Eythorsdottir, E., Mulsant, P., Lecerf, F., Hanrahan, J. P., Bradford, G. E. and Wilson, T. (2002). DNA tests in prolific sheep from eight countries provide new evidence on the origin of the Booroola (FecB) mutation. *Biology of Reproduction* **66:** 1869-1874.

Davis, M. E. and Bishop, M. D. (1992). Induction of multiple births with FSH: calving rate and subsequent performance. *Livestock Production Science,* **32:** 41-62.

Denning, C., Burl, S., Ainslie, A., Bracken, J., Dinnyes, A., Fletcher, J., King, K., Ritchie, M., Ritchie, W.A., Rollo, M., de Sousa, P., Travers, A., Wilmut, I., and Clark, A.J. (2001). Deletion of the α(1,3)galactosyl transferase (GGTA1) gene and the prion protein (PrP) gene in sheep. *Nature Biotechnology,* **19:** 559-562.

Dielemann, S.J., Hendriksen, P.J.M., Viuff, D., Thomsen, P.D., Hyttel, P., Knijn, H.M., Wrenzycki, C., Kruip, T.A.M., Niemann, H., Gadella, B.M., Bevers, M.M. and Vos, P.L.A.M. (2002). Effects of *in vivo* prematuration and *in vivo* final maturation on developmental capacity and quality of pre-implantation embryos. *Theriogenology,* **57:** 5-20.

Dingwall, W. S., Robinson, J. J., Aitken, R. P. and Fraser, C. (1987). Studies on reproduction in prolific ewes. 9. Embryo survival, early foetal growth and within litter variation in foetal size. *Journal of Agricultural Science (Cambridge),* **108:** 311-319.

Echternkamp, S. E. (1992). Fetal development in cattle with multiple ovulations. *Journal of Animal Science,* **70:** 2309-2321.

Eddy, R. G., Davies, O. and David, C. (1991). An economic assessment of twin births in British dairy herds. *Veterinary Record,* **129:** 526-529.

Ellis, M., Fowler, V. R., Franklin, M. F., Wood, J. D. and Varley, M. A. (1996). The influence of feeding regimen and lactation length on growth, carcass characteristics and meat quality of once-bred gilts. *Animal Science,* **62:** 561-571.

Eppleston, J. and Maxwell, W.M.C. (1993). Recent attempts to improve the fertility of frozen ram semen inseminated into the cervix. *Wool Technology and Sheep Breeding* **1:** 291-302.

Faustini, M., Torre, M. L., Stacchezzini, S., Norberti, R., Consiglio, A. L., Porcelli, F., Conte, U., Munari, E., Russo, V. And Vigo, D. (2004). Boar spermatozoa encapsulated in barium alginate membranes: a microdensitometric evaluation of some enzymatic activities during storage at 18°C. *Theriogenology,* **61:** 173-184.

Galli, C., Crotti, G., Notari, C., Turini, P., Duchi, R. and Lazzari, G. (2001).

Embryo production by ovum pick up from live donors. *Theriogenology,* **55:** 1341-1357.

Galloway, S. M., McNatty, K. P., Cambridge, L. M., Laitinen, M. P. E., Juengel, J. L., Jokiranta, T. S., McLaren, R. J., Luiro, K., Dodds, K. G., Montgomery, G. W., Beattie, A. E., Davis, G. H. and Ritvos, O. (2000). Mutations in an oocyte-derived growth factor gene (BMP15) cause increased ovulation rate and infertility oin a dosage-sensitive manner. *Nature Genetics*, **25:** 279-283.

Galmessa, U. and Prasad, S. (2003). Comparative study of PMSG treatment and different levels of concentrate supplements prior to mating on twinning rate of Horro ewes. *Indian Journal of Animal Sciences,* **73:** 851-853.

Gardner, D. K. (1999). Development of serum free media for the ruminant embryo and subsequent assessment of embryo viability. *Journal of Reproduction and Fertility Supplement* **54:** 461-475.

Garner, D.L. and Suh, T.K. (2002). Effect of Hoechst 33342 staining and laser illumination on the viability of sex-sorted bovine sperm. *Theriogenology,* **57:** 746 (abstr.)

Georges, M. and Massey, J. M. (1991). Velogenetics, or the synergistic use of marker assisted selection and germ-line manipulation. *Theriogeneology,* **35:** 151-159.

Gibbons, J., Arat, S., Rzucidlo, J., Miyoshi, K., Waltenburg, R., Repess, D., Venable, A. and Stice, S. (2002). Enhanced survivability of cloned calves derived from roscovitine-treated adult somatic cells. *Biology of Reproduction*, **66:** 895-900.

Golovan, S. p., Meidinger, R. G., Ajakaiye, A., Cottrill, M., Wiederkehr, M. Z., Barney, D. J., Plante, C., Pollard, J. W., Fan, M., Hayes, M A., Laursen, J., Hjorth, J. P., Hacker, R. R., Philips, J. P. and Forsberg, C. W. (2001). Pigs expressing salivary phytase produce low-phosphorus manure. *Nature Biotechnology*, **19:** 741-745.

Goodhand, K.L., Watt, R.G., Staines, M.E., Hutchinson, J.S.M. and Broadbent, P.J. (1999). *In vivo* oocyte recovery and *in vitro* embryo production from bovine donors aspirated at different frequencies or following FSH treatment. *Theriogenology,* **51:** 951-961.

Gordon, I., Williams, G. L. and Edwards, J. (1962). The use of PMS in the induction of twin pregnancy in the cow. *Journal of Agricultural Science, Cambridge* **59:** 143-198.

Grapes, L. and Rothschild, M. F. (2002). *BMP15* maps to the X chromosome in swine. *Animal Genetics*, **33:** 158-167.

Gregory, K. E., Bennett, G. L., Van Vleck, L. D., Echternkamp, S. E. and Cundiff, L. V. (1997). Genetic and environmental parameters for ovulation rate. Twinning rate, and weight traits in a cattle population selected for twinning. *Journal of Animal Science*, **75:** 1213-1222.

Guerra-Martinez, P., Dickerson, G. E., Anderson, G. B. and Green, R. D. (1990). Embryo-transfer twinning and performance efficiency in beef production.

Journal of Animal Science, **68:** 4039-4050.

Halbert, G.W., Dobson, H., Walton, J.S. and Buckrell, B.C. (1990). A technique for transcervical intrauterine insemination of ewes. *Theriogenology,* **33:** 977-992.

Hanrahan, J. P. (2003). Aspects of reproductive performance in small ruminants – opportunities and challenges. *Reproduction, Supplement* **61:** 15-26.

Hanrahan, J. P., Gregan, S. M., Muslant, P., Mullen, M., Davis, G.H., Powell, R. and Galloway, S. M. (2004). Mutations in the genes for oocyte-derived growth factors GDF9 and BMP15 are associated with both increased ovulation rate and sterility in Cambridge and Belclare sheep (*Ovis aries*). *Biology of Reproduction,* **70:** 900-909.

Hasler, J. F., Lane, M., Musser, J., Hasler, M. J. and Gardner, D. K. (2000). Culture of bovine embryos in the sequential media G1.2/G2.2. *Theriogenology,* **53:** 295 (abstr.).

Heyman, Y., Zhou, Q., Lebourhis, D., Chavatte-Palmer, P., Renard, J.P. and Vignon, X. (2002). Novel approaches and hurdles to somatic cloning in cattle. *Cloning and Stem Cells,* **4:** 47-55.

Hillard, M. A., Wilkins, J. F., Cummins, L. J., Bindon, B. M., Tsonis, C. G., Findlay, J. K. and O'Shea, T. (1995). Immunological manipulation of ovulation rate for twinning in cattle. *Journal of Reproduction and Fertility, Supplement,* **49:** 351-364.

Hinch, G. N., Kelly, R. W., Davis, G. H., Owens, J. L. and Crosbie, S. F. (1985). Factors affecting lamb birth weights from high fecundity Booroola ewes. *Animal Reproduction Science,* **8:** 53-60.

Hofmann, A., Kessler, B., Ewerling, S., Weppert, M., Vogg, B., Ludwig, **H.,** Stojkovic, M., Boelhauve, M., Brem, G., Wolf, E. and Pfeifer, A. (2003). Efficient transgenesis in farm animals by lentiviral vectors. *EMBO Reports,* **4:** 1054-1060.

Holt, W. V. and Watson, P. F. (2004). Role of new and current methods in semen technology for genetic resource conservation. In: F*arm Animal Genetic Resources*, BSAS publication No. 30. Eds G Simm, B Villanueva, KD Sinclair and S Townsend. P 191-205. Nottingham University Press.

Hunter, R. H. F. (2003). Advances in deep uterine insemination: a fruitful way forward to exploit new sperm technologies in cattle. *Animal Reproduction Science,* **79:** 157-170.

Johanson, J. M., Berger, P. J., Kirkpatrick, B. W. and Dentine, M. R. (2001). Twinning rates for North American Holstein Sires. *Journal of Dairy Science,* **84:** 2081-2088.

Johnson, L. A. (2000). Sexing mammalian sperm for production of offspring: the state-of-the-art. *Animal Reproduction Science,* **60-61:** 93-107.

Juengel, J. L., Hudson, N. L., Whiuting, L. and McNatty, K. P. (2004). Effects of immunization against Bone Morphogenetic Protein 15 and Growth Differentiation Factor 9 on ovulation rate, fertilization and pregnancy in ewes. *Biology of Reproduction,* **70:** 557-561.

Kasinathan, P., Knott, J.G., Moreira, P.N., Burnside, A.S., Jerry, J. and Robl, J.M. (2001). Effect of fibroblast donor cell age and cell cycle on development of bovine nuclear transfer embryos in vitro. *Biology of Reproduction*, **64:** 1487-1493.

Kästli, F. (1978). Cattle twins and freemartin diagnosis. *The Veterinary Record*, **102:** 80-83.

King, A. H., Jiang, Z, Gibson, J. P., Haley, C. S. and Archibald, A. L. (2003). Mapping quantitative traits loci affecting reproductive traits on porcine chromosome 8. *Biology of Reproduction,* **68:** 2172-2179.

King, M. E., McKelvey, W. A. C., Dingwall, W. S., Matthews, K. P., Gebbie, F. E., Mylne, M. J. A., Stewart, E. and Robinson, J. J. (2004). Lambing rates and litter size following intrauterine or cervical insemination of frozen/thawed semen with or without oxytocin administration. *Theriogenology* (in press).

Kreider, D. L., Rorie, R., Brown, D., Maxwell, C., Miller, F., Wright, S. and Brown, A. (2001). Ovulation rate and litter size in gilts immunized against androtenedione and 17 α-hydoxyprogesterone. *Journal of Animal Science*, **79:** 1691-1696.

Krueger, C., Rath, D. and Johnston, L.A. (1999). Low dose insemination in synchronized gilts. *Theriogenology*, **52:** 1363-1373.

Kumar, S., Millar, J. D. and Watson, P. F. (2003). The effect of colling rate on the survival of cryopreserved bull, ram, and boar spermatozoa: a caomparison of two controlled-rate cooling machines. *Cryobiology,* **46:** 246-253.

Kuroiwa, Y., Kasinathan, P., Choi, Y.J., Naeem, R., Tomizuka, K., Sullivan, E.J., Knott, J.G., Duteau, A., Goldsby, R.A., Osborne, B.A., Ishida, I. And Robl, J.M. (2002). Cloned transchromosomic calves producing human immunoglobulin. *Nature Biotechnology*, **20:** 889-894.

Land, R. B. and Hill, W. G. (1975). The possible use of superovulation and embryo transfer in cattle to increase response to selection. *Animal Production*, **59:** 465-468.

Leibo, S.P. and Rall, W.F. (1987). Increase in production of pregnancies by bisection of bovine embryos. *Theriogenology*, **27:** 245 (Abstr.).

Lewis, R.M. and Simm, G. (2000). Selection strategies in sire referencing schemes in sheep. *Livestock Production Science*, **67:** 129-141.

Lien, S., Karlsen, A., Klemetsdal, G., Vage, D. I., Olsaker, I., Klungland, H., aasland, M., Heringstad, B., Ruane, J. and Gomez-Raya, L. (2000). A primary screen of the bovine genome for quantitative trait loci affecting twinning rate. *Mammalian Genome*, **11:** 877-882.

Linville, R. C., Pomp, D., Johnson, R. K. and Rothschild, M. F. (2001). Candidate gene analysis for loci affecting litter size and ovulation rate in swine. *Journal of Animal Science*, **79:** 60-67.

Lohuis, M. M. (1995). Potential benefits of bovine embryo-manipulation technologies to genetic improvement programs. *Theriogenology,* **43:**

51-60.

Loi, P., Ptak, G., Barboni, B., Fulka Jr., J., Cappai, P. and Clinton, M. (2001). Genetic rescue of an emdangered mammal by cross-species nuclear transfer using post-mortem somatic cells. *Nature Biotechnology*, **19:** 962-964.

Lubritz, D. l., Eisen, E. J. and Robison, O. W. (1993). Effects of pregnant mare serum gonadotropin on reproduction in 4 genotypes of gilts. *Tierzuchtung und Zuchtungsbiologie,* **110:** 363-373.

MacPherson, R. M., Hovell, F. D. DeB. and Jones, A. S. (1977). Performance of sows first mated at puberty or second or third oestrus, and carcass assessment of once-bred gilts. *Animal Production*, **24:** 333-342.

Mapletoft, R. J. Steward, K. B. and Adams, G. P. (2002). Recent advances in the superovulation in cattle. *Reproduction Nutrition Development*, **42:** 601-611.

Martinez, E.A., Vazquez, J.M., Roca, J., Lucas, X., Gil, M.A., Parrilla, I., Vazquez, J.L. and Day, B.N. (2002). Minimum number of spermatozoa required for normal fertility after deep intrauterine insemination in non-sedated sows. *Reproduction*, **123:** 163-170.

Medrano, A., Watson, P. F. And Holt, W. V. (2002). Importance of cooling rate and animal variability for boar sperm cryopreservation: insights from the cryomicroscope. *Reproduction,* **123:** 315-322.

Merton, J. s., de roos, A. P. W., Mullaart, E., de ruigh, l., Kaal, L., Vos, P. L. A. M. And Dieleman, S. J. (2003). Factors affecting oocyte quality and quantity in commercial application of embryo technologies in the cattle breeding industry. *Theriogenology*, **59:** 651-674.

McCreath, k.J., Howcroft, J., Campbell, K.H.S., Colman, A., Schnieke, A.E. and Kind, A.J. (2000). Production of gene-targeted sheep by nuclear transfer from cultured somatic cells. *Nature*, **405:** 1066-1069.

McKelvey, W.A.C. (1999). AI and embryo transfer for genetic improvement in sheep: the current scene. *In Practice*, (April) 190-195.

McNatty, K. P., Juenge, J. L., Wilson, T., Galloway, S. M., Davis, G. H., Hudson, N. L., Moeller, C. L., Cranfield, M., Reader, K. L., Laitinen, M. P. E., Groome, N. P., Sawyer, H. R. and Ritvos, O. (2003) Oocyte-derived growth factors and ovulation rate in sheep. *Reproduction, Supplement* **61:** 339-351.

Mitchell, S. E., Robinson, J. J., King, M. E., McKelvey, W. A. C. and Williams, L. M. (2002). Interleukin 8 in the cervix of non-pregnant ewes. *Reproduction* **124:** 409-416.

Montgomery, G. W., Galloway, S. M., Davis, G. H. and McNatty, K. P. (2001). Genes controlling ovulation rate in sheep. *Reproduction,* **121:** 843-852.

Morris, C. A. (1984). A review of the genetics and reproductive physiology of dizygotic twinning in cattle. *Animal Breeding Abstracts*, **52:** 803-819.

Morris, C. A. and Wheeler, M. (2002). Genetic variation in twin calving incidence in herds with a high phenotypic mean. *New Zealand Journal of Agricultural Research*, **45:** 17-25.

Morris, D. G., Browne, D., Diskin, D. G. and Sreenan, J. M. (1997). Effect of peptide to carrier ratio on the immune and ovarian response to inhibin immunization in cattle. *Animal Reproduction Science*, **48:** 1-8.

Nagy, A. (2000). Cre Recombinase: The universal reagent for genome tailoring. *Genesis,* **26:** 99-109.

Nebel, R. L., Vishwanath, R., McMillan, W. H. and Saacke, R. G. (1993). Microencapsulation of bovine spermatozoa for use in artificial insemination: a review. *Reproduction Fertility and Development*, **5:** 701-712.

Nicholas, F. W. (1996). Genetic improvement through reproductive technology. *Animal Reproduction Science*, **42:** 205-214.

Niemann, H. and Kues, W. A. (2003). Application of transgenesis in livestock for agriculture and biomedicine. *Animal Reproduction Science*, **79:** 291-317.

Picton, H. M. (2004). Oocytes and assisted reproduction technology. In: F*arm Animal Genetic Resources*, BSAS publication No. 30. Eds G Simm, B Villanueva, KD Sinclair and S Townsend. P 207-221. Nottingham University Press.

Polejaeva, I.A., Chen, S.H., Vaught, T.D., Page, R.L., Mullins, J., Ball, S., Dal., Y., Boone, J., Walker, S., Ayares, D.L., Colman, A. and Campbell, K.H.S. (2000). Cloned pigs produced by nulear transfer from adult somatic cells. *Nature*, **407:** 86-90.

Ptak, G., Loi, P, Dattena, M., Tischner, M. and Cappai, P. (1999). Offspring from one-month-old lambs: studies on the developmental capability of prepubertal oocytes. *Biology of Reproduction*, **61:** 1568-1574.

Pursel, V.G., Wall, R.J., Solomon, M.B., Bolt, D.J., Murray, J.D. and Ward, K.A. (1997). Transfer of an ovine metallothionein-ovine growth hormone fusion gene into swine. *Journal of Animal Science*, **75:** 2208-2214.

Rhind, S. M., Robinson, J. J. and McDonald, I. (1980). Relationships among uterine and placental factors in prolific ewes and their relevance to variations in foetal weight. *Animal Production*, **30:** 115-124.

Ribeiro, E. L. de A., Nielsen, M. K., Bennett, G. L. and Leymaster, K. A. (1997a). A simulation model including ovulation rate, potential embryonic viability, and uterine capacity to explain litter size in mice: I. Model development and implementation. *Journal of Animal Science*, **75:** 641-651.

Ribeiro, E. L. de A., Nielsen, M. K., Leymaster, K. A. And Bennett, G. L. (1997b). A simulation model including ovulation rate, potential embryonic viability, and uterine capacity to explain litter size in mice: II. Responses to alternative criteria of selection. *Journal of Animal Science*, **75:** 652-656.

Robinson, J. S., Kingston, W. J., Jones, C. T. and Thornburn, G. D. (1980). Studies on experimental growth retardation in sheep. The effect of removal of endometrial caruncles on fetal size and metabolism. *Journal of Developmental Physiology*, **1:** 379-398.

Rothschild, M. F., Jacobson, C., Vaske, D., Tuggle, C., Wang, L., Short, T., Eckardt, G., Sasaki, S., Vincent, A., McLaren, D., Southwood, O., van der Steen, H., Mileham, A. and Plastow, G. (1996). The estrogen receptor locus is associated with a major gene influencing litter size in pigs. *Proceedings of the National Academy of Science*, USA, **93:** 201-205.

Ryan, D. P. and Boland, M. P. (1991). Frequency of twin births among Holstein-Friesian cows in a warm dry climate. *Theriogenology,* **36:** 1-10.

Saeki, K., Matsumoto, K., Kinoshita, M., Suzuki, I., Tasaka, Y., Kano, K., Taguchi, Y., Mikami, K., Hirabayashi, M., Kashiwazaki, N, Hosoi, Y., Murata, N and Iritani, A. (2004). Functional expression of α12 fatty acid desaturase gene from spinach in transgenic pigs. *Proceedings of the National Academy of Science*, **101:** 6361-6366.

Schnieke, A.E., Kind, A.J., Ritchie, W.A., Mycock, K., Scott, A.R., Ritchie, M., Wilmut, I., Colman, A. and Campbell, K.H.S. (1997). Human factor IX transgenic sheep produced by transfer of nuclei from transfected fetal fibroblasts. *Science*, **278:** 2130-2133.

Seidel, Jr. G. E. (2003). Sexing mammalian sperm – interwining of commerce, technology, and biology. *Animal Reproduction Science*, **79:** 145-156.

Seidel, Jr. G. E., Schenk, J. L., Herickhoff, L. A., Doyle, S. P., Brink, Z., Green, R. D. and Cran, D. G. (1999). Insemination of heifers with sexed sperm. *Theriogenology*, **52:** 1407-1420.

Simm, G., Villanueva, B., Sinclair, K.D. and Townsend, S. (2004) F*arm Animal Genetic Resources*, BSAS publication No. 30. Eds. Nottingham University Press.

Sinclair, K. D., Broadbent, P. J., Dolman, D. F., Watt, R. G. and Mullan, J. S. (1995a). Establishing twin pregnancies in cattle by embryo transfer. *Animal Science*, **61:** 25-33.

Sinclair, K. D., Broadbent, P.J. and Dolman, D. F. (1995b). In vitro produced embryos as a mean of achieving pregnancy and improving productivity in beef cows. *Animal Science*, **60:** 55-64.

Sinclair, K. D. and Broadbent, P. J. (1996). Increasing the efficiency of suckled calf production using embryo transfer technology. *Veterinary Record*, **139:** 409-412.

Sinclair, K. D., Young, L. E., Wilmut, I. and McEvoy, T. G. (2000). *In utero* overgrowth in ruminants following embryo culture: Lessons from mice and a warning to men. *Human Reproduction* 15, Supplement 5, 68-86.

Sinclair, K. D. and Young, L. E. (2002). Reprogramming the donor nucleus: Imprinting disorders following nuclear transfer. In: *Epigenetics in Human Disease. Proceedings of an International Symposium*, Bethesda, MD, USA (May, 2002), pp 7.

Sinclair, K. D., Garnsworthy, P. C. and Webb, R (2003a). Recent advances in reproductive technologies: Implications for livestock production and animal nutrition. In: *Recent Advances in Animal Nutrition 2003*. Ed: P

C Garnsworthy and J Wiseman. Nottingham University Press. pp 89-113.

Sinclair, K. D., Rooke, J. A. and McEvoy, T. G. (2003b). Regulation of nutrient uptake and metabolism in pre-elongation ruminant embryos. *Reproduction, Supplement* **61:** 371-385.

Singleton, W. L. (2001). State of the art in artificial insemination of pigs in the United States. *Theriogenology,* **56:** 1305-1310.

Sirard, M.A. (2001). Resumption of meiosis: Mechanisms involved in meiotic progression and its relation with developmental competence. *Theriogenology*, **55:** 1241-1254.

Taylor, St.C.S., Moore, A.J., Thiessen, R.B. and Bailey, C.M. (1985). Efficiency of food utilization in traditional and sex-controlled systems of beef production. *Animal Production*, **40:** 401-440.

Telfer, E., Webb, R., Moor, R. and Gosden, R. (1999). New approaches to increasing oocyte yield from ruminants. *Animal Science,* **68:** 285-298.

Thibier, M. and Wagner, H.G. (2002). World statistics for artificial insemination in cattle. *Livestock Production Science*, **74:** 203-212.

Thibier, M., Humblot, P. and Guerin, B. (2004). Role of reproductive biotechnologies: global perspective, current methods and success rates. In: F*arm Animal Genetic Resources*, BSAS publication No. 30. Eds G Simm, B Villanueva, KD Sinclair and S Townsend. P 171-189. Nottingham University Press.

Tregaskes, L. D., Broadbent, P. J., Hutchinson, J. S. M., Roden, J. A. and Dolman, D. F. (1996). Attainment of puberty and response to superovulation in performance-tested Simmental heifers. *Animal Science*, **63:** 65-71.

van Berkel, P. H., Welling, M. M., Geerts, M., van Veen, H. A., Ravensbergen, B., Salaheddine, M., Pauwels, E. K., Pieper, F., Nuijens, J. H., Nibbering, P. H. (2002). Large scale production of recombinant human lactoferrin in the milk of transgenic cows. *Nature Biotechnology*, **20:** 484-487.

Van Tassell, C. P., Van Vleck, L. D. and Gregory, K. E. (1998). Bayesian analysis of twinning and ovulation rates using a multiple-trait threshold model and Gibbs sampling. *Journal of Animal Science*, 76: 2048-2061.

van Rens, B. T. T. M., de Groot, P. N and van der Lende, T. (2002). The effect of estrogen receptor genotype on litter size and placental traits at term in F2 crossbred gilts. *Theriogenology*, **57:** 1635-1649.

Villanueva, B., Simm, G. and Wooliams, J. A. (1995). Genetic progress and inbreeding for alternative nucleus breeding schemes for beef cattle. *Animal Science*, **61:** 231-239.

Vishwanath, R. (2003). Artificial insemination: the state of the art. *Theriogenology*, **59:** 571-584.

Wall, R.J. (2002). New gene transfer methods. *Theriogenology*, **57:** 189-201.

Ward, K.A. (2000). Transgene-mediated modifications to animal biochemistry. *Trends in Biotechnology*, **18:** 99-102.

Webb, R., Land, R. B., Pathiraja, N. and Morris, B.A. (1984). Passive immunization against steroid hormones in the female. In: *Immunological Aspects of Reproduction*. Ed. D.B. Crighton, Chapt. 26, pp. 475-499, Butterworths, London.

Webb, R., Driancourt, M.A. & Hanrahan, J.P. (1998). Ovulation rate in the ewe: mechanisms underlying genetic variation. *Proceedongs 6th World Congress on Genetics Applied to Livestock Production* **27:** pp3-10.

Webster, A. J. F. (1989). Bioenergetics, bioengineering and growth. *Animal Production*, **48:** 249-269.

Wells, D. N., Misica, P. M. and Tervit, H. R. (1999). Production of cloned calves following nuclear transfer with cultured adult mural granulosa cells. *Biology of Reproduction*, **60:** 996-1005.

Wells, D. N. (2004). The integration of cloning by nuclear transfer in the conservation of animal genetic resources. In: F*arm Animal Genetic Resources*, BSAS publication No. 30. Eds G Simm, B Villanueva, KD Sinclair and S Townsend. P 223-241. Nottingham University Press.

Wheeler, M. B., Bleck, G. T. and Donovan, S. M. (2001). Transgenic alteration of sow milk to improve piglet growth and health. *Reproduction, Supplement*, **58:** 313-324.

Wilson, T., Wu, X-Y, Juengel, J. L., Ross, I. K., Lumsden, J. M., Lord, E. A., Dodds, K. G., Walling, G. A., McEwan, J. C., O'Connell, A. R., McNatty, K. P. and Montgomery, G. W. (2001). Highly prolific Booroola sheep have a mutation in the intracellular kinase domain of bone morphogenic protein IB receptor (ALK-6) that is expressed in both oocytes and granulosa cells. *Biology of Reproduction*, **64:** 1225-1235.

Woolliams, J. A. (1989). The value of cloning in MOET nucleus breeding schemes for dairy cattle. *Animal Production*, **48:** 31-35.

Wooliams, J. A. and Wilmut, I. (1989). Embryo manipulation in cattle breeding and production. *Animal Production*, **48:** 3-30.

Woolliams, J. A. (2004). Managing populations at risk. In: F*arm Animal Genetic Resources*, BSAS publication No. 30. Eds G Simm, B Villanueva, KD Sinclair and S Townsend. P 85-106. Nottingham University Press, Nottingham.

Yang, J., Ratovitski, T., Brady, J.P., Solomon, M.B., Wells, K.D. and Wall, R.J. (2001). Expression of myostatin pro domain results in muscular transgenic mice. *Molecular Reproduction and Development* **60:** 351-361.

5

ALLYING GENETIC AND PHYSIOLOGICAL INNOVATIONS TO IMPROVE PRODUCTIVITY OF WHEAT AND OTHER CROPS

SEAN MAYES*, MICHAEL HOLDSWORTH*, ALESSANDRO PELLEGRINESCHI[#] AND MATTHEW REYNOLDS[#]

**Division of Agricultural and Environmental Sciences, School of Biosciences, University of Nottingham, Sutton Bonington Campus, Loughborough, Leics., LE12 5RD, UK*

[#]CIMMYT, International maize and wheat improvement centre, Apdo. Postal 6-641 06600 Mexico, D.F., Mexico.

Introduction

Currently our understanding of the genetic basis for cultivar differences in yield potential in the post-green revolution period is extremely limited (e.g. Reynolds, Sayre and Rajaram, 1999). Section I of this paper establishes the state of the art regarding the physiological and biochemical limitations to yield potential, using wheat as the example, since this is probably the best studied crop from an agronomic point of view. Wheat is a very difficult genome with which to work, compared with other staple crops such as rice or maize whose genomes are a fraction of the size, and neither of which is a true polyploid. Even in these genomically simpler species, the genetic basis of cultivar differences in yield potential is not established. However with advent of new genetic tools, progress is inevitable and Section II addresses how these technologies may be applied to wheat and other crops.

Section I: The physiological and biochemical basis of crop productivity

INTRODUCTION

Yield potential can theoretically be increased by genetic modification of one of its main components expressed in the equation below

$$YP = LI \times RUE \times HI \qquad \text{(Equation 5.1)}$$

i.e. by increasing light intercepted (LI), radiation use efficiency (RUE), or harvest index (HI): In a high yield situation light interception of a well managed crop is close to 100% from canopy closure until the onset of senescence. Therefore the amount of radiation intercepted can only be increased by improving the rate of canopy establishment (Richards, 2000) or by delaying senescence in the final stages of grain-filling (Jenner & Rathjen, 1975). However, neither of these characteristics has been shown to be associated with genetic improvement of yield in favourable environments. Nor has a consistent relationship been observed between yield and biomass when comparing wheat cultivars released over time (Slafer, Satorre and Andrade, 1994; Calderini, Dreccer and Slafer, 1995; Calderini, Reynolds and Slafer, 1999) suggesting that neither light interception nor RUE have been significantly improved by breeding. In fact yield improvement in wheat is mainly associated with larger partitioning of total above-ground biomass to grain. This has come about as a result of the introgression of Rht genes (Gale & Youssefian, 1985; Calderini *et al.*, 1995) as well as empirical selection for yield in the post-green revolution period (Kulshrestha & Jain, 1982; Waddington, Ranson, Osmanzai and Saunders, 1986; Waddington, Osmanzai, Yoshida and Ransom, 1987; Calderini *et al.*, 1995; Sayre, Rajaram and Fischer, 1997; Reynolds *et al.*, 1999).

However, the most recent studies comparing genetic progress in HI over time in spring wheat indicate no additional progress since the mid 1980s (Sayre *et al.*, 1997), suggesting that the theoretical maximum of 60% (Austin, Bingham, Blackwell, Evans, Ford, Morgan and Taylor, 1980) may be overly optimistic. If yield cannot be raised any further through the traditional route of raising HI then the more difficult challenge of raising biomass must be tackled. Since a crop loses considerable light interception at the beginning of the cycle through incomplete ground cover, improvement in this trait seems an obvious target. Canopy establishment is a complex character, determined by a number of traits, including specific leaf area and embryo size, for which genetic diversity has been established (Richards, 2000). However, in experiments where canopy establishment has been manipulated agronomically with seed density or N application, yield responses have not been observed in a high yielding spring wheat environment (Sayre, unpublished data).

There is also genetic variation for the 'stay-green' trait, which can be manipulated through N applications during grain-filling. However, to date, no evidence has been presented to indicate source limitation in high yield environments, either during the latter stages of grain-filling when leaf senescence occurs, or prior to canopy closure. Therefore genetic modification of RUE may be a more promising approach for increasing crop biomass.

THEORETICAL LIMITS TO RUE

The theoretical limits to RUE were revised by Loomis and Amthor (1996). Theoretically, the minimum quantum requirement is 9 mol photons mol^{-1} CO_2 fixed as CH_2O (Nobel, 1991). Assuming 2.07 mol photons MJ^{-1} of solar radiation absorbed by the canopy (this allows for losses due to canopy reflection), and an energy content of 17 kJ mol^{-1} for crop dry matter, the maximum conversion efficiency, without any losses from respiration or any other cause, is approximately 12% of total solar radiation (approximately 25% of PAR). Quantum efficiency is reduced further by environmental factors, principally when CO_2 concentrations are inadequate to keep pace with photochemistry. Irrespective of quantum requirement (QR), there is a fixed energy cost for maintenance respiration, using approximately 20% of gross assimilates, plus the cost of growth respiration using approximately 30% of the remaining assimilates (Amthor, 1989). While estimates of QR for wheat typically range from 20-24 mol photons mol^{-1} CO_2 fixed as CH_2O (Fischer 1983), a value of 15 photons mol^{-1} CO_2 has been reported for a C3 crop species at 25°C and ambient CO_2 concentration in controlled situations (McCree, 1971). With values of QR from 15-24, RUE varies from 1.5 to 2.6g CH_2O MJ^{-1} solar radiation (Loomis and Amthor, 1996). These figures can then be used to estimate potential biomass in a given environment based on incident radiation.

In a typical spring wheat cycle in NW Mexico, incident radiation for the crop season is approximately 2,300 MJ m^{-2}. Allowing for incomplete (78%) light interception by the crop (Reynolds, van Ginkel and Ribaut, 2000), approximately 1,800 MJ m^{-2} of radiation are absorbed by photosynthetically active tissue between emergence and physiological maturity. Assuming a range of possible values of RUE from 1.5 to 2.6g CH_2O MJ^{-1} solar radiation (from previous paragraph), above-ground biomass would therefore range from 2,700 to 4,680 g m^{-2}. Irrigated wheat in this environment currently approaches an above ground biomass of 2,100 g m^{-2} (Reynolds *et al.*, 1999). This would suggest that if the cost of root growth and maintenance is not substantially higher than previously estimated for such an environment (i.e. 10% of above ground biomass, Weir, Bragg, Porter and Rayner, 1984) improvements in RUE in the field are theoretically possible, assuming lower values of QR can be achieved through genetic manipulation of photosynthesis from cellular to canopy levels.

APPROACHES FOR RAISING RUE

Genetic modification of RUE can be tackled at the following levels (i) cellular metabolism, (ii) leaf photosynthetic rate, (iii) canopy photosynthesis, (iv) optimisation of sink size to fully exploit photosynthetic capacity. At the cellular level RUE could be increased through reducing photorespiration by increasing the affinity of Rubisco for CO_2, thereby decreasing its oxygenase activity. Variation for CO_2 specificity has been found in land plants (Parry, Keys and Gutteridge, 1989; Delgado, Medrano, Keys and Parry, 1995). However, much higher values are reported in marine algae (Read and Tabita, 1994; Uemura, Anwarazzaman, Miyachi and Yokota, 1997). Molecular techniques may offer the possibility of genetically transforming wheat with these traits (see discussion later). If this were achieved, Austin (1999) predicts increases in leaf photosynthetic rate of perhaps 20%.

Evans and Dunstone (1970) and Austin, Morgan, Ford and Bhagwat (1982) observed superior maximum leaf photosynthetic rate (A_{max}) in certain wheat ancestors, with rates up to 35% higher for *Triticum urartu*. However, the trait is only expressed pre-anthesis, does not confer superior RUE, and in most studies appears to be linked to small leaf area, rendering it unsuitable for introgression into the cultivated species. Small leaf size is associated with smaller mesophyll cell size which itself is mechanistically involved in increasing A_{max} since it is associated with reduced resistance of the CO_2 diffusion pathway (Austin *et al.*, 1982). Significant genetic variation in A_{max} has been reported in *T. dicoccoides* that was not associated with leaf anatomy (Carver and Nevo, 1990).

At the canopy level, modification of leaf architecture may improve RUE by permitting a light distribution profile that reduces the number of leaves experiencing wasteful and potentially destructive supersaturated light levels, while increasing light penetration to canopy levels where photosynthesis responds linearly to light (Duncan, Jones and Wilson, 1971). There is some experimental evidence to support this (Innes & Blackwell, 1983; Angus, Reynolds and Acevedo, 1972; Araus *et al.*, 1993; Fischer, 1996). More erect leaf canopy types are characteristic of many of CIMMYT's best yielding wheat lines (Fischer, 1996) and genetic control of leaf angle is thought to be controlled by only two or three genes (Carvalho and Qualset, 1978).

Another way of improving canopy photosynthesis is to optimise the composition of the photosynthetic apparatus, as well as N distribution, throughout the canopy, so that leaf photosynthesis is equally efficient at different light intensities. Lucerne canopies showed a clear trend for reduced total leaf N at greater depth in the canopy (Evans, 1993). In addition, chlorophyll a:b ratios declined with depth (Figure 5.1a), indicating an increased ability to capture scarce light by increasing investment in chlorophyll associated with the light harvesting antennae, relative to the reaction centres. This was consistent with a lower total N to chlorophyll N ratio (Figure 5.1c), reflecting a smaller investment in soluble protein associated with CO_2 fixation. Consequently, lower leaves had a reduced overall photosynthetic capacity in normal light (Figure 5.1b), but equally efficient RUE per unit of N at the light intensities experienced towards the bottom of the canopy. Crop models (Dreccer, Slafer and Rabbinage, 1998) support the advantage of optimising vertical distribution of canopy nitrogen in wheat.

Although there is little evidence for increased RUE or biomass in most crops (Evans, 1993) increased yield and biomass has been reported quite recently in spring wheat cultivars carrying the 7DL.7Ag translocation (Reynolds, Calderini, London and Rajaram, 2001). Experimental data indicated that increased biomass was most strongly associated with a modified source:sink balance that favourably affected RUE in these backgrounds.

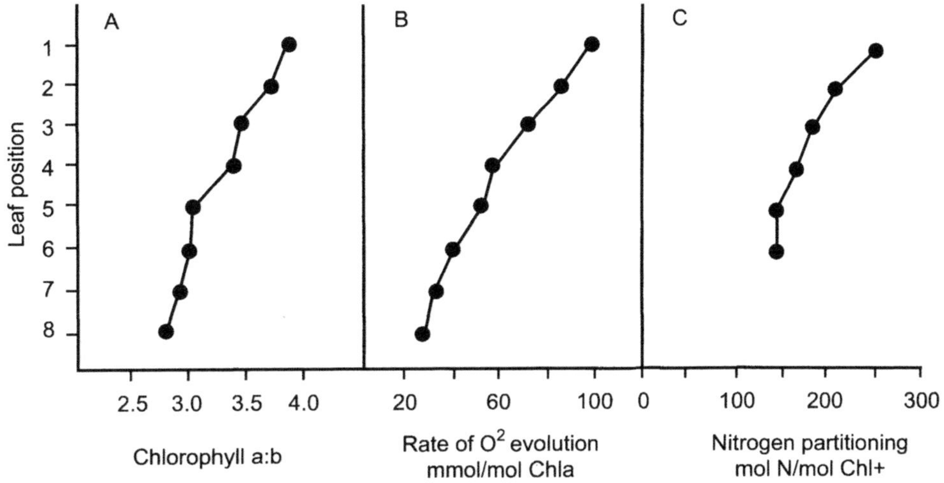

Figure 5.1 Profiles of characteristics of photosynthetic acclimation with position in the canopy. Chlorophyll *a:b* ratio (*A*), photosynthetic capacity per unit of chlorophyll (*B*) and nitrogen content per unit of chlorophyll (*C*). (Adapted from Evans JR, 1993).

EVIDENCE FOR EXCESS PHOTOSYNTHETIC CAPACITY AND SINK LIMITATION IN SPRING WHEAT

The 7Ag.7DL translocation from *Agropyron elongatum* was originally introgressed into wheat to introduce *Lr19* for leaf rust resistance. It had negligible influence on agronomic type and was found to be associated with increased yield and biomass in one background (Singh, Huerta-Espino, Rajaram and Crossa, 1998). In experiments with six spring wheat backgrounds carrying the 7DL.7Ag translocation, yield was increased on average by 13% while biomass was improved in five of the six backgrounds by 10% (Reynolds *et al.*, 2001). To explain the increased biomass, parameters relating to light interception, leaf and canopy photosynthesis, and source-sink balance were measured. There was no evidence for differences in light interception associated with the translocation (Reynolds *et al.*, 2001). However, total biomass measured shortly after flowering was 5% greater than controls, while final biomass was 9% greater (Table 5.1). In addition, grain number per spike and spike biomass at anthesis were increased by 11% and 15% respectively. This suggested that, while RUE may have been increased intrinsically prior to anthesis at the leaf or canopy level, there was a larger effect associated with a strong post-anthesis sink. This was supported by measures of leaf photosynthetic rate being 9% greater during booting, and 21% greater during grain-filling (Table 5.1).

It appears that the translocation is associated with increased spike fertility, and is manifested as a larger investment in spike mass at anthesis and a larger

Table 5.1 MAIN EFFECT OF 7DL.7AG TRANSLOCATION ON THE WHEAT CULTIVARS ANGOSTURA, STAR AND YECORA, SPRING WHEAT SEASON, NW MEXICO, 2000

	Yield ($g\ m^{-2}$)	*Bio-mass* ($g\ m^{-2}$)	*Harvest index* (%)	*Grains /spike*	*Kernel weight* (mg)	*Bio-mass Zadok -35* ($g\ m^{-2}$)	*Bio-mass at anth-esis* ($g\ m^{-2}$)	*Spike biomass at anth-esis* ($g\ m^{-2}$)	*Photo-synthesis booting* ($\mu mol\ m^{-2}s^{-1}$)	*Photo-synthesis grainfill* ($\mu mol\ m^{-2}\ s^{-1}$)
Controls	680	1,700	40.0	32.5	40.3	400	930	275	21.5	16.5
+ 7DL.7Ag	760	1,850	41.1	36.0	39.0	370	980	315	23.5	20.0
% effect	*12%*	*9%*	*3%*	*11%*	*-3.5%*	*7%*	*5%*	*15%*	*9%*	*21%*
LSD	*68*	*138*	*2.1*	*3.3*	*2.1*	*85*	*79*	*35*	*3.2*	*2.9*
P value	0.01	0.05	0.10	0.05	0.10	0.25	0.05	0.05	0.10	0.05

number of competent grains/spike, generating an increased demand for assimilates throughout grain-filling, associated with increased leaf photosynthetic rate and a greater final biomass. Although other studies have not generally shown increases in yield to be associated with increases in biomass, the results presented here are entirely consistent with a large body of literature indicating that yield is sink limited in wheat (Austin *et al.*, 1980; Kulshrestha & Jain, 1982; Fischer, 1993;1998; Slafer *et al.*, 1994; Slafer & Savin, 1994; Calderini *et al.*, 1995; Sayre *et al.*, 1997; Calderini *et al.*, 1999). The data suggest that photosynthetic rate per unit leaf area is apparently under-utilised in modern cultivars (under the relatively optimal conditions in which these experiments were conducted) and that if grain number can be improved, RUE during grain-filling can be increased in response to the need for more assimilates, permitting simultaneous increases in final biomass and yield. These hypotheses were tested in high yield wheat cultivars using a treatment that increased sink strength.

High yield wheat cultivars were grown in 4-row plots, and during booting stage a sink treatment was applied to increase grain number whereby light penetration to the inner rows was increased by bending outer rows away from inner ones during the day. At the end of the treatment rows naturally assumed a vertical position for the duration of grain-filling. The effect of the pre-anthesis sink treatment was to increase biomass at anthesis by 200 g m^{-2} and the proportion of assimilates invested in spikes of primary tillers by 5%. At maturity the sink treatment was associated with an extra 345 g m^{-2} of biomass and approximately four extra grains/spike, while average kernel weight and harvest index were not affected (Table 5.2). Leaf photosynthetic rate measured at different times during grain-filling were on average 10% higher in plots that had experienced the pre-anthesis sink treatment (Table 5.2). The data confirmed the hypothesis that photosynthetic rate is apparently under-utilised in modern cultivars and that RUE during grain-filling can be increased through increasing the number of grains per spike.

Table 5.2 EFFECT OF INCREASING LIGHT PENETRATION INTO A WHEAT CANOPY FOR 15 DAYS DURING BOOTING STAGE ON GROWTH TRAITS IN THE WHEAT CULTIVAR SERI 7DL.7AG, OBREGON, NW MEXICO, 2001

Treatment	*Grain Yield* ($g\ m^{-2}$)	*Biomass at maturity* ($g\ m^{-2}$)	*Harvest index* (%)	*Kernel weight* (mg)	*Grains /spike*	*Spike number* (m^{-2})	*Biomass at anthesis* ($g\ m^{-2}$)	*Tiller number at anthesis* (m^{-2})	*Spike: culm biomass at anthesis* (%)	*Photo-synthesis grainfill* ($\mu mol\ m^{-2}\ s^{-1}$)
Check	630	1555	40.5	35.4	36.3	495	950	600	27.7	21.1
Source-sink	785	1900	41.5	36.7	40.2	540	1150	710	29.1	23.2
% effect	25%	22%	2.5%	4%	11%	9%	21%	18%	5%	10%
LSD	152	345	1.7	3.1	4.0	149	206	284	0.74	2.0
P value	0.05	0.05	0.10	0.25	0.05	0.20	0.05	0.15	0.01	0.05

The evolutionary basis of this is not clear; under non-agronomically managed situations there would be a clear advantage in having an excess photosynthetic capacity at flowering to increase the chances of seed filling and maturation in an unpredictable environment. High temperatures, reduced solar radiation and low precipitation can all impact negatively on photosynthesis. In addition, biotic factors such as disease, pressure from herbivores and insects, and interplant competition for resources can also reduce photosynthetic capacity.

TRAITS FOR INCREASING SPIKE-FERTILITY

The evidence outlined in the previous sections clearly indicates a need to improve understanding of the genetic basis of spike fertility so that these genes can be targeted in both conventional and molecular approaches to breed for increased grain number. Breeders must incorporate these objectives into empirical breeding approaches using available genetic variation in traits that may influence spike fertility. One example is the multi-ovary (MO) trait (Chen, Skovmand, Rajaram and Reynolds, 1998) shown to increase grain number when introgressed into improved spring wheat backgrounds (Figure 5.2). Other traits associated with spike fertility (Richards, 1996; Slafer *et al.*, 1996) have been tested in spring wheat backgrounds by comparing trait expression with that in unselected progeny of three crosses whose parents contrasted in (i) growth rate during spike-growth stage, (ii) partitioning of biomass to spike at anthesis, (iii) relative duration of spike-growth stage and (iv) relative duration of pre-anthesis period. Preliminary data indicate that partitioning of biomass to the spike and a relatively longer pre-anthesis period were best associated with yield (Table 5.3). In addition to grain fecundity, increasing potential kernel weight independently of increases in grain number would be another avenue to increase sink demand during grain-filling (Calderini and Reynolds, 2000).

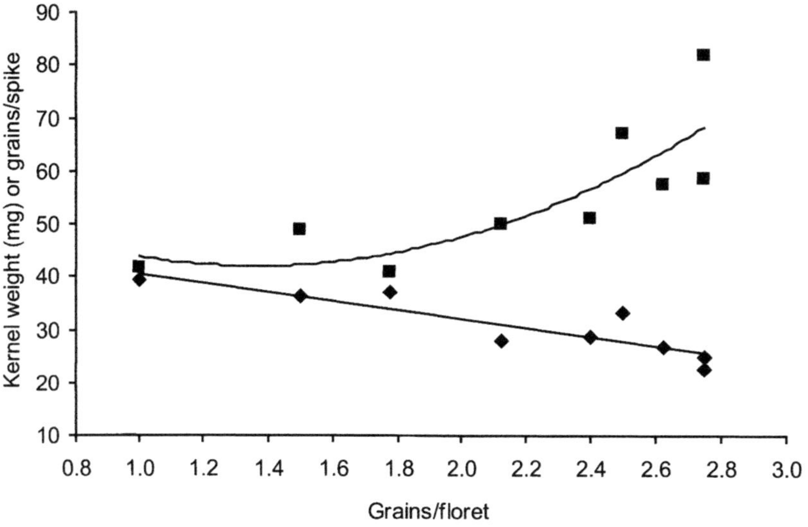

Figure 5.2 The relationship between the number of grains/floret and (i) kernel weight (◆) and (ii) grains/spike (■), in eight genotypes expressing the multi-ovary (MO) trait, and the average value of eight lines not expressing the MO trait, for progeny of crosses between an MO source and five elite spring bread wheat cultivars, NW Mexico, 2001. Fitted lines: Grains/spike = $12.9x^2 - 33.5x + 63.4$; $r^2 = 0.66$. Kernel weight = $8.6x + 49.2$; $r^2 = 0.79$.

Table 5.3 GENETIC CORRELATIONS BETWEEN SPIKE-FERTILITY RELATED TRAITS AND YIELD IN THREE POPULATIONS OF RANDOM INBRED SPRING WHEAT LINES, NW MEXICO, 2002

	Cross-1 (n=36)	*Cross-2 (n=25)*	*Cross-3 (n=25)*
Growth rate (spike-growth stage)	0.10	0.25	0.35
Partitioning to spike (at anthesis)	0.30	0.61	0.53
Relative spike-growth duration	0.08	0.04	0.65
Relative pre-anthesis duration	0.30	0.42	0.79

TRANSFORMATION OPPORTUNITIES TO INCREASE RUE

Although the genetic basis of variation in yield potential is poorly understood in crops, a number of physiological processes can be targeted for genetic improvement based on theory, although to date there are no candidate genes for increasing sink strength and only limited possibilities for increasing CO_2 fixation genetically. These include genes related to the C_4 photosynthesis (Hausler, Hirsch, Kreuzaler and Peterhansel, 2002), and those associated with improving the CO_2 specificity of ribulose-1,5-biphosphate carboxylase / oxygenase (Quick 1994; Uemura *et al.*, 1997; Austin, 1999; Hausler *et al.*, 2002).

To improve yield in C_3 crops, genes related to the C_4-cycle have been targeted (Hausler *et al.,* 2002) as a means to raise RUE. The rationale for focusing on the over-expression of the C_4 cycle enzymes is that concentrating CO_2 at the site of ribulose-1,5-biphosphate carboxylase / oxygenase (Rubisco) would be expected to increase RUE by reducing losses associated with photorespiration, as well as by permitting a reduction in stomatal conductance in response to mild soil water deficit without a dramatic decline in the rate of CO_2 assimilation (Drake & Gonzalez-Meler, 1997). The increased interest in this strategy is based on the fact that the anatomical division into mesophyll and bundle sheaths cells is not a prerequisite for the functioning of the CO_2 concentrating mechanism. Supporting evidence for this hypothesis includes the observation that submerged aquatic macrophytes can induce a C_4-like CO_2 concentrating mechanism in only one cell type when CO_2 becomes a limiting factor (Casati, 2000). However, because C_4–cycle enzymes have a distinct function in C_3 plants, an increase of their activity (over-expression) is likely to disturb the plant's metabolism or trigger compensational changes in the metabolic fluxes. An example of that is the introduction in potato of a PEPC from *Flaveria trinervia* (Rademacher, Hausler, Hirsch, Zhang, Lipka, Weier, Kreuzaler and Peterhansel, 2002). Both modifications resulted in enzymes with lowered sensitivity to malate inhibition and an increased affinity for PEP. Constitutive over-expression of PEPC carrying both N-terminal and internal modifications strongly diminished plant growth and tuber yield. Metabolite analysis showed that carbon flow was re-directed from soluble sugars and starch to organic acids (malate) and amino acids, which increased four-fold compared with the wild type. This and other observations are of interest in studying the regulatory interrelationship between different metabolic pathways, but for the moment, knowledge gained from such research is not applicable to the breeding process.

Another way to increase RUE via increased CO_2 fixation by Rubisco may be to increase the affinity of Rubisco for CO_2, which would also decrease its oxygenase activity and thereby reduce photorespiration. The ratio of carboxylase (Vc) to oxygenase (Vo) activity is known as the Rubisco specificity factor (expressed as relative units i.e. Vc/Vo). Rubisco specificity factor shows genetic variation in higher plants (Delgado *et al.*, 1995) and while wheat for example, already has a relatively favourable specificity factor in comparison with other crop species higher values are reported in marine algae (Read and Tabita, 1994; Uemura *et al.*, 1997). Molecular techniques may offer the possibility of genetically transforming Rubisco from its current specificity (i.e. Vc:Vo ratio) of 95 in wheat to values of 195 corresponding to thermophilic alga *Galderia partita* (Uemura *et al.*, 1997). If this were achieved, Austin (1999) predicts increases in leaf photosynthetic rate of 20% at 20°C and ambient CO_2, using a biochemical model for CO_2 assimilation presented by Farquhar, von Laemmerer and Berry (1980). Various Rubisco genes from pea, sunflower, and the cyanobacterium Synechococcus PCC6301 have been expressed in *Arabidopsis thaliana* and tobacco (Hausler *et al.,* 2002, Quick, 1994). Studies investigating this strategy demonstrate that the recombinant protein was expressed and able to assemble with the plant native subunits, however, the hybrid enzymes showed lower carboxylase activity at saturating substrate concentrations, suggesting that higher-plant Rubisco, unlike the cyanobacterial enzyme, has evolved species-specific interactions between S and the large subunit protein that are involved in carbamylation of the active site. These results indicate that further investigations are needed before transgenic lines with these genes can be used in the field.

YIELD POTENTIAL AND MOISTURE STRESS

Another strategy for improving aggregate yield potential is to improve yield stability in the face of environmental variation among sites and seasons. For example, moisture stress can occur in a number of high yield scenarios such as: (i) drier than average years in rain-fed environments, (ii) towards the end of irrigation cycles when plant roots may respond to moisture stress at certain soil profiles even when water is available at other depths (Davies and Zhang, 1991), (iii) when vascular capacity is exceeded by evaporative demand during periods of high vapour pressure deficit. The study of roots is fundamental to making progress in moisture-stressed environments. Field measurements have generally shown crops with more extensive root systems to be associated with improved performance under drought (Hurd, 1968; 1974; Boyer, 1996). Furthermore, when considering the genetic diversity found in root exploration capacity among crop species and wild plants (Taylor and Terrell, 1982), the potential benefits of genetic transformation are hard to ignore. Studies with trees suggest that the weakest link in the hydraulic flow path from roots to the atmosphere is in the roots, especially the small roots which are very vulnerable to cavitation (Hacke and Sauter, 1996). A recent modelling exercise of cereal crops by a UK group concluded that crop yield could be increased by increasing investment in fine roots at depth, at the expense of proliferation of surface roots. They estimate that the benefits to water capture would be especially important (King, Gay, Sylvester-Bradley, Bingham, Foulkes, Gregory and Robinson, 2003). Due to the logistical problems associated with direct measurement of root structure and function, this is also an area that is likely to benefit substantially from a sound understanding of the genetic basis of root characteristics and the identification of genetic markers for use in crop improvement.

CONCLUSIONS ON THE PHYSIOLOGICAL AND BIOCHEMICAL BASIS OF CROP PRODUCTIVITY

Physiological data collected in well managed high yield environments indicate that grain number m^{-2} (sink) is the current primary rate limiting step for yield potential in spring wheat, and that excess photosynthetic capacity (source) exists during grain-filling, even in the highest yielding cultivars. Improvement of RUE at the crop level should, therefore, be tackled both indirectly through increasing our understanding of the genetic and physiological basis of spike fertility, as well as directly at the level of canopy photosynthesis and photosynthetic metabolism. To reflect its complexity, a model describing yield potential needs to incorporate the fact that RUE varies over the crop cycle in response to both environment and source: sink balance. The following modification to equation 5.1 is suggested:

$$YP = HI \times \int_{e}^{m} (RUE \times LI \qquad \text{(Equation 5.2)}$$

where e and m represent emergence and maturity respectively. The equation derives from an analysis of dry matter accumulation presented by Monteith (1977) but, in Equation 5.2, dry matter is expressed as the product of LI and RUE integrated over the crop cycle, to recognise the fact that both RUE and LI have developmental components. The idea that the RUE of a genotype can be differentiated according to crop stage is extremely relevant to its genetic dissection. Physiological models tend to consider average effects of RUE (see Hunt, Reynolds, Sayre, Rajaram, White and Yan, 2003; S. Chapman, personal communication) overlooking evidence that genetic expression of yield and biomass are determined by interactions between phenological stage and environment (Fischer, 1975; Abbate, Andrade and Culot, 1995; Reynolds *et al.*, 2002). The theoretical upper limit to crop yield and biomass (Loomis & Amthor, 1996) implies a constant upper value for RUE which, though its achievement is difficult to imagine, nonetheless defines the ultimate goal for manipulation of the crop genome.

Section II: Genetic approaches for improving crop productivity

INTRODUCTION

Understanding the physiological basis for crop productivity is essential to identify traits which can be used in crop genetic improvement programmes. Section I dealt with the identification and assessment of such traits, particularly in wheat, and examined how they might be relevant to both transgenic and conventional breeding programmes. Section II surveys the tools which are currently available to identify genes that underlie such traits.

By the end of this year there will be two completely finished plant genome sequences available - *Arabidopsis thaliana* and rice. Many genomics-related resources have been or are being developed in *Arabidopsis*, including large insert genomic libraries, expressed sequence tags (ESTs), gene-chips, knock-out and knock-down mutation lines, enhancer-trap lines and Single Nucleotide Polymorphism (SNP) markers, to name but a few (see Nottingham Arabidopsis Stock Centre, http://arabidopsis.info/). While the resources available in rice are currently more limited, the production of a finished, publicly available genome sequence for this second most important cereal crop will clearly be a major step and of particular relevance to other cereals (Yu, Ricke, Lan and 52 others, 2002; Goff, Hu, Wang and 97 others, 2002).

These resources, and others, offer the possibility of using genomics to pursue genes which affect crop traits such as those described in the preceding section from two different directions: 1) Using molecular biology and transformation, 2) Using molecular genetics. The two molecular methodologies are complementary and non-exclusive; they approach trait dissection from different directions. The molecular biology approach focuses on a particular pathway and attempts to dissect its genes and their interactions, starting with single genes (often defined by mutations) and expanding this into networks of genes involved in the expression of a trait. The major limitations of this approach are that it is relatively inefficient with respect to trait complexity, (however, note transcriptomics and proteomics below) and considerable time is needed to fully characterise each gene. The molecular genetics approach involves the use of markers to tag the location of factors affecting a trait through genetic mapping. The major disadvantage of this approach is that many important classes of breeding traits are quantitative and multi-locus (some exceptions to this are disease resistance genes - e.g. Michelmore and Meyers, 1998, and specific traits such as the semi-dwarfing Rht genes- Gale & Youssefian, 1985; Calderini *et al.*, 1995). The potential of molecular genetics and associated marker-assisted breeding technologies has been recognised for the last twenty years or so, but how much of an impact has it made and how does the recent development of extensive genomics resources, both physical and informational, enhance its value?

This section of the review focuses primarily on the application of new breeding technologies to the genome challenge that crop scientists face. The review will consider genomics associated with crop species, but will not address current advances in the model *Arabidopsis*. It is considered that major investment in resources associated with crop genomics and, as importantly, crop physiological analysis, will be required before the promise of *Arabidopsis* as a true model for crop species can be realised. Wheat is taken as the example species here.

FINGERPRINTING AND DIVERSITY ANALYSIS

While fingerprinting may seem trivial, the lack of morphological markers in most crop species to distinguish between individuals or cultivars actually makes fingerprinting one of the most practical ways in which marker analysis can have an impact on plant breeding (e.g. Mohapatra, Manifesto, Schlaffer, Hopp, Suarez and Dublovsky, 2003; Manifesto, 2001; Warburton, Xianchun, Crossa, Franco, Melchinger, Frisch, Bohn and Hoisington, 2002; Huang, Boerner, Roeder and Ganal, 2002). Examination of diversity between cultivars, between landraces and between related species provides an idea of how individuals and populations are related. This is useful for maintaining efficient germplasm collections, selecting possible material for introgression into advanced material and for understanding how modern elite material is descended (e.g. Roussel, Koenig, Beckert and Balfaurier, 2004; Bandopadhyay, Sharma, Rustgi, Singh, Kumar, Balyan and Gupta, 2004; Queen, Gribbon, James, Jack and Flavell, 2004). This approach has developed recently to produce association mapping (see below). An example of a diversity analysis for CIMMYT wheat germplasm is given in Figure 5.3.

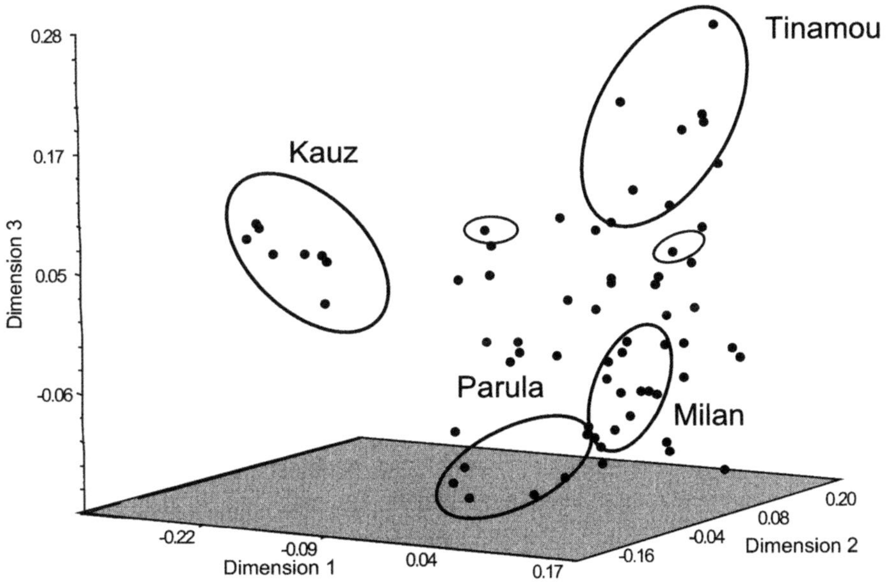

Figure 5.3 Principle coordinate analysis (PCoA) based on microsatellite markers showing relationships between CIMMYT breeding lines. Circled lines are sister lines closely related by pedigree.

GENETIC MAPPING

Genetic mapping can allow the dissection of simple traits and their assignment to a chromosomal location. Dense genetic maps are available for most crop species (see http://ukcrop.net/ and associated links). Dominant and multi-locus marker systems (such as RAPD, AFLP and most retrotransposon-based systems) have limited transferability between breeding crosses. However, co-dominant markers (initially RFLPs, more recently microsatellites and even SNPs) allow the same locus in the genome to be examined in many different crosses, and this permits links to be built between genetic maps constructed in different material. From these links, consensus maps can be constructed for a species as a whole. For RFLPs, this has even permitted cross-species comparisons and has led to the discovery of conserved synteny and the development of comparative mapping (see below) between species that diverged tens of millions of years ago.

MAPPING IN WHEAT

The first common mapping population that became available was the International Triticeae Mapping Initiative (ITMI) Opata85 x Synthetic cross, for which 115 Recombinant Inbred Lines (RILs) are available. There is also extensive marker data available for this cross, so it would be possible to assess further traits without the need to extensively re-map or even map at all (Leroy, Negre, Tixier and 13 others, 1997). The disadvantage of this cross is that it is extremely wide genetically and may not be relevant to breeding programmes.

The development of wheat doubled haploid (DH) populations (derived from haploid gametic tissue restored to diploid status by a treatment, such as colchicine, which 'doubles' this haploid complement) and their adoption as standards in map construction mean that marker density can be increased rapidly using multilocus systems, which are fully informative in DH populations. The lines themselves are also 'fully inbred'.

Hexaploid bread wheat has 3 genomes (6n = 3x = AABBDD) yet it behaves like a triple diploid in meiosis. This inhibition of homeologous recombination in favour of homologous recombination is controlled by the action of the Ph-loci. Of these, Ph1 on chromosome 5B is the most important and loss of activity leads to increased levels of homeologous recombination (see Moore, 2002 and Martinez-Perez, Shaw, Aragon-Aleable and Moore, 2003) that is usually detrimental to the offspring. Prevention of homeologous recombination permits genetic maps to be constructed for each homeologous chromosome-set and keeps the genetics essentially diploid. Another anomaly in wheat is the suppression of recombination in a large part of the centromeric region, with the majority of genetic maps representing only the distal third or so of the chromosome, compared to the physical maps (e.g. Faris, Fellers, Brooks and Gill, 2000). However, routine use of molecular markers to introduce single gene characters such as nematode resistance (*Cre1* and *Cre3*), barley yellow dwarf virus (BYDV) resistance, to detect the Sears deletion (*ph1b*) and to introduce stripe rust (*Yr17*) are routinely carried out by CIMMYT (William, Crosby, Trethowan, van Ginkel, Mujeeb-Kazi, Pfeiffer, Khairallale and Hoisington, 2003a) and other labs.

QTL ANALYSIS

Most traits show complex patterns of inheritance with a continuous distribution of trait values and these traits need to be approached as Quantitative Trait Loci. QTL analysis makes no assumptions as to the nature of the underlying gene or genes which are influencing the trait, it simply uses a variety of methods to try to explain phenotypic variation in terms of genetic factors along the chromosome groups. All of these methods rely on grouping trait data on the basis of genotype and then testing for significant differences between the groupings and all give broadly similar results, whether the method used is based upon Maximum Likelihood, multiple regression or even simple Students-t tests (Botstein, White, Skolnick and Davis, 1980; Kearsey and Pooni, 1998, for background)

Dissection of the overall genetic variation for a trait (segregating within a particular cross) into components localised to particular chromosomes allows a better understanding of the factors underlying a trait and permits the use of flanking markers to select for beneficial QTL and against detrimental ones. Two closely linked QTL may confound the analysis, potentially being mistaken for a single QTL, or, if the effect on the trait values of the QTL are in opposite directions, for the QTL to be missed entirely. Interactions between the QTL, such as epistasis may also confound the analysis (Jiang, Chee, Draye, Smith and Patterson, 2000). A further problem is pleiotropy versus independent QTL effects. A QTL may affect more than one trait. This may be the effect of the same gene on a linked pathway for the two traits (e.g. grain number and yield) or of two separate genes which are closely linked. Given that large numbers of locus / trait tests are carried out, deciding a suitable level of significance for a 'real' QTL can be debatable (Lander and Botstein, 1989; Lander and Kruglak, 1995; Churchill and Doarge, 1994).

Despite these obstacles, QTL analysis is popular and has been extensively used to dissect the genetics of agronomic traits. Table 5.4 lists some of the QTL for important traits which have been identified recently in particular crosses of wheat. This is not an attempt to produce a comprehensive list, but different traits have been illustrated.

Because wheat is hexaploid there is the added complication of three genomes contributing to QTL effects and the question of whether genes at the same position, but on different homeologous chromosomes, may multiply the contribution to a trait. The evidence so far suggests that this is not the case (however, note Tg and Tg2 in Jantasuriyarat, Vales, Watson and Riera-Lizarazu, 2004). It is not known whether this is because only one of the three copies is usually active, or because multiple copies confound the statistical analysis (as the genetics would essentially be hexaploid), or because there are two 'spare' copies on the other genomes which can compensate for any loss, so that the trait does not often segregate. Genome and transcriptome analysis may help to answer this question, as will the development of SNP markers from EST collections (Mochida, Yamazaki and Ogihara, 2003).

Table 5.4 EXAMPLES OF RECENT QTL STUDIES. POPN. = POPULATION USED FOR MAPPING; DH – DOUBLED HAPLOID LINES; RILS – RECOMBINANT INBRED LINES; SSD – DERIVED BY SINGLE SEED DECENT; F(X) INDICATES BREEDING GENERATION

Trait assessed	*Cross*	*Popn.*	*Major QTL detected, location and variation explained*	*Reference*
Flag leaf senescence: (GFLS% –14d and 35d post-anthesis; yield)	Beaver x Soissons	48DH	Effects (unirrigated): Chr 2D (25.7% – 32.9%) Effects (irrigated): Chr 2B (10.2% –11.4%)	Verma *et al.*, 2004
Free threshing Spike compactness	Opata85 x synthetic (ITMI)	115 RILs	Free threshing: Chr 2DS (17% - 44% associated traits; Tenacious glumes gene- Tg?) Chr 5AL (10% - 21% associated traits; 'Q' domestication gene Spike compactness: multiple)	Jantasuriyarat *et al.*, 2004
Cereal Cyst Nematode- Tolerance Resistance	Trident x Molineau	91 DH	Tolerance: Chr 6B (60%) Resistance: Chr 6B (64%)	William *et al.*, 2003
Seed dormancy on Chr 3A	Zen x Chinese Spring	125 RILs (SSD; F7)	Resistance to pre-harvest sprouting: Chr 3AS (23.3% - 38.2%) However, not Vp1 or red grain (RA-1) by position	*Osa et al.*, 2003
Grain hardness puroindoline-a (Pin-a) puroindoline-b (Pin-b) (Chr 5D)	Opata85 x synthetic (ITMI)	115 RILs	GH: Chr 5DS (63%-71%) Pin-a: Chr 5DS (72%-77%) Pin-b: Chr 5DS (25%-45%) May relate to *Ha* locus for soft grain	Igrejas *et al.*, 2002

FROM QTL TO GENES?

For QTL analysis, increasing marker density, hence decreasing marker intervals to less than 10cM, does little to improve the localisation of the QTL – the key factor for localisation is the number of meiotic events characterised, to identify rare recombination events which split closely linked markers away from the study trait. The recent cloning of wheat vernalisation genes illustrate the problems associated with the localisation of even simply assessed single genes, using positional cloning. A major constraint is the large numbers of meiotic events needed to localise the trait, due to the very high kilobase / centimorgan ratios in crops such as wheat (Yan, Loukoianov, Tranquill, Helguera, Fahima and Dubcovsky, 2003; Yan, Loukoianov, Bleckl, Tranquill, Ramakrishna, San Miguel, Bennetzen, Echenique and Dubcovsky, 2004). Both VRN1 and VRN2 from a wheat ancestor (*T. monococcum*) were cloned using such positional approaches. VRN2 was localised to a region within 0.04cM using 5,698 gametes. This represented a region of 328Kbp in *T. monococcum,* and contains 8 candidate genes and one pseudogene. Vernalisation requirement is a simple Mendelian trait, with a correspondingly relatively simple physiological screen.

Many QTL may have 95% confidence intervals which span 20 or 30cM and may be difficult to screen simply. The development of 'smart' physiological screens, to allow early or simple assessment of more complex traits would be a major step forward to permit the screening of large numbers of gametes. In terms of genes, a typical 'QTL' region would be expected to contain around 4,000 to 6,000 (using the gene density identified for VRN2 – results for VRN1 are similar, although slightly less gene-rich.) This is simply too many to be able to use the tools of molecular biology and transformation to evaluate each possible candidate gene. In order to clone the genes responsible for traits we need to find some way to reduce considerably this number of candidates (see below).

It is interesting to consider whether we actually need to know exactly which gene represents a particular QTL. Flanking markers can be used to introgress QTL into elite material. This is essentially a slightly more sophisticated approach to that taken by traditional plant breeders in wheat for the last century, with the introgression of alien chromosome segments carrying disease resistance genes. For example, the 7Ag.7DL translocation from *Agropyron elongatum* mentioned in Section I above was introduced into wheat to provide the Lr19 leaf rust resistance gene. It also carries factors that enhance spike fertility. This can essentially be viewed as a 'QTL', in which the two effects are probably due to different genes, rather than the pleiotropic effects of one gene. There may also be several thousand other *Agropyron* genes in the introgressed segment, depending upon the physical size of the segment. In different genetic backgrounds, or in different environments, these may act in a deleterious way. Even with the current introgression and results presented in this paper, the true value of the 'QTL' may be masked by associated deleterious genes. Once a wheat plant is homozygous for the introgression, the presence of the *Agropyron* 'QTL' is fixed and there is no opportunity to further reduce the introgressed segment size through recombination (if such recombination were to occur), as the *Agropyron* genes essentially act as a single unit. Cloning of the genes involved may therefore provide a route to allow the beneficial ones to be retained and the detrimental ones eliminated.

ASSOCIATION MAPPING

An important consideration with the QTL approach is that each detected QTL has a genomic 'context'. A detected QTL simply indicates which genetic factors affect the selected trait in a particular cross, in a particular environment, grown during a particular year. To be able to use QTL generically, it is necessary to test whether the QTL is still important if all three contexts are changed. QTL are also, as mentioned above, poorly localised. One attempt to identify whether there are more 'universal' loci involved that affect trait values, is association mapping. Rather than using a defined cross, or set of crosses, from a limited germplasm pool, attempts have been made to examine large numbers of individuals and to generate marker data across the genome for these. Trait data are then scored and the two data types are compared to identify any 'association' between alleles and traits. This approach relies on continued linkage disequilibrium between the marker and the locus defining the trait. As genes defining traits are introduced to a crop during its domestication and breeding, they will also carry flanking regions of the donor chromosome with them. These regions can only be broken up by the action of recombination during subsequent breeding. Linkage disequilibrium approaches are based on the assumption that this 'homogenisation' in most breeding material is incomplete. It also has potential advantages in polyploid crops where traditional analysis is complicated by multiple genomes.

One potential problem is that it is difficult to distinguish a marker linked to a trait from a marker which comes from the original genome donating the trait, but is not genetically linked to that trait, particularly with introgressions from alien species. Studies in sugar beet (Kraft, Hansen and Nilsson, 2000), maize (Thornsburry, Goodman, Doebley, Kresovich, Nielsen and Buckler, 2001) and potato (Gebhardt, Ballvosa, Walkemeier, Oberhagemann and Schuler, 2004) all indicate that the remaining linkage disequilibrium could be limited in extent, with 0.2cM, 1cM and 0.3cM being the extent of remaining disequilibrium association between trait and marker showing association, respectively. This is potentially also one of the strong points of association mapping. Any confirmed associations between marker and trait are at a much higher resolution than QTL analysis. However, it may be that large numbers of putative associations need to be tested by progeny screening to confirm authenticity. For cases such as the R1 resistance gene introgressed from *S. demissum* into potato, the original introgressed resistance gene is also no longer functional against *Phytophthora infestans*. Interestingly however, Gebhardt *et al.* (2004) found that closely linked resistance loci are still active.

Association mapping has also been used to test for alleles of genes which had been selected for in domestication of maize (Vigouroux, McMullen, Hittinger, Houchins, Schulz, Kresovich, Matsuoka and Doebley, 2002) and to examine the number of domestication events (Matsuoka, Vigouroux, Goodman, Sanchez, Buckler and Doebley, 2002).

DOMESTICATION GENES

As crops have been domesticated (for cereals see Salamini, Ozkan, Brandolini, Schäfer-Pregl and Martin, 2002) they have been selected for some common characteristics. A simple example is ear or seed-pod shattering. This is involved in the dispersal of seed following grain / pod ripening. For efficient collection of seed shattering is highly detrimental, so one of the initial changes that was made during domestication of many species was selection of non-shattering lines. Other changes were also important, for instance plant height and response to photoperiod (see Paterson, 2002).

In the history of maize domestication from the wild relative teosinte, there have been major morphological changes. These include the suppression of 2° lateral shoots in teosinte, the feminisation of the 1° shoots and the development of the multiple-rowed cobs. Using QTL analysis in a teosinte x maize cross, Doebley, Stec, Wendel and Edwards (1990) demonstrated that there were five key regions of the maize genome which accounted for most of the necessary differences between the domesticated and the wild versions. Considerable work has been undertaken to characterise these QTL and 3 major changes are closely localised; *teosinte branched, teosinte glume architecture* and *two ranked. Teosinte branched 1* has been cloned by insertional inactivation (Doebley, Stec and Hubbard, 1997; see deletion and insertional inactivation stocks below) and is involved in increasing apical dominance in maize compared to teosinte. Lev-Yadun, Abba and Doebley, 2002 recently suggested that once these genes have been cloned, transformation of temperate cereals (which resemble teosinte more than maize) with the equivalent genes, could lead to major breeding advances.

In tetraploid wheat, a recent QTL study crossing durum wheat (*T. durum*) with wild Emmer wheat (*T. diccocoides*) identified 7 'Domestication Syndrome Factors' (DSFs; Peng, Ronin, Fahima, Roder, Li, Nevo and Korol, 2002). This might indicate that relatively limited numbers of loci are needed for the development of 'modern' wheats (and 80% of DSF QTL were attributed to just 4 chromosomes out of 14), although whether these QTL represent single key genes or groupings of genes is unclear.

Recently, Faris *et al.* (2003) have identified a candidate gene in a BAC contig spanning the Q locus for 'free-threshing', which was an important step during domestication. The candidate identified is a MADS-box transcription factor with similarity to the *Arabidopsis* APETELA-2 gene for floral morphogenesis. If this proves to be the correct gene, it would be a second example where a transcription factor has been responsible for complex morphological change during domestication, as for *teosinte branched* (above).

COMPARATIVE MAPPING

RFLPs have proved invaluable in the production of comparative maps between different plant species and they have formed the basis of the study of conserved synteny between many related species. In the Poaceae, comparative mapping has led to the conclusion that in most cases gene order could be explained by rearranging 33 linkage blocks (Moore, 1995). While it is now clear that at the micro-synteny level, this relationship is often more complex [at least for maize (Bennetzen and Ramakrishna, 2002)], this was still a remarkable discovery, suggesting that tools developed in one species could be applied in another. However, it has also become clear that conserved synteny between the first sequenced plant, *Arabidopsis*, and the second, rice, is far more restricted (Paterson, Bowers, Chapman, Peterson, Rong and Wicker, 2004), as it is between *Arabidopsis* and maize (Brendel, Kurtz and Walbot, 2002). This has led to the adoption of *Arabidopsis* as the presumptive linkage model for dicotyledonous plants and rice for monocotyledonous plants. As more genomes are sequenced, more phylogenetically relevant comparisons will be made (for example, *Medicago truncatula* will act as a model species for many Legumes).

In the longer term, direct comparison of genome sequences will replace linkage-based approaches. We are still some time away from this for any of the major crop species, however, the approach is promising (Paterson, Lan, Amasino, Osborn and Quiros, 2001; Grant, Cregan and Shoemaker, 2000 and Mysore, Tuori and Martin, 2001). Detailed cross-species comparisons have been carried out in a number of cases for certain genome regions. In particular, in maize, the *alcohol dehydrogenase* (*adh1*), *shrunken2* (*sh2*) and *bronze* (*bz*) regions have all been compared to the gene order in the related species, sorghum and rice. These comparisons indicate that the genomes of the three species vary significantly for DNA content, gene content, gene orientation and repetitive element content within the regions under examination (see Bennetzen, 2000).

It had been assumed that gene order and content were likely to be relatively fixed within any given species. Fu and Donner (2002) have recently sequenced and compared 100Kbp of the *bz* region of two maize inbred lines, B73 and McC. Their results show that co-linearity is violated between the two lines for this region, that there are differences in the gene content and that even the repeat element blocks differ in both element types and position within the region. This major violation of intra-specific co-linearity presents a challenge to the applicability of conserved synteny and perhaps more importantly, using single genome sequence accessions to represent a species. It also intriguingly raises the possibility that heterosis in maize is due to the hybrid actually having a greater compliment of genes compared to the inbreds, with a significant number being effectively hemizygous in the hybrid (see Bennetzen and Ramakrishna, 2002).

However, maize (while having the greatest number of inter-species comparisons and the only major intra-species comparison available for cereal crops) may prove to be the exception rather than the rule. Maize is an ancient tetraploid, with a polyploidisation event estimated to have occurred around 11 million years ago. Since then, it appears to have been becoming progressively diploidised. This was illustrated by Ilic, San Miguel and Bennetzen (2003) who isolated the two 'homeologous' copies of the *adh1* region in maize, one derived from each of the original diploid genomes before the polyploidisation event. Compared to the relatively conserved gene content and order between rice and sorghum, both maize copies show the deletion of non-identical sets of genes, with only a single gene out of 12 present still producing a transcript in both regions. The Moore *et al.* (1995) paper on order of linkage blocks in maize suggested a significant number of translocations and rearrangements between the remnants of the original donor genomes in maize. This polyploidisation event may have resulted in rapid rearrangement and gene loss to allow functional meiotic recombination in the original tetraploid. However, it has focused a spotlight on the question of genotypic variation within species.

In hexaploid wheat, while some deletion of genes has been reported in newly synthesised hexaploids (Feldman, 1997) and by comparison with bread wheat ancestors (Akhunov, Akhunova, Linkiewicz and others, 2003), the situation may be different from that in maize, due to the action of the Ph-loci (see above). So, while there are reasons to think that such gross violations of conserved synteny and gene content and gene order may be less problematic in wheat than maize, we currently have little real data, particularly in hexaploid wheat. Cloning of the VRN2 region in rice, barley and wheat allowed a comparison between gene content, with barley and rice having significantly less DNA in the comparison region than wheat (328Kbp wheat, 26Kbp barley and 7Kbp; Yan *et al.*, 2004). While such comparisons are intrinsically useful in understanding and predicting possible candidate genes for a trait, there is no guarantee that the same genes will be identified in other cereal species. For the VRN2 gene, rice appears to have no equivalent. This might be predictable, as the natural habitat of rice would not experience vernalisation-like temperatures. In barley, however, vernalisation might be expected to have relevance, although no gene equivalent was found in the barley contig. The reason for this is relatively easy to explain as the BAC library used was made from the variety Morex, which is a Spring barley and does not have a vernalisation requirement. Once identified in wheat, it was possible to show that the gene is deleted in most spring barleys, including Morex, but present in winter barleys which have a vernalisation requirement. This latter study provided compelling evidence that the correct gene had been identified, but it did suggest that there are at least limited gene content differences intra-specifically in barley, as shown in maize.

CANDIDATE GENES

One way in which it might be possible to help predict which of a group of genes is a likely candidate for the one controlling a trait, is to try to predict what sort of gene might be expected from comparative studies in other organisms, or from the nature of the trait. A gene which causes a major morphological or developmental change is more likely to be a transcription factor. A good example already given is *teosinte branched 1* (*tb1*) in maize. This gene encodes a flowering–related MADS-box transcription factor and the mutant provides a similar phenotype to *cyclopeidia* in *Antyrrhinum majus.* Of course, similarity of phenotype can also be achieved by convergence of function, rather than through a common ancestry. The vernalization genes in *Arabidopsis*, for example, are not the same ones as those identified in wheat.

Another example where information in one species has been useful in identifying candidate genes in another is for the trait of Pre-Harvest Sprouting (PHS). Hexaploid bread wheat (*Triticum aestivum*) caryopses are characterised by relatively weak embryo dormancy and display PHS under cool moist conditions. The physiology of sprouted wheat is very similar to the phenotypes of *Arabidopsis* and maize mutants disrupted in proteins containing the plant-specific B2 and B3 DNA binding domains (*abi3*, *fus3*, *lec2*, and *vp-1*). These proteins are transcription factors that function both to activate gene expression involved in seed maturation and simultaneously inactivate genes involved in germination. Analysis of three of these wheat Vp-1homoeologues has shown that each gene produces a set of cytoplasmic mRNAs resulting from variant-splicing of the genes. Many of these transcripts do not have the capacity to encode full-length proteins, suggesting that the population of Vp-1 protein derived from variant-splicing may lead to susceptibility to sprouting. Alternative splicing is not the result of sequence evolution as a consequence of domestication of wheat, because Vp-1orthologues show splice variation in the presumed progenitors of wheat. However, alternative splicing is not detected outside the Triticeae (McKibben, Wilkinson, Bailey, Flintham, Andrew, Lasseri, Gale, Lenton and Holdsworth, 2002).

'Gene therapy' approaches have been used to complement endogenous wheat Vp-1 activity with *Avena fatua* Vp-1 (afVp-1). Results show that the presence of the afVP1 gene has a positive effect in transgenic lines; promoting resistance to sprouting. These results indicate that genes and genetic pathways identified in model species could have important applications for the manipulation of dormancy characteristics in crop species. As more genes are identified and their modes of action understood, this approach will become more powerful.

TRANSCRIPTOMICS AND PROTEOMICS

The development of gene microarrays and the ability to define protein components via micro-sequencing together offer one way in which it may be possible to help bridge the gap between QTL identification and the individual gene level. Microarrays (containing either microspots of amplified ESTs or the more powerful photo-lithographic (e.g. Affymetrix) process which allows over half-a-million 25-mer oligonucleotides to be built onto silicon), allow the expression patterns of very large numbers of genes to be assessed simultaneously, using mRNA derived from tissues of interest. For a particular trait, it may be possible to examine situations when the physiological expression of the trait is high or low and to examine whether there are corresponding differences in the expression patterns of any particular gene or sets of genes.

This approach has been used recently by Mitra, Gleason, Edwards, Hadfield, Downie, Oldroyd and Long (2004) who have cloned a nodulation signalling gene (DMI3) from *Medicago truncatula* by comparing gene expression levels in the wild with those in the mutant dmi3 type on a custom gene microarray (9,935 genes). Identification of mutant genes using microarrays may be aided by the fact that incomplete and frame-shifted transcripts are very unstable. A similar test to identify the already cloned RAR1 gene in barley using the barley Affymetrix chip, identified RAR-1 as the 11th best candidate. While not 'precise', the possibility of using this approach to significantly reduce the potential pool of candidate genes identified by QTL analysis is clear.

DELETION AND INSERTIONAL INACTIVATION STOCKS

For a number of species, insertional inactivation through the activity of transposons or T-DNAs (e.g. Maize - Mu; Hanley, Edwards, Stevenson, Haines, Hegartey, Schuch and Edwards, 2000; rice Tos17, Yamazaki, 2001; *Arabidopsis* T-tagged lines, Alonso, Stepanova, Leisse and 36 others, 2003) has been an invaluable tool for creating random mutation lines, which can then be screened for phenotypic effects of agronomic interest. The *teosinte branched 1* (*tb1*) gene (mentioned above) in maize was cloned following insertional inactivation, as was the *Antyrrinum* homologue. For wheat there is evidence for the presence of transposon-like sequences.

Its three genomes give wheat a unique advantage over many crop species. The use of deletion, addition and substitution lines in wheat has been heavily exploited during the last century to generate a series of invaluable cytogenetic stocks, such as ditelosomics, monosomics and nullitetrasomics, where part of one homeologous chromosome is missing and the genes lost on that section are compensated for by the remaining homeologous genes. The very fact that these can be made, suggests that there is at least some overlap of function between the homeologous genes. This has allowed many traits to be localised to particular chromosomes, for example, the Sears' deletion, which is on chromosome 5B originally defined the position of the Ph-1 locus, which maintains homologous recombination in the wild-type (Sears, 1976). Recently 3,977 ESTs have been localised into 159 chromosome bins in wheat through the use of hybridisation of such genetic stocks (Akhunov *et al.*, 2003; according to http://wheat.pw.usda.gov/index.shtml this number has now been extended to 16,000 EST sequences binned in wheat). With the impending release of an Affymetrix chip for wheat, and also the use of slide-based EST arrays, the placement of genes will become more comprehensive and more precise. Even with the relatively limited number of genes and bins above, a detailed assessment of homeologous conserved synteny was made. Four thousand neutron deletion lines in bread wheat are being developed using single seed decent by the John Innes Centre, together with EMS lines in a wheat ancestor at Rothampsted, as part of the Wheat Genetic Improvement Network (WGIN). In addition to providing a resource for the identification of agronomically interesting phenotypes, these lines could also be used to generate a dense and overlapping series of deletion sets for the entire genome, using the gene-chip hybridisation approach. Such a resource will be invaluable for reducing the list of candidates for QTL. If this were coupled with an EMS mutagenised population (EMS generally causes point mutations), changes in expression pattern for genes from wild-type, neutron mutagenised (deletion lines) and EMS lines (point mutations) could be examined to try to narrow the candidate gene pool.

General conclusions

Combination of data from physiology, molecular biology and molecular genetics still remains a challenge, largely due to the complexity of crops, but connection through computer modelling and bio-informatics may be important in handling and using the extensive datasets that will emerge. One of the major roles of physiology is to identify and characterise physiological and developmental processes that underlie the target traits that could make a real impact on plant breeding. The tools of molecular science can then be used to begin to dissect these highly defined traits, either from a forward (molecular genetics) or a reverse (molecular biology) direction.

High through-put systems for both marker data generation and for expression analysis will soon remove many of the practical constraints on the generation of molecular data, allowing the full potential of these approaches to be realised. However, it is only through a combination of approaches, particularly involving physiological assessment, that practical breeding progress will be made.

For wheat, the most promising trait for immediate progress appears to be RUE. While it would be theoretically possible to engineer plants to increase their affinity for CO_2 (decrease the oxygenase activity of Rubisco), other conventional approaches also show promise. The *Agropyron elongatum* introgression of the Lr19 resistance gene which also gives significant yield increases through increased spikelet fertility could be dissected through a combination of the approaches mentioned above, and the genes responsible for this trait could eventually be cloned, if this proves necessary and desirable. Reducing the number of candidate genes for a trait remains a problem at the moment, but as more becomes known over the next ten years about the interactions and their fundamental biology, it will become easier to select the most likely subset of candidates and to test their effects. An increase in the rate of meiotic recombination represents one long-term option to localise QTL with increasing accuracy or the use of overlapping deletion mutation sets for wheat could achieve the same end. The association mapping approach also needs to be examined, as it offers the potential to localise traits to within relatively small distances, with the generation of dense marker information unlikely to remain a major issue by the end of the decade. Comparative mapping can be used to inform our search for the gene responsible for a trait and to facilitate work in less developed crop species.

Major advances will be made in the next decade in fundamental plant science, the challenge is to ensure that these discoveries make it through into varieties, to ensure food quality and quantity, and environmental sustainability for their future.

References

Abbate, P.E., Andrade, F.H. and Culot, J.P. (1995) The effects of radiation and nitrogen on number of grains in wheat. *Journal of Agricultural Science*. (Cambridge) **14**: 351-360.

Akhunov., E.D., Akhunova, A.R., Linkiewicz, A.M., Dubcovsky, J., Hummel, D. and others. (2003) Synteny perturbations between wheat homoeologous chromosomes caused by locus duplications and deletions correlate with recombination rates. *Proceeding of the National Academy of Science (USA)* **100**: 19, 10836–10841

Alonso, J.M., Stepanova, Leisse and others (2003) Genome-Wide Insertional Mutagenesis of *Arabidopsis thaliana. Science* **301** (5633): 653-657

Amthor, J.S. (1989) *Respiration and Crop Productivity*. Springer-Verslag, New York.

Angus. J.F., Jones. R. and Wilson, J.H. (1972) A comparison of barley cultivars with different leaf inclinations. *Australian Journal of Agricultural Research* **23:** 945-957.

Araus, J.L., Reynolds, M.P. and Acevedo, E. (1993) Leaf posture, grain yield, growth, leaf structure, and carbon isotope discrimination in wheat. *Crop Science* **33:** 1273-1279.

Austin, R.B., Bingham, J., Blackwell, R.D., Evans, L.T., Ford, M.A., Morgan, C.L. and Taylor, M. (1980) Genetic improvement in winter wheat yields since 1900 and associated physiological changes. *Journal of Agricultural Science* **94**: 675-689

Austin R.B., Morgan C.L., Ford M.A., Bhagwat S.G. (1982) Flag leaf photosynthesis of *Triticum aestivum* and related diploid and tetraploid species. *Annals of Botany* **49**: 177-189

Austin, R.B. (1989) The climatic vulnerability of wheat. In *Climate and Food Security*. International Rice Research Institute and American Association for the Advancement of Science, Washington, DC.

Austin, R.B. (1999) Yield of wheat in the UK: Recent advances and prospects. *Crop Science* **39**: 1604-1610

Bandopadhyay, R., Sharma, S., Rustgi, S., Singh, R., Kumar, A., Balyan, H.S. and Gupta, P.K. (2004) DNA polymorphism among 18 species of Triticum-Aegilops complex using wheat EST SSRs *Plant Science*: **166**: no. 2, pp. 349-356.

Bennetzen, J.L. (2000) Comparative Sequence Analysis of Plant Nuclear Genomes: Microcolinearity and Its Many Exceptions. *The Plant Cell*, **12**: 1021–1029

Bennetzen, J.L. and Ramakrishna, W. (2002) Exceptional haplotype variation in maize. *Proceeding of the National Academy of Science (USA)* **99**: 14 9093–9095

Botstein D, White R, Skolnick M, Davis R. (1980) Construction of a genetic linkage map in man using restriction fragment length polymorphisms. *American Journal of Human Genetics* **32**: 314–331.

Boyer, J.S. (1996) Advances in drought tolerance in plants. *Advances in Agronomy* **56:** 187-218.

Brendel, V., Kurtz, F. and Walbot, V. (2002) Comparative genomics of *Arabidopsis* and maize: prospects and limitations *Genome Biology*, **3(3)**: reviews1005.1–1005.6

Calderini, D.F., Dreccer, M.F. and Slafer, G.A. (1995) Genetic improvement in wheat yield and associated traits. A re-examination of previous results and the latest trends. *Plant Breeding* **114:**108-112.

Calderini, D.F., Reynolds, M.P. and Slafer, G.A. (1999) Genetic gains in wheat yield and main physiological changes associated with them during the 20th century. In *Wheat: Ecology and Physiology of Yield Determination*. Edited by E.H. Satorre and G.A. Slafer. New York, Food Products Press.

Calderini, D.F, and Reynolds, M.P. (2000) Changes in grain weight as a consequence of de-graining treatments at pre- and post-anthesis in synthetic hexaploid lines of wheat (*Triticum durum* X *T. tauschii*). *Australian Journal of Plant Physiology* **27**: 183-191.

Canevara, M.G., Romani, M., Corbellini, M., Perenzin, M. and Borghi, B. (1994) Evolutionary trends in morphological, physiological, agronomical and qualitative traits of *Triticum aestivum* L. cultivars bred in Italy since 1900. *European Journal of Agronomy* **3:**175-185.

Carvalho, F.I.F. and Qualset, C.O. (1978) Genetic variation for canopy architecture and its use in wheat breeding. *Crop Science* **18:** 561–567.

Carver, B.F. and Nevo, E. (1990) Genetic diversity of photosynthetic characters in native populations of *Triticum dicoccoides*. *Photosynthesis Research* **25**:119-128.

Casati, P. (2000) Induction of a C_4-like mechanism of CO_2 fixation in *Egeria densa*, a submersed aquatic species. *Plant Physiology.* **123(4):**1611-1621, 2000.

Chen, T.Y., Skovmand, B.S., Rajaram, S. and Reynolds, M.P. (1998) Novel source of increased spike fertility in wheat multi-seeded flowers. *Agronomy Abstracts* (90th ASA, Oct.18-22, Baltimore, MD).

Churchill, G. and Doerge, R. (1994) Empirical threshold values for quantitative trait mapping. *Genetics* **138**: 963–971.

Davies, W.J. and Zhang, J. (1991) Root signals and the regulation of growth and development of plants in drying soil. *Annual Review of Plant Physiology and Plant Molecular Biology* **42**: 55-76.

Delgado, E., Medrano, H., Keys, A.J. and Parry, M.A.J. (1995) Species variation in Rubisco specificity factor. *Journal of Experimental Botany* **292:** 1775–1777.

D'Ennequin, M.L.T., Toupance, B., Robert, T., Godelle, B. and Gouyon, P. (1999) Plant domestication: a model for studying the selection of linkage. *Journal of Evological Biology* **12**: 1138–1147.

Doebley, J., Stec, A., Wendel, J. and Edwards, M. (1990) Genetic and morphological analysis of a maize-teosinte F2 population: Implications for the origin of maize. *Proceeding of the National Academy of Science (USA)* **87**: 9888-9892.

Doebley, J., Stec, A. and Hubbard, L. (1997) The evolution of apical dominance in maize. *Science* **286**: 485-488.

Drake B.G. and Gonzalez-Meler, M.A. (1997) More efficient plants: a consequence of rising atmospheric CO_2? *Annual Review of Plant Physiology and Plant Molecular Biology* **48:** 609–639.

Dreccer, M.F., Slafer, G.A. and Rabbinge, R. (1998) Optimization of vertical distribution of canopy nitrogen: an alternative trait to increase yield potential in winter cereals. *Journal of Crop Production* **1:** 47–77.

Duncan, W.G. (1971) Leaf angles, leaf area, and canopy photosynthesis. *Crop Science* **11:** 482-485.

Evans, L.T., and R.L. Dunstone. (1970) Some physiological aspects of evolution in wheat. *Australian Journal of Biological Science* **23**: 725–741.

Evans, J.R. (1993) Photosynthetic acclimation and nitrogen partitioning within a lucerne canopy. I. Canopy characteristics. *Australian Journal of Plant Physiology* **20**: 55–67.

Faris, J.D., Haen, K.M. and Gill, B.S. (2000) Saturation of a gene-rich recombination hotspot in wheat. *Genetics* **154:** 823–835

Faris, J.D., Fellers, J.P., Brooks, S.A. and Gill, B.S. (2003) A bacterial artificial chromosome contig spanning the major domestication locus Q in wheat and identification of a candidate gene. *Genetics* **164**: 311–321

Farquhar, G.D., von Caemmerer, S. and Berry, J.A. (1980) A biochemical model of photosynthetic CO_2 assimilation in leaves of C_3 species. *Planta* **149**:78–90.

Feldman, M., Liu, B., Segal, G., Abbo, S., Levy, A.A., and Vega, J.M. (1997) Rapid Elimination of Low-Copy DNA Sequences in Polyploid Wheat: A Possible Mechanism for Differentiation of Homoeologous Chromosomes

Genetics **147**: 1381-1387

Fischer, R.A. (1975) *Report on wheat physiology research at CIMMYT 1970-75.* Internal CIMMYT document. CIMMYT, Mexico, D.F.

Fischer, R. A. (1983) Wheat. In: Smith W H, Banta J J, eds. *Potential productivity of field crops under different growth conditions.* Los Baños, Philippines; IRRI, 129-154.

Fischer, R.A. (1993) Irrigated spring wheat, and timing and amount of nitrogen fertilizer. II. Physiology of grain yield response. *Field Crops Research* **33:** 57-80.

Fischer, R.A. (1996) Wheat physiology at CIMMYT and raising the yield plateau. In: *Increasing Yield Potential in Wheat: Breaking the Barriers.* Edited by M.P. Reynolds, S. Rajaram and A. McNab. Mexico: CIMMYT, 195-203.

Fischer, R.A., Rees, D., Sayre, K.D., Lu, Z-M., Condon, A.G. and Larque-Saavedra, A. (1998) Wheat yield progress associated with higher stomatal conductance and photosynthetic rate, and cooler canopies. *Crop Science* **38:** 1467-1475

Fu, H and Donner, H.K. (2002) Intraspecific violation of genetic colinearity and its implications in maize. *Proceedings of the National Academy of Science (USA)* **99** :14, 9573–9578

Gale, M.D. and Youssefian, S. (1985) Dwarfing genes in wheat. In *Progress in Plant Breeding.* Edited by E. Russel. London: Butterworth and Co. 1-35.

Gebhardt, C., Ballvora, A., Walkemeier, B., Oberhagemann, P. and Schüler, K. (2004) 'Assessing genetic potential in germplasm collections of crop plants by marker-trait association: a case study for potatoes with quantitative variation of resistance to late blight and maturity type.' *Molecular Breeding* **13**: 93-102.

Goff, S.A., Ricke, D., Lan, T-H., and 52 others (2002) A draft sequence of the rice genome (*Oryza sativa* L. ssp. japonica). *Science* **296**, 92–100

Grant, D., Cregan, P., and Shoemaker. R.C. (2000) Genome organization in dicots: genome duplication in *Arabidopsis* and synteny between soybean and *Arabidopsis*. *Proceeding of the National Academy of Science (USA)* **97:** 4168-4173.

Hacke, U. and Sauter, J.J. (1996) Drought induced xylem dysfunction in petioles, branches and roots of *Populus balsamifera* and *Alnus glutinosa* (L) Gaertn. *Plant Physiology* 111: 413-417

Hanley, S., Edwards, D., Stevenson, D., Haines, S., Hegarty, M., Schuch, W., Edwards, K.J. (2000) Identification of transposon-tagged genes by the random sequencing of Mutator-tagged DNA fragments from *Zea mays Plant Journal* **23**: 4, pp. 557-566

Hausler, R.E., Hirsh, H.J.F., Kreuzaler, F. and Peterhansel, C. (2002) Overexpression of C_4-cycle enzymes in transgenic C_3 plants: a biotechnological approach to improve C_3-photosynthesis. *Journal of Experimental Botany* **53:** 591-607.

Huang, X.Q., Boerner, A., Roeder, M.S. and Ganal, M.W. (2002) Assessing genetic diversity of wheat (*Triticum aestivum* L.) germplasm using microsatellite markers. *Theoretical and Applied Genetics* **105**: 5, 699-707

Hunt, L. A., Reynolds, M.P., Sayre, K.D., Rajaram, S., White, J.W. and Yan, W. (2003) Crop modelling and the identification of stable coefficients that may reflect significant groups of genes. *Agronomy Journal* **95**: 20-31

Hurd, E.A. (1968) Growth of roots of seven varieties of spring wheat at high and low moisture levels. *Agronomy Journal* **60**: 201-205

Hurd, E.A. (1974) Phenotype and drought tolerance in wheat. *Agricultural Meteorology* **14**: 39-55

Igrejas, G., Leroy, P., Charmet, G., Gaborit, T., Marion, D. and Branlard, G. (2003) Mapping QTLs for grain hardness and puroindoline content in wheat (*Triticum aestivum* L.) *Theoretical and Appied Genetics* **106**: 19–27

Ilic, K., San Miguel, P.A. and Bennetzen, J.L. (2003) A complex history of rearrangement in an orthologous region of the maize, sorghum, and rice genomes. *Proceedings of the National Academy of Science (USA)* **100**: 21, 12265–12270

Innes, P. and Blackwell, R.D. (1983) Some effects of leaf posture on the yield and water economy of winter wheat. *Journal of Agricultural Science* **101:** 367-376.

Jantasuriyarat, C., Vales, M.I., Watson, C.J.W. and Riera-Lizarazu, O. (2004) Identification and mapping of genetic loci affecting the free-threshing habit and spike compactness in wheat (*Triticum aestivum* L.) *Theoretical and Applied Genetics* **108:** 261–273

Jiang, C.P., Chee, X., Draye, P., Morrell, C.W., Smith, Paterson A.H. (2000) Multi–locus interactions restrict gene flow in advanced-generation interspecific populations of polyploid *Gossypium* (cotton). *Evolution* **54**: 798–814.

Jenner, C.F. and Rathjen, A.J. (1975) Factors regulating the accumulation of starch in ripening wheat grain. *Australian Journal of Plant Physiology* **2:** 311-322.

Kearsey, M.J. and Pooni, H.S. (1998) *The genetical analysis of quantitative traits.* Chapman and Hall, UK. ISBN: 0-7487-4082-1

King, J., Gay, A., Sylvester-Bradley, R., Bingham, I., Foulkes, J., Gregory, P. and Robinson, D. (2003) Modelling Cereal Root Systems for Water and Nitrogen Capture: Towards an Economic Optimum. *Annals of Botany* **91**: 383-390

Kraft T., Hansen M., Nilsson N.-O. (2000) Linkage disequilibrium and fingerprinting in sugar beet. *Theoretical and Applied Genetics* **101**: 323–326.

Kulshrestha, V.P. and Jain, H.K. (1982) Eighty years of wheat breeding in India: past selection pressures and future prospects. *Zeitschrift für Pflanzenzüchtung* **89:** 19-30.

Lander, E. and Botstein, D. (1989) Mapping Mendelian factors underlying quantitative traits using RFLP linkage maps. *Genetics* **121**: 185–199.

Lander, E. and Kruglyak, L. (1995) Genetic dissection of complex traits: Guidelines for interpreting and reporting linkage results. *Nature Genetics* **11**: 241.

Leroy, P., Negre, S., Tixier, M.H., Perretant, M.R., Sourdille, P., Gay, G., Bernard, M., Coville, J.L., Quetier, F., Nelson, C., Sorrells, M., Marino, C.L., Hart,

G., Friebe, B., Gill, B.S. and Roder, M. (1997) A genetic reference map for the bread wheat genome – *Triticum aestivum* L. em. Thell. In *Progress in genome mapping of wheat and related species*. Joint Proc 5th and 6th Public Workshops Int Triticeae Mapping Initiative. Edited by P.E. McGuire, C.O. Qualset. Report No. 18, University of California Genetic Resources Conservation Program, Davis, California, pp 134–140

Lev-Yadun, S., Abba, S. and Doebley, J. (2002) Wheat, Rye and Barley on the cob. *Nature Biotechnology* **20**: 337-338

Lin, Y., Schertz, K. and Paterson, A. (1995) Comparative analysis of QTLs affecting plant height and maturity across the Poaceae, in reference to an interspecific sorghum population. *Genetics* **141**: 391–411.

Loomis, R.S. and Amthor, J.S. (1996) Limits to yield revisited. In: *Increasing Yield Potential in Wheat: Breaking the Barriers*. Edited by M.P. Reynolds, S. Rajaram and A. McNab. Mexico: CIMMYT, 76-89.

Loomis, R.S. and Amthor, J.S. (1999) Yield potential, plant assimilatory capacity, and metabolic efficiencies. *Crop Science* **39:** 1584-1596.

Manifesto, M.M. (2001) Quantitative evaluation of genetic diversity in wheat germplasm using molecular markers. *Crop Science* **41**: 3, 682-690.

Martinez-Perez, E., Shaw. P., Aragon-Alcalde, L. and Moore, G. (2003) Chromosomes form into seven groups in hexaploid and Tetraploid wheat as a prelude to meiosis. *Plant Journal.* **36:** 21-29

Matsuoka, Y., Vigouroux, Y., Goodman, M.M., Sanchez, J.G., Buckler, E., and Doebley, J. (2002) A single domestication for maize shown by multilocus microsatellite genotyping. *Proceeding of the National Academy of Science (USA)* **99**: 9 6080–6084

McCree, K.J. (1971) The action spectrum, absorptance and quantum yield of photosynthesis in crop plants. *Agricultural Meteorology* **9:**191-216.

McKibbin, R.W., Wilkinson, M.D., Bailey, P.C., Flintham, J.E., Andrew, L.M., Lazzeri, P.A., Gale, M.D., Lenton, J.R., and Holdsworth, M.J. (2002) Transcripts of Vp- 1 homeologues are mis-spliced in modern wheat and ancestral species. *Proceedings of the National Academy of Science (USA)* **99:** 15 10203–10208.

Michelmore, R.W. and Meyers, B.C. (1998) 'Clusters of resistance genes evolve by divergent selection and a birth and death process.' *Genome Research* **8:** 1113 – 1130.

Mitra, R.M., Gleason, A.C., Edwards, A., Hadfield, J., Downie, J.A., Oldroyd, G.E.S and Long, S.R. (2004) A Ca^{2+} / Calmodulin-dependent protein kinase required for symbiotic nodule development: Gene identification by transcript cloning. *Proceeding of the National Academy of Science (USA)* Early Edition www.pnas.org/cgi/doi.10.1073/pnas 0400595101

Mochida, K., Yamazaki Y. and Ogihara, Y. (2003) Discrimination of homoeologous gene expression in hexaploid wheat by SNP analysis of contigs grouped from a large number of expressed sequence tags. *Molecular and General Genomics* **270**: 371-377.

Mohapatra, T., Manifesto, M.M., Schlatter, A.R., Hopp, H.E., Sua´ rez, E.Y., and Dubcovsky, J. (2003) STMS-based DNA fingerprints of the new plant type wheat lines. *Current Science*: **84**, no. 8, pp. 1125-1129

Monteith, J.L. (1977) Climate and the efficiency of crop production in Britain. *Phil Transactions of the Royal Society London*, **281**: 277-294

Moore, G. (1995) Cereal genome evolution: pastoral pursuits with 'Lego' *Current Opinions in Genetics and Development* **5**: 6 717-724.

Moore, G. (2002) Meiosis in allotetraploids – The importance of 'Teflon' chromosomes. *Trends in Genetics* **18**: 456-463

Mysore, K.S., Tuori, R.P. and Martin, G.B. (2001) *Arabidopsis* genome sequence as a tool for functional genomics in tomato. *Genome Biology* **2:** reviews1003.1-1003.4.

Nobel, P. (1991) *Biochemical and environmental plant physiology*. San Diego:Academic Press.

Parry, M.A.J., Keys, A.J. and Gutteridge, S. (1989) Variation in the specificity factor of C_3 higher plant Rubiscos determined by the total consumption of ribulose-P_2. *Journal of Experimental Botany* **40**: 317–320.

Paterson, A., Lander, E., Hewitt, J., Peterson, S., Lincoln, S. and Tanksley, S. (1988) Resolution of quantitative traits into Mendelian factors by using a complete map of restriction fragment length polymorphisms. *Nature* **335**: 721–726.

Paterson, A., Lin, Y., Li, Z., Schertz, K., Doebley, J., Pinson, S., Liu, S., Stansel, J. and Irvine, J. (1995) Convergent Domestication of Cereal Crops by Independent Mutations at Corresponding Genetic Loci. *Science* **269**: 1714–1718.

Paterson, A H., Lan, T., Amasino, R., Osborn, T.C. and Quiros, C. (2001) *Brassica* genomics: a complement to, and early beneficiary of, the *Arabidopsis* sequence. *Genome Biology:* **2**: reviews1011.1-1011.4.

Paterson, A.H. (2002) What has QTL mapping taught us about plant domestication? *New Phytologist* **154**: 591–608

Paterson, A.H., Bowers, J.E., Chapman, B.A., Peterson, D.G., Rong, J. and Wicker T.M. (2004) Comparative genome analysis of monocots and dicots, toward characterization of angiosperm diversity. *Current Opinion in Biotechnology* **15** (2) 120-125

Peng, J. Ronin, Y., Fahima, T., Roder, M.S., Li, Y., Nevo, E. and Korol, A. (2002) Domestication quantitative trait loci in *Triticum dicoccoides*, the progenitor of wheat. *Proceedings of the National Academy of Science (USA)*. **100**:5: 2489-2494.

Osa, M., Kato, K., Muri, M., Shindo, C., Torado, A and Miura, H. (2003) Mapping QTLs for seed dormancy and the VP1 homologue in Chr. 3A in wheat. *Theoretical and Applied Genetics* **106**: 1491-1496

Queen, R.A., Gribbon, B.M., James, C., Jack, P. and Flavell, A.J. (2004) Retrotransposon-based molecular markers for linkage and genetic diversity analysis in wheat. *Molecular Genetics and Genomics* **271**: 91–97

Quick, W.P. (1994) Analysis of transgenic tobacco plants containing varying amounts of ribulose-1,5-bisphosphate carboxylase/oxygenase. *Biochemical Society Transactions* **22:** 899-903

Read, B.A., and F.R. Tabita. (1994) High substrate specificity ribulose bisphosphate carboxylase/oxygenase from eukaryotic marine algae and properties of recombinant Cyanobacterial rubisco containing “algal” residue residue modifications. *Archives of Biochemistry and Biophysics*. **312**: 210–218.

Rademacher, T., Hausler, R.E., Hirsch, H.J., Zhang, L., Lipka, V., Weier, D., Kreuzaler, F. and Peterhansel, C. (2002) An engineered phosphoenolpyruvate carboxylase redirects carbon and nitrogen flow in transgenic potato plants *The Plant Journal* **32**: 25–39

Reynolds, M.P., Sayre, K.D. and Rajaram, S. (1999) Physiological and genetic changes of irrigated wheat in the post green revolution period and approaches for meeting projected global demand. *Crop Science* **39:** 1611-1621.

Reynolds, M.P., van Ginkel, M. and Ribaut, J.M. (2000) Avenues for genetic modification of radiation use efficiency in wheat. *Journal of Experimental Botany* **51:** 459-473.

Reynolds, M.P., Calderini, D.F., Condon, A.G. and Rajaram, S. (2001) Physiological basis of yield gains in wheat associated with the *LR19* translocation from *Agropyron elongatum*. *Euphytica* **119:** 137-141.

Reynolds, M.P., Trethowan, R., Sayre, K.D. and Crossa, J. (2002) Physiological factors influencing genotype by environment interactions in wheat. *Field Crops Research* **75:** 139-160

Richards, R.A. (1996) Increasing the yield potential of wheat: manipulating sources and sinks. In: *Increasing Yield Potential in Wheat: Breaking the Barriers*. Edited by M.P. Reynolds, S. Rajaram and A. McNab. Mexico: CIMMYT, 101-134.

Richards, R.A. (2000) Selectable traits to increase crop photosynthesis and yield of grain crops. *Journal of Experimental Botany* **51:**447-458.

Roussel, V., Koenig, J., Beckert, M., Balfourier, F. (2004) Molecular diversity in French bread wheat accessions related to temporal trends and breeding programmes. *Theoretical and Applied Genetics* **108**: 5, 920-930.

Salamini, F., Özkan, H., Brandolini, A., Schäfer-Pregl, R. and Martin, W. (2002) Genetics and geography of wild cereal domestication in the near east. *Nature Reviews Genetics* **3:** 429–441.

Sayre, K.D., Rajaram, S. and Fischer R.A. (1997) Yield potential progress in short bread wheats in northwest Mexico. *Crop Science* **37:** 36-42.

Sears, E.R. (1976) Genetic control of chromosome pairing in wheat. *Annual Review of Genetics* 31 -51.

Sharma, D. and Knott, D.R. (1966) The transfer of leaf-rust resistance from. *Agropyron* to *Triticum* by irradiation. *Canadian Journal of Genetics and Cytology* **8:** 137-143.

Singh, R.P., Huerta-Espino, J., Rajaram, S. and Crossa, J. (1998) Agronomic effects from chromosome translocations 7DL.7Ag and 1BL.1RS in spring wheat. *Crop Science* **38:** 27-33.

Slafer, G.A. and Savin, R. (1994) Sink-source relationships and grain mass at different positions within the spike in wheat. *Field Crops Research* **37:** 39-49.

Slafer, G.A., Satorre, E.H. and Andrade, F.H. 1994. Increases in grain yield in bread wheat from breeding and associated physiological changes. *Genetic Improvement of Field Crops* pp. 1-68. Edited by G.A. Slafer. New York: Marcel Dekker, Inc.

Slafer, G.A., Calderini, D.F. and Miralles, D.J. (1996) Yield components and compensation in wheat: Opportunities for further increasing yield potential. In: *Increasing Yield Potential in Wheat: Breaking the Barriers*. Edited by M.P. Reynolds, S. Rajaram and A. McNab. Mexico: CIMMYT, 101-134.

Taylor, H.M. and Terrell, E.E. (1982) Rooting patterns and plant productivity. In *Handbook of Agricultural Productivity* Vol. 1, pp. 185-200. Edited by M. Rechcigel, Jr. CRC press, Boca Raton, FL

Thornsburry, J.M., Goodman, M.M., Doebley, J., Kresovich, S., Nielsen, D. and Buckler, E.S. (2001) *Dwarf8* polymorphisms associated with variation in flowering time. *Nature Genetics* **28**: 286–289.

Uemura, K., Anwaruzzaman, S., Miyachi, S. and Yokota, A. (1997) Ribulose-1,5-bisphosphate carboxylase/oxygenase from thermophilic red algae with a strong specificity for CO_2 fixation. *Biochemistry and Biophysics Research Communications* **233:** 568-571.

Verma,V., Foulkes, M.J., Worland, A.J., Sylvester-Bradley, R., Caligari, P.D. S. and Snape, J.W. (2004) Mapping quantitative trait loci for flag leaf senescence as a yield determinant in winter wheat under optimal and drought-stressed environments. *Euphytica* **135**: 255–263

Vigouroux, Y., McMullen, M., Hittinger, H.T., Houchins, K., Schulz, L., Kresovich, S., Matsuoka, Y., and Doebley, J. (2002) Identifying genes of agronomic importance in maize by screening microsatellites for evidence of selection during domestication. *Proceedings of the National Academy of Science (USA)* **99:** 15 9650–9655

Villareal, R.L., Toro, E.D., Mujeeb-Kazi, A. and Rajaram, S. (1995) The 1BL/1RS chromosome translocation effect on yield characteristics in a *Triticum aestivum* L. cross. *Plant Breeding* **114:** 497-500.

Vos, P., Hogers, R., Bleeker, M., Reijans, M., van de Lee, T., Hornes, M., Frijters, A., Pot, J., Peleman, J., Kuiper, M. and Zabeau, M. (1995) AFLP: a new technique for DNA fingerprinting. *Nucleic Acids Research* **23**: 4407–4414.

Waddington, S.R., Ranson, J.K., Osmanzai, M. and Saunders, D.A. (1986) Improvement in the yield potential of bread wheat adapted to northwest Mexico. *Crop Science* **26:** 698-703.

Waddington, S.R., Osmanzai, M., Yoshida, M. and Ransom, J.K. (1987) The yield of durum wheats released in Mexico between 1960 and 1984. *Journal of Agricultural Science* **108:** 469-477.

Warburton, M.L., Xianchun, X., Crossa, J., Franco,J., Melchinger, A.E., Frisch, M., Bohn, M. and Hoisington, D. (2002) Genetic Characterization of

CIMMYT Inbred Maize Lines and Open Pollinated Populations Using Large Scale Fingerprinting Methods. *Crop Science* **42**: 1832–1840

Weir, A.H., Bragg., P.L., Porter, J.R. and Rayner, J.H. (1984) A winter wheat crop simulation model without water or nutrient limitations. *Journal of Agricultural Science, UK.* **102**(2):371-382, 1984.

William, H.M., Crosby, M., Trethowan, R., van Ginkel, M., Mujeeb-Kazi, A., Pfeiffer,W., Khairallah, M., and Hoisington, D. (2003a) Molecular Marker Service Laboratory - An Interface between the Laboratory and the Field. pp. 852-854. *Proceedings of the 10th International Wheat Genetics Symposium*, September 1 - 6, Paestrum, Italy, 2003.

William, K.J., Lewis, S.G., Bogacki, P., Pallotta, M.A., Willsmore, K.L., Kuchel, H and Wallwork, H. (2003) Mapping a QTL contributing to cereal cyst nematode tolerance and resistance. *Australian Journal of Agricultural Research* **54**: 731-737

Yamazaki, M., Tsugawa, H., Miyao, A., Yano, M., Wu, J., Yamamoto, S., Matsumoto, T., Sasaki, T and Hirochika, H. (2001) The rice retrotransposon Tos17, prefers low-copy marker sequence as a target. *Molecular Genetics and Genomics* **265**: 336-344

Yan, L., Loukoianov,A., Tranquilli, G., Helguera, M., Fahima, T., and Dubcovsky, J. (2003) Positional cloning of the wheat vernalization gene VRN1. *Proceedings of the National Academy Science (USA)* **100**: 10

Yan, L., Loukoianov, A., Blechl, A., Tranquilli, B., Ramakrishna, W., SanMiguel, P., Bennetzen, J.L., Echenique, Y. and Dubcovsky, J. (2004) The Wheat *VRN2* Gene Is a Flowering Repressor Down-Regulated by Vernalization. *Science* **303**: 6263–6268

Yu, J., Hu, S., Wang, J. and 97 others. (2002) A draft sequence of the rice genome (*Oryza sativa* L. ssp. indica). *Science* **296**, 79–92.

6

LIMITS TO YIELD OF FARM SPECIES: GENETIC IMPROVEMENT OF LIVESTOCK

G SIMM[1], L BÜNGER[1], B VILLANUEVA[1], W G HILL[2]
[1]*Sustainable Livestock Systems Group, SAC, West Mains Road, Edinburgh, EH9 3JG;* [2]*School of Biological Sciences, University of Edinburgh, King's Buildings, West Mains Road, Edinburgh, EH9 3JT*

Introduction

Genetic change in the performance of farm livestock has been brought about either by selection between breeds, crossbreeding or selection within breeds. In future, these methods may be augmented by the ability to remove or modify the expression of existing genes, or to add new ones - these techniques are not considered in this chapter, but they are discussed elsewhere in these proceedings.

Genetic improvement utilises genetic variation. Selection among breeds or crosses is a one-off, non-recurrent process: the best breed or breed cross can be chosen but further improvement can be made only by selection within the populations. Hence, interest in potential limits to genetic improvement in yield is primarily concerned with limits to selection within breeds or strains.

Selection within breeds of farm livestock is expected to produce annual genetic changes typically in the range 1 to 3% of the mean in the trait (or multi-trait index) concerned (see, for example, Smith, 1984). The higher rates of change are expected for traits with high genetic variability, in traits that are not age- or sex-limited, and in species, like pigs and poultry, with a high reproductive rate. These changes seem small when considered on an annual basis, but they are cumulative with continuous selection.

These expected rates of change have been achieved in practice over the last few decades in poultry and pig breeding schemes in several countries (McKay, Barton, Koerhuis and McAdam, 2000; Merks, 2000) and in dairy cattle breeding programmes in the US, Canada and New Zealand (Simm, 1998). Typically, rates of genetic change achieved in national beef cattle and sheep populations have been substantially lower than those theoretically possible, though they have been achieved in individual breeding schemes. Generally, rates of change achieved in practice in most species have increased over the last few decades. This is as a result of: better statistical methods for estimating the genetic merit (breeding value) of animals, especially best linear unbiased prediction (BLUP) methods; the wider use of reproductive technologies, especially artificial insemination; improved techniques for measuring performance (e.g. ultrasonic scanning to assess carcass composition *in vivo*); and more focussed selection on objective rather than subjective traits, such as milk yield rather than type. Developments in the statistical, reproductive and molecular genetic technologies available – some of which are discussed later, and some in other chapters in these proceedings – have the potential to increase rates of change further.

There has been a great deal of interest – from both practical agricultural and scientific viewpoints - in the possibility of limits to selection. In the first part of this chapter the evidence for limits to selection, from commercial and experimental populations, is reviewed. As it turns out, there is little evidence from either source of *genetic* limits to selection, but there are a few exceptions that we discuss. We then discuss recent advances in methods for increasing rates of change while simultaneously managing genetic variation (not because of compelling evidence of limits *per se*, but because more sustainable use of genetic variation offers opportunities to enhance responses to selection). Although there is little evidence of direct genetic limits to selection for yield, there is evidence of unfavourable correlated responses when selection is narrowly focused on 'yield' traits alone. In the final section the need to broaden selection objectives, to ensure that genetic improvement programmes remain both effective and acceptable to society at large, is considered.

Genetic and physiological limits to yield

BACKGROUND

Most experimental information on limits to selection comes from experiments in laboratory animals, albeit they are often undertaken with smaller populations than used in commercial livestock. In farm livestock, intense selection has been practised over many generations, particularly in poultry, and there is good information on the changes achieved. The meaning of 'physiological' limits *per se* is not clear, for physiological parameters such as food intake or metabolic rate are themselves under genetic control. There are, of course, absolute limits: fat content can not be less than 0%, and others are set by, for example, thermodynamic laws. It seems that, for example, there are lower limits to body size in mammals, of the order of 1-3 g, found in the Thai bat and the white toothed pygmy shrew mouse (Bünger, Renne and Buis, 2001); and the physiology of the chicken is such that one egg/day appears to be a limit.

Mendelian segregation generates new variation within families every generation. In an unselected population approximately half the variance is within families; in one undergoing extensive selection, which removes between family differences, more than half is within families. Useful genetic variation is lost in two ways: (i) increasing the frequencies of useful alleles towards fixation, the rate being unpredictable without knowledge of the distributions of effects and frequencies of genes affecting the traits of importance; and (ii) through random loss of variation by genetic drift, because the population is of finite size. For genes having little or no effect on the traits under selection, a proportion $1/2N_e$ of the genetic variance is lost each generation, where N_e is the effective population size. This is the number of breeding animals if they are used randomly with equal numbers of males and females; in practice N_e may be much less than the number of breeding individuals.

Genetic variation is also gained by *de novo* mutations arising each generation. The rate is not known for commercial traits of livestock, but estimates from experiments utilising highly inbred lines of laboratory animals and plants are remarkably consistent, the new variation contributing an increment in heritability of about 0.1% per generation. With such a rate of mutation, a population of size $N_e = 250$ would be sufficient to maintain a heritability of 1/3 (i.e. environmental variance σ_E^2 equals twice genetic variance σ_A^2) in the absence of selection, the loss by drift being $\sigma_A^2/(2 \times 250)$ and the increase by mutation $0.001\sigma_E^2$.

Rates of response to selection depend on the current levels of genetic variation and the accuracy of the selection practised. At its simplest, the rate (R) per generation is predicted by $R = ir\sigma_A$, where i is the selection intensity, r the accuracy with which superior genotypes are identified – this is influenced largely by the heritability of the trait(s) concerned and the number and class of relatives with performance records available - and σ_A^2 the additive genetic variance in the trait under selection, or in overall economic merit if selection is for a combination of traits. Reduction in genetic variation leads to a fall in both r, tempered by use of BLUP and family selection indexes, and particularly in σ_A^2. Current rates of response are maximised by selecting very intensely, but this leads to much smaller effective population sizes. However, the total response which can be achieved both from existing (Robertson, 1960) and mutational variance (Hill, 1982) is a function of N_e x the rate of response per generation. Hence there is a compromise to be made between immediate rate of progress and long term gain, for increasing selection intensity or use of relatives' information can reduce effective population size more than it increases response. There is no unequivocal way of deciding on the optimum point of the trade off: it depends on predictions of stability of markets and of businesses as long term responses have to be discounted, and it depends on the unknown relationship between N_e and long term response and on estimates of the loss of performance due to inbreeding. A common practice is to set a minimum level of rate of inbreeding, i.e. $1/2N_e$, and then optimize the structure. The theory for doing so is discussed later. Nevertheless it is quite clear that, to maximise long term response and limits, population size should be as large as can be afforded.

Unfavourable side effects of selection are potentially important in limiting long term responses to selection, but these are discussed in more detail in a later section.

There is no genetic prediction as to the extent to which new mutations that will provide useful variation can continue to be produced, nor, in general, knowledge to predict how rapidly useful variation will be lost due to selection, because these depend on unknown parameters. All one can say is that the loss is likely to be more rapid than $1/2N_e$, and is not likely to be reduced disproportionately less as N_e is increased. Some examples of very long term selection in livestock, animals bred for sporting purposes, and experimental animals, concentrating on species that have had the most intense and the longest selection history are considered below. There are some contradictions. Further data and discussions are given elsewhere (Bünger *et al.*, 2001; Hill and Bünger, 2004).

EVIDENCE FROM FARM LIVESTOCK AND SPORT ANIMALS

Broiler chickens

Excellent data have been published recently on rates of progress from both the public and commercial sector by comparing the performance of current broiler populations with those of unselected control populations founded many years ago. Data from Havenstein and co-workers comparing 1991 and 2001 commercial broilers with a 1957 origin control are summarised in Table 6.1 (Havenstein, Ferket, Scheideler and Larson, 1994; Havenstein, Ferket, Scheideler and Rives, 1994; Havenstein, Ferket and Qureshi, 2003a, b). These show that the modern broiler is over four times heavier at market ages than the 1957 bird. The comparisons shown in the table were between birds on the diets appropriate for the year, but each strain was grown on both diets. For example at 56 days of age, the body weights were as follows for birds reared on the 2001 diet: Select 3946 g, Control 886 g; and for birds reared on the 1957 diet: Select 2984 g, Control 809 g. Thus the great majority of the difference in growth was due to genetic change, but the modern broiler was able to make better use of the modern diet. Further comparisons between 1991 and 2001 results show that selection response continued at a high rate over that decade, at over 2%/year in weight for age, with substantial changes also in carcass yield and breast meat yield.

These genetic trends accord with those of McKay *et al.* (2000) from commercial broiler lines. For example, these authors reported annual rates of genetic improvement over the last few decades of around 1.2% for food conversion ratio, while the time taken for birds to reach 2 kg live weight fell from 63 days in 1976 to 36 days in 1999. These dramatic responses in broiler chickens indicate what can be achieved with long term and focused selection.

The heritability of growth rate in broilers is traditionally around 25%, as indicated from early studies (Siegel, 1963; Dunnington and Siegel, 1996). In the absence of new variation, some fall might have been anticipated. Albeit

Table 6.1 COMPARISON AT 56/57D OF MODERN BROILERS REARED ON A MODERN DIET VS. ACRBC BROILERS REARED ON A 1957 DIET IN TWO STUDIES CONDUCTED IN 1991 AND 2001[A] (SEXES AVERAGED)

	1991 Comparison		*1991*	*2001 Comparison*		*2001*	*Difference*
Strain	*Arbor Acres*	*ACRBC*	*Diff*	*Ross 308*	*ACRBC*	*Diff*	*in trait*
Diet	*1991*	*1957*		*2001*	*1957*		*2001-1991*
Body weight (g)	3108[b]	790	2318	3946	809	3137	819
Carcass weight (g)	2071[b]	498	1513	2815	480	2335	822
Carcass yield (%)	69.7[b]	61.2	8.5	74.4	60.8	13.6	5.1
Breast yield (%)	15.7[b]	11.8	3.9	21.3	11.4	9.9	6.0
Carcass fat (%)	15.3[b]	9.4	5.9	15.9	10.6	5.3	-0.6

[a]Data from Havenstein *et al.* (1994a,b, 2003a,b)
[b]Female data unavailable, but estimated using the 42 and 70 day data

using more sophisticated methods of analysis and probably better management and record keeping, recent estimates of heritability in commercial flocks show similarly high values, for example 32% and 27% in two Ross nucleus flocks (Koerhuis and Thompson, 1997); and recent analyses show heritabilities remain at similar levels (A. Koerhuis, personal communication). At this juncture, there are no obvious limits in sight, and the broiler chicken has a long way to go before it reaches the size of many other avian species.

Poultry for egg production

There were concerns expressed almost 50 years ago about limits, but there is good evidence of continued response. Jones *et al.* (2001) reared together stocks established as unselected controls from commercial stocks in 1950, 1958, 1972, and a commercial cross of 1993. Over the period 1972 to 1993, age at 50% lay was reduced from 166 to 155 days, hen day production increased from 64.2 to 73.4%, and egg weight increased from 61.0 to 63.6 g. Although body weight increased during this period from 1972 it was less than in 1950 stocks, such that feed conversion efficiency improved from 0.319 to 0.345 to 0.378 to 0.426 for 1950, 1958, 1972 and 1993 stocks, respectively. Thus, as a result of industry breeding programmes, birds come into lay earlier with adequate size eggs, more achieve one egg a day during peak, and they keep laying longer. It is hard to see that progress will not attenuate, but it seems not to have done so yet – bearing in mind that egg number is not the only objective.

Preisinger and Flock (2000) give estimates of heritabilities in modern nucleus flocks: both heritability and phenotypic variation were low at peak production when most birds lay daily, but earlier and later in the laying year heritability values were typically about 20%.

The obvious limiting factor with poultry is that they lay only one egg a day. Various experiments have been run to increase this laying frequency, either using continuous lighting or shortened (<24hr) lighting cycles, and in these laying frequency in the selection environment has been changed (e.g. Yoo *et al.*, 1986, changed oviposition interval using continuous light). Nevertheless, breeders do not use such methods. Presumably they consider insufficient response in overall egg production efficiency carries over into the commercial environment with normal lighting to justify the problems of running a selection programme. The issue is controversial: in a review, Gowe and Fairfull (1990) regarded the issue as not fully settled, while Sheldon (2000) argued that the promising results of this approach had been well publicised for at least a decade and there was no evidence that it did not work.

Pigs

It is well known that commercial pigs have been changed dramatically by selection in recent years, notably for increased growth and reduced fatness. In a review, Merks (2000) reported for the last decade annual changes in European pig breeding programmes of about 20 g/day increase (2.4%) in live weight gain, 0.2 piglets extra per litter (1.6%) and 0.5% extra lean meat (0.8% of the mean). Over the last century daily gain has increased by about 100%, while backfat depth has decreased by about 75% (see also Tribout, Caritez, Gogué, Gruand, Billon, LeDividich, Quesnel and Bidanel (2003), for a recent analysis of the positive trends in reproductive performance in French Large White pigs.) In pigs too there is evidence of unintended side effects of selection, which are discussed later.

Racehorses and Greyhound dogs

Selection for improved racing performance of horses and dogs has been conducted for centuries, race winners typically having been chosen for breeding. There are no direct estimates of genetic change using contemporary comparisons with unselected controls. However, there are data on the winning times in classic races. Gaffney and Cunningham (1988) pointed out that winning times for classic races in the UK had not improved much in recent decades, and we have recently presented data on the Kentucky Derby (Figure 1, from Hill and Bünger, 2004). There was improvement until about 1950, and since then none, and the record has been held by Secretariat since 1973. Indeed progress prior to 1950 may have been due in large part to better training, in the way that training has improved performance of athletes. A similar pattern appears for Greyhound dogs (Hill and Bünger, 2004): the winning time in the English Greyhound Derby improved rapidly from the mid 1920s to about 1950; since then there has been little or no progress. There has, however, been continued improvement in racing times of Swedish Standardbred trotters, albeit potentially attenuating (Arnason, 2001).

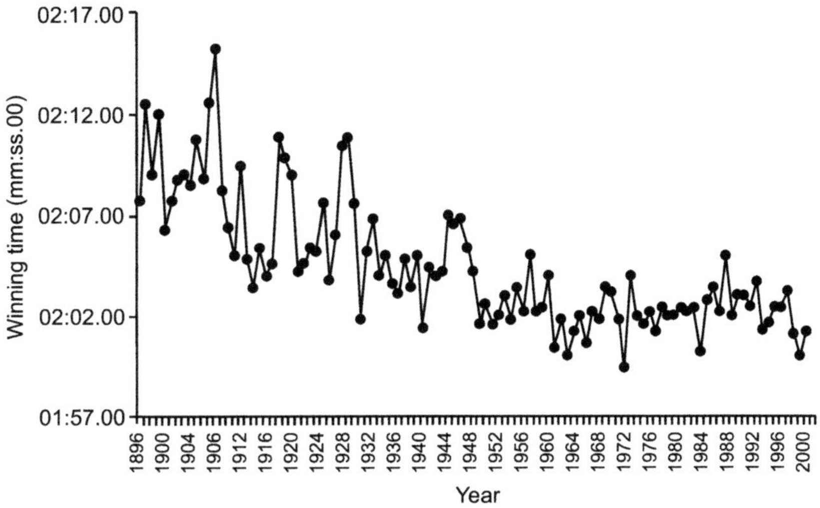

Figure 6.1 Winning times in the Kentucky Derby (Thoroughbred racehorses; data source: http://www.drf.com/home/crown2001/kd/derby_stats.html; Fig. from Hill and Bünger 2004)

The attenuation of speed for the horses is not fully understood. Although the Thoroughbreds have a very narrow genetic base, there have been many generations of mutation. Estimates of relative speed from Timeform data indicate moderately high heritabilities (Gaffney and Cunningham, 1988), although these may be biased by special treatment (would you send offspring from a stallion with a six figure stud fee to a second rate trainer?), and the racing community certainly believes in heritable differences and selects accordingly. One possible explanation is that mutations are indeed occurring, but that almost all are deleterious, either on speed, conformation such as leg strength, or viability/fertility.

EVIDENCE FROM LONG TERM EXPERIMENTS

Maize

The Illinois maize experiment was started over 100 years and 100 generations ago. Selection was practised for high and low kernel oil content - the initial level being about 5% - and, in separate lines, for high and low protein content

- initially about 11%. In each case, about 12 ears were picked as parents each generation. Response has attenuated for low oil, at effectively zero, and for low protein, at about 60g/kg - perhaps the lowest level at which the embryo is viable. But response for increased oil and protein has continued almost linearly, at least for 90 generations (Dudley and Lambert, 2004), current levels being around 200g oil/kg and 300g protein/kg. Although there has been little (oil) or no (protein) response in the last decade, there is still fairly clear evidence that genetic variation remains. The total responses upward in units of genetic (σ_G) standard deviation are $21\sigma_G$ for oil content in the high oil line, and $28\sigma_G$ in the high protein line. Lines in which the direction of selection was reversed after 48 generations have now reached levels close to those of the lines selected in the same direction throughout, indicating that there was substantial variability left even after almost 50 generations, but perhaps that limits are being approached in the high lines. The results show that large and continued genetic change can be made in commercial crops.

Poultry for meat

Selection for growth rate in broilers by Siegel and colleagues and in turkeys by Nestor and colleagues has shown continued responses over 40 generations (see Hill and Bünger, 2004 for a review). The longest running avian experiment has been conducted in Japanese quail, in which continued responses were made for over 90 generations in two lines selected for increased 4 week body weight. A total change of over $15\sigma_P$ or $30\sigma_G$ was achieved.

Mice

There have been many experiments in which selection for body weight has been practised in the laboratory mouse, using different kinds of base population and different population sizes, with duration up to 100 generations. Results have been reviewed elsewhere (Bünger *et al.*, 2001; Hill and Bünger, 2004). Response in all cases has exceeded $3\sigma_P$ or $7\sigma_G$, and there is a relationship between the long-term response and the limit (if selection was continued that long). The heaviest line, Dummerstorf, selected both longest (>110 generations to date) and with one of the largest population sizes (N_e about 60) now has a body weight over twice its initial value, showing a response of some $17\sigma_P$ or $24\sigma_G$. Typical of many selection experiments, whilst there have been indications of plateaux over periods of 10-30 generations, continued selection has realised further response. Plateaus turned out to be temporary and if and when they appear seems to be a function of time and N_e.

Other species

Selection experiments have been conducted for many traits in populations of very different sizes, Weber and colleagues selecting up to 1000 or so parents each generation in Drosophila using automatic scoring methods. Weber (2004) has summarised the relation between response and population size using data on both Drosophila (which have the extreme high values of N_e) and other species. He plotted response for 50 generations as a fraction of that in the first generation, R_{50}/R_1, against log N_e and found a sigmoid relationship, with values of R_{50}/R_1 equal to about 20, 30 and increasing up to about 40 for $N_e = 25$, $N_e = 30$ and $N_e > 250$, respectively (typical initial responses are less than 1 phenotypic standard deviation per generation). In view of the role of residual variation and mutation, proportionately larger further responses might be expected in later generations with larger populations. Overall, what selection experiments show is that essentially any trait can be changed, and that with selection in a sufficiently large population, very substantial changes can be made. Limits have been obtained in some experiments, albeit sometimes transient, and if so the levels depended on population size. There is also evidence from many experiments that if selection is relaxed, response may be lost, indicating that deleterious genes are segregating. In laboratory experiments, however, selection is usually for just a single trait, with no attempt to simultaneously select for fitness to eliminate harmful correlated changes (see e.g. Falconer and Mackay, 1996; Hill and Bünger 2004 for further discussion).

Advances in managing genetic variation

BACKGROUND

As discussed above, with a few exceptions, there is little evidence from either experiments or industry breeding programmes of limits to selection. However, the more sustainable use of genetic variation offers opportunities to enhance responses to selection. It has become apparent over the last couple of decades that techniques which can increase the accuracy and/or intensity of selection, such as the use of BLUP genetic evaluations, artificial insemination and embryo transfer, and the use of genetic markers, also potentially accelerate inbreeding and loss of genetic variation. For instance, Figure 2 shows how the rate of inbreeding in the US Holstein population has increased substantially over the last decade.

In addition to the concerns raised about the increases in inbreeding associated with past selection, there are scientific, heritage and aesthetic arguments for greater emphasis on conservation of farm animal genetic resources globally, and obligations on many governments to do so following the 1992 'Rio Convention' on biodiversity. The recent foot and mouth disease epidemic in the UK highlighted the need for a more coherent conservation strategy, not only for numerically small breeds but also for geographically isolated breeds. Although most publicity is given to loss of genetic variation at the species or breed level, there is potentially an equally serious loss of genetic variation within 'mainstream' breeds under intense selection.

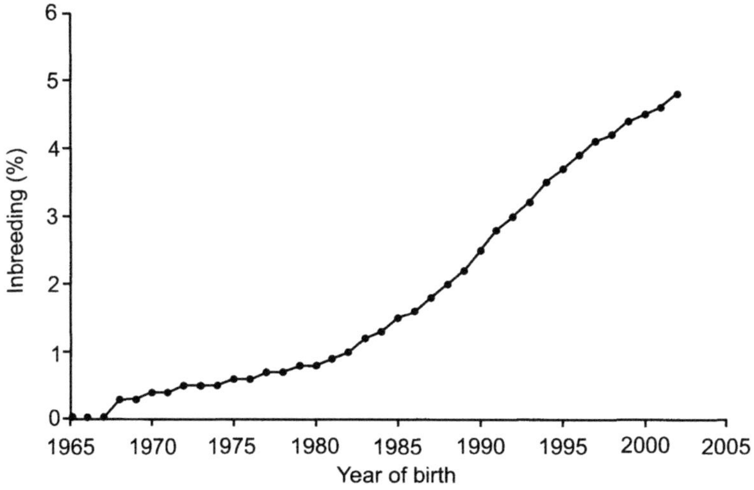

Figure 6.2 Trend of average inbreeding coefficient for the US dairy cattle population. (data source: http://www.aipl.arsusda.gov/dynamic/inbrd/current/kindx.html)

During the last few years there have been important developments for simultaneously managing rates of genetic gain and inbreeding and these are described below. Managing the rate of inbreeding (ΔF), or equivalently the effective population size ($N_e = 1/2\Delta F$), provides a general framework for managing genetic resources as it determines levels of neutral genetic variability and the relative effects of genetic drift and selection on non-neutral loci. The control of ΔF could avoid or alleviate the reductions in viability and fertility, which eventually limit further progress, despite the presence of genetic variance for traits under selection. Other processes influenced by ΔF such as increased probabilities of losing beneficial rare alleles (Caballero and Santiago, 1998) and risk of extinction (Saccheri, Kuussaari, Kankare, Vikman, Fortelius and Hanski, 1998) are also controlled.

OPTIMIZED SELECTION

Breeders have often applied strategies for controlling inbreeding (e.g. by avoiding matings between close relatives) but these strategies have been rather *ad hoc*, implying that the maximum achievable gains have not necessarily been achieved. The optimal way to restrict the loss of genetic variability in selection programmes is to assign differential contributions of selection candidates in a way that genetic gain is maximised, but at the same time the rate of inbreeding is restricted. There are now efficient methods for doing this and they lead to higher genetic gains at the same inbreeding or, alternatively, to lower inbreeding at the same genetic gains (Meuwissen, 1997; Grundy, Villanueva and Woolliams, 1998). These methods allow us to manage genetic gain and inbreeding simultaneously when making selection decisions, and to optimize the use of selected candidates.

This type of selection (referred to here as 'optimized selection') differs from the traditional method, usually referred to as 'truncation selection', in the information used when deciding which animals should be selected and how widely they should be used. With truncation selection, candidates are ranked according to the best estimates of breeding values available and only the candidates above (or below) a given value (the truncation point) are chosen as parents and they are expected to contribute equally to the next generation. Thus selection decisions are taken independently of the genetic relationships between candidates and therefore, in extreme cases, individuals selected can come from very few families. In contrast, optimized selection takes into account all genetic relationships, optimizes the number of individuals selected and allows them to contribute differently, resulting in higher gains at the same inbreeding, or to lower inbreeding at the same gain.

The application of these tools to manage rates of inbreeding only requires information that is commonly available in practical breeding programmes, i.e. pedigree information and estimates of breeding values (EBVs) for candidates for selection. EBVs are now routinely obtained using BLUP in most livestock species. A fundamental parameter in BLUP methodology is the additive relationship matrix (the **A** matrix) that describes the genetic variances and covariances of the complete population. The **A** matrix is usually computed from pedigree information. Optimized selection uses the **A** matrix not only to obtain accurate estimates of breeding values through BLUP but also for obtaining the desired rate of inbreeding. The restriction on ΔF is achieved by imposing the constraint $\mathbf{c}^{T}\mathbf{A}\mathbf{c} \leq C$, where $\mathbf{c}$ is the vector of solutions (i.e. contributions or proportions of total offspring left by each candidate) and C is the desired ΔF. With fully random union of gametes, $\mathbf{c}^{T}\mathbf{A}\mathbf{c}$ is twice the inbreeding coefficient of the next generation.

The benefits of optimized selection have been extensively proven in simulation studies. Also, the benefits of these techniques have been evaluated using data from actual livestock populations. These studies show that large improvements in rates of gain at a given ΔF are expected, compared to conventional truncation BLUP selection. Avendaño, Villanueva and Woolliams (2003) predicted extra gains of around 20% or more when the methods were applied to sheep and beef cattle populations. For instance, in the UK Aberdeen Angus beef cattle population they found that, for a restriction on ΔF of 0.2% (the value observed in the current population), the expected index gains (in index units) were 11.7, 15.2 and 21.1 for truncation selection, selection optimizing contributions of males only and selection optimizing contributions of both sexes, respectively. Valuable gains have also been predicted in dairy cattle, when optimized selection has been applied to groups of selection candidates whose offspring are potential young AI bulls (Kearney, Wall, Villanueva and Coffey, 2004).

Initially, optimization tools were described for selection programmes where the aim is to maximise the increase in performance for economically valuable traits but imposing restrictions on ΔF. However, these tools are also valid in a conservation context where the aim is to minimise ΔF but impose restrictions to avoid decreased performance in traits that make the breed valuable. The solutions from both types of formulations (minimise ΔF with a constraint on gain or maximise gain with a constraint on ΔF) are technically equivalent. The difference lies in the relative emphasis given to rates of gain and inbreeding (Villanueva, Pong-Wong, Woolliams and Avendaño, 2004).

OPTIMIZED SELECTION WITH MOLECULAR GENETIC INFORMATION

In the past, conventional selection programmes have been efficient in increasing genetic gains by making use of pedigree and performance information. Thus, the traditional approach has treated the genetic architecture of the trait itself as a black box, with no knowledge of the number of genes that affect the trait, or the effects of each gene or their locations in the genome. This has been mostly due to the fact that the individual properties of the genes affecting quantitative traits have been largely unknown. The standard model assumed has been the so called 'infinitesimal model', which assumes that there are many (strictly infinitely many) genes each of infinitely small individual effect. With more and more QTL (quantitative trait loci) of significant effect identified and firmer estimates of total gene number of thirty or so thousand from the genome projects, this model is clearly unrealistic, although it has been used highly successfully in breeding. Available information at the molecular level on genes that affect quantitative traits is being integrated with existing quantitative genetics techniques in new selection methods.

Information on QTL or on markers linked to them can be used to estimate breeding values with increased precision, leading to higher genetic gains. However, increased accuracies of evaluation can also lead to higher rates of inbreeding and higher rates of loss of genetic variation. Optimized selection described above has been extended to this situation where pedigree and performance records are jointly used with molecular information both on QTL affecting the trait under selection (Villanueva, Pong-Wong, Grundy and Woolliams, 1999) or on markers linked to QTL (Villanueva, Pong-Wong and Woolliams, 2002). Optimization tools for maximising gain while restricting the loss of genetic variability can be also applied to scenarios with multiple trait selection objectives (Avendaño, Visscher and Villanueva, 2002).

Another potential use of markers is to increase the accuracy of the estimates of genetic relationships between animals which are usually obtained from pedigree information. The pedigree-based relationship matrix gives only expectations of the proportion of genes that two particular animals have in common. However, there is variation around this expected value (Weir, Avery and Hill, 1980). For instance, full brothers and sisters may differ in their genetic relationship with each other, but the pedigree-based relationship assumes that full sibs have 0.5 of their genes in common (i.e. the expected value). Markers can be used to estimate the exact proportions with a high degree of precision and the genetic covariances may be re-calculated to reflect the true proportion of the genome in common (Christensen, Fredholm, Wintero, Jorgensen and Andersen, 1996). Efficient algorithms now exist to compute **A** matrices based on both pedigree and marker information (Pong-Wong, George, Woolliams and Haley, 2001); they are proving efficient in increasing the accuracy of evaluations and selection response (Villanueva, unpublished results). In principle, the control of diversity could be achieved by combining all the available information (markers and pedigree), appropriately weighted, in the optimization approach. Specific constraints could be applied to restrict ΔF and loss of variation at specific positions, genome regions or across the whole genome. Conceptually, different restrictions could be applied to different **A** matrices.

Finally, optimized selection could be also useful in breeding programmes specifically designed for eliminating deleterious identified alleles. Breeding schemes to eradicate scrapie (e.g. the National Scrapie Plan (NSP) in Britain) provide a good example. These rely on selection based on genotypes for the PrP locus and aim to reduce and eventually eliminate occurrence of TSEs (scrapie and risk of BSE) in sheep. Only rams of specific (i.e. less susceptible) PrP genotypes are allowed to be used for breeding. Whilst selective breeding for particular genotypes could speed up the control and eradication of scrapie, it is essential that selection for resistance to scrapie is seen in a wider context. In particular, in some rare and traditionally farmed sheep breeds, pursuing the strategy proposed in the NSP could lead to important losses in genetic variability. Existing optimization methods for maximising genetic progress while restricting the loss of genetic variability could be extended to optimize contributions when selection is applied in favour of desirable alleles (scrapie resistance) – whether or not this has an effect on the quantitative trait. In this case, the objective would be to fix the favourable gene without sacrificing large responses to selection for other important traits, and while maintaining genetic variability. An optimum balance between selection against scrapie and selection for other valuable traits could be achieved.

Health, welfare and fitness-related traits

BACKGROUND

There is now substantial evidence that selection narrowly focused on production traits often leads to unfavourable correlated responses in health-, welfare- or fitness-related traits. These are important in their own right, but they may also limit selection for yield or other characteristics. Rauw, Kanis, Noordhuizen-Stassen and Grommers (1998) reported over 100 examples of unfavourable correlated responses in a literature review, mainly in poultry, pigs and dairy cattle. Unfavourable responses were particularly common in reproductive performance, metabolic disease and functional fitness (e.g. leg weakness/gait disorders). However, it is important to note that reduced welfare is not a necessary consequence of selective breeding *per se* – rather these responses arise from a *narrowly-focused* breeding goal. Indeed animal breeding may have potential to enhance welfare (e.g. Jones and Hocking, 1999; Lawrence, Conington and Simm, 2004; Simm, Lawrence, Conington and Coffey, 2004).

Unfavourable correlated responses may arise as a result of pleiotropy or linkage. Pleiotropy occurs when a single gene affects two or more different traits of interest – in this case the trait under selection and one or more health, welfare- or fitness-related traits. Linkage occurs when genes are located close to each other on the same chromosome – in this case, genes affecting the trait under selection and one or more health-, welfare- or fitness-related trait. Unfavourable characteristics may also increase in frequency in selected populations by chance, or as a result of inbreeding leading to an increase in homozygosity. A rare deleterious condition or mutation may not be recognised until it reaches a high frequency. For example, the indirect effects of double muscling phenotypes in cattle seem not to have been extreme initially, until some constellation of modifying genes interacting with the mutation was reached (Arnold, Della-Fera and Baile, 2001; Bünger, Ott, Varga, Schlote, Rehfeldt, Renne, Williams and Hill, 2004).

MULTI-TRAIT SELECTION INCLUDING FITNESS-RELATED TRAITS

When there are unfavourable genetic correlations between 'production' and health-, welfare- or fitness-related traits, they can be addressed by some form of multi-trait selection. This may involve the use of 'independent culling levels' (i.e. animals selected for breeding must exceed threshold values for the fitness-related trait as well as other measures of performance), or the use of multi-trait selection indexes.

McKay *et al.* (2000) and McKay (2004) report that the incidence of tibial dyschondroplasia, and some other health and welfare-related conditions such as ascites, have been improving in breeding lines and commercial populations of broiler chickens as a result of the availability of improved techniques to assess them, and greater selection pressure against them. For example, as a result of selection for improved growth plate quality, the percentage of lesions was reduced rapidly, from ca. 7% to under 1% in male lines and, starting later, from 9% to ca. 1% in female lines, with most of the changes in each taking place over a three year period (McKay *et al.*, 2000).

In Scandinavian dairy cattle breeding programmes multi-trait selection indexes including health traits have been in use for several decades (see review of Simm, 1998). By accounting for genetic antagonisms between health and production in selection programmes, it is possible to improve both economic performance and animal welfare, compared to selection on production alone. For example, recent research in SAC and the University of Edinburgh has been expanding the current £PLI index used in the UK dairy industry. This index had milk, fat and protein yields and lifespan in the goal, but indicator traits for resistance to mastitis (somatic cell count) and lameness (locomotion score) have been added recently. Selection on the revised index is expected to benefit cow welfare, compared to selection on earlier versions, by slowing the expected unfavourable correlated responses in mastitis incidence (0.32 cases to 0.21 cases per cow). The additional benefit of selection on the expanded £PLI index in all parts of the breeding sector is predicted to be worth an extra 5.5% in economic returns per cow per year relative to selection on production only (Coffey, Stott and Brotherstone, 2004). This benefit is expected to rise further when fertility is included in £PLI later this year. We are continuing this approach by investigating additional indicators of 'robustness' in dairy cattle, including genetic variation in energy balance, in a range of production environments. We will also investigate whether selection on such indicators is likely to have any unforeseen adverse ethical or welfare consequences.

A similar approach has been followed in research at SAC and the Roslin Institute to develop broader breeding goals in hill sheep, with the emphasis on reducing mortality. In addition to being a production loss, lamb mortality is also a source of considerable suffering in the form of starvation and hypothermia. To tackle this we have developed indices that include ewe longevity and lamb survival in the selection objective alongside more conventional traits, such as lamb carcass production (Conington, Bishop, Grundy, Waterhouse and Simm, 2001). Importantly these indices take account of both the costs of higher lamb mortality and of ewes carrying a higher lamb burden. The indexes are expected to lead both to improved ewe and lamb survival and improved overall economic returns.

The advantage of the approaches currently used in multi-trait selection is that they transform all traits (production or welfare based) to a common currency allowing direct comparisons of costs and benefits. Currently the weights applied to traits reflect their expected economic value to the producer. This approach is likely to underestimate animal-based non-economic (moral) aspects such as the pain or discomfort associated with lameness, and new approaches are needed to more fully account for these moral values. The question of who should pay for the addition of these moral values to a breeding index remains open.

Conclusions

There is evidence from both commercial breeding, notably of poultry, and from many experiments in laboratory and farm animals, that responses to selection for most traits of interest continue over very long periods of time, and can be of a substantial size. Furthermore, there is evidence that genetic variation is being maintained in the lines, as would be expected from mutation. There is, however, a little understood exception where response does seem to have reached a plateau, namely speed in races. For most traits of interest in farm livestock there is no reason to expect direct genetic limits to yield. Indeed, long term responses to selection are expected to be higher than in the past as a result of the application of new methods for managing genetic variability, and new techniques (e.g. from molecular genetics) to increase accuracy of selection.

It is worth highlighting, though, that there are often unintended side effects of intense selection for productivity. These have been, are being, or could be, resolved in most cases through the use of broader multi-trait selection, including the health and welfare-related traits that are unfavourably correlated with production traits.

References

Arnason, T. (2001) Trends and asymptotic limits for racing speed in standardbred trotters. *Livestock Production Science*, **72**, 135-145.

Arnold, H.H., Della-Fera, M.A. and Baile, C.A. (2001) Review of myostatin history, physiology and applications. *LifeXY*, **1**, 1014-1022.

Avendaño, S., Villanueva, B. and Woolliams, J.A. (2003) Expected increases in genetic merit from using optimised contributions in two livestock populations of beef cattle and sheep. *Journal of Animal Science,* **81**, 2964-2975.

Avendaño, S., Visscher, P.M. and Villanueva, B. (2002) Potential benefits of using identified major genes in two trait breeding goals under truncation and optimal selection. *Proceedings of the 7th World Congress on Genetics Applied to Livestock Production*, Vol 33, pp 163-166.

Bünger, L., Renne, U. and Buis, R.C. (2001) Body weight limits in mice - Long-term selection and single genes. *Encyclopedia of Genetics*, pp 337-360. Edited by E.C.R. Reeve. Fitzroy Dearborn Publishers, London, Chicago.

Bünger, L., Ott, G., Varga, L., Schlote, W., Rehfeldt, C., Renne, U., Williams, J.L. and Hill, W.G. (2004) Marker assisted introgression of the *Compact* mutant *myostatin* allele: $Mstn^{Cmpt\text{-}dl1Abc}$ into a mouse line with extreme growth - effects on body composition and muscularity. *Genetical Research* (submitted)

Caballero, A. and Santiago, E. (1998) Survival rates of major genes in selection programmes. *Proceedings of the 6th World Congress on Genetics Applied to Livestock Production*, Vol 26, 5-12.

Christensen, K., Fredholm, M., Wintero, A.K., Jorgensen, J.N. and Andersen, S. (1996) Joint effect of 21 marker loci and effect of realized inbreeding on growth in pigs. *Animal Science*, **62**, 541-546.

Coffey, M.P., Stott, A. and Brotherstone, S. (2004) An update to the UK national profit index £PLI. *Proceedings of the British Society of Animal Science*, p 27.

Conington J., Bishop S.C., Grundy, B., Waterhouse, A. and Simm, G. (2001) Multi-trait selection indexes for sustainable UK hill sheep production. *Animal Science*, **73**, 413-423.

Dudley, J.W. and Lambert, R.J. (2004) 100 generations of selection for oil and protein in corn: people, progress, and promise. *Plant Breeding Reviews*, **24** (part 1), 79-110.

Dunnington, E.A. and Siegel, P.B. (1985) Long-term selection for 8-week body weight in chickens - direct and correlated responses. *Theoretical and Applied Genetics,* **71**, 305-313.

Dunnington, E.A. and Siegel, P.B. (1996) Long-term divergent selection for eight-week body weight in white Plymouth rock chickens. *Poultry Science,* **75**, 1168-1179.

Falconer, D.S. and Mackay, T.F.C. (1996) *Introduction to Quantitative Genetics, Ed 4.* Longman Group Ltd., Essex.

Gaffney, B. and Cunningham, E.P. (1988) Estimation of genetic trend in racing performance of thoroughbred horses. *Nature*, 332, 722-724.

Gowe, R.S. and Fairfull, R.W. (1985) The direct response to long-term selection for multiple traits in egg stocks and changes in genetic parameters with selection. In *Poultry Genetics and Breeding*, pp 125-146. Edited by W.G. Hill, J.M. Manson and D. Hewitt. Longman, Harlow, England.

Grundy, B., Villanueva, B. and Woolliams, J.A. (1998) Dynamic selection procedures for constrained inbreeding and their consequences for pedigree development. *Genetical Research,* **72**, 159-168.

Havenstein, G.B., Ferket, P.R. and Qureshi, M.A. (2003a) Growth, liveability, and feed conversion of 1957 vs. 2001 broilers when fed representative 1957 and 1991 broiler diets. *Poultry Science,* **82**, 1500-1508.

Havenstein, G.B., Ferket, P.R. and Qureshi, M.A. (2003b) Carcass composition and yield of 1957 vs. 2001 broilers when fed representative 1957 and 1991 broiler diets. *Poultry Science*, **82**, 1509-1518.

Havenstein, G.B., Ferket, P.R., Scheideler, S.E. and Larson, L.T. (1994a) Growth, livability, and feed conversion of 1957 vs. 1991 broilers when fed 'typical' 1957 and 1991 broiler diets. *Poultry Science*, **73**, 1785-1794.

Havenstein, G.B., Ferket, P.R., Scheideler, S.E. and Rives, R.D. (1994b) Carcass composition and yield of 1957 vs. 1991 broilers when fed 'typical' 1957 and 1991 broiler diets. *Poultry Science*, **73**, 1795-1804.

Hill, W.G. (1982) Predictions of response to artificial selection from new mutations. *Genetical Research*, **40**, 255-278.

Hill, W.G. and Bünger, L. (2004) Inferences on the genetics of quantitative traits from long-term selection in laboratory and domestic animals. *Plant Breeding Reviews,* **24 (part 2)**, 169-210.

Jones, R.B. and Hocking, P.M. (1999) Genetic selection for poultry behaviour: Big bad wolf or friend in need? *Animal Welfare*, **8**, 343-359.

Jones, D.R., Anderson, K.E. and Davis, G.S. (2001) The effects of genetic selection on production parameters of single comb White Leghorn hens. *Poultry Science*, **80**, 1139-1143.

Kearney, J.F., Wall, E., Villanueva, B. and Coffey, M.P. (2004) Inbreeding trends and application of optimised selection in the UK Holstein population. *Journal of Dairy Science* (in press).

Koerhuis, A.N.M. and Thompson, R. (1997) Models to estimate maternal effects for juvenile body weight in broiler chickens. *Genetics, Selection, Evolution*, **29**, 225-249.

Lawrence, A.B., Conington, J. and Simm, G. (2004) Breeding and animal welfare: practical and theoretical advantages of multi-trait selection. *Animal Welfare*, **13**, S191-196.

McKay, J.C., Barton, N.F., Koerhuis, A.N.M. and McAdam, J. (2000) The challenge of genetic change in the broiler chicken. In: *The Challenge of Genetic Change in Animal Production.* British Society of Animal Science Occasional Publication No 27. Edited by W.G. Hill, S.C. Bishop, B. McGuirk, J.C. McKay, G. Simm and A.J. Webb. British Society of Animal Science. (http://www.bsas.org.uk/publs/genchng/contents.pdf).

McKay, J. (2004) Breeding meat-type chickens for changing demands. *Proceedings of the British Society of Animal Science*, p 264.

Merks, J.W.M. (2000) One century of genetic change in pigs and the future needs. IIn: *The Challenge of Genetic Change in Animal Production.* British Society of Animal Science Occasional Publication No 27. Edited by W.G. Hill, S.C. Bishop, B. McGuirk, J.C. McKay, G. Simm and A.J. Webb. British Society of Animal Science. (http://www.bsas.org.uk/publs/genchng/contents.pdf).

Meuwissen, T.H.E. (1997) Maximizing the response of selection with a predefined rate of inbreeding. *Journal of Animal Science,* **75**, 934-940.

Pong-Wong, R., George, A.W., Woolliams, J.A. and Haley, C.S. (2001) A simple and rapid method for calculating identity-by-descent matrices using multiple markers. *Genetics, Selection, Evolution,* **33**, 453-471.

Rauw, W.M., Kanis, E., Noordhuizen-Stassen, E.N. and Grommers, F.J. (1998) Undesirable side effects of selection for high production efficiency in farm animals: a review. *Livestock Production Science,* **56**, 15-33.

Robertson, A. (1960) A theory of limits in artificial selection. *Proceedings of the Royal Society of London,* B153, 234-249.

Saccheri, I., Kuussaari, M., Kankare, M., Vikman, P., Fortelius, W. and Hanski, I. (1998) Inbreeding and extinction in a butterfly metapopulation. *Nature*, **392**, 491-494.

Sheldon, B.L. (2000) Research and Development in 2000: Directions and Priorities for the World's Poultry Science Community. *Poultry Science,* **78**, 147-158.

Siegel, P.B. (1963) Selection for body weight at eight weeks of age. 1. Short term response and heritabilities. *Poultry Science,* **42**, 954-962.

Simm, G. (1998) *Genetic Improvement of Cattle and Sheep.* CABI Publishing, Wallingford, UK.

Simm, G., Lawrence, A.B., Conington, J.E. and Coffey, M.P. (2004) Breeding and farm animal welfare. *Proceedings of the British Society of Animal Science*, p 263.

Smith, C. (1984) Rates of genetic change in farm livestock. *Research and Development in Agriculture,* **1**, 79-85.

Tribout, T., Caritez, J.C., Gogué, J., Gruand, J., Billon, Y., LeDividich, J., Quesnel, H. and Bidanel, J.P. (2002) Estimation of realised genetic trends in French Large White pigs from 1977 to 1988 for male and female reproduction traits using stored frozen semen. *Proceedings of the 7th World Congress on Genetics Applied to Livestock Production*, Vol 30, pp 79-82.

Villanueva, B., Pong-Wong, R., Grundy, B. and Woolliams, J.A. (1999) Potential benefit from using an identified major gene and BLUP evaluation with truncation and optimal selection. *Genetics, Selection, Evolution,* **31**, 115-133.

Villanueva, B., Pong-Wong, R. and Woolliams, J.A. (2002) Marker assisted selection with optimised contributions of the candidates to selection. *Genetics, Selection, Evolution*, **34**, 679-703.

Villanueva, B., Pong-Wong, R., Woolliams, J.A. and Avendaño, S. (2004) Managing genetic resources in commercial breeding populations. In *Farm Animal Genetic Resources*. British Society of Animal Science Occasional Publication No 30. Edited by G. Simm, B. Villanueva, K.D. Sinclair and S. Townsend. Nottingham University Press, Nottingham, UK.

Weber, K. (2004) Population size and long-term selection. *Plant Breeding Reviews*, **24 (part 1)**, 249-269.

Weir, B.S., Avery, P.J. and Hill, W.G. (1980) Effect of mating structure on variation in inbreeding. *Theoretical Population Biology,* **18**, 396-429.

Yoo, B.H., Sheldon, B.L. and Podger, R.N. (1986) Analyses of oviposition times and intervals in a wide-range of layer flocks under normal and continuous lighting regimes. *British Poultry Science*, **27**, 267-287.

7

GLOBAL WARMING AND AGRICULTURE

CYNTHIA ROSENZWEIG[1] AND JEREMY COLLS[2]
[1]NASA Goddard Institute for Space Studies, 2880 Broadway, New York, NY 10025, USA; [2]Division of Agricultural and Environmental Sciences, School of Biosciences, University of Nottingham, Sutton Bonington Campus, Loughborough, Leics., LE12 5RD, UK

Introduction

A changing climate due to increasing anthropogenic emissions of greenhouse gases will induce change in agricultural systems through a set of interactive processes. Both productivity and geographic distribution of crop species will be affected. The major climate factors contributing to these responses include increasing atmospheric carbon dioxide, rising temperature, and increasing extreme events, especially droughts and floods. These factors in turn will affect water resources for agriculture, grazing lands, livestock, and associated agricultural pests. Effects will vary, depending on the degree of change in temperature and precipitation and on the particular management system and its location. Several studies have suggested that recent warming trends in some regions may have already had discernible effects on some agricultural systems.

Climate change projections are fraught with uncertainty in regard to both the rate and magnitude of temperature and precipitation variation in the coming decades. This uncertainty arises from a lack of precise knowledge of how climate system processes will change and of how population growth, economic and technological developments, and land use patterns will evolve in the coming century (IPCC, 2000; 2001).

Nevertheless, three points regarding climate change can be made with some confidence (Figure 7.1). First, the natural presence of greenhouse gases is known to affect the planetary energy balance, causing the surface of the planet to be warmer than it would be otherwise. This natural warming amounts to 33°C, of which around 65% is due to CO_2. Second, greenhouse gas concentrations have increased progressively since the beginning of the Industrial Revolution. Such increases in greenhouse gases tend to enhance the natural 'greenhouse effect.' Third, the planet has indeed been warming over the last century, especially in the most recent two and a half decades. There is increasing evidence that this warming is due to the greenhouse gases. One of the most persuasive pieces of the evidence is from the marked global *cooling* which occurred between 1940 and 1970. During this period, sulphate particles that formed from gaseous sulphur dioxide (which was released with the CO_2) reflected solar radiation and more than offset the radiative warming. Inclusion of the particulate effect in the climate change models successfully reproduces this temporal variation.

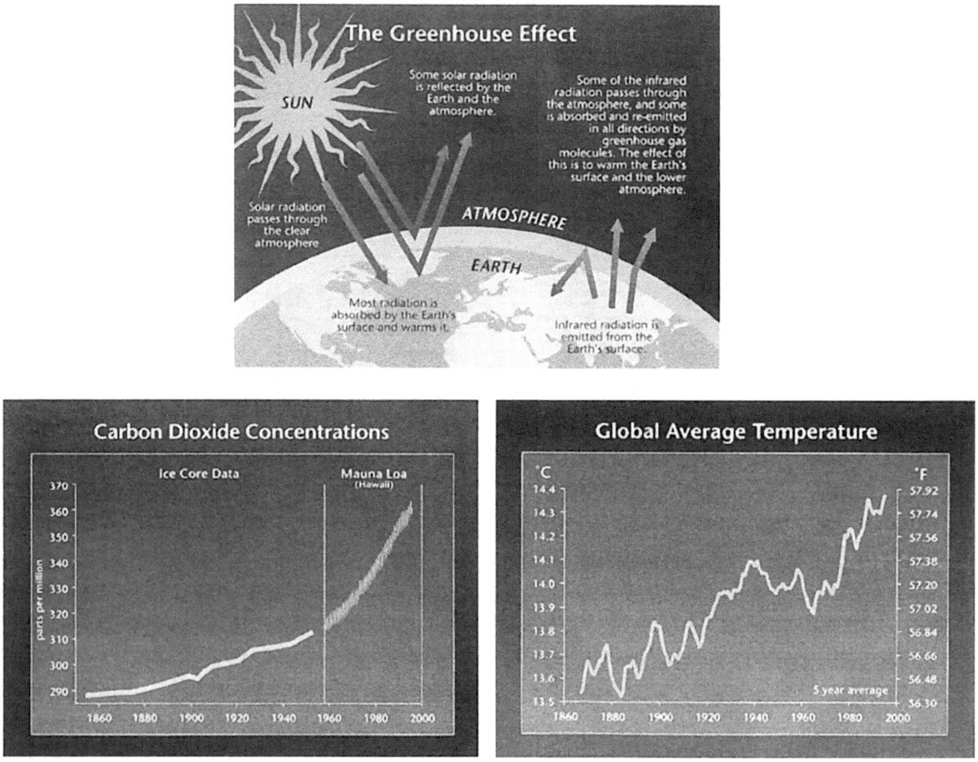

Figure 7.1 The three certainties of global climate change: (a) Atmospheric concentrations of greenhouse gases, 1860 to present; (b) The greenhouse effect and the planetary energy balance; (c) Mean global surface temperature, 1860 to present (OSTP, 1997).

The Intergovernmental Panel on Climate Change (IPCC) has attributed the observed warming over the last century to anthropogenic emissions of greenhouse gases, especially carbon dioxide (CO_2), methane (CH_4), and nitrous oxide (N_2O) (IPCC, 2001). Thus, anthropogenic emissions of greenhouse gases appear to be altering our planetary energy balance and to be manifested in an overall (though uneven) warming of the planet. If the measurable warming trend continues at the global scale, the association of greenhouse gas emissions, the greenhouse effect, and surface warming will acquire ever-greater certainty. The ultimate significance of the climate change issue is related to its global reach, affecting agricultural regions throughout the world in complex and interactive ways.

In this paper, we describe how climate processes affect agro-ecosystems in principle, summarise recent studies projecting how climate change will affect agriculture in the future, discuss the potential roles of mitigation and adaptation in responding to climate change, and suggest research pathways to achieve better understanding of how agriculture might be affected by global warming.

Climate/agro-ecosystem processes

A changing climate will affect agro-ecosystems in complex ways, with some potential benefits and some potential negative consequences (Figure 7.2). In any given agricultural region, varying factors will prevail at any given time. The factors that prevail may change over time, as gradual and possibly even abrupt climate change continues.

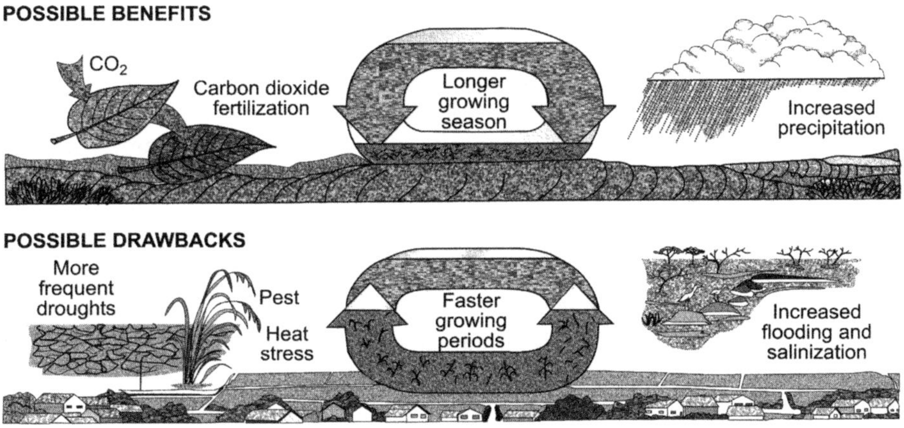

Figure 7.2 Agro-ecosystem processes and a changing climate (Bongaarts, 1994).

INCREASING ATMOSPHERIC CARBON DIOXIDE CONCENTRATION

Evidence from Antarctic ice-cores has shown that atmospheric CO_2 concentrations have varied between 200-300 ppm during the past 400,000 years. More recently, the concentration has risen from about 270 ppm before the Industrial Revolution (circa 1750) to 376 ppm in 2003. The current rate of increase is 1.8 ppm per year. Increasing atmospheric CO_2 concentration, in and of itself, is likely to influence crop yields and water use, and ultimately agricultural productivity, because photosynthesis, transpiration, and respiration are plant processes directly affected by CO_2 levels. More CO_2 in the atmosphere enhances the diffusion of CO_2 into the leaves, promoting its fixation into carbohydrates through photosynthesis. Experiments in controlled environments have demonstrated that photosynthesis in single leaves and whole plants is increased in CO_2-enriched atmospheres (Acock and Allen, 1985; Cure and Acock, 1986; Kimball, 1983; Poorter, 1993, Hsiao and Jackson, 1999). Free-air CO_2 enrichment (FACE) experiments (Hendrey, Lewin, and Nagy, 1993) generally confirm the positive results obtained in controlled environments. This influence of CO_2 enrichment on photosynthesis, which broadly amounts to an increase in biomass production of around 30% for a doubling of CO_2 concentration, is often called the *CO_2-fertilization effect*.

Plant species differ in their responses to CO_2 because of differences in their photosynthetic pathways. The two main types of photosynthetic pathway are referred to as C_3 and C_4 (so named because the precursor molecule in the C_3 pathway has three C atoms versus four C atoms in the C_4 pathway). The C_3 plants (legumes, small grains, cool-season grasses, and most trees) usually respond more positively than the C_4 plants (warm-season grasses, sorghum, corn, millet, and sugarcane). Responses also depend on other environmental variables (e.g. water and nutrient availability) and on genetics. There is some evidence that CO_2 enrichment is relatively more effective, at least in the short term, in plants that endure high-temperature, moisture, and salinity stress (Kimball, 1983; Kimball and Idso, 1983).

Another important physiological effect of CO_2 enrichment is the partial closure of stomata, the small openings in leaf surfaces through which CO_2 is absorbed and water vapour released. Accordingly, a rise in atmospheric CO_2 may inhibit transpiration even while stimulating photosynthesis. Water-use efficiency (biomass accumulation per volume of water transpired) is thereby improved (Allen, Boote, Jones, Jone, Valle, Acock, Rogers and Dahlman, 1987; Morison, 1985).

Whether the effects of enrichment demonstrated in controlled environments and in the limited number of FACE studies will prevail in farmers' fields remains uncertain, in part because of the unknown effect of enrichment on potential interactions with weed and insect pests (Rosenzweig and Hillel, 1998). For example, C_3 weeds may become more vigorous and hence more problematic in fields with C_4 crops.

RISING TEMPERATURE

Several studies have shown that changes in agricultural systems have already occurred because of rising CO_2 concentrations and temperatures in recent decades. Nicholls (1997) estimated the contribution of warming climate trends in Australia to the substantial increase in Australian wheat yields since 1952. Non-climate influences – such as new cultivars and changes in crop management practices – were taken into account by de-trending the wheat yield and climate variables and using the residuals to calculate the quantitative relationship between variations in climate and yield. Climate trends appear to be responsible for 30-50% of the observed increase in Australian wheat yields, with increases in minimum temperatures, and perhaps a decrease in frost frequency, being the dominant influence.

Temperature increases also appear to be extending the growing season in certain regions. Chmielewski, Muller and Bruns (2004) found that for the period of 1961-1990 the average annual air temperature increased by 0.36°C per decade ($P<0.01$) in Germany, resulting in a 1.4°C increase in temperature over the last forty years. As a result of this temperature change, over the same time period the beginning of the growing season has advanced by 2.3 days per decade ($P<0.10$). The beginning of stem elongation in winter rye advanced by 2.9 days per decade ($P<0.01$), the beginning of cherry tree blossom advanced by 2 days per decade ($P<0.05$), and the beginning of apple tree blossom advanced by 2.2 days per decade ($P<0.05$). All phenophases were well-correlated with the average air temperatures.

Rising temperatures may also be affecting yields in tropical regions. Peng, Huang and Sheehy (in review) analysed weather data from the International Rice Research Institute (IRRI) Farm in the Philippines from 1979 to 2002. They found that mean minimum temperature rose by 1.32°C in the dry season and by 0.79°C in the wet season. Mean radiation also rose during the same period. The authors concluded that rice grain yield declined by about 15% for each one-degree increase in growing-season mean temperature. Since there was no relationship between crop growth duration and minimum temperature, this effect was not associated with a change in growth duration. These recent results echo earlier relationships given by Monteith (1981), who showed that UK wheat yields had an inverse relationship with mean temperature during grain filling.

CLIMATE VARIABILITY AND EXTREME EVENTS

Climate change is likely to include changes in climate variability as well as in average conditions. Such changes are especially important during critical phases of crop development (e.g. more frequent or prolonged heat waves when maize is at anthesis, or late frost when wheat is at flowering). If temperature variability increases, crops growing near the top or bottom of their optimal temperature ranges could be affected adversely, as diurnal and seasonal temperature fluctuations may exceed the tolerance range for crop growth and development. If temperature variability diminishes, however, crops growing near their optimal ranges could benefit. Increases in daily temperature variability may reduce wheat yields through lack of cold hardening and resultant winter-kill. Extremes of precipitation (i.e. droughts or floods) are detrimental to the productivity of dry-land crops. Greater drought frequency would increase the need for irrigation, whereas greater incidence of flooding could cause damage due to soil waterlogging, crop lodging, and pest infestations (Rosenzweig, Iglesias, Yang, Epstein and Chivian, 2002a).

Africa's semi-arid Sahel has experienced a drying trend since the 1960s. Ben Mohamed, Van Duivenbooden and Abdoussalam (2002) found that, for millet in Niger, yields declined from 1967 to 1998, while total production increased in three study areas, evidently as a consequence of more land being cultivated for millet. Van Duivenbooden, Abdoussalam and Ben Mohammed (2002) found that groundnut production dropped from about 312,000 tons in the mid 1960s to as little as 13,000 tons in 1988 and rose again to 110,000 tons in 2000. A reduction in groundnut-cultivated area occurred because of the southward movement of isohyets, affecting villages that had previously been large producers (Ben Mohamed *et al.*, 2002). The reduced length of the vegetative period no longer allows the present varieties to complete their growth cycle. Increased length and frequency of dry spells has apparently contributed to these trends.

WATER RESOURCES

Climate change will most likely affect the availability of water supplies for irrigation, the requirements of crops and livestock for water, and the frequency and intensity of extreme events such as droughts and floods (Figure 7.3; Döll, 2002; Strzepek, 1999). Future availability of water for agriculture depends, in part, on possible changes in precipitation, potential and actual evaporation, and runoff, at the scale of watersheds and river basins. Warmer winters will induce loss of natural storage in mountain snowpacks and subsequent shrinking of stream flows in late summer and autumn (Gleick, 1987). This is already observed in some watersheds in California (Cayan, Kammerdiener, Dettinger, *et al.*, 2000). Crop water requirements will also be affected by increased evaporative demand. These and other changes may well affect the management of water resources, including reservoir operations, water allocations, and irrigation system development. Climate change also may increase the competition between industrial and domestic users, as well as the amount of water required to sustain non-agricultural ecosystems in riparian areas and watersheds.

The seasonality of water supply may also change as a result of global climate change, and thus it is likely that the intra-seasonal timing of water for crops will require increased attention. Global climate models project the potential for both increasing precipitation and intensifying hydrological variability. These projections cause concern regarding workability of soils in the spring and waterlogging damage to crops in the summer. Additional investment in drainage may be necessary if flooding of agricultural land occurs more frequently.

Finally, in agricultural regions close to the ocean, sea-level rise and associated saltwater intrusion and flooding can harm crops through impeded soil aeration and salination. This is likely to be most serious in countries such as Egypt and Bangladesh, which have major crop-growing areas in low-lying coastal regions.

GRAZING LANDS AND LIVESTOCK

Since a large percentage of total net primary productivity in grassland ecosystems occurs below-ground, grasslands may be more resilient to climate change than annual crops (Elliott, Hunt, and Walter, 1988; Rice, Todd, Blair, Seastedt, Ramundo and Wilson, 1998). Grassland productivity changes, due to decreased precipitation or increased temperature, could also be moderated by the enhanced water-use efficiency associated with increased atmospheric concentrations of CO_2 (CAST, 2004).

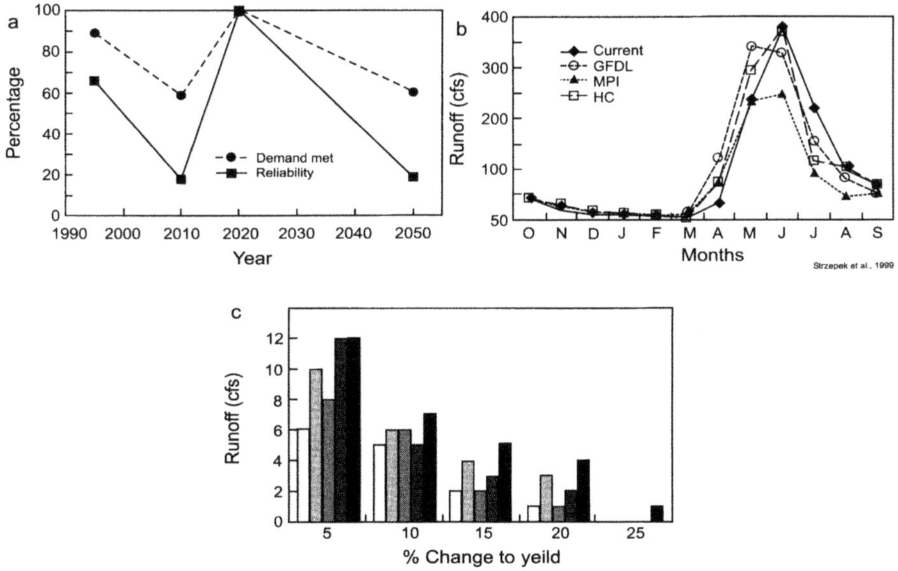

Figure 7.3 (a) Demand met (monthly average percentage of water demand met and reliability (percentage of years in which water demands are met) for the Lower Missouri Water Region for the present and for the Max Planck (MPI) climate change scenario for the 2010s, 2020s, and 2050s. (b) Runoff in the water regions supplying the U.S. Cornbelt for current climate, and the Geophysical Fluid dynamics Laboratory (GFDL), MPI, and Hadley Center (HC) climate change scenarios. (c) Number of events causing damage to maize yields due to excess soil moisture conditions, averaged over all study sites, under current baseline (1951-1998) and the HC and Canadian Climate Centre (CC) climate change scenarios. Events causing a 20% simulated yield damage are comparable to the 1993 US Midwest floods (a&b Stryzepek *et al.*, 1999; c Rosenzweig *et al.*, 2002).

Thus, the response of rangelands to land management and environmental change is a potentially significant component of the global carbon budget.

Modelling studies indicate that in the future, grasslands could function as either sinks or sources of carbon depending on management (e.g., Cole, Duxbury, Freney, Heinmeyer, Minami, Mosier, Paustian, Rosenberg, Sampson, Sauerbeck and Zhao, 1993; Hunt, Trlica, Redente, Moore, Detling, Kittel, Walter, Fowler, Klein and Elliott, 1991; Ojima, Parton, Schimel and Ownesby, 1990).

Just as with crops, the nature of direct and indirect effects of climate change on livestock will depend on the climate in which the livestock are located. In cooler areas, herds may benefit from warmer conditions, whereas in warmer areas herds may experience heat stress. Indirect effects of climate change on livestock production arise through impacts on the productivity and quality of forage from grasslands, quality of other feedstuffs available (e.g. grain and hay nutrient content), incidences of disease and pests, availability of water in grazing areas, and market-influenced input and output prices (CAST, 2004).

AGRICULTURAL PESTS

Climate affects not only agricultural crops but also their associated pests (Figure 7.4). Spatial and temporal distribution and proliferation of insects, weeds, and pathogens are determined to a great extent by climate, because their growth and development are driven by temperature, light, and water.

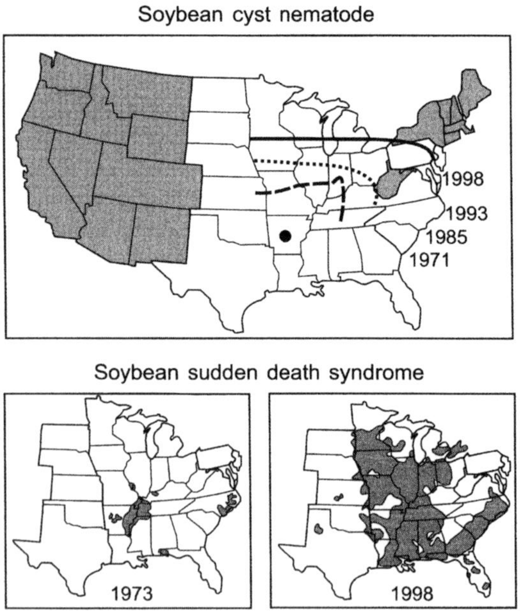

Figure 7.4 Spread of agricultural pests under current climate conditions: (top) Spread of soybean cyst nematode, 1971-1998; (bottom) Spread of soybean sudden death syndrome (Rosenzweig *et al.*, 2001).

Climate also affects the pesticides used to control pest outbreaks. For instance, rainfall intensity and timing influences pesticide persistence and efficiency. Through chemical alteration, temperature and light also affect pesticide persistence. Most analysts concur that, in a changing climate, pests will be able to expand their geographic ranges. An expansion of pest populations may necessitate increased use of agricultural chemicals, with consequent health, ecological, and economic costs (Rosenzweig *et al.*, 2002a; Chen and McCarl, 2001).

In a changing climate, pests may become more active, thus posing the threat of increased economic losses to farmers (Coakley, Scherm, and Chakraborty. 1999; IPCC. 1996). Whereas the majority of weeds are invasive species from temperate zones, certain weeds in temperate regions originated in tropical or subtropical regions, and in the current climate their distribution is limited by low temperatures. Such geographical constraints will be obviated under warm conditions. Warmer temperature regimes have been shown to increase the maximum biomass of certain grass weed species significantly. Crop weeds, insects, and diseases are projected to extend their geographic range to higher latitudes (Dahlsten and Garcia, 1989; Sutherst, 1990).

When temperatures are within their viable range, insects respond to higher temperature with accelerated rates of development and with shortened times between generations. Very high temperatures, however, decrease insect longevity. Warmer winters will decrease winterkill, and consequently there may be increased insect populations in subsequent growing seasons. With warmer temperatures occurring earlier in the spring, insect populations may become established and thrive during earlier and more vulnerable crop growth stages. Additional insect generations and greater populations encouraged by higher temperatures and longer growing seasons likely will require enhanced pest management efforts.

Increasing temperature, humidity, and precipitation favour the spread of plant diseases, because wet vegetation promotes germination of spores and proliferation of bacteria and fungi. Soil moisture also influences the life cycle of soil nematodes. In regions suffering from aridity, however, disease infestations may lessen, although certain diseases such as powdery mildews can thrive even in hot, dry conditions as long as there is dew. Interactions between crops and their associated pests in response to elevated levels of atmospheric CO_2 are not well understood and are difficult to predict. Because of the different growth responses of weeds and crops to elevated atmospheric CO_2, C_3 weeds are likely to become more aggressive (Patterson, 1993). Carbon dioxide enrichment also may modify insect-crop relations. Changes in C- and N-partitioning in crops grown under elevated CO_2 conditions may affect nutritional quality and attractiveness of foliage to various insects. For example, experiments have shown that higher CO_2 tends to increase the C:N ratio in crop leaves, stimulating the feeding of and the damage caused by certain insects (Lincoln, Sionit, and Strain, 1984; Salt, Brooks, and Whittaker, 1995).

Future effects

CHANGE OVER TIME

Due to all the agro-ecosystem processes described above, it is likely that agricultural regions will experience changes and that these changes will evolve through the coming decades. Shifts in crop zoning are likely to occur, with some crop types expanding and others contracting their ranges. Considering the imprecise temperature and precipitation projections from global climate models and the unknown degree of manifestation of direct CO_2 effects on crops growing in farmers' fields, however, the magnitudes and rates of these changes are yet uncertain.

The interactions between beneficial and detrimental agro-ecosystem processes are likely to vary over time for several reasons. First, as biophysical effects follow their characteristic temperature-response curves, crop responses to change in temperature may shift from positive to neutral, and then to negative (Figure 7.5). Another reason that climate change effects are likely to be transformed over time is the potential for decadal shifts in the hydrological cycle. While it is difficult to predict the direction of change in any specific agricultural region, global climate models do project increased decadal variability in hydrological regimes. Finally, as crop breeding and pest species evolve in the coming decades under changing climate conditions, there are likely to emerge new agro-ecosystem weeds, insects, and diseases, the adjustment to which may be costly.

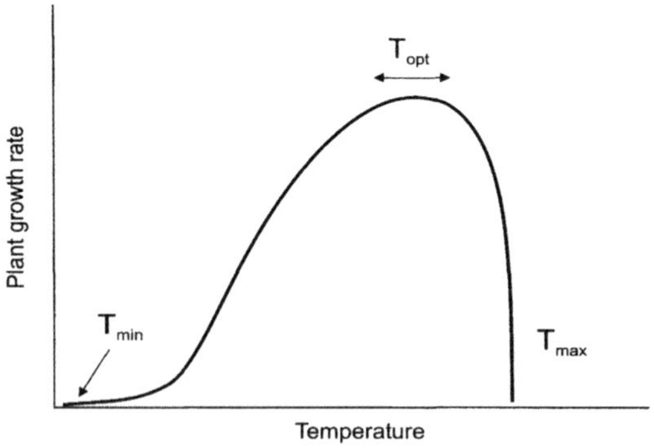

Figure 7.5 Temperature response curve for biological processes (Rosenzweig and Hillel, 1998).

HETEROGENEOUS EFFECTS

Global studies to date show that negative and positive effects will occur both within countries and across the world. In countries that are large such as the United States, Russia, Brazil, and Australia, agricultural regions will probably be affected quite differently (Figure 7.6). Some regions will experience increases in production and some declines (e.g. Reilly, Tubiello, McCarl *et al.*, 2003). At the international level, this implies possible shifts in comparative advantage for production of export crops. This also implies that adaptive responses to climate change will necessarily be complex and varied.

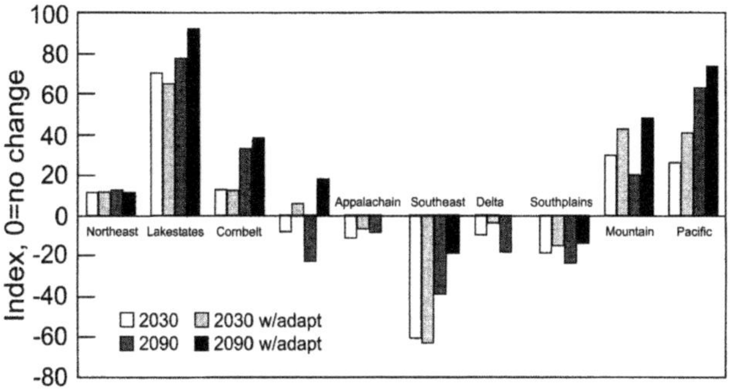

Figure 7.6 Simulated percentage changes in U.S. regional agricultural production, with adaptation, under the Canadian Climate Center scenario (Reilly et al., 2003).

VULNERABILITY IN DEVELOPING COUNTRIES

Despite general uncertainties about the rate and magnitude of climate change and about consequent hydrological changes, regional and global studies have consistently shown that agricultural production systems in the mid and high latitudes are more likely to benefit in the near term (to 2050), while production systems in the low-latitudes are more likely to decline (IPCC WG II, 2001). In biophysical terms, rising temperatures will likely push many crops beyond their limits of optimal growth and yield. Higher temperatures will intensify the evaporative demand of the atmosphere, leading to greater water stress, especially in semi-arid regions. Since most of the developing countries are located in lower-latitude regions (some of which are indeed semi-arid), while most developed countries are located in the more humid mid to high latitudes, this finding suggests a divergence in vulnerability between these groups of nations, with far-reaching implications for future world food security (Rosenzweig and Parry, 1994; Parry, Rosenzweig, Iglesias, Livermore and Fischer, 2004).

Furthermore, developing countries often have fewer resources with which to devise measures to meet changing agricultural conditions. The combination of potentially greater climate stresses and lower adaptive capacity in developing countries creates different degrees of vulnerability between rich and poor nations as they confront global warming. This difference is due in part to the potentially greater detrimental impacts of a changing climate in areas that are already warm (especially if such areas are also dry), and in part to the generally lower levels of adaptive capacity in developing countries.

LONG-TERM NEGATIVE EFFECTS

If the effects of climate change are not abated, even production in the mid and high latitudes is likely to decline in the long term (end of 21st century) (Figure 7.7). These results are consistent over a range of temperature, precipitation, and direct CO_2 effects tested, and are due primarily to the detrimental effects of heat and water stress as temperatures rise. While the beneficial effects of CO_2 may eventually level out, the detrimental effects of warmer temperatures and greater water stress are more likely to be progressive in all regions.

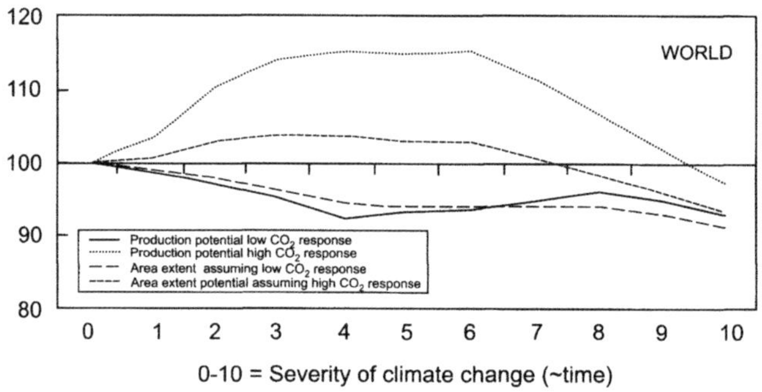

Figure 7.7 Generalized projection of world agricultural production potential and areal extent under low and high CO_2 responses for increasing severity of climate change. (Note: severity of climate change may be taken as a proxy for decadal-to-century timeframe.) (Fischer and Tubiello, personal communication).

A RECENT STUDY

Since climate is not the only driving force on agriculture, researchers now conduct scenario analyses that include linked sets of population projections, economic growth rates, energy technology improvements, land-use changes, and associated emissions of greenhouse gases. Parry *et al.* (2004) have analysed the global consequences of climate change on crop yields and production using scenarios developed from the HadCM3 global climate model, in connection with the Intergovernmental Panel on Climate Change Special Report on Emissions Scenarios (SRES) A1FI, A2, B1, and B2 (Table 7.1). Projected changes in yield were calculated using transfer functions derived from crop model simulations with observed climate data and projected data scenarios.

As has been found in previous studies (e.g. Rosenzweig and Parry, 1994; IPCC, 2001), aggregated results show that agricultural yields in developing countries are projected to be more vulnerable than in developed countries (Table 7.2). Maps of the crop yield results elucidate the complex regional patterns of projected climate

Table 7.1 IPCC SRES SCENARIO DESCRIPTIONS (IPCC, 2000)

Scenario	*Major underlying themes*	*Population*	*Economic development*	*Technology and energy*
A1FI	Convergence among regions, capacity building, and increased cultural and social interactions	Global population peaks in the mid-century and declines thereafter	Rapid economic growth and substantial reduction in regional differences in per capita income	Fossil-fuel intensive coal, oil, and gas continue to dominate energy supply for foreseeable future
A2	Self reliance and preservation of local identities	Continuously increasing global population	Regionally oriented, with per capita growth fragmented and slower than in other storylines	Fragmented and slower than in other storylines
B1	Emphasis is on global solutions to economic, social, and environmental sustainability, including improved equity, but without additional climate initiatives	Global population peaks in mid-century and declines thereafter	Rapid changes in economic structures toward a service and information economy, with reductions in material intensity	Introduction of clean and resource-free technologies
B2	Emphasis is on local solutions to economic, social, and environ-mental sustainability	Continuously increasing population at a rate lower than A2	Intermediate levels of economic development	Less rapid and more diverse technological change than in B1 and A1

Table 7.2 AGGREGATED DEVELOPING-DEVELOPED COUNTRY DIFFERENCES (PER CENT) IN AVERAGE CROP YIELD CHANGES FROM BASELINE FOR THE HADCM2 AND HADCM3 SCENARIOS (PARRY *et al.*, 2004).

	Had CM3 - 2080s							*Had CM2 - 2080s*	
Scenario	*A1F1*	*A2a*	*A2b*	*A2c*	*B1a*	*B2a*	*B2b*	*S550*	*S750*
CO_2 (ppm)	810	709	709	709	527	561	561	498	577
World	-5	0	0	-1	-3	-1	-2	-1	1
Developed	3	8	6	7	3	6	5	5	7
Developing	-7	-2	-2	-3	-4	-3	-5	-2	-1
Difference (%) Developed-Developing	10.4	9.8	8.4	10.2	7.0	8.7	9.3	6.6	7.7

variables, CO_2 effects, and agricultural systems that contribute to aggregations of global crop production. As expected, the A1F1 scenario, with its large increase in global temperatures, exhibits the most pronounced decreases in yields, both regionally and globally. The contrast between the predicted yield change in developed and developing countries is largest under the A2a-c scenarios. Under the B1 and B2 scenarios, developed and developing countries exhibit less contrast in crop yield changes, with future crop yield changes being slightly more favourable under the B2 than the B1 scenario.

Responses

MITIGATION

The practice of agriculture plays a major role in the global carbon cycle (Figure 7.8; Rosenzweig and Hillel, 2000). On a global scale, the process of photosynthesis by agricultural crops fixes about 2 Gt C year^{-1}, with about 1 Gt C year^{-1} in sustenance for the world's population that is respired back to the atmosphere as it is consumed. The additional 1 Gt C is returned to the soil annually as plant residues. Some of the latter carbon, however, is subsequently returned to the atmosphere as a consequence of soil microbial activity, and some is stored in the soil matrix. Furthermore, the fossil fuel that powers the machinery to sow, irrigate, harvest, and dry crops world-wide is responsible for atmospheric emissions of about 150 Mt C year^{-1}. Large amounts of fossil-fuel energy are used to produce fertilisers, especially nitrogenous compounds. Rice cultivation, livestock production, and soil processes are also responsible for considerable methane and nitrous oxide emissions (Rosenzweig and Hillel, 1998).

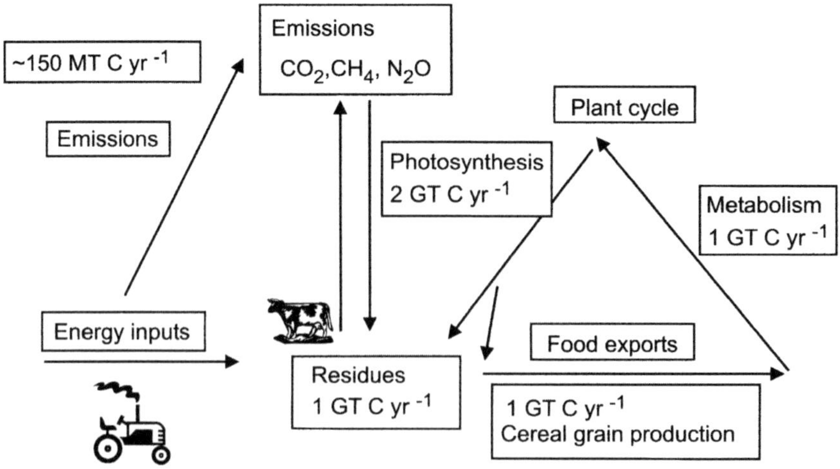

Figure 7.8 The agricultural carbon cycle (Rosenzweig and Tubiello, 2004).

The agricultural carbon cycle offers several entry points for mitigation of greenhouse gas accumulation in the atmosphere. An important one is the potential for agricultural soils to store carbon, particularly to the extent that its 'active' carbon stores had been depleted by past soil management practices (Rosenzweig and Hillel, 2000). Other ways that agriculture may help to mitigate the enhanced greenhouse effect are through the production of biofuels, the development of more efficient rice and livestock production systems, and the reduction of fossil-fuel use by farm machinery.

ADAPTATION

Adaptation is integral to the study of climate change impacts on agriculture. Social scientists have made a significant contribution to the field of climate change impacts by bringing forward this important point (IPCC, 2001; Smith, Klein and Huq, 2003). The task now is to integrate the findings and insights of economists, sociologists, political scientists, anthropologists, and psychologists in providing guidance to decision-makers so as to promote sectoral and international co-operation in minimising the potential negative impacts of, and maximising the opportunities for adjustment to, climate change.

Farmers have always had to adapt to the vagaries of weather, whether on weekly, seasonal, or annual timescales. They will undoubtedly continue to adapt to the changing climate in the coming decades, applying a variety of agronomic techniques, such as adjusting the timings of planting and harvesting operations, substituting cultivars, and – where necessary – changing the entire cropping system.

However, it is noteworthy that farming systems have never been perfectly adapted, even to the current climate (witness the recurrent effects of droughts and floods on various agricultural regions around the world). Hence it seems unreasonable to expect perfect adaptation in the future to changing climate conditions. Some adaptations will probably be successful (e.g. changing planting dates to avoid heat stress), while other attempted adaptations (e.g. changing varieties and breeds) may not always be effective in avoiding the negative effects of droughts or floods on crop and livestock production (Figure 7.9). There are numerous social constraints to adaptation, as well, some of which have been highlighted recently by social scientists (Smith *et al.*, 2003).

INTERACTIONS

Joint consideration of agricultural mitigation and adaptation is needed for several reasons. Our research shows that the carbon sequestration potential of agricultural soils varies under changing climate conditions (Figure 7.10). Thus, a changing climate clearly will affect the mitigation potential of agricultural

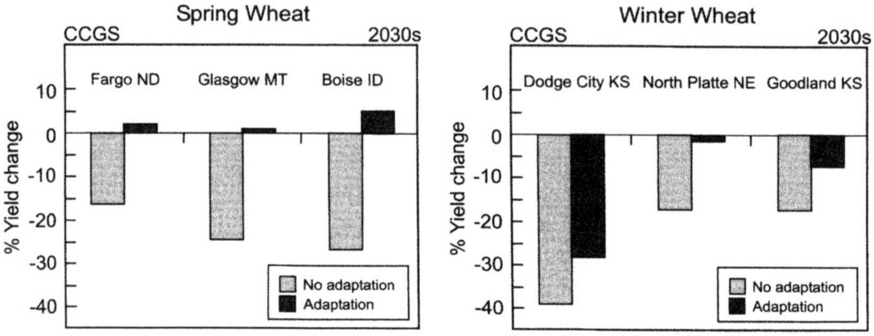

Figure 7.9 Percent yield changes with and without adaptation under the Canadian Climate Centre climate change scenario in the 2030s: (left) spring wheat with change of planting date; (right) winter wheat with change of cultivar (Rosenzweig and Tubiello, 2004).

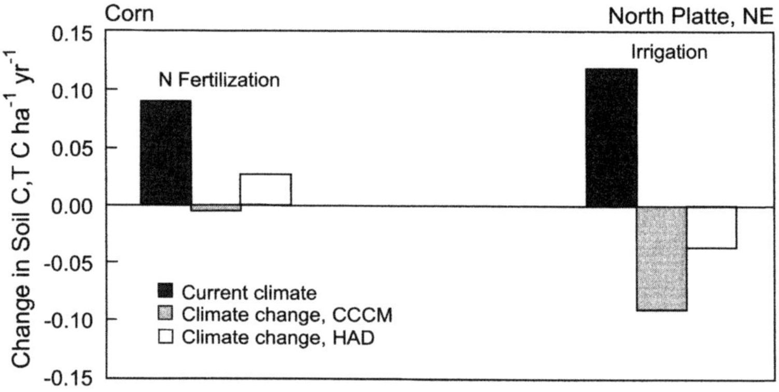

Figure 7.10 Change in soil carbon in corn production under nitrogen fertilization and irrigation under current climate and under the Canadian Climate Centre (CCCM) and Hadley Center (HAD) climate change scenarios (Rosenzweig and Tubiello, 2004).

practices. If changing climate is not taken into consideration, calculations such as those pertaining to 'carbon to be sequestered' may be in serious error.

On the other hand, mitigation practices can also affect the adaptation potential of agricultural systems. For example, by enhancing the ability of soils to hold soil moisture, carbon sequestration in agricultural soils may help crops to withstand droughts and/or floods, both of which are projected to increase in frequency and severity in a warmer climate. Additionally, sequestering carbon in soil supports larger and more diverse populations of microbes and other organisms that provide services to plants and indirectly to animals, such as producing root growth–promoting hormones. All these functions can contribute substantially to the sustainability of agricultural systems. Adaptation practices may in turn affect the mitigation potential. For example, irrigation and nitrogen fertilisation may greatly enhance the ability of soils in semi-arid regions to sequester carbon.

Finally, since it is likely that efforts to reduce the enhanced greenhouse effect (such as the Kyoto Protocol) will not be completely effective, farmers and others in the agricultural sector will be faced with the dual tasks of reducing carbon dioxide and other greenhouse gas emissions, while having to cope with an already changing climate.

Research pathways

To better address the interactions between climate change and sustainability of food and fibre production, we suggest the following areas for future research.

CLIMATE VARIABILITY AND CHANGE

A bifurcation in the field of climate impacts has occurred between research on responses to major systems of climate variability, such as the El Niño-Southern Oscillation, and responses to long-term global warming. The insights that have been gained from studies of agriculture in regard to these two timescales – seasonal-to-inter-annual versus decadal-to-century – need to be reconciled.

The work on seasonal-to-inter-annual climate forecasts has tended to focus on short-term decision-making in regard to predictions of El Niño and La Niña events, which are manifested in terms of climate extremes (Figure 7.11). The figure shows the normalised vegetative index (NDVI) for Uruguay, which tends to experience wet periods during El Niño events (e.g. February, 1998) and droughts during La Niña events (e.g. February, 2000). The role of local stakeholders is crucial at these shorter timescales, and responses are focused on adaptation.

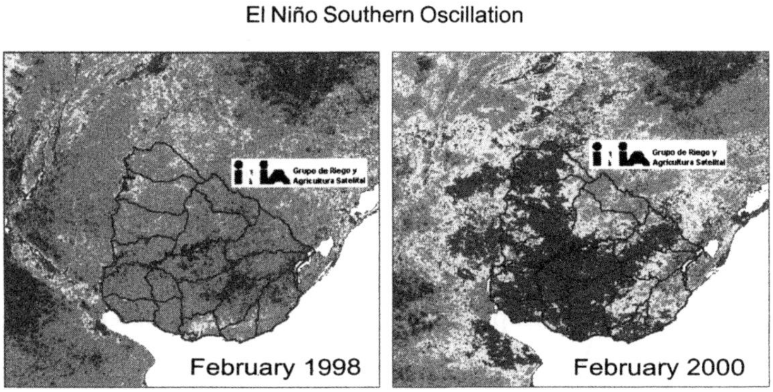

Figure 7.11 Observed temperature trends, 1970-2000 (NASA/GISS).

The work on the decadal-to-century timescale, on the other hand, has focused primarily on responses to mean changes and long-term decisions. The stakeholders for climate change impact studies have often been national policy-makers. The goal here has usually been to provide information needed to help these decision-makers to devise long-term strategies in regard to the climate change issue, in terms of both mitigation and adaptation.

New theoretical constructs are needed to link climate-agriculture interactions on the two timescales, as well as new ways to utilise analytic tools such as dynamic crop growth models and statistical analyses. The need is to move beyond the readily tractable projections of crop responses to mean changes, toward the more difficult yet highly relevant issue of how crops may respond to altered climate variability, such as changes in the frequency and intensity of extreme events.

Our predictions of future scenarios are underpinned by global climate models. Although the latter have steadily expanded in complexity as our understanding and computing power have increased, there remain major uncertainties. For example, 17% of incoming solar radiation is reflected from clouds, and a small change in the proportion of cloud cover would dominate the few W m^{-2} forcing that results from radiative warming. The models need to incorporate dynamically the types and heights of cloud droplets and ice crystals if they are to handle energy budgets realistically.

OBSERVED EFFECTS OF WARMING TRENDS

Analysis of temperature records from around the world shows that many regions have already begun to experience a warming trend, especially from the 1970s to the present. Warmer-than-normal springs have been documented in western North America since the late 1970s (Cayan *et al.*, 2001). In some areas of the world, there have also been recent episodic increases in floods (e.g. North America) and droughts (e.g. Sahel) (IPCC, 2001), with likely but as yet mostly undocumented effects on food production. The responses of agricultural systems to such changes need to be monitored and documented. Have farmers indeed switched to earlier planting dates? Have they changed cultivars? And are there any trends in yields that can be discerned in conjunction with the climate trends?

Such questions are difficult to answer because factors other than climate, including land-use change and pollution, have been changing simultaneously. But such questions are important for furthering our understanding of agricultural adaptation to climate, and for validating the many simulation studies done on potential climate change impacts in the future. These analyses will contribute to the IPCC Fourth Assessment now underway.

GLOBAL AND LOCAL SCALES

A final bifurcation that needs to be resolved is that between global and local/regional scales. Recent work has emphasised the importance of scale in estimating the impacts of climate variability and change on agriculture (Mearns, 2003). In order to understand how a changing climate will affect agriculture, we must find new ways to bring detailed knowledge at local and regional scales to bear on global analyses.

Agriculture in any one region is linked to other agricultural regions, and indeed to the world food system, both through trade and through the food-donor system. As a changing climate shifts the comparative advantage in one region versus another, various regions will inevitably be affected. Thus, in our research on agriculture and climate change, we need to link regional 'place-based' studies of vulnerability and adaptation, as well as mitigation, into a global synthesis. There are many reasons why the impact of the changing climate may be more complex than it seems at first. For example, warming in the northern hemisphere will nominally increase the area that is climatically suitable for temperate crops. However, much of this increase may be only theoretical in the medium term if soils for crop growth are not available or the terrain is impracticable. Also, other factors are changing that may offset potential benefits. For example, the long-term average ozone concentration is predicted to increase in parallel with CO_2, and may cause significant yield reductions for sensitive species.

Conclusions

Improving responses to climate variability and change must be a crucial requirement for future agricultural sustainability. The challenge for the study of climate change impacts on agriculture is to integrate insights from the physical, biophysical, and social sciences into a comprehensive understanding of climate-agriculture interactions at seasonal-to-inter-annual and decadal-to-century timescales, as well as at regional and global spatial scales. The final challenge is to apply this knowledge to 'real-world' agricultural practices and planning world-wide, so that long-term sustainability may be effectively enhanced.

References

Acock, B. and L. H. Allen, Jr. 1985. Crop responses to elevated carbon dioxide concentrations. Pp. 53–97. In B. R. Strain and J. D. Cure (eds.). *Direct Effects of Increasing Carbon Dioxide on Vegetation.* DOE/ERB0238. U.S. Department of Energy, Washington, D.C.

Allen, L. H., Jr., K. J. Boote, J. W. Jones, P. H. Jones, R. R. Valle, B. Acock, H. H. Rogers, and R. C. Dahlman. 1987. Response of vegetation to rising carbon dioxide: Photosynthesis, biomass, and seed yield of soybean. *Global Biogeochem* Cycles **1:**1–14.

Ben Mohamed, A., N. Van Duivenbooden, S. Abdoussalam, 2002. Impact of climate change on agriculture production on the Sahel – Part 1. Methodological approach and case study for millet in Niger. *Climatic Change* **54 (3):** 327-348.

Cayan, D.R., S.A. Kammerdiener, M.D. Dettinger, et al. 2001. Changes in the onset of spring in the western United States. B. *Am Meteorological Society* **8 (3):** 399-415.

Chen, C.-C. and B.A. McCarl. 2001. An investigation of the relationship between pesticide usage and climate change. *Climatic Change* **50:**475-487.

Chmielewski, FM, A. Muller, E. Bruns, 2004. Climate changes and trends in phenology of fruit trees and field crops in Germany, 1961-2000. *Agriculture and Forest Meteorology* **121 (1-2):** 69-78.

Coakley, S. M., H. Scherm, and S. Chakraborty. 1999. Climate change and plant disease management. *Annu Rev Phytopathology* **37:**399–426.

Cole, C. V., J. Duxbury, J. Freney, O. Heinmeyer, K. Minami, A. Mosier, K. Paustian, N. Rosenberg, N. Sampson, D. Sauerbeck, and Q. Zhao. 1997. Global estimates of potential mitigation of greenhouse gas emissions by agriculture. *Nutr Cycling Agroecosyst* **49:**221–228.

Council for Agricultural Science and Technology (CAST). 2004. Climate Change and Greenhouse Gas Mitigation: Challenges and Opportunities for Agriculture. Task Force Report. 133 pp.

Cure, J. D. and B. Acock. 1986. Crop responses to carbon dioxide doubling: A literature survey. *Agric For Meteorol* **38:**127–145.

Dahlsten, D. L. and R. Garcia (eds.). 1989. *Eradication of Exotic Pests: Analysis with Case Histories.* Yale University Press, New Haven, Connecticut.

Döll, P. 2002. Impact of climate change and variability on irrigation requirements: A global perspective. *Climatic Change* **54:**269-293.

Elliott, E. T., H. W. Hunt, and D. E. Walter. 1988. Detrital food web interactions in North American grassland ecosystems. *Agric Ecosyst Environ* **24:**41–56.

Fischer, G., M. Shah, H. Velthuizen, and F. O. Nachtergael. 2001. *Global Agro-ecological Assessment for Agriculture in the 21st Century. International Institute for Applies Systems Analysis.* IIASA Publications.Vienna.

Gleick, P. H. 1987. Regional hydrologic consequences of increases in atmospheric CO2 and other trace gases. *Clim Change* **10:**137–161.

Hendrey, G. R., K. F. Lewin, and J. Nagy. 1993. Free air carbondioxide enrichment: Development, progress, results. *Vegetation* **104:**17–31.

Hsiao, T. C. and R. B. Jackson. 1999. Interactive effects of water stress and elevated CO_2 on growth, photosynthesis, and water use efficiency. In:

Carbon Dioxide and Environmental Stress. Y. Luo and H. A. Mooney (eds.), Academic Press, New York. pp. 3-31.

Hunt, H. W., M. J. Trlica, E. F. Redente, J. C. Moore, J. K. Detling, T. G. F. Kittel, D. E. Walter, M. C. Fowler, D. A. Klein, and E. T. Elliott. 1991. Simulation model for the effects of climate change on temperate grassland ecosystems. *Ecol Model* **53:**205–246.

IPCC. 2001. *Climate Change 2001: Impacts, Adaptation, and Vulnerability.* Intergovernmental Panel on Climate Change. Cambridge University Press. Cambridge, UK. 1032 pp.

IPCC. 2000. *Special Report on Emissions Scenarios (SRES).* Nakicenovic, N., Alcamo, J., Davis, G., de Vries, B., Fenhann, J., Gaffin, S., Gregory, K., Grübler, A. *et al.*, (Eds.). Working Group III, Intergovernmental Panel on Climate Change. Cambridge University Press. Cambridge. 595 pp. http://www.grida.no/climate/ipcc/emission/index.htm

IPCC. 1996. R. T. Watson, M. C. Zinyowera, and R. H. Moss (eds.). *Climate Change 1995: Impacts, Adaptations and Mitigation of Climate Change: Scientific-Technical Analyses.* Chaps. 13 and 23. The Intergovernmental Panel on Climate Change Second Assessment Report, Vol. 2. Cambridge University Press, Cambridge, U.K. 878 pp.

Kimball, B. A. 1983. Carbon dioxide and agricultural yield: An assemblage and analysis of 430 prior observations. *Agron J* **75:**779–788.

Kimball, B. A. and S. B. Idso. 1983. Increasing atmospheric CO2. Effects on crop yield, water use and climate. *Agric Water Manage* **7:**55–72.

Lincoln, D. E., N. Sionit, and B. R. Strain. 1984. Growth and feeding response of *Pseudoplusia includens* (Lepidoptera: Noctuidae) to host plants grown in controlled carbon dioxide atmospheres.*Environ Entomol* **13:**1527–1530.

Morison, J. I. L. 1985. Sensitivity of stomata and water use efficiency to high CO_2. *Plant, Cell, Environ* **8:**467–474.

Mearns, L.O. (Ed.) 2003. *Issues in the Impacts of Climate Variability* and Change on Agriculture: Applications to the southeastern United States. Kluwer Academic Publishers. Dordrecht. 216 pp.

Monteith, J.L. 1981. Climatic variation and the growth of crops. *Quarterly Journal of the Royal Meterologiocal Society* **107**: 749-774.

Nicholls, N., 1997. Increased Australian wheat yield due to recent climate trends. *Nature*, **387:**484-485.

OSTP. 1997. *Climate Change: State of Knowledge.* Office of Science and Technology Policy. Washington, DC. 18 pp.

Ojima, D. S., W. J. Parton, D. S. Schimel, and C. E. Ownesby. 1990. Simulated impacts of annual burning on prairie ecosystems. Pp. 118–132. In S. L. Collins and L. L. Wallace (eds.). *Fire in North American Tallgrass Prairies.* University of Oklahoma Press, Norman.

Patterson, D. T. 1993. Implications of global climate change for impact of weeds, insects, and plant diseases. Pp. 273–280. In: *International Crop*

Science I. Crop Science Society of America, Madison, Wisconsin.

Peng, S., J. Huang, J.E. Sheehy, et. al. (In review). Rice Yields Decline with Higher Nightime Temperature from Global Warming. *Proceedings of the National Academy of Sciences.*

Parry, M.L., C. Rosenzweig, A. Iglesias, M. Livermore, G. Fischer. 2004. Effects of climate change on global food production under SRES emissions and socio-economic scenarios. *Global Environmental Change* **14:** 53-67.

Poorter, H. 1993. Interspecific variation in the growth response of plants to an elevated ambient CO2 concentration. *Vegetatio* **104/105:**77–97.

Reilly, J., F. Tubiello, B. McCarl, D. Abler, R. Darwin, K. Fuglie, S. Hollinger, C. Izaurralde, S. Jagtap, J. Jones, L. Mearns, D. Ojima, E. Paul, K. Paustian, S. Riha, N. Rosenberg, and C. Rosenzweig. 2003. U.S. Agriculture and climate change: New results. *Climatic Change* **57:**43-69.

Rice, C. W., T. C. Todd, J. L. Blair, T. Seastedt, R. A. Ramundo, and G. W. T. Wilson. 1998. Belowground biology and processes. Pp. 244–264. In A. K. Knapp, J. M. Briggs, D. C. Hartnett, and S. L. Collins (eds.). *Konza Prairie Long-term Ecology Research.* Oxford Press, New York, New York.

Rosenzweig, C. and M.L. Parry. 1994. Potential impacts of climate change on world food supply. *Nature* **367:**133-138.

Rosenzweig, C. and D. Hillel. 1998. *Climate Change and the Global Harvest: Potential Impacts of the Greenhouse Effect on Agriculture.*Oxford University Press, New York, New York. 324 pp.

Rosenzweig, C. and D. Hillel. 2000. Soils and global climate change: challenges and opportunities. Millennium Issue *Soil Science* 165(1):47-56.

Rosenzweig, C., A. Iglesias, X.B. Yang, P.R. Epstein, and E. Chivian. 2002. Climate change and extreme weather events: Implications for food production, plant diseases, and pests. *Global Change and Human Health* **2(2):**90-104.

Rosenzweig, C., F. N. Tubiello, R. Goldberg, E. Mills, and J. Bloomfield. 2002. Increased crop damage in the US from excess precipitation under climate change. *Global Environmental Change* **12:**197-202.

Rosenzweig, C. and F. Tubiello. 2004. Mitigation and adaptation in agriculture: An interactive approach. *Mitigation and Adaptation Strategies for Global Change* Special Issue. (in preparation).

Salt, D. T., B. L. Brooks, and J. B. Whitaker. 1995. Elevated carbon dioxide affects leaf-miner performance and plant growth in docks (*Rumex* spp.). *Global Change Biol* **1:**153–156.

Smith, J.B., R.J.T. Klein, and S. Huq. 2003. *Climate Change, Adaptive Capacity, and Development.* Imperial College Press. London. 347 pp.

Strzepek, K.M., D.C. Major, C. Rosenzweig, A. Ingesias, D.N. Yates, A. Holt, and D. Hillel. 1999. New methods of modeling water availability for agriculture under climate change: the U.S. Cornbelt. *Journal of the American Water Resources Association* **35:**1639-1655.

Sutherst, R. W. 1990. Impact of climate change on pests and diseases in Australia. *Search* **21:**230–232.

Van Duivenbooden, N., S. Abdoussalam and A. Ben Mohammed. 2002. Impacts of Climate Change on Agricultural Production in the Sahel – Part 2. Cast Study for Groundnut and Cowpea in Niger. *Climate Change* **54:**349-368.

8

LIMITS TO EFFICIENCIES OF PRIMARY PRODUCTION - CONSTRAINTS AND OPPORTUNITIES.

STEPHEN P. LONG[1], XINGUANG ZHU[1], SHAWNA L. NAIDU[1], CHRISTINE A. RAINES[2], DONALD R. ORT[3].

[1]Departments of Crop Science and Plant Biology, University of Illinois, Urbana, IL 61801, USA; [2]Department of Biological Sciences, University of Essex, Colchester, CO4 3SQ, USA; [3] Photosynthesis Research Unit, USDA-ARS, Urbana, IL 61801, USA

Introduction

As a process determining yield, photosynthesis in contrast to allocation, for example, is highly conserved. Thus, the process involves identical steps across green plants, with little variation in the major proteins. As a result, any means that are discovered to improve photosynthetic efficiency are likely to be applicable to all crops. Despite great advances in understanding of photosynthesis, photosynthesis research has contributed little to improvement of crop production in the past (Long, 1998). Does it have a future role in continuing the large rate of increase in the yield of farmed species that occurred in the last decades of the 20th Century?

Plant breeding over the past three decades has produced remarkable world-wide increases in the potential yields of many crops, most notably the improvements in the small grain cereals of the "green revolution" (Beadle and Long, 1985; Evans, 1993). Potential yield is defined as the yield that a genotype can achieve under optimal cultivation practice and in the absence of pests and diseases. What are the physiological bases of these increases? Following the principles of Monteith (1977) the potential yield (Y) of a crop and primary production (P_n) at a given location is determined by:

$$P_n = S_t.\varepsilon_i.\ \varepsilon_c/k \quad .... \text{(1a)}$$
$$Y = \eta/P_n \quad .... \text{(1b)}$$

Where S_t is the annual integral of incident solar radiation (MJ m^{-2}), ε_i the efficiency with which that radiation is intercepted by the crop; ε_c the efficiency with which the intercepted radiation is converted into biomass; η the harvest index or the efficiency with which biomass is partitioned into the harvested product; and k the energy content of the plant mass (MJ g^{-1}). S_t is determined by the site and year. Although k varies very little between vegetative organs, typically averaging 17 MJ/kg, grain with significant oil content may have significantly higher energy contents which need to be taken into account in computing yield from equation 1 (Roberts, Long, Tieszen and Beadle, 1993). P_n is the primary productivity, i.e. the total plant biomass produced over the growing season. Potential yield is therefore determined by the combined product of three efficiencies, each describing broad physiological properties of the crop: ε_i, ε_c, and η. ε_i is determined by the speed of canopy development and closure, and canopy longevity, size and architecture. ε_c is a function of the combined photosynthetic rate of all leaves within the canopy less crop respiratory losses. Since reported P_n for annual crops is commonly the total above-ground biomass, actual ε_c will also be lowered by the fact that some shoot tissue will have been shed prior to harvest and that some biomass will have been allocated to roots. These factors will lower apparent ε_c in the order of about 20% for annual grain crops.

In the context of equation 1, increase in potential yield over the past 40 years has resulted largely from increase in η. Increased yield potential has also resulted from increased ε_i through the development of larger leafed cultivars, whilst realized yields have improved through better fertilisation and improved disease protection, in turn increasing ε_i (reviewed: Beadle and Long, 1985; Evans, 1993; Hay, 1995).

With reference to equation 1, how can yield potential be increased further? Healthy crops of modern cultivars at optimised spacing intercept most of the available radiation within their growing season, limiting prospects for improving ε_i. However, a number of crops do not currently use the full potential growing season, i.e. period when there is adequate temperature and water for plant growth. So here, there is some room for further improvement in ε_i. For example, higher primary production in the perennial C4 grass *Miscanthus x giganteus* relative to silage maize (*Zea mays* L.) in southern England, corresponds to an additional 4 – 6 weeks in which >90% of the available sunlight is trapped by the canopy (cf. Beale and Long, 1995; Baker, East and Long, 1983).

Grain in the modern cultivars of cereals can represent 60% of the total biomass at harvest (Evans, 1993; Hay, 1995). Harvest index would therefore also appear to be approaching an upper limit, given that a minimum quantity of biomass that must remain in the plant body, to ensure that vital nutrients and reserves can be translocated into the grain or other harvested organ, and to account for cell wall materials that cannot be degraded. If η and ε_i are approaching an upper limit, further increase in potential yield can only be achieved by increase in ε_c which is determined by photosynthesis and respiration. In theory, ε_c depends on the efficiency with which the absorbed light can be transduced into biomass, i.e. the efficiency of photosynthesis corrected for respiratory losses. This review considers the limitations to and opportunities for increasing photosynthesis in crops. However, it is first necessary to establish whether photosynthesis limits crop production, and whether increase in photosynthesis will actually result in increased crop yields.

Why was leaf photosynthesis abandoned as a selection trait for crops?

The advent of transportable infra-red CO_2 analyzers opened the opportunity for selecting crop genotypes on the basis of leaf photosynthetic rates (Long, Farage and Garcia, 1996). However, some influential studies questioned the idea that leaf photosynthesis was limiting to crop production. Evans and Dunstone (1970) showed that modern bread wheats had lower leaf photosynthetic rates than their wild ancestors. This lack of correlation between crop yield and leaf photosynthetic rate has been noted frequently in other studies, and has been attributed to control by capacity to utilize additional photosynthate (sink strength) rather than photosynthetic capacity (Evans, 1993). The lack of correlation between leaf photosynthetic rate and yield in such studies should have been no surprise. Whilst it is implicit in Equation 1 that photosynthetic efficiency is critical to crop yield, this is the photosynthetic efficiency of the whole crop. Many surveys of leaf photosynthesis were based on the light-saturated rate of a single leaf at a single stage in crop development (Long, 1998). The relationship of single leaf measurements to the whole crop will be complex, and not intuitive. First, as much as 50% of crop carbon may be assimilated by leaves under light-limiting conditions where very different biochemical and biophysical properties determine photosynthetic rate (Long, 1993). Secondly, increase in leaf area may often be achieved by decreased investment per unit leaf area, thus photosynthetic capacity per unit area is commonly lower in species with lower leaf thickness (Beadle and Long, 1985).

If crop improvement has resulted in increased leaf area, mean leaf photosynthetic rate may decline because of increased self-shading and maximum leaf photosynthetic rates may decline because resources are spread more thinly across the larger leaf area (Evans, 1993). Contrary to the finding of Evans and Dunstone (1970), Watanabe, Evans and Chow (1994) showed a strong positive correlation between leaf photosynthetic rate and date of release of Australian bread wheat cultivars. This difference might be explained by the fact that the latter study was limited not only to a single species, but a narrow range of germplasm within that species. Here variability in leaf area per plant and its distribution would be smaller, and variation in leaf photosynthetic rate not confounded with large differences in total leaf area or specific leaf area. The potential of leaf photosynthetic rate in improving potential crop yield can only be evaluated when other factors, in particular leaf canopy size and architecture are held constant. Prior conclusions are based on correlation; can we increase canopy photosynthesis within a single genotype to test the question of whether increased yield can be achieved?

Fortuitously, the recent interest in the effects of rising atmospheric CO_2 concentration ([CO_2]) has provided this test. Increase in [CO_2] has two effects on C_3 plants: an increase in leaf photosynthesis and a decrease in stomatal aperture (reviewed: Long, Ainsworth, Rogers and Ort, 2004). Elevated [CO_2] increases net leaf photosynthetic rate primarily by competitive inhibition of the oxygenase activity of Rubisco and therefore photorespiration. At 25°C, increase in [CO_2] from the current 370 ppm to the 550 ppm forecast for the year 2050 would increase net photosynthesis by 15-25% by partial inhibition of photorespiration (Long *et al.*, 2004). In the first large scale open-field experiment to grow a food crop at 550 ppm [CO_2], bread wheat leaf photosynthesis and ε_c were increased throughout the growth of the crop. Grain yield was increased significantly by 10-12% in two consecutive growing seasons, yet there was no change in leaf area index, ε_i or η with $p$$CO_2$ treatment (Kimball, Pinter, Garcia, Lamorte, Wall, Hunsater, Wechsung, Wechsung and Kartschall, 1995, Pinter, Kimball, Garcia, Wall, Hunsaker and LaMoret, 1996). Similar effects have subsequently been seen with further FACE (Free-Air Carbon dioxide Enrichment) experiments with rice and soybean, but interestingly not with sorghum (Long *et al.*, 2004). Over a wide range of C_3 crops, an approximate doubling of the current [CO_2] in field or laboratory chambers caused no significant increase in leaf area, a 23-58% increase in leaf photosynthetic rate (Drake, Gonzalez-Meler and Long, 1997), and an average 35% increase in crop yield (Kimball, 1983). However, these increases could also result from decreased water loss and water stress, or/and from decreased respiration. Evidence that there is an independent increase due to increased leaf photosynthesis comes from two sources: 1) Large increases in yield occurred under elevated [CO_2] when wheat was irrigated in the field to the level required for maximum yield (Pinter *et al.*, 1996) and when lowland rice was grown in paddy conditions in field chambers (Baker, Allen and Boote, 1990); 2) C_4 plants show similar reductions in stomatal aperture to C_3 plants when grown at elevated [CO_2], but show no or little increase in net photosynthesis (Drake *et al.*, 1997; Long *et al.*, 2004). C_4 crops, by contrast to C_3 crops, grown under elevated $p$$CO_2$ show little or no increase in yield when grown under well watered conditions (Ghannoum, Van Laemmerer and Conroy, 2001; Long *et al.*, 2004). This is consistent with the expectation that C_4 photosynthesis is CO_2-saturated in the present atmosphere (Ghannoum *et al.*, 2001). Thus elevated [CO_2] does not directly increase leaf photosynthesis in C_4 plants but does, as in C_3 plants, decrease transpiration. In a mixed C_3/C_4 water-logged marsh community, elevated [CO_2] caused a large and sustained increase in the net carbon gain of stands of the C_3, but not the C_4, species (Drake *et al.*, 1997). These findings suggest that whilst the increased water use efficiency will contribute to increased primary production under elevated $p$$CO_2$, when plants are grown with ample water large production increases still result and are attributable to increased leaf photosynthesis. Manipulation of photosynthesis by elevating [CO_2] has also provided a test of the role of sink-size. Ainsworth, Rogers, Nelson and Long (2004) showed a sustained increase in photosynthesis in soybean cv. Williams grown in the field under open-air [CO_2] elevation. However, mutation at the *dt1* locus to make this line determinate, decreased potential reproductive sink-size. In normal air, photosynthesis of the two lines did not differ significantly, in elevated [CO_2] however there was a significant increase in non-structural carbohydrates in leaves of the determinate form during seed filling, which corresponded to a decline in photosynthesis, suggesting sink-limitation. When a normally determinate line cv. Elf was grown in elevated [CO_2] it showed a similar increase in yield to indeterminate Williams, and did not show any loss of photosynthetic capacity. An interpretation of these results is that, at least in soybean, conventional breeding has selected for a sink capacity, which normally exceeds photosynthetic capacity. If sink capacity limited yield then genetically decreasing potential reproductive axes in Williams would have decreased yield in normal air and the normal form of Williams could not have increased yield in response to elevated pCO_2.

In summary, the growth of crops at elevated [CO_2] shows that increased leaf and canopy photosynthesis results in large increases in crop yields. What opportunities are there to increase photosynthesis other than waiting for the atmospheric [CO_2] to rise.

What is the potential radiation use efficiency?

In this section we examine the limits to maximum conversion efficiency, providing a framework for discussing potential routes for improvement. While an ε_i of 0.9 and a η of 0.6 are high and probably maximal, the maximum ε_c reported is around 0.014 for C_3 crops and 0.020 for C_4 over a growing season, although higher efficiencies are observed in the short-term. For C_3 crops the highest short-term efficiencies are about 0.035 and for C_4 about 0.043 (Beadle and Long, 1985; Beale and Long, 1995; Piedade, Junk and Long, 1991). This section explores why these record numbers are apparently so low and whether photosynthesis in crops is as inefficient as we might at first assume from such numbers. About 50% of solar energy is in the near infra-red wavelengths (>700 nm). The energy of photons of >700 nm is too low to drive charge separation at the reaction centres of land plants, and therefore outside the photosynthetically active waveband (Table 8.1). Leaves scatter absorbed light, with the result that some photons will re-emerge as reflected light. The minimum photon requirement is 8 in C_3 plants, regardless of wavelength below 700 nm; i.e. a red photon has the same effect as a violet photon. However, a violet photon of 400 nm contains 75% more energy than a red photon of 700 nm. The additional energy of the violet photon is lost as heat, representing an intrinsic photochemical inefficiency (Table 8.1). Other pigments, such as anthocyanins in the epidermis will absorb some light, but cannot pass this energy on to photosynthesis, resulting in inactive absorption (Beadle and Long, 1985). One mole of photons of 690 nm wavelength, contains 173.3 kJ, yet when one mole CO_2 is released from carbohydrate it liberates 477 kJ. Since a minimum of eight moles of photons are required to convert one mole of CO_2 to carbohydrate, the synthesis of carbohydrate therefore has a maximum efficiency of $477/(8 \times 173.3) = 0.344$, a loss of about 66% of energy at this step. Because the C_4 pathway requires more ATP, carbohydrate synthesis here has a still lower maximum efficiency (Table 8.1). This is offset in C_3 plants by photorespiration, which re-oxidises a portion of this carbohydrate (Beadle and Long, 1985). Finally, mitochondrial respiration necessary for synthesis of new tissues and maintenance of existing tissues in all plants removes about 40% of the remaining energy (e.g. Gifford, 1995). In theory, a maximum ε_c of about 0.051 is possible in C_3 plants and 0.060 in C_4 plants (Table 8.1). Two possibilities for increasing ε_c are therefore increasing the portion of the growing season in which the theoretical maxima are obtained or increasing efficiencies of the different steps given in Table 8.1. The first has already been exploited to some extent by modifying plant canopies, as in the erect leaf cultivars of rice of the green revolution. Another factor lowering ε_c is the decrease in efficiency that develops to dissipate excess absorbed energy in full sunlight, but persists as lower light levels return. Increased efficiency of carboxylation at the enzyme Rubisco and increased capacity for regeneration of the primary CO_2 acceptor molecule ribulose-1:5-bisphosphate (RubP), would raise the radiant flux density at which photosynthesis is saturated, allowing more efficient use of radiation in direct sunlight. Below it is shown that all four of these approaches might be realized on a relatively short time scale (ca. 10 years). The second possibility, could be realized either by eliminating the oxygenase activity or Rubisco or by engineering C_4 photosynthesis in C_3 crops. As is shown below there are substantial barriers to both. These six possible routes to improving ε_c are discussed below.

Table 8.1 EFFICIENCY OF THE TRANSDUCTION OF INTERCEPTED SOLAR RADIATION INTO PLANT CARBOHYDRATE VIA PHOTOSYNTHESIS OF CROP LEAF CANOPIES. "% LOSS" SHOWS THE PROPORTION OF ENERGY LOST AT EACH STAGE FROM INTERCEPTION TO CARBOHYDRATE ACCUMULATION, EFFICIENCY AT EACH STAGE IS GIVEN IN PARENTHESIS. "% REMAINING" SHOWS HOW MUCH OF THE ENERGY REMAINS AT EACH STAGE ALONG THE TRANSDUCTION CHAIN. C3 CROPS (E.G. WHEAT, SOYBEAN) DIFFER FROM C4 (E.G. MAIZE, SUGAR CANE) THE LATTER LACKING PHOTORESPIRATION, BUT REQUIRING MORE ENERGY FOR CARBOHYDRATE SYNTHESIS, HENCE THE DIFFERENT OVERALL ε_c. ADAPTED FROM: (BEADLE AND LONG, 1985).

	% loss at each stage (efficiency at each stage)		*% remaining*	
Incident energy outside photosynthetically active wavebands	50 (0.5)	50		
Reflected and transmitted light	5 (0.9)	45		
Light absorbed by non-photosynthetic pigments	1.8 (0.96)	43.2		
Photochemical inefficiency	8.4 (0.8)	34.8		
Photosynthetic type	***C_3***	***C_4***	***C_3***	***C_4***
Carbohydrate synthesis	22.8 (0.34)	24.8 (0.29)	12	10.0
Photorespiration	3.5 (0.7)	0 (1.0)	8.5	10.0
Dark respiration	3.4 (0.6)	4 (0.6)	5.1	6.0
Resulting ε_c	*0.051*	*0.060*		

Modifying plant canopies to increase ε_c

Leaf photosynthesis responds non-linearly to increases in solar energy (Figure 8.1b). In C_3 crops, leaf photosynthesis is saturated at radiant flux densities of about one quarter of maximum full sunlight, therefore any solar energy intercepted above this level is wasted. A mature healthy crop may have three or more layers of leaves; i.e. a leaf area index of ≥3. If the leaves are roughly horizontal (Figure 8.1, Plant X), the upper-most layer will intercept most of the light, about 10% may penetrate to the next layer and 1% to the layer below that. With the sun overhead, the photosynthetically-active energy intercepted per unit leaf area by a horizontal leaf at the top of a plant canopy would be 900 J m^{-2} s^{-1}, or about 3 times the amount required to saturate photosynthesis (Figure 8.1b). Therefore at least two-thirds of the energy intercepted by the upper leaves is wasted. A better arrangement for these conditions would be for the upper layer to intercept a smaller fraction of the light, allowing more to reach the lower layers. This is achieved when the upper leaves are more vertical and lowermost leaves horizontal, as in the example of Plant Y (Figure 8.1) (Nobel, Forseth and Long, 1993). For a leaf with a 75° angle with the horizontal, the amount of light energy intercepted per unit leaf area would be 300 J m^{-2} s^{-1}, just sufficient to saturate photosynthesis, but the remaining direct light energy (600 J m^{-2} s^{-1}) would penetrate to the lower layers of the canopy. By distributing the energy among leaves in this way, in full-sunlight Plant Y would have over double the efficiency of solar energy use than Plant X (Figure 8.1c) (Ort and Long, 2003). Mathematical models have been developed to design optimum distributions of leaves to maximize efficiency, which have been used as a guide for selecting improved crops. This approach has been a major factor in improving the productivity of rice (Beadle and Long, 1985). Older varieties with more horizontal leaves such as plant X have been replaced by newer varieties that have been bred to have more vertical leaves in the top layer, such as Plant Y (Nobel *et al.*, 1993). The advantage of this change in canopy design would be greatest when the sun is overhead and diminishes progressively as sun angles decline and in diffuse lighting conditions.

Relaxing the photoprotected state more rapidly to increase ε_c

Figure 8.1b shows the typical non-rectangular hyperbolic response of photosynthesis to radiation flux density. As radiation increases photosynthesis saturates. However, the leaf continues to absorb photosynthetically active radiation. This additional energy exceeds the capacity for photosynthesis, and without some alternative mechanism to dissipate the energy will cause photo-oxidative damage to the photosynthetic apparatus. This is largely avoided by an induced increase in thermal dissipation of energy via the formation of epoxidated xanthophylls (Baroli and Niyogi, 2000; Havaux and Niyogi, 1999;

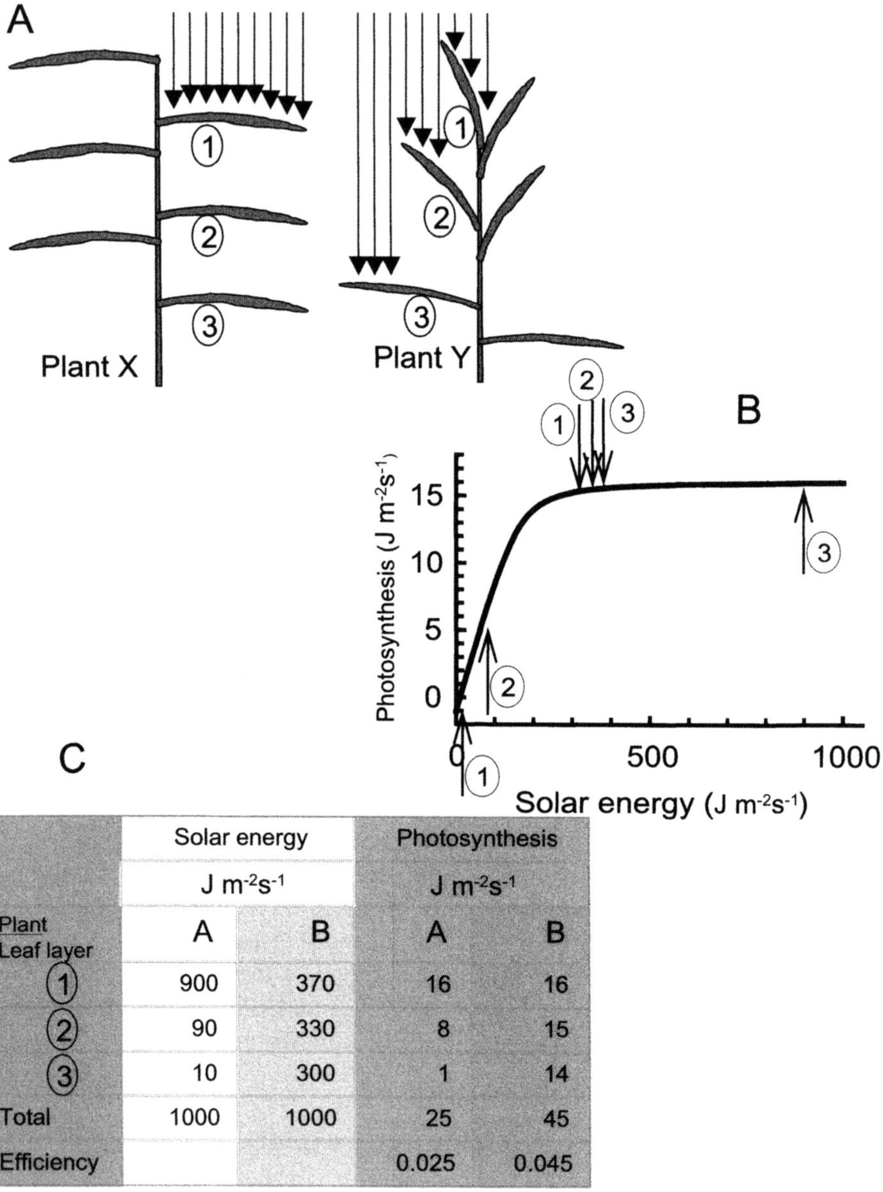

C

Plant / Leaf layer	Solar energy J m⁻²s⁻¹ A	Solar energy J m⁻²s⁻¹ B	Photosynthesis J m⁻²s⁻¹ A	Photosynthesis J m⁻²s⁻¹ B
①	900	370	16	16
②	90	330	8	15
③	10	300	1	14
Total	1000	1000	25	45
Efficiency			0.025	0.045

Figure 8.1 (a) Plant X has horizontal leaves, such that the upper layer ① will intercept most on the incoming solar energy, shading the lower layers; ② and ③. Plant Y has vertical leaves at the top, becoming more horizontal near the bottom. This arrangement spreads the solar energy more evenly between layers. (b) Photosynthesis, in terms of energy trapped in stored carbohydrate, for a leaf is plotted against solar energy. Arrows below the curve indicate the average amounts of solar energy at the three leaf layers of Plant X and arrows above the three leaf layers of Plant Y. (c) From the graph of B, the amount of solar energy and the photosynthesis for each leaf layer, and their totals, for the two Plants are given. Note that by spreading the same amount of solar energy more evenly among its leaves, Plant Y can achieve almost double the rate of photosynthesis of Plant X when the sun is directly overhead. Redrawn from Ort and Long (2003).

Long, Humphries and Falkowski, 1994b). This process increases thermal dissipation of absorbed light energy within the photosystem II (PSII) antenna system and protects PSII from damage in high light. The PSII reaction centre appears the most vulnerable part of the photosynthetic apparatus to damage under conditions of excess light. This reversible increase in thermal quenching, is termed photoprotection, and it decreases the maximum quantum yield of photosystem II (F_v/F_m) and CO_2 uptake (Φ_{CO2}), i.e. the initial slope of the response of photosynthesis to radiation (Figure 8.1b) (reviewed: Zhu, Ort, Whitmarsh and Long, 2004a). In addition it decreases the convexity (θ) of the non-rectangular hyperbolic response illustrated in Fig. 1B. At high light, decrease in Φ_{CO2} and θ has minimal impact on carbon gain, while the increased thermal energy dissipation protects PSII against oxidative damage. Light in leaf canopies in the field is continually fluctuating, but a finite period of time is required for recovery of Φ_{CO2} and θ when solar radiation drops, as for example when a cloud obscures the sun or change in sun-angle places one leaf in the shade of another. Low Φ_{CO2} and θ limit the rate of photosynthetic carbon uptake on transfer to low light, an effect prolonged by low temperature. What is the cost of this delayed recovery to potential CO_2 uptake by a canopy in the field? Zhu *et al.* (2004a) used a reverse ray-tracing algorithm for predicting light dynamics of 120 randomly selected individual points in a model canopy to describe the discontinuity and heterogeneity of light flux within the canopy. Because photoprotection is at the level of the cell, not leaf, light was simulated for small points of 10^4 µm rather than as an average for a leaf. The predicted light dynamics were combined with empirical equations simulating the dynamics of the light-dependent decrease and recovery of Φ_{CO2} and θ, and their effects on the integrated daily canopy carbon uptake (A_c'). The simulation was for a model canopy of leaf area index 3 with random inclination and orientation of foliage, on a clear sky day (latitude 44 °N, 120th day of year). The delay in recovery of photoprotection was predicted to decrease A_c' by 17-6.5 % at 30 °C and 32 – 12.5% at 10 °C. The lower figure is for a chilling tolerant species, the upper for a chilling susceptible species. Temperature is significant because it decreases photosynthetic capacity and rate of relaxation of photoprotection. However, this simulation suggests that the inability of leaves to recover efficiency on decrease in solar radiation, lowers ε_c by between 6.5 and 32%, average losses at typical temperatures for temperate crops would be of the order of ca. 15% (Zhu *et al.*, 2004a). Much larger losses result when photosynthesis is decreased by stresses (Long *et al.* 1994a).

Large gains in ε_c could be achieved if the lag in relaxation in photoprotection could be decreased or eliminated. Is this a possibility? Photoprotection fulfils a necessary function of decreasing the probability of oxidative damage to PSII, which in itself would lower photosynthetic efficiency, by photoinhibition, and would require repair and replacement of the protein before efficiency could be restored. In the longer term a continued excess of excitation energy would lead to irreversible photo-oxidation (reviewed: (Long *et al.*, 1994a). Could the loss found here be decreased without the risk of photoinhibition and photooxidation? Falkowski and Dubindky (1981) identified algae associated with corals that could withstand 1.5x full sunlight without evidence of loss of maximum photosynthetic efficiency or photoinhibition, showing that the loss of efficiency is not an intrinsic requirement of the photosynthetic apparatus. In higher plants, Wang, Zhang, Zhu, Lu, Kuang and Li (2002) have shown a close correlation between increased rate of recovery from the photoprotected state and increased biomass production in the "super-high yield" rice cultivars. This increased rate of recovery was associated with an increase in concentrations of the intermediates of the xanthophyll cycle. Across plant species higher rates of recovery have been associated with xanthophyll cycle capacity, including the epoxidation associated with recovery (Long *et al.*, 1994a). These findings suggest that up-regulation of capacity for recovery from photoprotection is feasible, and may already have been achieved in rice.

Photorespiration

About 30% of the carbohydrate formed in C_3 photosynthesis is lost via photorespiration. The amount increases with temperature so that photorespiration is a particularly significant inefficiency for C_3 crops in tropical climates and during hot summer weather (Figure 8.2). Photorespiration results from the apparently unavoidable oxygenation reaction of RuBP by Rubisco (reviewed: Long, 1998). Beyond this point the purpose of photorespiratory metabolism is to recover the carbon diverted into this pathway. Blocking photorespiratory metabolism downstream of Rubisco will simply result in this carbon entering a dead-end metabolic pathway. Indeed, mutants that lack any of the photorespiratory enzymes die, unless they are grown at low oxygen or at very high CO_2 to inhibit oxygenation of RuBP. The only remaining prospects for decreasing photorespiration then, is decreased oxygenation. Would decreased oxygenation result in higher yields? Photorespiratory metabolism can serve to dissipate excess excitation energy at high irradiances, involves the synthesis of serine and glutamate, and transfers reductive power from the chloroplast to the mitochondrion. This has led some to suggest that photorespiration has an essential role (e.g. Barber, 1998; Evans, 1998). However, xanthophylls provide a far more effective means of dissipating excess energy, and unlike photorespiration they do not dissipate energy and therefore impose decreased efficiency in limiting light. The photosynthetic cell has pathways beside photorespiration for amino acid synthesis and transfer of reductive energy to the cytosol (reviewed: Long, 1998). Further, because CO_2 is a competitive inhibitor of the oxygenase activity of Rubisco, photorespiration can be eliminated by growing plants in very high [CO_2]. Wheat can grow normally and complete its life cycle under these unusual conditions (Wheeler, Mackowiak, Sager, Knott and Berry, 1995). Commercial growers of some greenhouse crops increase [CO_2] to three or four times the outside concentration. This inhibits the oxygenation reaction of Rubisco, increasing photosynthetic efficiency and final yield. At the present time the global [CO_2] is rising and this too is diminishing photorespiration, but atmospheric change also includes many potentially negative effects for crops, including increased temperature, decreased soil moisture and an associated rise in phytotoxic tropospheric ozone (reviewed: Long *et al.*, 2004; Ort and Long, 2003). Healthy C_4 plants avoid photorespiration by concentrating CO_2 at the site of Rubisco.

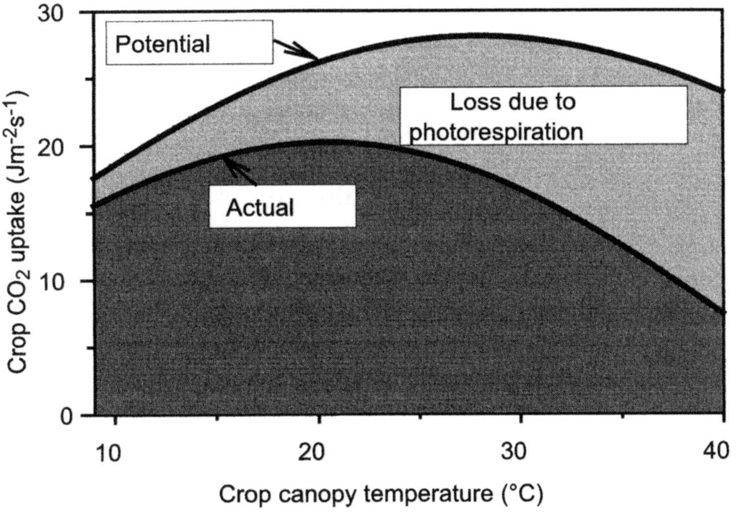

Figure 8.2 Calculated actual and potential rates of crop canopy photosynthesis versus temperature, where potential is defined as the rate in the absence of photorespiration. The difference represents the loss due to photorespiration. Calculation assumes a crop with a leaf area index of 3 and a photon flux above the canopy of 1800 µmol m^{-2} s^{-1}, i.e. full sunlight; parameters and method as detailed in Long (1991).

C_4 PHOTOSYNTHESIS A MEANS TO ELIMINATE PHOTORESPIRATION?

Terrestrial C_4 plants differ from C_3 in containing two distinct layers of photosynthetic tissue, one external to the other, each containing morphologically and functionally distinct chloroplasts. This cellular differentiation within the photosynthetic tissue is termed "Kranz" leaf anatomy. In this arrangement the mesophyll surrounds the inner photosynthetic bundle sheath, where Rubisco is localized. Only the mesophyll has intercellular air space and contact with the atmosphere. CO_2 is first assimilated into a C4 dicarboxylate via PEPcase (Phospho-enol pyruvate carboxylase) in the mesophyll. The dicarboxylate is transferred to the bundle sheath where it is decarboxylated releasing CO_2 at the site of Rubisco. The resulting pyruvate is transferred back to the mesophyll where it is phosphorylated to provide PEP, completing the C_4 cycle. The photosynthetic C_4 cycle is in effect a CO_2 pump which serves to concentrate CO_2 around Rubisco to ca. 10x current atmospheric concentrations (Hatch, 1987). It effectively eliminates photorespiration, but does require additional energy to operate the C_4 cycle (Table 8.1). C_4 photosynthesis in seed plants has evolved independently at least 31 times (Kellogg, 1999), the first appearances in the fossil record coinciding with the current low atmospheric $[CO_2]$, with respect to the geological record (Cerling, 1999). The repeated evolution of C_4 plants despite the complexity of the process suggests that there may be no other variability to exploit among land plants for decreasing photorespiration. If there were forms of Rubisco with improved ability to discriminate against oxygenation then it would surely have been a simpler route for evolution than selecting the complex syndrome of changes needed to provide functional C_4 photosynthesis. Table 8.1 shows that from theory C_4 plants will on average have a higher maximum ε_c than C_3. This difference increases with temperature due to the increase in photorespiration as a proportion of photosynthesis (Figure 8.2), such that advantage would be most pronounced in the tropics. Indeed the highest known productivity in natural vegetation is for a C_4 perennial grass in the central Amazon, which achieves a net production of 100 t (dry matter) ha^{-1} yr^{-1} (Long, 1999; Morison, Piedade, Muller, Long, Junk and Jones, 2000; Piedade *et al.*, 1991). Of our major food crops only maize and sorghum are C_4 (Long, 1998). Is there theoretical advantage in the C_4 process and could it be transferred to our major C_3 crops?

C_4 plants have the advantage of eliminating energy loss in photorespiration, but at the expense of additional energy, typically 2 ATPs per CO_2 assimilated. Because the specificity of Rubisco for CO_2 and solubility of CO_2 relative to O_2 decline with increase in temperature, photorespiration as a proportion of photosynthesis increases with temperature. In dim light, when photosynthesis is linearly dependent on the radiative flux, the rate of CO_2 assimilation will depend entirely on the energy requirements of carbon assimilation (Long, 1999; Long, Postl and Bolharnordenkampf, 1993). The additional ATP required for assimilation of one CO_2 in C_4 photosynthesis, compared to C_3 photosynthesis, increases the energy requirement in C_4 plants (Hatch, 1987). However, when the temperature of a C_3 leaf exceeds ca. 25°C, the amount of light energy diverted into photorespiratory metabolism in C_3 photosynthesis will exceed the additional energy required for CO_2 assimilation in C_4 photosynthesis (Hatch, 1992; Long, 1999). Below 25°C C4 photosynthesis is less efficient than C_3 photosynthesis under light-limiting conditions. However, total photosynthesis by a crop canopy reflects a combination of light-limited and light-saturated CO_2 assimilation. At light-saturation the maximum efficiency of solar-energy use does not determine actual efficiency, which is determined by the maximum rate of photosynthesis. Here the C_4 plant has an advantage, even below 25°C, since its maximum rate will be greater than that of an equivalent C_3 leaf because of the absence of photorespiration, as shown in Figure 8.2. Does a higher rate of light-saturated photosynthesis offset the lower rate of light-limited photosynthesis at the crop canopy level at temperatures below 25°C? In considering this question account has to be taken of the dynamic nature of the balance between light-limited and light-saturated photosynthesis within a canopy over the course of a day. By integrating established steady-state biochemical models of C_3 and C_4 leaf photosynthesis (Farquhar, von Caemmerer and Berry, 1980; Collatz, 1992) into canopy radiation transfer models the integrals of the diurnal course of photosynthesis can be calculated (Humphries and Long, 1995). Using this approach, Figure 8.3 shows that while the advantage of C_4 photosynthesis diminishes with temperature, there is still an advantage to the simulated daily integral of canopy CO_2 uptake even at down to 5°C. Thus, even at the cool growing season temperatures typical of the UK, and other cool temperate climates, some advantage could theoretically be gained from C_4 photosynthesis. That this can occur in practice is supported by the observation that the highest known dry matter productivity for the UK, is for the cold-adapted C_4 perennial grass *Miscanthus x giganteus* that has been shown to produce 29 t (dry matter) ha^{-1} in southern England with a measured ε_c of 0.039 (Beale and Long, 1995; Beale *et al.*, 1999). At the least, this suggests that with continued improvement in cold tolerance maize has the potential to out-yield C_3 crops even in cool climates, such as NW Europe.

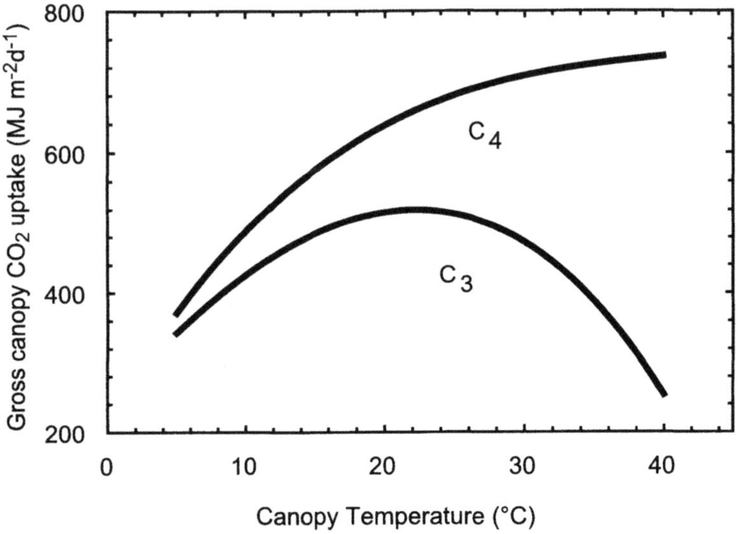

Figure 8.3 Predicted rates of gross canopy CO_2 uptake integrated over a diurnal course for a range of canopy temperatures. The simulation is for a leaf area index of three assuming a spherical distribution of foliar elements, on June 30th and with clear sky conditions (atmospheric transmittance = 0.75) at a latitude of 52°N. Equations and parameter from Humphries and Long (1995). Adapted from Long (1999).

Figure 8.3 shows that for a tropical C_3 crop such as rice, substantial gains in e_c might be gained by engineering the addition of the photosynthetic C_4 cycle into the crop. Genes coding for the enzymes of the photosynthetic C4 cycle have been isolated from maize and other C4 plants, and have been used to transform rice. It is now possible to transform C3 plants to express the C4 pathway enzymes to create C4 photosynthesis in a single cell (Miyao, 2003; Suzuki, Murai, Burnell and Arai, 2000). However, C4 plants differ not only in their use of the photosynthetic C4 cycle, but also in spatial separation of PEP carboxylase and Rubisco. In C4 plants there is a semi-impermeable barrier between the mesophyll and bundle sheath cells, which limits the diffusion of CO_2 released in the bundle sheath back into the mesophyll. Any CO_2 that does diffuse back must be re-assimilated, increasing the requirement of ATP and energy requirement per CO_2 molecule assimilated. Figure 8.3 assumes a leakage rate of 10%, i.e. 1 in 10 CO_2 molecules diffuses back into the mesophyll. If the entire mechanism was within a single cell as has been engineered in rice, i.e. PEPc in the cytoplasm and Rubisco in the chloroplast, then leakage of CO_2 would be very much higher. von Caemmerer (2003 shows from theory that such a single cell system would be very inefficient, because of the leakage of a large proportion of the CO_2 released at Rubisco. As such a single cell C4 system would allow a plant to maintain a positive carbon balance under high light and drought conditions, but would be very inefficient at low light or in dense canopies. Two naturally occurring C4 plants have been identified in which the process does occur within a single cell. However, these are elongated cells in which PEP carboxylase and Rubiscos are spatially separated by distance (Voznesenskaya, Franceschi, Kiirats, Artyusheva, Freitag and Edwards, 2002; Voznesenskaya, Franceschi, Kiirats, Freitag and Edwards, 2001). Both are slow-growing species of hot semi-arid environments consistent with the theoretical analysis of von Caemmerer (2003). Although, higher photosynthetic rates have been suggested to occur in rice transformed with the C4 cycle, this appears a result of increased stomatal aperture rather than increased capacity within the mesophyll (Ku, Cho, Li, Jiao, Pinto, Miyao and Matsuoka, 2001). The analysis of von Caemmerer (2003) shows that simple expression of the C4 enzymes in the mesophyll of C_3 crops will not be adequate to achieve the e_c advantages of C4 photosynthesis. This will require understanding of the integrated development of Kranz anatomy, localization of C_4 and C_3 enzymes, and necessary membrane transporters. Understanding of the development of C_4 photosynthesis is still too incomplete to determine the necessary transformations (Monson, 1999). It would appear, at the present time, that a more viable means to exploit C_4 photosynthesis is in improving the environmental range of existing C_4 crops, rather than in transforming C_3 crops. An alternative means of decreasing photorespiration is to decrease the oxygenation capacity of Rubisco, however as explained below this may come with the penalty of decreased carboxylation capacity.

Increasing the efficiency of Rubisco

In considering how to redesign plant canopies it was noted that photosynthesis at the leaf level is saturated by amounts of solar energy well below full sunlight (Figure 8.1b). Referring back to Figure 8.1b it can be seen that solar energy will exceed the amount needed to saturate photosynthesis for much of a sunny day. Are there other approaches to using this excess energy? The response of photosynthesis to solar energy describes a non-rectangular hyerbola, rising rapidly with increasing solar radiation at low levels, but saturating at about 25% of full sunlight. Why does this saturation occur? Several analyses suggest a co-limitation by Rubisco and by capacity for regeneration of RubP the primary substrate for CO_2 assimilation in C_3 leaves. So why not just increase the amount of Rubisco per unit of leaf area? Rubisco is already the most abundant protein in crop leaves, constituting about 50% of the soluble protein of the leaf. Calculations of volumes suggest there may not be physical capacity to add more (Pyke and Leech, 1987). Rubisco appears to carry a double penalty. First it catalyses the oxygenation of RubP leading to photorespiration. Second, the maximum catalytic rate (k_c^c) is remarkably slow compared to most plant enzymes, such that large amounts of the protein are required to achieve the photosynthetic rates necessary to support high productivities in C_3 crops. This inefficiency may explain why Rubisco is so much more abundant than any other protein in leaves.

It has long been recognized that genetic modification of Rubisco to increase the specificity for CO_2 relative to O_2 (τ) would decrease photorespiration and potentially increase C3 crop productivity. However, when the kinetic properties of Rubisco across photosynthetic organisms are compared it appears that forms with high τ have low maximum catalytic rates of carboxylation per active site (k^c_c) (Bainbridge, Madgwick, Parmar, Mitchell, Paul, Pitts, Keys and Parry, 1995). Theoretical considerations also suggest that increased τ might only be achieved at the expense of k^c_c. If a fixed inverse relationship between k^c_c and τ implied from measurements is assumed and if increased concentration of Rubisco per unit leaf area is not an option, will increased in τ result in increase leaf and canopy photosynthesis? Zhu, Portis and Long (2004b) used a mathematical model to explore these questions. From τ and k^c_c reported for Rubisco across diverse photosynthetic organisms an inverse relationship of k^c_c on τ was defined. Following the steady-state biochemical model of leaf photosynthesis of Farquhar *et al.* (1980) the C3 photosynthetic CO_2 uptake rate (A) is either limited by the maximum Rubisco activity ($V_{c,max}$) or the rate of regeneration of RubP, in turn determined by the rate of whole chain electron transport (J). If J is limiting, increase in τ will increase net CO_2 uptake because products of the electron transport chain will be partitioned away from photorespiration into photosynthesis. The effect of increase in τ on Rubisco-limited photosynthesis depends on both k^c_c and $[CO_2]$. As in the case of C4 photosynthesis, there are conflicting benefits at the level of the canopy. Increased τ would increase light-limited photosynthesis, while the associated decrease in k^c_c would lower the light-saturated rate of photosynthesis. Zhu *et al.* (2004b) simulated the consequences of variation in τ assuming an inverse change in k^c_c for carbon gain by crop canopies. Increase in τ results in an increase in leaf CO_2 uptake at low light, but decreases CO_2 uptake in high light. Over the course of a day total crop canopy CO_2 uptake results from significant amounts of both light-limited and light-saturated photosynthesis. A biophysical model of light transmission into leaf canopies was used to determine light flux at different leaves as sun-leaf geometry changes over the course of the day. The predicted light fluxes were then used to predict photosynthesis at the individual leaves from the biochemical model of Farquhar *et al.* (1980). Summing these yielded the canopy CO_2 uptake rate for the day. A leaf area index of 3 was assumed for a mid-summer day at mid-latitude and a canopy temperature of 25°C. The simulation was repeated for a wide range of atmospheric CO_2 concentrations. The results suggest that the present average τ found in C3 terrestrial plants is supra-optimal for the present atmospheric CO_2 concentration of 370 ppm, but would be optimal for 200 ppm a value remarkably close to the average of the last 400,000 years. This suggests that Rubisco has adapted to past CO_2 concentrations, but that the adaptation is slow and has failed to change in response to the rise in [CO_2] that has occurred since the start of the Industrial Revolution. The possibility that increased [CO_2] favours the selection of forms of Rubisco with increased k_c^c and decreased τ is consistent with the observation that Rubisco from C4 plants where the enzyme functions in a high [CO_2] typically has a higher k_c^c and lower τ than in C3 land plants (Sage, 2002; Seemann, Badger and Berry, 1984). Zhu *et al.* (2004b) showed that if Rubisco from the non-green algae *Griffithsia monilis* could be expressed in place of the present C3 crop Rubisco then canopy carbon gain could be increased by 27%. These simulations suggest that very substantial increases in crop carbon gain could result if exotic forms of Rubisco can be successfully expressed in C3 plants. However, if specificity can only be increased at the expense of catalytic rate, decrease in specificity rather than increase will increase C3 crop carbon gain. Ideally a crop would express a high k_c^c Rubisco in the upper canopy leaves exposed to full sunlight and a high τ Rubisco in the shaded lower canopy leaves. Most C3 annual crop canopies form leaves at progressively higher levels, so that leaves start life at the top of the canopy and then become progressively shaded as new leaves form above. Shading is sensed in plant leaves by the balance to red/far-red light via the phytochrome system (Gilbert, Jarvis and Smith, 2001). One possibility would be for this system to trigger the replacement of a high k_c^c Rubisco with a high τ form as the leaf acclimates to shade. Table 8.2 shows that such a system could increase ε_c by 31%, by comparison to an equivalent crop canopy with the current typical Rubisco of C3 crops. This would not be an increase over the theoretical maximum ε_c of 0.051 for C3 species, but an increase due to raising the point of saturation in the radiation response curve of photosynthesis (Figure 8.1b).

Regeneration of RubP

As noted previously at light-saturation photosynthesis in crop leaves is typically

Table 8.2 ESTIMATES OF THE DAILY CANOPY CARBON GAIN (A_C') AFTER ZHU *et al.* (2004b) AND ASSUMING THE HYPOTHETICAL REPLACEMENT OF THE AVERAGE FORM OF RUBISCO FROM C3 CROP SPECIES WITH RUBISCOS FROM OTHER SPECIES. REPORTED VALUES FOR K^C_C AND K^C_M OF THESE SPECIES (JORDAN AND OGREN, 1984; SEEMAN *et al.*, 1984; WHITNEY *et al.*, 2001). SINCE THE REPORTED K^C_M WAS FOR SOLUBLE [CO_2] AROUND RUBISCO IN SOLUTION, MESOPHYLL CONDUCTANCE WAS USED TO ADJUST K^C_M TO INTERCELLULAR CO_2 CONCENTRATION. THE FINAL ROW EXTENDS ON THE RESULTS OF ZHU *et al.* (2004b) TO SIMULATE THE GAIN THAT COULD BE ACHIEVED IF A FORM OF RUBISCO WITH A K^C_C (*AMARANTHUS EDULIS*) COULD BE EXPRESSED IN THE SUNLIT LEAVES AND A FORM WITH HIGH SPECIFICITY (CURRENT C3 AVERAGE) IN THE SHADE LEAVES.

Species	A_c' *(mmol m⁻² day⁻¹)*	A_c' *(%)*	A_{sat} *(μmol m⁻² s⁻¹)*
Current average C3 crop ($k^c_c = 2.5, \tau = 92.5$)	1040	100	14.9
Griffithsia monilis ($k^c_c = 2.6, \tau = 167$)	1430	127%	21.5
Amaranthus edulis *($k^c_c = 7.3, \tau = 82$)*	*1250*	*117%*	*28.3*
Amaranthus edulis/current ($k^c_c = 7.3, \tau = 82$) ($k^c_c = 2.5, \tau = 92.5$)	1360	131%	28.3

co-limited by capacity for carboxylation Rubisco ($V_{c,max}$) and capacity for regeneration of RubP, termed J_{max} in the context of the model of Farquhar *et al.* (1980), which has become the standard framework for analysing limitations to C_3 leaf photosynthesis. If the rate of carboxylation at Rubisco is increased, then J_{max} should also be increased to gain maximum benefit. By 2050 atmospheric [CO_2] will be about 50% higher than today. This change, without any modification of the protein, will increase the efficiency of Rubisco by partially inhibiting oxygenation. From kinetic data it may be calculated that as a result $J_{max}/V_{c,max}$ would need to increase by 30% to maintain an optimal distribution of resources (Long *et al.*, 2004). Interestingly, acclimation of soybean to growth under elevated [CO_2] in the field involves a significant increase in $J_{max}/V_{c,max}$, but of only 7%, so someway short of the theoretical change needed to maximize response (Bernacchi, Morgan, Ort and Long, 2004). Increase in J_{max} will therefore be necessary, simply to adapt plants to rising [CO_2].

Unlike $V_{c,max}$, regeneration of RubP does not depend on the amount and properties of a single protein, but on the complete photosynthetic electron transport chain and all the enzymes of the Calvin cycle, except Rubisco. Transgenic plants with small decreases in the quantities of specific proteins produced by anti-sense technology in tobacco suggest that two points in this chain limit J_{max}. Both the amount of the cytochrome b6/f complex in the electron transport chain and of sedoheptulose-1:7-bisphosphatase (SbPase) in the Calvin cycle have been shown to strongly control the rate of RubP synthesis (Harrison, Olcer, Lyloyd, Long and Raines, 2001; Price, van Caemmerer, Evans, Siebke,

Anderson and Badger, 1998, Raines, 2003). Of course, decreased photosynthesis due to a decrease in a specific protein, even in the absence of pleiotropic effects, is not absolute proof of limitation in the wild-type plant. However, transgenic tobacco plants over-expressing SbPase have now been produced and show a ca. 10% significantly greater light-saturated photosynthetic and growth rate than the wild-type from which they were derived (CA Raines, personal communication).

Conclusion

Theoretical considerations suggest that the maximum efficiencies of conversion of intercepted solar radiation into biomass (ε_c) are ca. 0.051 and 0.060 for C_3 and C_4 crops respectively. The C_3 maximum could be raised by decreasing photorespiration, either by identifying Rubisco with an increased specificity for CO_2 or by engineering C_4 photosynthesis into C_3 crops. At present higher specificity forms of Rubisco that have been found in other photosynthetic organisms carry the penalty of lower catalytic rates, and replacement of existing Rubisco with these forms would result in lower, not higher, rates of canopy carbon uptake. Theoretical analysis now shows that introduction of C_4 photosynthesis into a single cell will be energetically highly inefficient. Engineering C_4 photosynthesis into C_3 crops would therefore require not only the introduction of the C_4 photosynthetic cycle, but also the Kranz leaf anatomy and associated differential expression of photosynthetic proteins. Given that the basis of development of this differential expression is not as yet fully understood, such a complex transformation will probably not be possible for some time (Table 8.3). In the shorter term extending the environmental range of existing C_4 crops is likely to be a more successful route to higher total productivities.

Table 8.3 SUMMARY OF POSSIBLE INCREASES IN SOLAR RADIATION CONVERSION EFFICIENCY (ε_c) THAT MIGHT BE ACHIEVED, AND THE SPECULATED TIME-HORIZON FOR SUCH A CHANGE.

Change	*% Increase in e_c relative to current realized value*	*Speculated time horizon (years)*
Improved canopy architecture	10% (0-40%)	0-10
Rubisco with decreased oxygenase activity	30% (5-60%)	???
C4 photosynthesis engineered into C3 crops	18% (2-35%)	???
Increased rate of recovery from photoprotection of photosynthesis	15% (6-40%)	5
Introduction of higher catalytic rate foreign forms of Rubisco	22% (17-30%)	5-10
Increased capacity for regeneration of RubP via overexpression of SbPase	10% (0-20%)	0-5

The theoretical ε_c is not achieved, even by the most productive systems under optimum conditions. The major cause is likely the fact that in direct sunlight, significant parts of the crop canopy will be light saturated, because there is insufficient capacity to utilize all of the absorbed radiation. Here there are four areas, which could allow significant increases in ε_c. 1) Alteration of crop canopy architecture to improve the distribution of radiation and minimize the period over which any leaf is light saturated. Although this is an approach which has fallen from fashion, it is clear that for many crops, canopy architecture remains far from optimal. This could increase ε_c by as much as 40% at mid-day in full sunlight. Increased computational capacity now allows the use of complex reverse ray tracing algorithms to identify optimal architecture for different environments and identify selection criteria. This is also one area of improvement that can be approached by conventional plant breeding. 2) Increased recovery from photoprotection to increase efficiency of photosynthesis of leaves in the shade, could increase daily ε_c by ca. 15%, and by more at lower temperatures. Again there is variability among photosynthetic organisms that could be exploited. The molecular mechanism of photoprotection and its relaxation is understood, and might be improved by relatively simple transformations. 3) The amount of Rubisco is possibly already close to the capacity of photosynthetic cells and represents the largest single investment of nitrogen within most growing crops. Despite its high concentration it is a limitation to light saturated photosynthesis. Simulations show that replacement of current C3 crop Rubisco with forms with higher catalytic rates from other photosynthetic organisms could increase of daily ε_c by up to 31%. Because Rubisco is formed from two types of sub-unit, one coded for in the nucleus and one in the plastid genome, transformation is particularly challenging. However, such transformations have already been achieved with tobacco. Given the progress in transformation technology over the past ten years, such changes appear achievable on a ten-year time horizon. Even larger increases in ε_c would be possible if leaves could be engineered to express a high catalytic rate form of Rubisco initially and then replace this with a high specificity form during shade acclimation. 4) To gain full advantage of an engineered increase in Rubisco efficiency, or the increase in Rubisco efficiency that will simply result from rising [CO_2], increase in capacity for RubP regeneration will be necessary. Anti-sense transformations suggest that the cytochrome b6/f complex and sedoheptulose-1:7 bisphosphatase (SbPase) are the major limitations to RubP regeneration. Preliminary evidence suggests that engineered over-expression of SbPase could increase ε_c by 10%. Thus, this transformation is already realisable as a means to increase ε_c.

In conclusion, while opportunities to increase the maximum ε_c of C3 or C4 crops do not appear realizable on a 10 – 20 year time horizon, there are a number of opportunities to improve the ability of crops to approach the current theoretical maxima. Some increases may be achieved by conventional breeding, although most will require introduction of foreign genetic material.

All appear achievable within a 10-year time horizon (Table 8.3). Finally, both environmental stress and respiration have been beyond the scope of this analysis. Clearly there is a wide range of opportunities for improving the tolerance of ε_c to stress, although most are specific to the individual stress (Ort and Long, 2003). Respiration as a factor that might be decreased to increase ε_c has received very little attention, in part because knowledge of the limitations and full role of the process is far less complete than for photosynthesis. In particular, the role of cyanide-insensitive respiration which appears to impose a variable inefficiency on the net carbohydrate use. In the one example where respiration has been used as a selection criterion, substantial yield increases were achieved by selecting for decreased respiration rates (Wilson and Jones, 1982).

References

Ainsworth E.A., Rogers A., Nelson R. and Long S.P. (2004) Testing the "source-sink" hypothesis of down-regulation of photosynthesis in elevated CO2 in the field with single gene substitutions in Glycine max. *Agricultural and Forest Meteorology*, **122**, 85-94.

Bainbridge G., Madgwick P., Parmar S., Mitchell R., Paul M., Pitts J., Keys A.J. and Parry M.A.J. (1995) Engineering rubisco to change its catalytic properties. *Journal of Experimental Botany*, **46**, 1269-1276.

Baker J.T., Allen L.H. and Boote K.J. (1990) Growth and yield responses of rice to carbon-dioxide concentration. *Journal of Agricultural Science*, **115**, 313-320.

Baker N.R., East T.M. and Long S.P. (1983) Chilling damage to photosynthesis in young Zea mays. *Journal of Experimental Botany*, **34**, 189-197.

Barber J. (1998) What limits the efficiency of photosynthesis and can there be beneficial improvements? In: *Feeding a World Population of More Than Eight Billion People - A Challenge to Science* (eds J.C. Waterlow, D.G. Armstrong, L. FowdenandR. Riley), pp. 112-123. Oxford University Press, Cary, NC.

Baroli I. and Niyogi K.K. (2000) Molecular genetics of xanthophyll-dependent photoprotection in green algae and plants. *Philosophical Transactions of the Royal Society of London Series B-Biological Sciences*, **355**, 1385-1393.

Beadle C.L. and Long S.P. (1985) Photosynthesis - Is it limiting to biomass production. *Biomass*, **8**, 119-168.

Beale C.V. and Long S.P. (1995) Can perennial C-4 grasses attain high efficiencies of radiant energy-conversion in cool climates. *Plant Cell and Environment*, **18**, 641-650.

Beale C.V., Morison J.I.L. and Long S.P. (1999) Water use efficiency of C-4 perennial grasses in a temperate climate. *Agricultural and Forest Meteorology*, **96**, 103-115.

Bernacchi C.J., Morgan P.B., Ort D.R. and Long S.P. (2004) The growth of soybean under free air [CO2] enrichment (FACE) stimulates photosynthesis while decreasing in vivo Rubisco capacity. *Planta*, (in press).

Cerling T.E. (1999) Paleorecords of C4 Plants and Ecosystems. In: *The Biology of C_4 Photosynthesis* (eds R.F. Sage and R.K. Monson), pp. 445-469. Academic Press, San Diego.

Drake B.G., Gonzalez-Meler M. and Long S.P. (1997) More efficient plants: A consequence of rising atmospheric CO2? *Annual Review of Plant Physiology and Plant Molecular Biology*, **48**, (in press).

Evans L.T. (1993) *Crop Evolution, Adaptation and Yield.* CUP, Cambridge.

Evans L.T. (1998) Greater crop production: whence and whither? In: *Feeding a World Population of More Than Eight Billion People - A Challenge to Science* (eds J.C. Waterlow, D.G. Armstrong, L. Fowden and R. Riley), pp. 89-97. Oxford University Press, Cary, NC.

Evans L.T. and Dunstone R.L. (1970) Some physiological aspects of evolution in wheat. *Australian Journal of Biological Sciences*, **23**, 725-741.

Falkowski P.G. and Dubindky Z. (1981) Light shade adaption of Stylophora pistillata, a hermatypic coral from the Gulf of Eilat. *Nature*, **289**, 172-174.

Farquhar G.D., von Caemmerer S. and Berry J.A. (1980) A biochemical model of photosynthetic CO_2 assimilation in leaves of C_3 species. *Planta*, **149**, 78-90.

Ghannoum O., von Caemmerer S. and Conroy J.P. (2001) Plant water use efficiency of 17 Australian NAD-ME and NADP-ME C-4 grasses at ambient and elevated CO2 partial pressure. *Australian Journal of Plant Physiology*, **28**, 1207-1217.

Gifford R.M. (1995) Whole-plant respiration and photosynthesis of wheat under increased CO_2 concentration and temperature - long-term vs short-term distinctions for modeling. *Global Change Biology*, **1**, 385-396.

Gilbert I.R., Jarvis P.G. and Smith H. (2001) Proximity signal and shade avoidance differences between early and late successional trees. *Nature*, **411**, 792-795.

Harrison E.P., Olcer H., Lloyd J.C., Long S.P. and Raines C.A. (2001) Small decreases in SBPase cause a linear decline in the apparent RuBP regeneration rate, but do not affect Rubisco carboxylation capacity. *Journal of Experimental Botany*, **52**, 1779-1784.

Hatch M.D. (1987) C-4 photosynthesis - a unique blend of modified biochemistry, anatomy and ultrastructure. *Biochimica et Biophysica Acta*, **895**, 81-106.

Havaux M. and Niyogi K.K. (1999) The violaxanthin cycle protects plants from photooxidative damage by more than one mechanism. *Proceedings of the National Academy of Sciences of the United States of America*, **96**, 8762-8767.

Hay R.K.M. (1995) Harvest index - a review of its use in plant-breeding and crop physiology. *Annals Of Applied Biology*, **126**, 197-216.

Humphries S.W. and Long S.P. (1995) WIMOVAC: a software package for modelling the dynamics of plant leaf and canopy photosynthesis. *Computer Application in the Biosciences*, **11**, 361-371.

Jordan D.B. and Ogren W.L. (1984) The carbon dioxide/oxygen specificity of ribulose-1,5-bisphosphate carboxylase/oxygenase. *Planta*, **161**, 308-313.

Kellogg E.A. (1999) Phylogenetic Aspects of the Evolution of C4 Photosynthesis. In: *The Biology of C_4 Photosynthesis* (eds R.F. Sage and R.K. Monson), pp. 411-444. Academic Press, San Diego.

Kimball B.A. (1983) Carbon-dioxide and agricultural yield - an assemblage and analysis of 430 prior observations. *Agronomy Journal*, **75**, 779-788.

Kimball B.A., Pinter P.J., Garcia R.L., Lamorte R.L., Wall G.W., Hunsaker D.J., Wechsung G., Wechsung F. and Kartschall T. (1995) Productivity and water-use of wheat under free-air CO_2 enrichment. *Global Change Biology*, **1**, 429-442.

Ku M.S.B., Cho D.H., Li X., Jiao D.M., Pinto M., Miyao M. and Matsuoka M. (2001) Introduction of genes encoding C4 photosynthesis enzymes into rice plants: physiological consequences. In: *Rice Biotechnology: Improving Yield, Stress Tolerance and Grain Quality*, pp. 100-116.

Long S.P. (1991) Modification of the response of photosynthetic productivity to rising temperature by atmospheric CO_2 concentrations: has its importance been underestimated? *Plant, Cell and Environment*, **14**, 729-739.

Long S.P. (1993) The significance of light-limited photosynthesis to crop canopy carbon gain and productivity - A theoretical analysis. In: *Photosynthesis: Photoreactions to Plant Productivity* (eds Y.P. Abrol, P. Mohanty and Govindjee), pp. 547-560. Oxford & IBH Publishing, New Delhi.

Long S.P. (1998) Rubisco, the key to improved crop production for a world population of more than eight billion people? In: *Feeding a World Population of More Than Eight Billion People - A Challenge to Science* (eds J.C. Waterlow, D.G. Armstrong, L. Fowden and R. Riley), pp. 124-136. Oxford University Press, Cary, NC.

Long S.P. (1999) Environmental Responses. In: *The Biology of C_4 Photosynthesis* (eds R.F. SageandR.K. Monson), pp. 209-243. Academic Press, San Diego.

Long S.P., Ainsworth E.A., Rogers A. and Ort D.R. (2004) Rising atmospheric carbon dioxide: plants face their future. *Annual Reviews of Plant Biology*, **55**, 591-628.

Long S.P., Farage P.K. and Garcia R.L. (1996) Measurement of leaf and canopy photosynthetic CO_2 exchange in the field. *Journal of Experimental Botany*, **47**, 1629-1642.

Long S.P., Humphries S. and Falkowski P.G. (1994a) Photoinhibition of photosynthesis in nature. *Annual Review of Plant Physiology and Plant Molecular Biology*, **45**, 633-662.

Long S.P., Humphries S. and Falkowski P.G. (1994b) Photoinhibition of photosynthesis in nature. *Annual Review of Plant Physiology and Plant Molecular Biology*, **45**, 633-662.

Long S.P., Postl W.F. and Bolharnordenkampf H.R. (1993) Quantum yields for uptake of carbon-dioxide in C-3 vascular plants of contrasting habitats and taxonomic groupings. *Planta*, **189**, 226-234.

Miyao M. (2003) Molecular evolution and genetic engineering of C-4 photosynthetic enzymes. *Journal of Experimental Botany*, **54**, 179-189.

Monson R.K. (1999) The Origins of C4 Genes and Evoutionary Pattern in the C4 Metabolic Phenotype. In: *The Biology of C_4 Photosynthesis* (eds R.F. Sage and R.K. Monson), pp. 377-410. Academic Press, San Diego.

Monteith J.L. (1977) Climate and the efficiency of crop production in Britain. *Philosophical Transactions of the Royal Society of London*, **281**, 277-294.

Morison J.I.L., Piedade M.T.F., Muller E., Long S.P., Junk W.J. and Jones M.B. (2000) Very high productivity of the C-4 aquatic grass Echinochloa polystachya in the Amazon floodplain confirmed by net ecosystem CO2 flux measurements. *Oecologia*, **125**, 400-411.

Nobel P.S., Forseth I.N. and Long S.P. (1993) Canopy structure and light interception. In: *Photosynthesis and Production in a Changing Environment: A Field and Laboratory Manual* (eds D.O. Hall, J.M.O. Scurlock, H.R. Bolhàr-Nordenkampf, R.C. Leegood and S.P. Long), pp. 79-90. Chapman & Hall, London.

Ort D.R. and Long S.P. (2003) Converting Solar Energy into Crop Production. (eds M.J. Chrispeels and D.E. Sadava), pp. 240-269. American Society of Plant Biologists/Jones and Bartlett, Boston.

Piedade M.T.F., Junk W.J. and Long S.P. (1991) The Productivity of the C4 Grass Echinochloa-Polystachya on the Amazon Floodplain. *Ecology*, **72**, 1456-1463.

Pinter P.J., Kimball B.A., Garcia R.L., Wall G.W., Hunsaker D.J. and LaMorte R.L. (1996) Free-air CO_2 Enrichment: Responses of Cotton and Wheat Crops. In: *Carbon Dioxide and Terrestrial Ecosystems* (eds G.W. Koch and H.A. Mooney), pp. 215-249. Academic Press, San Diego.

Price G.D., von Caemmerer S., Evans J.R., Siebke K., Anderson J.M. and Badger M.R. (1998) Photosynthesis is strongly reduced by antisense suppression of chloroplastic cytochrome bf complex in transgenic tobacco. *Australian Journal of Plant Physiology*, **25**, 445-452.

Pyke K.A. and Leech R.M. (1987) Cellular-levels of ribulose 1,5 bisphosphate carboxylase and chloroplast compartment size in wheat mesophyll-cells. *Journal of Experimental Botany*, **38**, 1949-1956.

Raines C.A. (2003) The Calvin cycle revisited. *Photosynthesis Research*, **75**, 1-10.

Roberts M.J., Long S.P., Tieszen L.L. and Beadle C.L. (1993) Measurement of plant biomass and net primary production of herbaceous vegetation. In: *Photosynthesis and Production in a Changing Environment: A Field and Laboratory Manual* (eds D.O. Hall, J.M.O. Scurlock, H.R. Bolhàr-Nordenkampf, R.C. Leegood and S.P. Long), pp. 1-21. Chapman & Hall, London.

Sage R.F. (2002) Variation in the k(cat) of Rubisco in C-3 and C-4 plants and some implications for photosynthetic performance at high and low temperature. *Journal of Experimental Botany*, **53**, 609-620.

Seeman J.R., Badger M.R. and Berry J.A. (1984) Variations in the specificity activity of ribulose-1,5-bisphosphate carboxylase between species utilizing differing photosynthetic pathways. *Plant Physiology*, **74**, 791-794.

Suzuki S., Murai N., Burnell J.N. and Arai M. (2000) Changes in photosynthetic carbon flow in transgenic rice plants that express C4-type phosphoenolpyruvate carboxykinase from Urochloa panicoides. *Plant Physiology*, **124**, 163-172.

von Caemmerer S. (2003) C-4 photosynthesis in a single C-3 cell is theoretically inefficient but may ameliorate internal CO2 diffusion limitations of C-3 leaves. *Plant Cell and Environment*, **26**, 1191-1197.

Voznesenskaya E.V., Franceschi V.R., Kiirats O., Artyusheva E.G., Freitag H. and Edwards G.E. (2002) Proof of C-4 photosynthesis without Kranz anatomy in Bienertia cycloptera (Chenopodiaceae). *Plant Journal*, **31**, 649-662.

Voznesenskaya E.V., Franceschi V.R., Kiirats O., Freitag H. and Edwards G.E. (2001) Kranz anatomy is not essential for terrestrial C-4 plant photosynthesis. *Nature*, **414**, 543-546.

Wang Q., Zhang Q.D., Zhu X.G., Lu C.M., Kuang T.Y. and Li C.Q. (2002) PSII photochendstry and xanthophyll cycle in two superhigh-yield rice hybrids, Liangyoupeijiu and Hua-an 3 during photoinhibition and subsequent restoration. *Acta Botanica Sinica*, **44**, 1297-1302.

Watanabe N., Evans J.R. and Chow W.S. (1994) Changes in the photosynthetic properties of australian wheat cultivars over the last century. *Australian Journal of Plant Physiology*, **21**, 169-183.

Wheeler R.M., Mackowiak C.L., Sager J.C., Knott W.M. and Berry W.L. (1995) Proximate composition of celss crops grown in NASA's biomass production chamber. In: *Natural and Artificial Ecosystems*, pp. 43-47.

Whitney S.M., Baldett P., Hudson G.S. and Andrews T.J. (2001) Form I Rubiscos from non-green algae are expressed abundantly but not assembled in tobacco chloroplasts. *Plant Journal*, **26**, 535-547.

Wilson D. and Jones J.G. (1982) Effect of selection for dark respiration rate of mature leaves on crop yields of Lolium-Perenne Cv S23. *Annals of Botany*, **49**, 313-320.

Zhu X.G., Ort D.R., Whitmarsh J. and Long S.P. (2004a) The slow reversibility of photosystem II thermal energy dissipation on transfer from high to low light may cause large losses in carbon gain by crop canopies: a theoretical analysis. *Journal of Experimental Botany*, **55**, 1167-1175.

Zhu X.G., Portis A.R. and Long S.P. (2004b) Would transformation of C-3 crop plants with foreign Rubisco increase productivity? A computational analysis extrapolating from kinetic properties to canopy photosynthesis. *Plant Cell and Environment*, **27**, 155-165.

9

THE FUTURE OF CHEMICAL CROP PROTECTION

JOHN M CLOUGH
Syngenta, Jealott's Hill International Research Centre, Bracknell, Berkshire, RG42 6EY, UK

Introduction

This paper discusses the future of chemical crop protection. It starts by describing the reasons why crop protection (CP) chemicals are so widely used. The global market for these chemicals, of which herbicides, insecticides and fungicides are the most important, is then discussed. Worth $26.71 billion in 2003, its value has fallen sharply over the last five years, and the factors which have contributed to this decline are explored.

About 800 chemicals are currently registered for use in crop protection. New products have been introduced at an average of about 12 per annum since 1980, and the reasons why this rate is expected to be maintained over at least the next five years are described.

Although CP chemicals are available for the control of most commercially-important weeds, insects and fungi, it will be shown that there are still good reasons for the development of new ones, e.g. the onset of resistance to existing products and the loss of many older chemicals through re-registration programmes.

The ways in which new CP chemicals are invented is described. Methods have changed hugely in the last decade, e.g. high-throughput *in vivo* screening and combinatorial chemistry are now widely used.

Finally, the story of the strobilurin fungicides is outlined, because it illustrates the success that can be achieved in the market by a family of chemicals with a new mode of action, as well as the problems of resistance which can be encountered by chemicals which act at a single enzyme site.

Why use crop protection chemicals?

Farmers use CP chemicals because they provide a straightforward and cost-effective way of improving the yield and the quality of their crops. The leading classes of chemicals are herbicides, which reduce losses resulting from competition from weeds, and insecticides and fungicides, which protect crops from insects and fungi, both while they are growing and after harvest. Smaller amounts of other chemicals, e.g. nematicides, molluscicides, rodenticides and, although not strictly for crop protection, growth regulators, are also used. Importantly, the use of CP chemicals also helps the farmer to obtain consistently good yields from year to year (Avery, 1997; Evans, 1999; Foster, Atkinson and Burnett, 2003).

The losses caused by weeds, insect pests and fungal diseases were roughly quantified in a major series of studies reported 10 years ago. Thousands of data-points were collected in order to provide an accurate assessment of the global losses in eight principal food and cash crops. Reductions from the estimated maximum attainable yields varied from ~30% for wheat, barley and soybeans, to ~40% for potatoes, maize, coffee and cotton, and ~50% for rice. The use of various forms of crop protection, especially CP chemicals, reduced these losses, but not to zero. It was estimated that crop protection prevented losses totalling about 27% of the attainable yield: 16%, 7% and 4% from the control of weeds, insects and fungi, respectively (Oerke, Weber, Dehne and Schönbeck, 1994).

Comparisons have also been made between the yields obtained from crops grown conventionally and those grown according to traditional or organic systems in which inputs of CP chemicals are minimised. Of course, this is not precisely the same as measuring the effect of using CP chemicals, particularly because of different crop nutrition, but it does provide some idea of the yield benefits which they bring. In one example, a study of organic farming in seven European countries showed that organic yields are usually well below those achieved with conventional farming, e.g. ~30% lower for wheat and barley and ~40% lower for potatoes (Landell Mills, 1992). Other studies have given similar results (Mäder, Fließbach, Dubois, Gunst, Fried and Niggli, 2002; Phillips, Leifert, Santos, Juntharathep, Bodker, Tamm and Smit, 2002; Hewitt, 2004).

Finally, it is also significant that without the use of CP chemicals entire crops can be lost if weather conditions are conducive to the development of a fungal epidemic. High risk crops include potatoes and vines. This may affect an entire region, so yields, and therefore supplies, fluctuate much more without the use of CP chemicals (Tamm, 2000).

Figures such as those above, together with a knowledge of the cost of CP chemicals and the value of the crops, can be used to calculate that for each dollar the farmer spends on CP chemicals, he can expect a yield increase worth between five and 15 dollars (Jonathan Shoham, Syngenta, personal communication). On top of this, he enjoys season-to-season reliability, plus the relative simplicity that stems from the use of CP chemicals (e.g. crops can be harvested mechanically when weeds are properly controlled).

Attempts have also been made to calculate the benefits to national and global economies which accrue from the use of CP chemicals. These are difficult calculations, but one estimate suggests that the complete phase-out of CP chemicals in the USA (a model in which all animal feed would also be produced without CP chemicals) would cost the country $93-277 billion a year, equivalent to $320-950 per citizen per annum (Lomborg, 2003b).

The use of CP chemicals has made a key contribution to the steady increase in the yields of all the important crops over recent decades. Over the years 1961 to 2003, for example, average yields across the world in tonnes per hectare of soybean, maize and wheat rose by factors of 2.0, 2.3 and 2.5, respectively (FAO, website). Global output has increased correspondingly and, importantly, without a significant increase in the area of land used for its production (Evans, 1999).

Although the focus of this conference is yield, it is worth mentioning briefly that the use of CP chemicals can also improve the quality of crops. To pick just one example, the use of Syngenta's broad-spectrum strobilurin fungicide azoxystrobin leads to bigger grain size and improved malting properties in barley, bigger grain size and better milling quality in rice, increased tuber size in potatoes and bulb size in garlic, and increased soluble sugars and longer shelf life in tomatoes (Bartlett, Clough, Godwin, Hall, Hamer and Parr-Dobrzanski, 2002; Pragnell, 2003).

The market for crop protection chemicals

The global market for CP chemicals was valued at $26.71 billion in 2003. Herbicides, insecticides and fungicides accounted for 50.2%, 24.9% and 21.5% of this total, respectively, with CP chemicals of other types accounting for the remaining 3.4%. This represents a nominal growth of 6.2% in comparison to the market in 2002 but, in fact, once inflation and currency factors are taken into account, it can be shown that the real value of the market declined by 1.6%. Importantly, 2003 was the fifth consecutive year in which there was a decline in the value of this market and, in real terms, it has lost 18.6% of its value since it peaked in 1998 (Phillips McDougall, 2004b).

So the market for CP chemicals is mature. Its value rose steeply in the 1960s and 1970s, slowly in the 1980s, was overall flat in the 1990s, and has declined since the turn of the millennium (average growth rates per annum in the 1960s, '70s, '80s, '90s and 2000-2003 were 10.2%, 6.8%, 2.2%, 0.1% and –4.5%, respectively). Despite recent declines, a small degree of market growth, to the extent of 0.5% per annum, has been forecast for the next five years (Phillips McDougall, 2004b).

A variety of factors affect the value of the market for CP chemicals, and the important ones are outlined in turn in the following paragraphs. Where possible, their contributions to the recent decline in the value of the market are discussed and projections are made for the future.

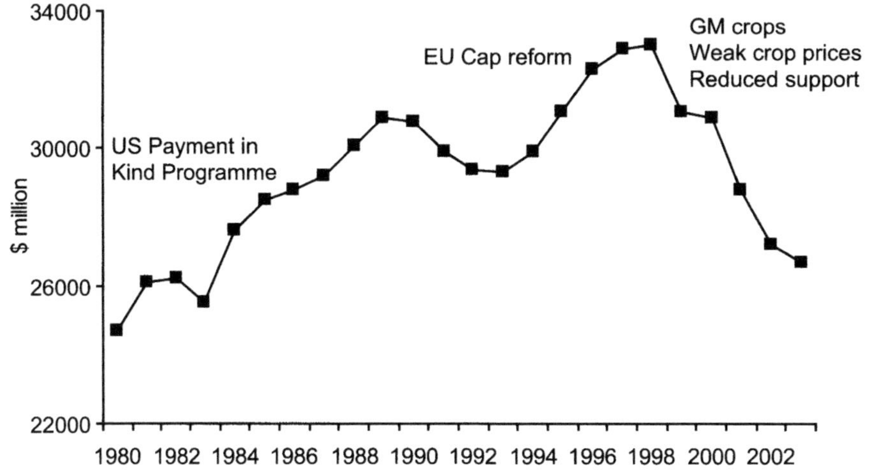

Figure 9.1 Value of the market for CP chemicals at constant 2003 dollars (courtesy of Phillips McDougall, 2004b).

GENETICALLY MODIFIED CROPS

The global uptake of GM crops, since they were first commercialised in 1995, has been patchy. In Europe, intense consumer resistance has meant that significant areas of GM crops are grown only in Spain and Romania (Phillips McDougall, 2003c and 2004a; Agrow, 2003). By contrast, growth in the use of GM crops in other parts of the world, especially in the USA, Argentina and Canada, has been explosive. Expansion into additional countries is continuing and, in 2003, seven million farmers in 18 countries grew a total of about 70 million hectares of GM crops, and the value of the market for GM seed (including technology fees) grew to $3.94 billion, an increase in 19.0% from the previous year (Phillips McDougall, 2004a and 2004b; Agrow, 2004a).

The significance of GM crops in this discussion is that they have taken at least $3 billion per annum out of the value of the market for CP chemicals by replacing selective herbicides with cheaper broad-spectrum herbicides, and by reducing the need for insecticides. This, then, is clearly one of the key contributors to the recent decline in the value of the market for CP chemicals (Blair, Benson, Burns, Heslop and Jacobs, 2002). How this has happened is described in the following paragraph.

The GM crops currently in use are of two main types, herbicide-tolerant crops (e.g. soybean and maize) and insect-resistant crops (e.g. maize and cotton). Herbicide-tolerant crops still require the use of a herbicide, but the selective herbicides which were used previously in the non-GM crop are replaced by a cheaper broad-spectrum herbicide, most often glyphosate. As the use of glyphosate has increased, and in the face of generic competition, its main producer, Monsanto, has aggressively reduced its price. Glyphosate is widely used in conventional applications as well as in GM crops, and it is an extremely valuable product with higher sales than those of any other CP chemical, so this decline in its price has had a significant negative effect on the total value of the market for CP chemicals (Agrow 2004a; Phillips McDougall, 2004b). Insect-resistant crops potentially provide a means of circumventing the use of chemical insecticides altogether. Although they have certainly reduced insecticide use, the spectrum of insecticidal activity afforded by the currently available insect-resistant crops means that the use of chemical insecticides is still required.

Over the next few years, the use of GM crops will continue to expand into further territories, especially in Asia. The established traits, such as glyphosate-resistant soybean, are now approaching maturity, but there will be further expansion through the introduction of new traits and new crop varieties in which genes coding for different traits are stacked. Although this will further erode the value of the market for CP chemicals, it is probable that the rate at which erosion is occurring has peaked (Blair *et al.*, 2002; Phillips McDougall, 2004a and 2004b).

Finally in this section, it is worth noting that genetic modification will only erode a part of the market for CP chemicals. It is not technically possible or cost effective to develop GM crops to replace all of the functions performed by CP chemicals. For example, there are no realistic alternatives to herbicides (the most valuable sector of the market for CP chemicals). In addition, genetic modification will have little impact on the valuable fruit and vegetable sector, partly because it is so fragmented and partly because of consumer resistance to genetic modification in directly consumed foods. Furthermore, the market for fungicides is also fragmented due to the variety of fungi which infect a wide range of crops, and it would not be economically viable to develop a host of different GM crops with resistance to each of these pathogens. Syngenta's broad spectrum fungicide azoxystrobin, for example, is used on more than 400 crop/disease systems (Bartlett *et al.*, 2002; Evans, 2000).

CROP PRICES, ECONOMICS AND THE FOOD SUPPLY CHAIN

Crop prices have always been volatile, driven, for example, by the weather, pest pressure and the global and national economies, but they have reached historic lows in recent years. Prices of many of the major crops, e.g. wheat, maize, soybean, rice and cotton, were about 40% lower in 2001 than they had been during the period 1995-1997, due to a succession of good harvests in Europe, North America and Latin America. When crop prices are low, farmers plant less of them, and it is not economical to spend so much on chemicals to protect what they plant. But the picture is now changing and the prices of many crops (but not of rice) have partially recovered during the last two years (Phillips McDougall 2002c, 2003g and 2004b).

Crops are then converted into the food products which consumers buy. The food supply chain, which links the farmer to the consumer, comprises a complex series of interacting businesses, each of which draws a profit. The farmer receives less than 20% of the value of the final food product and, importantly, this proportion has steadily decreased over the last 20 years. Most of the profit is made downstream, especially by the food processors and retailers. The major food companies are huge – the largest each have annual sales which are greater than the total global sales of CP chemicals, and they dwarf the individual crop protection companies (Evans, 1999; Harris, Kaufman, Martinez and Price, 2002).

Finally, recent economic problems in some parts of the world, notably in Brazil and Argentina, have affected the way in which the crop protection companies have been able to operate in these territories, and sales have been reduced. This was a major problem in 2002, but the situation eased in 2003 and now looks more promising (Phillips McDougall, 2004b).

FARM SUPPORT PROGRAMMES

In many parts of the world, farming is supported by subsidies from the taxpayer. In the EU, for example, the Common Agricultural Policy (CAP) was introduced in 1962 with the purpose of increasing food production in Europe, and farmers were subsidised on the basis of what they could produce. By the early 1990s, agriculture had been so successful that over-production had become an issue, and the CAP was distorting world trade, harming developing countries and proving to be very expensive for European taxpayers. Various reform packages, including set-aside programmes, have therefore been introduced, e.g. the so-called McSharry reform of 1992, in which crop prices were no longer supported at high levels and, as compensation, farmers received direct payments. The most recent, agreed in June 2003 and to be implemented from 2005, will fully decouple subsidies from production and, instead, farmers will receive payments provided they comply with certain standards relating to the environment, food safety and animal and plant health (Phillips McDougall, 2003c and 2004b; Mathisen, Cheung and Sirur, 2003).

GENERIC PRODUCTS

Once a CP chemical comes off patent in the majority of its key markets, it becomes a "generic", and there may or may not be immediate competition from "me-too" manufacturers. Of course, generics continue to be sold by the originating company, and a large proportion of the sales of the research and development companies are sales of generics (Uttley, 2004).

It is not straightforward for a would-be generic manufacturer to launch its own version of a CP chemical when it comes off patent. Not only does it need to be able to compete on price (and the original research and development company has had many years to refine its chemical processes), but it must prepare and submit its own registration dossier. Furthermore, the original company can implement various strategies to defend its product, e.g. by differentiating it in the market by introducing new and improved formulations, or by providing superior product support and customer service. It may also occasionally be able to switch to a single enantiomer form of the chemical which offers additional benefits and can be protected by newer patents. As a last resort, the company will reduce the price of the chemical. For these reasons, some important generic CP chemicals are still sold mainly by the original company. Monsanto's glyphosate is one example, and another is paraquat, where Syngenta still has ~95% of sales, despite the fact that the product was first sold more than 40 years ago (Blair *et al.*, 2002; Bateman, 2003).

So once a CP chemical becomes generic, there is a downward pressure on its price. This is beneficial to the farmer, but it erodes the value of the market for that chemical. The overall proportion of generic products has increased over recent years, and this has caused a reduction in the value of the global market for CP chemicals. Generics companies are also more important than they were in the past, and there are now five in the top 20 CP chemical companies (Copping, 2003; Sisson and Williams, 2001).

FOOD SAFETY AND THE ENVIRONMENT

For many years, consumers have been concerned about the effects of residues from CP chemicals in food. In the minds of many, these residues, where they exist, are linked to various health problems, especially cancer (Bates, 2002; Reynolds and Hill, 2002). One measure of the level of public interest is the fact that *The Guardian* published 81 articles about chemical residues in food during the period 1999 to 2002 (Foster *et al.*, 2003).

Although these concerns are easily understood (after all, we are all consumers), in reality they are broadly unjustified. First of all, CP chemicals are thoroughly tested for their safety before they can be registered for use, and wide safety margins are set by Government regulators regarding permissible levels of their residues in foods (Hamilton and Crossley, 2004). The fact that extremely small quantities of residues of these chemicals can be detected in some foodstuffs is a testimony to the power of modern analytical methods but not a reason to be concerned. In fact, almost all the pesticides in our food are the natural chemicals which plants produce to defend themselves against fungi, insects and other animal predators, and these have mostly not been tested for toxicity (Swirsky Gold, Slone and Ames, 2001; Wise and Findlay, 2003).

A recent analysis of all the best data shows that the number of cancers attributable to residues from synthetic CP chemicals in the diet is immeasurably small. In reality (and counter-intuitively), the use of CP chemicals reduces the incidence of cancer and therefore saves many lives each year because it keeps down the cost of fruit and vegetables, which are known to be key cancer-reducing components of the diet. It has been estimated that a decrease of just 10% in the consumption of fruit and vegetables in the USA, for example, would lead to 26 000 extra cancer deaths each year (Swirsky Gold *et al.*, 2001; Lomborg, 2003b).

Consumers are also increasingly interested in the effect that agriculture, and especially the use of CP chemicals, has on the environment. It was Rachel Carson who first drew attention to the problems of bioaccumulation associated with the persistent CP chemicals of her day in her beautifully written book *Silent Spring*, published in 1962 (Kidd, Pimentel and Wilkinson, 2002; Trewavas, 2004). Sophisticated modern CP chemicals are applied at lower rates and are biodegradable, so they do not accumulate in the environment. In fact, today's CP chemicals, when used properly, make an important contribution to the protection of the global environment because, by helping to maximise crop yields, they have minimised the number of the world's wild habitats which have had to be ploughed up for agricultural use (Avery, 1997).

However, the over-use of CP chemicals can damage the environment and, clearly, their unchecked use is highly undesirable. In the UK, various schemes are in place or are being developed to encourage farmers to reduce the use of CP chemicals to the minimum necessary for the effective control of pests. The so-called Voluntary Initiative, an industry-backed five-year programme of measures launched in April 2001, is one such scheme, and a national pesticides strategy is being developed by the Government regulator in the UK, the Pesticides Safety Directorate, to be introduced in the summer of 2004. If voluntary measures are not effective, the Government is likely to reconsider levying a tax on CP chemicals as a more powerful incentive to limit their use. Such a tax has been in place in Denmark since 1986 (Williams, 2003; Pesticide Outlook, 2003). Finally, in the EU (and, indeed, in other parts of the world), re-registration programmes are in hand to ensure that all the CP chemicals in use meet modern safety and environmental standards (see later).

Because of concerns about the effects of CP chemicals on food safety and on the environment, there has been a steep increase in the value of retail sales of organic food in the UK over the last 10 years, and the market continues to grow. However, the annual rate of growth peaked at 53% in 2000, and has fallen substantially since that time (31%, 11% and ~9% growth in 2001, 2002 and 2003, respectively). In the UK, the market for organic food is worth just below 1% of the value of the total market for groceries. About 4.3% of the agricultural land in the UK is either in organic production or is being converted. In Europe as a whole, about 2% of the total agricultural land is farmed organically, but the proportion is lower in other regions of the world, e.g. 0.3% and 0.07% in North America and

Asia, respectively. The increasing consumption of organic food has therefore had little effect so far on the value of the market for CP chemicals (Defra, web site; Organic Centre Wales, web site; Willer and Yussefi, 2004).

GLOBAL POPULATION AND PROSPERITY

Finally, one factor which will lead to a long-term growth in the demand for CP chemicals is the increasing global population. The number of people on earth is growing every day, and reached six billion in 1999. The massive growth in the world's population began in about 1950, mainly due to a fall in the death rate as a result of improved access to food, medicine, clean water and sanitation. But the rate of growth has already peaked and numbers will probably stabilise in about 2050. The United Nations forecasts a world population of between 7.4 and 10.6 billion in 2050, and about 9 billion in the year 2300 (Lomborg, 2003a; United Nations, 2003; United Nations Population Division, web site).

However, the global population is growing in prosperity as well as in number, and with this comes an increasing demand for meat and other animal products in the diet. The process of converting plant material into animal products is not an efficient one, and drives the need for even greater crop production. So the "dietary upgrading" associated with increasing wealth and standard of living also increases the demand for CP chemicals. This is an important consideration because global prosperity is increasing more quickly than population. Taking increases in both population and prosperity into account, the FAO forecasts that the global demand for agricultural products will be 60% higher in 2030 than it was in 2002 (Weber, 1994; Avery, 1997; Leaver, 2001; Bruinsma, 2002).

The number and rate of introduction of crop protection chemicals

There are about 800 CP chemicals currently in use. An analysis of the 13th edition of *The Pesticide Manual*, published in 2003, shows that this total comprises about 275 herbicides, 240 insecticides and 185 fungicides, with the remaining 100 or so chemicals having uses as pheromones, nematicides, molluscicides, rodenticides and so on (Tomlin, 2003). Importantly, however, many of these chemicals are being withdrawn in some parts of the world as re-registration programmes take place in Europe, the USA and Japan (see below).

There is a perception that the rate at which new CP chemicals have been introduced into the market has declined significantly in recent years, perhaps as a result of the distraction of industry consolidation and a refocus of research funds towards biotechnology. However, a proper analysis shows that this is not the case, and the number of new CP chemicals introduced during the 1980s, 1990s and the 2000s so far average 12.3, 12.6 and 12.0 per annum, respectively (Phillips McDougall, 2003b, 2003d and 2003e).

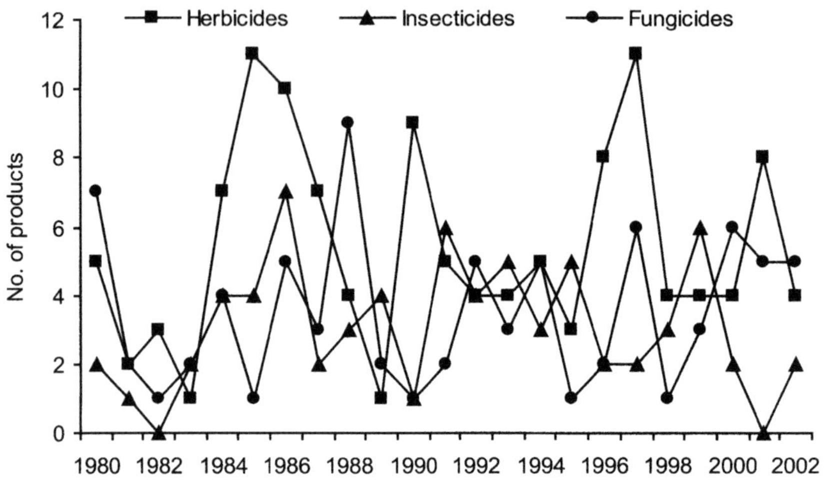

Figure 9.2 Number of herbicides, insecticides and fungicides introduced each year since 1980 (courtesy of Phillips McDougall, 2003d).

One could argue that insufficient time has passed for the effects of industry consolidation to be reflected in the number of new products being launched, because of the number of years it takes to develop a new CP chemical. However, looking to the future, the development pipelines within the crop protection industry are known to contain at least 54 new chemicals which are within five years of launch. While some of these will, no doubt, fall by the wayside, there are presumably others which have not yet been made public. These figures therefore suggest that the average rate of introduction of new chemicals will be maintained at close to 12 per annum over the next few years. The 54 chemicals are split almost equally between herbicides, insecticides and fungicides. Interestingly, many of them are aimed at specialty crop sectors and niche markets, e.g. fungicides for use on Oomycete diseases in fruit and vegetables (Phillips McDougall, 2003d).

A closer analysis of the 54 new chemicals known to be in development shows that only 20 of them (37%) are attributable to the six major multi-national companies (Bayer, Syngenta, BASF, Dow, Monsanto and DuPont), while 28 (52%) are from Japanese and South Korean companies (Phillips McDougall, 2003d and 2003e). These figures could be interpreted as being an effect of the consolidation which has led to the formation of several of the largest companies and has not distracted the far-eastern companies in the same way, together with the shift of research funds in the major companies towards biotechnology. The correct interpretation, however, probably relates to different company strategies. The major companies have been searching for "blockbuster" products, while the smaller companies have been satisfied with the development of smaller products, e.g. for use in Japan only. One observation which is consistent with this interpretation is that, although the rate of introduction of new CP chemicals has been maintained in recent years, chemicals launched since 1998 have had lower average sales values and there are few blockbusters (Phillips McDougall, 2003b).

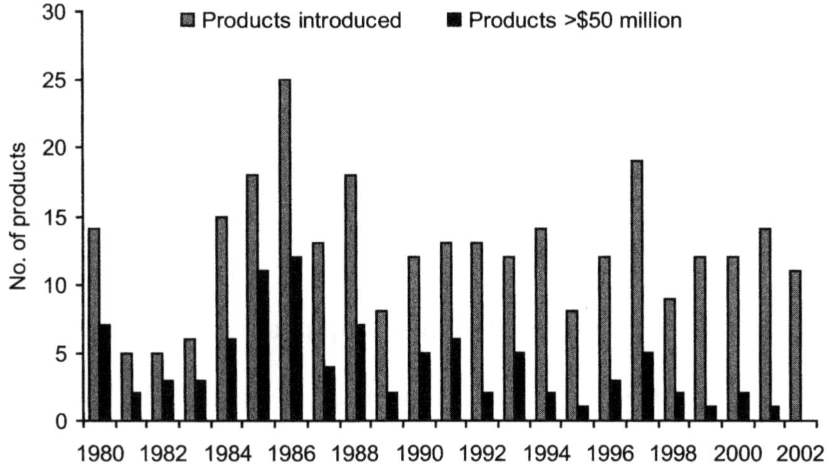

Figure 9.3 Number and size of new products introduced each year since 1980 (courtesy of Phillips McDougall, 2003b).

The increasing effectiveness of crop protection chemicals

New CP chemicals justify the cost of their development and are able to displace older products from the market because they are becoming more effective in various ways. The obvious way in which this has happened relates to their potency. Analysis shows that modern chemicals, broadly speaking, are effective at lower application rates than the older ones (although one has to look back over decades rather than years to see the trend). For example, the herbicide 2,4-D (introduced in 1942) is typically applied at a rate of ~1000g/ha, while the best of the modern compounds such as the sulfonylureas are exquisitely potent and are effective at rates as low as 10g/ha. The picture is similar for insecticides, e.g. DDT (also introduced in 1942) was used at rates in excess of 1000g/ha, while imidacloprid and the best of the pyrethroids are applied at rates of ~50g/ha and ~10g/ha, respectively. For fungicides, the older compounds such as mancozeb (introduced in 1961) and copper-based inorganics are applied at rates of at least 1000g/ha, while the triazoles and strobilurins are effective at rates in the region of 200g/ha (Dehne and Schönbeck, 1994; Bateman, 2003).

The modern CP chemicals are effective at lower rates because of higher intrinsic potency and better physical properties. Back-of-the-envelope calculations show that it should be possible to reduce rates much further still if methods can be found to deliver more of the chemical from the spray tank to the active site within the target organism.

Modern CP chemicals are also clearly more effective in terms of their safety and environmental profiles. The major re-registration programme currently being pursued in Europe, for example, has shown that some chemicals which fulfilled the registration criteria in the past fall short of the modern standards, and they are being de-registered.

The higher potency and selectivity of modern CP chemicals has been achieved by focusing on compounds which act at a single site (an enzyme or a receptor). As explained below, such compounds are prone to resistance, unlike some of the older ones, especially fungicides, which have multi-site modes of action. So from a resistance-management point of view, some modern CP chemicals are arguably less effective than the older ones (Hewitt, 2004)!

Drivers for the development of new crop protection chemicals

As stated above, the market for CP chemicals is mature. The industry has been highly successful in inventing effective products for most commercially-important problems associated with weeds, insects and fungi. Furthermore, the discovery and development of a new CP chemical costs more than ever before (about $184 million, according to the results of a recent survey – Phillips McDougall, 2003a). Despite these factors, there are very good reasons to invest in the search for new CP chemicals because there are still excellent prospects for new and innovative CP chemicals, especially those with new modes of action. For example, the onset of resistance to existing products, and the loss of many older products through the ongoing re-registration programmes in Europe and other parts of the world, present opportunities for the introduction of new CP chemicals. In addition, consolidation within the industry means that there are now fewer companies conducting research and development leading to new CP chemicals, which should improve the prospects for those which remain. These factors are discussed in turn in the following paragraphs.

THE ONSET OF RESISTANCE AND THE IMPORTANCE OF NEW MODES OF ACTION

An important driver for the development of new CP chemicals is that, sooner or later, many of the existing products encounter the problems associated with resistance. Most modern CP chemicals act at a single biochemical site and, indeed, it is this that leads to their high potency and selectivity. However, it also means that single genetic changes in the target organism can lead to resistance. Some of the older CP chemicals act at more than one site (and sometimes at many sites) which means they are generally less potent and less selective, but they are also less likely to be affected by resistance. Products do not have to be abandoned at the onset of resistance, but their continuing use needs to be carefully managed by mixing or alternating with other chemicals with different modes of action (Russell, 2003).

Resistance is a problem which affects fungicides, insecticides and herbicides. The industry has well-established "Resistance Action Committees" for each of these sectors, and it is their task to allocate new CP chemicals to appropriate resistance groups and to provide advice about resistance management to agronomists and farmers.

The problems associated with resistance mean that there is a continuing need for new CP chemicals with novel and resistance-breaking modes of action. It is true that new products with established modes of action, even with modes of action which have been known for decades, can be highly successful (Stetter and Lieb, 2000). However, in most cases, a "blockbusting" product (or family of products) has a new mode of action, which can lead to a new technical profile, differentiation in the market place, rapid market penetration, displacement of older products and substantial rewards for the innovator. The leading examples of such product families in recent years are the neonicotinoid insecticides, first sold in the early 1990s and taking an 8.5% share of the insecticide market by 2000, and the strobilurin fungicides, first sold in 1996 and rising to 10.9% of the fungicide market by 2000 (Phillips McDougall, 2002a).

The number of modes of action represented by the ~800 commercial CP chemicals is uncertain. About 50 have been determined [~20 for herbicides (Phillips McDougall, 2002b), ~10 for insecticides and ~20 for fungicides] and these cover most of the products. However, there are other products where the mode of action is only partially understood (e.g. the pathway on which the chemical acts has been determined, but not the specific enzyme) or is simply not known (and this applies even to some recently introduced compounds, e.g. BASF's fungicide metrafenone, launched in 2003 – Agrow, 2004b). What is certain is that there are many more modes of action waiting to be discovered – only a small fraction of the theoretically accessible biochemical targets which should be essential to the survival of plants, insects and fungi have been found (Ward and Bernasconi, 1999).

Not surprisingly, great importance is placed on the discovery of new modes of action in the crop protection industry. When a class of chemistry with a new mode of action is discovered, there is great excitement and fierce competition for patent property.

RE-REGISTRATION PROGRAMMES

Re-registration programmes are in hand in the EU, the USA and Japan to ensure that all CP chemicals currently in use meet modern safety and environmental standards. In the EU, this procedure is governed by the Council Directive 91/

414/EEC, which was enacted in 1991, came into force in 1993, and is now expected to be complete by 2008. If manufacturers choose to submit their products for re-registration, they are responsible for conducting the required health and safety testing. Of course, products of major commercial importance easily justify the cost of re-registration, but minor products may not, even if they are likely to comply with the safety and environmental standards of the Directive.

The effect of the re-registration programmes is therefore to remove products which are older, less safe and of lesser commercial importance. This is likely to have a disproportionately negative effect on many generic products which, by definition, are older chemicals, and growers will also lose access to certain smaller products for which there are no substitutes. For the crop protection industry, it means a substantial loss of sales, but it also represents a growth opportunity for the more modern products, as well as an opportunity for the introduction of new CP chemicals.

The number of CP chemicals registered for use in the EU is falling dramatically. By the end of 2003, about 450 chemicals had been withdrawn (more than half of those on the market at the beginning of the re-registration process!), and just 46 had been successfully re-registered (accepted into Annex 1 of the EU Directive). Others are still in evaluation (Phillips McDougall, 2003f; Bird, 2004; Rhodes, 2004).

INDUSTRY CONSOLIDATION

Finally, there has been considerable consolidation within the crop protection industry over the last 15 years. There are now just six global research and development companies with more than a 5% share of the market for CP chemicals and these (ranked in decreasing order of sales of CP chemicals) are Bayer, Syngenta, BASF, Dow, Monsanto and Dupont. These six companies, taken together, had an 85.6% share of the market in 2003. With fewer players, those that remain should have less competition and improved prospects for the introduction of new products (Copping, 2003; Phillips McDougall, 2004b).

The invention of new crop protection chemicals

It is not easy to invent a new CP chemical! It must have high but selective potency, in order to control a useful spectrum of weeds, insects or fungi when applied at low rates, without affecting non-target organisms. It must have appropriate chemical and metabolic stability, so that it survives in sunlight on a leaf surface and then within the target organism, but is not so stable that it persists in the environment. It requires physical properties which allow it to reach the enzyme or receptor at which it acts (systemic fungicides, for example, need to be able to penetrate the waxy surface of a leaf and then to move within the plant to the fungus). And finally it must be cost-effective, which usually means that it must be relatively cheap to manufacture, although chemicals with extremely high potency can support higher manufacturing costs. From a research investment point of view, it is important to understand that there is no guarantee that such a chemical can be found within one class of chemistry with a particular mode of action. Many man-years of work may be invested without finding a compound in which all the required properties are combined.

The way in which new CP chemicals are invented has changed a great deal in the last ten years. A huge amount of new technology is now being used, much of it developed by the pharmaceutical industry, which is also faced with the task of discovering new chemicals with useful biological properties, but which has a higher research budget than the crop protection industry.

In the 1990s, advances in robotics, automation and data handling made it possible to screen at much higher throughput than in the past. The industry seized the opportunity to do so because increased screening rates should provide higher numbers of lead compounds and therefore more choices of starting points for research programmes. Companies now screen in the order of 200k compounds per annum, a twenty-fold increase over the typical screening rate of ten years ago. Importantly, chemicals are still mainly screened *in vivo*, that is, directly against the organisms of interest, retaining the advantage that the crop protection industry has over the pharmaceutical industry which cannot, of course, test directly on the patient. However, the tests have been miniaturised so that the plants, insects or fungi are grown in 96-well microtitre plates which can be handled robotically, and sub-milligram quantities of the test chemicals are required.

This huge increase in screening capacity presented a challenge for the synthesis chemists, who were required to supply a stream of new chemicals for testing. Combinatorial chemistry was developed in response to this need (Smith, 2003). It allows "libraries" of hundreds or thousands of related compounds to be made in parallel using the same synthetic steps. In fact, it is not so difficult to make large numbers of compounds, and the real skill is to design libraries containing diverse sets of compounds with properties which are likely to lead to useful biological activity. After all, the driver for increasing screening throughput was the expectation that the hit rate would rise proportionately, and this will not be the case unless the quality of the input is maintained. Automation and robotics, together with advanced data manipulation, are central to the success of combinatorial chemistry.

The fact that screens require only very small quantities of material has also opened up opportunities for other sources of compound supply, e.g. purchased chemical libraries, sample exchange agreements with pharmaceutical companies, and natural product extracts.

When a new biologically-active chemical is identified by *in vivo* screening, it is useful to establish its mode of action. This is partly because of the commercial implications downstream, but it also enables the intrinsic activity of the compound to be determined using an appropriate enzyme assay. It is relatively easy to determine whether or not the mode of action is novel, simply by screening against a battery of enzymes representing each of the known modes of action, but elucidating a novel mode of action is much more difficult, and various biochemical and genomic methods are used.

A host of other miniaturised assays are also in place, allowing a wide variety of properties to be measured using tiny quantities of material, e.g. physical properties; specific biological effects such as systemicity, vapour activity and translaminar activity; metabolic stability; the rate of uptake into plants; and the potential for persistence and leaching in soil. These assays allow chemicals to be thoroughly characterised at an early stage of the research process. Chemicals with properties which make them unpromising as starting points for further synthesis can quickly be discarded, leaving the better ones on which to focus. Miniaturisation has therefore improved early decision-making and portfolio management.

Writing as one directly involved in the discovery of new CP chemicals, I will end this section by referring to the importance of establishing a research environment in which innovation and creativity can flourish. It is the nature of research that most ideas, even though scientifically rigorous, lead nowhere. Nevertheless, the value of these ideas should be recognised, and lessons must be learnt from the projects which fail.

A case study – the strobilurin fungicides

This section of the review focuses on one specific class of CP chemicals, the strobilurin fungicides, because their discovery, development and first eight years of use illustrate several of the issues mentioned in general terms in the sections above. In addition, the strobilurins are particularly interesting in the context of this conference because they deliver unexpected benefits in yield, over and above those which can be accounted for by their fungicidal activity.

The strobilurins are the most important class of agricultural fungicides to have been discovered during the last 25 years. Their invention was inspired by a family of fungicidal natural products. They have a novel mode of action, inhibiting mitochondrial respiration in fungi by binding at the so-called Qo site of cytochrome b, and thereby blocking the production of ATP and suppressing the energy cycle. The first of the strobilurins, launched in 1996, were azoxystrobin (from Syngenta) and kresoxim-methyl (from BASF). Azoxystrobin has outstanding broad-spectrum activity and it has risen rapidly to become the world's biggest selling fungicide, with sales now worth ~\$450 million per annum (Blair *et al.*, 2002). In a number of crops, strobilurins have led to significant changes in disease-control programmes. With azoxystrobin, for example, growers of grape-vines have, for the first time, a single chemical which controls both powdery and downy mildews (Bartlett *et al.*, 2002). Eight strobilurins are now in commercial use (Tomlin, 2003).

Yield and quality studies with the strobilurins have focused on wheat and barley, and data indicate clear and consistent benefits in terms of yield and grain size. However, of particular interest is the observation that cereal fungicide programmes based on strobilurins consistently give higher yields than those based on triazole fungicides, even though both give similar levels of visible disease control. This has been termed the strobilurin "greening effect" because it is closely associated with the ability of the strobilurins to maintain the green leaf area of a crop until late in the season, maximising the grain-filling period. At least two hypotheses have been proposed to explain this phenomenon. The first is that, as a result of their mode of action, strobilurins prevent the germination of spores of pathogenic, non-pathogenic and saprophytic fungi and thereby block the elicitation of energy-demanding responses from the plant in a way which triazoles do not. The second is that the strobilurins produce beneficial physiological effects on plants which are not related to disease control (and strobilurins can be shown to affect various physiological processes in plants). It is not clear whether either of these hypotheses is the correct explanation for the "unexpectedly good" yield benefits which follow from the use of strobilurins in wheat and barley, and perhaps elements of both contribute (Bartlett *et al.*, 2002; Reade, Milner and Cobb, 2003).

All the strobilurins act at the same Qo site on cytochrome b, so the Fungicide Resistance Action Committee (FRAC) has allocated them all to the same QoI (Qo Inhibitor) cross-resistance group. Since they act at a single site, it was anticipated that resistance would, in due course, be encountered in the field. In fact, the first signs of resistance emerged very quickly, as early as 1998, on *Erysiphe graminis tritici* (powdery mildew) in wheat, in northern Germany. Since that time, resistance has appeared in a variety of fungi throughout the world. Perhaps most significant from a commercial point of view has been the emergence of resistant strains of *Septoria tritici* (leaf spot) in wheat, initially in Ireland in 2002, and then spreading throughout Europe (Heaney, Hall, Davies and Olaya, 2000; Lucas, 2003).

For most of the pathogens in which QoI resistance has been detected, the major mechanism of resistance is the substitution of the amino acid glycine at position 143 of the protein chain of cytochrome b with alanine, the so-called G143A mutation. The amino acid residue at position 143 forms part of the Qo site, and the switch from glycine to alanine introduces an extra methyl group which protrudes into the site and prevents the strobilurins from binding. Importantly for the farmer and the crop protection industry, the G143A mutation leads to high resistance factors and little or no fitness penalties in the fungus (Gisi, Sierotzki, Cook and McCaffery, 2002).

In order to safeguard the future of the strobilurins, FRAC has issued guidelines about how they should be used. They recommend that the number of strobilurin applications each season should be limited, and alternated with applications of effective fungicides from alternative cross-resistance groups. In January 2004, Syngenta launched a mixture of azoxystrobin and chlorothalonil under the name "Amistar Opti" for use in cereals where strobilurin-resistant strains of *Septoria* are likely to be found. Chlorothalonil is a fungicide with a multi-site mode of action. No cases of chlorothalonil resistance are known, despite the fact that the chemical was first described 40 years ago and is now used on about 150 crops and 300 diseases (Market Scope Europe, 2004).

Conclusions and outlook

The crop protection industry has been highly successful, and chemicals are available for the control of most weeds, insects and fungi of commercial importance. The market for CP chemicals has declined in value over the last five years, partly because less chemicals are being used (some have been displaced by the use of GM crops, for example) and partly because the cost of many chemicals has been driven down by competition. In Europe, agriculture has been particularly successful, and subsidies to farmers are being decoupled from production.

Despite the decline over the last five years, a small growth in the value of the market for CP chemicals, to the extent of about 0.5% per annum, has been forecast for the next five years. The reasons for this optimism include increasing crop prices, recovery from drought in Europe and other parts of the world, and an improving economy in Latin America (Phillips McDougall, 2004b).

CP chemicals will continue to be used for the foreseeable future because of the outstanding benefits they bring. They provide major improvements in yield, higher quality and season-to-season reliability, and make an important contribution to the global economy. Although it is more expensive than ever before to invent, develop and launch a new CP chemical, innovative new products still bring substantial rewards.

Acknowledgements

I would like to thank my colleague Jonathan Shoham, Syngenta Business Intelligence, for his help in the preparation of this review. Figures 9.1, 9.2 and 9.3 are reproduced by kind permission of Phillips McDougall, Pathhead, Midlothian, Scotland.

References

Agrow (2003) GM crops to slash US pesticide sales. *Agrow* No. 438, December 19th 2003, p 17, PJB Publications Ltd., Richmond, Surrey, UK.
Agrow (2004a) Global GM crop plantings up 15% in 2003. *Agrow* No. 440,

January 23rd 2004, p 13, PJB Publications Ltd., Richmond, Surrey, UK.

Agrow (2004b) New active ingredients presented in France. *Agrow* No. 440, January 23rd 2004, p 21, PJB Publications Ltd., Richmond, Surrey, UK.

Avery, D.T. (1997) Saving the planet with pesticides, biotechnology and European farm reform. In *The 1997 Brighton Crop Protection Conference: Weeds – Volume 1*, pp 3-18. The British Crop Protection Council, Farnham, Surrey, UK.

Bartlett, D.W., Clough, J.M., Godwin, J.R., Hall, A.A., Hamer, M. and Parr-Dobrzanski, B. (2002) The strobilurin fungicides. *Pest Management Science*, **58**, 649-662.

Bateman, R. (2003) Rational pesticide use: spatially and temporally targeted application of specific products. In *Optimising Pesticide Use*, pp 131-159. Edited by M.F. Wilson. Wiley Series in Agrochemicals and Plant Protection, John Wiley and Sons, Chichester, West Sussex, UK.

Bates, R. (2002) Pesticide residues and consumer risk assessments. *Pesticide Outlook*, **13**, 142-147.

Bird, J. (2004) Dearth of new products causes concern. In *Agrow Review of 2003*, pp 9-14, PJB Publications Ltd., Richmond, Surrey, UK.

Blair, P., Benson, A., Burns, I., Heslop, M. and Jacobs, A. (2002) *Syngenta – Expanding the Pie*. Report by Schroder Salomon Smith Barney, 3 September 2002.

Bruinsma, J. (2002) *How to Feed 2 Billion More Mouths in 2030? Here are Some Answers*. Website of the Food and Agriculture Organization of the United Nations, http://www.fao.org/english/newsroom/news/2002/8280-en.html.

Copping, L. (2003) The evolution of crop protection companies. *Pesticide Outlook*, **14**, 276-278.

Defra, Organic Food and Farming, http://www.defra.gov.uk/farm/organic/introduction/.

Dehne, H.-W. and Schönbeck, F. (1994). Crop protection – past and present. In *Crop Production and Crop Protection – Estimated Losses in Major Food and Cash Crops*, pp 45-71. By E.-C. Oerke, H.-W. Dehne, F. Schönbeck and A. Weber. Elsevier, Amsterdam, The Netherlands.

Evans, D.A. (1999) How can technology feed the world safely and sustainably? In *Pesticide Chemistry and Bioscience – The Food-Environment Challenge*, pp 3-24. Edited by G.T. Brooks and T.R. Roberts. Royal Society of Chemistry, Cambridge, UK.

Evans, D.A. (2000) New era, new challenges, new solutions. In *The BCPC Conference: Pests and Diseases 2000 – Conference Proceedings, Volume 1*, pp 3-18. The British Crop Protection Council, Farnham, Surrey, UK.

FAO (Food and Agriculture Organization of the United Nations), http://www.fao.org/waicent/portal/statistics_en.asp.

Foster, G.N., Atkinson, D. and Burnett, F.J. (2003) Pesticide residues – better early than never? In *The BCPC International Congress – Crop Science and Technology 2003, Congress Proceedings – Volume 2*, pp 711-718. The British Crop Protection Council, Alton, Hampshire, UK.

Gisi, U., Sierotzki, H., Cook, A. and McCaffery, A. (2002) Mechanisms influencing the evolution of resistance to Qo inhibitor fungicides. *Pest Management Science*, **58**, 859-867.

Hamilton, D. and Crossley, S. (2004) Introduction. In *Pesticide Residues in Food and Drinking Water: Human Exposure and Risks*, pp 1-25. Edited by D. Hamilton and S. Crossley. Wiley Series in Agrochemicals and Plant Protection, John Wiley and Sons, Chichester, West Sussex, UK.

Harris, J.M., Kaufman, P.R., Martinez, S.W., and Price, C. (2002) *The U.S. Food Marketing System, 2002. Competition, Coordination, and Technological Innovations into the 21st century*. Agricultural economic report No. 811. Electronic report from the Economic Research Service of the United States Department of Agriculture, June 2002. http://www.ers.usda.gov/publications/aer811/.

Heaney, S.P., Hall, A.A., Davies, S.A. and Olaya, G. (2000) Resistance to fungicides in the QoI-STAR cross-resistance group: current perspectives. In *The BCPC Conference: Pests and Diseases 2000 – Conference Proceedings, Volume 2*, pp 755-762. The British Crop Protection Council, Farnham, Surrey, UK.

Hewitt, H.G. (2004) Do we need new fungicides? *Outlooks on Pest Management*, **15**, 90-95.

Kidd, H., Pimentel, D. and Wilkinson, C. (2002) Profile and viewpoints on the life of Rachel Carson. *Pesticide Outlook*, **13**, 204-208.

Landell Mills (1992) *Organic farming in seven European countries*. A study conducted by the Landell Mills research group for the European Crop Protection Association, Brussels, Belgium.

Leaver, C.J. (2001) Food for thought. In *The BCPC Conference: Weeds 2001 – Conference Proceedings, Volume 1*, pp 3-12. The British Crop Protection Council, Farnham, Surrey, UK.

Lomborg, B. (2003a). Measuring human welfare. In *The Skeptical Environmentalist: Measuring the Real State of the World*. Chapter 3, pp 45-49. Cambridge University Press, Cambridge, UK.

Lomborg, B. (2003b). Our chemical fears. In *The Skeptical Environmentalist: Measuring the Real State of the World*. Chapter 22, pp 215-248. Cambridge University Press, Cambridge, UK.

Lucas, J. (2003) Resistance to QoI fungicides: implications for cereal disease management in Europe. *Pesticide Outlook*, **14**, 268-270.

Mäder, P., Fließbach, A., Dubois, D., Gunst, L., Fried, P. and Niggli, U. (2002). Soil fertility and biodiversity in organic farming. *Science*, **296**, 1694-1697.

Market Scope Europe (2004) Chlorothalonil makes a comeback. In *Crop Protection Monthly by E-Mail*, Issue number 170, 31 January 2004, Market Scope Europe Ltd.

Mathisen, F., Cheung, K. and Sirur, G. (2003) EU adopts reform of Common Agricultural Policy. In *PhytoPhile*, June 2003, Agrochemical Service of Wood Mackenzie Ltd., Edinburgh, Scotland.

Oerke, E.-C., Weber, A., Dehne, H.-W. and Schönbeck, F. (1994) Conclusions and perspectives. In *Crop Production and Crop Protection – Estimated Losses in Major Food and Cash Crops*, pp 742-770. By E.-C. Oerke, H.-W. Dehne, F. Schönbeck and A. Weber. Elsevier, Amsterdam, The Netherlands.

Organic Centre Wales, The UK organic market in figures 2002-2003, http://www.organic.aber.ac.uk/statistics/uk2003.shtml.

Pesticide Outlook (2003) Regulatory News. In *Pesticide Outlook*, **14**, p 47.

Phillips, S.L., Leifert, C., Santos, J., Juntharathep, P., Bodker, L., Tamm, L. and Smit, A.B. (2002) Development of a systems approach for the management of late blight (*Phytophthora infestans*) in organic potato production: an update on the EU Blight-MOP project. In *The BCPC Conference: Pests and Diseases 2002 – Conference Proceedings, Volume 2*, pp 539-546. The British Crop Protection Council, Farnham, Surrey, UK.

Phillips McDougall (2002a) Agrochemical market development by chemistry. *Agrifutura*, No. 28, February 2002, Phillips McDougall, Pathhead, Midlothian, Scotland.

Phillips McDougall (2002b) Herbicide mode of action. *Agrifutura*, No. 32, June 2002, Phillips McDougall, Pathhead, Midlothian, Scotland.

Phillips McDougall (2002c) The global agrochemical market in 2002 – preliminary review. *Agrifutura*, No. 37, November 2002, Phillips McDougall, Pathhead, Midlothian, Scotland.

Phillips McDougall (2003a) *The Cost of New Agrochemical Product Discovery, Development and Registration in 1995 and 2000*. Final Report. A consultancy study for Crop Life America and the European Crop Protection Association. Phillips McDougall, April 2003, Pathhead, Midlothian, Scotland.

Phillips McDougall (2003b) Research and development; cost and return. *Agrifutura*, No. 43, May 2003, Phillips McDougall, Pathhead, Midlothian, Scotland.

Phillips McDougall (2003c) The EU agrochemical market. *Agrifutura*, No. 44, June 2003, Phillips McDougall, Pathhead, Midlothian, Scotland.

Phillips McDougall (2003d) Changing focus of R & D. *Agrifutura*, No. 47, September 2003, Phillips McDougall, Pathhead, Midlothian, Scotland.

Phillips McDougall (2003e) *AgriService, Research Section – 2003*. Phillips McDougall, September 2003, Pathhead, Midlothian, Scotland.

Phillips McDougall (2003f) EU re-registration issues. *Agrifutura*, No. 48, October 2003, Phillips McDougall, Pathhead, Midlothian, Scotland.

Phillips McDougall (2003g) The global agrochemical market in 2003 – preliminary review. *Agrifutura*, No. 49, November 2003, Phillips

McDougall, Pathhead, Midlothian, Scotland.
Phillips McDougall (2004a) Biotechnology in crop protection. *Agrifutura*, No. 51, January 2004, Phillips McDougall, Pathhead, Midlothian, Scotland.
Phillips McDougall (2004b) *AgriService, Industry Overview – 2003 Market*. Phillips McDougall, March 2004, Pathhead, Midlothian, Scotland.
Pragnell, M. (2003) Crop protection in the new millennium. In *Chemistry of Crop Protection – Progress and Prospects in Science and Regulation*, Chapter 1, pp 7-16. Edited by G. Voss and G. Ramos, Wiley-VCH, Weinheim.
Reade, J.P.H., Milner, L.J. and Cobb, A.H. (2003) Can picoxystrobin protect winter wheat from environmental stress? In *The BCPC International Congress – Crop Science and Technology 2003, Congress Proceedings – Volume 2*, pp 863-868. The British Crop Protection Council, Alton, Hampshire, UK.
Reynolds, S. and Hill, A. (2002) 'Cocktail' effects – stirred, not shaken...yet. *Pesticide Outlook*, **13**, 209-213.
Rhodes, J. (2004) EU agrochemical review moves up a gear. In *Agrow Review of 2003*, pp 15-19, PJB Publications Ltd., Richmond, Surrey, UK.
Russell, P. (2003) Taking the path of least resistance. *Pesticide Outlook*, **14**, 57-61.
Sisson, A.M. and Williams, R.O. (2001) The benefits of generic products to the market. In *The BCPC Conference: Weeds 2001 – Conference Proceedings, Volume 1*, pp 107-116. The British Crop Protection Council, Farnham, Surrey, UK.
Smith, S. (2003) Combinatorial chemistry in the development of new crop protection products. *Pesticide Outlook*, **14**, 21-25.
Stetter, J. and Lieb, F. (2000) Innovation in crop protection: trends in research. *Angewandte Chemie International Edition*, **39**, 1724-1744.
Swirsky Gold, L., Slone, T.H. and Ames, B.N. (2001) Natural and synthetic chemicals in the diet: a critical analysis of possible cancer hazards. In *Food Safety and Quality*, pp 95-128. Edited by R.E. Hester and R.M. Harrison, Royal Society of Chemistry, Cambridge, UK.
Tamm, L. (2000) The impact of pests and diseases in organic agriculture. In *The BCPC Conference: Pests and Diseases 2000 – Conference Proceedings, Volume 1*, pp 159-166. The British Crop Protection Council, Farnham, Surrey, UK.
Tomlin, C.D.S., Editor (2003). *The Pesticide Manual*, Thirteenth Edition, British Crop Protection Council, Alton, Hampshire, UK.
Trewavas, A. (2004) The true price of progress. *Chemistry and Industry*, 19 January 2004, pp 12-13.
United Nations (2003). World population in 2300 to be around nine billion. http://www.un.org/esa/population/publications/longrange2/longrange2.htm.
United Nations Population Division, World population prospects: The 2002

revision population database, http://esa.un.org/unpp/.
Uttley, N. (2004) Is the future generic? *Speciality Chemicals Magazine*, **24**, April 2004, pp 13-15.
Ward, E. and Bernasconi, P. (1999) Target-based discovery of crop protection chemicals. *Nature Biotechnology*, **17**, 618-619.
Weber, A. (1994) Population growth, agricultural production and food supplies. In *Crop Production and Crop Protection – Estimated Losses in Major Food and Cash Crops*, pp 1-44. By E.-C. Oerke, H.-W. Dehne, F. Schönbeck and A. Weber. Elsevier, Amsterdam, The Netherlands.
Willer, H. and Yussefi, M. (2004) The world of organic agriculture – statistics and emerging trends 2004. International Federation of Organic Agriculture Movements, Bonn, Germany. http://orgprints.org/00002555/.
Williams, D.P.E. (2003) Developing a national pesticide strategy. In *The BCPC International Congress – Crop Science and Technology 2003, Congress Proceedings – Volume 1*, pp 163-170. The British Crop Protection Council, Alton, Hampshire, UK.
Wise, C.J.C. and Findlay, A. (2003) The growers' perspective on strategies for the minimisation of pesticide residues in food. In *The BCPC International Congress – Crop Science and Technology 2003, Congress Proceedings – Volume 2*, pp 723-726. The British Crop Protection Council, Alton, Hampshire, UK.

10

PLANT BREEDING IN THE 21ST CENTURY

P. STEPHEN BAENZIGER[1], KEN RUSSELL[1], BRIAN BEECHER[1], AND ROBERT A. GRAYBOSCH[2]
[1]Department of Agronomy and Horticulture, University of Nebraska, Lincoln, NE 68583-0915, USA; [2]USDA-ARS and Department of Agronomy and Horticulture, University of Nebraska, Lincoln, NE 68583-0915, USA

> "We recently advanced our knowledge of genetics to a point where we can manipulate life in a way never intended by nature. We must proceed with the utmost caution in the application of this new found knowledge."
>
> *Review of Luther Burbank on plant breeding, 1906*

> "It is well for people who think to change their minds occasionally in order to keep them clean. For those who do not think, it is best at least to rearrange their prejudices once in a while."
>
> *Luther Burbank, 1849 - 1926*

> ". . It is impossible to estimate the wealth he has created. It has been generously given to the world. Unlike inventors, in other fields, no patent rights were given him, nor did he seek a monopoly in what he created. Had that been the case, Luther Burbank would have been perhaps the world's richest man. But the world is richer because of him. In this he found joy that no amount of money could give."
>
> *Eulogy to Luther Burbank by Judge Ben Lindsey, April 14, 1926*

The need for plant breeding and crop management

All plant breeders are familiar with the formula, Phenotype (P) = Genotype (G) + Environment (E) + Genotype by Environment Interaction (GEI). Because the environment consists of both random weather effects and controlled management effects such as, tillage, pesticide use, fertiliser, and irrigation, this formula has been modified to:

P = G + E +Management (M) + Genotype by Management Interaction (GMI) + GEI + Genotype by Management by Environment Interaction (GMEI).

When the interactions of genotype with controlled and uncontrolled environmental effects are pooled together under GEI, then the formula becomes: P = G + M + E + GEI. These formulae are useful in considering the future role for plant breeding in sustainable agricultural productivity. Plant breeders develop new genotypes which interact with the environment and crop management systems. Cropping-system specialists work with management systems and how they interact with other environmental factors and with genotypes. The need for greater agricultural productivity is clear, due to expected population growth, increased wealth, and a concomitant diet of greater calories and diversity. The fundamental question is the relative importance of the plant breeder and the cropping-system specialist in the future. Historically, increased crop productivity was due roughly half to genetic improvements and half due to cropping system improvements (Specht and Williams, 1984), though genetic improvement can be higher in some crops (Duvick, 1984). Will the past predict the future? We believe that this depends on the crop. In maize (*Zea mays* L.) and rice (*Oryza sativa* L.) there may be relatively little improvement left for maximum grain yield in very high yielding environments (our definition of yield potential; Cassman, 1999; Duvick and Cassman, 1999). The breeders' future contribution to increased productivity in these crops may largely come through improving stress tolerance, though the definition of stress will change as random and controlled environmental variables change over time. In wheat (*Triticum aestivum* L.), however, the maximum grain yield in high yielding environments does not appear to have been reached and genetic progress for both actual yield and yield potential is possible.

Similarly, many of the cropping-system practices have captured much of their benefit. For example, once you have a fully irrigated or well-fertilised and pest-controlled field, can you expect to achieve large future gains by additional irrigation, fertiliser, or pest control? Also, it will be difficult to allocate more of world's resources, where already it is estimated that 80% of all fresh water consumption is used in irrigation and 50% of the world's reactive nitrogen is used in agriculture (Cassman, Dobermann, Walters and Yang, 2003).

Though there are large parts of the world where these improved management practices still need to be adopted, the ability to increase agricultural productivity through improved management practices becomes less with their adoption. Also, long-time use of many cropping systems that were previously regarded as successful, have degraded soil quality (Dobermann, Dawe, Roetter and Cassman, 2000). Hence, much of the future research into crop production practices will need to address resource-use efficiency and sustainability, and to reverse the soil degradation that is reducing crop yields (e.g. Cassman, 1999; Dobermann *et al.*, 2000; Cassman *et al.*, 2003). This research should allow agricultural productivity to remain constant or be slightly improved, but it is unlikely to bring large improvements. This conclusion is essentially analogous to the outcome of breeding for pest resistance, where biotic changes in pests dictate that breeding effort is sustained in order to maintain the same level of resistance (and crop yield) over time.

The current state of plant breeding technology

Plant breeding technologies and methodologies are determined by breeding objectives (including those determined by the consumer and market), which are tempered by crop biology. The biology of the crop includes whether or not it is self-pollinated (e.g. wheat), cross-pollinated (e.g. maize), or clonally propagated (e.g. bananas [*Musa* spp.]). In addition, the crop may be grown as a hybrid (usually a cross between two inbred lines, e.g. hybrid maize or rice), a population (e.g. alfalfa [*Medicago sativa* L.]), an inbred line (e.g. wheat or barley [*Hordeum vulgare* L.]), or as a clone (e.g. bananas, sugar cane [*Saccharum officinarum* L.], potatoes [*Solanum tuberosum* L.], and apomictic grasses). Unadjusted for moisture, the top ten crops of the world, based on global production by weight, are sugar cane, maize, rice, wheat, alfalfa, potatoes, sugar beets (*Beta vulgaris* L.), soybeans (*Glycine max* (L.) Merr.), cassava (*Manihot esculenta* Crantz), and barley (FAOSTATdata, 2004). These crops represent the diversity of pollination mechanisms, propagules, and ploidy levels that are found in flowering plants (angiosperms).

This paper will concentrate on plant breeding methodologies for 21st century of those crops that can be sexually hybridised. This decision is not to diminish the importance of clonally propagated crops, but rather recognises that various aspects of sexually hybridised crops can be used for clonally propagated crops and that most breeding efforts are concentrated in sexually hybridised crops.

Traditional crop breeding has a long and highly successful history of improving agricultural productivity. Hybrid maize and the green revolution in wheat and rice have greatly added to the world's stores of food. Hence, as we look to the 21st century, the first question is: "Will traditional plant breeding as practised in the 20th century be sufficient to meet the world's growing needs for food for humans, feed for livestock, and fibre in a sustainable manner or will new technologies be needed?" It should be remembered that in the 20th century, plant breeding was greatly changed by the rediscovery of Mendelian genetics, and by the birth of statistics which introduced the ideas of replication and randomisation, hence various forms of family-based selection procedures. In addition to these highly successful procedures, plant breeders tried, with less success, increases in ploidy level, and mutation breeding, including variation induced through tissue culture. In the 21st century, plant breeding will have similar new technologies that will greatly affect the way the crop is improved, and other technologies that incrementally affect our understanding of crop breeding. For example, much new research focuses on single genes. This may be a necessary stage that plant breeding will pass through to achieve the real target which is increased understanding of complex biological systems that evolve from interactions of gene networks and multiple environmental stimuli.

These new technologies may allow plant breeders, and agriculture in general, to meet the world's needs for food, feed, and fibre in a sustainable manner. However, it must be recognised that conventional plant breeding and agriculture have the capability today to feed the millions that go hungry every night, yet humanity as a whole chooses not to use its resources and technology to feed them. One should not think that conventional or new technologies will change human nature.

Plant breeders will only change their breeding methods for two reasons: 1. they see an ability to do something that they previously were unable to do (novelty), or 2. they see an ability to carry out an existing process more efficiently than they could previously (efficiency). As we predict what plant breeding will be like in the 21st century, the changes we suggest will be justified by these two causes.

The phases of plant breeding

INTRODUCING GENETIC VARIATION

The three main phases in plant breeding are the identification and incorporation of genetic variation, selection of the best variants (often accompanied by inbreeding), and finally the evaluation of elite lines, populations, or hybrids for possible release. As mentioned above, genetic variation is most commonly introduced through a cross or sexual hybrid (Fehr and Hadley, 1980). Sexual hybrids are used because no cultivar is perfect and none is generally lacking in only one trait. Sexual hybrids allow numerous genes to segregate for those traits controlled by many genes (quantitative traits) and for those traits controlled by a few genes (qualitative traits). In all cases, standard breeding methods [with the exception of back-crossing (Baenziger and Peterson, 1992; Fehr, 1987)] are working with multiple traits and multiple genes.

One of the most difficult aspects facing a plant breeder is predicting which breeding crosses are best (Baenziger and Peterson, 1992). In most crops, it is relatively easy to make a cross and often hundreds or thousands of crosses are made each year. However, only one or a few cultivars are released each year from a successful breeding program, so most of the crosses and their progeny are used to create new parents for future crosses or are discarded completely. The crosses that are the most obvious to make are those that involve transferring a single gene or trait from a donor to a recipient parent (e.g. adding disease or insect resistance, height or maturity genes, etc.). With a better understanding of genes that control useful traits (Ho, McCouch and Smith, 2002; Liu and

Anderson, 2003) it may be possible to design better crosses (Cooper, Podlich, Micallef, Smith, Jensen, Chapman and Kurger, 2002). For example, traditionally a breeder wanting to increase drought tolerance has crossed two of the best drought tolerant parents. But as specific genes that control drought tolerance are identified, greater focus will shift to crossing parents that truly have complementary drought-tolerant gene complexes. These may not be the same as the parents with the highest levels of drought tolerance.

For some species that are apomictic or clonally propagated, crosses may not be preferred, or even possible. In these cases, genetic variation is introduced by mutations and potentially by transformation. In this paper, we use the term plant transformation for the introduction of DNA (a transgene) into the plant usually by micro-projectile bombardment or with the aid of *Agrobacterium tumefaciens*. Others describe this process as genetic engineering and the resultant plant as a genetically modified organism (GMO). However, as plant breeders have always striven to genetically modify crops by conventional methods, GMO is an ambiguous or misapplied term. Even in those crops where sexual hybrids are easily made, genetic variation is supplemented by mutation and transformation (Ahloowalia and Maluszynski, 2001; Gepts, 2002). The techniques for mutagenesis have been developed for all major crops (Maluszynski, 1990) as, in general, have the techniques for transformation.

Crop transformation remains a controversial issue, though it is our belief that breeders of the world's major crops will be able eventually to use appropriate transgenes and plant transformation technology. The sheer level of effort, the increasing need for enhanced agricultural productivity, its benefits to the environment and to sustainability, and the increasing acceptance of transgenic crops will lead to its greater use. The argument is not if, but when. It is estimated that in 2003, 67.7 million hectares were planted to transgenic crops (James, 2003). The countries with the largest transgenic crop area are the United States (42.8 million ha), Argentina (13.9 million ha), Canada (4.4 million ha), Brazil (3 million ha) and the People's Republic of China (2.8 million ha). It is estimated that 7 million farmers in 18 countries now produce transgenic crops. The main transgenic crops are soybeans (55% of the global crop is now transgenic), maize (11% transgenic), cotton (*Gossypium spp.*; 21% transgenic), and canola (*Brassica napus* L.; 16% transgenic); the main traits are herbicide tolerance and insect resistance (James, 2003).

Interestingly, two of the world's great cereal crops, rice and wheat, have no commercial transgenic production despite both having transformation capability (Cheng, Lowe, Spencer and Armstrong, 2004). Golden rice and rice with improved iron content are being developed (Lucca, Hurrell and Potrykus, 2001; Ye, Babili, Kloti, Zhang, Lucca, Beyer and Potyruks, 2000) for commercialisation. In wheat, herbicide tolerant types have been developed (Obert, Riddley, Schneider, Riodran, Nemeth, Trujillo, Breeze, Sorbet and Astwood, 2004; Zhou, Berg, Blank, Chay, Chen, Eskelsen, Fry, Hoi, Isakson, Lawton, Metz, Rempel, Ryerson, Sansone, Shook, Starke, Tichota and Valenti,

2003) but not commercialised, due to lack of market acceptance and concerns of growers about losing export markets. Because rice and wheat are primarily food grains and because the latter is widely exported in highly competitive markets, current consumer attitudes dictate the acceptability of transformation technology. While breeders will change their methods to achieve increased novelty or efficiency, consumers will change their buying habits on the basis of how the food, feed, or fibre directly benefits them. Benefiting the consumer explains the wide acceptance of medical biotechnology and the concern with agricultural biotechnology, especially among the well fed and well clothed.

Plant transformation will be used to incorporate one or a few genes, either in a sense form (adding a function to the plant) or an anti-sense form (taking an undesirable function away from the plant). As such, plant transformation is expected to have its greatest impact on qualitative traits. Additionally, it may be able to change regulatory events and affect cascades of genes, thus affecting more complex traits. Transformation will also affect how traits are manipulated, in that the transgene insert consists of linked genes that segregate as a single cluster (Baenziger and Peterson, 1992); it is much easier to select for a cluster of genes than to select for unlinked genes that segregate independently (see example below in selection strategies).

Transgenics have already allowed the development of crops for truly novel uses, such as pharmaceutical production or production of renewable raw materials (Kern, 2002). Both of these uses involve novel genetic variation that was not available by conventional plant breeding. While both uses are attractive as ways of reducing production costs, great concern remains that the products of these transgenic plants must be separated in the market so that grains developed for non-food uses are not mixed with grains developed for food or feed. Some of these concerns could be removed by judicious choice of pharmaceutical or renewable raw material product and of plant species used for the transformation (Peterson and Arntzen, 2004), but this may also increase the cost of "manufacturing" the transgenic product.

As can well be imagined, effects of transgenic crops in the environment are important in considering their release. In most countries, all transgenic crops are heavily regulated and evaluation of their safety reflects the attitudes in the country. What is known is that transgenic crops can reduce exposure of workers and beneficial organisms to pesticide use, drift, residues, and run-off (Carpenter, Felsot, Goode, Hammig, Onstad and Sankula, 2002; Gianessi, Silvers, Sankula, and Carpenter, 2002). Whilst there are concerns about transgene 'flow' to wild and weedy plants and about natural selection for pest tolerance, proper stewardship and monitoring should reduce the risks of transgenic crop use. It should also be remembered that agriculture has a long history of pollen flow and pest selection before the advent of transgenic crops. There are additional concerns that crop production may become too weed free, due to herbicide tolerant crops, resulting in loss of bio-diversity in cultivated fields. These concerns reflect a particular societal view in favour of weed biodiversity in cultivated fields and against improved technology or successful weed control.

SELECTION

In the second phase, selecting useful variants, selection protocols are often dictated by the large numbers of plants in a plant-breeding program, the information that is known about a line, and the value of a trait. In many crops, the populations are large in the early generations and the only cost efficient mechanism is visual selection in carefully planned selection nurseries that allow good separation of the desirable types. A selection nursery is different from an evaluation nursery (or trial) because the selection nursery has the goal of best separating lines for the trait of interest and may not be representative of the environments where the cultivar will eventually be grown (Baenziger and Peterson, 1992). For example a selection nursery might be placed where a crop is rarely grown because the disease pressure is too high for economic production, but this allows the breeder to determine which lines have good disease resistance. An evaluation nursery (see below) is representative of the environments for which the breeder is selecting improved lines; the information gained from evaluation nurseries is used to determine in which of those environments the elite line is adapted and can be successfully produced.

There is also a hierarchy of selection, with the less costly and easily measured traits selected for first, and the more complex, difficult, and costly traits selected later. The rationale is that, unless the difficult traits are the only essential traits for the cultivar, there is no reason to select for the difficult traits until the easily selected and important traits have been selected. For example, assuming that a breeder wanted to select in a population for five traits, A, B, C, D, and E, and assuming each is controlled by an unlinked, single dominant gene (hence in an F_2 population ¾ of the population would express each trait: ¼AA:½Aa:¼ aa, or ¾A_:¼ aa), the total number of plants that would have all five dominant traits is: $(3/4)^5 = 0.24$ or about 24% of the population. In a population of 1000 F_2 plants about 240 would be selected. In practice, if one trait were very costly to select, the breeder would not screen all 1000 plants for the trait, but rather would screen for the first four traits [e.g. $(3/4)^4 = 32$ %] and would then screen those 320 plants for the fifth trait, thus saving ²/₃ of the cost of screening for the fifth trait. In crops where inbred lines are created either as cultivars or as parents for hybrids, inbreeding will increase the level of homozygosity (e.g. ½AA:½ aa in later generations). If selection can be done sequentially or can be postponed until later generations, fewer plants would need to be screened for the costly traits. For example in the F_6, the frequency of A_ is 33/64 (and the frequency of aa is 31/64). With no prior selection, the frequency of lines with the four easily selected traits would be $(33/64)^4 = 0.07$ or 7% of the population (about 70 plants out of 1000). Only these 70 plants would need to be screened, of which 33/64 would have the dominant allele for the fifth trait. Waiting until later generations, greatly reduces the cost of screening lines and is one of the reasons that plant breeders use the single seed descent breeding method or the conceptually related doubled haploid breeding method (Baenziger, 1996). This saving is very important because most plant breeders work with numerous populations, often more than 300 (hence more than 300,000 plants) at the same time. In the University of Nebraska wheat-breeding program, we work with about 700 F_2 populations (over 1,000,000 plants) and slightly fewer F_3 populations (over 2,000,000 plants). However, waiting to select in the later generations does lengthen the time required for selection. In a commercial world, if "time is money", the cost:benefit ratio will determine at which generation the lines are screened.

The reason why it is important to understand the power of visual selection, and the importance of good selection nurseries, and the size and scope of breeding programs, is that these become critical aspects of how modern technologies will be used in the 21st century. While some researchers (Stuber, Polacco and Senior, 1999) have described marker assisted selection (MAS) as being very important to future plant breeders, the plant numbers with which breeders must work, as well as their genetic understanding, must constrain the use of MAS. Currently, some commercial breeding companies are giving each of their breeding programs 60,000 molecular data points to assist with their breeding effort. The breeders decide how to allocate these assays. For example, they could screen 60,000 plants for one marker, 12,000 plants for 5 markers, 1,000 plants for 60 markers, or 1,000 plants for 12 traits, where each trait is controlled by 5 markers. It would be rare that more than a few traits (less than 10) would be screened at a time because population sizes work against selection (in the F_2, $(3/4)^{10} = 6\%$ of the population; in the F_6 $(33/64)^{10} = 0.1\%$ or one plant in a 1000), as does our ability to understand the complex genetics of traits (Bernardo, 2001). Breeders also decide in which generation the tests should be completed. Because there are only a limited numbers of assays that can be done, the selection or information hierarchy again becomes important, but in this case the hierarchy is often reversed because the cost of each assay is the same. If each trait were equally valuable, the most difficult traits for visual selection would be the ones that would be given the highest priority for molecular marker assays. Marker assisted selection is very applicable to back-cross breeding, where a few genes are added to an existing elite line that is currently deficient in the traits controlled by those genes. It becomes extremely powerful when the target gene(s) is recessive, hence is masked in the heterozygous stages.

It is hard to predict how new technology will drive MAS in plant breeding. With automated laboratories, the cost per data point is becoming progressively less expensive. In addition, as more genes are identified or marked, it is possible to develop a better understanding of gene action and how they interact with other genes and with the environment. A greater ability to understand the genes that affect productivity and to predict their resultant phenotypes in different environments will provide powerful support for plant breeders in selection, as well as in choosing parents (Chapman, Hammer, Podlich and Cooper, 2002; Cooper *et al.*, 2002).

Although MAS is an area of great activity and interest, other technologies are being developed that will also have great use for future plant breeding.

Two examples concern the evaluation of grain for specific traits, and automated seed sorting based on near infrared reflectance (NIR) or transmission (NIT) technology. NIR can be used to identify kernels with differing colour (Wang, Dowell and Lacey, 1999a; Wang, Dowell and Lacey, 1999b), starch properties (Delwiche and Graybosch, 2002), rye translocations (Delwiche, Graybosch and Peterson, 1999), disease (Dowell, Boratynski, Ykema, Dowdy and Staten, 2002; Dowell, Ram and Seitz, 1999), different protein contents (Bramble, Herrman, Loughin and Dowell, 2002), end-use quality attributes (Delwiche, Graybosch and Peterson, 1998), or vitreousness (Dowell, 2000). The advantages of NIR and NIT technologies are that multiple assays can be performed on the same sample and, with NIT, the grain sample is intact and would not need to be destroyed by grinding. With both technologies, the cost of sample preparation is relatively inexpensive and quick. Hence, the assays can be done for crops where there is a short turn-around time between harvest and planting and in early generations where seed supply is limited. With the ability to sort kernels on the basis of these traits, plant breeders will be able to enhance segregating populations for the traits that they desire, again between harvest and planting, even when there is little time.

One of the interesting dichotomies within the major crops, at least within the United States, was the relative interest in MAS vs. the interest in plant transformation. In maize, many breeders believed that there was ample genetic variation for plant breeding, and therefore the best use of new molecular technologies was to improve the understanding of genetic variation and to improve selection efficiency. This interest may have arisen in maize because this species is one of the earliest crops to have plentiful good molecular markers and defined heterotic germplasm pools, and because it is a somewhat recalcitrant species for plant transformation. In addition, maize being a hybrid crop, the ability and requirement to manipulate inbred lines may have channelled much of the breeding effort into MAS. In other crops, such as wheat, plant transformation to increase genetic variation had a greater interest, perhaps because initially molecular markers were difficult and there were clear genetic limitations (e.g. winter survival, etc.). This dichotomy may be changing as wheat breeders become frustrated with their inability to deploy transgenic crops and the development of very good molecular markers. In maize, much of the molecular information will be validated by transformation; hence maize geneticists now have a greater interest in maize transformation.

EVALUATION

In the final phase of plant breeding, evaluation of elite lines, plant breeders will still want to have as many evaluation sites as they can afford. Nothing will change regarding the desire to have as much field data as possible to evaluate and describe a new cultivar. The biggest changes in line evaluation will concern data collection and analysis. In wealthy countries where capital is relatively inexpensive and labour is expensive, mechanisation will increase the efficiency of planting, harvesting, data collection and data analysis. Currently grain-harvesting combines are able to cut two plots at a time, weigh them, and sub-sample (i.e. store grain from) those that were predetermined (e.g. check cultivars) or that meet the selection criteria. Unselected samples are harvested and discarded into a bulk storage collection bin. How the data will be analysed will also change, so that more information can be obtained from the same number of experimental units (e.g. Landes, 2002; Stroup, Baenziger and Mulitze, 1994), and the environments will be better understood, so improving prediction of how lines might perform in other environments. In understanding our evaluation environments and how they may relate to future production environments, crop modelling may play a major role (Chapman *et al.*, 2002). Plant breeders will want to become smarter in their evaluation processes so they can get the maximum amount of information from each trial. Models may help create a better understanding of GEI, hence may allow testing to be more predictive. Regardless of the number of evaluation sites in which breeders can test their materials, sites are always insufficient to represent the diverse sites in which the cultivar will be grown.

Who will be the plant breeders in the 21st century?

As described above, the three phases of plant breeding are the creation of novel genetic variation, selection of useful variants (often coupled with inbreeding), and the evaluation of elite lines for agronomic performance and other traits of economic value. Plant breeding is an expensive enterprise and is becoming more expensive (Frey, 1996; Duvick, 2003). For crops where returns can be made on investment, the private sector, as is to be expected, is a dominant force. Duvick (in press) suggested that 43% of the world's maize crop is planted with private sector cultivars (27% in the developing world and 99% in the industrialised world). For soybeans, private sector cultivars are grown on 45–75% of the world's land (70–90% in the industrialised countries and 30-60% in the developing world). For wheat, private sector cultivars are grown on 9% of the world's land (27% in the industrialised countries and 3% in the developing countries). For rice, less than 1% of the world's land (developing or industrialised) is planted to private sector lines. The importance of the private sector in plant breeding and the ability of the private sector to invest in crop breeding can be seen in the last census of plant breeders in the U.S. (Frey, 1996). In 1994, there were approximately 900 cereal breeders in the U.S. Of those, 600 breeders were involved in maize breeding and 93% of the 600 were private sector plant breeders. Basically there were twice as many maize breeders as there were breeders for all of the other cereal crops grown in the U.S. and virtually all of the maize breeding was done in the private sector. As plant breeding costs increase due to new technology (genomics, machinery, and information technology) and to intellectual property rights, including proprietary information, it is becoming more difficult for new companies to start up, and for public sector and private sector breeders in smaller companies to continue. These costs are a barrier to entry for new businesses and for the continuation of smaller private and public sector breeding research. Our understanding of the current state of plant breeding is that in Europe, much of the plant breeding is done in the private sector or in institutes that have commercial ties. In Australia and in India, much of their plant breeding is also becoming more commercially orientated.

For plant breeders in smaller companies and in the public sector, there are strategies for continued success. As mentioned previously, most plant breeders continue to develop new cultivars and inbred lines by making a cross, selecting, and undertaking careful evaluation. The advantages of size and the requirement for large resources are often less in conventional breeding than in the adoption of new molecular technologies, especially when considering the regulatory costs of bringing a transgenic cultivar to market. In addition, conventional plant breeding works with thousands of genes that are not very amenable to molecular manipulation. Hence smaller companies and public sector breeders can build the genetic platforms for new cultivars and work with the larger seed or biotechnology companies for access to their traits and to technologies to create the finished cultivar.

As plant breeding becomes more privatised, the education of future plant breeders will evolve. In many crops, university programs no longer have the size, scope and breadth of private plant breeding organisations. As such, many new plant breeders educated at universities work initially with established plant breeders to learn the company protocols before they become independent or senior plant breeders. In many ways, these early years with a company are similar to an apprenticeship where the new employee can learn plant breeding on the grand scale. In the future, educational institutions will probably have closer ties to companies so that the students can be educated in programs with the scale and scope in which they may eventually work.

A second aspect of the increasing privatisation of plant breeding is that it will become increasingly global in its efforts. The shuttle breeding that made CIMMYT famous and the transfer of germplasm from Mexico to Pakistan and India, may well be a harbinger of how plant breeding will happen in future. For similar agro-ecological zones, plant breeding can be located where there is a critical mass of plant breeders and where breeding is resource efficient. Already, it is believed that considerable vegetable breeding is undertaken in India where there are numerous vegetarians and a huge market for improved vegetables. Materials from this activity can then be transferred to similar regions outside of India for testing and for potential release. While the evaluation of cultivars is site-specific, the creation of genetic variation and of the initial selections can be done in facilities, where it is most cost efficient, within major eco-geographic regions but representative of similar regions elsewhere.

Critical needs for plant breeding in the 21st century

Plant breeders in the 21st century must know whether the crop they are breeding has reached its genetic maximum yield in high yielding environments, or if further genetic improvements can be made. Maximum yield potential may be a circular idea. With current cultural practices or management systems, the maximum grain yield in high yielding environments may have been reached but with different management systems a higher yield might be achieved. Certainly plant breeders cannot select for maximum grain yield in high yielding environments if they cannot test their lines in such environments.

Similarly, plant breeders need better definitions of the world's eco-geographic zones so that they can identify germplasm for specific regions that would contribute to a global crop improvement effort. We also need the better tools to describe our selection and evaluation environments so that we extrapolate our data better from the limited number of evaluation sites to the multitude of potential production sites.

There also needs to be a common consensus of the role of technology in crop improvement. It is somewhat surprising that, globally, transgenic soybeans are the predominant form of the crop, while there is no transgenic wheat or rice marketed commercially. Is this an unwitting design to reduce the genetic improvement of wheat and rice, thus making two of the world's staple crops "technology orphans"? Many small-acreage crops currently lack the 'critical mass' necessary to provide for competitive improvement; these are already considered "orphaned" crops, thus decreasing diversity and choice of economic crops.

Finally, we need to learn how to partner public sector resources with private sector resources so as to ensure the human and intellectual capital for future plant breeding needs. Educational institutions need to make sure that they are educating students to meet future needs, perhaps in partnership with both the large, global seed companies and the smaller seed companies. In addition, there are certain major questions in plant breeding that are too long-term for the private sector, but are appropriate for public sector investment. There are two ready examples in the development of hybrid wheat and non-hybrid maize.

Acknowledgement

Partial funding for the senior author is from USDA- IFAFS competitive grant 2001-04462 and USDA, NRICGP 00-353000-9266

References

Ahloowalia, B. S. and Maluszynski, M. (2001). Induced mutations—a new paradigm in plant breeding. *Euphytica*, **118**(2), 167-173.

Baenziger, P. S. (1996). Reflections on doubled haploids in plant breeding. *In Vitro Haploid Production in Higher Plants*, **1**, 35-48.

Baenziger, P. S. and Peterson, C. J. (1992). Genetic variation: its origin and use for breeding self-pollinated species. In *Plant Breeding in the 1990s*, eds. H. T. Stalker and J. P. Murphy, C.A.B. International, Wallingford, UK, pp. 69-92.

Bernardo, R. (2001). What if we knew all the genes for a quantitative trait in hybrid crops. *Crop Science*, **41**(1), 1-4.

Bramble, T., Herrman, T. J., Loughin, T. and Dowell, F. (2002). Single kernel protein variance structure in commercial wheat fields in western Kansas. *Crop Science*, **42**(5), 1488-1492.

Carpenter, J., Felsot, A., Goode, T., Hammig, M., Onstad, D. and Sankula, S. (2002). *Comparative Environmental Impacts of Biotechnology-derived and Traditional Soybean, Corn, and Cotton Crops*. Council for Agricultural Science and Technology, Washington, D.C. USA.

Cassman, K. G. (1999). Ecological intensification of cereal production systems: yield potential, soil quality, and precision agriculture. *Proceedings of the National Academy of Science U S A*, **96**(11), 5952-5959.

Cassman, K. G., Dobermann, A., Walters, D. T. and Yang, H. (2003). Meeting cereal demand while protecting natural resources and improving environmental quality. *Annual Review of Environment Resources*, **28**, 315-58.

Chapman, S. C., Hammer, G. L., Podlich, D. W. and Cooper, M. (2002). Linking biophysical and genetic models to integrate physiology, molecular biology, and plant breeding. In *Quantitative Genetics, Genomics, and Plant Breeding*, ed. M. S. Kang, CABI Publishing. New York, New York, USA, pp. 400.

Cheng, M., Lowe, B. A., Spencer, T. M., Ye, X. and Armstrong, C. A. (2004). Factors influencing *Agrobacterium*-mediated transformaton of monocotyledonous species. *In Vitro Celluar Developmental Biology Plant*, **40**, 31-45.

Cooper, M., Podlich, D. W., Micallef, K. P., Smith, O. M., Jensen, N. M., Chapman, S. C. and Kruger, N. L. (2002). Complexity, quantitative traits and plant breeding: a role for simulation modeling in the genetic improvement of crops. In *Quantitative Genetics, Genomics, and Plant Breeding*, ed. M. S. Kang, CABI Publishing. New York, New York, USA, pp. 400.

Delwiche, S. R. and Graybosch, R. (2002). Identification of waxy wheat by near-infrared reflectance spectroscopy. *Journal of Cereal Science*, **35**, 29-38.

Delwiche, S. R., Graybosch, R. A. and Peterson, C. J. (1999). Identification of wheat lines possessing the 1AL.1RS or 1BL.1RS wheat-rye translocation by near-infrared reflectance spectroscopy. *Cereal Chemistry*, **76**(2), 255-260.

Delwiche, S. R., Graybosch, R. A. and Peterson, C. J. (1998). Predicting protein composition, biochemical properties, and dough-handling properties of hard red winter wheat flour by near-infrared reflectance. *Cereal Chemistry*, **75**(4), 412-416.

Dobermann, A., Dawe, D., Roetter, R. P. and Cassman, K. G. (2000). Reversal of rice yield decline in a long-term continuous cropping experiment. *Agronomy Journal*, **92**, 633-643.

Dowell, F. E. (2000). Differentiating vitreous and nonvitreous durum wheat kernels by using near-infrared spectroscopy. *Cereal Chemistry*, **77**(2), 155-158.

Dowell, F. E., Boratynski, T. N., Ykema, R. E., Dowdy, A. K. and Staten, R. T. (2002). Use of optical sorting to detect wheat kernels infected with Tilletia indica. *Plant Disease*, **86**(9), 1011-1013.

Dowell, F. E., Ram, M. S. and Seitz, L. M. (1999). Predicting scab, vomitoxin, and ergosterol in single wheat kernels using near-infrared spectroscopy. *Cereal Chemistry*, **76**(4), 573-576.

Duvick, D. N. (2003). The current state of plant breeding: how did we get here? *Summit on Seeds and Breeds for the 21st Century*, Washington, D.C.

Duvick, D. N. (1984). Genetic contributions to yield gains of U.S. hybrid maize, 1930 to 1980. In: *Genetic Contributions to Yield Gains of Five Major Crop Plants*. (ed. Fehr W.R.) Madison, WI: ASA and CSSA, 15-47 b CSSA Spec. Publ. 7.

Duvick, D. N. and Cassman, K. G. (1999). Post-Green Revolution trends in yield potential of temperate maize in the North-Central United States. *Crop Science*, **39**(6), 1622-1630.

FAOSTATdata. (2004). Vol. February 2004.

Fehr, W. E. (1987). *Principles of Cultivar Development*. MacMillan Publishing Company, New York, USA.

Fehr, W. E. and Hadley, H. H. (1980). *Hybridization of Crop Plants*. American Society of Agronomy and Crop Science Society of America, Madison, WI.

Frey, K. J. (1996). *National Plant Breeding Study-1 Human and Financial Resources Devoted to Plant Breeding Research and Development in the United States in 1994*. Iowa Agriculture and home Economics Experiment Station, Ames, IA.

Gepts, P. (2002). A comparison between crop domestication, classical plant breeding, and genetic engineering. *Crop Science*, **42**(6), 1780-1790.

Gianessi, L., Silvers, C., Sankula, S. and Carpenter, J. (2002). *Plant Biotechnology - Current and Potential Impact for Improving Pest Management in US Agriculture. An Analysis of 40 Case Studies*. National Center for Food and Agricultural Policy.

Ho, J. C., McCouch, S. R. and Smith, M. E. (2002). Improvement of hybrid yield by advanced backcross QTL analysis in elite maize. *Theoretical and Applied Genetics*, **105**(2/3), 440-448.

Kern, M. (2002). Food, feed, fibre, fuel and industrial products of the future: challenges and opportunities. Understanding the strategic potential of plant genetic engineering. *Journal of Agronomy and Crop Science*, **188**(5), 291-305.

Kong, Q., Richter, L., Yang, Y. F., Arntzen, C. J., Mason, H. S. and Thanavala, Y. (2001). Oral immunization with hepatitis B surface antigen expressed in transgenic plants. *Proceedings of the National Academy of Science U S A*, **98**(20), 11539-11544.

Landes, R. D., K. M. Eskridge, P. S. Baenziger, D. B. Marx. (2002). Are Spatial Models needed with Adequately Blocked Field Trials? *Applied Statistics in Agriculture,* p. 234-246. G.A. Milliken, Ed. Kansas State University, Manhattan, KS.

Liu, S. and Anderson, J. A. (2003). Marker assisted evaluation of Fusarium head blight resistant wheat germplasm. *Crop Science*, **43**(3), 760-766.

Lucca, P., Hurrell, R. and Potrykus, I. (2001). Genetic engineering approaches to improve the bioavailability and the level of iron in rice grains. *Theoretical and Applied Genetics*, **102**(2/3), 392-397.

Maluszynski, M. (1990). Induced mutations—an integrating tool in genetics and plant breeding. *Stadler Genetics Symposium* (19th), 127-162.

Obert, J. C., Riddley, W. P., Schneider, R. W., Riordan, S. G., Nemeth, M. A., Trujillo, W. A., Breeze, M. L., Sorbet, R. and Astwood, J. D. (2004). The composition of grain and forage from glyphosate tolerant wheat Mon 71800 is equivalent to that of conventional wheat (*Triticum aestivum* L.). *J. Agric. Food Chem.*, in press.

Peterson, R. K. and Arntzen, C. J. (2004). On risk and plant-based biopharmaceuticals. *Trends in Biotechnology*, **22**(2), 64-6.

Specht, J. E. and Williams, J. H. (1984). Contribution of genetic technology to soybean productivity—retrospect and prospect.

Stroup, W. W., Baenziger, P. S. and Mulitze, D. K. (1994). Removing spatial variation from wheat yield trials: a comparison of methods. *Crop Science*, **34**(1), 62-66.

Stuber, C. W., Polacco, M. and Senior, M. L. (1999). Synergy of empirical breeding, marker-assisted selection, and genomics to increase crop yield potential. *Crop Science*, **39**(6), 1571-1583.

Wang, D., Dowell, F. E. and Lacey, R. E. (1999a). Predicting the number of dominant R alleles in single wheat kernels using visible and near-infrared reflectance spectra. *Cereal Chemistry*, **76**(1), 6-8.

Wang, D., Dowell, F. E. and Lacey, R. E. (1999b). Single wheat kernel size effects on near-infrared reflectance spectra and color classification. *Cereal Chemistry*, **76**(1), 34-37.

Ye, X., Al Babili, S., Kloti, A., Zhang, J., Lucca, P., Beyer, P. and Potrykus, I. (2000). Engineering the provitamin A (beta-carotene) biosynthetic pathway into (carotenoid-free) rice endosperm. *Science*, **287**(5451), 303-305.

Zhou, H., Berg, J. D., Blank, S. E., Chay, C. A., Chen, G., Eskelsen, S. R., Fry, J. E., Hoi, S., Hu, T., Isakson, P. J., Lawton, M. B., Metz, S. G., Rempel, C. B., Ryerson, D. K., Sansone, A. P., Shook, A. L., Starke, R. J., Tichota, J. M. and Valenti, S. A. (2003). Field efficacy assessment of transgenic Roundup Ready wheat. *Crop Science*, **43**(3), 1072-1075.

11

FUTURE WHEAT YIELDS: EVIDENCE, THEORY AND CONJECTURE

ROGER SYLVESTER-BRADLEY[1], JOHN FOULKES[2] AND MATTHEW REYNOLDS[3]
[1] *ADAS Centre for Sustainable Crop Management, Boxworth, Cambridge, CB3 8NN, UK;* [2] *Division of Agricultural and Environmental Sciences, School of Biosciences, University of Nottingham, Sutton Bonington Campus, Loughborough, Leics., LE12 5RD, UK;* [3] *CIMMYT, Apartado Postal 6-641, 06600 Mexico, D.F., Mexico*

Introduction

Progress in improving the productivity of wheat (*Triticum aesivum* L.) has probably been more successful and sustained than for any other farmed species. A tripling of UK wheat yields (Figure 11.1) has come to exemplify the enormous agricultural progress in temperate countries over the past half century, and has been used both to laud and denigrate the technological and economic agents responsible: high output agriculture is seen to be both good and bad.

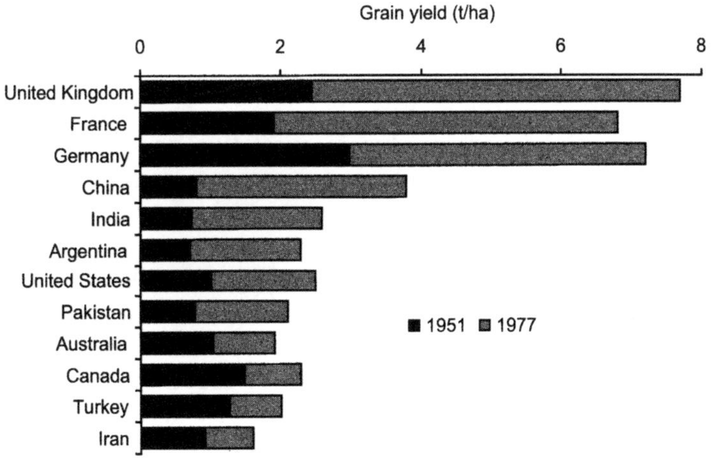

Figure 11.1 Estimated wheat yields in 1951 and 1997 for the 12 countries producing more than 10Mt in 1997, sorted according to their increase in yield. Data from Pingali (1999).

Much has been written about the causes of this 'progress', but attempts to extrapolate into the future are rare. This paper aims to account for changes in wheat's productivity in terms that enable us to form some tentative expectations of trends in wheat productivity over the coming century. Certainly in technological terms the high yields of modern wheat crops can be considered a resounding success. It was and is the main aim of technologists concerned with breeding wheat, developing its mechanisation, agrochemical use and fertilisation, and its on-farm husbandry, to raise its yield as far as can be justified economically. Advances in grain quality and control of diseases, pests and lodging have accompanied those in yield, but they have been secondary in extent.

Wheat is discussed elsewhere in these proceedings in relation to advances in technology (Legg), genetics (Mayes, Holdsworth, Pellegrineschi and Reynolds) and breeding (Baenziger, Russell, Beecher and Graybosch). The approach taken here is primarily physiological. What have been the major drivers of the massive heist in yield, and how can we expect these to change in future? What do we know about the limits to growth of annual species such as wheat in the UK? What have been the changes in form and function of the wheat crop that have been associated with its increased yields? How much further can we expect to be able to change the wheat phenotype, through plant breeding and crop management, such that it's productivity approaches its potential? What will be the technological challenges in working towards this potential, and what repercussions would be entailed for resource use, and for the environment in which wheat is grown? In seeking answers to these questions, the UK environment will be our prime focus, but reference will be made to evidence from other wheat-growing regions.

Yield trends past and future

Of the world's major wheat producers, west European countries stand out as having achieved the largest yield improvements over the past half century (Figure 11.1). Considering the extent of yield improvements by all the major wheat producers, it appears that success has not so much been determined by their state of technological sophistication as by suitability of their environment for wheat growth. Of the countries at the forefront of technological innovation throughout the period, yields in Western Europe have largely been constrained by available light energy, whereas those in the United States, Canada and Australia have been largely constrained by available water. Larger improvements have been achieved in China and India, where irrigation has been extended, than in the North America or Australia. Clearly it has proved easier to overcome the challenge of light limitation than that of water limitation; progress in overcoming water limitation will be a crucial consideration when imagining prospects for future yields.

The UK has achieved the largest improvement in wheat yield of the major producing countries, and UK data for the whole of the 20th century are presented in Figure 11.2. Whilst there is obvious short-term variation which must largely be accounted for by weather, and which will be considered later, there is an obvious underlying long-term trend which must, at its simplest, be seen as curvilinear. In previous analyses of yield trends, whether genetic or phenotypic, it has been conventional to consider these as linear (e.g. Calderini, Reynolds and Slafer, 1999). However, a linear analysis is not satisfactory if we are to extrapolate over significant periods outside the 20th century. Our approach in Figure 11.2 has therefore been to take four possible socio-economic scenarios, all leading to stable wheat yields, but ranging from extreme pessimism to extreme optimism.

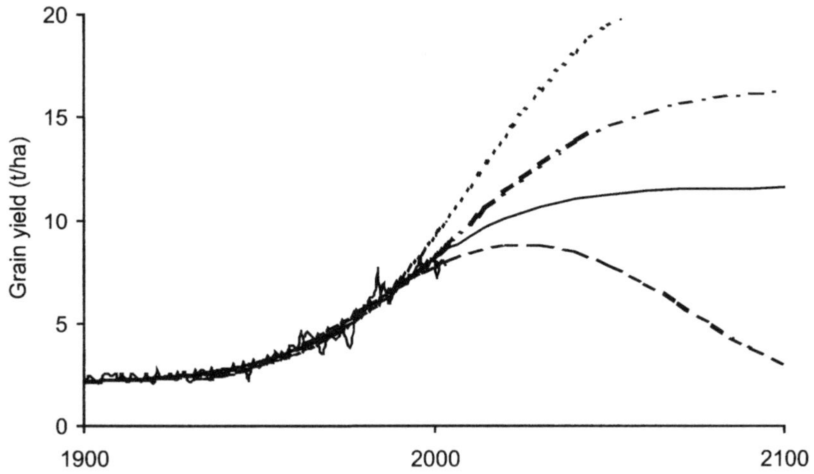

Figure 11.2 Mean grain yields of wheat in the UK over the 20th century, and four projections for the 21st century based on a sine curve and three logistic curves giving maxima of 8.8, 11.6, 16.3 and 21.9 t/ha in 2020, 2080, 2090 and 2100 respectively. For all curves, the mean of the residuals for all 20th century data is zero (sine curve from 1930).

The pessimistic scenario is that the global scientific, technological and economic revolutions that have gave rise to yield increases in the 20th century are temporary phenomena: two world wars and subsequent aspirations for global communication and interdependence gave rise to conditions which encouraged high output cropping but which depended on unsustainable exploitation of resources, and were associated with unstable political, economic and climatic conditions. We are reaching a turning point when science becomes distrusted, global communications are disrupted by terrorism, transportation costs become prohibitive, the weather becomes too unstable for reliable crop growth, and there are ever strengthening 'green' socio-political pressures to localise trade and social contacts. By the end of the 21st century we can expect to have reverted to wheat growing practices somewhat similar to those of the 19th and early 20th centuries, and best exemplified by current 'organic' production systems where average yields, taking the associated crop failures and fallowing into account, are probably of the order of 3 t/ha/year (Lamkin, Measures and Padel, 2004; Poudel, Horwath, Lanini, Temple and van Bruggen, 2002). The rate at which these changes might take place is debatable; if we set them as symmetrical with the yield trend since 1930 then we can expect yields to maximise in 2020 and start decreasing after 2030, reaching 3 t/ha by 2100.

On the other hand, the optimistic scenario is that an era of global peace and prosperity emerges in which wheat demand is driven ever higher by expanding global bread consumption, and increased use as a feedstock for global production of white meat, biofuels and renewable materials. The synergies between science and technology, and between plant breeding and crop husbandry are increasingly recognised and are given full reign, benign genetic solutions are found to the problems now solved with unsustainable chemistry, winters become milder, summer rainfall becomes less erratic, increases in carbon dioxide enhance photosynthesis and wheat yields increase rapidly until they begin to meet the physical limits of available light energy and water, and the physiological limits of energy conversion, water use and harvest index. Given unconstrained scope for genetic modification, these limits may only apply because the crop is losing its integrity as *Triticum aestivum*! Again, the rates at which such changes could take place are very uncertain, but the physical and physiological limits, hence the most probable of the three positive scenarios in Figure 11.2, are subject to more precise estimation.

Morris, Audsley and Rickard discuss global scenarios such as these in more detail elsewhere in this volume, so they will not be developed further here. Rather, the prime task for this paper is to define physical and physiological limits to wheat productivity in the UK, and then to consider the technological issues that might be raised in attempting to reach such limits. The impacts of increasing carbon dioxide and changing climate will largely be ignored so that the discussion retains its main focus on technological issues. It is possible that, by tracing the technological path to achievement of maximum yields, a picture may emerge of the rate at which such changes could take place. With this in mind, it will be pertinent to contrast the challenges of the future with those that have been addressed successfully by plant breeding and husbandry over the past half century (Silvey, 1986; 1994; Ingram, Macleod and McCall, 1997; see also Legg, this volume).

Two world wars, and an intervening agricultural depression, must account for the static wheat yields from 1900 until about 1950. With low world prices and little government support for agriculture until the Second World War, there was little stimulus to invest in farming. However, this changed during the '40s. In a push for national self-sufficiency, price support brought greater market stability, subsidies were introduced for land improvements such as drainage, hedge removal and liming, and successive governments invested heavily in research and technology, including plant breeding. Selective hormone herbicides were introduced in the '40s, the rate of variety introduction escalated in the '50s, increased use of phosphate and potash fertilisers reduced crop deficiencies, myxomatosis reduced rabbit populations by 90% and, by 1964, wheat yields had doubled from the pre-war average to 4.6 t/ha. A downturn in yields towards the end of the '60s was attributed to a breakdown in resistance to yellow rust (*Puccinia striiformis*) in the widely grown variety Rothwell Perdix and, as detailed in the Strutt Report (MAFF, 1971), a deterioration in soil conditions after a series of wet autumns. The introduction of effective fungicides in the late '60s and their more widespread use through the '70s, together with the introduction of semi-dwarf varieties and heavier use of nitrogen (N) fertilisers, combined to bring a fast period of yield increase, interrupted by the 1976 drought, but culminating with the exceptionally high yields of the 1984 harvest (7.7 t/ha; 2.9 t/ha above the linear trend). More recent progress has been slower. Breeders have continued to increase yield at about 1 t/ha per decade as well as making big improvements in quality, but advances in husbandry were more concerned with improving efficiency: reducing labour by further mechanisation, reducing nutrient leaching by improved fertiliser timing, introducing agrochemicals with better efficacy, adjusting to the ban on burning straw, and reducing costs of inputs such as seeds.

Since the turn of the 21st century, mean UK wheat yields have fluctuated around 8 t/ha. It seems unlikely that this represents the turning point of a long-term logistic curve of yield improvement (Figure 11.2). The low yields from the 2001 harvest can be accounted for by seasonal factors, particularly autumn rain that seriously affected crop establishment. Trials of new varieties show a genetic component of yield progress that remains as strongly positive as ever (Legg, this volume). Also, it does not appear that the 2 t/ha difference is diminishing between average yields in these trials (managed to express full potential of the site and season by minimising diseases, weeds and lodging, and with supra-optimal fertiliser N) and national average yields. It thus seems doubtful that we are yet approaching an upper yield ceiling.

Potential wheat yield in the UK

POTENTIAL CROP GROWTH

Consideration of prospects for UK crops at a Nottingham Easter School would not be complete without reference to Monteith's work. His much-quoted paper on efficiency of British crops (Monteith, 1977) estimated the potential for total growth in the UK to be 46 t/ha[1] on the basis of average insolation and a

[1] It is not entirely clear at what point Monteith's analysis made allowance for root dry matter, but we choose to do this by reducing the 54 t/ha in his Table 2 for 'DM equivalent of net photosynthesis' to 46 t/ha, which is then equivalent to ε being 2.8 g/MJ PAR, as adopted elsewhere in his paper.

conversion efficiency (ε) equivalent to 2.8 g/MJ photosynthetically active radiation (PAR). The shortfall shown by the best crop yields at that time was attributed in part to inhibition of photosynthesis by low temperatures in winter and by droughts in summer, but largely to the short duration of full light interception (-25 t/ha), inability to harvest or utilise straw and chaff (-10 t/ha), and uncontrolled stresses (-6.5 t/ha). In reviewing this analysis today, the main deficiency appears to concern availability of water; the constraint of available rainfall was not separated out. In attempting to consider rainfall and light limitation together, the following analysis counts all of annual rainfall as available. This is on the grounds that local water is already used occasionally for irrigation, but that the inherent status of crops as staple feedstocks will never justify long-distance transfers of water (such as occurred when the Central Valley Project was developed for the emerging Californian citrus industry in the late 19th century).

Long-term average rainfall in the UK varies from 550 mm in the east to more than 1000 mm in the south west, exceeding spatial variation in long-term average solar radiation which ranges from 1.6 GJ/m^2/year PAR in the north to 1.8 GJ/m^2/year in the south and west. Thus a regional analysis is required when considering crop growth in relation to both light and rainfall. Taking ε at 2.8 g/MJ PAR and water-use efficiency at 5 g/litre (or 5 g/m^2/mm; Green, Vaidyanathan and Hough, 1983; French and Schultz, 1984), crude estimates of potential annual growth in above-ground dry matter are presented in Figure 11.3. It is clear that, assuming full light interception throughout the year, and availability of all annual rainfall, the area in the east where arable cropping currently predominates would first be limited by rainfall rather than by incident solar radiation. The distribution shows considerable geographic variation, with greatest potential growth in the Southwest, and least in central East Anglia, where a significant area has estimated average potential growth of between 30 and 35 t/ha.

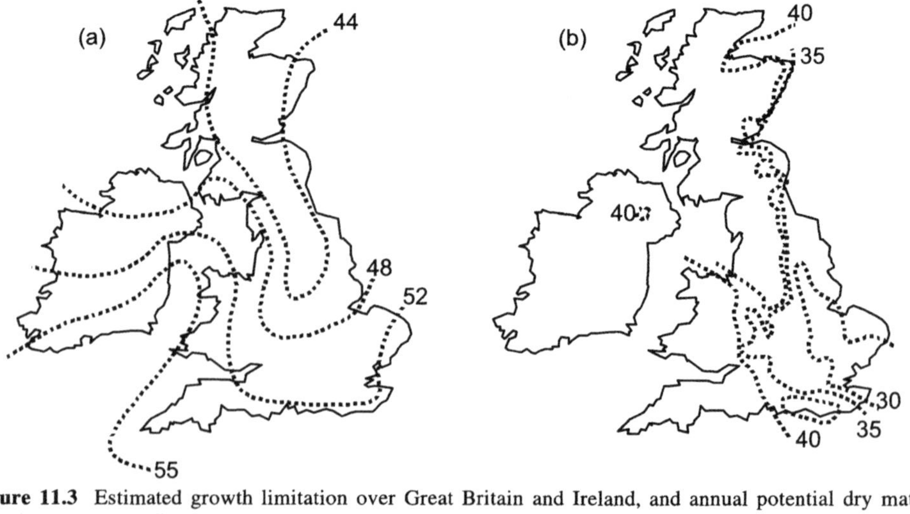

Figure 11.3 Estimated growth limitation over Great Britain and Ireland, and annual potential dry matter growth assuming all light energy (a) and all annual rainfall (b) are used at dry matter conversion efficiencies of 2.8 g/MJ PAR and 5.0 g/l rainfall respectively.

For annual crops such as wheat it is not possible to maintain light interception for a full year: an interval must occur between canopy death of one crop and canopy closure of the next. Climatic requirements for reliable annual re-establishment (canopy senescence, grain ripening, harvest, soil cultivation, re-sowing, and canopy expansion) dictate that this interval must include at least some of the period of high insolation. For irrigated spring wheat it has been estimated that 28% of annual insolation could not ever be used for photosynthesis (Reynolds, van Ginkel and Ribault, 2000), whereas for wheat in the cooler, duller, unirrigated conditions of western Europe we estimate that about 40% of annual insolation is 'unusable'.

A belief in the stability of ε, encouraged by Monteith (1977) and Gallagher and Biscoe (1978), has proved enormously valuable, by focussing attention of the more tractable aspects of crop improvement. However, through numerous subsequent studies, there is now good evidence of variation in ε for field-grown wheat, and some recent genetic improvement is clear for pre-anthesis ε in the UK (e.g. Shearman, Sylvester-Bradley and Foulkes, in press). Looking at genetic advances elsewhere in the world, e.g. those reported for CIMMYT spring wheat releases, it appears that they have been associated with changes in pre-anthesis and post-anthesis growth rates (Sayre, Rajaram and Fischer, 1997) involving increased ε. Higher yielding lines showed increased sink capacity and higher rates of photosynthesis during grain-filling (Fischer, Rees, Sayre, Lu, Condon and Larque Saavedra, 1998), and lines containing the 7Ag.7DL translocation showed on average a 10% increase in final biomass compared to check lines (with the same background and no translocation). These were associated with an increased partitioning of biomass to spikes at anthesis (sink strength) and larger values of ε post-anthesis compared to checks, as indicated by larger growth rates and leaf photosynthetic rate (Reynolds, Calderini, Condon and Rajaram, 2001; Reynolds, Pellegrineschi and Skovmand, 2005). As far as crop management is concerned, ε can be influenced by leaf N content (Bindraban 1999; Dreccer, van Oijen, Schapendonk, Pot and Rabbinge 2000) and plant density (Whaley, Sparkes, Foulkes, Spink, Semere and Scott, 2000). It therefore seems reasonable to expect that the mean value for ε for the period of grand growth could change from typical current values of 2.2 g/MJ to eventually reach Monteith's value of 2.8 g/MJ. Indeed, values of this order have already been observed, on occasion, in modern wheat crops (e.g. Foulkes, Scott and Sylvester-Bradley, 2001).

With potential annual light interception at 60% and ε at 2.8 g/MJ, potential long-term average above-ground growth of wheat in the UK must range from about 27 t/ha in the north east to 31 t/ha in the south and west. Final total (above-ground) dry matter of wheat under trial conditions ranges between 13 and 21 t/ha (Spink, Foulkes, Gay, Bryson, Berry, Sylvester-Bradley, Semere, Clare, Scott, Kettlewell and Russell, 2000) so, even in East Anglia, there appears to be ample scope for greater growth before any limit is reached. It is now necessary to translate this picture of potential growth, into a potential for grain yield. Essentially, the issue is the determination of harvest index but, in contrast to the approach Gallagher and Biscoe took in 1978, it must now be accepted that harvest index varies widely, for example from 35 to 55% for one variety over 18 site-seasons (Sylvester-Bradley, Scott, Clare and Duffield, 1998), so our approach is to look more closely at the utilisation of dry matter in determination of grain yield.

DETERMINATION OF WHEAT YIELD

Physiologists have reached a strange dichotomy in their explanation of wheat yields. Those concerned with simulation modelling have generally emphasised climatic, edaphic and cultural effects, and have come to consider that grain growth is source-driven (e.g. Jamieson, Semenov, Brooking and Francis, 1998), whilst those concerned with explaining genetic effects have come to consider grain growth as being sink-driven (Evans, 1978; Calderini, Dreccer and Slafer, 1995; Shearman *et al.* in press; Reynolds, Pellegrineschi and Skovmand, 2005). Probably the best resolution here is to treat yields as being ultimately source-limited, but recognising a continuous inter-play between sources and sinks such that the level of source at any point in development will influence subsequent sink size, and *vice versa*. Unfortunately there is an implication here of a need for rather sophisticated calculations which, even if each component is considered proven, when integrated into computerised models, may produce inherently imprecise or unstable predictions (Passioura, 1996). It is only in climates where yield variation is large, and generally attributable to one dominant factor, such as available water, that physiological explanations have proved generally acceptable for use by industry (e.g. in Australia; French and Schultz, 1984). In the UK, it is currently the case that quantitative physiological models of variation in wheat yields are still too imprecise to have been adopted for practical use. It is therefore necessary to be cautious in assuming an ability to extrapolate on physiological grounds.

In an attempt to explore the dominant climatic influences on UK wheat yields, a new empirical analysis of seasonal yield variation was undertaken here. We took annual deviations from the linear trend in national average wheat yields between 1978 and 2002, and related these to weather data for each cropping year (September to August) obtained from a site in central East Anglia (Marham, Norfolk) which is representative of the main wheat growing area in the UK. The aim was to find the periods during wheat growth when weather had most effect on yield variation and, by identifying the weather variables of most influence, to infer the growth processes that should be addressed in any yield prediction exercise. We used the 'Window Pane' data-mining approach of Coakley and Line (1978), developed by Pietraville, Shaw, Parker and van den Bosch (2003), whereby correlations between the deviations in yield were calculated iteratively for a series of weather parameters (minimum, maximum and mean air temperatures, rainfall, sun hours, total solar radiation, photo-thermal quotient, wind-speed, and potential soil moisture deficit), for a series of dates during the season (from 1 September to 31 July in 5 day increments), and for a series of periods (10, 20 and 30 days). Deviations in grain yield were initially taken for the 25 years since 1977, when semi-dwarf varieties were grown, fungicides and PGRs were widely used, and there was no sign of any short-term pattern in the data. The analysis was subsequently repeated for the 40-year period over which data were available. This approach identified 12 weather variables as being potentially correlated with seasonal deviations in grain yield (Table 11.1). Individually these explained no more than 30% of the variation, but the best combinations of parameters in a multiple regression could explain about 70% of variation for the 25 year period and 60% of variation for the 40-year period (Figure 11.4). Only four or five variables were required to maximise the proportion of variation explained in either the 25- or the 40-year dataset.

Table 11.1. WEATHER PARAMETERS RESULTING FROM WINDOW-PANE ANALYSIS OF UK WHEAT YIELDS: PARAMETERS HAVING SIGNIFICANT INDIVIDUAL CORRELATIONS WITH DEVIATIONS FROM THE LINEAR TREND IN AVERAGE YIELDS OVER THE LAST 25 OR 40 YEARS.

Weather parameter and summary period				*Correlation coefficient (r)*	
		Start	*Duration (days)*	*1978-2002*	*1960-2002[1]*
a	Minimum temperature (grass)	15 September	30	0.39	0.26[1]
b	photo-thermal quotient (PTQ)	20 October	10	0.55	0.30[1]
c	wind-speed	25 October	30	-0.49	-0.32[2]
d	Rainfall	4 November	20	-0.51	-0.49
e	Sun	8 January	20	0.24	0.32
f	wind-speed	8 January	20	0.47	0.24[2]
g	Minimum temperature (grass)	2 April	30	-0.47	-0.24[1]
h	wind-speed	2 April	30	-0.40	-0.39[1]
i	photo-thermal quotient (PTQ)	2 April	30	0.53	0.10[1]
j	mean temperature	1 May	30	-0.39	-0.20[1]
k	Minimum temperature (air)	21 June	20	-0.26	-0.54[1]
l	wind-speed	16 July	10	-0.49	-0.35[2]

[1] omitting four seasons from 1967 to 1970 when data from Marham were unavailable
[2] omitting 11 seasons from 1960 to 1970 when wind data were unavailable

Of course this empirical approach to yield variation may just give results unique to the years chosen, and some of the relationships may well be spurious. However, the results were corroborated by parallel examination of independent data from ADAS Boxworth (not reported here). As expected, there was no relationship with soil moisture deficit (range at the end of July 76-421 mm) but, surprisingly for a crop-environment that is understood to be light limited, neither was there a relationship with sunshine hours or solar radiation in summer (ranges for May 256-586 MJ/m^2, and June 358-638 MJ/m^2).

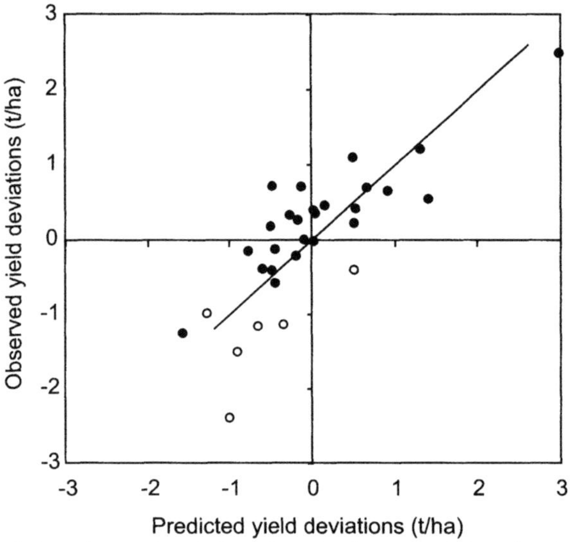

Figure 11.4. Best prediction of annual deviations from the linear trend in national wheat yields based on five aspects of weather listed in Table 11.1: 4.35 + 1.905**b* - 0.326**d* - 0.123**j* - 0.081**k* - 0.2876**l*. Closed symbols are for harvests in 1978-2002 (r^2=0.73); open symbols add in the years 1972-77 (r^2=0.63).

Of the several surprises in the parameters identified, there appears to be no explanation for a benefit from bright, windy weather during January. The only significant and interpretable parameter involving radiation was the photo-thermal quotient (PTQ); a relationship in late October seems rather tenuous, and perhaps it is just concerned with fewer late drillings, but PTQ during April might work by affecting initial sink size (and then sink-source synergy) for the period of stem extension, hence the rate of grand growth.

The five most independent parameters, hence the most useful as combined predictors, were those listed as *b, d, j, k* and *l* in Table 11.1. The negative influence of rainfall during November (range 8-87 mm) is assumed to arise through reducing plant establishment. The negative effect of mean temperatures during May (range 9-14°C) must be concerned with prolonging a crucial developmental phase when shoot numbers, spikelet fertility (Fischer 1985), water soluble carbohydrates (WSC) in stems and nodal rooting are all being determined. Low minimum temperatures in late June (range 8.6 to 14.2) are likely to signify reduced dark respiration at a time when crop biomass is reaching a maximum; this was a particular feature of 1984, the year giving the greatest positive deviation in yield. Lastly the negative effect of wind-speed in late July (range 2.7-5.3 m/s) is clearly concerned with lodging: widespread lodging was observed in most of the years with high mean wind-speeds during this period, and not in years with less wind.

Comparing this approach with yield prediction by simulation modelling, it is interesting to note that quite different aspects of seasonal weather are identified as being influential. Also, despite their perceived complexity, simulation models apparently are still not complex enough! Few simulation models of wheat are able to deal adequately with the dynamics of topsoil hydrology, hence effects of rainfall on seedling establishment, few consider sink effects (except perhaps indirectly by adopting a radiation-use efficiency which reflects feed-back inhibition by sinks on photosynthesis), few consider respiration independently of photosynthesis, and few consider lodging!

Thus, in seeking a realistic basis on which to analyse current wheat yields and move towards a prediction of wheat's ultimate potential, we have not resorted to simulation modelling. We choose to retain a belief that UK wheat yields generally depend upon interception of radiation, but that we must augment this with thoughts that sink determination and lodging resistance are also important.

POTENTIAL YIELD, AND CONSEQUENT CHANGES TO THE WHEAT PHENOTYPE

If potential growth of wheat in the UK is taken as 27 t/ha, the minimum of the range discussed earlier, the main question concerns how little of this dry matter must be invested in un-harvestable crop components? Since all post-flowering growth goes to grain, this quantity is best derived from a knowledge of the functions of the non-grain components at flowering. The optimum green area index (GAI) for a high yielding crop (using the approach of Sylvester-Bradley, Scott, Stokes and Clare, 1997, to optimise between extra cost of fertiliser N and return from extra grain DM) is about 7, with 4.5 being lamina, 2 as leaf-sheath and 0.5 as ear. Specific weight for culm leaves of wheat is relatively consistent at around 40 g/m^2 (Fisher *et al.*, 1998; Beed, Sylvester-Bradley and Paveley, unpublished), giving a leaf DM requirement of 2.6 t/ha.

The function of stem DM is in physical support, and storage of fructans. All fructans can be considered available for redistribution to grain (Foulkes *et al.,* 2002), so un-harvestable stem DM requirements are just for support. Berry, Sterling, Spink, Baker, Sylvester-Bradley, Mooney, Tams and Ennos (2004) suggest how a lodging-proof wheat ideotype might be designed in mechanical terms. Using this approach, the base bending moment on each stem created by a crop with a dry weight of 27 t/ha, with say 600 stems per m^2 and a height of 0.6 m is estimated to be 390 Nmm. Assuming the best stem material strength yet recorded for current UK germplasm (43 Mpa; Berry, Spink, Gay and Craigon, 2003), this requires a stem base diameter of 4.7 mm, a stem wall width of 0.9 mm, hence a DM requirement per unit length at the stem base of 3.0 mg/mm (the material density of 0.27 is taken from Mulder, 1954).

Given the reduction in bending moment with height, stem DM can diminish by 0.2% per mm, hence the DM requirement for structural stem is 6.0 t/ha. The equations used to make this estimate indicate that the penalty in DM, hence in potential yield, of increasing crop height is about 2 t/ha per 100 mm, so it will be important to balance this disadvantage of tallness with the advantages seen in terms of better light distribution, and in escape from disease inoculum (Lovell, Parker, Hunter, Royle and Coker, 1997).

The remaining above-ground crop component at flowering is the ear. Numerous studies show a relatively consistent requirement for floral DM of 8.5 mg/grain (e.g. Sylvester-Bradley *et al.*, 1997; Whaley *et al.*, 2000). Weight per grain will be discussed more thoroughly below but, if maximum average weight per grain is taken to be 60 mg, the minimum requirement for non-grain ear DM will be 14% of grain weight. Total potential grain DM will thus be

$$(27.0 - 2.6 - 6.0) \text{ x } (1.00 - 0.14) = 16.4 \text{ t/ha}$$

equivalent to a potential grain yield of 19.2 t/ha at 85% DM. The extent of changes needed through plant breeding and crop management to create a 'maxi-wheat' of this potential yield will now be considered by comparison with the growth and yield of current wheat crops.

DESIGNING A 'MAXI-WHEAT' PHENOTYPE

Wheat growth and yield were monitored closely (Gay, Stokes, Weightman and Sylvester-Bradley, 1998) over 27 site-seasons, using cv. Mercia from harvest 1993 to 1996 (Sylvester-Bradley *et al.*, 1998) and cv. Consort from 1996 to 1998 (Spink *et al.*, 2000). Sites in these studies were chosen to represent the range of climates and soils in which wheat is grown in the UK, and the husbandry regime was almost entirely successful in eliminating observable weeds, disease, pests and lodging. Medians taken from the data for Mercia were adopted as 'benchmarks' by Sylvester-Bradley, Watson, Dewes, Clare, Scott and Dodgson (1997); the benchmark yield was 9.1 t/ha, about half of the potential yield estimated above.

Benchmark patterns of light interception, above-ground growth and grain growth are shown in Figure 11.5. A GAI of 3 was reached on 1st May and grand growth then continued for 75 days at a rate of 0.2 t/ha/d. Set against incident PAR, and assuming an extinction coefficient (K) for the canopy of 0.46 (Thorne, Pearman, Day and Todd, 1988), this growth was equivalent to ε being 2.2 g/MJ PAR, with partitioning to grain resulting in a harvest index (HI) of 47%. Benchmark values can now be used to indicate the changes that must be made to achieve a 'maxi-wheat' phenotype for the UK, and to suggest whether these may most easily come about through plant breeding or crop management.

It appears eminently feasible that crops could be managed consistently to reach GAI 2.5 in mid April, GAI 4.0 on 1st May and that the maximum GAI reaches 7. This can largely be achieved by judicious use of seed rate and fertiliser N. Patterns such as this are commonly observed in current crops. As for canopy senescence, although there is less variation in this than in canopy closure, significant genetic variation has been shown (Miralles, Dominguez and Slafer, 1996; Verma, Foulkes, Caligari, Sylvester-Bradley and Snape, in press), and significant delays of 5-8 days have been achieved by use of strobilurin fungicides and urea (Ruske, Gooding and Jones, 2003). It therefore seems feasible that canopy senescence could be delayed by the requisite 8 days (Figure 11.5) so that, overall, the period of grand growth can be extended from 75 to 90 days.

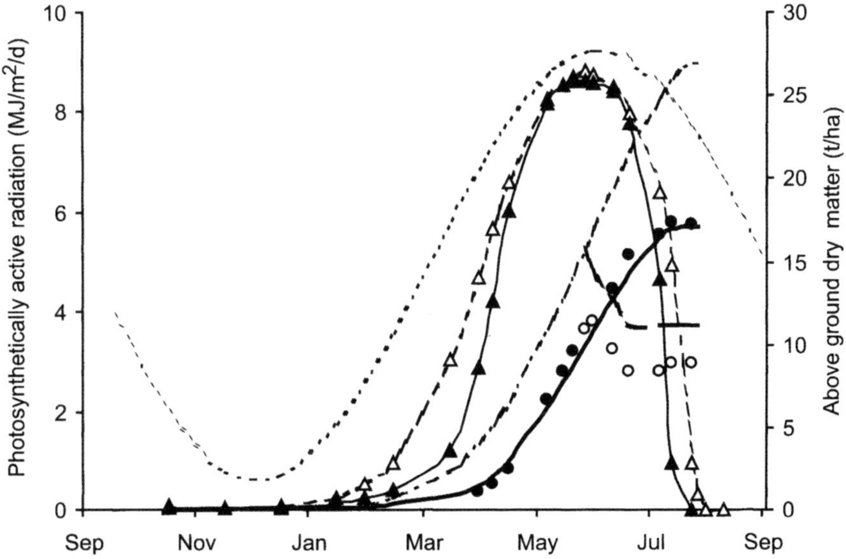

Figure 11.5. Physiological changes needed to double wheat yields in the UK. The seasonal time-course shows mean incident PAR (dotted), intercepted PAR (closed triangles, assuming a canopy extinction coefficient of 0.46), total above ground dry matter (closed circles), non-grain dry matter (open circles), and predicted total above ground dry matter (bold line, assuming PAR use efficiency of 2.2 g/MJ) for the UK 'benchmark' crop of wheat cv. Mercia yielding 9.1 t/ha grain (Sylvester-Bradley *et al.*, 1997). Dashed lines for a 'maxi-wheat' crop show intercepted PAR (open triangles) and predicted total above ground dry matter (assuming PAR use efficiency of 2.8 g/MJ; bold) and non-grain dry matter consistent with redistributed dry matter of 3.4 t/ha and a grain yield of 19 t/ha. PAR interception in 'maxi-wheat' crop assumes: GAI 2.5 in mid April, GAI 4.0 on 1st May, maximum GAI of 7, and canopy senescence delayed 8 days compared to the 'benchmark' crop.

Improvement of ε from 2.2 to 2.8 g/MJ PAR certainly appears feasible for growth pre-flowering (Shearman *et al.*, in press), but this will be a greater challenge after flowering, particularly during the latter stages of grain filling. Recent cases where ε has been estimated at this level have apparently been associated with varieties having the 1BL.1RS wheat–rye translocation

(Shearman *et al.,* in press) or the 7DL.7Ag wheat-*Agropyron elongatum* translocation (in spring wheat in Mexico, taking light saturated net CO_2 exchange rate of the flag leaf as an indicator of ε: Reynolds *et al.,* 2001), low plant populations (Whaley *et al.,* 2000), and high N nutrition (Dines, 1998). In general, it appears that high ε in light-limited conditions is brought about by having high specific leaf N at the top of the canopy (Evans, 1989), and high sink capacity (Evans and Wardlaw, 1996). If grain yield is to be doubled, sink capacity is bound to be enhanced at the end of the grand growth phase, however, if sink capacity is to be enhanced at the beginning of grand growth, it will probably be necessary to breed for an earlier start to stem extension, as has been proposed by Slafer, Abeledo, Miralles, Gonzalez and Whitechurch (2001). Overall, it seems that potential growth of 27 t/ha DM could be achieved, through combining husbandry and breeding, to jointly extend the duration of grand growth, and to maximise ε.

A number of mechanisms have been suggested for genetic improvement of ε (Reynolds *et al.*, 2000). On the source side, these include genetic transformation of Rubisco, manipulation of leaf angle (which is under relatively simple genetic control), and manipulation of leaf-N distribution within the canopy. On the sink side, recent evidence suggests that increased assimilation rate during the spike-growth stage, and a relatively larger partitioning of assimilates to the developing spike at anthesis were associated with higher yield in populations of recombinant inbred lines (Reynolds, Pellegrineschi and Skovmand, 2005). However, balancing source- and sink-strength is a complex genetical challenge. A crop will switch between source and sink limitation with phenological stage, and as conditions vary week by week, day by day, and even during the day. Thus any genetic improvement in source-sink balance must be robust to seasonal changes in environment; further analysis of the physiological basis of genotype by environment interactions may be needed to indicate the best avenues for genetic improvement.

Turning to DM partitioning, apart from increasing the duration of grain filling, there are two routes to enhancement of grain weight: (i) increasing the part of the growing season that is taken up by grain filling, by bringing flowering earlier in the life of the crop, or (ii) increasing the amount of pre-flowering assimilation that is redistributed to the grain, by enhanced deposition of fructans pre-flowering. Essentially these two routes are of equal utility in terms of grain DM accumulation, but advanced flowering has the disadvantages of attracting greater risk of frost damage in spring (Spink, Kirby, Frost, Sylvester-Bradley, Scott, Foulkes, Clare and Evans, 2000), and of dictating that leaves must live longer through grain filling and that grain filling must be prolonged beyond current experience. Possibly these are the reasons that there has been no detectable advance in flowering dates of UK varieties over the past 30 years, whilst there has been a significant trend for greater WSC in stems (Shearman *et al.,* in press). However, disadvantages in enhancing fructan deposition further may be that this will reduce availability of DM for concurrent formation of fertile florets (Blum, 1998). Current UK varieties have a 6-day range of flowering around the 12th June, with the exception of the French-bred variety Soissons, which possesses the *Ppd-D1* gene for photoperiod insensitivity, and flowers at the end of May. The success of Soissons in the UK encourages the belief that there is scope for advancing flowering to early June without incurring undue frost risk; it was only when Soissons reached stem extension in early March that significant frost damage has occurred (Whaley, Kirby, Spink, Foulkes and Sparkes, 2004). The pattern of total DM accumulation for the 'maxi-wheat' crop (Figure 11.5) indicates that about 14.5 t/ha DM would be accumulated by 4th June. This is sufficient for 3.5 t/ha WSC formation, in addition to the DM requirements for leaf, structural stem and ear already discussed, and is similar to the larger quantities of WSC in current varieties (Foulkes *et al.,* 2002). Thus, we suggest that the potential yield could be achieved, either by advancing flowering to early June, or by holding flowering in mid-June whilst seeking further enhancement of stem WSC to 5.5 t/ha by breeding.

Enhancing stem WSC is probably preferable to advancing flowering because it imposes less on leaf longevity. The flag leaf generally emerges (benchmark date: 24th May) about two phyllochrons before flowering. Its senescence is a little delayed but largely concurrent with that of the second leaf and the third leaf (counting back from the ear), so the benchmark pattern of canopy senescence (with GAI 3 on 13th July and GAI 1 on 25th July) indicates that current green duration for these leaves is 55 to 75 days. If the date of flowering remains on 12th June the need to prolong canopy life by 8 days appears realistic. However, further extension of leaf life may be more difficult because photosynthetic efficiency tends to diminish with leaf age (Dreccer, van Oijen, Schapendonk, Pot and Rabbinge, 2000) and because, in other environments, redistribution of stem DM has been associated with earlier senescence (Blum, 1998).

The choice between large numbers of shoots with small leaves and small ears, or small numbers of shoots with large leaves and large ears does not appear to be particularly crucial in terms of achieving potential yield. For instance, when yield and yield components were compared for the highest yielding CIMMYT spring wheat cultivars released between 1982 and 1996, yield ranged from 9.2 to 10.1 t/ha (10%), but spikes per m^2 ranged from 440 to 615 (40%), grains per spike from 40 to 48 (22%) and weight per grain from 34 to 50 mg (46%) (Reynolds, Sayre and Rajaram, 1999). It is possible that there may be subtle advantages in having many shoots because this would increase the area of leaf sheaths (which show greater disease resistance and longevity than leaf laminae, and are more erect) as a proportion of total green area. However, the more important issue regarding yield components concerns the balance between grain number per m^2 and weight per grain; current UK varieties show a wide range in mean weight per grain from 45 to 56 mg (Home-Grown Cereals Authority, 2004). There is an obvious inverse correlation between these traits, so the main consideration appears to be that, as grain weight increases, a smaller proportion of ear DM is required for non-grain components. For this reason (as well as for quality considerations) it seems that large grains are to be preferred. We can thus conclude that the approximate best structure for the 'maxi-wheat' crop yielding 19.2 t/ha at harvest will be to have 600 shoots per m^2, bearing 53 grains per ear, each weighing on average 51 mg DM. The harvest index of such a crop would be 60%.

The issue of N nutrition must now be addressed. An optimum canopy of GAI 7 predicted for 'maxi-wheat' will contain approximately 210 kg/ha N (Sylvester-Bradley *et al.,* 1997). If the large grain sink were not to cause early redistribution of N, hence premature canopy senescence, and if final N harvest index were 80%, the final grain protein concentration would be only 7.3% DM. It therefore seems that N supplies must be augmented to encourage further late uptake, both to ensure it is useful for bread-making, where this is intended, and to prevent premature redistribution of canopy protein to the grain. About 160 kg/ha extra N uptake would be required if the grain protein is to meet the current UK threshold for bread-making (2.2% N, DM basis); implications that this has for fertiliser requirements are discussed below.

Considering both the structure and composition of this 'maxi-wheat' crop, very few individual traits have been described that are beyond common experience in current UK crops. Therefore the main challenge for physiologists, agronomists and, in particular, for wheat breeders, will be in their combination: negative interdependencies must be targeted and broken, for instance between ear fertility and stem growth (Slafer, Araus and Richards, 1999; Slafer *et al.,* 2001), between grain number and grain size, or between rate and duration of grain filling (Spiertz, 1977). This must be a protracted enterprise in which breeders may find some success through the coming decades. However, most rapid progress is likely to come through interdisciplinary collaborations; crop management regimes must be developed by agronomists that enable expression and selection of 'maxi-wheat' traits by breeders. The 'maxi-wheat' phenotype envisages (i) prolonging stem extension which must depend on combining photoperiod responses with particular sowing dates, (ii) increasing ε and minimising unproductive tillering which can probably only be achieved at particular plant populations and levels of nutrition, (iii) increasing partitioning to the developing ear, which depends on setting up particular shoot numbers, and (iv) prolonging leaf longevity and grain filling which must depend on establishing late N availability and fungicidal activity. Thus the rate at which wheat yields are enhanced will tend to be governed by the level of technical support to, hence the level of investment in, wheat breeding. The wisdom of investing in enhancement of wheat yields will be determined by the anticipated repercussions, environmental as well as economic, of realising such large yields, compared to other options for land use.

Repercussions of realising potential yields

So far we have considered the feasibility of breeding a 'maxi-wheat' crop. It remains to consider the associated effects on grain quality, and the demands on husbandry, farming logistics and the environment, hence to predict the significant technological and other challenges in seeking to maximise wheat yields.

Taking the evidence of national trials series such as those supporting the Recommended List of varieties (Home-Grown Cereals Authority, 2004), where particular care is taken to minimise levels of weeds, diseases, pests, nutrient deficiencies and lodging, for most of the UK's arable land and in most seasons, it is currently possible to grow wheat crops that are ostensibly free of biotic and nutritional stresses. Of course, the costs of achieving this in management skill, time and inputs is uneconomic on a farm scale. But, in the main, it is possible to conclude that the non-genetic issues associated with maximising UK wheat yields are more economic and environmental than technological.

Economic determination of input levels

Any impression that requirements for agronomic inputs relate directly to the level of output is not supported in the literature. From a physiological point of view, the feedstocks for crop growth and yield are carbon dioxide, light energy and water, rather than the machinery, seeds, nutrients and agro-chemicals in which the grower must invest; the latter are used to initiate, maintain and capitalise on the production process, rather than to fuel it. Farming statistics tend to support this thesis: farms producing the best yields generally use *less* inputs than poorer performing farms (e.g. Renwick and Rush, 2001). Inputs are used, with varying skill, with the intention of optimising between extra input cost and additional returns in yield and quality. High (i.e. super-optimal) use of inputs is, by definition, counter-productive.

Rather, the relationships between individual agronomic inputs and grain yield generally show diminishing responses, which can largely be attributed to *direct* effects on canopy size causing a *diminishing* effect on light capture, and thus on yield. Example analyses have been published recently for seed rates (Spink *et al.,* 2000; Whaley *et al.,* 2000), N fertiliser amounts (Sylvester-Bradley *et al.,* 1997), and fungicide doses (Paveley, Sylvester-Bradley, Scott, Craigon and Day, 2001). Thus the association between high output cropping and large applications of N or fungicides is indirect: as the response to an input becomes amplified, a little more of that input becomes affordable, but the increase is relatively less than the yield increase.

To illustrate the relationships between grain yield and input requirements, Table 11.2 compares a typical organic wheat crop producing 4.5 t/ha grain

Table 11.2. TYPICAL COSTS AND RETURNS FROM CURRENT UK ORGANIC AND CONVENTIONAL SYSTEMS OF WHEAT PRODUCTION, AFTER NIX (2003) AND LAMPKIN *et al.* (2004), COMPARED TO A 'MAXI-WHEAT' SYSTEM YIELDING ABOUT TWICE THE CURRENT CONVENTIONAL SYSTEM.

Wheat system: Materials (M) & application (A) costs:	*Organic '04*			*Conventional '04*			*'Maxi-Wheat'*		
	M	*A*	*Total*	*M*	*A*	*Total*	*M*	*A*	*Total*
Inputs				***equivalent t grain /ha***[1]					
ploughing	0.00	0.52	0.52	0.00	0.52	0.52	0.00	0.52	0.52
seedbed cultivations	0.00	0.53	0.53	0.00	0.53	0.53	0.00	0.53	0.53
seed[2]	0.83	0.23	1.06	0.55	0.23	0.78	0.55	0.23	0.78
phosphate and potash fertiliser				0.32	0.10	0.42	0.64	0.10	0.74
slug pellets				0.03	0.07	0.10	0.03	0.07	0.10
herbicides for grass weeds				0.63	0.09	0.71	0.63	0.09	0.71
herbicides for broad-leaved weeds				0.14	0.09	0.23	0.14	0.09	0.23
insecticide for virus and grub control				0.07	0.09	0.16	0.07	0.09	0.16
animal manures	0.00	0.10	0.10						
growth regulator to shorten stems				0.07	0.09	0.16	0.21	0.09	0.29
N fertiliser for shoot production				0.22	0.10	0.32	0.24	0.10	0.34
N fertiliser for canopy expansion				0.65	0.15	0.80	0.73	0.15	0.88
N fertiliser for canopy survival							1.38	0.23	1.60
fungicide for lower leaves				0.33	0.09	0.41	0.33	0.09	0.41
fungicide for flag leaf and ear				0.33	0.09	0.41	0.33	0.09	0.41
fungicide for canopy survival							0.33	0.09	0.41
combining and carting	0.00	1.16	1.16	0.00	1.16	1.16	0.00	1.74	1.74
grain drying[3] and storage		0.53	0.53	0.00	1.06	1.06	0.00	2.75	2.75
Total variable costs	0.83	3.06	**3.90**	3.33	4.44	**7.76**	5.60	7.02	**12.62**
Output as grain			4.50			9.10			19.20
Margin			**0.60**[4]			**1.34**			**6.58**

[1] Assuming all conventional and organic inputs are costed at £72 per tonne conventional grain.
[2] Organic or conventional with fungicide for bunt and smut.
[3] Drying required for 50% crops in 2004, and 75% of 'maxi-wheat' crops.
[4] Note organic grain has a value about twice that of conventional grain.

with a conventional benchmark crop yielding 9.1 t/ha, and then with a 'maxi-wheat' crop, producing 19.2 t/ha. For organic and conventional crops, levels of inputs and costs of application have been taken from standard statistics. Interestingly the predicted margins of the two systems are much the same, taking into account that the value of organic wheat is about double that of conventional wheat. Beside these are set levels of inputs for 'maxi-wheat', predicted assuming that nutrients will be needed to provide for a slightly enlarged canopy and to replace the nutrients exported in grain, that pest and weed control will be little different from conventional (indeed high yielding crops may suppress weeds better, hence need less herbicide), and that affordable levels of growth regulators and fungicides would be increased a little. Significantly, the financial margin after doubling yield is predicted to increase five-fold. Hence, unless land values or rents (the only costs omitted from this analysis) were affected by yield achievement, it is clear that there are considerable economic incentives to the improvement of wheat yields: in other words, crops with doubled yields would produce grain at 20% less cost.

Of the inputs used to grow wheat (both those termed 'fixed' and 'variable' by economists), it is only drying that is directly proportional to yield. The costs of establishing a high yielding crop should be little different from that for a conventional crop, although note that increasing annual light interception would tend to shorten the interval between harvest and sowing, hence increase demands on machinery in that period. Other inputs would increase either less than in proportion to yield (e.g. fungicides) or more than in proportion to yield (e.g. N), with consequent implications for the environment.

Given the main uses of wheat, the important criteria governing grain quality, hence value per tonne, are bulk density (or specific weight, kg/hl), Hagberg falling number, and protein content. Taking each of these in turn, it seems likely that the enlarged grains from 'maxi-wheat' crops would, if anything, confer better bulk density. The prolonged grain-filling period and consequent delayed ripening period would confer a slightly greater risk of low Hagberg falling numbers. However, most significantly, the large increase in grain carbohydrate production, with only a slightly enlarged canopy (hence little extra redistributable protein) would lead to dramatically reduced grain protein concentrations.

ENVIRONMENTAL REPERCUSSIONS OF HIGH WHEAT YIELDS

It is not possible to predict environmental implications of very high yielding wheats without some further specification of the contexts in which these would be grown. If they do not much alter the demand for wheat or the distribution of wheat production, it appears that high yields would allow land to be relinquished for other purposes and that, overall, agro-chemical use would be reduced. Undoubtedly, *within the cropped area*, the competitive ability of the crop and the somewhat greater use of agrochemicals would further reduce diversity of macro-flora and some fauna such as ground-nesting birds (see Johnson and Hope, this volume). However, the net effect on the biodiversity of the landscape as a whole would probably be positive because (a) intensively cropped areas already have low biodiversity, and (b) uncropped land could be increased. As for the soil, large crops would contribute more fresh organic matter than small crops, so it is likely that the activity of micro-flora and fauna would be improved.

On the other hand, the large increases in nutrient requirements of 'maxi-wheat' have serious implications for the environment. With present uptake and redistribution efficiencies it is predicted that crops yielding 19.2 t/ha would have fertiliser requirements exceeding 400 kg/ha N, as well as high requirements for phosphate. The net increase in phosphate input to the environment might not be huge, depending on the reduction in cropped area, and on whether minimum soil phosphate levels would have to be increased, to provide for the increased crop demand. However, high yielding systems using large N applications would have considerably increased carry-over of residual N from one crop to the next. Extrapolating from long-term experiments where carry-over effects have been assessed (e.g. Bhogal, Rochford and Sylvester-Bradley, 2000) we predict that over-winter mineral N levels in soil would increase by about 50%; depending on soil texture, a portion this would leach and a portion would be available to the succeeding crop. Breeding for increased yields of wheat in recent years has also inadvertently increased wheats capacity to recover fertiliser N (Foulkes, Sylvester-Bradley and Scott, 1998), so there is hope of moderating these effects. However, as breeders seek ever-higher wheat yields, it is likely that they will have to be more conscious of improving the efficiency with which the crop acquires and uses N, and it is possible that, to avoid excessive nutrient loading of soils, greater use will have to be made of foliar fertilisers, and of gluten enhancement of flours for bread-making (e.g. Robertson and Cao, 2001).

Table 11.3. CRUDE ESTIMATES OF WATER USE BY WHEAT UNDER THREE GROWING SYSTEMS IN EASTERN ENGLAND (WITH AVERAGE PRECIPITATION OF 600 mm) AND THE REPERCUSSIONS FOR DRAINAGE. CROP WATER USE EFFICIENCY OF 5 g/l HAS BEEN ASSUMED IN EACH CASE.

	Units	*Organic*	*Conventional*	*'Maxi-wheat'*
Mean grain yield	t/ha	4.5	9.1	19.2
Harvest index	ratio	0.35	0.47	0.60
Total crop biomass	t/ha	10.9	16.5	27.2
Crop water use	mm	219	329	544
Drainage	mm	381	271	56

Possible impacts of 'maxi-wheat' on water in the environment are, if anything, more serious than the predicted effects of nutrients. There is little evidence that efficiencies of water use by crops have increased as yields have been increased by breeding, except to the extent that harvest index has improved (Foulkes, Sylvester-Bradley and Scott 2001, 2002; Sylvester-Bradley and Foulkes, 2003). In temperate environments such as the UK where soil evaporation represents a small part of total evapo-transpiration, early vigour and extending canopy life can only have a small effect on overall crop transpiration. Assuming a fixed water-use efficiency (WUE, ratio of above-ground DM production to concurrent evapo-transpiration), Table 11.3 provides estimated annual water balances for the three crop types described in Table 11.2, and shows that for the average year in eastern England, water draining from cropped land could be reduced drastically, to about one fifth of current amounts as a result of future yield enhancement. Of course this is an extreme case where it is assumed that artificial water storage and irrigation are widely available. However, it illustrates that, depending on uncropped land and other changes in land use, there is a clear risk that breeding for improved crop yields will increasingly reduce both the flows of surface waters and the recharge of aquifers.

Currently, for the 12% of UK wheat fields located on drought-prone sandy or shallow soils, yield-limiting droughts occur about two years in three (Foulkes *et al.*, 2001), and for all UK wheat fields it has been estimated that the loss in yield potential due to drought over a run of years is in the region of 10-20% (Foulkes *et al.*, 2002). Taking into account the new environmental constraints on extension of irrigation (Department for Environment, Food and Rural Affairs, 2004), we have to conclude that it is likely there will be much land where rainfall becomes the main determinant of yield, and that there will be many seasons when water supplies are inadequate everywhere. Although root biomass may increase in proportion to aerial biomass at anthesis in the 'maxi-wheat' crop, this can only improve recovery of soil-stored winter rainfall, not annual water supply.

It is possible that genetic increases in WUE could ameliorate the potential influence of UK droughts. However, success depends upon approach: under drought in Australia Rebetzke, Condon, Richards and Farquhar (2002) found that increases in transpiration efficiency (TE; ratio of above-ground DM production to crop transpiration), after selection for reduced carbon isotope discrimination, gave greater crop biomass and greater grain yield. On the other hand, using the same selection approach, Araus, Slafer, Reynolds and Royo (2002) found reduced stomatal conductance gave lower intercellular levels of CO_2 and decreased photosynthesis and biomass. It appears that the better result comes from seeking increased WUE through increasing photosynthesis (perhaps through an increasing sink), rather than through directly reducing stomatal conductance. Seeking improvement of ε from 2.2 to 2.8 g/MJ could possibly translate into improvements in WUE. However, in practice, stomatal conductance would most probably increase under these circumstances, tending to maintain intercellular levels of CO_2 and moderating or negating potential improvements in WUE. Furthermore, in the 'maxi-wheat' phenotype, a greater proportion of seasonal growth occurs in June and July under conditions of high evaporative demand on the leaf (i.e. when the vapour pressure deficit is greatest), and this will tend to decrease season-long WUE. In summary, although small improvements in WUE through breeding may be feasible, it is unlikely that they will be sufficient to alleviate the anticipated limitation to growth and grain yield of water availability. Thus, irrespective of any effects of climate change, we anticipate that wheat improvement in the UK will increasingly be influenced by drought, and the rate of yield progress is likely to slow. We estimate, assuming no breeding improvement in WUE, that a potential yield, unrestricted by water, of 19.2 t/ha reduces to 14 t/ha if irrigation remains unavailable in wheat-growing regions. This estimate equates to about 12 t/ha, or 150% of the current national average wheat yield, if the current difference between average trial yields and average farm yields is applied.

Conclusions

Taking the UK as a case study for wheat production in temperate environments, advances in national average yields are still apparent, and further advances are eminently feasible. As in the last century, advances are more likely to result from economic incentives and political will than from scientific breakthrough. Physiological analysis indicates that current average UK yields of 8 t/ha grain could be increased by at least 50%. Economic analysis, which shows a possible quintupling of profit margins, indicates that the direct incentives for yield improvement are almost as great as they have ever been; however, wider political will is less certain. Agricultural policies in the EU and other developed economies take an increasingly holistic view of the productivity of agriculture, and it appears that crop production will have to contend with increasing environmental constraints. Thus, in addition to the economic forces encouraging yield enhancement, plant breeders will probably have to contend with a set of increasingly disparate objectives. New varieties will need to be less dependent on agrochemicals, use nutrients with greater efficiency, be better suited to market requirements (despite constraints on fertiliser use), and be better able to contend with water-limitation. With current breeding methods, such an extension of objectives will inevitably cause slower progress in yield. Meanwhile, public support for new research and development will tend to favour encouragement of more benign farming methods, rather than increased productivity. However, if breeders can harness new genomic information, and physiologists can show some success in crop design, it should prove possible to embrace these broader aims, and still maintain some yield progress, albeit at slower rates than in recent decades. We estimate that the most likely yield trajectory for UK wheat yields is somewhere between the second and third scenarios in Figure 11.2, giving national average yields of approximately 9.5, 10.7 and 13.1 t/ha in 2010, 2020 and 2050 respectively.

From a technological view-point, seed and agrochemical input requirements predicted here for a 'maxi-wheat' crop, are surprisingly modest compared to current use. The main demands on new technologies will arise primarily through the much increased demands for water and N, and the requirement for machinery that enables reliable re-sowing within a few weeks of harvest.

Progress in mechanisation still continues apace, and there are grounds for optimism that N requirements of wheat could be moderated where it is not used for bread-making. The crop develops significant stores of N (Martre, Porter, Jamieson and Triboi, 2003) that could be reduced without affecting its value as a high energy feedstock, and little or no attention has yet been paid to this by breeders. However, other technological solutions will be needed to moderate N requirements for bread-making wheats. Although some inherent association is expected between the efficiencies with which wheat uses radiation and water, we conclude that the tractability or otherwise of water availability and water use efficiency will increasingly dictate rates of yield improvement in this important crop species.

Acknowledgements

We would like to thank CJ Dyer for undertaking the Windowpane analysis, PM Berry for checking the section on lodging, J Spink for helpful criticism of the manuscript, and the UK Department for Environment Food and Rural Affairs for financial support (Project IS0210).

References

Araus, J.L., Slafer, G.A., Reynolds, M.P. and Royo, C. (2002). Plant breeding and drought in cereals: what should we breed for? *Annals of Botany* **89**, 925-940.

Berry, P.M., Spink, J.H., Gay, A.P. and Craigon, J. (2003). A comparison of root and stem lodging risks among winter wheat cultivars. *Journal of Agricultural Science* **141**, 191-202.

Berry, P.M., Sterling, M., Spink, J.H., Baker, C.J., Sylvester-Bradley, R., Mooney, S.J., Tams, A.R. and Ennos, A.R. (2004). Understanding and reducing lodging in cereals. *Advances in Agronomy* **84**, 215-269.

Bhogal, A., Rochford, A. D. and Sylvester-Bradley, R. (2000). Net changes in soil and crop nitrogen in relation to the performance of winter wheat given wide-ranging annual nitrogen applications at Ropsley, UK. *Journal of Agricultural Science* **135**, 139-149.

Bindraban, P.S. (1999). Impact of canopy nitrogen profile in wheat on growth. *Field Crops Research* **63**, 63-77.

Blum, A. (1998). Improving wheat grain filling under stress by stem reserve mobilisation. *Euphytica* **100**, 77-83.

Calderini, D.F., M.F. Dreccer and G.A. Slafer. (1995). Genetic improvements in wheat yield and associated traits. A re-examination of previous results and latest trends. *Plant Breeding* **114**, 108-112.

Calderini, D.F., Reynolds, M.P. and G.A. Slafer. (1999). Genetic gains in wheat yield and associated physiological changes during the twentieth century. p. 351–377. In: *Wheat: Ecology and Physiology of Yield Determination.* Eds. E.H. Satorre and G.A. Slafer. The Haworth Press, Inc., New York

Department for Environment, Food and Rural Affairs (2004). *Water Resources and Supply: Agenda for Action.* www.defra.gov.uk/environment/water/resources/wrs/summary.htm

Dines, L. (1998). *Processes Limiting the Response of Winter Wheat to Applied Nitrogen Fertiliser.* MSc Thesis, University of Nottingham. 268 pp.

Dreccer, M.F., van Oijen, M., Schapendonk, A.H.C.M., Pot, C.S. and Rabbinge, R. (2000). Dynamics of vertical leaf nitrogen distribution in a vegetative wheat canopy. Impact on canopy photosynthesis. *Annals of Botany* **86**, 821-831.

Evans, J.R. (1989). Photosynthesis – the dependence on nitrogen partitioning. pp. 159-174 In: *Causes and Consequences of Variation in Growth Rate and Productivity of Higher Plants.* Eds. H. Lambers, M.L. Cambridge, H. Konnings and T.L. Pons. The Hague: SPB Academic Publishing.

Evans, L.T. (1978). The influence of irradiance before and after anthesis on grain yield and its components in microcrops of wheat grown in a constant daylength and temperature regime. *Field Crops Research* **1**, 5-19.

Evans, L.T. and Wardlaw, I.F. (1996). Wheat. In: *Photoassimilate Distribution in Plants and Crops: Source-sink Relationships.* Eds. Zamski, E. and Schaffer, A. Marcel Dekker, Inc. pp. 501-518.

Fischer, R.A. (1985). Number of kernels in wheat crops and the influence of solar radiation and temperature. *Journal of Agricultural Science* **105**, 447-461.

Fisher, R.A., Rees, D., Sayre, K.D., Lu, Z.-M., Condon, A.G. and Larque Saavedra, A. (1998). Wheat yield progress associated with higher stomatal conductance and photosynthetic rate, and cooler canopies. *Crop Science* **38**, 1467-1475.

Foulkes, M.J., Sylvester-Bradley, R. and Scott, R.K. (1998). Evidence for differences between winter wheat cultivars in aquisition of soil mineral nitrogen and uptake and utilization of applied fertilizer nitrogen. *Journal of Agricultural Science, Cambridge* **130**, 29-44.

Foulkes, M.J., Scott, R.K. and Sylvester-Bradley, R. (2001). A comparison of the ability of wheat cultivars to withstand drought in UK conditions: resource capture. *Journal of Agricultural Science* **137**, 1-16.

Foulkes, M.J., Scott, R.K. and Sylvester-Bradley, R. (2002). The ability of wheat cultivars to withstand drought in UK conditions: formation of grain yield. *Journal of Agricultural Science* **138**, 153-169.

French, R.G. and Schultz, J. (1984). Water use efficiency in a Mediterranean environment I. The relationship between yield, water use and climate. *Australian Journal of Agricultural Research* **35**, 734-64.

Gallager, J.N. and Biscoe, P.V. (1978). Radiation absorption, growth and yield of cereals. *Journal of Agricultural Science* **91**, 47-60.

Gay, A.P., Stokes, D.T., Weightman, R.M. & Sylvester-Bradley, R. (1998). How to run a reference crop. In *Assessments of Wheat Growth to Support its Production and Improvement.* HGCA Research Report No. 151 Volume 2. Eds R. Sylvester-Bradley, R.K. Scott, R.W. Clare & S.J. Duffield. London: Home-Grown Cereals Authority. 78 pp.

Green, C.F., Vaidyanathan, L.V. and Hough, M.N. (1983). An analysis of the relationship between potential evapotranspiration and dry-matter accumulation for winter wheat. *Journal of Agricultural Science* **100**, 351-358.

Home-Grown Cereals Authority. (2004). *Recommended List for Cereals and Oilseeds 2003/04.* London: HGCA.

Ingram, J. Macleod, J. and McCall, M.H. (1997). The contribution of varieties to the optimisation of cereal production in the UK. *Aspects of Applied Biology* 50, *Optimising Cereal Inputs: Its Scientific Basis* 31-38.

Jamieson, P.D., Semenov, M.A., Brooking, I.R. and Francis, G.S. (1998). Sirius: a mechanistic model of wheat response to environmental variation. *European Journal of Agronomy* **8**, 161-179.

Lampkin, N., Measures, M. and Padel, S. (2004) *Organic Farm Management Handbook (6th ed.)* Elm Farm Research Centre and the Welsh Institute of Rural Studies, Aberystwyth.

Lovell, D.J., Parker, S.R., Hunter, T., Royle, D.J. and Coker, R.R. (1997). The influence of growth habit and architecture on the risk of epidemics of *Mycosphaerella graminicola (Septoria tritici)* in winter wheat. *Plant Pathology* **47**, 126-138.

MAFF. (1971) *Modern Farming and the Soil: Report of the Agricultural Advisory Council on Soil Structure and Soil Fertility.* Chairman N. Strutt. HMSO, London.

MAFF. (2000) *Climate Change and Agriculture in the United Kingdom.* MAFF, London.

Martre, P., Porter, J.R., Jamieson, P.D. and Triboi, E. (2003). Modeling grain nitrogen accumulation and protein composition to understand the sink/source regulations of nitrogen remobilization for wheat. *Plant Physiology* **133**, 1959-1967.

Miralles, D.J., Dominguez, C.F. and Slafer, G.A. (1996). Relationship between grain growth and postanthesis leaf area duration in dwarf, semidwarf and tall isogenic lines of wheat. *Journal of Agronomy and Crop Science* **177**, 115-122.

Monteith, J.L. (1977). Climate and efficiency of crop production in Britain. *Philosophical Transactions of the Royal Society, London, Series B.* **281**, 277-294.

Mulder, E.G. (1954). Effect of mineral nutrition on lodging in cereals. *Plant and Soil* **5**, 246-306.

Nix J. (2003) *Farm Management Pocketbook - 34th edition.* Imperial College at Wye.

Pietravalle, S. Shaw M.W., Parker, S.R. and van den Bosch, F. (2003). Modeling of relationships between wetaher and Septoria tritici epidemics on winter wheat: a critical approach. *Phytopathology* **93**, 1329-1339.

Passioura, J.B. (1996). Simulation models, snake oil, education, or engineering? *Agronomy Journal* **88**, 690-694.

Paveley, N.D., Sylvester-Bradley, R., Scott, R.K., Craigon, J. and Day, W. (2001). Steps in the prediction of yield on fungicide dose. *Phytopathology* **91**, 708-716.

Pingali, P.L. (ed.) (1999) *CIMMYT 1998-9 World Wheat Facts and Trends. Global Wheat Research in a Changing World: Challenges and Achievements.* Mexico, D.F.: CIMMYT.

Poudel, D.D., Horwath, W.R., Lanini, W.T., Temple, S.R. and van Bruggen, A.H.C. (2002). Comparison of soil N availability and leaching potential, crop yields and weeds in organic, low-input and conventional farming systems in northern California. *Agriculture Ecosystems and Environment* **90**, 125-137.

Rebetzke, G.J., Condon, A.G., Richards, R.A. and Farquhar, G.D. (2002). Selection for reduced carbon isotope discrimination increases aerial biomass and grain yield of rainfed bread wheat. *Crop Science* **42**, 739-745.

Renwick, A. and Rush, C. (2001). Economic and technical performance in wheat production in England. *Farm Management* **11**, 177-192.

Reynolds, M.P., Sayre, K.D. and Rajaram, S. (1999). Physiological and Genetic Changes in Irrigated Wheat in the Post Green Revolution Period and Approaches for Meeting Projected Global Demand. *Crop Science* **39**, 1611-1621.

Reynolds, M.P, van Ginkel, M. and Ribault, J-M. (2000). Avenues for genetic modification of radiation-use efficiency in wheat. *Journal of Experimental Botany* **51**, 459-473.

Reynolds, M.P., Calderini, D.F., Condon, A.G. and Rajaram S. (2001). Physiological basis of yield gains in wheat associated with the *LR19* translocation from *A. elongatum. Euphytica* **119**, 137-141

Reynolds, M.P., Pellegrineschi, A. and Skovmand, B. (in press). Sink-limitation to yield and biomass: a summary of some investigations in spring wheat. *Annals of Applied Biology*.

Robertson, G.H. and Cao, T.K. (2001). Farinograph responses for wheat flour dough fortified with wheat gluten produced by cold-ethanol or water displacement of starch. *Cereal Chemistry* **78**, 538-542.

Ruske, R.E., Gooding, M.J. and Jones, S.A. (2003). The effects of adding picoxystrobin, azoxystrobin and nitrogen to a triazole programme on disease control, flag leaf senescence, yield and grain quality of winter wheat. *Crop Protection* **22**, 975-987.

Sayre, K.D., Rajaram, S. and Fischer, R.A. (1997). Yield potential progress in short bread wheats in northwest Mexico. *Crop Science* **37**, 36-42.

Shearman, V., Sylvester-Bradley, R. and Foulkes, M.J. (in press). Physiological processes associated with wheat yield progress in the UK. *Crop Science.*

Silvey, V. (1986). The contribution of new varieties to cereal yields in England and Wales UK between 1947 and 1983. *Journal of the Institute of Agricultural Botany,* **17**, 155-168.

Silvey, V. (1994). *Plant Breeding in Improving Crop Yield and Auality in Recent Decades.* Paper presented at the Agribex Symposium, February 1994.

Slafer, G.A., Araus, J.L and Richards, R.A. (1999). Physiological traits that increase the yield potential of wheat. pp. 379-416. In: *Wheat: Ecology and Physiology of Yield Determination.* Eds. Satttore, E.H. and G.A. Slafer. New York, Food Products Press.

Slafer, G.A., Abeledo, L.G., Miralles, D.J., Gonzalez, F.G. and Whitechurch, E.M. (2001). Photoperiod Sensitivity During Stem Elongation as an Avenue to Raise Yield Potential in Wheat. In: *Wheat in a Global Environment*, 487-496. Eds. Bedo, Z. and Lang L.

Spiertz, J.H.J. (1977). The influence of temperature and light intensity on grain growth in relation to the carbohydrate and nitrogen economy of the wheat plant. *Netherlands Journal of Agricultural Science* **25**, 182-197.

Spink, J.H., Semere, T., Sparkes, D.L., Whaley, J.M., Foulkes, M.J., Clare, R.W. and Scott, R.K. (2000). Effect of sowing date on the economic optimum plant density of winter wheat. *Annals of Applied Biology* **137**, 179-188.

Spink, J.H., Kirby, E.J.M., Frost, D.L., Sylvester-Bradley, R, Scott, R.K., Foulkes, M.J., Clare, R.W. and Evans, E.J. (2000). Agronomic implications of variation in wheat development due to variety, sowing date, site and season. *Plant Varieties and Seeds* **13**, 91-105.

Spink, J.H., Foulkes, M.J., Gay, A., Bryson, R.J., Berry, P., Sylvester-Bradley, R., Semere, T., Clare, R.W., Scott, R.K., Kettlewell, P.S. and Russell, G. (2000). *Reducing Winter Wheat Production Costs Through Crop Intelligence Information on Variety and Sowing Date, Rotational Position, and Canopy Management in Relation to Drought and Disease Control.* HGCA Project Report No. 235. pp. 135. London: Home-Grown Cereals Authority.

Sylvester-Bradley, R., Scott, R.K., Stokes, D.T. and Clare, R.W. (1997) The significance of crop canopies for nitrogen nutrition. *Aspects of Applied Biology* 50, *Optimising cereal inputs: its scientific basis,* 103-116.

Sylvester-Bradley, R., Watson, N.A., Dewes, M.E., Clare, R.W., Scott, R.K. and Dodgson, G. Eds. (1997). *The wheat growth guide.* Home-Grown Cereals Authority, London. 32 pp.

Sylvester-Bradley, R., Scott, R.K., Clare, R.W. and Duffield, S.J. (1998) Eds. *Assessments of Wheat Growth to Support its Production and Improvement. Volume 3 The Dataset.* HGCA Research Report No. 151.

Sylvester-Bradley, R. and Foulkes, M.J. (2003). Wheat varieties and diminishing UK water supplies. *Journal of the UK Irrigation Association* **31**, 9-11.

Thorne, G.N., Pearman, I., Day, W. and Todd, A.D. (1988). Estimation of radiation interception by winter wheat from measurements of leaf area. *Journal of Agricultural Science* **110**, 101-108.

Verma, V., Foulkes, M.J., Worland, A.J., Sylvester-Bradley, R., Caligari, P.D.S. and Snape, J.W. (2004). Mapping quantitiative trait loci for flag leaf senescence as a yield determinant in winter wheat under optimal and drought-stressed environments. *Euphytica* **135**, 255-263.

Whaley, J.M., Sparkes, D.L., Foulkes, M.J., Spink, J.H., Semere, T. and Scott, R.K. (2000). The physiological response of winter wheat to reductions in plant density. *Annals of Applied Biology* **137**, 165-177.

Whaley, J.M., Kirby, E.J.M., Spink, J.H., Foulkes, M.J. and Sparkes, D.L. (in press). Frost damage to winter wheat in the UK: The effect of plant population density. *European Journal of Agronomy*.

12

THE RISE AND FALL OF GRAIN LEGUMES AND THEIR PREDICTED YIELDS IN UK AGRICULTURE

R. M. WEIGHTMAN
ADAS Boxworth, Battlegate Road, Boxworth, Cambridge CB3 8NN, UK

Introduction

In 1984, the 40th Easter School in Agricultural Science took place, with the title 'The Pea Crop – A Basis For Improvement' (Hebblethwaite, Heath and Dawkins, 1985). During the 1980's the area of peas and beans increased in response to EEC subsidies (Heath, 1987). It is timely therefore 20 years on, to review the progress which has been made in genetic improvement of grain legume crops, and contrast this with yield trends on-farm, in the UK. At the present time, we are considering the role of grain legumes in the context of a more sustainable agriculture and superficially, peas and other legumes, with their ability to fix atmospheric nitrogen (N), would appear to be ideal candidates to take a place in a sustainable cropping system.

This paper addresses principally spring-sown, white flowered combining peas (*Pisum sativum* L), as these are most instructive in teaching us both about yield potential in temperate grain legumes, and also constraints to progress in the UK. Vining peas are grown for direct human consumption, and are not grown to maturity. In the latter, quality is of more importance than yield *per se*, and much of the breeding effort has been focussed on producing vining peas, which meet the needs of the end-user in terms of flavour and texture. Moreover, as the wrinkled-seed types used for vining pea production have an inherent yield penalty, compared against round seeded peas, they are less relevant. The paper refers to faba beans (*Vicia faba* L) where applicable but, as will be seen, both genetic improvement and improvement in performance on-farm are currently very difficult to detect, or quantify, in this species.

The pea crop for many years was the model plant species of choice for physiologists and biochemists (e.g. Sutcliffe 1977). The work that led up to the 40th Easter School built on a broad science base, which spanned from molecular biology through to field-based crop physiological disciplines. The pea is now no longer the model crop of choice for plant researchers, having been supplanted by *Arabidopsis* and *Medicago*. In 2000/01, of the UK government-funded projects surveyed for a review of UK plant genetic research, Caligari, Lobley, Weightman, Sylvester-Bradley and Temple (2002) showed that £719k was spent on pea research, out of a total of £9M on arable crops. This compared against £4M on horticultural crops and £26M on general plant genetic research on model systems. There is currently no UK government funded research on faba beans, although there is some collaborative work on genomics funded by the EU (EU-FABA; Fred Stoddard, personal communication). However, the fact that research activity twenty years ago on peas or beans was greater than it is now, does not necessarily mean that yields will no longer be increasing. The time lag from basic research to farm yields has been estimated at 18 to 19 years in the case of winter wheat (Thirtle, Bottomley, Palladino, Schimmelpfennig and Townsend, 1998), and is of the order of 16 years in peas (Caligari *et al.*, 2002). Thus, we should be able to detect, over the past ten years, continued improvement in pea yields as the result of ongoing commercial plant breeding activity, and application of the results of research carried out 20-30 years ago.

UK yield trends

Yields of peas were reviewed by Knott (1996), who concluded that over an 18 year period (1976-1993), the yield of dry peas increased by 23.4% for England and Wales. Of that yield increase, on average, 72% of the improvement was considered to be due to genetic factors. Knott (1996) has given an excellent summary of the changes to morphological traits of pea genotypes over that period, and the reader is directed there for further information. The purpose of this paper is to review the wider trends in pea yields over a longer period, and to consider whether UK yields are following a different pattern to those in the rest of the world, and more importantly whether on-farm yields follow the same trends as those recorded in UK Recommended List (RL) trials (currently our best measure of genetic improvement of peas).

YIELD TRENDS FROM RECOMMENDED LIST TRIALS

Assessment of genetic improvement in grain legumes can be carried out with some confidence, because one variety has generally been used consistently in all RL trials. Therefore assuming that there has been no genetic drift in that variety, its yield can be used as a reference point. For the period 1984-2002, the reference varieties are Maro (spring peas; 1985-2002), Bourdon (winter beans; 1987-1998) and Maris Bead (spring beans; all years). Tables 12.1 and 12.2 for peas and beans respectively show the annual yields of the control varieties used by the National Institute of Agricultural Botany (NIAB), compared against the reference variety yields in those years (Anon, 1984-2002). These data clearly show detectable genetic improvement in pea yields. In 2002, the average yield of the control varieties was 1.25 t/ha (or 30%) higher than that of Maro. The data for field beans contrast with those for peas, in that for winter beans, there has been no discernable yield increase above that of Bourdon, whereas for spring beans genetic improvement has led to an increase of 0.66 t/ha (18%) over Maris Bead. The greater progress in peas reflects a greater number of commercial breeders and a more competitive commercial breeding environment. It also reflects that peas 'travel well': improvements made from an initial cross in any one country can be exploited across Europe, which is perhaps not the case with a crop like winter beans which, it is argued, is UK-specific.

Table 12.1 MEAN YIELDS (AT 85% DM) OF CONTROL VARIETIES, AND THE REFERENCE VARIETY MARO, IN UK RECOMMENDED LIST TRIALS OF SPRING COMBINING PEAS BETWEEN 1985 AND 2002.

Year	*Control varieties*				*No. years in average*	*Mean yield of control varieties (t/ha)*	*Yield of reference variety (t/ha)*
1985	Birte	Maro	Progreta		1	4.55	4.37
1986	Birte	Maro	Progreta		1	4.51	4.33
1987	Birte	Maro	Progreta		1	4.63	4.44
1988	Birte	Countess	Progreta		1	4.65	4.28
1989	Birte	Countess	Progreta	Solara	1	4.84	4.26
1990	Birte	Countess	Solara		1	4.59	3.95
1991	Countess	Orb	Solara		1	4.79	4.17
1992	Baroness	Orb	Solara		1	4.85	3.98
1993	Baroness	Orb	Solara		1	5.01	4.36
1994	Baroness	Orb	Solara		1	5.00	4.40
1995	Baccara	Baroness	Orb	Solara	1	5.18	4.45
1996	Baccara	Baroness	Orb	Solara	1	5.16	4.39
1997	Baccara	Eiffel	Orb	Solara	5	5.18	4.61
1998	Baccara	Carrera	Eiffel	Elan	5	5.38	4.30
1999	Baccara	Carrera	Eiffel	Elan	5	5.08	4.01
2000	Baccara	Carrera	Eiffel	Espace	5	5.15	3.91
2001	Baccara	Croma	Eiffel	Espace	5	5.38	4.20
2002	Arrow	Croma	Eiffel	Espace	5	5.42	4.17

The control yields in a particular year do not necessarily represent the best varieties available to growers. Figure 12.1 shows the yield trends for the top varieties of peas and beans on the RL in each year. Previous reviewers e.g. Knott (1996) have used 'smoothed' data, of rolling five year means, but the individual data are presented here. As expected, there is considerable variation from year to year, and some of the reasons for this variability are considered

Table 12.2 MEAN YIELDS (AT 85% DM) OF CONTROL AND REFERENCE VARIETIES IN UK RECOMMENDED LIST TRIALS OF FIELD BEANS (*V. FABA* L.) BETWEEN 1985 AND 2002.

Year	*Control varieties used*	*Number of years for average*	*Control yield (t/ha)*	*Yield of reference variety (t/ha)*
Winter beans				(Bourdon)
1984	Banner, Bulldog, Maris Beagle, Throws MS	1	3.78	*
1985	Banner, Bulldog, Maris Beagle, Throws MS	1	4.00	*
1986	Banner, Bulldog, Maris Beagle, Throws MS	1	4.06	*
1987	Banner, Bulldog, Maris Beagle, Throws MS	1	4.19	4.48
1988	Banner, Bourdon, Bulldog	1	4.33	4.46
1989	Banner, Bourdon, Bulldog	1	4.55	4.73
1990	Banner, Bourdon, Bulldog	1	4.60	4.78
1991	Banner, Bourdon, Boxer	1	4.82	4.92
1992	Banner, Bourdon, Boxer	1	4.93	5.03
1993	Bourdon, Boxer, Punch	1	5.02	5.07
1994	Bourdon, Boxer, Punch	1	4.97	4.97
1995	Bourdon, Boxer, Punch	1	4.98	4.98
1996	Bourdon, Punch	1	4.87	4.82
1997	Bourdon, Punch	7	4.97	4.92
1998	Bourdon, Punch	7	4.92	4.92
1999	Punch, Striker, Target	7	4.68	*
2000	Clipper, Punch, Target	7	4.76	*
2001	Clipper, Punch, Target	7	4.52	*
2002	Clipper, Target	7	4.56	
Spring beans				(Maris Bead)
1984	Blaze, Danas, Herra, Maris Bead	1	3.85	3.73
1985	Blaze, Danas, Herra, Maris Bead	1	4.04	3.92
1986	Blaze, Danas, Herra, Maris Bead	1	4.08	4.00
1987	Blaze, Danas, Herra, Maris Bead	1	3.99	3.99
1988	Alfred, Blaze, Maris Bead, Troy	1	4.26	4.05
1989	Alfred, Maris Bead, Troy	1	4.51	4.28
1990	Alfred, Corton, Troy	1	4.57	4.30
1991	Alfred, Corton, Troy	1	4.51	4.24
1992	Alfred, Corton, Troy, Victor	1	4.55	4.23
1993	Gobo, Troy, Victor	1	4.61	4.20
1994	Caspar, Gobo, Troy, Victor	1	4.74	4.31
1995	Caspar, Gobo, Vasco, Victor	1	4.83	4.25
1996	Gobo, Scirocco, Vasco, Victor	1	4.54	3.95
1997	Maya, Scirocco, Vasco, Victor	7	4.45	3.83
1998	Maya, Scirocco, Vasco, Victor	7	4.50	3.92
1999	Maya, Quattro, Scirocco, Victor	7	4.56	3.97
2000	Maya, Quattro, Scirocco, Victor	7	4.47	3.80
2001	Maya, Quattro, Scirocco, Victor	7	4.36	3.71
2002	Maya, Quattro, Victor	7	4.39	3.73

below. Simple linear regression for pea yields over time suggest an average increase in yield of 0.05 t/ha/yr, approximately half the value estimated for winter wheat (Sylvester-Bradley *et al.*, 2002). In contrast, spring and winter bean yields increased by 0.09 and 0.13 t/ha/year respectively in the period 1984-1994, but subsequently decreased by 0.04 and 0.07 t/ha/yr in the period 1995-2002. These data mirror those in Table 12.2, indicating no net yield improvement in beans. This is particularly the case for winter beans where, with only one UK breeder, a limited number of new varieties enter the RL system each year, and so the control varieties are often also the top three recommended varieties.

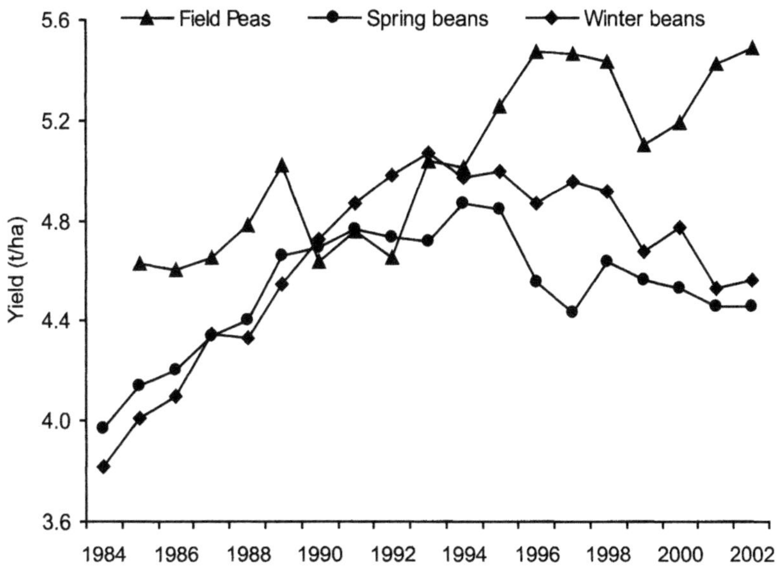

Figure 12.1 Average grain yields of the best three performing winter bean, or four pea and spring bean varieties. Source: NIAB Recommended List trials, 1984-2002.

FARM AND NATIONAL YIELDS

Heath and Hebblethwaite (1985a) showed that UK farm yields of peas were increasing at a rate of 0.04 t/ha/yr over the period 1940-1980, approximately half of that seen in wheat during the same period. Considering the next twenty years, pea yields (ex. farm) are presented in Figure 12.2 for the period 1962-2002 (FAO statistics). UK national pea yields have increased on average by 0.02 t/ha/yr, a rate considerably less than the 0.1 t/ha/yr noted for winter wheat (Sylvester-Bradley, Lunn, Foulkes, Shearman, Spink and Ingram, 2002). More importantly, over the last 10 years, yields have decreased by 0.07 t/ha/yr. This is a worse position than the 0.006 t/ha/yr increase estimated previously by Caligari *et al.* (2002) for the period 1989-1999, and which led them to conclude that the rate of return on the investment in genetic improvement (the sum of public research plus private breeding efforts) of combining peas was negative and low.

The simplest explanation for the higher yields in RL trials is that of 'edge effects': it has been known for some time that RL trials will out-perform farm crops, and this was thought to be particularly the case with peas, where tall, vigorous varieties benefit from the design of the trialling system (Heath and Hebblethwaite, 1985b). However, changes in stature of modern pea varieties would tend to mean that shorter, less competitive varieties are now being tested. Thus, one might expect recent RL trials to be less likely to overestimate yields. The more important question is why the national farm yield of combining peas in the UK is decreasing, whereas the RL trials demonstrate that yields of the top pea varieties are increasing at *c.* 0.05 t/ha/yr ? This is a key question if we are to predict future pea yields successfully.

Figure 12.2 also shows yields for France and the world, the former being particularly interesting. Having started with lower pea yields in the late 1960's, French pea yields overtook those of the UK during the early 1970's, and now outperform them by more than 1 t/ha. Over the period 1961-2002 the average rate of yield increase in French pea crops has been 0.08 t/ha/yr, four times the rate estimated for the UK.

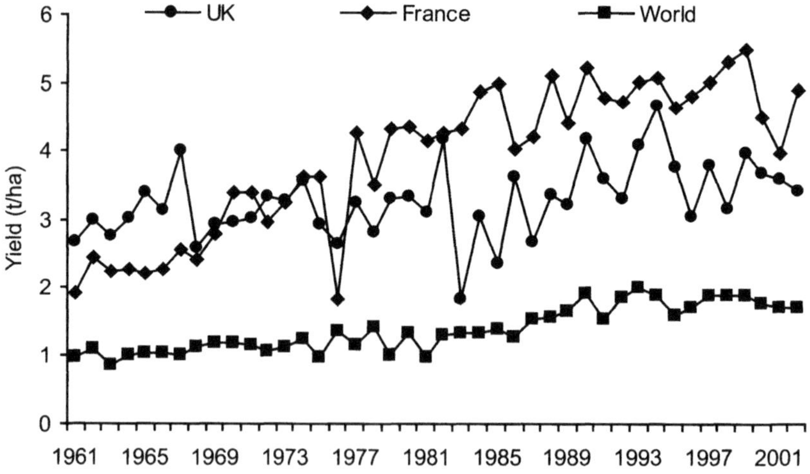

Figure 12.2 Average national pea yields for world pea crop, UK and France for the period 1961-2002. Source: FAO statistics.

It is well documented that peas have gained in popularity in France, and have benefited from targeted agronomic advice, built on a broad research programme. France also has larger areas of uniform soil types, whereas in the UK, soils are perhaps more variable. The area sown to combining peas in France increased during the 1980's and 1990's (Table 12.3) to a peak of 727,000 ha in 1993, but has since declined to approximately half this area. This reduction in area appears to be due to cropping too frequently in the rotation, and build up of diseases such as *Aphanomyces* in certain areas, but changes to CAP reform and support prices will also have played a part. In contrast, the increase in area of peas in the UK has been less dramatic, with the peak area of 102,400 ha recorded in 2001. These figures underline the fact that peas in France are treated as a serious crop. The typical agronomic advice for pea growing in France was outlined by Plancquaert (1991). It is clear from the French experience that much fine-tuning of pea agronomy can be achieved: optimum sowing dates are chosen based on long term 30 year meteorological records for each region, in 1990, 80,000 ha (12%) of pea crops were irrigated and, particularly in conventional leaved-types, controlled lodging was carried out in order to improve harvestability. Conventional wisdom suggests that these methods would not be cost-effective for pea crops in the UK, but there appears to be no published data to support this. What is certain however, is that as more growers become involved with grain legume crops, and with greater levels of research and technology transfer, there is potential to increase their yields, as demonstrated in France.

Table 12.3 HARVESTED AREAS OF COMBINING PEAS IN FRANCE AND THE UK FROM 1962 TO 2002 (SOURCE: FAO STATISTICS).

Year	*France*	*UK*
	Area (ha)	
1962	9,200	12,545
1967	6,289	16,187
1972	9,128	21,853
1977	12,900	37,000
1982	104,000	21,000
1987	451,000	116,600
1992	700,000	79,000
1997	606,000	97,800
2002	338,000	85,000

The decline in pea yields is not unique to Europe. For example, Peck and McDonald (1998) have also reported pea yield decline in good pea growing areas in Australia. A large increase in the area of field peas sown in the 1980's was accompanied by yield decreases of 24-38%, with Southern Australia being significantly affected. The conclusion, perhaps mirroring the position in France, was that peas have been sown too closely in the rotation. Steps have now been taken in Australia to widen the rotation and to advise on sowing peas once in every five years (Glenn McDonald, personal communication), as well as avoiding sowing too early in the spring. It is noted that disease control in peas, rather than yield improvement *per se*, has been the focus of agronomic advice in Australia.

Before considering the limitations of peas, and the improvements which could be made to agronomic advice in the UK, the physiology of the crop must be considered, and how this relates to those major arable species against which peas compete.

Physiological constraints

This section presents peas and beans as 'typical' arable crops whose productivity is determined by the duration of the growing period, the maximal rate of crop growth during the grand phase of growth, and the partitioning of a portion of the total dry matter into grain (harvest index). This analysis was developed by Monteith and Scott (1982).

CROP GROWTH RATES AND DRY MATTER ACCUMULATION

Of key interest in the productivity of an arable crop is its ability to intercept light. A number of studies have compared conventional, leafless and semi-leafless peas (Heath and Hebblethwaite, 1995c). All these have shown that there are no differences between leaf-types in terms of the relationship between biomass accumulated and light energy intercepted, and that the figures for peas of 2.94 g/MJ PAR (Pyke and Hedley, 1985) and 2.25-3.43 g/MJ PAR (Heath and Hebblethwaite, 1985c) are similar to the value of 2.9 g/MJ PAR (1.4 g/MJ total radiation) derived for several British crops (Monteith, 1977), and to the values for modern wheat varieties (2.8-3.9 g/MJ PAR; Gillett, Crout, Stokes, Sylvester-Bradley and Scott, 2001). In contrast, some studies on faba beans report higher radiation use efficiencies of up to 3.56 g/MJ PAR (see Table 3. in Stützel and Aufhammer, 1991), and values of 2.36-5.54 g/MJ PAR for winter beans (Pilbeam, Hebblethwaite, Nyongesa and Ricketts, 1991a).

The higher values of radiation use efficiency in field beans indicate greater potential gross productivity, but there is little evidence that breeding *per se* has, or will, increase yield potential of beans further through this route. Thus, assuming that there have been no changes in radiation use efficiency (RUE) associated with more recent pea or bean genotypes, it is likely that most improvements in productivity will come from changes in the rates and durations of crop growth, and the efficiency with which the total dry matter can be partitioned into the grain. However, it is accepted that changes in RUE may come through changes in plant stature, and lowering the incidence of lodging in farm crops, at some point in the future.

Figure 12.3 shows total dry matter production for three varieties of combining peas at two sowing dates in the spring. The data for the linear phase of growth was estimated, by fitting a linear regression to the data. The crop growth rates, final yields and harvest indices are summarised in Table 12.4. The average growth rate of 18.1 g/m^2/d for peas agrees with the values reported by other workers (e.g. 19.6 g/m^2/d by Sibma, 1968; 18.6 g/m^2/d by Greenwood, Cleaver, Loguens and Niendorf, 1977). These estimates are within the range of growth rates for a winter wheat crop (mean, 19.8 g/m^2/d; range, 16.1-23.6 g/m^2/d) when data from 18 crops of Mercia winter wheat were estimated similarly (based on data of Sylvester-Bradley, Scott, Clare and Duffield, 1998a). Spring beans have similar growth rates (Stützel and Aufhammer, 1991), but winter beans have reported growth rates somewhat higher than the other crops (19.9-29.0 g/m^2/d, Pilbeam *et al.*, 1991a), although the available data are limited.

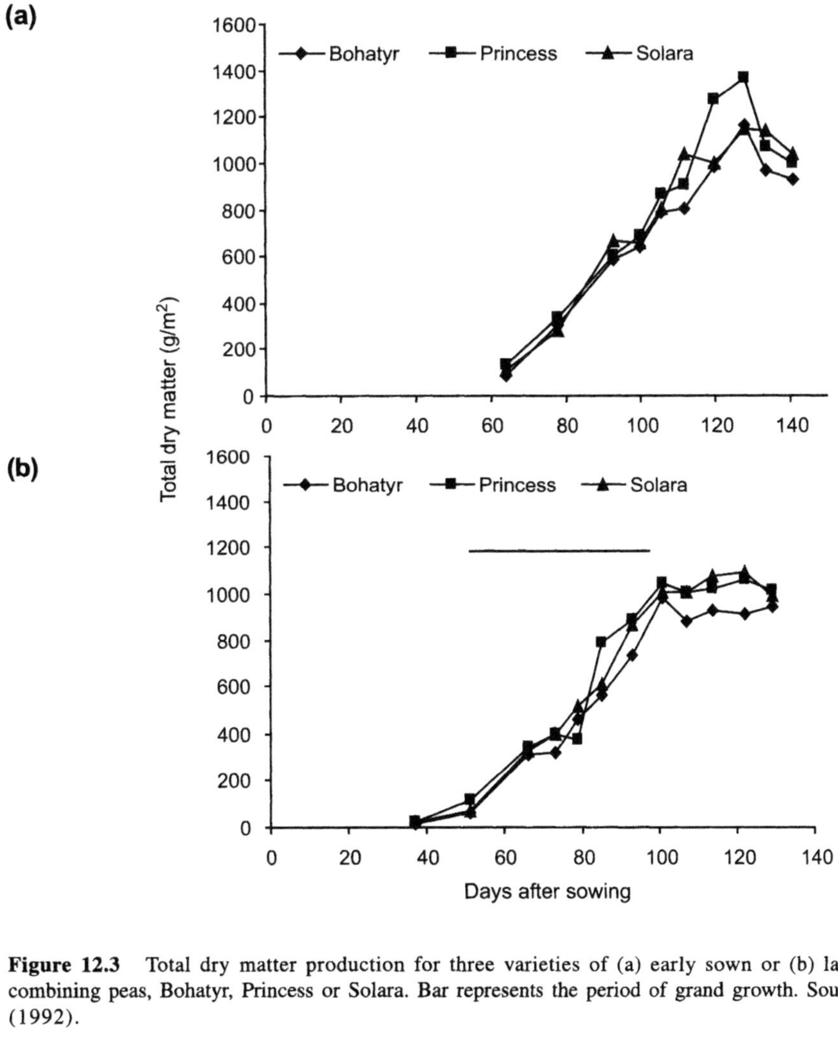

Figure 12.3 Total dry matter production for three varieties of (a) early sown or (b) late sown spring combining peas, Bohatyr, Princess or Solara. Bar represents the period of grand growth. Source: Weightman (1992).

Table 12.4 CROP GROWTH RATES DURING THE PHASE OF 'GRAND GROWTH', AND COMBINE HARVEST YIELDS (AT 85%DM) AND HARVEST INDICES FOR EARLY (28TH MARCH) AND LATE (24TH APRIL) SPRING SOWN PEA CROPS AT COCKLE PARK, UK IN 1990. SOURCE: WEIGHTMAN (1992).

Variety	*Crop growth rate ($g/m^2/d$)*	*Combine harvested yield (t/ha)*	*Harvest index*
		Early sowing	
Bohatyr	16.4	5.92	50.1
Princess	19.7	6.73	44.8
Solara	16.8	5.58	54.0
		Late sowing	
Bohatyr	17.7	5.17	53.0
Princess	19.3	5.91	48.7
Solara	18.8	5.24	59.1

Harvest indices for the pea crops described above are close to the upper ranges of published data for other carbohydrate storers including winter wheat (0.43-0.54), maize (0.42-0.49) and barley (0.43-0.57) summarised by Hay (1995). Thus genetic improvement in peas has certainly produced crops which are able to partition a high proportion of accumulated dry matter into harvestable yield. The short-stature, semi-leafless varieties such as Solara can therefore exhibit very high harvest indices (Weightman, 1992).

In contrast, spring field beans tend to have lower harvest indices of 0.24-0.43 (Stützel, Aufhammer and Löber, 1994), although Pilbeam, Duc and Hebblethwaite (1990) reported higher harvest indices (0.35 to 0.55) in some field experiments. For two cultivars of winter beans grown in two seasons in the UK, harvest indices of 0.39-0.55 were recorded (Pilbeam, Hebblethwaite, Ricketts and Nyongesa, 1991b). Field beans tend to produce larger canopies than peas, but suffer from higher levels of competition for assimilate between reproductive structures, leading to pod abortion. As pods maintain a large pool of N, and contribute significantly to photosynthesis during seed filling, this investment in carbon and N within structures which subsequently do not translate to seed yield is probably the main reason for the lower progress in bean yields. In the 1980's, attempts to increase bean yields using determinate genotypes (by introducing a terminal inflorescence) did not succeed because the determinate beans had fewer pods and therefore lower yield potential. Although branching was greater than in indeterminate lines, the side branches were shown to be reproductively inferior (Pilbeam *et al.*, 1990). Changing the components of yield (pods/node, seeds/pod, etc.) through breeding has proved notoriously difficult in beans, because of the inherent plasticity of these components.

If efficiency of light interception, rates of above-ground dry matter accumulation, and efficiency of partitioning dry matter into harvestable yield are comparable with other crops, then the major limitation to the yield potential of peas is likely to be the duration of the grand growth phase. For the crops in Figure 12.3, durations of grand growth were 50 and 64 days for the late and early sown crops respectively. Stützel and Aufhammer (1991) reported maximum durations of *c.* 50 days of growth for the main phase dry matter accumulation in spring-sown field beans, and Pilbeam *et al.*, (1991a) reported durations of *c.* 90 days for winter beans. These compare against an average duration of the main growth phase of 91 days for Mercia winter wheat (range 76-109 days, based on data of Sylvester-Bradley *et al.*, 1998a). An example of the growth of Mercia winter wheat is shown in Figure 12.4. Thus, based on the duration of growth, we would expect the productivity of spring peas or spring beans to be approximately half that of winter wheat and winter beans in the UK. The key to increasing the yield potential of peas is therefore to increase the duration of growth. This amounts to achieving canopy closure earlier in the season, maintaining the canopy above the critical leaf area index which gives 95% interception of the incident radiation, and delaying the onset of senescence (Monteith and Scott, 1982).

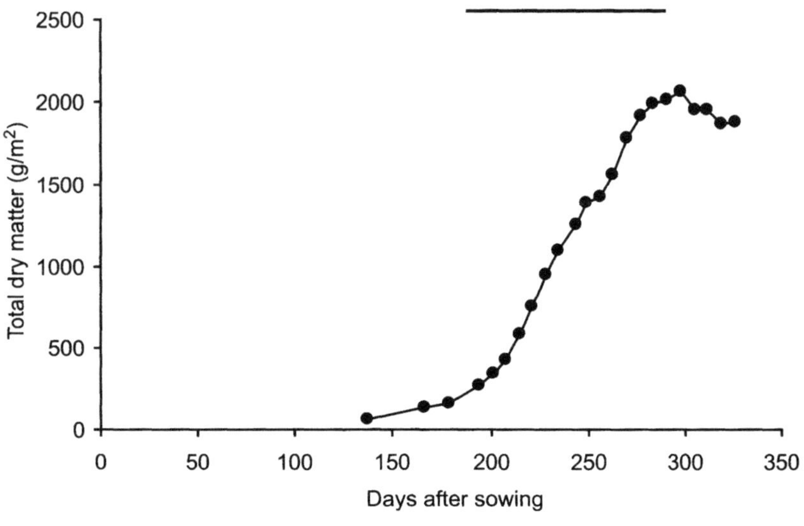

Figure 12.4 Total dry matter production for a crop of Mercia winter wheat sown at Boxworth, Cambridgeshire, UK in the 1992/93 growing season. Bar represents the period of grand growth. Source: Sylvester-Bradley *et al.* (1998a).

THE DURATION OF GROWTH

Autumn-sown peas

Peas are restricted to the spring-sown types in UK, as autumn-sown pea varieties are considered unpredictable in their ability to survive over-winter conditions and yield well (Silim, Hebblethwaite and Heath, 1985). Heath (1987) described the historical unsuccessful attempts to introduce autumn-sown (or 'winter') peas into the UK. Despite a number of companies listed as agents for winter peas (Anon. 2002), there is apparently little commercial activity, and UK RL yields are not available for winter peas. The main problems with winter peas are cited as a high risk of total crop loss over winter and an unpredictable harvest date. Clearly, autumn-sown peas can potentially increase the duration of growth by allowing the crop to develop a larger canopy earlier in the season. There may be disadvantages in bringing flowering forward, as this may increase the exposure of reproductive organs to frost damage, but it could extend the accumulation of N (see discussion below) and allow the canopy to close earlier in the season. Silim *et al.* (1985) demonstrated very high peak leaf area indices for the winter variety Frimas sown in the autumn, but there was no yield advantage over spring types. However, commercial practice tends to indicate that the main benefit of autumn sown peas would be in providing an earlier harvest.

Spring-sown peas

Peas will not emerge below 3-5°C although emerged spring peas will tolerate temperatures of –6°C (Plancquaert, 1991). Date of sowing is the first critical decision to be made when sowing peas in spring, and is a key factor in determining the length of the growing season. In the UK it has been established for many years that, for each week's delay after the first week in March, yield falls by approximately 0.125 t/ha (Anon. 2002). In France the optimum sowing dates in Brittany and the Paris basin are some two weeks earlier than the UK, with earliest sowing dates around the end of January.

Pyke and Hedley (1985) showed that, for field-sown peas, conventional and semi-leafless types had achieved leaf area indices of over 3 by 45-50 days from sowing (Figure 12.5). For a pea crop sown in mid-March, this means that canopy closure should be achieved by early May. This is not particularly limiting in comparison with winter cereals in the UK, for which the benchmark date to reach a crop green area index of 3 is estimated as 1st May (Sylvester-Bradley *et al.*, 1998b).

The next determinant of the duration of the closed canopy, is the length of time at which the canopy is above the critical leaf area index of 3. This is driven by the peak leaf area achieved, for a variety of given maturity. In comparison with the extensive knowledge of the growth of modern wheat varieties, there are few data on peak leaf area indices in peas. In the data of Nichols, Ragan and Floyd (1985), vining peas sown at 100 plants/m^2 achieved a leaf area index of *c.* 7.5 in New Zealand, comparable with that for a well-fertilised wheat crop. Silim *et al.* (1985) found photosynthetic area indices in the range 4 to 12 for spring sown peas grown under trial conditions, a similar range to the green area indices of 3 to 9 have been found in winter wheat

(Sylvester-Bradley *et al.*, 1998b). For semi-leafless peas, Heath and Hebblethwaite (1985c) reported peak photosynthetic area indices of *c.* 2.5 and 6 at two contrasting sites in the UK.

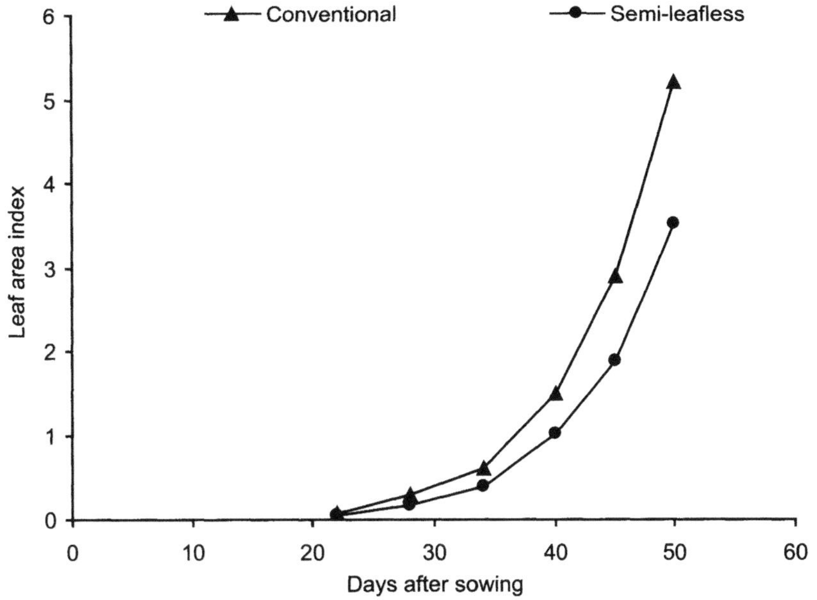

Figure 12.5 Leaf area indices for conventional and semi-leafless peas during early crop growth. Data from Pyke and Hedley (1985).

Production of more foliage and vegetative dry matter in peas, is contrary to the generally accepted pea ideotype in the UK (i.e. a short statured plant, which matures quickly). Excess canopies are undesirable as they cause crops to lodge, and encourage the build up of pests and diseases. Modern practice for other crops tends to involve avoidance of excess canopy production, but postponement of senescence in order to maintain green leaf area for longer; the aim is to consider the factors that influence persistence of an efficient, photosynthesising canopy in the field, rather than to generate more canopy *per se*. This draws us on to consider nutrition, hence the root and soil environment.

NITROGEN ECONOMY

The arguments have been made for extending the period of growth of peas through delaying the onset of senescence, hence maintaining canopy activity. However, as well as physiological maturity, which is genetically driven, persistence of the canopy in most crop species is determined by N availability. A pea seed of average thousand seed weight (250 g) contains approximately 25% protein. When drilled to establish 70 plants/m^2, the seeds themselves can be estimated to provide 7.5 kg/ha of N to the developing crop. Assuming that N is used in the same way as other crops in order to build leaf area, this gives a starting green area index of approximately 0.25, and provide a photosynthetic surface which can intercept light. Most data suggest that rhizobial N fixation is very efficient, and the fact that peas can generate over three units of leaf area within 45 days of sowing suggests rapid accumulation of N at 2 kg/ha/d (assuming that *c.* 90kg of N is needed to generate three units of leaf area, as is the case for wheat). Gaillard and Duchene (1995) suggested that a high yielding spring pea crop will have accumulated 129 kg/ha of N in the aerial parts by the start of flowering, and this again fits with such rates of N accumulation. A high yielding pea crop of 6.7 t/ha yield at maturity (all yields will be quoted at 85% dry matter unless otherwise stated), with a crude protein concentration of 22 g/100g dry matter (N x 6.25) will, assuming a N harvest index of 0.77, have accumulated *c.* 260 kg/ha of N above ground by crop maturity, comparable with uptakes recorded in winter wheat (Sylvester-Bradley *et al.*, 1998b). The crop therefore has to accumulate a further 130 kg/ha of N during the grand phase of growth. Assuming that this phase lasts 50-65 days, uptakes of N during this period of *c.* 2-2.6 kg/ha/d are indicated. Indeed, very high rates of N uptake are possible in peas: Aman and Graham (1987) demonstrated that, for a pea crop in Sweden, N uptake during the main phase of growth was of the order of 5 kg/ha/d N.

Clearly a high yielding pea crop has to accumulate N at a very high rate. It must do so with a shallow rooting system; it does not have the time to develop as extensive a root system as an autumn sown crop. Low temperatures in the spring are probably not significant as *Rhizobium* activity shows a broad temperature optimum of 5-30°C (Pate, 1977). However, any factors which limit root exploration or nodulation with *Rhizobium* will limit N capture by peas, for instance poor soil structure (compaction) and waterlogging early in the season, deficiencies in mineral nutrition, particularly sulphur (Zhao, Wood and McGrath, 1999), and drought late in the season. These factors are considered further (below) when the discussion returns to on-farm yields, and considers why pea yield increases are currently negligible in the UK. It should be noted that, in France, fertiliser N is often used in spring to give pea crops a boost, but this is not standard practice in the UK.

As described above, a modern high yielding pea crop (yield *c.* 5.5 t/ha; crude protein *c.* 22%) stores *c.* 200 kg/ha N in the seed. Given the typical amounts of N accumulated prior to flowering, then continued N fixation is essential after the start of flowering, to supply the amount of N required to satisfy seed demand. This continued fixation depends on provision of energy through photosynthesis during flowering and early grain filling, and availability of sufficient water for effective root nodule activity. Anything which limits N fixation at this stage (such as drought) means that the pea plant will literally 'self destruct', translocating much of the N in vegetative parts to the seed. This in turn means increased rate of loss of photosynthetically active green area, which will serve to halt growth, and further limit N fixation. Maintaining N fixation for as long as possible is therefore critical to achieving the highest yields of peas. Smartt (1990) suggested that in selecting for more convenient morphology, we have selected against efficient N fixation, by reducing the time span over which it can occur effectively.

Site and location effects on yield

Peas rely heavily on symbiotically fixed N, but can also rely on soil mineral N uptake. At low concentrations of mineral N, peas can take up nitrate-N, and also fix N_2 through rhizobial activity. For example, in pot-grown plants with 10-60 mg/l nitrate-N in the rooting medium, nodule numbers were not reduced (Pate, 1977). Moreover nitrate benefited nodule growth relative to those plants receiving zero N. However, very high concentrations of soil mineral N will inhibit fixation. One interesting set of data, which suggests that N could play a significant role in pea yields, is based on analysis of regional RL trials data. The site that has consistently produced the highest pea yields in RL trials is on fen peat, near Chatteris, Cambridgeshire (Table 12.5).

Table 12.5 ANNUAL COMBINING PEA YIELDS (AT 85%DM) FROM REGIONAL LOCATIONS USED AS RECOMMENDED LIST TRIAL SITES, FOR THE 10 YEAR PERIOD, 1992-2002. SOURCE: PROCESSORS AND GROWERS RESEARCH ORGANISATION.

Location	*Location mean (t/ha)*	*SE*	*n*	*1992*	*1993*	*1994*	*1995*	*1997*	*1998*	*1999*	*2000*	*2001*	*2002*
Chatteris, Cambs.	6.32	0.267	10	7.30	7.10	5.08	6.37	6.30	6.28	7.57	6.25	5.13	5.77
Cambridge, Cambs.	4.97	0.286	8	5.22	6.11	5.06	*	5.88	3.70	4.97	4.66	4.14	*
Thornhaugh, Cambs.	4.49	0.355	8	5.50	4.84	2.65	*	*	5.72	4.81	4.19	4.61	3.58
Littlebourne, Kent	5.72	0.395	4	*	*	*	*	*	6.09	6.34	5.89	*	4.57
Wye, Kent	5.54	0.377	4	*	5.94	6.02	4.42	5.79	*	*	*	*	*
Codford, Wiltshire	5.85	0.520	2	*	*	6.37	5.33	*	*	*	*	*	*
Wiltshire	5.28	0.192	5	*	*	*	*	5.34	5.42	*	5.69	5.39	4.55
Headley Hall, Yorks.	5.11	0.459	5	5.84	5.23	5.93	3.38	5.19	*	*	*	*	*
North Yorkshire	4.35	0.401	3	*	*	*	*	*	4.21	*	*	5.10	3.73
Shropshire	4.76	0.431	9	4.52	5.17	4.68	2.16	7.00	*	5.03	4.36	4.23	5.69
Hampshire	4.74	0.451	9	3.58	4.96	6.91	4.15	2.72	6.24	*	3.81	5.72	4.55
Norfolk	4.72	0.421	7	3.13	*	3.58	4.95	6.57	*	4.97	4.91	4.94	*
Annual mean				5.01	5.62	5.14	4.39	5.60	5.38	5.62	4.97	4.91	4.63

Fen soils are well known to mineralise and supply appreciable quantities of N such that, for many crops, application of fertiliser N shows no yield response. Heath *et al.* (1991) carried out a series of plant population experiments with peas across a range of soil types including an organic fen soil (also at Chatteris). It was shown that sowing peas on the organic soil was a less risky strategy than sowing on mineral soils, when peas were established at low plant populations (e.g. target density 40 plants/m^2). On the organic soil, the peas were more able to meet their yield potential by increasing yield/plant, particularly at low plant populations. Factors other than N availability may be significant, as the site was also reported to have a relatively high water table, which would also prolong the duration of growth. Also the fen soil suffers less from compaction than a clay soil, and is likely to enable better root development.

WATER AVAILABILITY AND LODGING

Because of the limitations of rooting depth in peas and beans, and the requirements of these crops for transpiration/photosynthesis, and also to support N mobilisation from roots to aerial parts via the phloem, water availability is key. It has been known for many years that grain legumes are very responsive to water at critical phases during growth. Spring bean yields at ADAS Boxworth, Cambridgeshire are shown in Figure 12.6, illustrating the seasonal variability in yields, and how low rainfall in spring and summer can be a limiting factor (slope of the line represents a 0.81 t/ha yield increase for each 100mm of rain). Similarly, Hebblethwaite *et al.* (1991) reported that field beans gave average yield responses to irrigation of 0.66 t/ha for each 100mm of water, although responses of individual crops could be much higher.

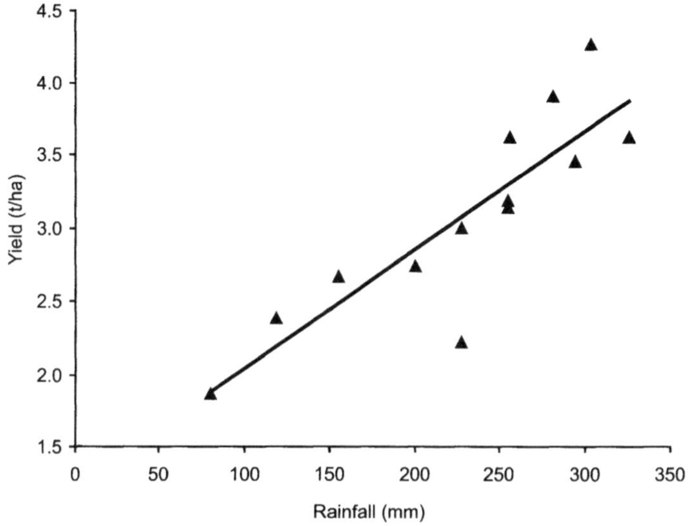

Figure 12.6 Annual yields of spring field beans plotted against the total rainfall between 1st April and 31st August, for crops grown on chalky boulder clay (Hanslope series) at ADAS Boxworth, Cambridgeshire, UK. Source: S Cook.

In peas, there is a notion of a fine balance between restricting water in order to make the crop determinate (and thus avoiding excess canopy growth), and applying water during flowering to ensure that pod set is not restricted. This has been turned into real advice in France (see earlier discussion). However advice in the UK is still somewhat contradictory – it is suggested that peas are grown on lighter land, because on heavier soils, with greater soil moisture availability, peas will produce excessive haulm, and the risk of lodging and more difficult harvests is increased. Yet evidence in the NIAB RL leaflet (Anon. 2002) shows that the mean control yield in the wet years 1997, 1998, 1999 and 2000 was 0.69 t/ha higher than mean yield for the drier years 1992, 1995, 1996 and 2001. This mean masks the fact that the older varieties (Maro, Supra, Princess and Samson) performed better in dry years (perhaps because they could be harvested more easily), whereas varieties Nitouche, Arrow, Venture and Espace performed better in wet years (when their greater yield potential was expressed, but harvesting was still possible due to the superior growth habit). Of course these are simplistic inferences that may not always translate to the farm situation, because it may be possible to harvest RL trials in wet seasons, whereas harvesting would be much harder at the larger farm scale. Nevertheless, these limited data suggest that modern pea varieties could be better matched to soil type and water availability.

Thirty years ago, peas may have lodged every year, whereas now, severe lodging may only take place one year in four. The modern semi-leafless varieties have much improved stature and so lodging incidence has been reduced. However, the data in Figure 12.7, demonstrating an increasing variability in yield, may be due in part to the fact that in dry years, modern short statured varieties are less productive, because they produce less canopy, intercept less light, and senesce earlier, giving less time for maximal growth and limiting N fixation. In contrast, in wetter years, biomass production and yield are maximised, but the semi-leafless crop is much easier to harvest, and so the full yield potential is realised. These points suggest that the improved crop ideotypes developed by breeders have not been optimised to soil type, and further work is needed to match peas to the varied growing locations found in the UK.

SOIL MANAGEMENT

Matching a legume crop to soil type is not only about matching with soil water availability. Some soils will simply not grow peas, due to *Fusarium* and foot rot complexes. This is partly the reason for the historic advice, referred to above, that peas should be grown on lighter soils, as these tend to be less susceptible to root diseases. Moreover, growing peas too close in the rotation will encourage a build up of pathogens. Because peas are so shallow rooting, any pathogens that attack the root or stem base will quickly affect N fixation, water uptake and standing ability.

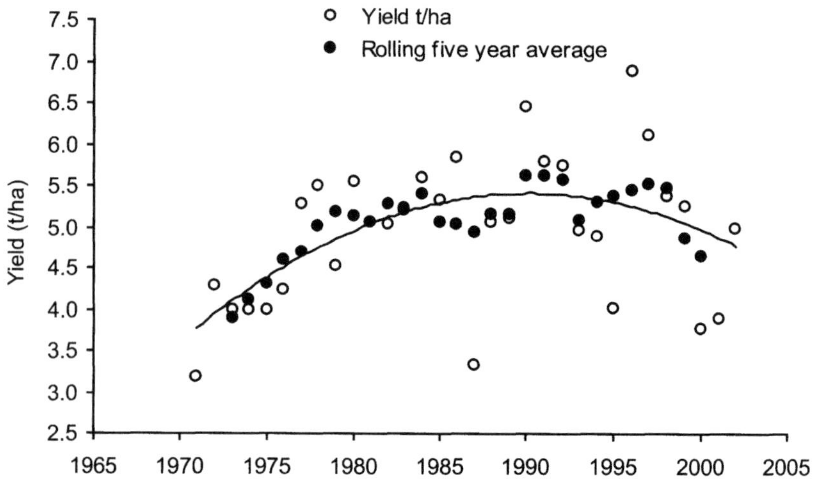

Figure 12.7 Annual or rolling five-year average yields of spring peas, from Weeley Lodge Farm, Essex, UK during the period 1971-2002. Varieties grown: 1971-1976, Maro/Vedette; 1975-1985, Birte; 1986-1998, Bohatyr, Solara, Elan, Baccara; 1999-2002, Espace, Nitouche. Line represents a 2nd order polynomial equation fitted to the rolling five-year average data. Source: Processors and Growers Research Organisation.

It has been well established that peas are very sensitive to soil compaction. The general effects of soil compaction are production of smaller plants with shallow and less branched root systems. In vining peas, compaction caused by wheelings was shown to reduce yields by up to 70% (Dawkins and McGowan, 1985). Minimum tillage regimes like autocasting may leave a subsoil which will suit cereals, but may over time become too consolidated for legumes. In recent years we have seen a change in agricultural practice to greater use of contractors. This indicates a tendency towards operators who are less familiar with the soil than a resident grower, the use of bigger farm machinery, and a tendency to move onto land at sub-optimal times, for example, sowing earlier in the spring, when soils are likely to be wetter. A combination of all these factors may have caused an increasing limitation on yield through damage to soil structure. Against this, the change to larger machinery may mean fewer passes per unit area of soil, and lower pressures from larger tyres. Data would need to be gathered before any conclusive statements could be made concerning changes in soil structure over the past 10 years, which may have contributed to lower yields.

SULPHUR

Sulphur is an essential component of proteins, hence is associated with N metabolism, and is an important nutrient in all crop plants. Deficiencies can be detected through determination of N:S ratios in leaf tissues. Sulphate behaves very much like nitrate in the soil, and leaches over winter. However, many soils do not receive sufficient quantities of sulphur, because industrial emissions and deposition on land have been reduced in recent years. For many cereals and oilseed rape crops now, on lighter soils, and particularly in the east of the country, sulphur is routinely applied. In legumes, there appears to be an additional requirement for sulphur in the root nodules. Zhao *et al.* (1999) demonstrated with pot-grown plants, that sulphur deficiency severely restricted N fixation in peas. Data for sulphur use on peas and beans is not routinely published but, for 2002, the evidence suggests that only 5% of combining pea crops received any sulphur (British Survey of Fertiliser Practice: fertiliser use on farm crops for crop year 2002, courtesy of the BSFP Authority). It is likely therefore that sulphur deficiency is limiting pea yield, particularly as peas tend to be grown on lighter soils. However, there has been no extensive experimental programme on fertiliser requirements of peas or beans in recent years in the UK, so there is little evidence on which to base advice. Most growers, because of the necessity to minimise operations in the spring, and the fact that peas are seen as a 'low input' or 'cheap' crop to grow, will tend to avoid putting fertiliser onto pea crops in the spring (P and K levels can be managed through maintenance applications at other times in the rotation).

PLANT POPULATIONS AND SEED

The data of Nichols *et al.* (1985) are interesting because they show the change in leaf area index with time at different planting densities. Above 100 plants/m^2 the LAI increases linearly with plant density, by approximately 0.2 leaf area units for every increase of 10 plants/m^2. However, pea seed is expensive and the pressure on the economics of farming would suggest that the trend will be towards lower plant populations. As growers try to save costs, they do so on seed dressings and by saving farm-produced seed. Discussions with breeders suggest that around 30% of seed sown on farm has been farm-saved. There is little evidence that use of farm-saved seed has led to a build up of any catastrophic diseases such as bacterial blight in peas, however, stem nematode in beans is a serious problem that poses real risks if growers do not get seed tested. Assuming that seed stocks are routinely tested, then vigour should not decline. Perhaps more significant in the case of peas is that growers may save money through using cheaper seed dressings, whereas a triple dressing will give much better protection against pests and diseases. Therefore changes to use of farm-saved seed and cutting costs in seed testing and dressings may have contributed to the lack of improvement in yields.

CONSTRAINTS TO FARM YIELDS IN THE UK

Figure 12.7 shows a set of data for commercial pea crops grown over a 31 year period on a single farm in Essex, UK. Soil type will have varied from year to year, and the varieties changed, as the grower adopted new improved cultivars. The data illustrate firstly the very variable yields; this is typical of the situation for most UK growers. What is perhaps more important is the apparent increase in variability of yield. When the yields are replotted as rolling five year averages, a curve can be fitted to the data which suggests, at best, that yields have plateaued, and at worst that they are declining. These data support on an individual farm, the UK national figures described in Figure 12.2.

This paper has reviewed the current limits to productivity in peas and beans. As indicated above, the lack of improvement on farm is probably due to a number of factors, all contributing small effects. However, given the current lack of research on crops in the field, it has been difficult, or even impossible in some cases, to gain objective data documenting decisions made on farms. Moreover, we do not know what constitutes either a low or high yielding pea or bean crop in the UK. Wheat, a crop about which a lot is now known, has 'benchmarks' for crop growth; such information is lacking for peas in the UK. In France, Gaillard and Duchene (1995) have described a benchmarking exercise on the variety Baccara from nine growing locations in 1994 (Table 12.6). The lack of benchmarks and measurements made under commercial conditions in the UK means that it is difficult to explain the poor performance of legumes in the field, and difficult to be sure about the potential of these crops.

Table 12.6 OPTIMAL PLANT POPULATIONS, YIELD COMPONENTS AND GROWTH RATES FOR A HIGH YIELDING PEA CROP OF VARIETY BACCARA GROWN IN 1994 ON LOAMY SOILS IN WESTERN FRANCE. SOURCE: GAILLARD AND DUCHENE (1995).

Parameter	*Value*
Plant density	65-70 /m^2
Stem density	95-105 /m^2
Biomass at beginning of flowering	3.0 t/ha
Nitrogen content of biomass at beginning of flowering	4.3 g/100g
Growth rate during the 250 day degrees following flowering	1.6 g/m^2/d
Number of fertile nodes/stem	6
Number of seeds/stem	25
Number of seeds per unit area	2450 /m^2
Thousand seed weight	285 g
Yield (at 85% DM)	7.0 t/ha
Seed protein content (N x 6.25)	24.8 g/100g

Yield predictions

POTENTIAL YIELDS OF PEAS

Notwithstanding the want of good evidence, this section looks towards the future, by seeking an estimate of the maximum theoretical yield that might be achieved in the UK for spring-sown combining peas. The average UK environment has been considered using the same data for incident solar radiation during grain filling as Spink and Berry (this volume), and the resultant theoretical potential pea yields are compared against the best yields seen in RL trials, and against typical farm yields (Table 12.7).

Table 12.7 CROP CHARACTERISTICS OF SPRING-SOWN COMBINING PEAS IN THE UK, FOR FARM, RECOMMENDED LIST (RL) AND A THEORETICAL HIGH YIELDING CROP.

	Units	*Average farm crop*	*RL crop*	*Maximum yielding crop**
Duration of initial growth period	(d)	40	40	40
Duration of period from canopy closure until flowering	(d)	27	30	30
Average rate of N uptake prior to flowering	(kg/ha/d)	1.2	1.6	2.0
Biomass at start of flowering	(t/ha)	2.5	3.1	3.5
Duration of period from flowering until end of grain filling	(d)	29	30	42
Average rate of N uptake during flowering and grain filling	(kgN/ha/d)	2.5	3.1	3.8
Incident PAR during flowering	($MJ/m^2/d$)	9	9	9
Proportion of light intercepted during grain filling	(%)	75	90	98
Radiation (PAR) Use Efficiency	(g/MJ)	2.20	2.36	2.70
Carbohydrate assimilated during grain filling	(t/ha)	4.5	5.7	10.0
Proportion of DM accumulated during grain filling which is translocated to seed	(%)	65	65	75
Grain yield (at 85% DM)	(t/ha)	3.6	4.8	8.9
Grain protein (at 100% DM)	(g/100g)	24.6	24.7	17.3

* estimated for an average UK environment

The model adopted for peas focuses on N uptake and allocation, and assumes that a modern semi-leafless pea genotype is grown. Typical values for rates of N uptake during early growth have been discussed earlier. In the UK it is assumed that there can be no advance from current sowing dates, and that the pre-flowering period could not be shortened, because of the increased risk of frost sensitivity during flowering. However, the highest yielding pea crops will make use of optimal soil conditions, by matching sowing date to soil type, to begin N fixation earlier. By improving rooting and N fixation early on (possibly including application of some fertiliser N at sowing) total N uptake by flowering would be of the order of 140 kg/ha N in the potential crop (compared to 112 and 80 kg/ha N in RL and farm crops respectively). RL trials are assumed to have higher N uptake than farm crops, as sites are most likely to be chosen with a longer break from peas, and will not be limiting in terms of key nutrients such as sulphur. Of the N accumulated pre-flowering in the RL crop it is assumed that 85% is translocated to the grain, whereas in the potential pea crop only 80% would be translocated, leaving more in vegetative structures to maintain photosynthesis. Typically crude protein left in the vegetative dry matter of peas at harvest is 5-7% (Aman and Graham, 1987), but it is proposed that this increases to *c.* 11% in the high yielding pea crop. By flowering, the potential pea crop will have accumulated 3.5 t/ha of above ground biomass compared to 3.1 and 2.5 t/ha for RL and farm crops respectively, and this additional biomass will have been allocated to stronger stems. Improvement in standing ability will, during the main phase of growth, cause improved light interception and RUE. The maintenance of an effective photosynthetic area for longer will also lead to improved light interception (Table 12.7).

Maintenance of green area in the potential pea crop would increase the duration of the main phase of growth by 12 days compared to RL crops, and by 16 days compared to current farm crops. The longer duration of grand growth, and the increased light interception and RUE during grain filling is therefore responsible for a much higher allocation of post anthesis DM assimilation to grain, leading to a potential grain yield of 8.9 t/ha for the potential crop (compared to 4.8 and 3.6 t/ha for RL and farm crops respectively). The main consequence of the higher biomass allocation to the seed, accompanied by higher N residues in vegetative structures, is a lower seed protein content in the potential pea crop. This theoretical yield (*c.* 9 t/ha) for the UK is realistic when we consider the best average yields for farm crops achieved in France (see Figure 12.2) and the high yields occasionally recorded in the UK (see Table 12.5 and Figure 12.7). However, it anticipates greater determinacy in UK pea crops, either by breeding or with more consistently dry summers (one possible scenario with expected climate change), and at the same time, an ability to irrigate to maintain growth and N fixation.

POTENTIAL YIELDS OF BEANS

This section estimates the maximum theoretical yields achievable in the UK for winter-sown field beans. Potentially the situation is different to that for peas, because winter beans already have durations of growth comparable to those for winter wheat, and the overall rate of biomass accumulation is high (see discussion earlier). Given the high canopy sizes achieved, and the higher N residues left for succeeding crops by beans (compared to peas), it is likely that N accumulation is, in most cases, not limiting. Total biomass production of 20 t/ha has been demonstrated experimentally in winter beans in the UK

(Pilbeam *et al.*, 1991a) and therefore, given a potential harvest index of 50% (harvest indices of 39-55% have been reported in winter beans by Pilbeam *et al.*, 1991a), a grain yield of 10 t/ha of winter beans should theoretically be possible. It is likely that winter beans will not reach the same high harvest indices as winter wheat because of their inability to re-translocate large quantities of biomass accumulated pre-anthesis to the growing seed. However, higher harvest indices have been recorded in spring beans (60-65%; Hebblethwaite, Batts and Kantar, 1991), where there is less investment in pre-anthesis biomass than winter beans.

Pilbeam *et al.* (1991a) reported yields in field experiments of up to 6.6 t/ha for winter beans in the UK, and these should be achievable in RL trials. However, even in RL trials, such grain yields are not currently realised (see Figure 12.1 and Table 12.2). In a collaborative EEC study, the highest yields of beans (>4.5 t/ha) were only achieved where seed growth rates of greater than 5 $g/m^2/d$ were maintained during the reproductive period (Bond, Lawes, Hawton, Saxena and Stephens, 1985). UK average yields for beans from Defra statistics (spring and winter beans combined) were only 3.7 t/ha for the period 1998-2003. One main reason for the low farm yields, as for peas, must be that the majority of the carbon assimilated pre-flowering cannot be re-translocated to the grain. The seed must therefore rely on photosynthesis from flowering onwards, in order to satisfy its demands for carbohydrate. The processes involving accumulation of N and C and partitioning into seed would be similar to those described for peas (Table 12.7). In beans (more so than in peas) the growing apex or side branches compete with the developing reproductive structures for assimilate (up to 30 pods/plant; Pilbeam *et al.*, 1991a) during this period. Thus, taking a simple approach, it can be seen that beans accumulate more N than peas, but they only allocate similar amounts of carbohydrate to grain, leading to the higher protein contents of beans (*c.* 35%) than in peas. Despite the better opportunities for rooting, the winter bean crop is also affected by drought during the later stages of growth (Hebblethwaite *et al.*, 1991) so, as with peas, the canopy senesces rapidly, and dry matter accumulation often ceases abruptly.

Bond *et al.* (1985) noted that most breeders at the time had a desired ideotype which typically would possess a strong stem, few side branches, a concentrated region of pods on the stem, the first pod not less than 20 cm from the ground, and little vegetative growth after the last flower. Such an ideotype is still desirable today in order to maximise the allocation of assimilate to seeds, and to avoid wasteful allocation to growing tips, side branches, and pods (which later abort), rather than to harvestable yield. It is believed that most of the genetic variation required to select for such attributes exists in beans, through various mutants including types with terminal inflorescences, synchronous flowering, and an independent vascular supply to each floret. Significant progress was made at the Plant Breeding Institute (Cambridge, UK) in altering crop canopy structure and short-stature beans, also known as 'semi-determinate' types (such as the spring variety 'Alfred'), are now available. However, no research into such physiological traits is now undertaken. For example, relatively little is known of the genetics involved with straw shortening in *V. faba* (D. Bond, cited by Heath, Pilbeam, McKenzie and Hebblethwaite, 1994) in contrast with recent advances in understanding lodging and dwarfing in cereals. Since there is now no public research programme on field beans in the UK, it seems unlikely that this or any of the other important physiological traits of beans will be further advanced in the foreseeable future; hence little further genetic improvement seems likely.

Detrimental aspects

As discussed in the opening paragraph, temperate grain legumes such as peas and beans are thought to be desirable in a sustainable cropping system. However, this view can be challenged. Peas in particular are a weedy crop, and progress since the 1960's has, in part, been due to the ability to reduce row width, dispense with mechanical weeding, and use herbicides instead. Despite their beneficial effects on soil fertility, peas do not fit well within organic systems because of their lack of competitiveness with weeds. The list of available herbicides for peas reduces every year. This returns the focus of technology to mechanical weeding, and asks whether new advances in agricultural engineering will allow effective weed control, while maintaining narrower row spacing. Currently there is little research on peas but, with cereals, current research into mechanical weeding tends to favour wide rows (22-25cm; N. Saunders, personal communication). Given the discussion on ground cover and leaf areas above, this move can only be seen as likely to reduce yields compared to those achieved with herbicides. If mechanical weeding were to be applied to semi-leafless peas, further research would be required into effects on the tendrils which cause plants in adjacent rows to meet and intertwine, and which help support the crop. Repeated passes with a mechanical weeder would break this supporting network, and may increase the incidence of lodging. Thus it is likely that continued chemical control of weeds will be required for the foreseeable future, and loss of available products will be detrimental to yields.

Although most UK soils possess ample levels of *Rhizobia*, and peas and beans are not artificially inoculated in the UK, there has been much commercial activity to produce improved *Rhizobium* strains through genetic modification. However, the use of GM *Rhizobia* itself poses a number of environmental questions (Hirsch, 2002), and the various environmental lobby organisations have already raised public awareness of this technology (see van Aken, 2000).

Summary

This paper has evaluated crop growth rates and dry matter partitioning in peas and beans, and has highlighted areas that need addressing in order to increase yields into the future. The evidence also shows that, for peas on-farm, yields lag somewhat behind those recorded for the best varieties in RL trials. Thus, despite tremendous improvements to the pea crop through the efforts of plant breeders, the benefits of continued genetic improvement are not being realised. For beans, yields appear to be static or declining at both farm and RL level, and there is little breeding activity now taking place in the UK. Lack of progress on-farm cannot be related to any one particular factor, but lack of research into physiological traits and their genetic control, lack of benchmarking of crops, and lack of new ideas being promoted through agronomic advice, suggest that yields are unlikely to rise above their current plateau. The current economic climate, which is leading farmers to cut costs, is probably not compatible with gaining the highest yields. Variability is still a major issue in the yields of peas and beans, and there is some evidence that variability has increased in recent years with short-stature varieties, which are probably not well matched to growing location or soil type. Therefore, it is predicted that average farm yields in 2010 will probably be little different from those in 2003 (*c.* 3.6 t/ha for both peas and beans). This will remain the case unless there is a run of seasons with favourable climatic conditions similar to those in the Mediterranean, where peas or beans originate, and an economic climate which justifies higher inputs including, importantly, water through irrigation.

Acknowledgments

Assistance is gratefully acknowledged from Dr Steve Wilcockson of the University of Newcastle-upon-Tyne, Mr Geoffrey Gent and staff of the Processors and Growers Research Organisation for helpful discussions, and provision of long term yield data, and from Mr Nigel Simpson and Dr Martin Heath of ADAS Boxworth for helpful comments on the manuscript.

References

van Aken, J. (2000). *Genetically Engineered Bacteria: US Lets Bad Gene Out of the Bottle.* Paper by Greenpeace International, Genetic Engineering Campaign, Chausseestr. 131, 10115 Berlin, Germany.

Aman, P. and Graham, H. (1987). Whole crop peas I. Changes in botanical and chemical composition and rumen in vitro digestibility during maturation. *Animal Feed Science and Technology, **17***; 15-31.

Anon (1985-2002). NIAB Recommended Lists of field peas and field beans. National Institute of Agricultural Botany, Huntingdon Road, Cambridge, UK.

Bond, D. A., Lawes, D. A., Hawtin, G. C., Saxena, M. C. and Stephens, J. H. (1985). *Faba bean (*Vicia faba *L.).* In: Grain Legume Crops. Eds. R. J.

Summerfield and E. H. Roberts. pp 199-265. Collins, London, ISBN 0-00-383037-3.

Caligari, P.D.S., Lobley, K., Weightman, R.M., Sylvester-Bradley, R. and Temple, M. (2002). *The Role of Future Public Investment in the Genetic Improvement of UK Grown Crops.* Final report on project ST0158 for the Department for Environment, Food and Rural Affairs. September 2002.

Dawkins, T. C. K. and McGowan, M. (1985). The influence of soil physical conditions on the growth, development and yield of vining peas. In: *The Pea Crop – A Basis for Improvement.* Eds. P. D. Hebblethwaite, M. C. Heath and T. C. K. Dawkins, pp 153-162. Butterworths, London. ISBN 0-407-00922-1.

Gaillard, B. and Duchene, E. (1995). Crop diagnosis, attending to pea crop management. *Proceedings of the 2nd European Conference on Grain Legumes, 9-13 July, 1995, Copenhagen, Denmark.* A.E.P., Paris, France, p153.

Gillett, A. G., Crout, N. M. J., Stokes, D. T., Sylvester-Bradley, R. and Scott, R. K. (2001). An approach to modelling the effect of environmental and physiological factors upon biomass accumulation in winter wheat. *Journal of Agricultural Science, Cambridge,* ***136***; 369-381.

Greenwood, D. J., Cleaver, T. J., Loquens, S. M. H. and Niendorf, K. B. (1977). Relationship between plant weight and growing period for vegetable crops in the United Kingdom. *Annals of Botany,* ***41***; 987-977.

Hay, R. K. M. (1995). Harvest Index: a review of its use in plant breeding and crop physiology. *Annals of Applied Biology,* ***126***; 197-216.

Heath, M. C. (1987). Grain legumes in UK agriculture. *Outlook on Agriculture,* ***16*** (1); 2-7.

Heath, M. C. and Hebblethwaite, P. D. (1985a). Agronomic problems associated with the pea crop. In: *The Pea Crop – A Basis for Improvement.* Eds. P. D. Hebblethwaite, M. C. Heath and T. C. K. Dawkins, pp 19-30. Butterworths, London. ISBN 0-407-00922-1.

Heath, M. C. and Hebblethwaite, P. D. (1985b). Evaluation of field-plot estimates for pea varieties. In: *The Pea Crop – A Basis for Improvement.* Eds. P. D. Hebblethwaite, M. C. Heath and T. C. K. Dawkins, pp 105-114. Butterworths, London. ISBN 0-407-00922-1.

Heath, M. C. and Hebblethwaite, P. D. (1985c). Solar radiation interception by leafless, semi-leafless and leafed pea (*Pisum sativum*) under contrasting field conditions. *Annals of Applied Biology,* ***107***; 309-318 .

Heath, M. C., Knott, C. M., Dyer, C. J. and Rogers-Lewis, D. (1991). Optimum plant densities for three semi-leafless pea (*Pisum sativum*) cultivars under contrasting field conditions. *Annals of Applied Biology,* ***118***; 671-688.

Heath, M. C., Pilbeam, C. J., McKenzie, B. A. and Hebblethwaite, P. D. (1994). Plant architecture, competitive ability, and crop productivity in food legumes, with particular emphasis on pea (Pisum sativum L.) and faba bean *(Vicia faba* L.*)*. In: *Expanding the Production and Use of Cool Season Food Legumes*. Eds. F. J. Muehlbauer and W. J. Kaiser, pp 771-790. Kluwer Academic publishers, London. ISBN 0-7923-2535-4.

Hebblethwaite, P. D., Heath, M. C. and Dawkins, T. C. K. (1985). *The Pea Crop – A Basis for Improvement*. Eds. P. D. Hebblethwaite, M. C. Heath and T. C. K. Dawkins. Butterworths, London. ISBN 0-407-00922-1.

Hebblethwaite, P. D., Batts, G. R. and Kantar, F. R. (1991). Influence of sowing date, irrigation, plant growth regulator use and flower colour on yield of faba beans. *Aspect of Applied Biology, **27***; pp77-84.

Hirsch, P. R. (2002). Fate of GM rhizobial inoculants; lessons from Europe and elsewhere. *Proceedings of the 7th International Symposium on the Biosafety of GMO's*. Beijing, China, October 10-17, 2002, pp 243-250.

Knott, C. M. (1996). The contribution of plant breeding to improvement in yield, agronomic and quality characters for field peas (*Pisum sativum* L.). *Plant Varieties and Seeds, **9***; 167-180.

Monteith, J.L. (1977). Climate and the efficiency of crop production in Britain. *Philosophical Transactions of the Royal Society of London, **B***; 277-294.

Monteith, J. L. and Scott, R. K. (1982). Weather and yield variation in crops. In: *Food Nutrition and Climate*. Eds. K. Blaxter and L. Fowden. pp 127-153. Applied Science Publishers, London.

Nichols, M. A., Ragan, P. and Floyd, R. M. (1985). Temperature and plant density studies with vining peas. In: *The Pea Crop – A Basis for Improvement*. Eds. P. D. Hebblethwaite, M. C. Heath and T. C. K. Dawkins, pp173-184. Butterworths, London. ISBN 0-407-00922-1.

Pate, J. S. (1977). Nodulation and nitrogen metabolism. In: *The Physiology of the Garden Pea*. Eds J. F. Sutcliffe and J. S. Pate, pp 349-383. Academic Press, London. ISBN 0-12-677550-8.

Peck, D. M. and McDonald, G. K. (1998). Field pea yields are declining in medium to high yielding areas. *Proceedings of the 9th Australian Agronomy Conference*, Wagg Wagga, Australia, July 1998.

Pilbeam, C. J., Duc, G. and Hebblethwaite, P. D. (1990). Effects of plant population density on spring-sown field beans (*Vicia faba*) with different growth habits. *Journal of Agricultural Science, Cambridge, **114***; 19-33.

Pilbeam, C. J., Hebblethwaite, P. D., Nyongesa, T. E. and Ricketts, H. E. (1991a). Effects of plant population density on determinate and indeterminate forms of winter field beans (*Vicia faba*). 2. Growth and development. *Journal of Agricultural Science, Cambridge, **116***; 385-393.

Pilbeam, C. J., Hebblethwaite, P. D., Ricketts, H. E. and Nyongesa, T. E. (1991b). Effects of plant population density on determinate and indeterminate forms of winter field beans (*Vicia faba*). 1. Yield and yield components. *Journal of Agricultural Science, Cambridge, **116***; 375-383.

Plancquaert, P. (1991). Peas in France. *Aspects of Applied Biology, **27** Production and Protection of Legumes*; 189-198.

Pyke, K. A. and Hedley, C. L. (1985). Growth and photosynthesis of different pea phenotypes. In: *The Pea Crop – A Basis for Improvement*. Eds. P. D. Hebblethwaite, M. C. Heath and T. C. K. Dawkins, pp 297-306. Butterworths, London. ISBN 0-407-00922-1.

Sibma, L. (1968). Growth of closed green crop surfaces in The Netherlands. *Netherlands Journal of Agricultural Science, **16***; 211-216.

Silim, S. N., Hebblethwaite, P. D. and Heath, M. C. (1985). Comparison of the effects of autumn and spring sowing date on growth and yield of combining peas (*Pisum sativum* L.) *Journal of Agricultural Science, Cambridge, 104*; 35-46.

Smartt, J. (1990). The role of grain legumes in the human economy. In: *Grain Legumes - Evolution and Genetic Resources*, pp 9-29. Cambridge University Press. ISBN 0-521-30797-X.

Stützel, H. and Aufhammer, W. (1991). Light interception and utilization in determinate and indeterminate cultivars of *Vicia faba* under contrasting plant distributions and population densities. *Journal of Agricultural Science, Cambridge, **116***; 395-407.

Stützel, H. Aufhammer, W and Löber, A. (1994). Effect of sowing technique on yield formation of *Vicia faba* as affected by population density, sowing date and plant type. *Journal of Agricultural Science, Cambridge, **122***; 255-264.

Sutcliffe, J. F. (1977). History of the use of pea in plant physiological research. In: *The Physiology of the Garden Pea*. Eds J. F. Sutcliffe and J. S. Pate, pp 1-19. Academic Press, London. ISBN 0-12-677550-8.

Sylvester-Bradley, R., Scott, R.K., Clare, R.W. and Duffield, S.J. (1998a) (Eds.) *Assessments of Wheat Growth to Support its Production and Improvement. Volume III: The Dataset.* Research Report No. 151. Home Grown Cereals Authority, Caledonia House, London.

Sylvester-Bradley, R., Scott, R.K., Clare, R.W. and Duffield, S.J. (1998b). (Eds.) *Assessments of Wheat Growth to Support its Production and Improvement. Volume I: The Wheat Growth Digest.* Research Report No. 151. Home Grown Cereals Authority, Caledonia House, London.

Sylvester-Bradley, R., Lunn, G., Foulkes, J., Shearman, V., Spink, J. & Ingram, J. (2002). Management strategies for high yields of cereals and oilseed rape. In: *Agronomic Intelligence: The Basis for Profitable Production.* Home-Grown Cereals Authority, R & D Conference, 8.1-8.17.

Thirtle, C., Bottomley, P., Palladino, P., Schimmelpfennig, D and Townsend, R. (1998). The rise and fall of public sector plant breeding in the United Kingdom: a causal chain model of basic and applied research and diffusion. *Agricultural Economics, **19***; 127-143.

Weightman, R. M. (1992). *Nutritional Changes During Maturation of the Dry Pea Crop.* PhD thesis, University of Newcastle upon Tyne, UK.

Zhao, F. J., Wood, A. P. and McGrath, S. P. (1999). Effects of sulphur nutrition on growth and nitrogen fixation of pea (*Pisum sativum* L.). *Plant and Soil, **212***; 209-219.

13

PHYSIOLOGICAL AND TECHNOLOGICAL LIMITS TO YIELD IMPROVEMENT OF POTATOES

E J ALLEN[1], M F ALLISON[1] AND D L SPARKES[2]
[1]*Cambridge University Farm, Cambridge, CB3 0LH;* [2]*Division of Agricultural and Environmental Sciences, School of Biosciences, University of Nottingham, Sutton Bonington Campus, Loughborough, Leics., LE12 5RD, UK.*

Introduction

Potatoes are produced to meet a wide range of consumer requirements, tuber size, appearance, freedom from disease, dry-matter content and chemical composition. Thus, crops are defoliated after quite different growing periods. This must be considered when trends in national yields are discussed, because the types of crops grown change from year to year: utilisation of crops may not be as originally intended (due to market forces) and, over time, there have been major changes in the types of crops grown. The latter is well illustrated by the expansion in salad potato production (tubers <40 mm diameter) over the last 20 years. The UK and parts of Europe are unusual in this diversity.

All growers aim to achieve the highest yield from their specific crops but the challenges and methods used can be quite different. In most sectors, yield *per se* is no longer the main driver of economic success; the achievement of demanding quality criteria is paramount. In future, the objective will be to achieve quality without prejudicing yield, rather than increasing yield without affecting quality. Not many growers have truly recognised this change of emphasis.

Yields

NATIONAL YIELDS

As Figure 13.1 shows, average yields increased until the nineties. Subsequently, it is difficult to establish whether there is a continuing trend or a plateau. It has been argued (Allen and Scott, 2001) that the main drivers of these increasing yields were straightforward: nutrients, irrigation and agro-chemicals, and consequently, the utilisation of these inputs must be improved, or other limitations must be addressed, if yields are to continue to increase. Certainly, the last 25 years have seen increased use of these three inputs; for agro-chemicals this has been in response to emerging yield-limiting pests and diseases e.g. potato cyst nematodes (PCN), blight and black scurf.

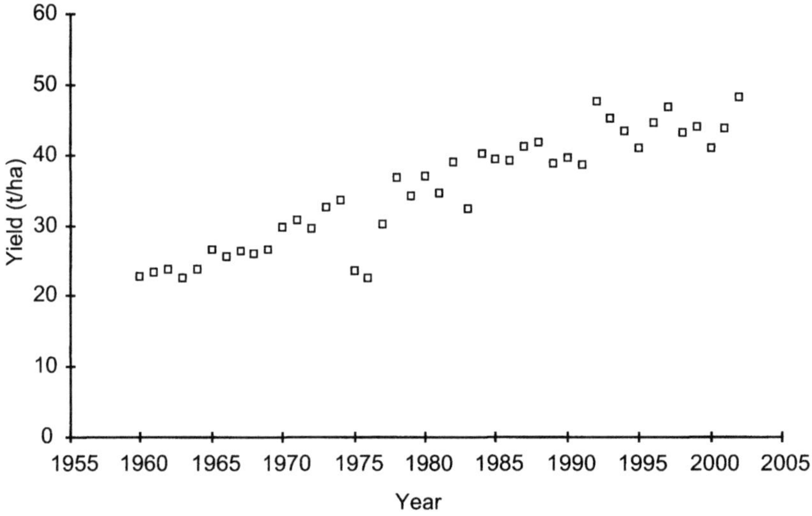

Figure 13.1 National potato yields 1960-2002 (British Potato Council, 2003).

In contrast to many crops, especially cereals, the increases in potato yields have not been achieved through continuous introduction of higher-yielding varieties. New varieties have been introduced over the last 50 years, notably from breeding programmes in Holland (Desiree, Estima, Wilja, Marfona, Lady Rosetta, Saturna) and the UK (Pentland Crown, Pentland Dell, Pentland Squire, Maris Peer, Maris Piper) but some have already disappeared, and only Estima, Marfona, Maris Peer, Maris Piper, Lady Rosetta and Saturna remain as major varieties. Of these, Maris Peer owes its survival to the emergence of the salad potato sector and Maris Piper has the desirable combination of PCN resistance and cooking/processing quality. Two varieties have been successfully introduced from North America: Russet Burbank and Shepody, but the former was already over 100 years old when introduced. Most of these varieties are no more efficient in total dry-matter production than the varieties which they replaced, and several, e.g. Estima, owed their success to higher tuber fresh-weight yields resulting from lower dry-matter contents (Table 13.1). In addition, the Dutch varieties had a growth pattern which was much earlier than traditional UK varieties, being effectively second earlies; this had major practical advantages for achieving good tuber appearance with earlier harvesting. Currently, almost all major varieties grown in the UK were bred between 30 and 130 years ago (Table 13.2) and in all sectors there appears to be no prospect of a new higher-yielding variety. This may also be the case in North America where Russet Burbank still dominates both the 'table' and 'French fry' sectors; this variety is challenged principally on quality not yield.

Table 13.1 EFFECT OF VARIETY ON TUBER DM CONCENTRATION (%) IN EXPERIMENTS AT CUF, 1997-2003.

	Estima	*Cara*	*Hermes*
1997	20.0	22.1	22.9
1998	19.4	17.5	23.5
1999	18.1	19.2	22.7
2000	20.3	20.8	25.1
2001	19.2	19.9	23.5
2002	18.9	21.7	22.8
2003	17.9	22.1	22.1
Mean	19.1	20.5	23.2

Table 13.2 AREA GROWN (ha) AND YEAR OF FIRST LISTING FOR THE TWELVE MOST POPULAR VARIETIES IN GREAT BRITAIN (BRITISH POTATO COUNCIL, 2003).

Variety	*Area grown in 2003*	*First listed*
Russet Burbank	2944	1874
King Edward	3590	1902
Marfona	3880	1975
Hermes	3986	1973
Nadine	4808	1987
Desiree	4929	1962
Saturna	5259	1964
Pentland Dell	5455	1961
Maris Peer	5787	1962
Lady Rosetta	5944	1988
Estima	10840	1973
Maris Piper	29275	1964

For the UK, it would seem reasonable to take *c.* 50 t/ha as the current national yield for full-season crops (Figure 13.1). Such yields have been documented in experimental plots for more than 60 years and have been reported increasingly from farm crops over this period. Over the same period, plot yields in excess of 100 t/ha have been reported (Table 13.3) and documented farm yields of >80 t/ha are well established. Thus, in a very broad sense, national yields have the potential to be very much higher without any new knowledge. A similar situation is found in other developed potato industries. In the major areas of the USA, e.g. Washington, plot yields of >130 t/ha and farm yields close to 100 t/ha are documented, with an average yield of 70 t/ha (Thornton, Washington State University, *personal communication*).

Table 13.3 MEAN TREATMENT YIELD AND MAXIMUM PLOT YIELD IN EXPERIMENTS CONDUCTED BY CUF.

Year	*Variety*	*Experiment*	*Mean treatment yield (t FW/ha)*	*Largest plot yield (t FW/ha)*
1997	Estima	Fulcrum	85	121
1998	Maris Piper	Variety & N	88	108
1998	Hermes	Variety & N	85	102
1999	Maris Piper	Sulphur	101	125
2002	Russet Burbank	Variety & N	81	102
2003	Estima	Variety & N	88	100

YIELD FORMATION

In potatoes, dry-matter production is directly proportional to the amount of radiation absorbed by the canopy (Figure 13.2) provided nutrient and water needs are met. Therefore, for all potato crops, rapid generation of complete ground cover and its maintenance for the allotted growing season is the primary objective in order to maximise radiation absorption. Two factors restrict generation and maintenance of potato canopies so that commercial yields are less than potential. First, the crop is susceptible to frost at the beginning and end of the season, and secondly, defoliation of the canopy is usually essential prior to harvest, especially to ensure successful storage.

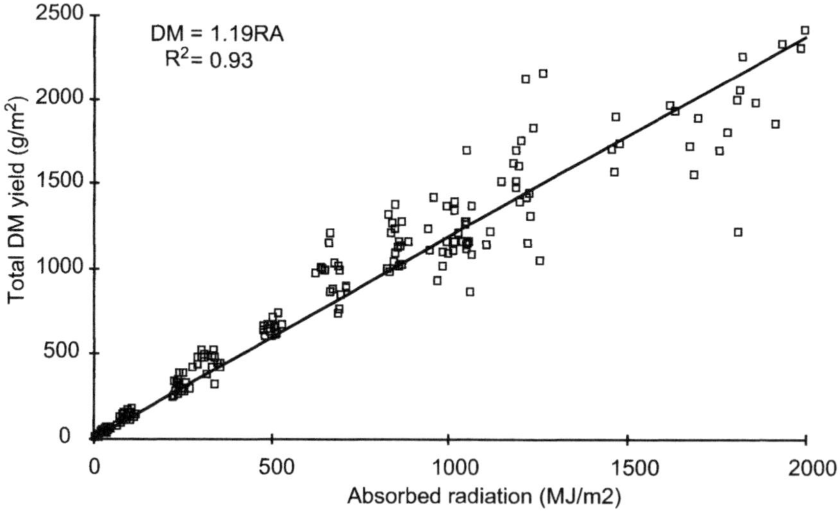

Figure 13.2 Relationship between total dry matter yield and radiation absorption estimated from ground cover for Estima and Cara at CUF 2003 (Allison and Allen, 2004).

The first factor could be moderated by breeding; the research group of Professor L Frusciante at the University of Naples claim to have clones which show resistance to 1-2°C of frost. It is, however, unlikely that such characteristics will appear in commercial varieties in the near future and the speed of their introduction will be affected by attitudes to genetic modification. There are also risks associated with planting in colder conditions, for emergence is delayed and the canopy morphology changes; internodes and petioles become shorter (Figure 13.3) and leaf expansion is restricted so that planting densities, which produce complete ground cover in warmer conditions, fail to do so in colder conditions. This is illustrated for dates of planting in Figure 13.4.

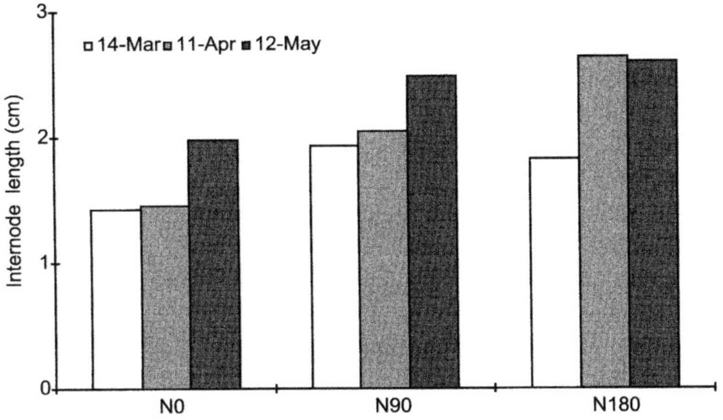

Figure 13.3 Effect of planting date and N application rate (kg N/ha) on internode length (Firman, 1987).

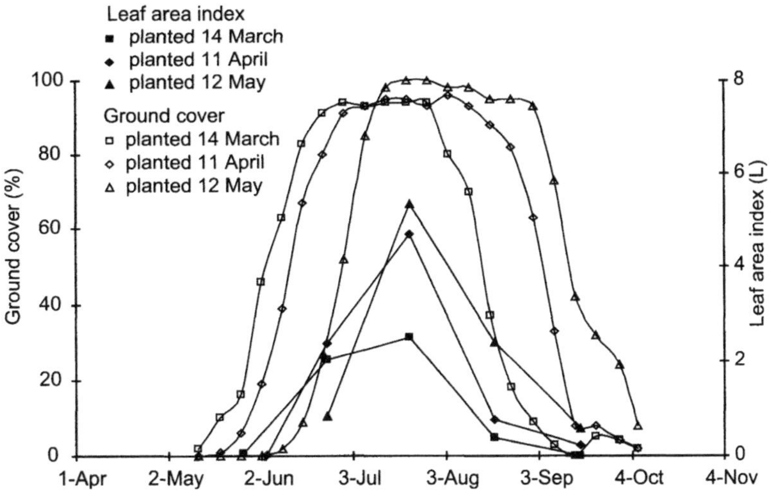

Figure 13.4 Effect of planting date on leaf area index (closed symbols) and ground cover (open symbols) for Estima given 180 kg N/ha (Firman, 1987).

The second factor is also unlikely to be overcome as the destruction of the canopy is considered essential for skin-set and a pre-requisite for harvesting. In order to minimise the risk of bruising, harvesting should occur in warm soil conditions (10-18°C), so defoliation normally occurs before any serious risk of frost, and this effectively limits the life of the crop.

Potato varieties vary greatly in their leaf growth characteristics and knowledge in this area is now extensive. All varieties conform to a basic morphology, illustrated in Figure 13.5. Each mainstem produces a relatively fixed number of leaves (according to variety) before an inflorescence is initiated. Subsequently, sympodial branches continue the elongation of the main axis and axillary branches are formed and both produce leaves. Each sympodial branch carries fewer leaves than the initial mainstem and ends in an inflorescence. Some varieties, e.g. Estima, and some agronomic treatments, e.g. low N or extensive sprouting, restrict the production of sympodial branches (and axillaries) and the growth habit is described as determinate. In contrast, other varieties (e.g. Cara, Russet Burbank) and other agronomic treatments (e.g. high N availability or delayed planting) enhance the production of sympodial branches (and axillaries) and the resultant growth habit is called indeterminate. As a consequence, the morphology, extent and longevity of a potato crop with complete ground cover is extremely variable. It can vary by 30-60 cm in height and 4 or 5 units of leaf area (i.e. 3 to 8), and it may be actively producing leaves or it may have completely stopped leaf production. Crops with such varied canopies may all maximise radiation absorption but give differing tuber yields, due to differences in dry-matter partitioning. For future improvements in commercial yields, knowledge of canopy growth and its relation to tuber bulking must be exploited more fully.

The balance between leaf and tuber growth has always been thought important (Ivins and Bremner, 1965) but the pre-occupation with tuber initiation as the central event in the developmental life of the crop led to the acceptance that once initiation occurred, tuber bulking would begin and dominate subsequent growth and development. It is now clear (O'Brien, Allen and Firman, 1998) that tuber initiation occurs early in the life of the crop (15-25 days post 50% emergence) and has no obvious impact on leaf initiation, leaf emergence and expansion, LAI or ground cover (Figure 13.6). This is not surprising, as the tubers are merely an integral part of the developing stem. However, it is now clear that, whilst few factors affect the timing of tuber initiation, the onset of bulking can be delayed and the extent of partitioning altered. The major agronomic factor is N nutrition and its effect is much influenced by variety. In determinate varieties, a considerable amount of N is needed to ensure sufficient leaf area is generated to last through the available growing season, as few sympodial and axillary branches are produced. Similar amounts of N applied to indeterminate varieties produce a profusion of branches which continue growth throughout the season and in so doing delay the onset and restrict the rate of bulking.

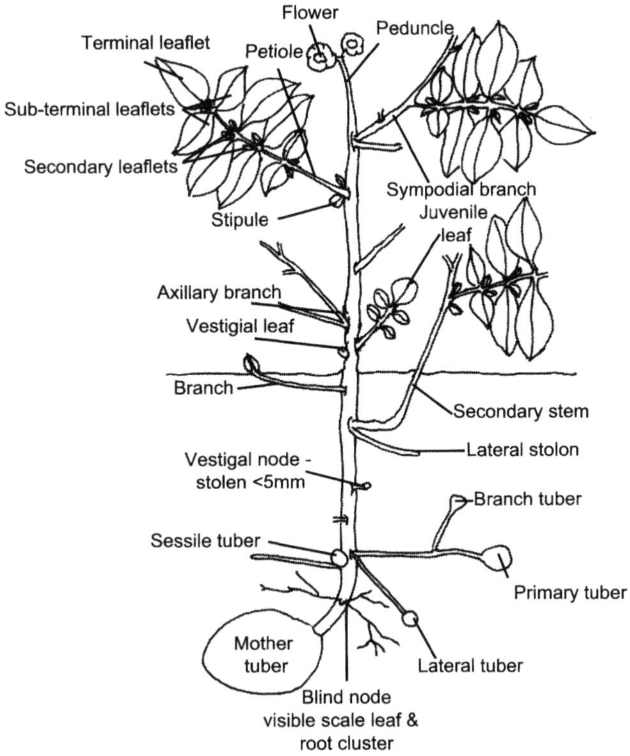

Figure 13.5 Structure of the potato plant detailing stem, leaf and tuber hierarchy (incomplete stem and not to scale), (Allen and Scott, 2001).

The greater persistence of such canopies provides limited compensation, as ambient radiation levels are decreasing at this time and the growth period is ended by defoliation. Table 13.4 shows effects in Estima and Cara grown with and without fertiliser N and demonstrates the enormous production of haulm when Cara has access to N through to the end of the season. In principle, much of the dry matter in this haulm could have been diverted to the tubers. Thus, managing the canopy for increased yields depends on linking the developmental predisposition of the variety and the availability of N from soil and fertiliser in a more effective way (see below).

POTENTIAL YIELD

The biological yield potential of the potato can be estimated by considering an 'ideal' potato crop that exploits resources to the maximum. Taking estimates from Allen and Scott (1980) for average fractional interception, approximately 2000 MJ/m^2 solar radiation could be intercepted by a full season canopy in the UK. Work in the UK (Allen and Scott, 1980; Khurana and McLaren, 1982;

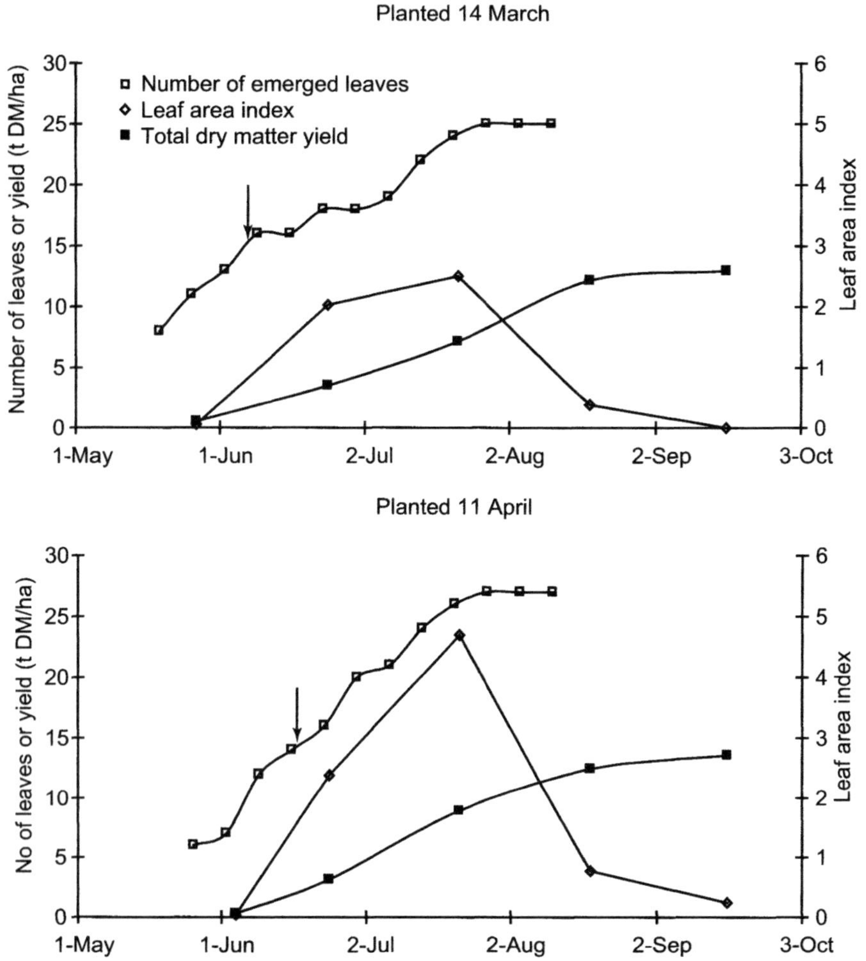

Figure 13.6 Relationship between number of emerged leaves, leaf area index and total dry matter yield in relation to tuber initiation (arrow) for Estima planted on 14 March or 11 April and given 180 kg N/ha. (Firman, 1987).

Jefferies and MacKerron, 1989) has shown radiation use efficiencies in the range 1.32 to 1.81 g/MJ. Taking the maximum value of 1.81 g/MJ, this equates to 3,620 g/m^2 biomass. Assuming a harvest index of 0.8, tuber dry matter would be 2,896 g/m^2. Tuber dry matter content is not constant, but assuming 20% this equates to a maximum fresh tuber yield of 14,480 g/m^2 or *c.* 145 t/ha. This compares with maximum recorded plots yields of 125 t/ha (Table 13.3) but average farm yields of only 50 t/ha.

The considerable gap between the potential yields of UK environments and average yields suggests that there are serious limitations to yield formation in most potato crops which, if corrected, would lead to significant increases in yield. These limitations are likely to have been present for some time, which suggests that utilisation of knowledge within the industry is a constraint on improvement of yields.

Table 13.4 EFFECTS OF VARIETY AND N APPLICATION RATE (kg/ha) ON HAULM, TUBER AND TOTAL DRY WEIGHT YIELDS (CUF 2003).

Total dry		*Estima*		*Cara*		
weight t/ha	*Mean*	*0*	*300*	*0*	*300*	*S.E.*
29 May						
Haulm	0.29	0.24	0.25	0.34	0.34	0.018
Tuber	0.00	0.00	0.00	0.00	0.00	-
Total	0.29	0.24	0.25	0.34	0.34	0.018
13 June						
Haulm	1.75	1.45	1.72	1.88	1.97	0.194
Tuber	0.34	0.63	0.56	0.15	0.03	0.040
Total	2.10	2.08	2.28	2.03	2.01	0.199
27 June						
Haulm	3.46	2.36	2.81	3.75	4.93	0.131
Tuber	2.07	3.17	2.85	1.68	0.58	0.078
Total	5.53	5.53	5.67	5.43	5.51	0.163
17 July						
Haulm	4.75	2.45	3.85	4.68	8.03	0.268
Tuber	5.93	8.23	8.26	5.10	2.11	0.193
Total	10.68	10.69	12.11	9.77	10.13	0.371
8 August						
Haulm	4.85	1.72	2.49	5.48	9.70	0.349
Tuber	8.84	10.18	11.16	9.03	5.01	0.490
Total	13.69	11.90	13.65	14.51	14.72	0.731
7 October						
Haulm	2.85	0.81	1.26	3.86	5.47	0.236
Tuber	13.59	10.54	14.45	14.62	14.76	0.874
Total	16.44	11.35	15.70	18.48	20.23	0.969

Limits to yield formation

SOIL COMPACTION

Potatoes are grown in soil types ranging from the lightest sands to quite heavy clay loams although there has been a major movement towards the lighter soils in the last 30 years. A similar move to lighter soils has occurred in Europe, e.g. Portugal. It is widely accepted that soils with a significant clay or silt content are easily compacted but only recently has the extent and severity of compaction on sandy soils become apparent. Figure 13.7 shows the change in resistance down the profile created by cultivation of wet, sandy soil and the dramatic difference created by delaying cultivation by only days. Such changes in resistance delay root penetration and, where more severe, may prevent further penetration. As a result, stems and petioles become shorter and growth of ground cover is restricted, thereby reducing yield (Figure 13.8). Restricted root penetration reduces water and nutrient uptake and, unless the crop can eventually penetrate the obstruction, a severe yield penalty results which cannot be avoided by use of more fertiliser N, although this strategy has been suggested in the USA (Westermann and Sojka, 1996). Use of water can exacerbate the problem as drainage is also impeded by compaction and with shallow compaction (<35mm) serious waterlogging can result which will shorten the life of the canopy and create a further yield penalty (Rosenfeld, 1997).

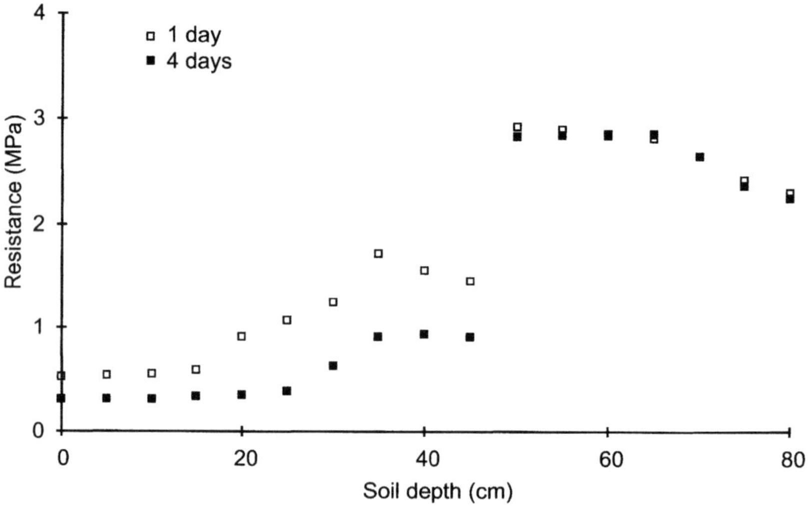

Figure 13.7 Effect of delaying planting 1 day or 4 days after heavy rain on soil resistance in a sandy loam soil, (Stalham, unpublished).

The occurrence of compaction on sandy soils results from their perceived suitability for early planting and the type of equipment now used. The development of potato production in many areas of sandy soils was facilitated by the availability of irrigation; this combination offered a lengthy growing season which would lead to higher yields. In order to minimise machinery costs, very large areas were planted with a single machine, necessitating an early start to soil cultivations (frequently at the end of February). The standard cultivation and stone separation system used by most growers operates above the plastic limit of the soil, leaving compaction at the depth of the main share and a surface soil of excellent appearance.

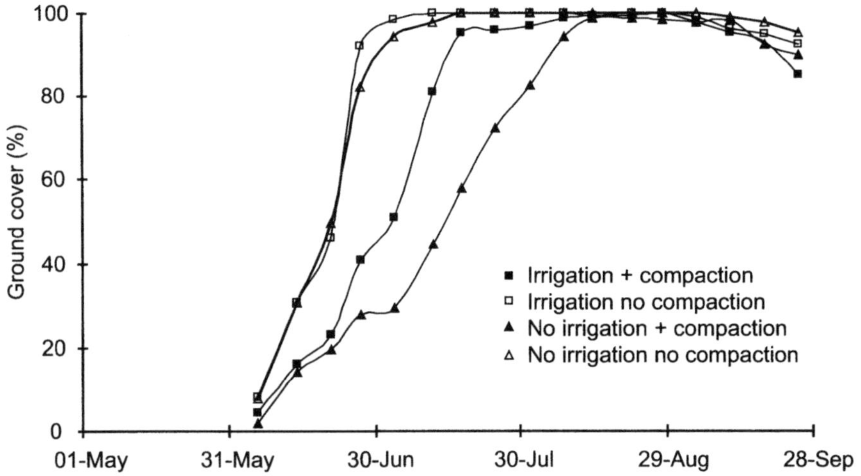

Figure 13.8 Effect of compaction and water supply on ground cover of Maris Piper (Rosenfeld, 1997).

Even on soils without compaction, there is accumulating evidence that soil type affects the rate of increase in ground cover. Figure 13.9 shows the time-course of ground cover of two plantings of Cara made on light and heavy soils approximately 250 m apart within the same field. The emergence and rate of increase in ground cover of the crop on the heavy soil was slower than on the light soil especially at the first planting. The fineness of the soil particles almost certainly affected the root-soil interface and thereby water and nutrient uptake. Such effects would normally be much smaller than the effects of compaction but suggest that soil type and cultivations may also affect yield potential.

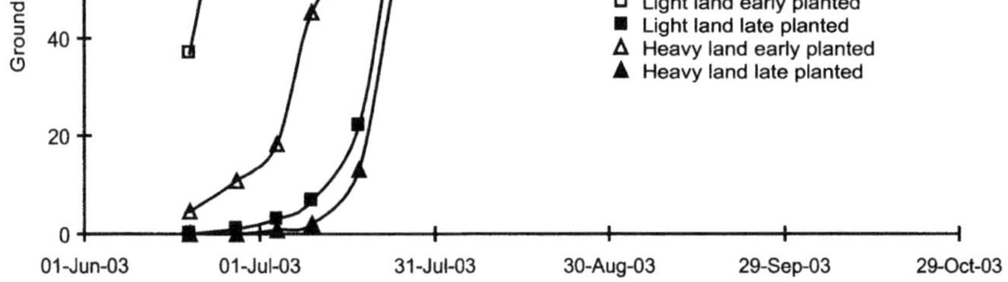

Figure 13.9 Effect of soil texture and planting date on pattern of ground cover development in Cara. (Allison and Allen, 2004).

NUTRIENTS

Increased availability of the major nutrients was a significant contributor to the yield increases until the 1980s by extending canopy duration. Since then few crops have received insufficient nutrients so lack of any nutrient is no longer a limitation to yields. Indeed, many soils have accumulated considerable residues of P and K and reductions in fresh applications are possible. Paradoxically, N may now be limiting yields through excess availability, especially in indeterminate varieties. As indicated above, such varieties respond to excess N by delaying and restricting tuber bulking and maintaining extensive haulm growth for the whole season. This shift in dry-matter partitioning reduces tuber yields (Table 13.4; Allen and Scott, 2001). In determinate varieties the canopy can be complete with only mainstems and a single sympodial tier of leaves and for the remaining part of the growing season all products of photosynthesis pass to the tubers. Such varieties require considerable amounts of fertiliser N to achieve this relatively small canopy and are, by definition, earlies, which cannot produce a full season canopy even with unlimited N e.g. FL 1953 (Courlan). It seems clear that managing a canopy to achieve the potential yield requires knowledge of varietal response and all sources of N and their availability. Even where applications of N do not reduce yield, there may be scope to reduce N applications so that, in many cases, the amount of N in the environment could be reduced. The incentives to apply current knowledge are, therefore, considerable.

In addition to the excessive application of fertiliser N, there are other contributors to excessive availability of N which are poorly understood. These relate principally to the release of N from soil organic matter and organic manure and the effect of cultivations on this process. Potatoes are grown on mineral soils with a wide range of organic matter contents (1-6%) and on much of the remaining area of organic soils. Many of these soils receive applications of organic manures. Cultivations for potatoes disturb soil to a depth of 60 cm and sift most of that depth in preparation for planting. These cultivations aerate the soil and can be carried out over a period of months when soil temperatures range from close to freezing initially to 20°C for final separation. In combination, an enormous volume of soil is moved in differing temperatures and soil water contents and the implications for N mineralisation and ultimately fertiliser N requirement have gone largely unrecognised. Their importance is illustrated by the variation of optimum rates found in experimental series apparently conducted in the same way (Table 13.5). There are no adequate ways of predicting N release from soil for potatoes as yet and therefore fertiliser applications inevitably do not take this source into account. Recent research at Cambridge University Farm (CUF) has shown that the timing of cultivations has a major effect on the release of N from the soil (Figure 13.10) and the quantity released over the whole season is usually more than the total uptake needed for full yields. This work has also shown that, in contrast to cereals, soil analysis for ammonium and nitrate has no utility in potatoes in improving the accuracy of fertiliser use. This should not be surprising, as cereals are largely grown in less disturbed soil conditions. Use of organic manures on potato land further complicates this issue; there is accumulating evidence that current predictions of N release underestimate the true values (Allison and Allen, 2004). The importance of precise N use is under-appreciated but is crucial if potatoes are to be produced without risk to the environment. Progress in this area should ensure both increasing yields and reductions in fertiliser N use, hence minimising N residues in the environment.

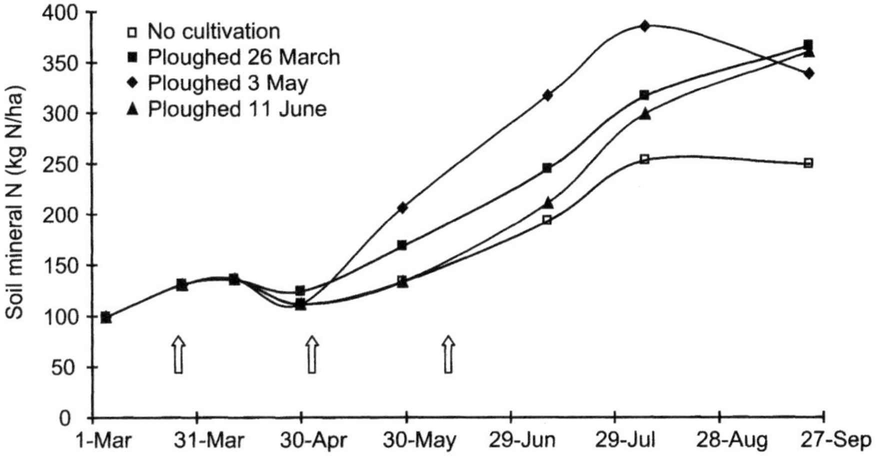

Figure 13.10 Effect of timing of cultivation on amount of soil mineral N in soil, arrows indicate cultivation state if crop present. (Allison and Allen, 2003).

Table 13.5 VARIATION IN OPTIMUM N RATE AND CORRESPONDING TUBER YIELD FOR ESTIMA AND CARA AT CUF 1986-2003 (ALLEN, ALLISON AND FOWLER, UNPUBLISHED).

	Estima		*Cara*	
Year	*(kg N/ha)*	*(t FW/ha)*	*(kg N/ha)*	*(t FW/ha)*
1986	180	49	90	56
1987	90	52	90	68
1991	180	78	120	91
1992	60	54	60	51
1993	240	81	120	92
1994	180	58	60	71
1995	120	46	240	72
2000	300	56	200	59
2001	100	51	100	48
2002	160	85	80	79
2003	120	69	0	73
Mean	157	62	105	69

WATER

Potato growers are the major users of irrigation in the UK and even current yield levels are dependent on the continuing availability of irrigation. Further increases in yield will require the use of more water, which must largely come from irrigation. Justification for its use is therefore essential.

Inadequacies in water use limit the achievement of potential yields either because crops are not irrigated and soil water is inadequate for potential yield, or because irrigation is inefficient. Inefficiency has two main components: first, most growers do not have sufficient capacity to fully irrigate their crops in the majority of years and this problem is still increasing as the renting of land is dispersing the production base over a greater area. This inevitably leads to more and longer movements of equipment and less time spent actually applying water. Secondly, over- and under-application of water are common and both can reduce yields (and quality). Over-watering can impair root growth and penetration and may lead to disease outbreaks, e.g. potato blight. Future improvements in yield will require more efficient use of available water, assisted by further research.

PESTS

Potato yields are severely affected by the widely distributed potato cyst nematodes *Globodera rostochiensis* and *Globodera pallida,* unless nematicides are applied to the soil. Distribution of this pest has increased over the last 25 years and is now endemic in most of the light soils in the UK. This amounts to considerably more than half the land base of the industry and for individual sectors, e.g. processing, even more. Its extent, its effects on yield and the opportunities for control revealed by improved knowledge of its population dynamics are all well documented (Trudgill, 1986; Trudgill, Elliot, Evans and Phillips, 2003). The pest is erratically distributed and, as initial infestations develop, it is easily 'missed' in sampling, so it is almost certain that its distribution is more extensive than surveys suggest. Moreover, the widespread renting of land frequently restricts good management; records are in different hands, so longer-term responsibilities are ignored and opportunities for spread from farm to farm are increased. As a consequence, it is hard to exaggerate the importance of nematodes to the future of the industry. Yields without nematodes are higher than can be achieved after their control with nematicides, so it is likely that future yields will be increasingly restricted by the pest, irrespective of improved knowledge and practices in all other areas of agronomy, as more and more land has become infested. The situation will continue to deteriorate unless nematicides are used more effectively, in particular by controlling large populations and restricting the multiplication of small ones.

The UK is fortunate in not suffering from leaf-eating pests such as Colorado potato beetle which affect much of Europe and North America. The stringent exclusion of this pest is very valuable, especially as its suitability will increase with any warming of our climate. If Colorado beetle were to become established in the UK there are good models which predict its life-cycle in relation to weather in USA and these should aid its control.

Other pests, such as aphids, which both affect leaf growth and transmit viruses, are effectively controlled by insecticides. Where aphid feeding is a problem, the continuing availability of these chemicals will be essential to maintaining productivity, whereas the significance of virus transmission in restricting productivity is small.

DISEASES

The generation and maintenance of a complete canopy in potatoes can be threatened by a range of diseases – fungal, viral and bacterial – which damage leaf tissue and, in some cases, impair photosynthetic efficiency. Leaf growth can be restricted by the loss of whole plants, starting before emergence right through almost to final defoliation. Blackleg (*Erwinia atroseptica* subsp. *atroseptica*) is a seed-borne, bacterial disease throughout the world, which causes rotting of stems and whole plants. Its epidemiology is still poorly understood, and disease expression is unpredictable but often severe. There are no chemical control measures. It is a particular problem in some varieties, e.g. Estima, and greater understanding combined with improved diagnostics should lead to a reduction in its effect on overall production. Another bacterial disease, brown rot (*Ralstonia solanacea*), causes severe plant loss in warm environments and can result in premature crop death. The bacterium has been found only rarely in UK and stringent tests on tubers and waterways (where the bacterium survives on other *Solanaceae*) have been in place since 1992. However, its presence in a shipment of seed potatoes delivered to Lancashire in February 2004 re-emphasised the potential danger. Although regarded as a disease of warmer climates it is clearly now established in NW Europe and the increased movement of potatoes within Europe increases the chance of spread. The same can also be said of ring rot (*Clavibacter michiganensis*) which was found for the first time in UK in 2003. This disease can cause plant loss in the field but is primarily a cause of serious breakdown during storage. It has now been found in several EU countries and must be regarded as a serious potential threat.

Leaf (and stem) tissue can be damaged by infection with several fungal diseases of which early and late blight are the most important. The former (*Alternaria solani*) is favoured by higher temperatures than normally prevail in UK, but it is a major problem in the Mediterranean and the USA. Late blight

(*Phytophthora infestans*) is the more significant disease in NW Europe and it remains a major potential restriction on productivity. Infection occurs via the seed tuber but almost all crop infections result from aerial spores produced on surviving potato debris and dumps from the previous year. (In the USA some transference of infection can occur during the cutting operation). The initial infection can occur at any time from stem emergence and thus the disease can severely curtail the length of the growing season. Unfortunately all varieties can become infected and many commercially important varieties are highly susceptible e.g. Russet Burbank (Table 13.6). As a result, control of late blight is dependent on rigorous spray programmes using a range of fungicides. The success of this approach is influenced by the weather conditions, for in wet weather the risk of infection and disease development is much increased and the wetter conditions can reduce the timeliness and efficiency of the spray operations. It is not therefore surprising that in most seasons some serious blight epidemics occur.

Table 11.6. AREA GROWN (HA) AND FOLIAGE AND TUBER BLIGHT RESISTANCE (9 = HIGH RESISTANCE) OF THE TWELVE MOST POPULAR VARIETIES IN GREAT BRITAIN (BRITISH POTATO COUNCIL/NATIONAL INSTITUTE OF AGRICULTURAL BOTANY, 2003).

Variety	*Area grown in 2003*	*Foliage*	*Tuber*
Russet Burbank	2944	3	1
King Edward	3590	3	4
Marfona	3880	4	4
Hermes	3986	3	4
Nadine	4808	6	4
Desiree	4929	4	5
Saturna	5259	4	5
Pentland Dell	5455	6	5
Maris Peer	5787	4	4
Lady Rosetta	5944	4	4
Estima	10840	4	5
Maris Piper	29275	4	5

Given the large risk of late blight reducing yield and quality, spraying every seven days is not uncommon. Thus, the use of fungicides is intense and resistance to some classes of fungicide has been found, e.g. phenylamides. For the foreseeable future (say 15 years) control of late blight will be dependent upon the extensive use of agro-chemicals. There is a world-wide search for resistance to blight e.g. the PICTIPAPA program based in the Toluca Valley in Mexico (e.g. Flier, Grunwald, Kroon, Starbaum, Van den Bosch, Garay-Serrano, Lozoya-Saldana, Fry and Turkensteen, 2003) and there are some reasons for optimism. However, none of the most promising sources of resistance has yet been combined with the commercial characteristics that are essential for the creation of improved genetic backgrounds. Thus, until about 2020 blight control will depend upon the use of fungicides and these will be major contributors to the overall use of agro-chemicals in potato production. Thus, over this period, it is essential that all other aspects of blight control be improved. The starting point must be better farm hygiene to reduce the inoculum burden carried over from season to season. Secondly, the management of fungicide use in relation to weather, crop characteristics, and application technology can be enhanced, leading to improved decision-support systems that minimise fungicide use.

Infection of potato plants with a range of viruses can lead to both restricted leaf expansion and reduced photosynthetic efficiency. Seed-borne diseases (severe mosaic, leaf roll, mild mosaic) are less important than in the past as improvements in seed production have reduced their incidence. Resistance to some of these viruses exists in commercial varieties and in breeding sources so it is unlikely that viruses will be a major limitation to future productivity.

In addition to diseases which affect the leaf canopy, potatoes are affected by a range of diseases which may affect the plant but which also, more importantly, infect the tuber – black scurf, silver scurf, black dot, powdery scab, common scab, gangrene, dry rot. The individual (and sometimes collective) effect of these diseases in determining tuber quality can negate all improvements in overall productivity by rendering the crop unsuitable for its intended use. For many of these diseases, our understanding of the epidemiology is inadequate and this inhibits effective control. More crucially, many of these diseases are seed and soil borne, so they are becoming endemic in many production areas. This we see as the major limitation on future potato productivity in the UK, unless the industry moves its land base.

UPTAKE OF NEW INFORMATION

This paper has highlighted a number of areas where recent research has provided improved knowledge and hence has increased the potential yield of commercial potato crops. However, in many cases, management of the potato crop has not changed to reflect this new understanding.

One example is the variation in nitrogen (N) requirement of the potato crop depending on variety: the inclusion in RB209 (MAFF 2000) of varietal characteristics as a determinant of N requirement should have made most growers aware of current knowledge, but fertiliser statistics show no evidence of decreasing use of N. Another important area is that of PCN which, as discussed earlier, now infests over 50% of the UK potato area. There can be no doubt that growers underestimated the consequences of failure to properly manage fields containing PCN, and they were frequently unaware of increasing understanding of its population dynamics. In this, the almost complete absence of advisors cognisant of current understanding was central to the failure to manage PCN better. This is a crucial lesson for the future. The successful application of better understanding requires the transfer of new information into the practical planning and operations of growers, but unless the researchers who advance knowledge are directly involved in this transfer there is no impetus for change. This is because the information to be transferred is not a single definitive protocol, but rather a complex of interacting knowledge that must be interpreted in the context of the individual grower. Few growers (or their advisors) can become fully conversant with the latest understanding, e.g. about PCN, therefore the best advice requires that researchers become more intimately involved in transfer of the knowledge. Where this does not happen, the rate of crop improvement is inevitably slowed.

The examples of N and PCN management illustrate the ineffective transfer of research results to practice, which has been a major constraint to improvement of potato production in the UK. Significant yield improvements could be made on-farm if current knowledge were applied appropriately.

Potential for genetic improvement

As seen in Table 13.2, the major potato varieties in current use in the UK were all introduced more than 30 years ago. Whilst new varieties have been introduced over the past 30 years, these have not been taken up by the major processors and thus they have taken only a small market share.

There has been considerable progress in breeding for pest and disease resistance (notably PCN and blight) but, as yet, the resultant varieties have not been accepted by the industry. For example, the recently introduced varieties Lady Balfour and Eve Balfour have good resistance to potato blight and partial resistance to both *G. pallida* and *G. rostochiensis* (Table 13.7). These varieties are being targeted for the organic sector where chemical control of these problems is not possible.

Table 13.7 RESISTANCE OF LADY BALFOUR AND EVE BALFOUR TO BLIGHT AND PCN. RESISTANCE IS SCORED ON SCALE OF 1-9, WITH 9 BEING COMPLETELY RESISTANT. FOR PCN 1, 2 ARE CLASSED AS SUSCEPTIBLE, WITH 3-9 INDICATING INCREASING RESISTANCE (NIAB POCKET GUIDE TO POTATO VARIETIES, 2004).

	Resistance to blight		*Resistance to PCN*	
Variety	*Foliage*	*Tuber*	*G. pallida*	*G. rostochiensis*
Lady Balfour	8	7	pr 4	pr 4
Eve Balfour	8	9	pr 5	pr 5

To incorporate resistance into existing varieties (with proven processing quality) requires a GM approach. As consumers have not accepted this technology, there is no short-term prospect of GM varieties. However, the technology is in place, as demonstrated by 'Newleaf' Russet Burbank developed in the USA, which is resistant to Colorado potato beetle (Coombs, Douches, LiWen Bin, Gratius and Pett, 2003). In the UK, H. Atkinson's group at Leeds University has produced transgenic lines of Desiree and Sante with good resistance to *G. pallida* (Urwin, Troth, Zubko and Atkinson, 2001).

In summary, the tools for genetic improvement of the potato crop are in place. Advances in conventional plant breeding have been made but as yet pest and disease resistance have not been combined with acceptable quality characteristics. If GM technology were accepted, then transgenic lines of cultivars, already accepted by the industry, could be produced with resistance to the major pests and diseases. Given the demands of the industry, genetic improvement of the potato is being targeted to improved quality and pest and disease resistance, rather than yield *per se.* However, complete resistance to PCN and late blight may increase yield of many UK crops.

Summary

The average yield of potato crops in the UK has shown no positive trend for the past 10 years. This contrasts with wheat where farm yields have continued to increase by c. 0.13 t/ha/year (Defra, 2003). The major reason for the contrast with wheat appears to be the emphasis of the potato industry on quality. Over recent years, the market for potatoes has diversified (e.g. salad potatoes, bakers, chips, crisps), and each sector has its own set of quality criteria (e.g. tuber size, shape, skin finish, dry matter content). Therefore the emphasis for both breeders and growers has been on achieving the necessary quality standards, rather than on increasing yield *per se.*

It is clear that the average UK yield is less than half of the biological potential. The key limitations to yield are soil compaction, through its effect on water and nutrient uptake; N fertilisation due to poor understanding of N mineralisation and because N fertiliser strategies are not tailored to the needs of individual varieties; poor irrigation management; and a complex range of pests and diseases of which potato cyst nematodes and late blight are the most serious. Soil compaction, N fertiliser application and irrigation can all be addressed through better crop husbandry, and the control of PCN and blight could also be improved using current knowledge. Improved transfer of knowledge from researchers to advisors and growers would facilitate this. In addition, a greater understanding of N dynamics in the potato crop is needed to predict the contribution of soil-derived N to the crop. The control of PCN and blight relies on the input of relatively large amounts of agro-chemicals. If tighter controls on the use of agro-chemicals are introduced (notably on nematicides) then this may further limit potato yield. Genetic modification of the potato crop to provide commercially acceptable varieties with good resistance to PCN and blight would help to overcome these limitations, but this seems unlikely in the short to medium term. Due to the constraints on yield mentioned above, and the continuing emphasis on quality, rather than yield *per se*, we therefore predict no significant increase in the yield of potatoes in the UK until after 2010.

Acknowledgements

The results reported in this paper are from experimental programmes supported by the British Potato Council, Cambridge University Potato Growers Research Association, Pepsico International, Syngenta Crop Protection UK Ltd. and Bayer CropScience plc.

References

Allen, E.J. and Scott, R.K. (1980). An analysis of growth of the potato crop. *Journal of Agricultural Science, Cambridge* **94**, 583-606.

Allen, E.J. and Scott, R.K. (2001). *The Agronomy of Effective Potato Production - British Potato Council Research Review*. British Potato Council: Oxford.

Allison, M.F. and Allen, E.J. (2003). *Evaluation of the Soil Nitrogen Supply System – Opportunities for Further Improvement to the Nitrogen Economy of the GB Potato Crop.* British Potato Council Project 807/228. British Potato Council: Oxford.

Allison, M.F. and Allen, E.J. (2004). *Evaluation of the Soil Nitrogen Supply System – Opportunities for Further Improvement to the Nitrogen Economy of the GB Potato Crop.* British Potato Council Project 807/228. British Potato Council: Oxford.

Coombs, J.J., Douches, D.S., Li WenBin, Grafius, E.J. and Pett, W.L. (2003) Field evaluation of natural, engineered, and combined resistance mechanisms in potato for control of Colorado potato beetle. *Journal of the American Society for Horticultural Science*, **128**, 219-224.

Defra (2003). http://statistics.defra.gov.uk/esg/publications/auk/2003/

Firman, D.M. (1987). *Leaf Growth and Senescence of the Potato.* PhD thesis, University of Cambridge.

Flier, W.G., Grunwald, N.J., Kroon, L.P.N.M., Sturbaum, A.K., Van den Bosch, T.B.M., Garay-Serrano, E., Lozoya-Saldana, H., Fry, W.E. and Turkensteen, L.J. (2003). The population structure of *Phytophthora infestans* from the Toluca Valley of central Mexico suggests genetic differentiation between populations from cultivated and wild Solanum spp. *Phytopathology* **93**, 382-390.

Ivins, J.D. and Bremner, P.M. (1965). Growth, development and yield in the potato. *Outlook on Agriculture* **4**, 211-217.

Jefferies, R.A. and MacKerron, D.K.L. (1989). Radiation interception and growth of irrigated and droughted potato (*Solanum tuberosum*). *Field Crops Research* **22**, 101-122.

Khurana, S.C. and Mclaren, J.S. (1982). The influence of leaf area, light interception and season on potato growth and yield. *Potato Research* **25**, 329-342.

MAFF (2000). *Fertiliser Recommendations for Agricultural and Horticultural Crops (RB209).* The Stationary Office, London.

NIAB (2004). *Pocket Guide to Varieties of Potatoes.* NIAD Association, Cambridge.

O'Brien, P.J., Allen, E.J. and Firman, D.M. (1998). A review of some studies into tuber initiation in potato (*Solanum tuberosum*) crops. *Journal of Agricultural Science, Cambridge* **139**, 251-270.

Rosenfield, A.B. (1997) *Effects of Nitrogen and Soil Conditions on Growth, Development and Yield in Potatoes.* PhD thesis, University of Cambridge.

Stalham, M.A. and Allen, E.J. (2004). Water uptake in the potato (*Solanum tuberosum*) crop. *Journal of Agricultural Science* **137**, 251-270.

Trudgill, D.L. (1986). Yield losses caused by potato cyst nematode: a review of the current position in Britain and prospects for improvement. *Annals of Applied Biology* **108**, 181-198.

Trudgill, D.L., Elliott, M.J., Evans, K. and Phillips, M.S. (2003). The white potato cyst nematode (*Globodera pallida*) – a critical analysis of the threat in Britain. *Annals of Applied Biology* **143**, 73-80.

Urwin, P.E., Troth, K.M., Zubko, E.I. and Atkinson, H.J. (2001). Effective transgenic resistance to *Globodera pallida* in potato field trials. *Molecular Breeding*. 2001. 8: 1, 95-101.

Westermann, D.T. and Sojka, R.E. (1996). Tillage and nitrogen placement effects on nutrient uptake by potato. *Soil Science Society of America Journal* **60**, 1448-1453.

14

YIELD OF UK OILSEED RAPE: PHYSIOLOGICAL AND TECHNOLOGICAL CONSTRAINTS, AND EXPECTATIONS OF PROGRESS TO 2010

J. SPINK[1] AND P. M. BERRY[2]

[1] *ADAS Rosemaund, Preston Wynne, Hereford, HR1 3PG, UK;* [2] *ADAS High Mowthorpe, Duggleby, Malton, North Yorkshire, YO17 8BP, UK*

Introduction

Oilseed rape (*Brassica napus*) is grown on 4-500,000 ha of UK arable land (DEFRA, 2004) and is the most important non-cereal combinable crop, worth £360 million to UK agriculture. The seed is an important source of home-grown oil, with lower levels of saturated fatty acids than soya, sunflower, and olive oils and accounts for about 60% of UK oilseeds crushing. The meal provides high protein animal feeds and the crop also has potential for industrial uses, with over 100,000 ha grown on set-aside for industrial oils and speciality uses. Oilseed rape is a very important break crop in mainly cereal rotations, especially for heavy land farmers where there are few alternatives.

This paper considers recent yield trends for oilseed rape in the UK and then, having examined their underlying physiology, it estimates the maximum yield that might be achieved with this species, identifying likely physiological, physical and technological constraints, and any detrimental repercussions that might arise.

Yield trends

Oilseed rape began to be grown on an extensive scale in the UK in the 1970s. Average farm yields are available for the UK from 1980 onwards and these show that seed yields have remained static, at about 3 t/ha, during the last two decades (Figure 14.1). Similarly static yields have occurred in the main oilseed rape producing countries of the EU (France and Germany), where yields also averaged 3 t/ha between 1982 and 1998. A survey of Canadian farm yields of canola (*B. napus* and *B. rapa*) showed that yields have increased from 0.9 t/ha in 1956 to 1.3 t/ha in 1994 (Daun and DeClerg, 1995). In Australia, field experiments were used to measure the yield of oilseed rape varieties introduced between 1966 and 1992. This showed that, in the presence of stem canker

(*Leptosphaeria maculans*), yields of new varieties were about 25% greater than the oldest. However, no yield differences were observed in the absence of stem canker (Potter, Salisbury, Balliinger, Wratten and Mailer, 1995).

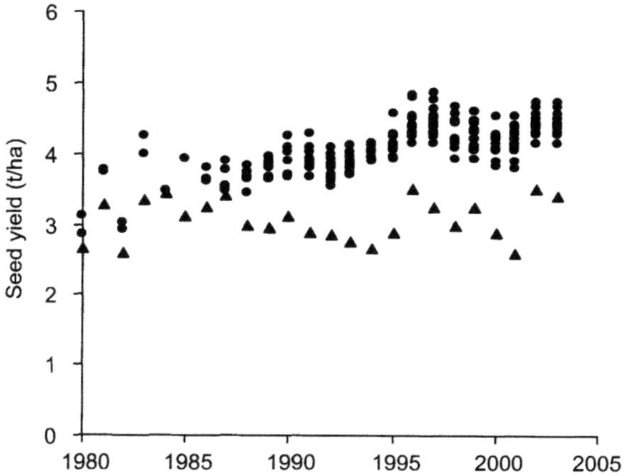

Figure 14.1 Average UK farm yields (s) (no significant regression) and UK Recommended List (RL) (l; y=0.053x - 101; R^2 = 0.58). UK farm yield data are calculated from UK Oilseed Rape production and area figures (Defra, 2004). UK RL data include the treated yield of each variety grown in a particular year, averaged across all UK RL trials.

In contrast to the UK farm averages, yields from the UK Recommended List (RL) variety trials (averaged across 20 sites around the UK) have increased by about 0.5 t/ha/decade since 1980 (Figure 14.1). This trend is impressive when it is considered that the breeding programmes suffered a major interruption to breed for low glucosinolate in the mid 1980s. Surveys have shown that uptake of new oilseed rape varieties is rapid (Sylvester-Bradley, Lunn, Foulkes, Shearman, Spink and Ingram, 2002), so the varieties used to calculate the RL yields will approximate to the same varieties grown on farms. The difference between the farm and RL trends is therefore surprising. Over the same period seed oil content of varieties in the RL increased from about 40% to 43% (Figure 14.2), representing a further increase in output. Almost all of this improvement occurred between 1987 and 1995. The first challenge in assessing the physiological and technological constraints on yield in oilseed rape must be to explain the RL yield trends and explain why UK farm yields have not increased at a similar rate.

Physiology that accounts for yield trends

FACTORS THAT MAY ACCOUNT FOR THE YIELD IMPROVEMENTS IN THE RL TRIALS

No studies have specifically investigated the physiological basis for the yield

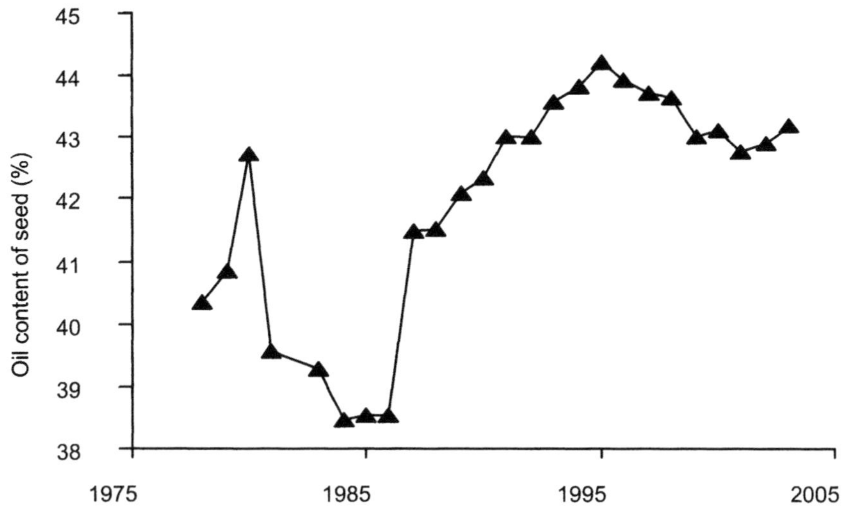

Figure 14.2 Oil content (%) of seed (at 9% moisture) averaged across all varieties described in each year of the UK Recommended List (RL).

improvements of oilseed rape observed in the RL trials. Analysis of the scores for various plant traits recorded in the RL over the past 25 years provide some insight as to what changes the breeders have made to oilseed rape varieties. However, interpretation of the trait scores is not straightforward. Time trends for the mean score of a particular trait averaged across varieties are not necessarily indicative of breeding progress for two reasons. Firstly, the method of scoring a trait may change, e.g. a score of 7 for stem shortness may correspond to a height of 160 cm in 1980, and 150 cm in 2000. Secondly, G x E interactions may cause trait scores to change, e.g. a breakdown of disease resistance. Therefore, it is also important to investigate whether the trait scores of individual varieties also change with time to quantify the influence of non-breeding factors. If these changes are equivalent to the time trends for the scores averaged across all varieties, then it may be concluded that breeding has not altered this trait. On the other hand, if the trait scores of individual varieties do not change, or the change does not fully account for the time trends averaged across all varieties, then it may be concluded that breeding has made a contribution.

The scores for seven traits (lodging resistance, stem shortness, earliness to flowering and ripening and resistance to stem canker (*Leptosphaeria maculans* (*Phoma lingam*)), light leaf spot (*Pyrenopeziza brassicae*) and downy mildew (*Peronospora parasitica*) recorded in the RL between 1978 and 2003 have been analysed to investigate whether they have been altered by breeding. The variety scores of each trait have been averaged for each year and these means have been plotted against time. Linear regression has then been used to calculate the mean annual change for each score (Table 14.1). This analysis showed strong increases in the scores for earlier flowering and greater downy mildew resistance, and a weaker increase in the score for earlier ripening. Scores for stem shortness and stem canker resistance decreased, with non-significant changes in lodging resistance and light leaf spot resistance.

Table 14.1 TRENDS IN THE RL TRAIT SCORES BETWEEN 1978 AND 2003.

Trait	*Annual change for the mean trait score (average across varieties)*	*Annual change to the mean trait score caused by non-breeding factors*	*Annual change to the mean trait score caused by breeding*
Lodging resistance	0	0	0
Shortness of stem	-0.030	0	-0.030
Earliness of flowering	+0.079	+0.046	+0.033
Earliness of ripening	+0.029	+0.040	-0.011
Resistance to stem canker	-0.046	-0.060	+0.014
Resistance to light leaf spot	0	-0.102	+0.102
Resistance to downy mildew	+0.108	-0.050	+0.158

An increasing score represents increased resistance to lodging or diseases, earlier flowering/ripening and shorter stems.

The annual change in each trait score was also calculated individually for each of the 45-55 varieties that have been present in the RL for three or more years. A CHI squared test was used to estimate whether the frequency of positive and negative changes was significantly different from what could be expected by chance ($P<0.05$). The mean change across all varieties is presented in Table 14.1 for the statistically significant traits. Non breeding factors are assumed to cause these changes. Light leaf spot, downy mildew and stem canker scores decreased. The light leaf spot scores of some varieties such as Madrigal, Lipton and Synergy dropped by two during their seven to eight year duration in the RL. These changes may be caused by a breakdown of the resistance genes. The earliness scores may have increased due to a change in the scoring system or an interaction with the environment.

The breeding contribution to the time trends for the mean trait scores can be estimated as the difference between the time trends for the mean trait scores and the annual changes that have been estimated to be caused by non-breeding factors. The annual changes estimated to be caused by breeding are described in the final column of Table 14.1. This indicates that the greatest breeding advances have been to increase the resistance to light leaf spot and downy mildew. This has increased the resistance score for downy mildew by more than two since 1978. In the case of light leaf spot, the breeding improvements appear to have been just sufficient to counter the breakdown of resistance by new strains of the pathogen. The small improvements in stem canker appear to have been insufficient to keep pace with resistance breakdown. Breeding has also produced earlier flowering varieties. Importantly, the date of ripening does not appear to have been altered by breeding, which indicates that breeders have extended the period from flowering to ripening. Surprisingly, and contrary to received wisdom, this analysis indicates that breeders have selected for slightly taller varieties. Kightley (2001), who measured height of three commonly grown varieties, gives some support for this finding. Variety Apex was introduced in 1993 with a stem shortness score of 7 and was 160 cm tall. Madrigal and Pronto were introduced in 1997-98 with a stem shortness score of 6 and were 170-174 cm tall. This may in part be caused by the introduction of hybrid varieties which tend to be taller than conventional varieties. The current top yielding variety (cv. Winner) is also one of the tallest (HGCA, 2003), which further supports the finding that yield and height may not be very well related. The absence of significant changes in height may also explain the lack of breeding progress found for lodging resistance.

There have also been improvements in agronomy during the last 25 years which may have contributed to the yield improvements observed in the RL trials. The introduction of triazole fungicides during the early 1990s improved control of major diseases including light leaf spot, phoma leaf spot and canker, altenaria (*Altenaria brassicae*) and sclerotinia stem rot (*Sclerotinia sclerotiorum*). The growth regulatory properties of triazole fungicides (tebuconazole and metconazle) have also been shown to improve the yield of crops with large canopies, but can reduce yield in crops with small canopies (Lunn, Spink, Wade and Clare, 2003). Changes in disease control practice have led to a greater number fungicides applied in the autumn (P. Gladders, *pers. comm.*), which is likely to have improved disease control. No other significant agronomic changes appear to have occurred, with seed rates remaining at 120 seeds/m^2 and 70 seeds/m^2 for conventional and hybrid varieties respectively and sowing date targeted at around the first week of September.

PHYSIOLOGICAL MECHANISMS FOR THE YIELD IMPROVEMENTS IN THE RL TRIALS

The previous section indicates that yield improvement through breeding may have been brought about by lengthening the seed filling period and improving downy mildew resistance, and by improved chemical control of diseases. The mechanisms by which these factors improve yield are likely to have been in the processes that determine the two main yield components; seed number/m^2 (which determines the sink size) and post-anthesis assimilate supply (which determines the size of the source for filling the seeds). The next three paragraphs describe the processes that determine sink and source, then explain how the genetic and agronomic changes described above may have affected these processes.

The number of seeds per metre squared is the primary determinant of the size of the sink. This character is the most important yield component and has been shown to account for 85% of yield variation (Mendham, Shipway an Scott, 1981). The number of seeds per metre squared is determined during a critical phase for pod and seed abortion lasting for 200-300°Cd after anthesis (Leterme, 1988). In most UK situations this translates to about 13-25 days. Pod and seed survival have been shown to be related to the amount of radiation intercepted by photosynthetic tissue per flower and per pod respectively during this critical period (Leterme, 1988; Mendham *et al.*, 1981). The amount of radiation that is usefully intercepted at this time is severely reduced by the layer of flowers, which have been shown to reflect and absorb 60-80% of the incoming radiation (Mendham *et al.*, 1981; Yates and Steven, 1987; Fray, Evans, Lydiate and Arthur, 1996). In addition, Keiler and Morgan (1988) showed that the apical regions of the inflorescence were a stronger sink for photo-assimilate than young pods, hence some abortion occurred due to lack of assimilate. Thus, where potential pod number (flower number) is high, the final pod canopy is likely to consist of a greater proportion of later formed pods, which are generally characterised by smaller seed (Milford, Fieldsend, Spink and Child, 1989) and lower yield potential. Therefore any agronomic or genetic changes that reduce the size of the flower layer and increase light transmission are expected to increase seed number, seed size and overall yield.

The post-anthesis assimilate supply (source) is mainly determined by current photosynthesis by the pods and stems. During early pod development, the pods and stems intercept 84% and 72% of the light in thick and thin canopies respectively, with the leaves intercepting the remaining light (McWilliam, Stafford, Scott, Norton, Stokes and Sylvester-Bradley, 1995). By the time pods have attained their maximum hull weight and area, Norton, Bilsborrow and Shipway (1991) showed that pods and stems were responsible for 95% of crop photosynthesis, due to their position in the crop and loss of green leaf area. The type of tissue responsible for photosynthesis during seed fill is of significance as it is generally recognised that pods and stems have a lower rate of photosynthesis than leaves. Rode, Gosse and Chartier (1983) estimated that the photosynthetic rate of stems is 25% less than leaves and the photosynthetic rate of pods is 75% less than leaves. However, Mendham (1995) reported that photosynthetic rate was least for stems, with pods intermediate between leaves and stems. These differences in photosynthetic rate correlated with the frequency of stomata on these tissues (Major, 1977). Radiation use efficiency (RUE) of oilseed rape during seed filling (g of seed dry matter per m^2 per MJ PAR) has been estimated at 0.5 (Justes, Denoroy, Gabrielle and Gosse, 2000), 0.8 (Habekotte, 1997) and 1.5 in pot experiments under shelters (Dreccer, Schapendonk, Slafer and Rabbinge, 2000). The latter experiment may have calculated a high RUE because the plants were beneath polycarbonate and would have experienced more diffuse light. A low RUE may have been estimated by Justes *et al.* (2000) because the seed yields were only 1.8 t ha^{-1}.

The RUE estimate of Habekotte (1997) was based on yields of 3.1 to 3.4 t ha^{-1}. When it is considered that the synthesis of oil rich dry matter (seed) requires 50% more energy than pre-flowering biomass (Dreccer *et al.*, 2000), then the RUE during seed filling is equivalent to 0.75-1.9 g of pre-flowering dry matter/m^2. The average RUE just prior to flowering, which is primarily due to leaf photosynthesis, has been measured at 2.7 g/MJ/m^2 (Habekotte, 1997; Justes *et al.*, 2000).

The amount of assimilate contributed by pre-flowering photosynthesis as a result of relocation from the stems has been estimated as negligible (Stafford, 1996) or up to 10% (Mendham, 1995). Habekotte (1993) estimated that relocation of assimilate from the pod walls and stems accounts for about 12% of the seed yield. However, as pod walls are predominantly produced post flowering this is likely to represent the relocation of reserves accumulated after flowering. The source of photosynthates for yield formation is, therefore, predominantly from current photosynthesis post flowering, the duration of which has been shown to be about 40 days (Mendham *et al.*, 1981; Habekotte, 1993).

This paper has indicated that increasing RL yields are associated with earlier flowering, longer seed filling, better downy mildew resistance and improved chemical control of diseases. Analysis of yield formation in wheat has shown that earlier flowering is expected to lengthen the seed filling period by ensuring that this phase of growth occurs in cooler temperatures. A similar effect would also be expected in oilseed rape. However, earlier flowering also causes an increased risk of frost coinciding with flowering. This can reduce flower viability and potentially limits sink size. Improved fungicide efficacy is likely to prolong the photosynthetic activity of the leaves, stems and pods after flowering. Together these changes would be expected to increase the size of the source available for filling the seeds. This would be expected to result in heavier seeds and would increase the number of seeds that reach a harvestable weight. Improved fungicide efficacy may also increase the leaf area at flowering and may improve RUE during this critical period when the number of seeds per unit area is set. The growth-regulatory activity of triazole fungicides may also increase the number of seeds set by restricting the production of dense flowering layers (Lunn *et al.*, 2003). It is likely that breeders have altered oilseed rape in other ways that have not yet been detected and that these have also contributed to the yield increases. For example, the architecture of the canopy may have been altered, which may have affected RUE. Rooting may have been improved and varieties may have become more shatter proof, so increasing the proportion of yield recovered. Without any formal analysis of the phenotypic and the physiological changes that have occurred through oilseed rape breeding it is impossible to fully account for yield improvement. Past and present varieties need to be characterised more completely if we are to fully understand how breeders have achieved such significant yield improvements during the past 25 years.

EXPLAINING THE DIFFERENCE BETWEEN RL AND FARM YIELDS

The difference in yield trends between farm and RL is unlikely to be due to the use of inferior varieties because UK farmers have been quick to grow new oilseed rape varieties soon after they are tested. About 10% of the UK area of oilseed rape is spring sown, which is inherently lower yielding than the winter crop, and is included in the average farm yields. This does not, however, explain the yield gap between farm and RL yields; if we assume that spring oilseed rape crops yield on average 2.0 t/ha then removing them from the analysis only increases the average yield of autumn sown farm crops by 0.1 t/ ha. Therefore, the smaller farm yield must be due to crop management practices which are restricting the crops ability to achieve its full yield potential. Industrial oilseed rape is grown on set-aside land and accounts for 10 to 20% of the oilseed rape crop. These crops are probably managed less intensively than oilseed rape grown for human consumption. However, this also does not appear to account for the difference as industrial crop yields have averaged just 0.2 t/ ha less than conventional crops over the last six years (DEFRA, 2004). Several differences between the management of farm crops and RL trials can be identified, many of which have resulted from pressure to reduce the costs of growing farm crops. Compared with RL trials, farm oilseed rape crops tend to: be grown in shorter rotations, be established using minimal cultivations, be drilled earlier and at higher seed rates, and have fewer fungicide applications, less sulphur and possibly less nitrogen.

The frequency with which oilseed rape is grown in arable rotations has increased because it is commonly the second most profitable crop to winter wheat and is an effective break for wheat. Rotations with oilseed rape grown one year in three had increased from 8% in 1990 to 15% in 1994 (Turner and Hardwick, 1995), and then to 23% in 2003 (J. Turner, *pers. comm.*). Shorter rotations increase the likelihood of diseases and disease-associated yield loss. Sieling and Christen (1997) showed that oilseed rape grown every 2 years yielded about 0.5 t/ha less than when it was grown once every 3 years.

Many farm crops are 'autocast' directly into the preceding cereal crop (HGCA 2002), direct drilled or established after minimum tillage, as compared with ploughing used in the RL trials. Direct drilling or autocasting have been shown to result in lower yields compared with ploughing, but yields from crops established using minimal cultivations are similar to those established with ploughing (Sauzet, Reau and Palleau, 2003; Bowerman, Chambers and Jones, 1995). Conversely, Christensen, Hofmann and Bischoff (2003) found no detrimental effect of direct drilling on yield. It appears direct drilling and autocasting are less effective when there are large amounts of poorly chopped straw from the previous crop, wet conditions or wet and poor soil structure. Non-burial of the residues of the previous crop has also been shown to increase lodging and *Phoma* infection (Sauzet *et al.*, 2003) and increase slug damage, which may cause sub-optimal plant populations.

Many farm crops are established two to three weeks earlier and at high seed rates compared with the RL trials. The reduction in winter barley area may result in this sowing date difference diminishing as more farm crops must be sown after wheat. High seed rates provide insurance against poor emergence and pest damage which are difficult to predict. Early sowing and high plant populations have commonly been shown to reduce seed yields (Mendham *et al.*, 1981; Leach, Darby, Williams, Fitt and Rawlinson, 1994; Jenkins and Leitch, 1986; Leach, Stevenson, Rainbow and Mullen, 1999). Early sowing also increases the risk of pests such as cabbage root fly (*Delia radicum* L.), light leaf spot aphids and viruses. Extensive trials work in Germany showed that drilling two weeks earlier (late August compared with early September) reduced yields by 0.1 to 0.2 t/ha (Baer and Frauen, 2003). This is thought to be due to the production of an over thick canopy from early drilled crops which reduces the number of seeds set via the mechanism described above.

Oilseed rape crops yielding 3 t/ha require about 50 kg of sulphur per hectare (McGrath and Zhao, 1996). In the past, atmospheric depositions have largely met this requirement but, between 1970 and 2002, SO_2 emissions decreased by 82% (source: NET CEN). The percentage of farm crops which had sulphur applied increased from 10 to 30% between 1993 and 1996, then remained constant at 30% until 2001 (Figure 14.3). Since 2001 the percentage of crops receiving sulphur has risen sharply to 50%. The average application rate has risen steadily from less than 15 kg /ha S in 1993 to 30 kg /ha S in 2003 (Figure 14.3).

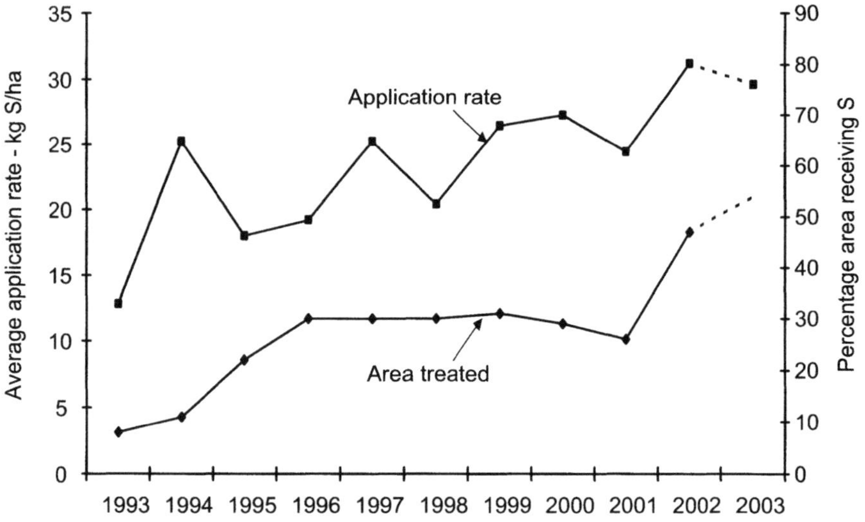

Figure 14.3 Average application rate of S (kg /ha) (n) and percentage of the oilseed rape area receiving sulphur (u) in the UK. Data for 2003 are provisional. Source: British Survey of Fertiliser Practice.

Many oilseed rape crops are responsive to S fertiliser, with yield increases of 0.7 to 1.6 t/ha recorded in response to 40 kg/ha S on a sandy soil (McGrath and Zhao, 1996). An extra application of sulphur began to be applied to RL trials during the early 1990s (S. Kightley, *pers. comm.*) and a minimum of 30 kg /ha S is currently recommended. Therefore, it is probable that inadequate sulphur applications on commercial crops has contributed to the increasing yield differences between the farm and RL yields during the past decade.

Nitrogen fertiliser applications to commercial oilseed rape crops increased from 250 kg to 280 kg/ha N between 1980 and 1984, then decreased steadily to 180 kg/ha N by 1993. Since then the application rates have risen slowly to current levels of about 200 kg/ha N (British Survey of Fertiliser Practice). The reduction in N fertiliser during the late 80s and early 90s may have been the result of new research into the optimum fertiliser rates and falling oilseed prices. Nitrogen fertiliser applications to the RL consist of 30 kg/ha N in the autumn plus those of the farm on which the trial is situated. Autumn applications of nitrogen to commercial farms are becoming less common; 65% of farms received autumn N in 1988 compared with 40% in 2002 (BSFP). So it seems likely that the RL trials receive more early nitrogen fertiliser than many farm crops.

Ongoing HGCA (research project 2890) has correlated low farm yields with years when the incidence of *Phoma* canker, light leaf spot and dark pod spot (*Altenaria brassicae*) were high. Between 1987 and 1994, less than 50% of commercial crops received a fungicide in autumn or spring to control light leaf spot or *Phoma* and between 25 and 50% received a fungicide at or post flowering to control stem rot (*Sclerotinia scleretiorum*) and dark pod spot (Fitt, Gladders, Turner, Sutherland, Welham and Davies, 1997). Since 1998, the number of crops receiving a fungicide has stabilised at 90% (Turner, Elcock, Walters and Wright, 2002). Disease induced yield losses increased steadily between 1997 and 2000 despite the use of fungicides (Turner *et al.*, 2002) and it is clear that dose and timing of the fungicide is not always optimal (Fitt *et al.*, 1997; Turner *et al.*, 2002). The increase in disease has been attributed to warmer and wetter weather and the use of more susceptible varieties (although our analysis indicates that greater incidence of disease is probably due to the breakdown of varietal resistance). It is probable that RL yields will be affected less by diseases due to a more robust programme of fungicides compared with what is used on commercial crops. The RL trial protocol includes three compulsory fungicides (in the autumn, at the beginning of stem extension and at flowering), with a further three fungicide applications possible depending on disease pressure. This contrasts with an average of just 1.8 fungicide sprays applied to farm crops in 2002 (Garthwaite, Thomas, Dawson and Stoddart, 2002).

Estimate of the maximum yield in an average UK environment

Given the yield trends just described, it now becomes possible to attempt estimates of the maximum possible oilseed rape yield that could be achieved in the current average UK environment. The average UK environment has been estimated from the long-term weather for the main arable areas in England and Scotland. This shows that the photosynthetically active radiation during June and July varies from 8.5 to 9.5 MJ/m^2/day (a mean of 9 MJ/m^2/day is used for this exercise) and the annual rainfall is 550 to 1000 mm. Currently the very best yielding farm crops can produce 4.5-5 t/ha (at 100% dry matter) in seasons with favourable weather. Seed weight tends to be fairly conservative at 4 to 5 mg so, if we assume a seed weight of 4.5 mg, then these crops must produce 100,000 seeds/m^2. There is a strong inverse relationship between the numbers of pods and the numbers of seeds per pod. Given that the number of seeds per pod can vary from 8 to 28 (Sylvester-Bradley *et al.*, 2002), the required number of seeds per m^2 can be achieved with pod densities varying from 4,000 to 12,000 pods/m^2. If the relationship between the number of pods and seeds per pod can be broken, for example by improving RUE at flowering, then there is the potential to greatly increase the number of seeds/m^2. Currently, the flowering layer reflects at least 50% of the incoming radiation. If this can be halved by growing more open canopies or using apetalous varieties, then it is possible that the number of seeds/m^2 could be increased to 150,000.

The challenge then lies in filling the seeds. This should not be underestimated due to the extra energy costs of synthesising lipid-rich seeds. Dreccer (2000) estimated that oilseeds require about 50% more energy per gram of dry matter than pre-flowering oilseed rape biomass or cereal grains. At the beginning of seed filling 95% of crop photosynthesis has been shown to be carried out by pods and stems (Norton *et al.*, 1991). Therefore, we estimate that over the entire seed filling period of typical farm crops almost all of the photosynthesis is carried out by the pods and stems. Habekotte (1997) estimated that the RUE of crops yielding just over 3 t/ha was 0.8 g of seed/m^2/MJ$_{PAR}$. It therefore seems reasonable to assume this RUE for the seed filling period of an average farm crop. We also assume a seed filling period of 35 days and an average daily incident radiation of 9 MJ/m^2 (all of which is intercepted). These assumptions lead to a yield estimate at 91% dry matter of 2.8 t/ha (Table 14.2). The leaves of a RL crop, or a well managed farm crop, would be expected to intercept a proportion of the radiation due to the more open structure of RL crops, arising from their later drilling and better disease control giving longer leaf duration. Leaves of thin canopies that are subjected to several fungicide applications can intercept 30% of light at the beginning of seed filling (McWilliam *et al.*, 1995) and can persist for 50-75% of the seed filling period (Stafford, 1996). If we assume a leaf RUE for producing carbohydrate of 2.7 g/m^2/MJ$_{PAR}$ (Justes *et al.*, 2000) then the RUE for producing lipid rich seeds would be 1.8 g/m^2/MJ$_{PAR}$. The RUE of the pods would also be expected to increase slightly compared with an average farm crop due to the more open canopy and better protection from diseases. These calculations estimate a yield of 4.1 t/ha, which is typical of a good farm crop or RL yield (after accounting for the plot edge effect). More radical changes must be assumed, to estimate the maximum yield for the average UK environment. Currently the rate of seed filling is severely limited by the RUE of the pods. If this can be increased to the level expressed under diffuse light conditions (Dreccer *et al.*, 2000), the RUE of leaves increased to the maximum that has been observed during pre-anthesis growth (Justes *et al.*, 2000), the duration of the green leaf and pod tissues extended and the proportion of light intercepted by the leaves can be increased to 25% (averaged over the seed filling period), then oilseed rape could yield 7.6 t/ha (at 91% dry matter). Yield could even top 8 t/ha if a proportion of the pre-flowering biomass can be relocated to the seed. However, it must be recognised that the literature is inconsistent over whether this is possible, with estimates for the proportion of yield originating from pre-anthesis assimilate ranging from 0 to 10%. A 7.6 t/ha (6.9 t/ha at 100% dry matter) crop would require 150,000 seeds/m^2 each weighing 4.6 mg, which should be possible assuming the pod number and seed number per pod relationship can be broken. What these calculations do show very clearly is that improvements to both the source availability and sink capacity must be made to realise maximum yield.

Table 14.2 CROP CHARACTERISTICS DURING THE SEED FILLING PERIOD AND ESTIMATES OF YIELD

	Average farm crop		*RL crop*		*Maximum yielding crop**	
	Pod/stem	*Leaf*	*Pod/stem*	*Leaf*	*Pod/stem*	*Leaf*
Mean daily incident radiation (MJ_{PAR})	9	9	9	9	9	9
Mean percentage of PAR intercepted	100	0	85	15	75	25
RUE (g of seed/m^2/MJ_{PAR})	0.8	-	1.0	1.8	1.5	2.6
Duration during seed filling (days)	35	-	40	25	45	40
Seed yield at 100% dm (t/ha)	2.8	0	3.1	0.6	4.6	2.3
Total seed yield at 91% dm (t/ha)				2.8	4.1	7.6

* estimated for an average UK environment

Physiological and physical constraints to achieving the maximum yield

The most obvious potential constraint concerns improvement of RUE, especially during the critical phase of pod and seed determination. The fact that some crops utilise as little as 20% of the radiation at this time, due to absorption and reflection by the flowers, illustrates that there is great scope for improvement. Other potential constraints are associated with improving RUE of the pod canopy, prolonging the duration of green tissues during seed filling and relocating a proportion of the pre-flowering assimilate to the seed. The greater biomass of heavy yielding crops would also be expected to increase lodging risk. Lodging has been shown to reduce the yield of oilseed rape by 16 to 50%

(Bayliss and Wright, 1990; Armstrong and Nichol, 1991). Lodging compresses the canopy which reduces the transmission of radiation through the pod layer. This must reduce RUE, due to light saturation of the upper layer of the pod canopy during pod expansion and seed growth, which will reduce both the number of seeds/m^2 and the supply of assimilate for filling the seeds.

In France, dry years have been associated with low oilseed rape yields (Champolivier and Merrien, 1995). Our data also indicate a trend for dry summers to be associated with low yields in the UK (Figure 14.4). The amount of water required to produce a crop with a maximum yield of 6.9 t/ha (at 100% dry matter) can be estimated by assuming the commonly-observed water use efficiency of 5 g of carbohydrate dry matter per litre of water (Green, Vaidyanthan and Hough, 1983). Taking into account the higher energy demand for the production of oil it may be assumed that this translates to 3.4 g/l for oilseed (Penning, Jansen, ten Berge and Bakema, 1989). If we assume that the stem and leaf biomass of the maximum yielding crop remains the same as current crops at 7 t/ha, and the pod biomass is 50% of the seed weight, then the total amount of non-seed biomass would be 10.5 t/ha. This crop would require 413 mm of rainfall or soil water storage. The majority of this water will be required between the onset of stem extension in mid March and canopy senescence in mid July. If we assume that the crop biomass in mid March is 3 t/ha, then the crop will require about 350 mm during the following 4 months. Assuming that the soil is at field capacity in mid March, then the available water capacity of the soil to a depth of 1.8m would vary between 252 and 396 mm depending on soil type. Using information about the root length density of oilseed rape (Barraclough, 1989) together with relationships linking root length density with water extraction for cereals (King, Gay, Sylvester-Bradley, Bingham, Foulkes, Gregory and Robinson, 2003), we estimate that oilseed rape roots may extract about 68% of available water to a depth of 1.5m. This means that oilseed rape crops could extract 143 to 224 mm of water from the soil, leaving a shortfall of 126 to 207 mm which must be supplied by rainfall between mid March and mid July. Average monthly rainfall varies between 45 mm in the east of the UK and 83 mm in the south and west. Therefore, in a year with average rainfall, the water supply would prevent the maximum yield from being attained in crops grown on loamy sands, sandy loams and clays in the east. Assuming that the water restriction began during seed filling then the maximum yield on these soil types would be 5.8, 6.9 and 7.4 t/ha respectively. At this point it must be recognised that oilseed rape is mainly grown on heavy soils, so it must be concluded that a maximum yield of close to 7.6 t/ha would be possible in most crops with average rainfall. However, it is also clear that moderate changes in the weather would lead to a significant proportion of crops being water limited. For example, a reduction in rainfall between March and July of 30% would prevent the maximum yield of 7.6 t/ha from being achieved on any soil type in the east. Oilseed rape is also sensitive to water logging and this may represent a further obstacle to high yields.

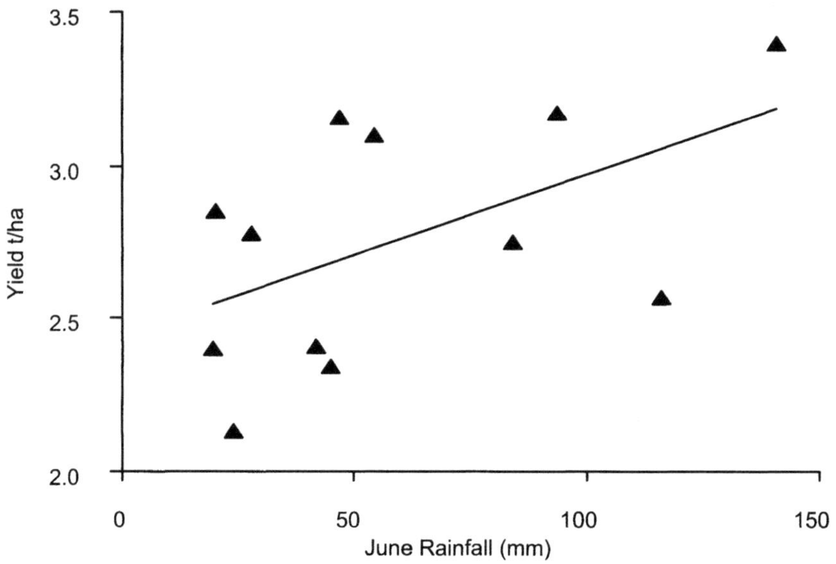

Figure 14.4 Farm oilseed rape yields and June rainfall at ADAS Boxworth between 1991 and 2002.

The realisation of the maximum yield could be prevented by the breakdown of disease resistance. The proportion of crops sown with a resistance rating to *Phoma* of 5 or less has increased from 20% in 1997 to around 50% in 2001. The less frequent use of varieties that are resistant to light leaf spot has been even more dramatic. In 1997, zero crops had a resistance rating to light leaf spot of 5 or less, but this had risen to 40% in 2001 (Turner *et al.*, 2002). This may be evidence of a breakdown in disease resistance. The general reduction of disease resistance scores for individual varieties over their lifetime in the RL (Table 14.1) may be further evidence of breakdown to disease resistance.

A review of harvest losses in oilseed rape (Ogilvy, Milford, Evans and Freer, 1992) concluded that under favourable ripening conditions the pre-harvest treatments, swathing, desiccation, direct combining or the use of a pod sealant had little effect on yield. In these cases, harvest losses were 0.05 t/ha or less. Losses of 0.05 to 0.15 t/ha occurred in less than ideal harvesting conditions. Standing crops on exposed sites were most prone to losses. In these situations losses of over 0.2 t/ha occurred one year in three.

The achievement of high yield will be strongly influenced by non-biological factors such as government policy, economics, consumer demand for safer foods, quality and environmental concerns. These will influence the level of management and inputs that farmers use, and this will ultimately determine the rate at which new crop management, varieties and agro-chemicals are developed. In 2005, the decoupling of farm payments from production is likely to focus oilseed rape on to the most profitable land, hence accelerate yield improvement, however this effect may be countered by an increase in the frequency of oilseed rape within the rotation, hence increased disease.

Technology that might be used to overcome the constraints

The physiological constraints described above may be distilled into a series of changes in plant characteristics that are required to 1) increase seed set at flowering, 2) increase post-anthesis assimilate and 3) improve water use efficiency (Table 14.3). We now consider how these target traits could be achieved through breeding, agronomy and chemistry.

Table 14.3 TARGET PLANT TRAITS TO ACHIEVE HIGH YIELDS

Increase seed set	*Increase post-anthesis assimilate*	*Increase water use efficiency*
Early flowering	Erect pods	Early flowering
Fewer flowers	Persistent green canopy	Deeper rooting
Retention of leaves	Retention of leaves	More root branching
Long pods	More assimilate relocated from non seed organs	
Less branching	Greater pod photosynthesis	
Lodging resistant	Lodging resistance	

BREEDING

An obvious method of reducing PAR reflection during flowering is the use of apetalous varieties. In the UK, an apetalous variety increased the amount of radiation transmitted beyond the flowering layer by 70% (Fray *et al.*, 1996), but it did not increase seed yields. This may have been due to the production of more pods and branches rather than the anticipated increase in seeds per pod. In Australia, 30% extra light interception was recorded with apetalous lines (Rao, Mendham and Buzza, 1991). This increased the number of seeds/m^2 by 12% compared with a conventional variety already shown to have many seeds per pod. Seed size was also increased and overall seed yield was 40% greater. Another potential advantage of apetalous lines is a reduction in *Sclerotinia* stem rot fungus through avoiding petals (which can carry inoculum) adhering to stems.

The selection of varieties with long pods has been shown to increase the average number of seeds per pod from 12 to 17 (Chay and Thurling, 1989a). However, no yield advantage was observed because of a negative relationship with the number of pods per plant. Further studies (Chay and Thurling, 1989b) showed that the long pod character is determined by two dominant genes. They also showed that the negative association could be broken by intercrossing and that seed yield could be improved by selecting genotypes with long pods. Thus, as with apetalous lines, it seems that further work with long pods may pay dividends. A variety with erect pods (20-25° closer to the vertical) has been shown to increase the amount of PAR reaching the base of the pod layer by 26% (Fray *et al.*, 1996). This resulted in more fertile pods containing a greater weight of seed in the lower horizons of the pod canopy. The results showed that erectophile pod characteristics should increase yield most when introduced into genotypes producing a deep pod canopy, where the increased PAR transmission could maintain higher seed yields at the base of the canopy.

In oilseed rape, lodging can occur by either anchorage failure or stem buckling/breakage (Goodman, Crook and Ennos, 2001). Lodging risk can be effectively reduced with shorter crops, stiffer stems and stiffer longer tap roots (Goodman *et al.*, 2001). Breeders could exploit semi-dwarfs to reduce lodging risk, but there may also be genetic variation in stem stiffness and tap root properties, as found in wheat.

Recent advances in genetics, molecular biology and genomics should assist in accelerating oilseed rape breeding programmes. *Brassica napus* relates more closely than other arable crops to the cruciferous weed *Arabidopsis thaliana*, whose genome has now been sequenced. A wealth of information about the physiological effects of *A. thaliana* mutants exists in *Arabidopsis* and other *Brassicase* that may also be useful in locating genes affecting canopy structure. For example, mutants (*ein*) in the phytochrome system affecting seedling etiolation, internode elongation and shade avoidance responses exist in *B. rapa* (McCormac, Whitelam, Boylan, Quail and Smith, 1992; Robson, Whitelam and Smith, 1993; Devlin, Rood, Somers, Quai and Whitelam, 1992; Devlin, Somers, Quail and Whitelam, 1997). Recessive terminal flower (*tfl*) mutants which convert indeterminate wild-types of *Arabidopsis* to determinate types have been identified (Alvarez, Guli, Yu and Smyth, 1992; Bradley, Ratcliffe, Vincent, Carpenter and Coen, 1997). Mutants (*hy*) affecting the branching pattern of *Arabidopsis*, which appear to relate to the phytochrome system, have also been reported (Ballare and Scopel, 1997; Pigliucci and Scmitt, 1999; Schmitz and Theres, 1999). Mutations leading to increased branching are also known for pea (7 loci), *Petunia* (3) and *Antirrhinum* (2). Characterisation of these mutants and gene functions has led to major improvements in the models to explain the genetic control of branching in plants. QTLs have also been found for branching traits in *Brassicae*. Thus genetic information for locating genes conferring desirable flowering and branching determinacy traits in *Brassica napus* could be generated through reference to existing information. It should be emphasised that this process will not be straightforward, due to the amphidiploid nature of oilseed rape, arising from a relatively recent hybridisation between cabbage and turnip. Genome triplication occurred during evolution of cabbage and turnip from the common ancestor with *Arabidopsis*.

Oilseed rape thus has six copies of genes in the *Arabidopsis* genome, some of which may have ceased to function or which may have adopted novel functions.

AGRONOMY

RUE during flowering may be improved by avoiding the production of excessively large canopies, which have many flowers. Habekotte (1997), reinterpreting the work of Mendham *et al.* (1981), showed that a leaf area index (LAI) of 2 units before flowering was necessary for high yields. Lunn, Spink, Stokes, Wade, Clare and Scott (2001) estimated that a maximum green area index of 3.5 was required for maximum yield. Farm crops can reach a LAI of 6 or 7, so there is much potential to reduce canopy size. This can be achieved by delaying drilling until September, establishing fewer plants and by managing nitrogen supply (Lunn *et al.*, 2003). It must be recognised that some of these practices will be difficult to achieve due to pressures to drill oilseed rape before cereals and to avoid sub-optimal plant establishment caused by poor emergence or pest damage. Lunn *et al.* (2003) also showed that flowers of crops with smaller canopies reflected and absorbed less radiation, and the small canopies also aborted fewer pods, which would be expected to further improve RUE. Tommey and Evans (1992) showed that seed yield was increased when pod abortion was reduced by artificially removing flowers from all the branches that developed after the 4th branch. Excessive production of pods and flowers also reduces the survival of leaves and bracts at the base of the pod layer due to limitation of light penetration (McWilliam *et al.*, 1995), resulting in a canopy that consists almost entirely of pods and branches. Therefore, smaller canopies at flowering are likely to have greater RUE, due to the higher photosynthetic efficiency of leaves compared to pods and branches.

Delaying nitrogen applications could be used to prolong canopy greenness and improve late season RUE (Justes *et al.,* 2000). More widespread application of sulphur to farm crops should avoid a limitation on nitrogen uptake (McGrath and Zhao, 1996). Sulphur deficiency has also been linked to restricted root growth (Helal and Schnug, 1995), with evidence from glass-house experiments indicating that that adequate S levels are essential to maintain root integrity, prevent root mortality, and improve root efficiency.

CHEMISTRY

New fungicides will be required to combat diseases, particularly if the rate of breakdown of varietal resistance increases. The development of fungicides that prolong leaf life beyond the flowering phase would significantly increase the post-anthesis supply of assimilates. The more frequent use of triazole fungicides with growth regulating properties could be used to effect yield improvements in crops with large canopies. An autumn applied fungicide (metconazole) has been shown to increase root biomass (BASF data). In Germany, autumn applications of metconazole have been promoted for both winter hardiness, and to improve the root/shoot balance, based on lab experiments.

Detrimental aspects of increasing yield

Given the high energy cost associated with the production of oil it may be expected that increased yields may be associated with reduced oil content. So far, there has been no indication that heavier yields are associated with a smaller proportion of oil in the seed. In fact oil percentage has increased with seed yield. However, oil content and protein content are negatively correlated, so any improvements in oil content may reduce the quality of oilseed rape meal as a livestock feed. Increasing nitrogen applications have been shown to increase protein content at the expense of oil (Asare and Scarisbrick, 1995). This study also showed that applying 80 kg of sulphur per hectare increased seed glucosinolate levels from 8.9 umol/g seed to 11.8 umol/seed. Increasing N applications from 120 to 240 kg/ha increased seed glucosinolate concentration from 14.4 to 19.5 umol/g seed in a droughty year, but had no effect in the following year.

The pursuit of the maximum yield is not expected to be significantly detrimental for the environment. Much of the yield advance is likely to come from conventional breeding to produce canopies which use light more efficiently. Genetic modification may be exploited in the long-term, but it is uncertain what impact this may have on the environment. If additional chemical applications are needed they are likely to be fungicides. At a landscape scale, higher yields may mean that a smaller proportion of arable land is cropped with oilseed rape, more crop species will be grown, or more land will be used for environmental schemes, both of which would benefit biodiversity. High yielding crops may require more nitrogen fertiliser, which will increase nitrate pollution, but this effect would be slight.

Summary

OBSTACLES PREVENTING THE MAXIMUM YIELD FROM BEING ACHIEVED

Obstacles preventing the maximum yield from being realised in an average UK environment can be divided into two areas: use of sub-optimal inputs and crop management, and physiological constraints that must be overcome through breeding. The first of these is undoubtedly the more important, as it threatens to nullify the second. Management intensity and input use are both quite sensitive to input and output prices, so the position is by no means fixed. Also, the optimum level of a given input is often determined from experiments containing single factor comparisons. It seems possible that a co-optimisation study would show larger individual optimums.

Of the physiological constraints, the most important is increasing the supply of assimilate during seed filling, as currently this appears to be the most yield limiting step. This will involve improving RUE, lengthening the duration of the seed filling phase, reducing disease and, if possible, relocating pre-anthesis stem reserves. Other physiological constraints that must be overcome include increasing the number of seeds set, increasing lodging resistance and increasing water acquisition and water use efficiency. Breeding offers the greatest potential for overcoming these constraints in the long term. However, many of the traits are either poorly understood in oilseed rape, difficult to screen or require the introduction of novel germplasm. Therefore, improvements are likely to be slow.

YIELD PREDICTIONS

This paper argues that the increasing yield potential of new varieties is not being realised on farms because the levels of inputs and management are too low. This has happened because low oilseed rape prices and relatively high input costs have meant that profits were best maximised by reducing the cost of growing the crop. It seems likely that this trend will continue and as a result it is difficult to predict farm yields rising much above 3 t/ha before 2010. Decoupling subsidies from production may cause yields to rise in the short term, with oilseed rape production ceasing on low yielding and least profitable land. However, this outcome is far from certain as oilseed rape provides a valuable break between winter wheat crops and some of its value lies in the improved performance of the following first wheat crop.

Acknowledgements

The authors would like to thank Defra for funding this work, Chris Dawson for supplying data from the British Survey of Fertiliser Practice and Judith Turner of the Central Science Laboratory for supplying survey information about oilseed rape diseases. RL data were taken from trials funded by the British Society of Plant Breeders and the Home-Grown Cereals Authority; full results can be found at www.hgca.com.

References

Alvarez, J., Guli, C.L., Yu, X.H. and Smyth, D.R. (1992) Terminal flower: a gene affecting inflorescence development in *Arabidopsis thaliana. Plant Journal*, **2**, 103-116.

Armstrong, E.L. and Nicol, H.I. (1991) Reducing height and lodging in rapeseed with growth regulators. *Australian Journal of Experimental Agriculture,* **31**, 245-250.

Asare, E. and Scarisbrick, D.H. (1995) Rate of nitrogen and sulphur fertilizers on yield and yield components and seed quality of oilseed rape (*Brassica napus* L.). *Field Crops Research,* **44**, 41-46.

Baer, A. and Frauen, M. (2003) Yield response of winter oilseed rape hybrids to sowing date, sowing density, nitrogen supply and triazol input. In *Proceedings of the 11th International Rapeseed Congress Volume 3*, pp 887- 889. Edited by H. Sprensen. Copenhagen, Denmark.

Ballare, C.L. and Scopel, A.L. (1997) Phytochrome signalling in plant canopies: testing its population-level implications with photoreceptor mutants of *Arabidopsis. Functional Ecology*, **11**, 441-440.

Barraclough, P.B. (1989) Root growth, macro-nutrient uptake dynamics and soil fertility requirements of a high yielding winter oilseed rape crop. *Plant and Soil,* **119**, 59-70.

Bayliss, D. and Wright, T.J. (1990) The effects of lodging and a paclobutrazol – chlormequat chloride mixture on the yield and quality of oilseed rape. *Annals of Applied Biology,* **116**, 287-295.

Bowerman, P., Chambers, B.J. and Jones, A.E. (1995). Winter oilseed rape establishment methods on clay soils. In *Proceedings of the 9th International Rapeseed Congress Volume 2,* pp 220- 222. Edited by D. Murphy. Cambridge, UK.

Bradley, D., Ratcliffe, O., Vincent, C., Carpenter, R. and Coen, E. (1997) Inflorescence commitment and architecture in *Arabidopsis. Science*, **275**, 80-83.

Champolivier, L. and Merrien, A. (1995). Effect of water stress applied at different growth stages to Brassica napus L. var. Oleifera: I – Effect on yield and yield components. In *Proceedings of the 9th International Rapeseed Congress Volume 2,* pp 494- 496. Edited by D. Murphy. Cambridge, UK,

Chay, P.M. and Thurling, N. (1989a) Variation in pod length in spring rapeseed (*Brassica napus* L.) and its effect on seed yield and yield components. *Journal of Agricultural Science,* Cambridge, **113**, 139-147.

Chay, P.M. and Thurling, N. (1989b) Identification of genes controlling pod length spring rapeseed (*Brassica napus* L.) and their utilization for yield improvement. *Plant Breeding*, **103**, 54-62.

Christen, O., Hofmann, B. and Bischoff, J. (2003) Oilseed rape in minimum tillage systems. In *Proceedings of the 11th International Rapeseed*

Congress Volume 3, pp 762- 764. Edited by H. Sprensen. Copenhagen, Denmark.

Daun, J.K. and DeClerq, D.R. (1995) Interrelationships between quality factors and yield in canadian canola from harvest surveys, 1956 to 1994. In *Proceedings of the 9th International Rapeseed Congress Volume 1,* pp 336- 338. Edited by D. Murphy. Cambridge, UK.

Defra (2004). http://statistics.defra.gov.uk/esg/statnot/osrsur.pdf

Devlin, P.F., Rood, S.B., Somers, D.E., Quali, P.H. and Whitelam, G.C. (1992) Photophysiology of the elongated internode (ein) mutant of Brassica rapa ein mutant lacks a detectable phytochrome B-like polypeptide. *Plant Physiology*, **11**, 1442-1447.

Devlin, P.F., Somers, D.E., Quail, P.H. and Whitelam, G.C. (1997). The *Brassica rapa* elongated internode (ein) gene encodes phytochrome B. *Plant Molecular Biology*, **43**, 537-547.

Dreccer, M.F., Schapendonk, A.H.C.M., Slafer, G.A. and Rabbinge, R. (2000) Comparative response of wheat and oilseed rape to nitrogen supply: absorption and utilisation efficiency of radiation and nitrogen during the reproductive stages determining yield. *Plant and Soil,* **220**, 189-205.

Fitt, B.D.L., Gladders, P., Turner, J.A., Sutherland, K.G., Welham, S.J. and Davies, J.M.L. (1997) Prospects for developing a forecasting scheme to optimise use of fungicides for disease control on oilseed rape in the UK. *Aspects of Applied Biology,* **48**, 135-141.

Fray, M.J., Evans, E.J., Lydiate, D.J. and Arthur, A.E. (1996) Physiological assessment of apetalous flowers and erectophile pods in oilseed rape (*Brassica napus*). *Journal of Agricultural Science.* Cambridge, **127**, 193-200.

Garthwaite, D. G., Thomas, M. R., Dawson, A. and Stoddart, H. (2000) *Pesticide Usage Survey*. DEFRA Report No. 187. Department for the Environment Food and Rural Affairs, London.

Goodman, A.M., Crook, M.J. and Ennos, A.R. (2001) Anchorage mechanics of the tap root system of winter sown oilseed rape (*Brassica napus* L.). *Annals of Botany,* **87**, 397-404.

Green, C.F., Vaidyanathan, L.V. & Hough, M.N. (1983) An analysis of the relationship between potential evapotranspiration and dry matter accumulation for winter wheat. *Journal of Agricultural Science, Cambridge,* **103**, 189-199

Habekotté, B. (1993) Quantitative analysis of pod formation, seed set and seed filling in winter oilseed rape (*Brassica napus* L.) under field conditions. *Field Crops Research,* **35**, 21-33.

Habekotté, B. (1997). Identification of strong and weak yield determining components of winter oilseed rape compared with winter wheat. *European Journal of Agronomy* **7**, 315-321.

Helal, H.M. and Schnug, E. (1995) Root development and nutrient utilisation by *Brassica napus* as affected by sulphur supply. In *Proceedings of the*

9th International Rapeseed Congress Volume 1, pp 556- 558. Edited by D. Murphy. Cambridge, UK.

Home-Grown Cereals Authority (2002). *Establishing Oilseed Rape Using Autocast.* Topic Sheet No. 59. Home-Grown Cereals Authority, London.

Jenkins, P.D. and Leitch, M.H. (1986) Effects of sowing date on the growth and yield of winter oilseed rape (*Brassica napus*). *Journal of Agricultural Science Cambridge*, **107**, 405-420.

Justes, E., Denoroy, P., Gabrielle, B. & Gosse, G. (2000) Effect of crop nitrogen status and temperature on the radiation use efficiency of winter oilseed rape. *European Journal of Agronomy,* **13**, 165-177.

Keiller, D.R. and Morgan, D.G. (1988) Effects of pod removal and plant growth regulators on the growth, development and carbon assimilate distribution in oilseed rape *(Brassica napus, L). Journal of Agricultural Science, Cambridge*, **111**, 357–362.

Kightley, S. (2001) Lodging control in winter rape. *Plant Varieties and Seeds,* **14**, 171-180.

King, J., Gay, A., Sylvester-Bradley, R., Bingham, I., Foulkes, J., Gregory, P. and Robinson, D. (2003). Modelling cereal root systems for water and nitrogen capture: Towards an economic optimum. *Annals of Botany*, **91**, 383-390.

Leach, J.E., Darby, R.J., Williams, I.H., Fitt, B.D.L. and Rawlinson, C.J. (1994) Factors affecting growth and yield of winter oilseed rape (*Brassica napus*), 1985-89. *Journal of Agricultural Science,* Cambridge, **122**, 405-413.

Leach, J.E., Stevenson, H.J., Rainbow, A.J. and Mullen, L.A. (1999) Effects of high plant populations on the growth and yield of winter oilseed rape (*Brassica napus*). *Journal of Agricultural Science,* Cambridge, **132**, 173-180.

Leterme, P. (1988). Modelisation du fonctionnement du peuplement de colza d'hiver en fin de cycle: elaboration des composantes finales du rendement. In *Colza: Physiologie et Elaboaration du Rendement CETIOM*, pp 124-129. Paris.

Lunn, G.D., Spink, J., Stokes, D.T., Wade, A., Clare, R.W. & Scott, R.K. (2001). *Canopy Management in Winter Oilseed Rape.* HGCA Project Report No. OS49, 86 pp. Home-Grown Cereals Authority, London.

Lunn, G.D., Spink, J., Wade, A. and Clare, R.W. (2003) *Spring Remedial Treatments to Improve Canopy Structure and Yield in Winter Oilseed Rape.* HGCA Project Report No. OS64, 97 pp. Home-Grown Cereals Authority, London.

Major, D.J. (1977) Influence of seed size on yield and yield components of rape. *Agronomy Journal*, **69**, 541-543.

McCormac, A.C., Whitelam, G.C., Boylan, M.T., Quail, P.H., and Smith, H. (1992) Contrasting responses of etiolated and light-adapted seedlings to red:far-red ratio: a comparison of wild type, mutant and transgenic plants has revealed differential functions of members of the phytochrome family. *Journal of Plant Physiology*, **140**, 707-714.

McGrath, S.P. and Zhao, F.J. (1996) Sulphur uptake, yield responses and interactions between nitrogen and sulphur in winter oilseed rape (*Brassica napus*). *Journal of Agricultural Science,* Cambridge, **126**. 53-62.

McWilliam, S.C., Stafford, J.A., Scott, R.K., Norton, G., Stokes, D.T. and Sylvester-Bradley, R. (1995) The relationship between canopy structure and yield in oilseed rape. In *Proceedings of the 9th International Rapeseed Congress Volume 2,* pp 485- 490. Edited by D. Murphy. Cambridge, UK.

Mendham, N.J. (1995). Physiological basis of seed yield and quality in oilseed rape. In *Proceedings of the 9th International Rapeseed Congress Volume 2,* pp 491- 493. Edited by D. Murphy. Cambridge, UK.

Mendham, N.J., Shipway, P.A. and Scott, R.K. (1981) The effects of delayed sowing and weather on growth, development and yield of winter oilseed rape (*Brassica napus*). *Journal of Agricultural Science,* Cambridge, **96**, 389-416.

Milford, G.F.J., Fieldsend, J.K., Spink, J. and Child, R.D. (1989) Between and within plant variation in seed glucosinolate concentrations in the winter oilseed rape cv. Ariana. *Aspects of Applied Biology*, **23,** 173-176.

Norton, G., Bilsborrow, P.E. and Shipway, P.A. (1991) Comparative physiology of divergent types of winter rapeseed. In *Proceedings of the 8th International Rapeseed Congress,* pp. 578-582. Edited by D.I. McGregor. Saskatoon, Canada.

Ogilvy, S.E., Milford, G.F.J., Evans, E.J. and Freer, J.B.S. (1992) *Effects of Pre-harvest Treatment on the Yield and Quality of Winter Oilseed Rape.* HGCA Research Review No. OS7, 76 pp. Home-Grown Cereals Authority, London.

Pigliucci, M . and Scmitt, J. (1999) Genes affecting phenotypic plasticity in *Arabidopsis*: pleiotropic effects and reproductive fitness of photomorphogenic mutants. *Journal of Evolutionary Biology*, **12**, 551-562.

Potter, T.D., Salisbury, P.A., Balliinger, D.J., Wratten, N. and Mailer, R.J. (1995) Comparison of historical varieties of rapeseed and canola in Australia. In *Proceedings of the 9th International Rapeseed Congress Volume 1,* pp 365- 367. Edited by D. Murphy. Cambridge, UK.

Penning de Vries, F.W.T., Jansen, D.M., ten Berge, H.F.M. and Bakema, A. (1989) *Simulation of Ecophysiological Processes of Growth in Several Annual Crops*. Pudoc, Wageningen.

Rao, M.S.S., Mendham, N.J. and Buzza, G.C. (1991) Effect of apetalous flower character on radiation distribution in the crop canopy, yield and its components of oilseed rape (*Brassica napus*). *Journal of Agricultural Science,* Cambridge, **117**, 189-196.

Robson, P.R.H., Whitelam, G.C. and Smith, H. (1993) Selected components of the shade-avoidance syndrome are displayed in a normal manner in mutants of *Arabidopsis thaliana* and *Brassica rapa. Plant Physiology,* **102**, 1179-1184.

Rode, J.J.C., Gosse, G. and Chartier, M. (1983) Vers une modelisation de la production de graines de colza de printemps. *Information Techniques* CETIOM 82, 10-20.

Sauzet, G., Reau, R. and Palleau, J. (2003) Evaluation of oilseed rape crop managements with minimum tillage. In *Proceedings of the 11th International Rapeseed Congress Volume 3*, pp 863- 864. Edited by H. Sprensen. Copenhagen, Denmark.

Schmitz, G. and Theres, K. (1999) Genetic control of branching in Arabidopsis and tomato. *Current Opinion in Plant Biology,* **2**, 51-55.

Sieling, K. and Christen, O. (1997) Effect of preceding crop combination and N fertilization on yield of six oilseed rape cultivars (*Brassica napus* L.). *European Journal of Agronomy,* **7**, 301-306.

Stafford, J.A. (1996). *The Effects of Prochloraz on the Growth and Yield of Oil Seed Rape.* University of Nottingham, Ph.D. thesis.

Sylvester-Bradley, R., Lunn, G., Foulkes, J., Shearman, V., Spink, J. and Ingram, J. (2002) Management strategies for high yields of cereals and oilseed rape. In *Proceedings of HGCA Conference "Agronomic Intelligence: The Basis for Profitable Production"*, pp9.1-8.17. Home-Grown Cereals Authority, Coventry.

Tommey, A.M. and Evans, E.J. (1992) Analysis of post-flowering compensatory growth in winter oilseed rape (*Brassica napus*). *Journal of Agricultural Science,* Cambridge, **118**, 301-308.

Turner, J.A. and Hardwick, N.V. (1995) The rise and fall of *Sclerotinia sclerotiorum*, the cause of stem rot of oilseed rape in the UK. In *Proceedings of the 9th International Rapeseed Congress Volume 2,* pp 640- 642. Edited by D. Murphy. Cambridge, UK.

Turner, J.A., Elcock, S.J., Walters, K.F.A. and Wright, D.M. (2002). A review of pest and disease problems in winter oilseed rape in England and Wales. In *Proceedings of the BCPC Conference – Pests & Diseases*, pp555-562. Brighton, UK.

Yates, D.J. and Stevens, M.D. (1987) Reflection and absorption of solar radiation by flowering canopies of oilseed rape (*Brassica napus* L.). *Journal of Agricultural Science,* Cambridge, **109**, 495-502.

15

PRODUCTIVITY, BIODIVERSITY AND SUSTAINABILITY

BRIAN JOHNSON AND ANNA HOPE
English Nature, Roughmoor, Bishop's Hull, Taunton TA1 5AA

Introduction

Arable and grassland agricultural systems have been developed over thousands of years from the first agrarian societies in the late Neolithic period in the Middle East. Some arable systems and modified grasslands are older than the heathland communities that evolved in Northern Europe. Arable agriculture is essentially a disturbed ground habitat, and so it is no coincidence that arable crops have been developed from plant species that originated in disturbed ground habitats such as cliff tops, erosion areas and river banks. As arable systems were refined and developed, wild plants from disturbed habitats colonised them and co-evolved with the arable habitat. Many of the weeds inhabiting early arable fields adapted to the rotational cycles of tillage, growing and harvesting, and may themselves have been important food plants for humans (such as Fat Hen, *Chenopodium album*) or were later developed into crops (such as the mustards and other brassicas) by hybridisation and artificial selection. Until the nineteenth century arable fields were often very weedy with up to 20 species of arable plant within the crop, forming a substantial proportion of the total biomass in the fields. These weeds and their seed-rain supported substantial diversity and abundant higher trophic levels of insects and farmland birds.

The expansion of arable agriculture was primarily into areas of disturbed ground and grassland that could be easily tilled, destroying a large part of the natural habitat and forcing surviving non-crop biodiversity to adapt to farmland. Some species of farmland flora and fauna became almost entirely dependent on farmland landscapes at crucial stages in their life cycles, all but disappearing from natural habitats. With around 60% of Britain under farming in the nineteenth century a large part of Britain's wildlife in the landscape was adapted to farmland.

In the early part of the twentieth century methods were developed that allowed farmers to divert significantly more field biomass towards the crop at the expense of non-crop species, primarily by using cultivations and rotations to control weeds and artificial fertilisers to increase crop growth and competitiveness. The utilised metabolisable energy output per hectare was estimated to have increased by 1.8 per cent per year from 1950 to 1970 (Wilkins 2000). The process of reducing weed competition and enhancing crop growth accelerated in the mid-twentieth century with the introduction of herbicides, insecticides, fungicides and new crop varieties. There was also a switch from spring to winter cropping, enabled by novel varieties and agrochemicals, resulting in the virtual eradication of winter stubbles from the arable landscape. Figure 15.1 illustrates how some of these changes have affected yields of winter wheat in the Broadbalk experiment. This emphasises the importance of fertiliser or manure in raising yields (increases of two to three times compared to the unfertilised plots of continuous wheat) and the rapid improvements in yield on the fertilised/manured plots following the introduction of herbicide applications in the mid-1960s (yields approximately doubled within 10 years).

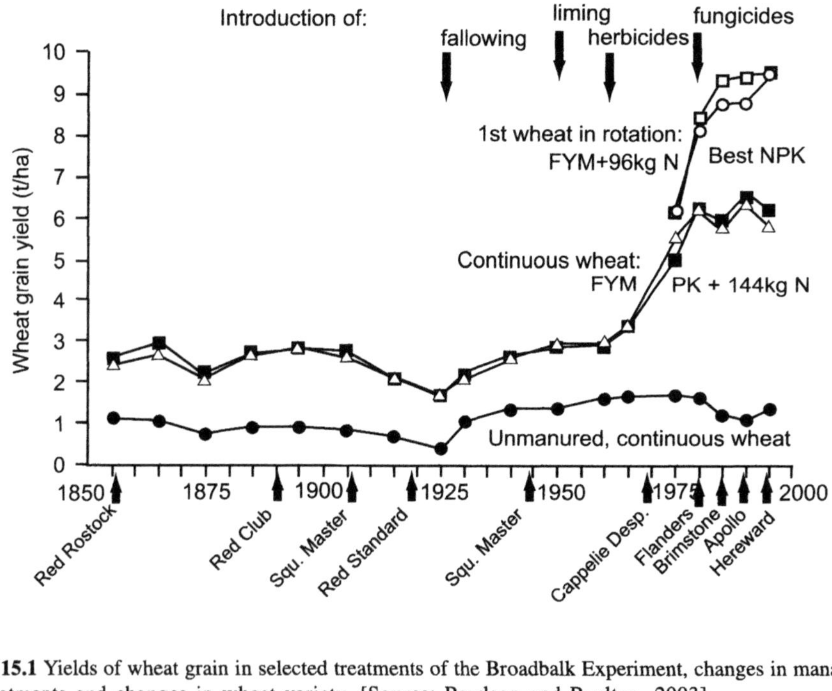

Figure 15.1 Yields of wheat grain in selected treatments of the Broadbalk Experiment, changes in management and treatments and changes in wheat variety. [Source: Powlson and Poulton, 2003].

In a similar way, grasslands have been modified from early grazed and mown natural systems such as chalk grassland and valley hay meadows that were very rich in herbaceous plants, insects and birds, to the species-poor artificially sown grass leys used for high productivity grazing and silage-making in modern livestock systems. Most natural grasslands have also been substantially modified using artificial fertilisers and manure, coupled with selective herbicides to remove broad-leaved plants. These modified natural grasslands are usually species-poor, and may be as productive as sown leys, with similar nutritional and digestibility qualities. Specialised livestock such as Friesian/Holstein cattle and lowland sheep have been developed to take advantage of the more productive swards. Between 1940 and 1990 increases in yields of meat and milk were very large, with output of beef and lamb doubling, milk production trebling, and numbers of stock increasing, whilst the total area of grassland fell (Figures 15.2a and 15.2b; Hopkins and Hopkins, 1994).

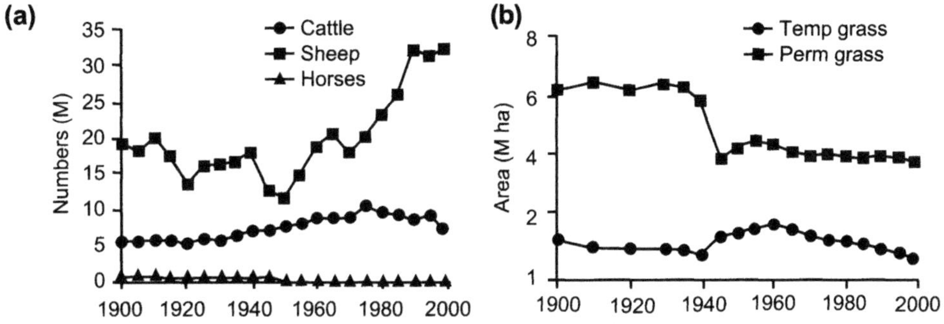

Figure 15.2a Numbers of cattle, sheep and agricultural horses in England and Wales [Source: Wilkins 2000]
Figure 15.2b Area of temporary and permanent grassland in England and Wales [Source: Wilkins 2000]

RELATIONSHIPS BETWEEN INCREASES IN YIELD AND DECLINES IN BIODIVERSITY

These developments in agriculture were aimed at maximising yield and minimising biotic stresses, including pests and diseases, hence minimising non-crop biodiversity. They were very successful in agronomic terms, so that substantial reductions in biodiversity correlated with yield increases. For example, it is estimated that weed seedbanks on UK arable farms have declined on average by 3% per year since the 1940s (Robinson and Sutherland, 2002). In the long-term Broadbalk Experiment carried out at Rothamsted, between 10 and 20 weed species occur on each arable plot that has never received herbicides, while in the stubble of the cleanest herbicide-treated plots there may be fewer than five species, represented by only one or two plants of each (Rothamsted Research website). The highest trophic levels are most sensitive to reductions in non-crop primary production, making birds that specialise on farmed habitats particularly sensitive indicators of change.

If population trends in farmland birds are compared with trends in arable yields for the past 40 years a clear pattern emerges, with a strong negative correlation between cereal yield and bird numbers (Figure 15.3). Another analysis found that, excluding seabirds and pheasants, bird biomass in Great

Britain fell by 29 per cent between 1968 and 1988, indicating a decrease in energy flux into the trophic levels occupied by birds (Dolton & Brooke, 1999). Analysis of 27 years of suction trap data and the BTO Common Birds Census in Scotland also reveals a trend of declines in numbers of farmland birds and invertebrates (Benton, Bryant, Cole and Crick, 2002).

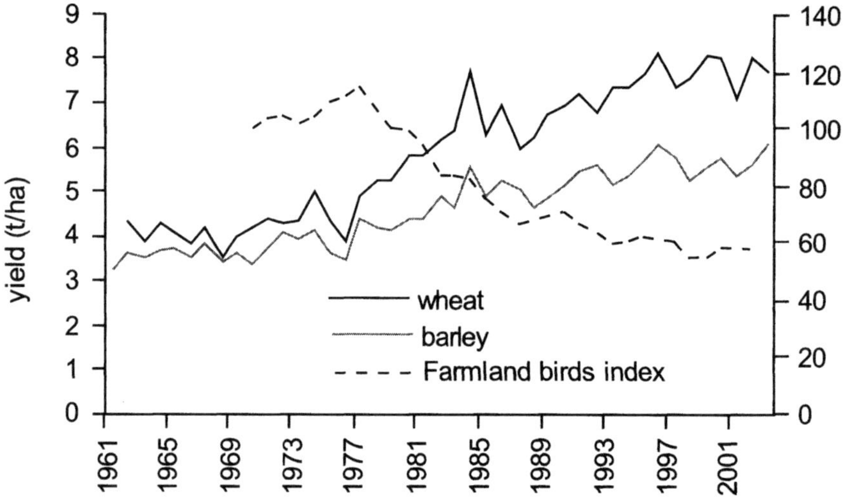

Figure 15.3 Recent trends in yields of wheat and barley and farmland bird populations in the UK. [Source: FAO; Defra]

However, research investigating relationships between arable productivity and non-crop biodiversity is remarkably thin on the ground. Some general relationships between farming intensity and populations of farmland birds (in both arable and grassland situations) in the UK has been established through research by the Game Conservancy, BTO and the RSPB (Fuller, Gregory, Gibbons, Marchant, Wilson, Baillie and Carter, 1995 and Siriwardena, Baillie, Buckland, Fewster, Marchant and Wilson, 1998, Ewald and Aebischer 1995, Potts 1997). This scenario of declining farmland wildlife correlating with increased farming intensity is typical over large parts of Europe (Krebs, Wilson Bradbury and Siriwardena, 1999) and elsewhere where intensive chemical-dependent agriculture is practised (McLaughlin and Mineau, 1995, Freemark and Boutin, 1995).

Yields of milk and meat from UK livestock have increased dramatically over the past few decades. For example, milk production per cow has increased by more than 17 per cent since 1995 (Defra Statistics website). The application of fertilisers to grasslands has had a major effect on plant productivity and therefore yields of animal products. For example, Kirkham and Wilkins (1994) reported an increase in yield from 4.7 t ha^{-1} to 10.5 t ha^{-1} in a herb rich meadow in Somerset when 200 kg nitrogen, 75 kg phosphate and 200 kg of potash per ha^{-1} were added without herbicide applications. The switch from hay to silage as winter feed has also been a major factor in increasing yields, aided by increased fertiliser applications and the use of new, highly competitive grass and clover varieties that contain the optimum balance of nutrients. These developments have been accompanied by animal breeding programmes to increase the "genetic potential" of modern livestock to maximise outputs of milk and quality/quantity of meat.

These yield increases correlate with large reductions in non-grass biodiversity. Large applications of nitrogen, phosphorus and potassium, whilst increasing productivity, have large negative effects on plant species diversity (e.g. Mountford, Talowin, Kirkham and Lakhani, 1994). Silvertown, Dodd, McConway, Potts and Crawley (1994) used data from the Rothamsted Park Grass Experiment to demonstrate a strong negative relationship between sward productivity and floristic composition, with the grass component of the sward increasing at the expense of other less competitive vegetation. Between 1960 and 1980, management changes had a major impact on grasslands. The average use of nitrogen fertiliser on grassland trebled, while the area under silage production expanded at a rate of around 200,000 hectares annually (Hopkins and Hopkins, 1994). The switch from hay to silage production, with associated increases in fertiliser inputs, ploughing and reseeding, herbicide applications and frequency of cutting, has probably been the single most important factor in reducing grassland biodiversity in the UK.

Ecological research has demonstrated that species diversity tends to be highest at intermediate levels of nutrient availability, since at very low levels plants are under severe stress and few can survive (although these may be of high conservation value because they are rare and specialised), whereas at very high nutrient levels competition for light becomes the most important factor, leading to dominance by a few highly competitive species (Grime, 1979, Vermeer and Berendse, 1983).

In arable crops, there seems to be a general trend that increases in yield result in deceases in non-crop biodiversity, but because there may not be a clear relationship between grain yield and crop biomass, it is difficult to assign causality in terms of biomass diversion from non-crop organisms to crops. For example, some of the adverse impacts on biodiversity resulting from agricultural intensification are only indirectly caused by increases in yield, but the cause and effect chain is nonetheless important if we are to understand how to reverse these declines. For example, the demand for increased yields has led to increased regional specialisation in the UK, with arable farming becoming more concentrated in the eastern counties of England while livestock farming is more dominant in Wales, Scotland and the west of England. This has adversely affected those species which require a mosaic of arable and grassland habitats to complete their life cycles (especially breeding birds such as Lapwings *Vanellus vanellus* that need a combination of crops and grassland for success), but the effects seem to result from a lack of habitat diversity rather than a lack of food resources on a per field basis. In this respect, organic farming may have clear benefits over conventional practice, since it tends to promote longer and more diverse rotations which more frequently include both arable and grassland elements. There is a need for more research comparing yields and biodiversity over the course of whole rotations, to augment the research into the impacts of farming intensity on breeding farmland birds in Southern England (Potts, 1997; Brickle, Harper, Aebischer and Cockayne, 2000).

To begin to understand the underlying causes of such biodiversity reductions, it is necessary to look at research into the relationship between increases in yield and crop management.

Increases in yield enabled by agricultural intensification can have both direct and indirect effects on biodiversity. Broadly speaking, the direct effects result from:

a) management interventions to increase nutrient levels in the field (fertilisers), combined with the use of high-yielding crop varieties that can out-compete most weeds under these conditions (e.g. by rapid canopy closure)
b) management interventions aimed at removing species that could compete with the crop for water, nutrients or light (i.e. herbicide and pesticide applications; seed cleaning, planting density)
c) management to extend the growing season of the crop (i.e. autumn sowing) at the expense of the non-crop biodiversity that would normally occupy stubble fields during the winter months
d) grazing regimes: these can have a positive or negative impact, depending on the grazing intensity and type of grassland (e.g. heavy grazing can increase biodiversity on chalk grassland but decrease it on acid grassland).

Indirect effects include:

a) "collateral damage" to non-target and beneficial species of plants, animals, fungi and bacteria (e.g. through tillage, harvesting, hedgerow removal, drainage, toxicity of broad-spectrum agrochemicals and diffuse pollution of watercourses)
b) changes in community composition through increases in soil fertility, changes in sowing date, etc. (e.g. highly fertile soils eliminate plant species that are competitive in conditions of low fertility; autumn sown crops reduce abundance of spring-germinating weeds)
c) changes in crop diversity, whether within a rotation or within a landscape (e.g. regional specialisation, continuous cropping) can adversely impact species requiring a mixture of habitats and food resources to complete their life cycles.

In practice it is often difficult to separate out each of the effects listed above, since they are highly interdependent (e.g. a large-scale switch to autumn sown crops was enabled by effective herbicides and fungicides). However, the following sections attempt to explore some of these interactions in more detail.

IMPACTS OF FERTILISERS ON YIELD AND BIODIVERSITY

Farmyard manure has probably always been used as a fertiliser on both arable crops and grasslands. Some primitive agrarian systems brought animals from outfield pastures to dung on infield cropping areas, and manure from livestock over-wintering enclosures was used (as now) as a manure on cropped areas. These systems essentially removed nutrients from grazed areas, transferring them to cropped land. This led to nutrient-impoverishment of grazed grasslands (probably with an increase in biodiversity) and nutrient enrichment of arable soils. Until the late eighteenth century in Britain, grasslands were rarely artificially fertilised either with manure or inorganic chemicals. At that time artificial phosphate-rich chemicals became available either as industrial by-products or from mining guano. These were enthusiastically applied to both grasslands and arable crops with remarkable increases in yield. With the advent of industrially produced nitrate fertilisers and potash, the late nineteenth century saw the introduction of routine applications of artificial fertilisers to arable crops and lowland grasslands, gradually allowing for greater crop yields and the development of grassland leys capable of supporting high densities of livestock (Stoate, 1995). This latter development continued in the twentieth century; increases in livestock carrying capacity were supported by increased yields of conserved grass, initially in hay production, but from the mid-twentieth century in silage-making, using repeated cuts of highly productive swards. Arable crop yields started to show perceptible increases from 1900 to 1930 as a result of increased fertiliser use. During this time weeds were controlled using cultivation (ploughing, harrowing and hoeing) and increasingly ingenious crop rotations that often included fallow grasslands on mixed farms.

Research has shown that the application of fertilisers alone to grasslands reduces the biodiversity associated with these systems. Both manure and artificial fertilisers increase grass species at the expense of low-growing broad-leaved herbs (Kirkham, Mountford and Wilkins, 1996), with a consequent reduction in invertebrates. On hay meadows such reductions still allow viable populations of birds to thrive, especially those that nest within the fields (such as partridges, skylarks and waders) and seed eating species such as buntings and finches. However, silage production has a damaging impact on all ground-nesting birds, because the fields are cut up to three times during the growing season, greatly reducing seed production and return because the grass is harvested before seed is set. Lowering productivity by withdrawing fertilisers can to some extent reverse these changes, but as the Somerset Levels experiments showed, reversion rates are very slow even on peat-based soils that do not hold nutrients very well. Reversion rates on clay and chalk soils may be even slower.

On arable cropped land the application of fertilisers has led to weed shifts favouring species that respond rapidly to high nutrient levels. Artificial fertilisers reduce weed diversity by increasing growth rate of the crop to the point where canopy closure occurs early in the growing season, suppressing weed germination, growth and development. Wilson (1992) describes the effects of applying different amounts of nitrogen on the numbers of weed plants present in two different cereal crops. Of 13 rare species sown, nine decreased significantly with increasing levels of nitrogen, and of these, very few plants of lamb's succory, weasle's snout, mousetail, and rough poppy survived when nitrogen was applied at a level similar to that used on modern cereals. On the other hand, species such as cleavers and blackgrass grow even better than modern crop varieties under highly fertilised conditions. Under highly fertile conditions, weeds may make up less than 2 per cent of the total above ground biomass, so competition between weed plants is probably negligible compared to interspecific competition with the crop (Jørnsgård, Rasmussen, Hill and Christiansen, 1996). However, there is some evidence from the Broadbalk experiments that even with long-term fertilisation of cereals, arable weeds can remain diverse and abundant, and certainly the biomass of weeds in fertilised fields can reach very high levels if not controlled by cultivation or herbicides.

THE IMPACTS OF HERBICIDES AND PESTICIDES

Although crop rotations, tillage and grassland management could allow increases in yield from fertiliser application, weeds were still a major factor limiting productivity and pests and diseases could have adverse impacts on yield. The agrochemicals industry began in the 1930s to develop herbicides such as MPCA, capable of selectively controlling broad-leaved herbs in grassland, and broad spectrum herbicides such as paraquat that could be used to control early flushes of weeds prior to the emergence of arable crops. Herbicides had an immediate and dramatic impact on yield, not only suppressing weed competition, but allowing the development of simpler arable rotations, where cereals could be grown for several years without a break crop, winter cropping that increased yield substantially, and longer lasting grass leys where weeds could be selectively controlled within the sward. Coupled with the application of higher fertiliser levels, insecticides and fungicides, and new crop varieties that were less competitive (allowing closer planting and more metabolic resources for grain production), yields rose to present day levels.

Pesticides have had both direct and indirect impacts on farmland biodiversity. Herbicides produce immediate and significant reductions in both grassland

and arable biodiversity by removing most of the non-crop primary trophic level that in turn leads to rapid reductions in invertebrate populations and in higher level populations such as birds. The evolution of herbicide use in the UK over the last 40 years has led to a concomitant change in the frequency of arable weeds. Research has identified a number of rare arable plants whose decline has been largely brought about by herbicides, including some species that were once common weeds, such as cornflower *Centaurea cyanus* and shepherd's needle *Scandix pecten-veneris* (Cooke and Burn, 1995). However, for many species the major effect of herbicide usage in arable fields has been to reduce rather than eliminate populations (Fryer and Chancellor, 1970). Herbicide use has been shown to be a key factor in reductions in bird populations partly because reductions in weeds lead to invertebrate biomass reductions that cannot support developing chicks (Potts, 1997; Brickle *et al* ., 2000). Herbicides also reduce seed return, with recent estimates of a 3% per annum reduction in seed return to arable soils over the past 50 years (Robinson and Sutherland, 2002). This in turn reduces weed abundance and diversity in the arable landscape. Insecticides and fungicides may also contribute to reductions in biodiversity but may not have such long-lasting impacts as herbicides because they are applied periodically and usually have short-lived effects. The organochlorine pesticides, which accumulate in tissues and increase in concentration through the food chain causing harm to predatory birds and other non-target fauna, have now been banned. However, pesticides can still have significant immediate effects on numbers of non-target invertebrates (e.g. Rands and Sotherton, 1986) and can reduce the food resource for higher predators such as birds and mammals, particularly if applied during the breeding season (Campbell, Avery, Donald, Evans, Green and Wilson, 1996). There is some evidence that reductions in non-crop biodiversity may increase the need for chemical control of pests and diseases due to the removal of predators and parasitoids that provide 'natural' pest and disease control (Gurr, Wratten and Luna, 2003; Cardinale, Harvey, Gross and Ives, 2003). Many high-yielding arable fields and grasslands in Northern Europe are largely occupied for most of the year by crop biomass, with virtually no non-crop biomass present, and no over-winter stubbles or fallows. The consequence has been dramatic reductions in plant, invertebrate, and bird populations to the point where many may be no longer viable, in the sense that they would be unlikely to survive extreme climatic events such as severe cold or drought.

THE ROLE OF ROTATIONS AND FARMING SYSTEMS

Some of the adverse impacts on farmland biodiversity have resulted from the reduction of crop diversity, and therefore habitat diversity within farmed landscapes. Simplifying crop rotations can significantly increase yield, particularly on soils that are suitable for a particular type of crop, and can also cut costs. In the mid-20th century mixed farming with grass leys as part of the rotation were still the dominant type of farming in most of the UK, however today mixed farms only account for a small percentage of farm businesses in Europe as a whole, and this is in rapid decline – between 1987 and 1997 the proportion of mixed farms in Europe fell from 27% to 18% (European Environment Agency, 2002). A far higher proportion of organic farms are mixed farms, since organic farming relies on the application of animal manure, making it economically efficient to keep livestock on the farm.

Organic farming provides a good demonstration of the importance of comparing yields and biodiversity across a whole rotation and not on a single year basis. Annual crops are not managed in isolation, and decisions on weed management are based on knowledge of the agronomy of previous and following crops. Yields of individual organic arable crops are almost always lower than yields of conventionally farmed crops, although the difference may not be great. However, when the yields are assessed over the course of the whole rotation, the total saleable yield of an organic farm is much lower than a conventional farm, because organic rotations tend to be longer and include grass leys for perhaps two out of seven years, which do not produce a yield (Ian Alexander, pers. comm.). Organic rotations also tend to include a wider variety of crops, whereas most conventional rotations are largely based on wheat. This diversity of crops and grass is one of the major reasons for the increased biodiversity that has been recorded on organic farms in the UK and elsewhere.

A study comparing crop and weed biomass in 38 matched pairs of organic and conventional cereal fields in Denmark found that organic crops had on average 25% lower crop biomass than conventional crops (presumably translated into considerably lower yield), but five times more weed biomass. Interestingly, the total biomass (crop + weed) in organic cereal fields was 16% lower than in conventional fields, perhaps because of lower levels of fertility (Hald, 1999).

Small pockets of arable farming within grassland-dominated areas of the UK have been demonstrated to be of great importance to a range of declining farmland birds (Robinson, Wilson and Crick, 2001). An increase in mixed farming in areas currently dominated by grassland or arable farming is one of the few options that is insufficiently catered for under existing agri-environment measures (Aebischer, Bradbury, Eaton, Henderson, Siriwardena and Vickery, 2003).

Are high output/low biodiversity systems sustainable?

The evidence above demonstrates that agricultural productivity has increased and biodiversity has been lost because of a drive for increased efficiency leading to changes on at least three levels: the field scale (the crop), the farm scale (the rotation and uncropped habitats), and the landscape scale (regional cropping patterns). In terms of the survival and viability of farmland biodiversity, modern intensive systems are clearly not sustainable at any of these levels. Constant reductions in non-crop biodiversity will continue to the point where Europe will fail to deliver the statutory commitments inherent in legislation such as the Birds Directive and the Habitats and Species Directive. High output/low biodiversity systems also rely on inputs of finite resources such as fossil fuels, phosphates and freshwater, and produce unacceptably high levels of atmospheric and water pollution. High yielding intensive agriculture, both cropping and grazing, can also produce serious soil degradation, although there have been some attempts (such as the introduction of minimal tillage) to alleviate this in the past decade. Soil is a finite resource and may have functions in maintaining above ground biodiversity that we do not yet fully understand. Globally, some 562 million hectares of agricultural land are now classified as degraded, of which around 27% have been degraded by agricultural activities (UNEP, 1992), and 10 million hectares of cropped soils are being lost each year to degradation by inappropriate arable cultivation (World Bank/UN estimates at http://lnweb18.worldbank.org).

A clear distinction needs to be made between mitigation, (providing alternative wildlife habitats and food resources outside the main cropped area) and adaptation of whole farming systems to allow biodiversity to coexist with the crop. Mitigation measures such as conservation headlands and "skylark scrapes" can undoubtedly play an important role in halting and reversing declines in many farmland species, but they will only be effective if adopted at a sufficient scale and targeted appropriately, and even then they may need to be augmented by organic and other low-input farming systems (Vickery, Bradbury, Henderson, Eaton and Grice, 2004).

Our view is that to produce more sustainable farming systems, there needs to be change at all three scales identified above, which might entail some drop in productivity within individual crops, over whole rotations and at the scale of total UK agricultural production.

Can biodiversity be increased and sustainability improved by lowering yields?

In arable systems the main factor impacting on non-crop biodiversity appears not to be yield or crop biomass *per se,* but the chemicals used to suppress primary weed production, mostly herbicides, and agrochemicals used to control pests and diseases. As many arable soils are now so depleted of weed seeds, simply reducing yields by reducing fertiliser inputs and reverting to spring cropping, without reducing the impacts of pesticides, would probably not have a significant impact on biodiversity (the use of unsprayed field margins in agri-environment schemes produces significant increases in biodiversity despite the fact that fertility remains high). The key issue is how to reduce herbicide and pesticide inputs to the point where arable non-crop biomass increases, but crop yields are economically viable. Development of crop plants that defend themselves against pests and diseases could greatly reduce pesticide inputs to the point where non-target predators and parasitoids control pests, and where fungal attack is avoided, resisted or tolerated by the crop. Recent advances in our understanding of the genetic control of plant architecture could also result in the development of crops that could tolerate weeds better, and there is also the prospect of harnessing allelopathic mechanisms (Grundy, Bond, Burston and Jackson, 1999; Chou, 1999), introducing these traits to crops to suppress weed growth and germination in the immediate vicinity of crop root systems. Herbicide tolerance traits, introduced into crops either by conventional breeding or by transgenic means, might also offer novel methods of controlling weeds in ways that allow insects and birds to develop at critical periods of their growth, and it may even be possible to increase autumn seed return to the soil by spraying herbicides early in the growing season and not later.

For agronomic reasons, reducing fertiliser levels on grasslands is already a trend in UK livestock production, although yield reductions have been marginal. Reversion experiments on the Somerset Levels (Kirkham and Tallowin, 1995) have shown that substantial reductions in fertilisers are needed, coupled with cessation of herbicide use, before any substantial increase in plant biodiversity is observed. Wildlife conservation organisations have shown that herb-rich swards can be re-established on some soils by using techniques including reseeding, phosphate-stripping using high nitrogen input and intensive harvesting for several years, and in extreme cases topsoil removal or burial. These techniques are expensive and would not be suitable for some heavy UK agricultural soils, so a return to lower fertilisation and hay-based livestock enterprises appears to be the most feasible option to increase biodiversity. Establishment of more diverse swards using these techniques has been shown to be a lengthy process, but they can be both productive and nutritious, both in conventional and organic systems (Mountford *et al.,* 1994). Low input, hay-based grasslands should be more sustainable systems than current silage based livestock production, with less use of fossil fuels, less water pollution, and probable gains in animal health. Such systems are used in organic livestock and milk production, which also use livestock varieties more suited to the conditions. The recent CAP reforms offer an opportunity to provide incentives to encourage these systems.

Conclusions

Lowering yields in arable cropping by using less fertiliser would not necessarily give higher biodiversity. As field margin agri-environment schemes have shown, there is also a need to reduce the impacts of herbicides and other pesticides. This does not necessarily equate with reducing inputs. Research, including the datasets obtained during the Farm Scale Evaluations of herbicide tolerant crops, has shown that the type of herbicide, dose rates, application frequency and timing have a greater impact on non-crop biodiversity than the amount of active ingredient applied. Reductions in the impacts of agrochemicals would by themselves reduce yield, especially if modern crop varieties, which are poorly competitive in the absence of high fertiliser inputs, are used. There is a need for new arable crop varieties that defend themselves against biotic and abiotic stress, and a need for crop rotations that include a greater frequency of break crops between cereals. There is also a need to address the issue of low weed seed banks in arable soils. Traits needed to enable more sustainable arable systems may become available via marker assisted breeding and transgenic techniques. More sustainable methods of managing arable rotations that support higher biodiversity are emerging from research into modern organic agriculture. These systems could be more widely adopted into 'quasi-organic' agriculture in areas where manure is not readily available, if novel crops were developed that had high resistance to abiotic and biotic stresses.

Development of lower yielding more sustainable grasslands supporting higher biodiversity will require lower fertiliser inputs and reversion to hay-based livestock systems. The latter factor is of great importance not only in terms of increased biodiversity but also in contributing to greater whole system sustainability, if pollution, use of fossil fuels and consumption of finite resources are also taken into account. Lower yields would be an inevitable consequence of adopting such systems but overall profitability of low input systems could be maintained in the new agronomic climate resulting from CAP reform.

Biodiversity gains from using lower input lower yielding cropping and grassland systems would be greatly enhanced by increasing the proportion of mixed farms in the landscape, together with the introduction of more diverse crop rotations.

References

Aebischer, N.J., Bradbury, R., Eaton, M., Henderson, I.G., Siriwardena, G.M. and Vickery, J.A. (2003) *Predicting the Response of Farmland Birds to Agricultural Change.* BTO Research Report No. 289. British Trust for Ornithology, Norfolk, UK.

Benton, T.G., Bryant, D.M., Cole, L. and Crick, H.Q.P. (2002) Linking agricultural practice to insect and bird populations: a historical study over three decades. *Journal of Applied Ecology* **39**, 673-687.

Brickle, N.W., Harper, D.G.C., Aebischer N.J., and Cockayne S.H. (2000) Effects of agricultural intensification on the breeding success of corn buntings *Miliaria calandra. Journal of Applied Ecology* **37**, 742-755.

Campbell, L.H., Avery, M.I., Donald, P., Evans, A.D., Green, R.E. and Wilson,

J.D. (1996) *A Review of the Indirect Effects of Pesticides on Birds*. A report to the DoE, JNCC and English Nature prepared by the RSPB.

Cardinale, B.J., Harvey, C.T., Gross, K. and Ives, A.R. (2003) Biodiversity and biocontrol: emergent impacts of a multi-enemy assemblage on pest suppression and crop yield in an agroecosystem *Ecology Letters* **6** (9), 857-865.

Chou, C-H (1999) Roles of allelopathy in plant biodiversity and sustainable agriculture. *Critical Reviews in Plant Sciences* **18** (5), 609-636.

Cooke, A.S. and Burn, A.J. (1995) The environmental impact of herbicides used in intensive farming systems. *Brighton Crop Protection Conference - Weeds - 1995*, **6B-2**: 603-612.

Dolton, C.S. and Brooke, M. de L. (1999) Changes in the biomass of birds breeding in Great Britain, 1968-88. *Bird Study*, **46**, 274-278.

European Environment Agency (2002) *Environmental Signals*. EEA, Copenhagen.

Ewald, J. A., and Aebischer, N.J. (1995) Pesticide Use, Avian Food Resources and Bird Densities in Sussex. *Joint Nature Conservation Committee*, Peterborough, UK

Freemark, K. and Boutin, C. (1995) Impacts of agricultural herbicide use on terrestrial wildlife in temperate landscapes: a review with special reference to North America. *Agriculture, Ecosystems and the Environment* **52** 67-91

Fryer, J.D. and Chancellor, R.J. (1970) Herbicides and our changing weeds. In: *The Flora of a Changing Britain*, FH Perring (ed)., Faringdon: Classey, pp. 105-118.

Fuller, R .J., Gregory, R.D., Gibbons, D.W., Marchant, J.H., Wilson, J.D., Baillie, S.R. and Carter, N. (1995) Population Declines and range Contractions among Lowland Farmland Birds in Britain *Conservation Biology* **9**: No 6 pp 1425-1441.

Grime, J.P. (1979) *Plant Strategies and Vegetation Processes*. Wiley, Chichester.

Grundy, A.C., Bond, W., Burston, S. and Jackson, L. (1999) Weed suppression by crops. *Proceedings of the BCPC Conference - Weeds – 1999*, 957-962.

Gurr, G.M., Wratten, S.D. and Luna, J.M. (2003) Multi-function agricultural biodiversity: pest management and other benefits. *Basic and Applied Ecology* **4** (2), 107-116.

Hald, A.B. (1999) Weed vegetation (wild flora) of long established organic versus conventional cereal fields in Denmark. *Annals of Applied Biology* **134**, 307-314.

Hopkins, A. and Hopkins, J.J. (1994) UK grasslands now: agricultural production and nature conservation. In *Grassland Management and Nature Conservation*. Ed. Haggar RJ and Peel S pp 10-19. British Grassland Society, Occasional Symposium 28.

Jørnsgård, B., Rasmussen, K., Hill, J. and Christiansen, J.L. (1996) Influence of nitrogen on competition between cereals and their natural weed populations. *Weed Research* **36**, 461-470.

Kirkham, F.W. and Wilkins, R.J. (1994) The productivity and response to inorganic fertilisers of species-rich wetland hay meadows on the Somerset Moors: nitrogen response under hay cutting and aftermath grazing. *Grass and Forage Science* **49**, 152-162.

Kirkham, F.W. and Tallowin, J.R.B. (1995) The influence of cutting date and previous fertiliser treatment on the productivity and botanical composition of species-rich hay meadows on the Somerset Levels. *Grass and Forage Science* **50**, 365-377.

Kirkham, F.W., Mountford, J.O. and Wilkins, R.J. (1996) The effects of nitrogen, potassium and phosphorus addition on the vegetation of a Somerset peat moor under cutting management. *Journal of Applied Ecology* **33**, 1013-1029.

Krebs, J.R., Wilson, J.D., Bradbury, R.B. and Siriwardena, G.M. (1999) The second Silent Spring? *Nature* **400** 611-612

McLaughlin, A. and Mineau, P. (1995) The impact of agricultural practices on biodiversity. *Agriculture, Ecosystems and Environment* **55**, 201-212.

Mountford, J.O., Talowin, J.R.B., Kirkham, F.W. and Lakhani, K.H. (1994) Effects of inorganic fertilisers in flower-rich hay meadows on the Somerset Levels. In *Grassland Management and Nature Conservation.* Ed. Haggar RJ and Peel S pp 74-85. British Grassland Society, Occasional Symposium 28.

Potts, G.R. (1997) Cereal farming, pesticides and grey partridges, in *Farming and Birds in Europe*, Chapter 6 Academic Press

Powlson, D.S. and Poulton, P.R. (2003) Long-term experiments in the 21st century – continuity or change? *Proc. of the NJF's 22nd Congress: Nordic Agriculture in Global Perspective, July 1-4, 2003, Turku, Finland.*

Rands, M.R.W. and Sotherton, N.W. (1986) Pesticide use on cereal crops and changes in the abundance of butterflies on arable farmland in England. *Biological Conservation* **36**, 71-82.

Robinson, R.A., Wilson, J.D. and Crick, H.Q.P. (2001) The importance of arable habitat for farmland birds in grassland landscapes. *Journal of Applied Ecology* **38**, 1059-1069.

Robinson, R. A. and Sutherland, W.J. (2002) Post-war changes in arable farming and biodiversity in Great Britain. *J Appl. Ecol.* **39**, 157-176

Silvertown, J., Dodd, M.E., McConway, K., Potts, J. and Crawley, M. (1994) Rainfall, biomass variation, and community composition in the park grass experiment. *Ecology* **75**, 2430-2437

Siriwardena, G.M., Baillie, S. R., Buckland, S.T., Fewster, R.M., Marchant, J.H. and Wilson, J.D. (1998) Trends in the abundance of farmland birds: a quantitative comparison of smoothed Common Birds Census indices. *Journal of Applied Ecology* **35**: 24-43.

Stoate, C. (1995) The changing face of lowland farming and wildlife. Part 1: 1845-1945. *British Wildlife* **6**, 341-350.

UNEP (1992) *World Atlas of Desertification.* London, Arnold

Vermeer, V.G. and Berendes, F. (1983) The relationship between nutrient availability, shoot biomass and species richness in grassland and wetland communities. *Vegetation* **53**, 121-126.

Vickery, J.A., Bradbury, R.B., Henderson, I.G., Eaton, M.A. and Grice, P.V. (2004) The role of agri-environment schemes and farm management practices in reversing the decline of farmland birds in England. *Biological Conservation* **119**, 19-39.

Wilkins, R.J. (2000) Grassland in the 20th Century. *IGER Innovations*, 25-33.

Wilson, P.J. (1992) Britain's arable weeds. *British Wildlife* **3** (3), 149-161.

16

PHYSIOLOGICAL LIMITATIONS, NUTRIENT PARTITIONING

ROBERT J. COLLIER[1], LANCE H. BAUMGARD[1], ADAM L. LOCK[2], AND DALE E. BAUMAN[2]
[1]*Department of Animal Science, University of Arizona, Tucson, AZ 85721, USA and* [2]*Department of Animal Science, Cornell University, Ithaca, NY 14853, USA*

Introduction and background

An important goal in animal agriculture is to achieve both economic and sustainable improvements in productive efficiency. Global prosperity continues to increase and affluent nations traditionally demand and consume more foods of animal origin (Roche and Edmeades, 2004). Thus, worldwide dairy product consumption is increasing and meeting this demand requires improving productive efficiency. Impressive gains in food production efficiency have occurred during the second half of the 20th century as is illustrated by the enhanced annual US and UK milk yield/cow (Figure 16.1; www.usda.gov/nasss; www.defra.gov.uk). Reasons for these gains in dairy production include a better understanding of nutrient requirements, improvements in diet formulation, utilizing artificial insemination, applying more accurate genetic selection methods, improved milking management practices, mastitis control and the effective use of herd health programs to prevent disease. Furthermore, new technologies and management tools such as bovine somatotropin, estrus synchronization and pregnancy detection have enhanced the production potential of dairy cows and allowed them to achieve that potential.

Performance is the best indicator of a dairy cow's well-being, and gains in performance and productive efficiency over the last half-century are truly remarkable. Of special importance is to identify the biological processes altered to achieve these advancements. Bauman, McCutcheon, Steinhour, Eppard and Sechen (1985); Gordon, Patterson, Yan, Porter, Mayne and Unsworth (1995) and more recently Reynolds (2004) reviewed the biological basis for such improvements and found that digestion and nutrient absorption, maintenance requirements and the partial efficiency of nutrient use for productive functions (i.e. milk synthesis) were only minor sources of the variation among animals and contributed little to gains in performance and productive efficiency. In contrast, animal differences in nutrient partitioning represented the major source of improvement. For example, feed intake is largely dependent on signals relating to tissue nutrient utilization and high producing cows consume more, and direct these nutrients toward milk synthesis rather than excessive fattening/body condition. Maintenance requirements for dietary energy and protein are a substantial portion of total requirements, but they are constant regardless of milk production levels. Thus, high producing cows have a greater energy and nutrient intake in order to support additional milk production, but they also have a higher productive efficiency due to a larger portion of total energy and nutrient intake being used to synthesize milk. This is often referred to as "dilution of maintenance" and represents the basis for remarkable gains in productive efficiency (Bauman *et al.*, 1985).

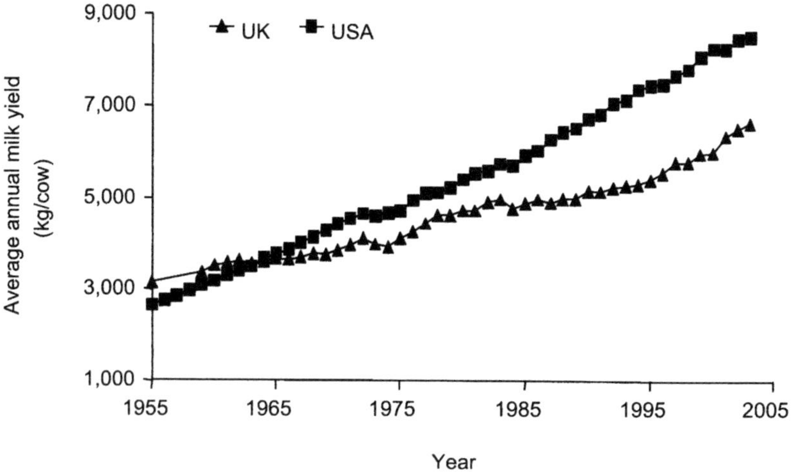

Figure 16.1 Annual milk yield/cow in the United States and the United Kingdom.

With each advance in dairy production over the last century, some have commented that cows are being pushed too far, thereby causing metabolic stress and compromising cow health and well-being. This concern was articulated over 50 years ago by Sir John Hammond when he pointed out that "considerable doubts have been expressed as to whether we are not now pressing high production in our farm animals too far, thereby undermining their constitution and so shortening their life" (Hammond, 1952). Hammond's review of lactational physiology provided no support for this opinion, and he concluded, "the physiological limits to intensive milk production are...only limited by our knowledge concerning the specific nutrients required for milk production" (Hammond, 1952). Subsequent re-evaluations of the biological limits of production by Bauman *et al.* (1985); Knight, Alamer, Sorensen, Nevison, Flint and Vernon (2004) and Reynolds (2004) reached similar conclusions. Indeed, increased productivity has continued at an amazingly consistent rate during the last half-century and there is no evidence that a plateau is nearing (Figure 16.1). Hammond (1952) described Manningford Faith Jan Graceful, a cow of that era that by the end of her third lactation had produced more milk than what 15 average contemporaries would have achieved. It has been the selection of such animals and an improved understanding of dairy cow nutrition and management that has allowed for this continual increase in average milk yield over the last 50 years. Today some herds produce over 15,000 kg/cow, a value similar to the production of Manningford Faith Jan Graceful, and modern day record cows, such as Muranda Oscar Lucinda, have an annual milk yield of over 30,000 kg (www.holsteinusa.com).

Two main arguments are often proposed for reducing the rate or even stopping gains in farm animal efficiency: 1) that high production and productive efficiency are contradictory to the concept of 'sustainable' agriculture; and 2) that increasing yield is contrary to the welfare and well-being of animals. However, when these arguments are examined more closely, it is apparent that in fact the opposite is true. Sustainable agriculture has become synonymous with protecting the environment at all costs; some even suggest that food production efficiency and environmental sustainability are mutually exclusive terms (Roche and Edmeades, 2004). However, it can be argued that for any system to be truly sustainable it must, by definition, be efficient. Similarly, in a review of dairy cow welfare, Broom (1999) suggested that it might be necessary to stop using genetic selection and some feeding/nutrition methods to increase milk yield because these practices have resulted in stressed cows in which their normal biological controls are overtaxed. However, proponents of this idea fail to understand that genetic selection and management improvements are successful because they have altered the biological controls in a coordinated manner. Rather than the biological controls being at discord with increased performance, it is the improvements in the biological control systems which are responsible for the increases in milk yield and gains in productive efficiency. In fact, it is only when the coordination of energy and nutrient use is inadequate or an imbalance occurs that animal well-being is compromised.

The objective of the chapter is to outline understanding of productive efficiency and how it may be affected by genetic selection and management practices. In the following sections, concepts involving homeostasis and homeorhesis and their application to the regulation of metabolism in different physiological situations will be reviewed. Of special interest will be lactation and both well-managed high producing cows and metabolic regulation in management systems where cows are chronically underfed will be considered. In addition, animal acclimation to environmental stress will be examined and how the concept of homeorhesis can be applied to this physiological situation. Finally, there will be speculation on what the future might hold and the limits of production efficiency in dairy cows.

Concepts of regulation

The regulation of physiological processes has the overall goal of maintaining the animal's well-being regardless of the physiological situation or environmental challenges that are encountered. From a conceptual basis, this involves two types of regulation, homeostasis and homeorhesis. Cannon (1929) described homeostasis as the condition of relative uniformity, which results from the adjustments of living things to changes in their environment. From an operational standpoint, homeostatic controls operate on an acute basis so that despite challenges from the external environment, different tissues and organs are "all working cooperatively" to maintain physiological equilibrium. There are many well-established examples of the multiple compensatory mechanisms functioning to maintain physiological equilibrium. One example involving nutrient partitioning is the maintenance of circulating glucose. Glucose supply is of critical importance for many tissues and physiological processes. Over the short term, homeostatic controls, primarily insulin and glucagon, maintain a relatively constant supply of glucose to peripheral body tissues by promoting glucose storage following a meal and subsequent mobilization from glycogen stores during the post absorptive period (Bauman and Elliot, 1983; Vernon and Sasaki, 1991). Thus, acute regulation of plasma glucose concentration by the reciprocal actions of insulin and glucagon ensures the proper balance in glucose supply and utilization by body tissues.

The second type of control is homeorhesis, which was defined as the "orchestrated changes for priorities of a physiological state" (Bauman and Currie, 1980). The concept originated from considering how physiological processes are regulated during pregnancy and lactation (Bauman and Currie, 1980), but application of the general concept has been extended to include different physiological states, nutritional and environmental situations and even pathological conditions (Table 16.1). Thus, homeorhetic regulation involves a coordination of physiological processes in support of a dominant physiological state or chronic situation.

Table 16.1. PARTIAL LIST OF PHYSIOLOGICAL SITUATIONS WHERE THE GENERAL CONCEPT REPRESENTING HOMEORHETIC REGULATION HAS BEEN APPLIED.[1,2]

Lactation	Hibernation
Pregnancy	Premigration / Migration
Growth	Egg laying
Puberty	Incubation anorexia
Aging	Seasonal cycles
Exercise	Environmental limitations[3]
Chronic undernutrition	Management limitations[3]
Chronic illness	Physiological limitations[3]

[1]Adapted from Bauman (2000).

[2]References include Bauman and Currie, 1980; Bauman, Eisemann and Currie, 1982; Dilman, 1982; Nicolaïdis, 1983; Mrosovsky, 1990; Wade and Schneider, 1992; Vernon, 1998; Chilliard, 1999; Kuenzel, Beck and Teruyama 1999.

[3]Application developed in subsequent sections of this chapter.

Key features of homeorhetic controls are: its chronic nature, hours and days versus seconds and minutes required for most examples of homeostatic regulation; its simultaneous influence on multiple tissues and systems that results in an overall coordinated response, which is mediated through altered responses to homeostatic signals (Bauman and Elliot, 1983; Vernon, 1989; Bell and Bauman, 1997).

Lactation is perhaps the best characterized example of homeorhesis. The successful transition from pregnancy to lactation is dependent on an extensive series of physiological adaptations that include many, perhaps most, body tissues and involve all nutrient classes (Table 16.2). The role of somatrotropin (ST) and other homeorhetic controls in the regulation of these physiological adaptations have been reviewed elsewhere (Bauman and Elliot, 1983; Vernon and Sasaki, 1991; Bell and Bauman, 1997; Chilliard, 1999). Again, metabolic adaptations related to glucose metabolism serve as an example. At the onset of lactation there is a marked increase in mammary utilization of glucose, primarily for lactose synthesis. Overall, the total glucose turnover in a high producing cow can exceed 3 kg/d with up to 85% of this being utilized by the mammary gland (Bauman and Elliot, 1983; Beever, Hattan, Reynolds and Cammell, 1999). If there is an imbalance between the availability and requirement for glucose, ketosis may result and well-being is compromised. To ensure an adequate glucose supply to support lactation, the biological regulation involves a series of orchestrated changes and these include increased hepatic rates of gluconeogenesis, decreased glucose uptake, reduced utilization and oxidation by adipose tissue and muscle, and a shift in whole body nutrient oxidation so less glucose is used as an energy source. Somatotropin is a key homeorhetic control in these adaptations and a central component in the mechanism involves attenuating tissue responses to insulin. At the onset of lactation, the ability of insulin to stimulate glucose disposal by adipose and muscle is reduced and insulin is less effective at inhibiting hepatic gluconeogenesis. The net effect is that these glucose-related metabolic alterations are coordinated with the increase in milk synthesis and substrate use by the mammary gland. Thus, homeostatic controls still function on a minute-by-minute basis to maintain steady state conditions during lactation, but on a chronic basis homeorhetic controls have orchestrated a series of adaptations in both physiological processes and extramammary metabolism to ensure the mammary gland is provided with an adequate quantity and pattern of substrates to support milk synthesis.

Management limitations

What if a high producing dairy cow is placed in a management system that is substantially less than ideal, a management system where the cow is unable to achieve her productive/genetic potential? Does the cow's genetic programming

Table 16.2 PARTIAL LIST OF PHYSIOLOGICAL ADAPTATIONS THAT OCCUR IN LACTATING DAIRY COWS.

Process or tissue	*Response*
Mammary tissue	Increased number of secretory cells
	Increased nutrient use
	Increased blood supply
Food intake	Increased quantity
Digestive tract	Increased size
	Increased absorptive capacity
	Increased rates of nutrient absorption
Liver	Increased size
	Increased rates of gluconeogenesis
	Increased glycogen mobilization
	Increased protein synthesis
Adipose tissue	Decreased de novo fat synthesis
	Decreased preformed fatty acid uptake
	Decreased fatty acid reesterification
	Increased lipolysis
Skeletal muscle	Decreased glucose utilization
	Decreased protein synthesis
	Increased protein degradation
Bone	Increased Ca and P mobilization
Heart	Increased cardiac output
Plasma hormones	Decreased insulin
	Increased somatotropin
	Increased prolactin
	Increased glucorticoids
	Decreased thyroid hormones
	Decreased IGF-I

Adapted from Bauman and Currie, 1980; Bauman and Elliot, 1983; Vernon, 1989, 1998; Chilliard, 1999.

for high lactational performance result in her becoming stressed and thus experience compromised health when faced with adverse management conditions? The belief that this occurs is often stated (for example see: Broom, 1995, 1999; FAWC, 1997; Rauw, Kanis, Noordhuizen-Stassen and Grommers, 1998), but does knowledge of biological regulation provide support for this idea? Knight *et al.* (2004) experimentally evaluated these questions by using cows differing in genetic merit and subjected them to three management challenges that increased their metabolic burden; 1) milking four times daily, 2) treatment with recombinant bST, and 3) thyroxine administration. Each of these management challenges increased milk yield in both the high and low genetic merit groups indicating neither group was at their metabolic ceiling, and when used in combination the increases in milk yield (+30%) were similar for both groups (Figure 16.2). No health problems were encountered, thus these results offer no support for the belief that selecting dairy cows for high milk yield has rendered them more susceptible to stress or metabolic problems. Thyroxine causes increases in heart rate and the metabolic rate of all cells, and the results observed with thyroxine treatment were particularly insightful. As a result of the increased rate of metabolism that occurred during thyroxine treatment, intake was inadequate to meet energy and nutrient requirements, and by the 3rd and 4th week on treatment body fat reserves were exhausted and cows in both genetic merit groups had markedly reduced milk yield (Knight *et al.*, 2004). Thus, in this experiment, cows responded to an inadequate energy and nutrient supply by down-regulating milk synthesis rather than developing a pathological deterioration that would compromise health, and this shift in the priority in energy and nutrient use occurred regardless of whether cows had a high or low genetic potential for milk production.

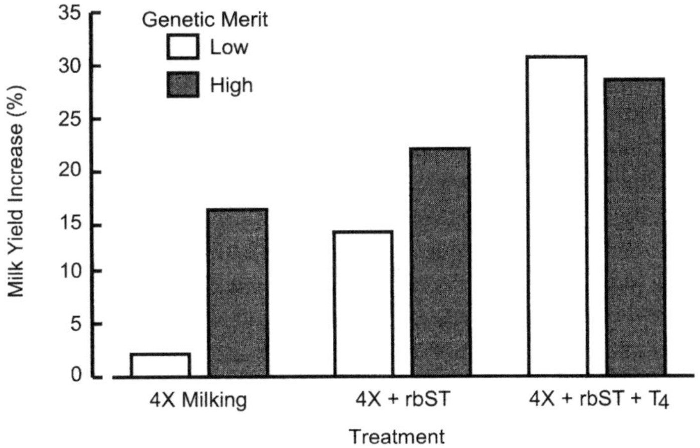

Figure 16.2 Milk yield response of high and low genetic merit cows to combinations of galactopoietic stimuli: frequent milking (4x), growth hormone (rBST) and thyroxine (T_4). Based on Knight (1999) and Knight *et al.* (2004).

Results from Knight *et al.* (2004) demonstrated that the same exquisite coordination of physiological processes that occurs at the onset of lactation in high producing dairy cows is equally important to maintain well-being when cows are subjected to an experimentally-induced metabolic burden. Do these coordinated biological responses also occur when management quality is less than ideal in commercial dairy production or has the selection for high milk yield put the well-being of these cows at greater risk? In fact, most commercial dairy operations are not optimally managed and in some management systems the cow's ability to achieve her genetic potential is substantially limited. An example is pasture-based management systems where cows generally have an adequate protein supply, but are chronically underfed in energy, especially when pasture is the only feed source (Arriaga-Jordan and Holmes, 1986; Kolver and Muller, 1998; Buckley, Dillon, Rath and Veerkamp, 2000). Kolver and Muller (1998) demonstrated the impact of this in high yielding dairy cows in a study where the lactational performance was compared between two dietary management systems; one where the latest dynamic model (Cornell Net Carbohydrate and Protein System) was used to formulate a total mixed diet ideally balanced to meet nutrient needs versus one where grazing of spring pasture supplied the only nutrients. Milk yield on the TMR diet averaged 44.1 kg/d, but was decreased by 33% when cows were fed only pasture. Although cows had the genetic potential for high milk yield, the chronic under-nutrition imposed by the pasture management system did not lead to stress or create a metabolic disease state; rather homeorhetic control mechanisms resulted in a shift in the priority for energy and nutrient use and a decrease in lactational performance, thereby maintaining cow health and well-being.

When management quality is less than optimal, individual differences in genetic merit will still be expressed, but all cows will be producing less than their genetic potential. By extension, technologies that enhance milk production may still have application under these conditions, although the magnitude of response will be related to the quality of management, and can be substantially less than if the animal was in an optimum environment (Figure 16.3; Bauman, 1992). An example is provided by Hoogendoorn, McCutcheon, Lynch, Wickham and MacGibbon (1990) in a study that examined bST-induced milk responses in pasture-fed cows. Not only was nutritional management less than ideal because the only feed source was pasture, but there was also seasonal variation in pasture growth over the 26-week treatment period. The milk response to bST was greatest (+18%) in the spring and early summer, when adequate pasture was available, declined to zero during the summer drought when pasture growth was low, and then increased again when autumn rains revived pasture growth (Figure 16.4). During the drought season when pasture growth was low, milk yield declined over 40% and production was similar between groups. Thus, even though bST treatment increased milk yield, there was no milk response during the summer drought period when nutrient supply was inadequate. A portion of the mechanism by which ST regulates lactation involves insulin-like growth factor (IGF). Key to the ability of nutrition to modulate performance is its effect on the ST/IGF axis and when nutrition is inadequate this axis is uncoupled (McGuire, Vicini, Bauman and Veenhuizen, 1992). Thus, nutritional regulation of the ST/IGF system appears to be a key component signaling the appropriate energy and nutrient utilization thereby preserving animal health and well-being (Bauman and Vernon, 1993; McGuire and Bauman, 1997).

Figure 16.3 Effects of management quality on predicted production responses to new technology.

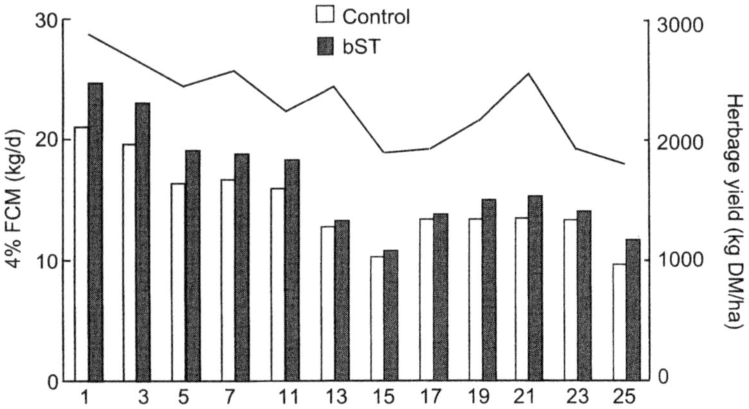

Figure 16.4 Effects of bST and nutrient availabiltiy on production responses in grazing cows. Kolver and Muller, 1998.

Environmental limitations

Environmental factors such as ambient temperature, solar radiation and humidity have direct and indirect effects on animals (Collier, Beede, Thatcher, Israel and Wilcox, 1982) and impact worldwide livestock production (Neinaber, Hahn and Eigenberg, 1999). Direct effects include altered production, reproduction and resistance or susceptibility to disease. Indirect effects involve changes in pathogen and pest populations as well as animal behavior that can cause increases in exposure to pathogens and pests, such as lying in muddy, wet areas to cool themselves when heat-stressed. Heat stress effects on post-absorptive metabolism and nutrient partitioning/utilization are not well understood even though the effect of hyperthermia on the endocrine system have been characterized (Collier and Beede, 1985). When an animal's microenvironment ventures outside its thermoneutral zone, a portion of the metabolizable energy typically used for production must be diverted to assure thermal balance. Therefore, selection for thermal tolerance to environmental stress has traditionally resulted in reduced productivity. For example, cattle breeds adapted to heat stress take longer to reach maturity and have lower levels of milk production (McDowell, 1972; Collier and Beede, 1985). Thus, it has generally been faster and easier to increase production by altering the environment. However, environmental modification comes at a high cost and in many cases these expenses cannot be economically justified. Therefore, economic advantages to producers exist if improved thermotolerance could be accomplished without adversely effecting production. Key to achieving this goal is an improved understanding of the physiological mechanisms of acclimation to environmental stressors.

Acclimation is a phenotypic response developed by the animal to an individual stressor within the environment (Fregley, 1996). However, under natural conditions it is rare for only one environmental variable to change over time. Acclimatization is the process by which an animal adapts to several stressors within its natural environment (Bligh, 1976). Acclimation and acclimatization are therefore not evolutionary adaptations or natural selection, which are defined as changes allowing for preferential selection of an animal's phenotype and are based on a genetic component passed to the next generation. The altered phenotype of acclimated animals will return to normal if environmental stressors are removed, which is not true for animals which are genetically adapted to their environment. Acclimatization is a process that takes several weeks to occur and close examination of this process reveals that it occurs via homeorhetic and not homeostatic mechanisms. As described by Bligh (1976), there are three functional differences between acclimatory responses and homeostatic or "reflex responses". First, the acclimatory response takes much longer to occur (days or weeks versus seconds or minutes). Second, the acclimatory responses generally have a hormonal link in the pathway from the central nervous system to the effector cell. Third, the acclimatory effect usually alters the ability of an effector cell or organ to respond to environmental change. These acclimatory responses are characteristic of homeorhetic mechanisms as described earlier and the net effect is to coordinate metabolism to achieve a new physiological state. Thus, the seasonally adapted animal is different metabolically in winter than in summer.

Adaptations to heat stress-hormonal control

Acclimation to heat stress is also a homeorhetic process, which involves changes in hormonal signals that affect target tissue responsiveness to environmental stimuli. Understanding of this process will lead to improved genetic selection of heat stress resistant genotypes. Hormones known to be homeorhetic regulators are also implicated in acclimatory responses to thermal stress. These include thyroid hormones, prolactin (PRL), ST, glucocorticoids and mineralcorticoids. The thyroid hormones, thyroxine (T_4) and triiodothyroxine (T_3), provide a major mechanism important for acclimation and have received considerable research attention. As a general observation, it is well known that heat acclimation decreases endogenous levels of thyroid hormones (in an attempt to reduce endogenous heat production) and that mammals adapted to warmer climates follow this pattern (Johnson and Vanjonack, 1976; Collier *et al.*, 1982a; Horowitz, 2001).

Prolactin, which is increased during thermal stress (Mueller, Chen, Dibbet, Chen and Meites, 1974; Wetteman and Tucker, 1979; Collier *et al.*, 1982a), may play an important role in acclimation through improved insensible heat loss and sweat gland function regulation (Beede and Collier, 1986). Bromocryptine, a blocker of PRL secretion, treatment during heat acclimation in humans affects sweat gland function by preventing PRL-induced increases in sweat gland discharge and decreases Na^+ output from the gland (Kaufman, Mills, Hughson and Peake, 1988). There is also a seasonal rhythm in PRL concentration that appears to be involved in animals acclimating to seasonal changes in temperature and day length (Leining, Bourne and Tucker, 1979).

Another important hormone within the galactopoetic complex of signals is ST. In hot environments the most common methods cows use to dissipate additional heat is through increasing respiration rate and/or sweating rate (McDowell, Hooven and Camoens, 1976). Lactating Holstein cows treated with bST during the summer increased heat production by 19 and 25% under thermoneutral and heat stressed conditions, respectively (Table 16.3, Manalu, Johnson, Li, Becker and Collier, 1991). However, heat dissipation in bST treated cows was increased 36 and 24% under thermoneutral and heat stressed conditions (Manalu *et al.*, 1991). In the same experiments replicated in the winter, heat production increased 18 and 10% while heat dissipation increased only 8 and 0.3% compared to controls under thermoneutral and cold stress conditions (Manalu *et al.*, 1991). In summer-acclimated cattle, their ability to carry an increased thermal load was improved by bST while winter-acclimated cows improved their ability to withstand cold stress when treated with bST (Manalu *et al.*, 1991). In addition, the treatment of ST-deficient human patients with hST restored IGF-I concentrations to physiological normality and increased sweating rates (Hasan, Cowen, Barnett, Elliot, Coskeren and Bouloux, 2001). Using immunohistochemistry, sweat gland motor neurons increased expression of acetylcholinesterase and vasoactive intestinal polypeptide (Hasan *et al.*, 2001), which are important for neural induced sweat gland activation. Somatotropin treatment in these two cases increased sweating rates demonstrating the beneficial role of ST in regulating heat dissipation of heat-acclimated animals.

Table 16.3 EFFECT OF HEAT STRESS ON WATER TURNOVER IN COWS TREATED WITH BOVINE SOMATOTROPIN (bST).[1,2]

	Thermoneutral		*Heat*		*Main Effect*[3]		
H_2O Source	*Control*	*bST*	*Control*	*bST*	*SEM*	*B*	*T*
	Kg/day						
Intake	73.5	112.2	94.5	110.2	12.1	0.04	0.4
Food[4]	12.2	14.1	9.6	1.4	0.5	0.003	0.0001
Metabolic[5]	3.8	4.5	3.7	4.7	0.1	0.0001	0.4
Total Intake	89.5	130.8	107.8	126.3	12.2	0.02	0.6
Skin Vaporization	12.9	17.6	19.3	24.5	1.4	0.002	0.001
Resp. Vaporization	3.1	3.7	4.3	5.0	0.2	0.002	0.001

[1]Manalu *et al.*, 1991.
[2]Data represents 6 cows for each treatment in switchback design.
[3]Abbreviations used B= bST effect, T= temperature effect.
[4]Calculated from moisture content of diet.
[5]Estimated using 1.23 kg water produced per 41.84 MJ (10 Mcal) heat production.

The hypothalamic-pituitary-adrenal axis including corticotropin releasing hormone, adrenocorticotropic hormone or corticotropin (ACTH), cortisol, and aldosterone are also altered by thermal stress and are involved in acclimatory responses to thermal stress (Collier *et al.*, 1982a; Beede and Collier, 1986). Corticotropin releasing hormone stimulates somatostatin release from the hypothalamus, which can inhibit ST and TSH secretion from the pituitary (Riedel, Layka and Neeck, 1998), and down regulate the thermogenic effects of both ST and thyroid hormones. In dairy cattle, the glucocorticoids decrease during acclimation at 35°C (Alvarez and Johnson, 1973) and are lower in thermal acclimated animals compared to controls (Collier *et al.*, 1982a; Beede and Collier, 1986).

Whole body integration

There is evidence for a biphasic pattern of heat acclimation divided into periods based on time. Short-term heat acclimation (STHA) is the phase where changes begin to take place within cellular signaling pathways (Horowitz, Kaspler, Marmary and Oron, 1996). These changes create disturbances in cellular homeostasis and begin to reprogram cells to combat the deleterious effects of heat stress (Horowitz, 2001). Full expression of the STHA is obtained when the decline in T_3 and T_4 reaches 30-40% (Horowitz, 2001). When plasma thyroid hormones concentrations decline, the regulation of gene transcription is altered transforming myosin from type V1 to a slower, more efficient form, type V3 (Horowitz, Peiser and Muhlrad, 1986). In lactating dairy cattle, STHA is characterized by an initial overcompensation to heat exposure resulting in a decrease in milk production (Johnson and Vanjonack, 1976). These STHA adaptations allow animals to compensate for increased heat stress before more permanent adaptations can be made.

When all of the changes occurring during STHA are complete and the heat-acclimated phenotype is expressed, long-term heat acclimation (LTHA) has occurred (Horowitz, 2001). In LTHA animals, heat shock protein (HSP) 72 is expressed at 200-fold compared to non-acclimated controls (Maloyan and Horowitz, 2002). During 30 d acclimation (34°C), rats grew at a slower rate than controls while rats treated with thyroxine or ß-adrenergic receptor antagonist did not deviate from controls (Maloyan and Horowitz, 2002). Maximal response to heat stress (41°C) occurred with heat acclimated rats treated with ß-adrenergic receptor antagonist (Maloyan and Horowitz, 2002). From these experiments, a model for HSP response to acclimation has been proposed where T_3 and T_4 exert a negative feedback on HSP production at the cellular level. When ß-adrenergic receptors are activated under conditions of elevated temperature, HSP production increases allowing for HSP synthesis to be controlled by both the decrease in thyroid hormones and ß-adrenergic receptor activation. The positive effects that T_3 and T_4 create on adrenergic receptors has yet to be explained in the model context, but it could allow for increased adrenergic receptors at the initiation of heat acclimation in order to provide more signaling cascades to activate HSP production while T_3 and T_4 are still decreasing (Maloyan and Horowitz, 2002). The muscarinic signal transduction pathway is important for activating the evaporative cooling system in the rat (Horowitz *et al.*, 1996). An expanded list of endocrine changes that occur in response to LTHA are presented in Table 16.4. Most of these examples deal with decreased heat production within the body and the animal's increased ability to dissipate heat obtained from the environment. In lactating cows, when LTHA has occurred, the low-producing cow's milk output will reach a level comparable to what she should produce while not under heat stress, but in the higher producing cow it could still be below the milk production possible in a thermoneutral environment (Johnson and Vanjonack, 1976). Adaptations during this phase of acclimation allow the animal to persist in an environment without experiencing the negative effects of acute heat stress. The bST examples demonstrate that LTHA can be improved by increasing heat dissipation capability of animals.

Heat stress and bioenergetics

Heat stress effects on feed intake, post-absorptive metabolism, nutrient partitioning, energy requirements, nutritional physiology and production variables have been well documented (Collier *et al.*, 1982a,b; Collier and Beede, 1985; West, 2003).

Table 16.4. A PARTIAL LIST OF ENDOCRINE ADAPTATIONS MADE DURING HEAT ACCLIMATION IN CATTLE

Tissue	*Response*	*Reference*
Adrenal	Reduced aldosterone secretion	Collier *et al.*, 1982a
	Reduced glucocorticoid secretion	Collier, Doelger, Head, Thatcher and Wilcox, 1982
	Increased epinephrine secretion	Alvarez and Johnson, 1973
	Increased progesterone secretion	Collier *et al.*, 1982b
Pituitary	Increased prolactin secretion	Wetteman and Tucker, 1979
	Decreased somatotropin secretion	McGuire, Beede, Collier, Buonomo, Delorenzo, Wilcox, Huntington and Reynolds, 1991
Thyroid	Decreased thyroxine secretion	Collier *et al.*, 1982b
Placenta	Decreased estrone sulfate secretion	Collier *et al.*, 1982b

Reductions in energy intake during heat stress result in a majority of lactating cows entering into negative energy balance (NEBAL, ≤10 MJ/d; Moore, Hafliger, Mendivil, Collier and Baumgard, 2004), and this is probably stage of lactation independent. Essentially, because of reduced feed and energy intake, the dairy cow is putting herself in a bioenergetic state, similar to the NEBAL observed in early lactation. The NEBAL associated with the early postpartum period is coupled with increased risk of metabolic disorders and health problems (Goff and Horst, 1997; Drackley, 1999), decreased milk yield and reduced reproductive performance (Lucy, Staples, Thatcher, Erickson, Cleale, Firkins, Clark, Murphy and Brodie, 1992; Beam and Butler, 1999; Baumgard, Moore and Bauman, 2002).

Heat stress-induced NEBAL mediates reductions in milk yield and reproductive performance/efficiency and is thus a costly industry issue. Typical management procedures are to increase dietary energy density via supplementing grains and/or fats. However, decreased milk production and impaired reproductive performance remain costly during heat stress. We have hypothesized that reducing milk fat synthesis would minimize the negative effects of heat stress on NEBAL via providing extra energy that could be utilized for the synthesis of milk and alternative milk components and would supply a signal allowing for improved reproduction (Bauman, Corl, Baumgard and Griinari, 2001; Baumgard *et al.*, 2002). It has recently been demonstrated that heat stress-induced NEBAL can be improved by more than 12.5 MJ/d (Moore *et al.*, 2004a, b) by utilizing dietary conjugated linoleic acid (CLA) to reduce milk fat synthesis (for more information on how CLA regulates milk fat synthesis see reviews by Bauman, Baumgard, Corl and Griinari, 2000; Bauman *et al.*,

2001). However, the impact of this improved energy balance on production and reproduction during heat stress has not yet been examined.

Interestingly, the NEBAL caused by heat stress and early lactation has different effects on the somatotropic axis. Animals in positive energy balance have a coupled ST axis, meaning pituitary-derived ST causes hepatic IGF-I synthesis and secretion (for reviews on the ST axis see: McGuire *et al.*, 1992; McGuire and Bauman, 1997). However, during periods of undernourishment or stages of lactation when animals experience NEBAL (i.e. early lactation) the somatotropic axis uncouples resulting in increased circulating ST levels and decreased IGF-I concentrations (McGuire *et al.*, 1992). As discussed earlier, nutritional modulation of the ST/IGF axis is a homeorhetic mechanism that allows for a shift in nutrient partitioning towards survival/vital processes rather than growth and lactation (McGuire and Bauman, 1997). However, heat stressed animals have or tend to have reduced ST levels (Mitra, Christison and Johnson 1972; McGuire *et al.*, 1991). In agreement, heat-acclimated rodents release less NEFA and glycerol when stimulated with epinephrine (a lipolytic signal; Rabi, Cassuto and Gutman, 1977). However, it is well known that increased ST levels enhance adipocyte sensitivity to lipolytic agents (Bauman and Vernon, 1993). Heat stress-induced changes in the somatotropic axis have not been well characterized and are ill-defined, but it is possible that reduced ST levels during heat stress may reduce NEFA oxidation and thus lower heat production in response to sympathetic stimulation.

Physiological limitations

With increases in milk yields over the last 50 years, two major concerns regarding dairy cow physiology have often been discussed; increased use of body reserves, and decline in fertility. Typically in dairy cows, milk yield peaks at 3-4 wk postpartum whereas energy intake does not peak until several weeks later and thus during this interval cows are in NEBAL. The energy deficit is met by mobilizing body fat reserves and in high producing cows the contribution of reserves will often be energetically equivalent to over one-third of the milk produced (Bauman and Elliot, 1983; Knight *et al.*, 2004). This extensive use of body fat reserves in early lactation is viewed by some as evidence that cows are stressed and may be over taxed (Broom, 1995, 1999; Rauw *et al.*, 1998). This is misleading at best and fails to recognize that selection for high milk performance includes the ability to use body energy reserves as one of the genetic components. Successful lactation initiation is essential for survival and all mammals are programmed to use body fat reserves during this period. The extent to which reserves are utilized varies across species and some animals, such as whales and seals, rely on body fat reserves to such a degree that in early lactation they will lose 40-50% of body weight. Thus, the role and magnitude at which body fat reserves are utilized in high producing dairy cows is only modest among mammals (see review by Bauman, 2000).

The use of body fat reserves in dairy cows has recently been addressed by comparing the extent of energy mobilization in cows that were selected for low or high genetic potential. Annual milk yields differ between the two selection lines by more than 4500 kg, with estimated ME intake and milk energy output greater for the high genetic line. Despite a large difference in milk output, tissue energy balance between the two genetic lines reached the same nadir of -64 MJ/d at 7 d postpartum, with both remaining in NEBAL until 60 d in milk (Crooker, Weber, Ma and Lucy, 2001). Reviewing these data, it is apparent that cows optimally managed for adequate energy intakes are not necessarily required to mobilize more body tissue reserves than lower-yielding animals. Indeed, for the early lactating animal the use of body reserves as a source of energy is logical. Energetically, mobilization confers approximately 80% efficiency, whilst digestion is only about 60%, so at the time of greatest demand there is an energetic advantage to utilize adipose derived energy (although the total energetic cost of depositing fat and then mobilizing it is obviously much greater). There is also less need for physical activity (foraging), and gut proliferation does not need to occur so rapidly, all of which confers more energetic efficiency (Knight, 2001).

The most documented change associated with increased milk yield potential has been a decline in fertility (Butler, 2000; Lucy, 2001). Milk yield increases achieved in the last 50 years have provided evidence for an antagonistic relationship between high milk production and reproductive success (Taylor, Beever and Wathes, 2004). This is not a new phenomenon since in 1929 there was a report of an association between ‘high’ milk yield (relative to breed averages) and decreased fertility (Hansen, 2000). As milk yield has increased, there has been a corresponding decrease in the fertility of dairy cows in both UK and US dairy herds (Butler and Smith, 1989; Royal, Darwash, Flint, Webb, Woolliams and Lamming, 2000). This decline has been associated with increased genetic capability for milk production, changes in nutritional management and larger herd sizes (Butler, 1998). Although there is ample evidence that increasing genetic merit for milk yield reduces fertility (Pryce and Veerkamp, 1999; Veerkamp, Beerda and van der Lende, 2003), heritabilities for measures of cow fertility have been shown to be very low, 3% or less; the majority of variation in cow fertility (>97%) is due to factors other than genetics (Hansen, 2000). However, establishing a direct cause and effect between high milk yields and reduced fertility has not been universally demonstrated. Some argue that reduced reproduction efficiency in high producing cows is evidence of metabolic stress or “pushing” cows too far. Again, this concept is misleading. The reduction in fertility is based on a producer imposed time constraint (for example a goal of calving every 365 days) and not an actual reduction related to an aberration in the cow’s biological control.

The decline in reproductive performance is especially evident when animals are experiencing NEBAL, with greater severity and longer NEBAL duration prolonging return to estrus. NEBAL has been shown to effect ovarian activity, reproductive efficiency and oocyte quality, with nutrition, especially undernutrition, arguably having the greatest impact on a whole range of fertility parameters (see review by Taylor *et al.*, 2004). In this regard, dairy cows are not unique. One of the first investigators to observe and comment on the link between food supply and fertility was Charles Darwin in '*The Origin of Species*' (Darwin, 1859). In an extensive review Wade, Schneider and Li, (1996) pointed out that either limited food availability or extraordinary energy demands unaccompanied by compensatory increases in energy intake diminishes fertility in a vast array of mammalian species, ranging from shrews to whales. Consequently, reproductive function need not be compromised by very high milk output, as long as the nutritional inputs are adequate to maintain a balance between input and output (Webb, Garnsworthy, Gong, Robinson and Wathes, 1999). During the first 3-4 weeks postpartum NEBAL has been shown to be highly correlated with both milk yield and the interval to first ovulation (Butler, Everett and Coppock, 1981); with the number of ovulatory cycles prior to breeding influencing conception rates (see review by Butler, 1999). The effects of NEBAL on reproduction are clearly demonstrated by the fact that when cows are dried at d 1 of lactation (thus experiencing a less severe NEBAL), days to first ovulation are markedly reduced (Canfield and Butler, 1991). It is important to realize that NEBAL is not indicative of high-producing dairy cows per se; rather it is a result of the difference between energy intake and output, and can therefore occur in cows of any yield potential (as shown in the preceding section). The failure of nutrient intake to increase in tandem with milk yields has resulted in some animals having longer and more severe periods of NEBAL. Thus, the fact that some high producing cows have a delay in the time to first ovulation is not evidence of metabolic stress, but rather indicates that the biological controls are coordinating physiological processes similar to what occurs in all mammals and what is homeorhetically predictable. Therefore in many species, including the dairy cow, it is normal for reproduction to be compromised during early lactation to a degree that is related to milk output, so that the energetic burdens of lactation and gestation are separated in time (Webb *et al.*, 1999). One could therefore argue that infertility induced by high metabolic turnover is a predictable physiological response rather than a pathophysiological failure (Wade *et al.*, 1996; Knight, Beever and Sorensen, 1999).

As discussed above, energetic status is a major determinant of delays in cyclicity resumption and normal reproductive function. However, on a within-herd basis a wide variation in reproductive function has been observed (Butler and Smith, 1989). The question as to why some animals within a herd are unable to conceive in early lactation whilst others can has perplexed reproductive biologists for decades. It is therefore important to determine how some cows manage to produce high milk yields and simultaneously conceive and maintain their pregnancy whilst others cannot (Taylor *et al.*, 2004) and this is an active research area.

A delay in return to normal reproductive function may be inevitable in many dairy cows. In the future, the greatest opportunity for addressing infertility during early lactation appears to be either to delay the initiation of a re-breeding programme and/or reduce the extent and duration of NEBAL. The concept of a 365 d calving cycle will likely become less common in the future, apart from herds where utilizing feed resources at certain times of the growing season is essential (e.g. pasture based system). The increased calving interval in high producing dairy cows is already evident and is expected due to the prioritization of nutrient partitioning to support high yields rather than reproductive processes. However, increasing calving intervals should be a conscious management decision not a result of poor conception rates. For extended calving intervals to be successful it is crucial for lactation persistency to be increased so that cows can maintain high yields for longer periods of time. Lactation persistency is dependent on maintaining the number and activity of milk secreting cells with increasing days in milk (see reviews: McFadden, 1997; Capuco, Ellis, Hale, Long, Erdman, Zhao and Paape, 2003). Certainly in the US, extended calving intervals are a possibility with the use of frequent milking and bST, which results in an overall improvement in herd life, animal well-being and dairy farm profitability (Van Amburgh, Galton, Bauman and Everett, 1997). However, since bST use is not a management option in some countries, maintaining lactation persistency will be more difficult.

As stated earlier, recovery of energy balance from its nadir to a more positive balance provides an important signal for resuming ovarian activity (Webb *et al.*, 1999). As a consequence reducing the extent of NEBAL in early lactation offers potential as a means to improve fertility in early lactation. Traditionally, this has been attempted by enhancing energy intake through increasing dry matter intake and energy concentration of the diet (Grummer, 1995; Drackley, 1999). An alternative approach is to reduce energy output. Since milk fat production is the major energetic cost of milk synthesis, methods specifically to reduce milk fat content in early lactation may improve energy balance and have a positive effect on reproductive performance. It has recently been demonstrated that energy balance parameters can be improved by supplementing CLA (Moore, Hafliger, Mendivil, Sanders, Bauman and Baumgard, 2004; Kay, Roche and Baumgard, 2004) in early lactation (in both TMR and pasture-based systems) and there is preliminary data suggesting that dietary CLA may improve reproductive indices during this stage of lactation as well (Bernal-Santos, Perfield, Barbano, Bauman and Overton, 2003).

Future challenges and opportunities

The regulation of biological processes is exquisitely coordinated and it is this coordination that allows for high performance. As discussed in previous sections, mechanisms involve homeostatic and homeorhetic regulation and when animals are placed in less advantageous environments and/or management situations, they shift priorities to maintain animal well-being by reducing energy and nutrient use for milk production. It is important to remember that there is no better indicator of overall animal well-being than performance (e.g. lactation, growth, etc.). Top dairy cows do not become stressed because they achieve a high yield; rather they achieve a high yield because they are not stressed. Consequently, there is little or no support for the notion that high lactational performance results in 'stressed' dairy cows with compromised health and welfare as suggested by some (e.g. see Broom, 1995, 1999; FAWC, 1997; Rauw *et al.*, 1998). Rather, stress occurs when regulation is out of balance, not because the genetic merit of the cow is too high. Problems allied to metabolic load are not restricted to high genetic merit cows, nor are they a consequence of absolute input or output within the range of commercial practice (Knight, 1999); rather, it is an imbalance between input and output and the inability to adjust metabolism quickly enough to meet these needs that frequently results in subclinical or acute metabolic disorders in farm animals (Bauman and Currie, 1980; Bauman and Elliot, 1983). For example, epidemiological studies have shown no relationship between milk yield and risk of disease in US commercial herds (Grohn, Eicker and Hertl, 1995) and no evidence of increased metabolic abnormalities in high-yielding cows in commercial UK herds (Ward and Parker, 1999). Similarly, Logue, Berry, Offer, Chaplin, Crawshaw, Leach, Ball and Bax (1999) found that high yielding cows managed by high or low input systems had remarkably few health disorders.

What are the limits to an individual animal's production in the future? For the last half-century the record-setting performance of top-producing cows from the recent past has become the production of today's average cows. It is expected that this trend will continue and certainly there is no evidence for a slow-down in the increase in average milk yields (Figure 16.1). There must be an upper limit where the biological controls in the regulation of milk synthesis are optimized, but again there is no evidence that this plateau is approaching (Figure 16.1). As cited earlier, there are cows capable of producing in excess of 30,000 kg per lactation so it is known that the biological control systems will allow for increases in milk yields to at least the level of current record holders. Will such averages ever be reached? This will be critically dependent on being able to select and breed suitable animals and our ability to optimally manage and feed such animals. Genetic selection and proper management are effective because we are affecting the 'natural' biological control systems.

Techniques to increase milk yield and productive efficiency involve altering many biological components involved in coordinating energy and nutrient utilization by the animal's tissues during its productive lifespan (Bauman *et al.*, 1985). New genomic technologies offer the opportunity to further delineate these controls and this includes identifying specific genes which are associated with sensitivity or resistance to stress, or which confer improved adaptation to physiological and environmental challenges. Thus, as pointed out by Sir John

Hammond more than 50 years ago, progress in dairy cattle as in other farm species will be critically dependent on an ability to understand more completely the control of energy and nutrient partitioning (Hammond, 1952).

References

Alvarez, M.B. and Johnson, J.D. (1973) Environmental heat exposure on cattle plasma catecholamine and glucocorticoids. *Journal of Dairy Science*, **56**, 189-194.

Arriaga-Jordan, C.M. and Holmes, W. (1986) The effect of concentrate supplementation on high-yielding dairy cows under two systems of grazing. *Journal of Agricultural Science*, **107**, 453-461.

Bauman, D.E. and Currie, W.B. (1980) Partitioning of nutrients during pregnancy and lactation: a review of mechanisms involving homeostasis and homeorhesis. *Journal of Dairy Science*, **63**, 1514-1529.

Bauman, D.E., Eisemann, J.H. and Currie, W.B. (1982) Hormonal effects on nutrients for tissue growth: role of growth hormone and prolactin. *Federation Proceedings*, **41**, 2538-2544.

Bauman, D.E. and Elliot, J.M. (1983) Control of nutrient partitioning in lactating ruminants. In *Biochemistry of Lactation,* pp 437-468. Edited by T.B. Mepham. Elsevier Science Publishers B.V., Amsterdam, The Netherlands.

Bauman, D.E., McCutcheon, S.N., Steinhour, W.D., Eppard, P.J. and Sechen, S.J. (1985) Sources of variation and prospects for improvement of productive efficiency in the dairy cow: a review. *Journal of Animal Science*, **60**, 538-592.

Bauman, D.E. (1992) Bovine somatotropin: review of an emerging animal technology. *Journal of Dairy Science*, **75**, 3432-3451.

Bauman, D.E. and Vernon, R.G. (1993) Effects of exogenous bovine somatotropin on lactation. *Annual Review of Nutrition*, **13**, 437-461.

Bauman, D.E. (2000) Regulation of nutrient partitioning during lactation: homeostasis and homeorhesis revisited. In *Ruminant Physiology: Digestion, Metabolism, Growth, and Reproduction*, pp 311-327. Edited by P.B. Cronje. CAB Publishing, New York, NY.

Bauman, D.E., Baumgard, L.H., Corl, B.A. and Griinari, J.M. (2000) Biosynthesis of conjugated linoleic acid in ruminants. *Proceedings of the American Society of Animal Science*, 1999. Available at: http://www.asas.org/jas/symposia/proceedings/0937.pdf.

Bauman, D.E., Corl, B.A., Baumgard L.H. and Griinari, J.M. (2001) Conjugated linoleic acid (CLA) and the dairy cow. In: *Recent Advances in Animal Nutrition*, pp 221-250. Edited by P. C. Garnsworthy and J. Wiseman. Nottingham University Press, Nottingham, UK.

Baumgard, L.H., Moore, C.E. and Bauman, D.E. (2002) Potential application of conjugated linoleic acids in nutrient partitioning. Proceedings of the Southwest Nutrition Conference, pp 127-141. http://animal.cals.arizona.edu/SWNMC/2002/Baumgard_Moore_Bauman2002.pdf

Beede, D.K. and Collier, R.J (1986) Potential nutritional strategies for intensively managed cattle during thermal stress. *Journal of Animal Science*, **62**, 543-554.

Beam, S.W. and Butler, W.R. (1999) Effects of energy balance on follicular development and first ovulation in postpartum dairy cows. *Journal of Reproduction and Fertility*, **54**, 411-424.

Beever, D.E., Hattan, A., Reynolds, C.K. and Cammell, S.B. (1999) Nutrient supply to high-yielding dairy cows. In *Fertility in the High-Producing Dairy Cow*, pp 119-131. Edited by M. G. Diskin. Occasional Publication No 26 - British Society of Animal Science.

Bell, A.W. and Bauman, D.E. (1997) Adaptations of glucose metabolism during pregnancy and lactation. *Journal of Mammary Gland Biology and Neoplasia,* **2**, 265-278.

Bernal-Santos, G., Perfield II, J.W, Barbano, D.M., Bauman, D.E. and Overton, T.O. (2003) Production responses of dairy cows to dietary supplementation with conjugated linoleic acid (CLA) during the transition period and early lactation. *Journal of Dairy Science*, **86**, 3218-3228.

Bligh. J. (1976) Introduction to acclamatory adaptation-including notes on terminology. In *Environmental Physiology of Animals*, pp 219-229. Edited by J. Bligh, J.L. Cloudsley-Thompson and A.G. Macdonald. John Wiley and Sons, New York.

Broom, D.M. (1995) Measuring the effects of management methods, systems, high production efficiency and biotechnology on farm animal welfare. In *Issues in Agricultural Bioethics*, pp 319-334. Edited by T.B. Mepham, G.A. Tucker and J. Wiseman. Nottingham University Press, Nottingham, UK.

Broom, D.M. (1999) The welfare of dairy cattle. In *Future Milk Farming. FIL-IDF 25th International Dairy Congress*, pp 32-39. Edited by K. Aagaard. Danish National Committee. IDF. Aarhus, Denmark.

Buckley, F., Dillon, P., Rath, M. and Veerkamp, R.F. (2000) The relationship between genetic merit for yield and live weight, condition score, and energy balance of spring calving Holstein Friesian dairy cows on grass based systems on milk production. *Journal of Dairy Science*, **83**, 1878-1886.

Butler, W.R., Everett, R.W. and Coppock, C.E. (1981) The relationships between energy balance, milk production and ovulation in postpartum Holstein cows. *Journal of Animal Science*, **53**, 742-748.

Butler, W.R. and Smith, R.D. (1989) Interrelationships between energy balance and postpartum reproductive function in dairy cattle. *Journal of Dairy Science*, **72**, 767-783.

Butler, W.R. (1998) Review: Effect of protein nutrition on ovarian and uterine

physiology in dairy cattle. *Journal of Dairy Science*, **81**, 2533-2539.
Butler, W.R. (1999) Nutritional effects on resumption of ovarian cyclicity and conception rate in postpartum dairy cows. In *Fertility in the High-Producing Dairy Cow-BSAS Occasional Publication No. 26,* pp 133-145. Edited by M. G. Diskin. BSAS, Edinburgh, UK.
Butler, W.R. (2000) Nutritional interactions with reproductive performance in dairy cattle. *Animal Reproduction Science*, **60**, 449-457.
Canfield, R.W. and Butler, W. R. (1991) Energy balance, first ovulation and the effects of naloxone on LH secretion in early postpartum dairy cows. *Journal of Animal Science*, **69**, 740-746.
Cannon, W.B. (1929) Organization for physiological homeostasis. *Physiological Reviews* IX, 399-431.
Capuco, A.V., Ellis, S.E., Hale, S.A., Long, E., Erdman, R.A., Zhao, X. and Paape, M.J. (2003) Lactation persistency: insights from mammary cell proliferation studies. *Journal of Animal Science*, **81 (Suppl. 3)**, 18-31.
Chilliard, Y. (1999) Metabolic adaptations and nutrient partitioning in the lactating animal. In *Biology of Lactation,* pp 503-552. Edited by J. Martinet, L.M. Houdebine, and H.H. Head. INRA Editions, Paris, France.
Collier, R.J., Beede, D.K., Thatcher, W.W., Israel, L.A. and Wilcox, C.J. (1982a) Influences of environment and its modification on dairy animal health and production. *Journal of Dairy Science*, **65**, 2213-2227.
Collier, R.J., Doelger, S.G., Head, H.H., Thatcher, W.W. and Wilcox, C.J. (1982b) Effects of heat stress on maternal hormone concentrations, calf birth weight, and postpartum milk yield of Holstein cows. *Journal of Animal Science*, **54**, 309-319.
Collier, R.J. and Beede, D.K. (1985) Thermal stress as a factor associated with nutrient requirements and interrelationships. In *Nutrition of Grazing Ruminants*, pp 59-71. Edited by L. McDowell. Academic Press, New York, NY.
Crooker, B.A., Weber, W.J., Ma, L.S. and Lucy, M.C. (2001) Effect of energy balance and selection for milk yield on the somatotropic axis of the lactating Holstein cow: endocrine profiles and hepatic gene expression. In *Proceedings of the 15th Symposium on Energy Metabolism in Animals*, pp 345-348. Edited by A. Chwalibog and K. Jakobsen. Wageningen Pers, Wageningen, The Netherlands.
Darwin, C.R. (1859) The Origin of Species. Penguin Books Ltd, New York, N.Y.
Dilman, V.M. (1982) Endocrinology and diseases of aging. In *Regulatory Processes in Clinical Endocrinology*, pp 123-161. Edited by W.B. Essman. Spectrum Publications, Inc., New York, N.Y.
Drackley, J.K. (1999) Biology of dairy cows during the transition period: the final frontier? *Journal of Dairy Science*, **82**, 2259-2273.
FAWC (1997) Report on the Welfare of Dairy Cattle. FAWC, Surbiton KT6 7NF, UK. MAFF Crown Publication. Available at: www.fawc.org.uk

Fregley, M.J. (1996) Adaptations: some general characteristics. In: *Handbook of Physiology*, Section 4: Environmental Physiology. pp 3-15 Vol I. Edited by M.J Fregley and C. M. Blatteis. Oxford University Press.

Goff, J.P. and Horst, R.L. (1997) Physiological changes at parturition and their relationship to metabolic disorders. *Journal of Dairy Science*, **80**, 1260-1268.

Gordon, F.J., Patterson, D.C., Yan, T., Porter, M.G., Mayne, C.S. and Unsworth, E.F. (1995) The influence of genetic index for milk production on the response to complete diet feeding and the utilization of energy and nitrogen. *Animal Science*, **61**, 199-210.

Grohn, Y.T., Eicker, S.W. and Hertl, J.A. (1995) The association between previous 305-day milk yield and disease in New York State dairy cows. *Journal of Dairy Science,* **78,** 1693-1702.

Grummer, R.R. (1995) Impact of changes in organic nutrient metabolism on feeding the transition dairy cow. *Journal of Animal Science*, **73**, 2820-2833.

Hammond, J. (1952) Physiological limits to intensive production in animals. *British Agricultural Bulletin,* **4,** 222-224.

Hansen, L.B. (2000) Consequences of selection for milk yield from a geneticist's viewpoint. *Journal of Dairy Science*, **83**, 1145-1150.

Hasan, W., Cowen, T., Barnett, P.S., Elliot, E., Coskeren, P. and Bouloux, P.M. (2001) The sweating apparatus in growth hormone deficiency, following treatment with r-hGH and in acromegaly. *Autonomic Neuroscience*, **89**, 100-109.

Hoogendoorn, C.J., McCutcheon, S.N., Lynch, G.A., Wickham, B.W. and MacGibbon, A.K.H. (1990) Production responses of New Zealand Friesian cows at pasture to exogenous recombinantly derived bovine somatrotropin. *Animal Proceedings*, **51**, 431-439.

Horowitz, M., Peiser, M.J. and Muhlrad, A. (1986) Alterations in cardiac myosin distribution as an adaptation to chronic heat stress. *Journal Molecular and Cellular Cardiology*, **18**, 511-515.

Horowitz, M., Kaspler, P., Marmary, Y. and Oron, Y. (1996) Evidence for contribution of effector organ cellular responses to biphasic dynamics of heat acclimation. *Journal of Applied Physiology*, **80**, 77-85.

Horowitz, M. (2001) Heat acclimation: phenotypic plasticity and cues underlying the molecular mechanims. *Journal of Thermal Biology*, **26**, 357-363.

Johnson, H.D. and Vanjonack, W.J. (1976) Effects of environmental and other stressors on blood hormone patterns in lactating animals. *Journal of Dairy Science*, **59**, 1603-1617.

Kaufman, F.L., Mills, D.E., Hughson, R.L. and Peake, G.T. (1988) Effects of bromocriptine on sweat gland function during heat acclimatization. *Hormone Research*, **29**, 31-38.

Kay, J.K., Roche, J.R and Baumgard, L.H. (2004) Effects of dietary CLA on production parameters in pasture-fed transition dairy cows. *Journal of Dairy Science*, **87 (supplement 1)**, 95.

Knight, C. (1999) Metabolic stress unravelled. In: *Research Reviews, Hannah Yearbook,* pp 32-37. Edited by E. Taylor. Hannah Research Institute, Ayr.

Knight, C.H., Beever, D.E. and Sorensen, A. (1999) Metabolic loads to be expected from different genotypes under different systems. In: *Fertility in the High-Producing Dairy Cow-BSAS Occasional Publication No. 24*, pp 27-35. Edited by M. G. Diskin. BSAS, Edinburgh, UK.

Knight, C. (2001) Lactation and gestation in dairy cows: flexibility avoids nutritional extremes. *Proceedings of the Nutrition Society*, **60**, 527-537.

Knight, C.H., Alamer, M.A., Sorensen, A., Nevison, I.M., Flint, D.J. and Vernon, R.G. (2004) Metabolic safety-margins do not differ between cows of high and low genetic merit for milk production. *Journal of Dairy Research*, **71**, 1-13.

Kolver, E.S. and Muller, L.D. (1998) Performance and nutrient intake of high producing Holstein cows consuming pasture or a total mixed ration. *Journal of Dairy Science*, **81**, 1403-1411.

Kuenzel, W.J., Beck, M.M. and Teruyama, R. (1999) Neural sites and pathways regulating food intake in birds: A comparative analysis to mammalian systems. *Journal of Experimental Zoology,* **283**, 348-364.

Leining, K.B., Bourne, R.A. and Tucker, H.A. (1979) Prolactin response to duration and wavelength of light in prepubertal bulls. *Endocrinology*, **104**, 289-294.

Logue, D.N., Berry, R.J., Offer, J.E., Chaplin, S.J., Crawshaw, W.M., Leach, K.A., Ball, P.J.H. and Bax, J. (1999) Consequences of metabolic load for lameness and disease. In: *Metabolic Stress in Dairy Cows*, pp 83-98. Occasional Publication No. 24 British Society of Animal Science.

Lucy, M.C., Staples, C.R., Thatcher, W.W., Erickson, P.S., Cleale, R.M., Firkins, J.L., Clark, J.H., Murphy M.R. and Brodie, B.O. (1992). Influence of diet composition, dry matter intake, milk production and energy balance on time of postpartum ovulation and fertility in dairy cows. *Animal Production*, **54**, 323-331.

Lucy, M.C. (2001) Reproductive loss in high- producing dairy cattle: Where will it end? *Journal of Dairy Science*, **84**, 1277-1293.

Maloyan, A. and Horowitz, M. (2002) ß-Adrenergic signaling and thyroid hormones affect HSP72 expression during heat acclimation. *Journal of Applied Physiology*, **93**, 107-115.

Manalu, W., Johnson, H.D., Li, R.Z., Becker, B.A. and Collier, R.J. (1991) Assessment of thermal status of somatotropin-injected Holstein cows maintained under controlled-laboratory thermoneutral, hot, and cold environments. *Journal of Nutrition*, **121**, 2006-2019.

McDowell, R.E. (1972). Improvement of Livestock Production. In: *Warm Climates*. Edited by W.H. Freeman and Co., San Francisco.

McDowell, R.E., Hooven, N.W. and Camoens. J.K. (1976). Effects of climate on performance of Holsteins in first lactation. *Journal of Dairy Science*, **59**, 965-973.

McFadden, T.B. (1997) Prospects for improving lactational persistency. In: *Milk Composition, Production and Biotechnology*. pp 319-340. Edited by R.A.S. Welch, D.J. W. Burns, S.R. Davis, A.I. Popay and C.G. Prosser. CAB International, Wallinford, Oxfordshire, UK.

McGuire, M.A., Beede, D.K., Collier, R.J., Buonomo, F.C., Delorenzo, M.A., Wilcox, C.J., Huntington, G.B. and Reynolds, C.K. (1991) Effect of acute thermal stress and amount of feed intake on concentrations of somatotropin, insulin-like growth factor I (IGF-I) and IGF-II and thyroid hormones in plasma of lactating dairy cows. *Journal of Animal Science*, **69**, 2050-2056.

McGuire, M.A., Vicini, J.L., Bauman, D.E. and Veenhuizen, J.J. (1992) Insulin-like growth factors and binding proteins in ruminants and their nutritional regulation. *Journal of Animal Science*, **70**, 2901-2910.

McGuire, M.A. and Bauman, D.E. (1997) Regulation of nutrient use by bovine somatotropin: the key to animal performance and well-being. In: IX *International Conference on Production Diseases in Farm Animals,* pp 125-137. Edited by M. Martens. Free University of Berlin.

Mitra, R., Christison, G.I. and Johnson, H.D. (1972) Effect of prolonged thermal exposure on growth hormone (GH) secretion in cattle. Journal of Animal Science, **34**, 776-786.

Moore, C.E., Hafliger III, H.C., Mendivil, O.B., Collier, R.J. and Baumgard, L.H. (2004a) Effects of dietary CLA on production parameters and milk fatty acid variables in Holstein and Brown Swiss cows during heat stress. *Journal of Dairy Science*, **87 (supplement 1)**, 307.

Moore, C.E., Hafliger III, H.C., Mendivil, O.B, Sanders, S.R., Bauman, D.E. and Baumgard, L.H. (2004b) Increasing amounts of conjugated linoleic acid (CLA) progressively reduce milk fat synthesis immediately postpartum. *Journal of Dairy Science*, **87**, 1886-1895.

Mrosovsky, N. (1990) *Rheostasis: The Physiology of Change.* Oxford University Press, Inc., New York.

Mueller, G.P., Chen, H.T., Dibbet, J.A., Chen, H.J. and Meites, J. (1974) Effects of warm and cold temperatures on release of TSH, GH, and prolactin in rats. *Proceedings of the Society for Experimental Biology and Medicine*, **147**, 698-700.

Neinaber, J.A., Hahn G.L. and Eigenberg, R.A. (1999) Quantifying livestock responses for heat stress management: a review. *International Journal of Biometeorology*, **42**, 183-188.

Nicolaïdis, S. (1983) Physiologie du comportement alimentaire. In: *Physiologie Humaine,* pp 1059-1076. Edited by P. Meyer. Flammarion Médecine-Sciences, Paris.

Pryce, J.E. and Veerkamp, R.F. (1999) The incorporation of fertility indices in genetic improvement programmes. In: *Fertility in the High-Producing Dairy Cow-BSAS Occasional Publication No 26*, pp 237-249. Edited by M. G. Diskin. BSAS, Edinburgh, UK.

Rabi, T., Cassuto, Y. and Gutman, A. (1977) Lipolysis in brown adipose tissue of cold- and heat-acclimated hamsters. *Journal of Applied Physiology*, **43**, 1007-1011.

Rauw, W.M., Kanis, E., Noordhuizen-Stassen, E.N. and Grommers, F.J. (1998) Undesirable side effects of selection for high production efficiency in farm animals: a review. *Livestock Production Science*, **56**, 15-33.

Reynolds, C.K. (2004) Metabolic consequences of increasing milk yield-revisiting Lorna. In: *UK Dairying: Using Science to Meet Consumers' Needs*, pp 73-83. Edited by E. Kebreab, J. Mills and D. Beever. Nottingham University Press, Nottingham.

Riedel, W., Layka, H. and Neeck, G. (1998) Secretory pattern of GH, TSH, thyroid hormones, ACTH, cortisol, FSH, and LH in patients with fibromyalgia syndrome following systemic injection of the relevant hypothalamic-releasing hormones. *Zeitschrift für Rheumatologie*, **57 (supplement 2)**, 81-87.

Roche, J.R. and Edmeades, D.C. (2004) The paradigm of efficiency and sustainability - a dairying perspective. *South African Journal of Animal Science*, **34 (supplement 2)**, 8-16

Royal, M.D., Darwash, A.O., Flint, A.P.E., Webb, R., Woolliams, J.A. and Lamming, G.E. (2000) Declining fertility in dairy cattle: changes in traditional and endocrine parameters of fertility. *Animal Science*, **70**, 487-501.

Taylor, V.J., Beever, D.E. and Wathes, D.C. (2004) Physiological adaptations to milk production that affect the fertility of high yielding dairy cows. In: *UK Dairying:Using Science to Meet Consumers' Needs,* pp 37-71. Edited by E. Kebreab, J. Mills and D.E. Beever. Nottingham University Press, Nottingham, UK.

Van Amburgh, M.E., Galton, D.M., Bauman, D.E. and Everett, R.W. (1997) Management and economics of extended calving intervals with use of bovine somatotropin. *Livestock Production Science*, **50**, 15-28.

Veerkamp, R.F., Beerda, B. and van der Lende, T. (2003) Effects of genetic selection for milk yield on energy balance, levels of hormones, and metabolites in lactating cattle, and possible links to reduced fertility's. *Livestock Production Science*, **83**, 257-275.

Vernon, R.G. (1989) Endocrine control of metabolic adaptation during lactation. *Proceedings of the Nutrition Society,* **48**, 23-32.

Vernon, R.G. and Sasaki, S. (1991) Control of responsiveness of tissue to hormones. In: *Physiological Aspects of Digestion and Metabolism in Ruminants*. pp 155-182. Edited by T. Tsuda, Y. Sasaki and R. Kawashima. Academic Press, Inc., San Diego, CA.

Vernon, R.G. (1998) Homeorhesis. In: *Research Reviews, Hannah Yearbook,* pp 64-73. Hannah Research Institute, Ayr.

Wade, G.N. and Schneider, J.E. (1992) Metabolic fuels and reproduction in female mammals. *Neuroscience and Biobehavioral Reviews,* **16**, 235-272.

Wade, G.N., Schneider, J.E. and Li, H. Y. (1996) Control of fertility by metabolic cues. *American Journal of Physiology-Endocrinology and Metabolism,* **33**, E1-E19.

Ward, W.R. and Parker, C.S. (1999) Field evidence of metabolic stress in dairy cows. In: *Metabolic Stress in Dairy Cows*, pp 21-26. Occasional Publication No 24-British Society of Animal Science.

Webb, R., Garnsworthy, P.C., Gong, J.G., Robinson, R.S. and Wathes, D.C. (1999) Consequences for reproductive function of metabolic adaptation to load. In *Metabolic Stress in Dairy Cows,* pp 99-112. BSAS Occasional Publication 24.

West, J.W. (2003) Effects of heat-stress on production in dairy cattle. *Journal of Dairy Science*, **86**, 2131-2144.

Wetteman, R.P. and Tucker, H.A. (1979) Relationship of ambient temperature to serum prolactin in heifers. *Proceedings of the Society for Experimental Biology and Medicine*, **146**, 909-911.

17

LIVESTOCK YIELD TRENDS: IMPLICATIONS FOR ANIMAL WELFARE AND ENVIRONMENTAL IMPACT

P.C GARNSWORTHY
Division of Agricultural and Environmental Sciences, School of Biosciences, University of Nottingham, Sutton Bonington Campus, Loughborough, Leics., LE12 5RD, UK

Introduction

The objective of this chapter is to consider whether the yield trends over the next 10 to 50 years, which are predicted for individual livestock species in other chapters, will have implications for animal welfare or for the environmental impact of animal production systems. These issues, driven by public opinion, are becoming of increasing concern to policy makers, legislators and producers. It is likely, therefore, that development of future animal production systems could be constrained more by the need to comply with legislation on welfare and environmental impact than by physiological or technological limitations.

Agricultural policy, from the 1940s until the 1960s, was to produce food as cheaply and efficiently as possible, using home-produced resources and reliance on Government support. This encouraged the selection of stock and production systems that favoured higher yields per unit of input. Units of input include feed, land, labour, machinery and capital, so efficiency of a production system can be measured by comparing yield with the most limiting input. In virtually every case, this comparison shows that intensive systems are more efficient.

From the 1960s onwards, the UK general public gradually forgot about food rationing, which ended in the 1950s, and increasing affluence allowed them to consider other parameters of food, such as taste, quality, additives and safety, in addition to price. Over the same period, there was increasing awareness of animal welfare and the impact of agricultural practices on the environment. However, numerous studies have shown that price is still the main driver of food purchasing decisions for all but a minority of the public. This means that producers and food processors have to take all these aspects on board, and still produce food that can compete with imports and leave a profit to make their businesses sustainable. Apart from niche markets, where specifically-defined products command a premium price or subsidy, economic sustainability usually still involves intensification and higher yields. The alternative extensive, low-input systems can only provide sufficient income without subsidies if they are performed on a large scale.

As will be seen in subsequent sections, intensification and high yields *per se* are not necessarily bad for animal welfare or environmental impact. The important thing is that more intensive systems require a greater level of management, planning and attention to detail. A holistic approach is needed that encompasses the total animal production system; one that balances yields, inputs, animal welfare and waste disposal.

Animal welfare

Animal welfare can be defined in several ways and the implications of yield trends for animal welfare vary with each definition. One definition is that an animal's welfare is a measure of its ability to cope with its environment (Broom, 1993). In this context both yield and environment are important, and the two need to change in harmony. Environment covers all non-genetic aspects of the phenotype, including housing or climatic environment, nutrition, and management. If yields increase without appropriate changes in environment, animal welfare will be compromised. For example, animals selected for larger size need more space, they need to be supplied with more energy and nutrients, and they will produce more excreta that must be removed efficiently. This definition can include mental well-being because if an animal cannot cope with its environment it will suffer stress and may exhibit inappropriate behaviour patterns.

Webster *et al.* (2004) use an operational definition of animal welfare as "Fit and Feeling Good". They explained that, for production animals, "Fitness" implies the capacity to sustain health and vigour throughout an effective working life, and "Feeling good" implies a lack of suffering plus provision of comfort, companionship and security. They expanded the scope of fitness to include the Darwinian concept of transmitting genes to the next generation. It is obvious that under this definition animals suffering from disease or lameness will not be fit and, therefore, will have bad welfare. However, the Darwinian definition of fitness rules out husbandry practices such as castration not on the grounds of pain or suffering, but because they deprive animals of the ability to transmit their genes. On the other hand, the Darwinian definition firmly encompasses fertility as a welfare issue, even though the inability of a female to conceive does not, in itself, cause pain or suffering.

Fraser (2003) identified three different views about how animal welfare should be judged. The first view is that animals should be raised under conditions that promote good biological functioning in the sense of health, growth and reproduction; this view is often held by those involved in animal production. The second view is that animals should be raised in ways that minimise suffering and promote contentment; this view is often held by humanitarians concerned about animal welfare. The third view is that animals should be allowed to lead relatively natural lives; this view is often held by radical consumers who reject modern society (Fraser, 2003). It is not surprising that most animal scientists tend to favour the first view for judging welfare because biological function can be measured scientifically, hypotheses can be formed, and systems can be compared objectively. Most scientists probably accept the second view as well, but find it more difficult to measure objectively either suffering or contentment, except in extreme cases. Scientifically, the third view has to be rejected as a subjective judgement that is impossible to test because nobody can quantify a 'natural life' for a domesticated animal. Animals have been domesticated for over ten thousand years and modern animals bear little resemblance to either their wild ancestors or to non-domesticated species. It is, however, a matter of concern that the public tends towards the third view even if they don't know what is natural for an animal.

There is a widespread belief amongst the general public, encouraged by the popular press, that intensive animal production is bad for animal welfare and that extensive systems are good. This simplistic view appears to be driven mainly by the perception that natural is best, and results from urbanisation of the population and their dissociation from primary food production. However, there are many examples where extensive systems have elements that are worse for animal welfare than their intensive counterparts.

Hill sheep production is the most extensive animal production system in the UK. With stocking rates as low as 0.5-2.0 ewes per hectare, and very little contact with humans, hill sheep are completely free to express their natural behaviour. However, the incidences of starvation, exposure, lameness, mineral deficiencies, parasitic infections and death are high, so only 77 lambs are reared per 100 ewes mated (SAC, 2003). The same ewes, when moved to upland farms at 4 years of age and stocked at the rate of up to 7 ewes per hectare (i.e. more intensive and less 'natural') rear 135 lambs per 100 ewes mated (SAC, 2003). Much of the improvement in performance is due to reduced mortality of lambs (10-30 % in hill flocks; Waterhouse, 1996) and ewes (8 % in hill flocks; SAC, 2003). In terms of the three views of animal welfare, the productionist would say that the intensive system is better because biological performance has doubled; the humanitarian would say that the intensive system is better because there is less suffering; the radical consumer, however, would argue that the extensive system is better because the animals lead natural lives, even if those lives are shorter.

This example illustrates the main problem with any consideration of animal welfare – that different people have different perceptions of what constitutes 'good' and 'bad' welfare. This makes it difficult to assess the potential impacts of yield trends on animal welfare because any conclusion will be a compromise. One solution is to use the widely accepted set of ideals for animal welfare, which is the "Five Freedoms" published by the Farm Animal Welfare Council (FAWC, 2004):

1. FREEDOM FROM HUNGER AND THIRST - by ready access to fresh water and a diet to maintain full health and vigour.
2. FREEDOM FROM DISCOMFORT - by providing an appropriate environment including shelter and a comfortable resting area.
3. FREEDOM FROM PAIN, INJURY OR DISEASE - by prevention or rapid diagnosis and treatment.
4. FREEDOM TO EXPRESS NORMAL BEHAVIOUR - by providing sufficient space, proper facilities and company of the animal's own kind.
5. FREEDOM FROM FEAR AND DISTRESS - by ensuring conditions and treatment which avoid mental suffering.

The Farm Animal Welfare Council is an independent advisory body established by the UK Government in 1979. Its terms of reference are to keep under review the welfare of farm animals on agricultural land, at market, in transit and at the place of slaughter; and to advise the Government of any legislative or other changes that may be necessary. In developing the five freedoms, the aim was to define ideal states rather than standards for acceptable welfare. FAWC believes that they form a logical and comprehensive framework for analysis of welfare within any system together with the steps and compromises necessary to safeguard and improve welfare within the proper constraints of an effective livestock industry (FAWC, 2004).

At first sight, the five freedoms appear to be concerned only with environmental factors (the sub clause of each freedom deals with provision of conditions that assure that freedom). However, as stated previously, an animal's environmental requirements are related to its genotype, so the environment provided for one type of animal might not be appropriate for another type. Therefore, assessment of the impact of yield trends on welfare must consider multiple interactions for each freedom. The questions to be addressed are:

- To what extent is the yield trend a result of changing genotype?
- To what extent is the yield trend a result of changing technology independently of genotype?
- To what extent does the yield trend depend on altering the interaction between genotype and environment?
- Can advances in technology (and knowledge) keep pace with yield changes so that welfare is not compromised?

Rauw *et al.* (1998) reviewed the effects of genetic selection for production efficiency on metabolic, reproduction and health traits in farm animals. They found that animals in a population that have been selected for high production efficiency seem to be more at risk from behavioural, physiological and immunological problems. This was ascribed to a breakdown in the ability of selected animals to maintain homeostasis. In other words, the production process consumes such a large proportion of available resources that the remaining resources are insufficient to allow an adequate response to a stressor.

In the process of natural selection, animals that could not maintain homeostasis would die out, but artificial selection of animals for artificial environments means that the balance can become upset. This is particularly true in species where breeding goals have been narrow and selection has concentrated on a single trait. The problem is that negative associations between traits are generally not known when breeding goals are set. For example, leg weakness in poultry would not have shown up until after many generations of selection for carcass size.

Once a potential welfare problem has been identified, steps can be taken to redress the balance. Short term solutions might involve changing energy and nutrient allowances, housing, husbandry or other environmental factors; long term solutions might involve altering breeding goals and weighting of individual traits in selection indices. These solutions might slow the rate of increase for yield trends, but are unlikely to reverse them.

POULTRY

Bone weakness in poultry has been a serious concern in commercial production of meat and eggs for several years. Calcium homeostasis can be upset due to selection for high growth rate or high egg production, without appropriate changes in nutrition (Rauw *et al.*, 1998). This can lead to fractures of legs or wings during the normal production cycle and during depopulation. The problem is exacerbated by systems where exercise is limited so birds do not develop strong bones. Changes in calcium and vitamin D supply provide short-term relief, but the long term solution is likely to be genetic. Breeding companies have already started selecting birds for skeletal quality, which has now improved in commercial broilers (McKay, 2004). This claim is supported by a survey of commercial broiler flocks, which found that the incidence of leg problems was 2.5% in 2000, and that the incidence is currently decreasing at the rate of 20% per year (Pfeiffer and Dall'Aqua, 2002). It appears, therefore, that yield trends in meat and egg production can continue without adversely affecting bone strength in poultry, provided this factor is retained in selection indices.

Ascites and sudden death syndrome are two non-infectious conditions that affect rapidly-growing broilers (Maxwell, 1998). Both are caused by problems with the heart and circulatory system that result in hypoxia and damage to other organs and general metabolism, leading to death. There is a strong genetic component in both syndromes, since different strains vary in their susceptibility (Maxwell,1998) and recent breeding strategies have reduced their incidence (McKay, 2004).

Feather pecking and cannibalism are undesirable behaviour patterns found in poultry that result from the natural desire to establish a dominance hierarchy or 'pecking order'. The incidence is related to colony size and stocking density, which influence the frequency with which a bird has to re-establish its dominance, and can be reduced by dim lighting. The problem is much worse in free-range flocks, where the only effective control is beak trimming, which is clearly undesirable from a welfare point of view (FAWC, 2004). A long-term genetic solution to the problem is to keep birds in colony cages and to select on group performance rather than individual performance, which always favours dominant and aggressive birds (Muir, 2004). In laying hens, group selection improved livability by over 800%; in Japanese quail, mortality was reduced by 32% using group selection, compared with a 400% increase in mortality amongst birds selected on an individual basis (Muir, 2004).

In broiler production, growing birds are normally fed *ad libitum*, and diet composition is varied to ensure they consume the balance of energy and nutrients necessary for optimum growth (Filmer, 2002). Broiler breeders, however, must not be overfed during the growing phase or they will develop hypertrophic ovaries, erratic ovulation and defective eggs; growth is severely restricted in comparison with bird potential by giving restricted amounts of food only once per day (Fisher and Willemsen, 1999). This is one of the few instances where withholding the first freedom (hunger and thirst) is essential in a production system to ensure adequate performance and to alleviate other potential welfare problems, such as internal laying leading to peritonitis, and misshapen and poorly shelled eggs leading to deformed offspring. The yield trend in broiler production, for increased growth rates and body size, will probably lead to greater proportional restrictions of energy and nutrient intake and growth rate in broiler breeders. In terms of freedom from hunger, the only solution is to design diets with lower energy and nutrient concentrations that can be fed at higher rates. A genetic solution is to use broiler parent birds with a recessive dwarfing gene, such as the Ross PM3 strain (Aviagen, Midlothian, UK), which has lower energy and nutrient requirements because of its small size, but produces heterozygous offspring that perform as well as conventional broilers.

Although the increase in popularity of free range systems of poultry production has been ascribed to consumer desire to enhance welfare, free range systems also have negative implications for welfare. In terms of the five freedoms, the major gain from free range systems is the freedom to express normal behaviour, although this can also be provided in well-designed indoor systems. Freedom from hunger and thirst is equal in both types of system, but freedoms from discomfort (2), from pain, injury and disease (3), and from fear and distress (5) are all at risk in free range systems. The environment is not controlled, so lighting, heating and shelter are not optimal; pecking and aggression are increased; parasitic diseases are increased; and mass panic can be caused by responses to perceived predators, such as wild birds, aeroplanes and animals (FAWC, 2004).

PIGS

Leg weakness was identified as a major concern in pigs intensively selected for performance in the 1970s (Webb *et al.*, 1983). Records were examined for nearly 24,000 boars in central testing stations and adverse genetic and phenotypic correlations were found between back fat depth and leg weakness. As in poultry, one of the principal causes of leg weakness is breakdown of homeostasis, leading to poor cartilage growth and osteochondrosis, but reduced oxidative capacity of muscles in fast-growing pigs might also be a contributory factor (Rauw *et al.*, 1998). Locomotor problems account for 15% of premature culls in breeding sows (Hughes and Varley, 2003), so there would be an economic benefit, as well as welfare improvement, if leg problems could be reduced. Lopez-Serrano *et al.* (2000) reported negative genetic relationships between leg scores and performance traits (daily gain and backfat thickness), but the heritability of 0.13 for leg score indicates that response to selection would be expected. Due to the longer generation interval in pigs, genetic improvements would take longer than in poultry, but a QTL for osteochondrosis has been identified that might speed the process (Andersson-Eklund *et al.*, 1998).

There has been a rapid increase in the number of outdoor pig breeding herds in recent years, and it is estimated that over 30% of herds are now housed outdoors (SAC, 2004). The main driver for this trend has been economic – outdoor units require a significantly lower capital investment. However, as with free-range poultry, animal welfare is perceived to be improved because outdoor pigs can express more 'natural' behaviours. Again, as with free-range poultry, there are also negative welfare aspects associated with outdoor pig herds. Because outdoor pigs are not in a controlled environment, they are at the mercy of the elements and need protection from rain, wind, snow, sunburn and extremes of temperature. Piglet mortality is generally higher in outdoor units, mainly due to crushing. A survey by SAC found pre-weaning mortality of piglets born alive was 13.2% for indoor systems and 16.7% for outdoor systems (SAC, 2004), and a survey of pig herds participating in the MLC Pigplan scheme recorded piglet mortality rates of 16.3% for indoor herds and 19.6% for outdoor herds. Edwards (2002) quoted two surveys that found 45% of piglet deaths were due to crushing in outdoor herds, compared with 20% for indoor herds. Genotypes for outdoor pig production need to be hardier, which usually means slower growth rates and fatter carcasses. Thus, the yield trends towards faster growth rates and leaner carcasses might be partially offset if the trend towards outdoor production continues.

Probably the most controversial area of pig production has been the housing of the sow during pregnancy and lactation. Sow tether stalls were banned in the UK in 1999 (tethering in stalls will be banned in the EU in 2006 and stalls themselves by 2013), but there is still much debate about the use of farrowing crates which are designed to protect piglets from crushing by the sow and also to provide supplementary heat and food for the piglets. However, protection of the piglets is achieved by restricting the movements of the sow, which some people find unacceptable. Much of the reduction in piglet mortality for indoor herds compared with outdoor herds can be attributed to the use of farrowing crates indoors, but even when comparing only indoor systems the effects can be dramatic. Kavanagh (1995) established a 500-sow unit with free-access farrowing nests and monitored performance for six months following a 12-month establishment period. The farrowing nests were replaced by farrowing crates and performance was monitored for a further six months. Pre-weaning mortality was 19.2% for farrowing nests, with most of the deaths being due to starvation and crushing; mortality was only 6% for farrowing crates. Sow mortality also decreased from 9.7% in nests to 3.5% in crates. Judged against the five freedoms, Kavanagh (1995) concluded that the farrowing nests failed on 3-4 freedoms for piglets and 2-4 freedoms for sows, whereas the farrowing crates satisfied all the freedoms for piglets and only failed on Freedom 5 (natural behaviour) for sows. Edwards (2002) suggested that a piglet mortality rate of 10-20% was normal in pigs and represents an optimal evolutionary strategy (where success of the species is attributable to the large numbers born, even though mortality is high). The trade-offs between sow welfare and piglet welfare, and between freedoms for sows, therefore remains. However, any decision to prohibit the use of farrowing crates will have serious consequences for productivity and yield trends of breeding units.

DAIRY COWS

Mastitis and lameness continue to pose significant welfare problems and, in conjunction with fertility problems, are the principal reasons for premature culling from the dairy herd.

Whitaker (2002) compared the incidence of mastitis in over 300 herds in 1998 and 2000; somatic cell counts increased from 88 700 to 140 000 cells/ml and the incidence of clinical mastitis increased by 36.6%, often in conjunction with an increase in herd size. However, in another study Esslemont and Kossabati (2002) reported the incidence of disease from 1988/89 to 1998/99 in herds recorded under the DAISY system. They found no trend in the incidence of mastitis, calf mortality, retained placenta or vulval discharge. Moreover, the incidence of several health problems had decreased: milk fever from 7% to 4%; lameness per 100 cows from 29 cases to 22 cases; assisted calving from 9-11% to 7%. This might be taken to suggest disease incidence had been unaffected or reduced as average milk yields had increased. However, since the incidence of many diseases increases with age, premature culling for fertility could be masking other effects on disease incidence.

High milk yields are detrimental to the welfare of dairy cows if they are accompanied by increased susceptibility to mastitis and udder damage. These can be caused by increased pressure on udder tissue and increased weight of milk. A cow with an annual milk yield of 6,000 litres will produce about 30 litres per day at peak lactation. If the cow is milked twice each day, with 12 hours between milkings, the maximum weight of milk that has to be supported by the suspensory ligaments of the udder is 15 kg. Milking intervals are usually not equal, however, because of the need to restrict the number of unsocial hours worked by stockmen. An interval of 15 hours would increase the maximum weight of milk to 20 kg.

Cows yielding 10,000 litres per year, however, yield 50-60 litres per day at peak lactation and have to carry up to 30-40 kg in the udder with twice-daily milking. This would be expected to cause increased udder pressure, stretched ligaments and increased incidence of mastitis. In addition, because milking speed does not increase with milk yield, cows will spend longer waiting in collecting yards and stockmen will have longer working hours.

In the USA herds with high annual milk yields are usually milked three times daily, which reduces the pressure on the udder. This practice is not widespread in the UK because of economic and social pressures on farm labour. However, automatic milking systems, which have recently been introduced to the UK, could provide a solution.

Automatic milking systems, popularly called robotic milking, allow cows to choose when to milk themselves. Cows typically attend for milking up to six times per day in early lactation and the frequency drops to twice per day in late lactation. The main incentive for a cow to enter the stall is not the desire to be milked, but the reward of some concentrates. In addition to the welfare benefits of relieving udder pressure and weight, thereby decreasing the incidence of mastitis, there are many other welfare improvements in the automatic milking system. Cows can choose when they want to be milked instead of following a timetable imposed for human convenience; individual cows show preferences, including some that like to attend at 2:00 am; cows are not forced into collecting yards where they have to wait for milking, often in close proximity to other cows; bullying is reduced or eliminated, and feeding or resting times are increased. Freeing the stockman from the drudgery of milking cows in a parlour allows extra time for other tasks, leading to better quality of life for the stockman and better stockmanship.

Higher milk yields have also been associated with increased incidence of lameness in dairy cows. Again, it is not milk yield *per se* that causes lameness, but an imbalance between resources and requirements. Lameness is a symptom that results from a complex series of interacting factors, including injury, disease and nutrition. Increases in milk yield over the past 20 years have been accompanied by selection of larger cows, which require larger cubicles; incremental increases in frame size might go unnoticed until a serious differential between cow size and cubicle size has been established, which predisposes cows to injury. Increased herd size and reduced labour inputs per cow can increase the likelihood of injury and infectious transmission, and could delay detection and treatment. The use of pedometers, which monitor cow activity for the detection of oestrus, can assist with the detection of lameness because lame cows are reluctant to move. However, there is no substitute for high levels of stockmanship.

Genetic correlations between milk yield and reproductive measures in dairy cows are unfavourable. This suggests that successful selection for higher yields may have led to a decline in fertility. There is also evidence that an imbalance of energy and nutrients, in either high genetic merit cows or those fed diets not matched to their performance, leads to poorer reproductive performance (Pryce *et al.*, 2004). However, it is debatable whether infertility is a welfare problem in dairy cattle. As already suggested, the inability of a female to conceive does not, in itself, cause pain or suffering, so infertility is only detrimental in terms of the inability to pass on genes to the next generation. However, some conditions that result in reduced fertility, such as assisted calving, retained foetal membranes, endometritis and vulval discharge, do have welfare implications, as might repeated veterinary intervention and artificial insemination. Discussions of causes of infertility, and possible remedial actions, have been reviewed by Garnsworthy and Webb (1999), Pryce *et al.* (2004), and Webb *et al.* (2004).

SHEEP AND BEEF

Due to the extensive nature of most sheep production systems, sheep have escaped the close attention of animal welfare campaigners. As noted earlier, however, the extreme environmental conditions found in some systems of hill sheep production have severe welfare implications. In addition, sheep at all altitudes are notorious for the number of diseases to which they are susceptible. Some of these diseases, such as foot rot and fly strike, can cause long term pain and suffering, particularly if diagnosis and treatment are delayed when topography makes daily inspection of stock impossible.

Yield improvements for hill sheep can only be achieved by selection for improved conformation and disease resistance within hardy breeds. Overall productivity cannot be increased by higher stocking rates or greater litter sizes because nutrition and other environmental resources impose absolute limits. Therefore, any future yield trends are likely to be accomplished by improvements in welfare. Because 70 % of the UK breeding ewe population is in less favoured areas, proposals to alter support payments and control stocking rates to protect the environment in these areas (Ashworth *et al.*, 2000) will have the greatest impact on yields. These changes are likely to be neutral or beneficial for the welfare of hill sheep.

In upland and lowland sheep production, yield improvements are predicted to result from selection for improved live-weight gain and conformation. Predicted changes are unlikely to have implications for animal welfare.

There are few welfare issues associated with yield in beef production. The possible exception is the production of purebred Belgian Blue calves. The Belgian Blue breed has a double-muscling gene that enhances the growth of lean tissue. This gene also results in relatively large calves, which frequently have to be delivered by caesarean section in purebred animals or when crossed with a relatively small dam.

CONCLUSIONS

In all species, good stockmanship is the key to good animal welfare. There is evidence that some welfare problems have been associated with trends for increasing yields, and these usually result from an imbalance between resources and requirements. In most cases, potential welfare problems have been addressed by changing nutrient allowances, housing, husbandry or other environmental factors in the short term, and altering breeding goals over the longer term.

Assessment of animal welfare is subjective, and we can no longer equate welfare solely with performance. Some consistency can be obtained by considering the five freedoms, but opinions differ as to which is the most important freedom. Most animal production systems involve a trade off between freedoms for individual animals. Some also involve conflict between freedoms for different generations.

Continuation of current yield trends is unlikely to impose welfare problems beyond those already identified, and steps have already be taken to address major problems. Future legislation purporting to improve welfare in some systems, particularly pigs and poultry, will undoubtedly had negative welfare implications for some animals within each system and will slow the overall rate of gain in performance. However, it can be concluded that animal welfare will not be a major limitation to continued increases in yields.

Environmental impact

The main environmental impact factors affected by animal production are gaseous emissions in the form of methane, ammonia and nitrous oxide, and pollution of groundwater by nitrate and phosphate runoff and leaching. Pollution incidents, such as spillage of silage effluent or chemicals, might arise from animal production enterprises, but they will not be discussed here because they are avoidable and are not related to yield level. Similarly, odour emissions are not related to yield levels and can be minimised by appropriate manure handling protocols.

Yield levels influence national pollution levels by affecting animal numbers, daily emissions by individual animals, and the number of days that individual animals take to reach slaughter condition. National inventories of greenhouse gas emissions by animals are calculated on a per head basis, using standard factors agreed by the Intergovernmental Panel on Climate Change (IPCC), which are multiplied by animal populations given in the annual census. The standard factors can be either Tier 1 factors, which use an average emission factor for each category of animal, or Tier 2 factors, which vary according to feed intake and reflect differences in production systems found between countries (IPCC, 1997; 2000). In the UK, Tier 2 factors are only applicable to dairy cows because of their greater variation in feed intake and their relative contribution to methane emissions.

Total UK emissions of methane were estimated to be just over 2 million tonnes in 2002. Enteric fermentation contributed 38.4 % of this total, and fermentation of animal wastes contributed 4.7 %; other major sources of methane were landfill sites (20 %), gas leakage (16 %) and coal mines (11 %). Emissions by different types of livestock are shown in Table 17.1.

Table 17.1. METHANE EMISSIONS FROM ENTERIC FERMENTATION AND FERMENTATION OF ANIMAL WASTES IN THE UK FOR 2002

	Emission factors (kg/hd/yr)		*Number of animals (millions)*	*Total methane (kT/yr)*	*% of livestock total*
	Enteric	*Manure*			
Dairy cattle	118.8	13.37	2.2	291	30.4
Other cattle[1]	48	6	8.4	395	41.2
Sheep[1]	8	0.19	33.1	216	22.5
Pigs	1.5	3	8.0	36	3.8
Poultry	-	0.078	160.3	13	1.4

Source: Defra (2003)

[1]Emission factors are for adult non-dairy cattle and sheep, but animal numbers and total emissions include weighted allowances for young animals (see source for details).

Methane is produced by methanogenic Archaea, which are anaerobic organisms that live in mud at the bottom of lakes and swamps, and in the intestinal tracts of animals. They primarily ferment cellulose and release methane as a waste product of cellular metabolism. Enteric fermentation in the rumens of ruminant animals is the main source of methane produced by agriculture, and emissions are proportional to body weight and the proportion of grass in the diet (Table 17.1). Some enteric fermentation also occurs in the hind-gut of ruminants and non-ruminants.

Total UK emissions of ammonia were estimated to be 306,000 T in 2002, with 70 % coming from animal wastes, 13 % from non-livestock agriculture, and the rest from non-agricultural animals, road transport and disposal of human waste (Defra, 2003). Ammonia pollution is a problem because ammonia nitrogen can be deposited in rainfall on nitrogen-sensitive habitats, thus destroying the natural balance in ecosystems. Potential ammonia emissions from animal wastes depend upon nitrogen excretion by animals, but actual emissions are affected mainly by storage and handling methods. For example, Misslebrook and Smith (2002) estimated that ammonia emissions from cattle wastes were 34 % during housing, 9 % during storage, 48 % during spreading, and 9 % during grazing.

Nitrous oxide emissions also arise from animal wastes and are related to nitrogen excretion, but animals only account for 3.3 % of total UK emissions, with 64.4 % coming from non-animal agriculture and 32.3 % from non-agricultural sources (Defra, 2004).

Pollution of groundwater by nitrates is a serious problem in the UK because it causes eutrophication and could also cause methaemaglobinaemia (blue baby syndrome) in young babies if high nitrate concentrations occurred in drinking water. Nitrate gets into groundwater through leaching, runoff, denitrification and soil erosion, and the amount lost depends on rainfall, nitrogen applied in animal manures and inorganic fertilizer, residual soil nitrogen, soil type, and nitrogen uptake by plants. In the current context, the most important factors affected by yield trends will be nitrogen excretion per animal and animal numbers, although some changes in production systems might also have an impact on nitrate pollution. Nitrogen excretion factors and total excretion by different types of livestock are shown in Table 17.2. Lord *et al.* (2002) calculated that the mean UK nitrogen surplus in 1995 was 115 kg N/ha, made up of 51 kg N/ha for arable land, 140 kg N/ha for grassland (excluding rough grazing) and 14 kg N/ha for agricultural land from pig and poultry units.

Table 17.2 NITROGEN EXCRETION BY DIFFERENT UK ANIMAL TYPES IN 2002

	kg/head/yr	*n ('000)*	*kT N*	*% total*
Dairy cows	93.8	2,227	209	23
Beef cows	60.0	1,657	99	11
Cattle 1-2 yr	47.0	3,595	169	19
Cattle <1 yr	11.8	2,867	34	4
Ewes	9.2	17,630	162	18
Lambs	3.4	17,310	58	6
Sows & gilts	14.3	557	8	1
Growing pigs	8.6	10,062	87	10
Broilers	0.5	105,137	52	6
Layers	0.6	28,778	17	2

Sources: IPCC (2000); Defra (2004)

POULTRY

The UK slaughtered 852 million broilers in 2003 at an average weight of 2.25 kg. The yield projection for broilers is that slaughter weight will remain the same, but the number of days to slaughter will reduce from the current 40 days to 34 days by 2025. Nitrogen excretion factors are currently calculated on a per head basis, with nitrogen retention being 30% of nitrogen intake. Because increases in growth rate have been paralleled by increases in feed intake, the reduction in days to slaughter is unlikely to have a significant impact on nitrogen emissions per bird. If national production is limited by housing capacity, the reduction in days to slaughter will allow greater annual throughput. This would increase the number of broilers produced to 1,000 million birds per year, increasing emissions by 18%. If, on the other hand, national production is determined by the demand for home-reared broilers, there will be no change in emissions unless there is a change in feed conversion efficiency.

Due to physiological limitations on egg production, the ultimate yield of eggs is predicted to be 360 eggs per bird per year, which could be attained between 2020 and 2030. This is an increase of 24% from the average 290 eggs per bird in 2003. If national egg production remains the same, the national flock could be reduced to 27 million birds, which would reduce nitrogen excretion by 24%.

Any change from cages to litter-based systems will result in an increase in nitrous oxide emissions because the emission factors are 0.005 kg N_2O-N/kg N excreted for systems where manure is excreted on a floor without bedding and birds do not walk on the waste, and 0.02 kg N_2O-N/kg N excreted for systems where manure is excreted on a floor with bedding and birds walk on the waste (IPCC, 2000). This 400% difference represents a potential increase of 1.54 kT/yr in N_2O emissions, which is 33 % of total emissions from poultry waste handling (Defra, 2003).

PIGS

In growing pigs, the predicted trends are for faster growth rates and leaner carcasses, with improved feed conversion efficiencies but greater feed intakes (Wiseman *et al.*, 2005). In 2050, growth rates are predicted to be 28% faster, so the time taken to grow a pig from 25 to 95 kg live weight will be reduced by 22% (from 110 days to 86 days). This will reduce the nitrogen excreted per pig, but the effect will be partially offset by the 9% increase in daily feed intake. Assuming a nitrogen retention of 30%, and a diet containing 160 g CP/kg, nitrogen excretion per pig raised will decrease by 14% from 3.4 to 2.9 kg. The current estimate of annual nitrogen excretion (11.3 kg N/pig/year) is close to the value given by Verstegen and Tamminga (2002) for pigs in the Netherlands (11.7 kg N/pig/year), but annual estimates on a per pig basis give a false impression of changes in nitrogen excretion. Because it will take less time for a pig to reach slaughter weight, annual nitrogen excretion will apparently increase by 9% per pig. A more appropriate indicator is to look at the national excretion of nitrogen produced by raising 10 million pigs per year; this would decrease by 14% from 34 to 29 Gg/year.

It is likely that improvements in growth rate and feed conversion efficiency in growing pigs will be accompanied by improvements in the precision of nutrition. Verstegen and Tamminga (2002) reviewed the potential for using adjustments in feed composition to reduce environmental pollution. Removal of safety margins from protein requirements would reduce nitrogen excretion, but would also be expected to reduce growth rate; using dietary ingredients with highly digestible nitrogen contents could reduce nitrogen excretion by 5%; optimising amino acid balance, particularly with the inclusion of synthetic amino acids, can reduce nitrogen excretion by up to 40% (Verstegen and Tamminga, 2002), which is equivalent to 13.6 Gg/year.

In the breeding herd, the optimum production level is assumed to be around 26-28 piglets reared per sow per year (Wiseman *et al.*, 2004), so there is unlikely to be any significant reduction in the number of sows required to produce 10 million pigs per year. However, there are suggestions that replacement rate might be increasing due to reproductive failure, particularly of animals in parity 1 and 2 (Hughes and Varley, 2003). The annual replacement rate was 43% in 2003, compared with a rate of 36% in 1980. With a sow population of 500,000, the increase in replacement rate represents an increase of 17% in the number of gilts reared per year, which is equivalent to an extra 360 tonnes of nitrogen excreted per year. In the breeding herd, as for growing pigs, opportunities exist to reduce nitrogen excretion by optimising amino acid balance and reducing dietary nitrogen concentrations.

The increasing trend towards keeping breeding herds outdoors has implications for nitrogen losses to the environment through leaching and volatilisation. Outdoor herds consume 17% more feed (MLC, 2000) to compensate for lack of controlled environment and increased activity, which will result in an equivalent increase in nitrogen excretion. A greater problem is the lack of control over the time and spatial distributions of excretion in outdoor herds. Whereas excreta from indoor herds can be stored and spread on land when weather conditions and plant uptakes minimise losses, excreta from outdoor herds are deposited on the land throughout the year and tend to be concentrated in certain areas of the paddock. Williams *et al.* (2000) measured nitrogen losses from outdoor pig systems with stocking rates of 25 (current commercial practice), 18 and 12 dry sows per hectare and compared these with an arable control. Over a two-year period, losses of nitrogen by leaching were between two and four times control levels, and losses by ammonia emissions were similar to leaching losses (Table 17.3); losses of nitrogen as nitrous oxide were relatively minor and were confined to urine patches. Nitrogen accumulation in the soil was between ten and twenty times control values, which was expected to exacerbate leaching losses in subsequent seasons (Williams *et al.*, 2000). In a similar study, Eriksen (2001) found relatively low nitrate leaching (25 – 30 kg N/ha) whilst sows were grazing paddocks for six months, but nitrate leaching was considerably increased (320 kg N/ha) during the following autumn and winter, even though an arable crop had been grown after the pigs were removed; leaching was highest (500 kg N/ha) 10m from the area where the sows had been fed.

Table 17.3 NITROGEN LOSSES (KG N/HA) FROM OUTDOOR PIG SYSTEMS AT DIFFERENT STOCKING RATES (DATA FROM WILLIAMS *et al.*, 2000)

	Stocking rate[1] (dry sows/ha)			
	25	*18*	*12*	*0*
Leaching	192	169	126	52
NH_3	200	144	96	0
N_2O	10	7	5	0
Soil N accumulation	576	398	265	27

[1] In addition to stocking rate (SR), ground cover was varied: treatments were stubble for SR 25, stubble undersown with grass for SR 18, and established grass for SR 12. The arable control (SR 0) was two crops of winter barley.

Losses of nitrogen after spreading of manure from indoor pigs depend on application rate and application method. Lord *et al.* (1999) recorded nitrogen losses due to leaching in the winter following cereal crops of 51 kg N/ha where no manure had been spread; 88 kg N/ha where manure had been applied to the cereal crop at up to 175 kg N/ha; 99 kg N/ha where 175 – 350 kg N/ha was applied; and 455 kg N/ha where 350 – 1000 kg N/ha was applied. Leaching losses from all crops in a survey of farms in nitrogen-sensitive areas were reduced from 71 kg N/ha to 47 kg N/ha following the introduction of control measures, which included reduced application rates, prohibition of application during autumn, and planting cover crops (Lord *et al.*, 1999). Losses of nitrogen as ammonia were reduced by 84% from 12.8 to 2.0 kg N/ha when pig slurry was injected 50 – 70 mm into growing crops compared with surface application (Pahl *et al.*, 2001); odour emissions were also reduced. Nitrogen losses from controlled application of slurry from indoor pigs are, therefore, considerably lower than reported values for outdoor pig units.

DAIRY COWS

Dairy cows account for 30 % of methane emissions and 23 % of nitrogen excretion from livestock agriculture. The trends over the past 40 years have been for a linear increase in average milk yield per cow, and a linear decrease in the total cow population; these trends are predicted to continue, certainly over the next ten years and probably for the next fifty years.

As mentioned previously, methane emissions by dairy cows are calculated using Tier 2 calculations, so they are based on energy intake calculated from live weight and milk yield (IPCC, 2000) and emissions are assumed to be 6 % of gross energy intake. This methane factor has been used for many years in energy evaluation and was calculated by Blaxter and Clapperton (1965) from studies of over 2,500 sheep and cattle performed in the 1950s and 1960s. However, its applicability to the modern dairy cow is questionable because methane production varies with level of feeding and with diet composition. These facts were acknowledged in the original publication (Blaxter and Clapperton, 1965); firstly, it was observed that methane production, as a proportion of gross energy intake, fell as level of feeding increased; secondly, methane production increased as digestibility increased when animals were at maintenance and twice-maintenance levels of feeding, but methane production decreased as digestibility increased when animals were fed at three times maintenance. Modern dairy cows are fed at levels above twice maintenance for most of the year, some at over four times maintenance, and even the average daily milk yield used in the calculation of the national methane inventory (17 l/d; Defra, 2003) requires a feeding level greater than twice maintenance. Furthermore, the trend away from diets based predominantly on grass and grass silage towards maize silage and concentrates will decrease the proportion of dietary energy lost as methane.

Yates *et al.* (2000) analysed data from 140 energy balance studies performed with dairy cows given a wide range of diets in calorimeters. They found that methane emissions were positively related to dry matter intake and dietary fibre content, and negatively related to the proportion of concentrates in the diet, according to the formula:

$$\text{Methane (MJ/d)} = 1.36 + 1.21\ \text{DMI} - 0.825\ \text{CDMI} + 12.8\ \text{NDF}$$

Where DMI = dry matter (DM) intake (kg/d)
CDMI = concentrate intake (kg DM/d)
NDF = neutral-detergent fibre concentration (kg/kg DM)

Using this formula, in the model of Garnsworthy (2004), it can be estimated that annual methane emissions from a herd of cows with 1 million litres of milk quota will be approximately halved (-47 %) if annual average milk yield increases from the current national average of 6,000 litres/cow to the predicted average for 2030 of 9,000 litres/cow. If methane emissions are calculated simply as 6 % of energy intake, with no allowance for change in diet composition to support higher milk yields, the predicted decrease would be only 13 %. Both of these predictions assume that milk quotas will limit national milk production and, therefore, decrease cow numbers as milk yield per cow increases.

Total nitrogen excretion by dairy cows is a function of cow numbers and excretion per cow, both of which are responsive to yield trends. Although dairy cows use nitrogen more efficiently than other ruminants, nitrogen output in milk is only 20 – 25 % of consumed nitrogen, so the remaining 75 – 80 % is excreted in dung and urine. Pollution potential depends mainly on the proportion of time spent grazing, the housing system and systems of storing, handling and spreading manure (IPCC, 2000; Misslebrook and Smith, 2002). These all affect the proportions of excreted nitrogen converted to ammonia or nitrous oxide, or lost to groundwater. Manure management systems are independent of milk yield, but the same principles apply as those discussed in the previous section on pigs.

The predicted increase in milk yield will increase total nitrogen excretion per cow. However, it is likely that the proportional increase in excretion will be smaller than the increase in milk yield because the number of cows will be reduced, so nitrogen excretion due to maintenance requirements will form a smaller proportion of the total. Also, grassland management is likely to improve and there will be a decrease in the proportion of milk produced from grazing, which can only support milk yields of up to 25 litres per day in spring and 5 litres per day in late season (Leaver, 1983). Grazed grass and legumes contain high ratios of rumen degradable protein to fermentable carbohydrates (AFRC, 1993), resulting in excess rumen ammonia that is excreted as urea in the urine. Külling *et al.* (2003) found that ammonia production from manure was two to three times greater when cows were fed on grass than when they were fed on a winter ration of hay plus concentrates. Similarly, Valk (1994) found that when grass was partially replaced by maize silage, total nitrogen excretion was reduced by 35 %, consisting of a 10 % increase in faecal nitrogen, but a 50 % reduction in urine nitrogen. Furthermore, urine and dung are not spread evenly on grazed pastures, but are voided on approximately 15 % of the field area and produce local concentrations of up to 1000 kg N/ha. This can lead to high losses of nitrogen through leaching and ammonia production, compared with spreading manure evenly under controlled conditions.

Apart from changes in grazing efficiency and forage sources, composition of supplements is also likely to change. Higher milk yields will only be achieved if energy and protein nutrition of dairy cows is more carefully balanced than in the past. If cows are to be able to consume sufficient nutrients within their dry matter intake limits, ration formulation programmes will have to produce diets with higher concentrations of energy from greater proportions of starch and fats, and higher concentrations of digestible undegradable protein. There is already a suggestion that the new "Feed Into Milk" ration formulation system (Offer *et al.*, 2002) has resulted in a marked decrease in protein concentration of commercial diets for dairy cows (D. Allen, Personal Communication). The increase in concentrate to forage ratio required to support higher milk yields will also have a beneficial effect on methane emissions.

The predicted decreases in methane emissions and nitrogen excretion might not be achieved if fertility in dairy cows continues to decline in association with increased genetic potential for milk production. Royal *et al.* (2000) estimated that conception rate to first service has declined by 1 % per year over the past 20 years to a current rate of 38 %. As fertility declines, replacement rate has to increase to maintain herd size. These extra replacements will be raised mainly on grass-based diets and will increase total emissions and excretions by dairy herds. Garnsworthy (2004) modelled the effect of fertility on methane and ammonia emissions by dairy herds; at current levels of fertility, replacements contribute 27 % of methane emissions and 15 % of ammonia emissions in a high-yielding herd, but these could be reduced to 15 % and 8 % if fertility was restored to ideal levels. Fortunately, current nutritional strategies designed to restore fertility (Garnsworthy and Webb, 1999) should also have beneficial effects on reducing methane and ammonia.

SHEEP AND BEEF

As previously mentioned, 70 % of the UK sheep breeding flock is found in hill and upland areas and there are proposals to alter support payments and control stocking rates to protect the environment in these areas (Ashworth *et al.*, 2000). This form of environmental protection is to maintain the physical appearance of the landscape, rather than to reduce emissions or nitrogen pollution. However, because beef cattle and sheep produce 60 % of methane emissions and 50 % of nitrogen excreted, reduced numbers of cattle and sheep will also reduce national methane emissions and nitrogen excretion.

Overgrazing in hills and uplands can lead to the destruction of heather and dwarf shrubs, which are replaced by poor quality grass that might not have the same soil retention properties to prevent erosion. Undergrazing can lead to the development of scrubland on the hills and to the destruction of marshland areas in the lowlands. In some areas, mixed grazing of cattle and sheep is desirable to maintain the balance of plant species. Because of the complexity of relationships between grazing pressure and herbage growth, the ideal carrying capacity of the land needs to be determined at the local level and stocking rates adjusted accordingly.

CONCLUSIONS

In all species, the predicted trends for increased yields, accompanied by increases in efficiency, are expected to reduce environmental impact of animal agriculture. This is mainly through reductions in animal numbers required to produce the same national levels of output.

Indoor systems have greater potential to reduce emissions because animal wastes can be stored and handled under controlled conditions. Indoor systems also give greater control over nutritional inputs, which can be matched more precisely to the animals' requirements, thereby reducing nitrogen excretion.

Grazing systems have historically relied on increasing levels of nitrogen application to increase output per hectare by growing more grass. There is a strong environmental case for limiting applications in future to curb both nitrogen pollution and methane emissions. For dairy cows in particular, future yield potentials will not be supported by grass alone and it is likely that nitrogen efficiency will be improved by application of animal wastes to crops that can be used subsequently for supplementary feeding.

There will still be a place for low input systems based mainly on grazing hill and upland areas, if only for aesthetic reasons. In future, these will require continued economic support through subsidies or through price differentials when supplying niche markets. However, strict controls will need to be placed on stocking rates to avoid overgrazing or undergrazing of fragile habitats.

Overall conclusions

In general, the yield trends predicted for the next fifty years are not likely to introduce any new welfare problems and are likely to be beneficial in terms of reduced environmental impact. Extensification of animal production systems is claimed to be better for animal welfare and the environment because it is more 'natural'. However, outdoor systems can actually reduce the welfare of some animals and are certainly detrimental to the environment in terms of methane emissions and lack of control over nitrogen pollution, compared with indoor systems.

Overall, environmental impact from animal production is predicted to decrease because of a reduction in animal numbers as yields increase. This assumes that total demand for animal products remains static in the UK. Demand is mainly a function of price, and globalisation tends to smooth out fluctuations by allowing import levels to change. Therefore, demand for home production, which determines profitability, yield levels, animal numbers and environmental impact in the UK, is more likely to be influenced by World prices and levels of imports than during recent years of price support.

References

AFRC (1993) *Energy and Protein Requirements of Ruminants.* CAB International, Wallingford.

Andersson-Eklund, L., Marklund, L., Lundstrom, K., Haley, C.S., Andersson, K., Hansson, I., Moller, M. and Andersson, L. (1988) Mapping quantitative trait loci for carcass and meat quality traits in a wild boar x Large White intercross. *Journal of Animal Science*, **76**, 694-700.

Ashworth, S., Palmer, M. and Northen, J. (2000) *An Economic Evaluation of the Sheepmeat Regime as Applied in the United Kingdom*. http://statistics.defra.gov.uk/esg/evaluation/shpmtreg

Blaxter, K.L. and Clapperton, J.L. (1965) Prediction of the amount of methane produced by ruminants. *British Journal of Nutrition*, **19**, 511-522.

Broom, D.M. (1993) A usable definition of animal welfare. *Journal of Agricultural & Environmental Ethics*, **6**, Special Supplement 2, 15-25.

Defra (2003) *UK Greenhouse Gas Inventory: 2003 Submission*. http://www.defra.gov.uk/environment/statistics/index.htm

Defra (2004) *e-Digest of Environmental Statistics* http://www.defra.gov.uk/environment/statistics/index.htm

Edwards, S.A. (2002) Perinatal mortality in the pig: environmental or physiological solutions? *Livestock Production Science*, **78**, 3-12.

Eriksen, J. (2001) Implications of grazing by sows for nitrate leaching from grassland and the succeeding cereal crop. *Grass and Forage Science*, **56**, 317-322.

Esslemont, D. and Kossabati, M. (2002) The Costs of Poor Fertility and Disease in UK Dairy Herds. *Daisy Research Report No. 5*, Intervet UK Limited, Milton Keynes.

FAWC (2004) Farm Animal Welfare Council. http://www.fawc.org.uk/

Filmer, D. (2002) Putting poultry nutrition into practice: experience from the field. In *Recent Advances in Animal Nutrition – 2002* (Eds. P.C. Garnsworthy and J. Wiseman) pp. 239-268. Nottingham University Press, Nottingham.

Fisher, C. and Willemsen, M.H.A. (1999) Nutrition of broiler breeders. In *Recent Advances in Animal Nutrition – 1999* (Eds. P.C. Garnsworthy and J. Wiseman) pp. 165-166. Nottingham University Press, Nottingham.

Fraser, D. (2003) Assessing animal welfare at the farm and group level: the interplay of science and values. *Animal Welfare*, **12,** 433-443.

Garnsworthy, P.C. (2004) The environmental impact of fertility in dairy cows: a modelling approach to predict methane and ammonia emissions. *Animal Feed Science and Technology*, **112**, 211-223.

Garnsworthy, P.C. and Webb, R. (1999) The Influence of nutrition on fertility in dairy cows. In *Recent Advances in Animal Nutrition - 1999* (Eds P.C. Garnsworthy and J. Wiseman), pp. 39-57, Nottingham University Press, Nottingham.

Hughes, P.E. and Varley, M.A. (2003) Lifetime performance of the sow. In *Perspectives in Pig Science* (Eds. J. Wiseman, M.A. Varley and B. Kemp) pp. 333-355. Nottingham University Press, Nottingham.

Intergovernmental Panel on Climate Change (IPCC) (1997) *Revised 1996 IPCC Guidelines for National Greenhouse Gas Inventories*, J.T. Houghton *et al.*, IPCC/OECD/IEA, Paris, France.

Intergovernmental Panel on Climate Change (IPCC) (2000) *Good Practice Guidance and Uncertainty Management in National Greenhouse Gas Inventories,* IPCC/OECD/IEA, Paris, France.

Kavanagh, N.T. (1995) A comparison between free-access farrowing nests and farrowing crates on a 500-sow unit. *Pig Journal*, **35**, 10-19.

Külling, D.R., Menzi, H., Sutter, F., Lischer, P. and Kreuzer, M. (2003) Ammonia, nitrous oxide and methane emissions from differently stored dairy manure derived from grass- and hay-based rations. *Nutrient Cycling in Agroecosystems*, **65**, 13–22.

Leaver, J.D. (1983) *Milk Production: Science and Practice.* Longman, Harlow.

López-Serrano, M., Reinsch, N., Looft, H. and Kalm, E. (2000) Genetic correlations of growth, backfat thickness and exterior with stayability in Large White and Landrace sows. *Livestock Production Science*, **64**, 121-131.

Lord, E.I., Johnson, P.A. and Archer, J.R. (1999) Nitrate sensitive areas: a study of large scale control of nitrate loss in England. *Soil Use and Management*, **15**, 201-207.

Lord, E.I., Anthony, S.G. and Goodlass, G. (2002) Agricultural nitrogen balance and water quality in the UK. *Soil Use and Management*, **18**, 363-369.

Maxwell, M.H. (1998) Ascites in broilers. In *Recent Advances in Animal Nutrition – 1998* (Eds. P.C. Garnsworthy and J. Wiseman) pp. 33-48. Nottingham University Press, Nottingham.

McKay, J. (2004) Breeding meat-type chickens for changing demands. *Proceedings of the British Society of Animal Science 2004*, 264.

Misselbrook, T. and Smith, K. (2002) Ammonia emissions from cattle farming. In *Ammonia in the UK*. pp.40-47 DEFRA Publications, London.

MLC (2000) *Pig Yearbook.* Meat and Livestock Commission, Milton Keynes.

Muir, W.M. (2004) Breeding and animal welfare: threats and opportunities. *Proceedings of the British Society of Animal Science 2004*, 265.

Offer, N.W., Agnew, R.E., Cottrill, B.R., Givens, D.I., Keady, T.W.J., Mayne, C.S., Rymer, C., Yan, T., France, J., Beever, D.E. and Thomas, C. (2002) Feed into milk - an applied feeding model coupled with a new system of feed characterisation. In *Recent Advances in Animal Nutrition – 2002* (Eds. P.C. Garnsworthy and J. Wiseman) pp. 167-194. Nottingham University Press, Nottingham.

Pahl, O., Godwin, R.J., Hann, M.J. and Waine, T.W. (2001) Cost-effective pollution control by shallow injection of pig slurry into growing crops. *Journal of Agricultural Engineering Research*, **80**, 381-390.

Pfeiffer, D.U. and Dall'Aqua, F. (2002) *An Analysis of an Industry National Broiler Chicken Leg Weakness Study in the United Kingdom.* British Poultry Council, London.

Pryce, J.E., Royal, M.D., Garnsworthy, P.C. and Mao I.L. (2004) Fertility in the high producing dairy cow. *Livestock Production Science*, **86**, 125-135.

Rauw, W.M., Kanis, E., Noordhuizen-Stassen, E.N. and Grommers, F.J. (1998) Undesirable side effects of selection for high production efficiency in farm animals: a review. *Livestock Production Science*, **56**, 15-33.

Royal, M.D., Darwash, A.O., Flint, A.P.F., Webb, R., Woolliams, J.A. and Lamming, G.E. (2000) Declining fertility in dairy cattle: changes in traditional and endocrine parameters of fertility. *Animal Science*, **70**, 487-501.

SAC (2003) *Hill Sheep Breeding Project.* http://www.sac.ac.uk/envsci/external/Hill&Mountain/projects/hsbpdesc.htm

SAC (2004) *Pigcare Campaign.* Http://www.sac.ac.uk/animal/External/AnimalWelfare/Pig/Pigcare/PCAREpage2.asp

Valk, H. (1994) Effects of partial replacement of herbage by maize silage on N-utilization and milk production of dairy cows. *Livestock Production Science*, **40**, 241-250.

Verstegen, M. and Tamminga, S. (2002) Feed composition and environmental pollution. In *Recent Advances in Animal Nutrition – 2002* (Eds. P.C. Garnsworthy and J. Wiseman) pp. 45-65. Nottingham University Press, Nottingham.

Waterhouse, A. (1996) Animal welfare and sustainability of production under extensive conditions - A European perspective. *Applied Animal Behaviour Science*, **49**, 29-40.

Webb, R., Garnsworthy, P.C., Gong, J-G. and Armstrong, D. G. (2004) Control of follicular growth: local interactions and nutritional influences. *Journal of Animal Science*, **82**, E63–E74.

Webster, A.J.F., Main, D.C.J. and Whay, H.R. (2004) Welfare assessment: indices from clinical observation. *Animal Welfare*, **13**: S93-98.

Whitaker, D.A. (2002) Clinical mastitis in British dairy herds. *Veterinary Record*, **151**, 248.

Williams, J.R., Chambers, B.J., Hartley, A.R., Ellis, S. and Guise, H.J. (2000) Nitrogen losses from outdoor pig farming systems. *Soil Use and Management*, **16**, 237-243.

Wiseman, J., Varley, M.A., Knowles, A. and Waters, R. (2004) Livestock yields now, and to come: case study pigs. In: *Yields of Farmed Species*. pp 495-518. Nottingham University Press, Nottingham.

Yates, C.M., Cammell, S.B., France, J. and Beever, D.E. (2000) Prediction of methane emissions from dairy cows using multiple regression analysis. *Proceedings of the British Society of Animal Science 2000*, 94.

18

EPIDEMIOLOGY OF LIVESTOCK DISEASES

J.A. (ARJAN) STEGEMAN
Department of Farm Animal Health, Faculty of Veterinary Medicine, Utrecht University, The Netherlands

Introduction

Ever since livestock were domesticated infectious diseases have played a major role. Stories about huge epidemics go back long before the beginning of the modern era (see for example the Biblical sixth plague of Egypt). Despite the increased possibilities of therapy and prevention, however, infectious diseases in livestock are still currently important. This chapter will focus on the epidemiology and future control of livestock diseases that are classified as list A diseases according to the criteria of the Office Internationale des Epizootie (OIE). These diseases cause severe clinical signs, can spread quickly irrespective of borders and can have huge socio-economic impact in countries and are thus of major importance to livestock yields now and to come. These diseases include, amongst others, foot and mouth disease (FMD), classical swine fever (CSF) and avian Influenza (AI). Livestock in countries of the European Union are generally free of these diseases. The risk of new introductions of the causative agents is, however, constantly present. The reasons for this risk are that wildlife may be endemically infected, e.g. waterfowl with AI virus, and wild boar with CSF virus (Fritzemeier *et al.*, 2000). Moreover, the causative agents are also present in livestock populations in large parts of the world. Consequently, importation of farm animals or products from such infected regions creates a high risk for re-introduction of these infections into free countries. Therefore, one should constantly be aware of new outbreaks of list A diseases and control strategies should be kept up-to-date, depending on scientific developments, developments in the livestock industry and socio-economic changes.

In recent years, outbreaks of list A diseases have been controlled by eradicating infected herds, and other veterinary and zoö-technical measures like tracing, screening and a transport ban in the surveillance zone. Additional measures such as pre-emptive culling of suspected or contiguous farms are allowed and are necessary in densely populated areas. Vaccination is also a measure that can be applied to stop epidemics. However, vaccination hampers international trade and, as a result, the negative economic consequences of vaccination may be huge. An exception is made for Newcastle disease (ND) in poultry, for which preventive vaccination is allowed.

Since the non-vaccination policy became official in the EU, several epidemics of list A diseases have occurred. Outbreaks of CSF occurred in Germany, the Netherlands, Belgium, Spain, United Kingdom (UK) and Italy. Outbreaks of FMD occurred in France, The Netherlands, UK and Ireland; outbreaks of AI in Italy, The Netherlands, Belgium and Germany and outbreaks of ND in Denmark, UK, Ireland and The Netherlands.

The consequences of some epidemics were devastating. The outbreak of CSF in 1997 and 1998 in The Netherlands lasted over a year and the costs mounted to 2.5 billion Euro (Meuwissen *et al.*, 1999). The epidemics of FMD in the UK and the Netherlands in 2001 cost approximately 3.1 billion and 250 million Euro, respectively (Source: Defra and Dutch Ministry of Agriculture, respectively). The epidemics also had a disturbing effect on the local society, and the slaughter and destruction of millions of, mostly, uninfected animals was considered unethical.

Since new outbreaks can be expected in the EU, the conventional control of these diseases by eradication has to be reconsidered, especially the slaughter of uninfected animals. An alternative may be the application of preventive or emergency vaccination as an additional measure. The main arguments against vaccination have been that infected animals cannot be distinguished from vaccinated animals (Van Oirschot, 2001). However, the development of marker vaccines and discriminating tests provides authorities with an opportunity to reconsider the non-vaccination policy.

A risk of vaccination for list A diseases can be that the infection may spread unnoticed, because vaccinated animals do not show signs of the disease anymore, but remain infectious. An example of this can be found in a paper about a vaccination against bovine herpes virus 1 (BHV1) (Bosch, 1997). Another risk is the occurrence of carriers which have developed an immune response and have recovered from disease. Despite the immune response, the live virus may remain in the animal and the animal is persistently infected. This risk is mainly described for FMD in cattle, although it has not been demonstrated that vaccinated carriers can transmit FMD virus. In cases of CSF, sows that become infected during pregnancy may produce a persistently infected litter. This is a phenomenon that has been described by various authors, e.g. Terpstra (1987) and Dahle and Liess (1992), and recently by Depner *et al.* (2001) and Dewulf *et al.* (2003).

At this moment the question is still raised whether emergency vaccination can prevent further outbreaks and whether transmission of the infections can be stopped. Moreover, the quality of the discriminating tests is still disputed. In the following section, the effectiveness of vaccines against various list A diseases and the characteristics of the discriminating tests will be discussed. In the last section, the possibilities of emergency vaccination with conventional or marker vaccines, with respect to the effectiveness, costs and risk of FMDV-carriers will be considered.

Vaccination and the epidemiology of livestock diseases

In general, the effectiveness of vaccines against list A diseases can only be determined in specialized laboratories under high containment conditions. Until the mid 1990s, vaccines were primarily evaluated in vaccination-challenge experiments, sometimes followed by field trials to determine antibody levels in livestock populations in the field. In those vaccination-challenge experiments, vaccinated animals were inoculated with wild type virus. Subsequently, the antibody response, virus excretion and clinical signs were determined and compared to the outcomes of a non-vaccinated control group. In this way, the immunity of individual animals was determined. Examples can be found, amongst others, in Terpstra *et al.* (1990), Biront *et al.* (1987), Doel (2003) and Stone (1987).

However, when vaccines are applied during an epidemic, the aim of vaccination is not to induce clinical protection, but to prevent further spread of the virus in the population. Therefore, the degree of immunity of individual animals is not of primary interest, but the degree of herd-immunity induced by vaccination. Herd immunity is defined as 'the resistance of a group of individuals to attack by a disease to which a large proportion of the member are immune, thus lessening the likelihood of a patient with a disease coming into contact with a susceptible individual' (Fine, 1993).

A key parameter to be determined is the reproduction ratio R, that is the average number of secondary cases caused by one infectious individual in a completely susceptible population during its entire infectious period (Diekman *et al.*, 1990). R has a threshold value of one. This implies that an infection may spread when $R > 1$, resulting in a major outbreak. On the other hand the infection will fade out when $R < 1$, resulting in a minor outbreak (Diekman *et al.*, 1990). As a consequence, vaccination can effectively stop an epidemic when it can reduce R to a value below one.

Whether vaccination can achieve this goal can be established in transmission experiments. In such experiments, infectious (I) and susceptible (S) animals are housed together in an experimental unit. Subsequently, the spread of the infection from I to S is determined from the number of contact infections. Next, R can be estimated from the final size of an outbreak, i.e. the total number of contact-infected animals at the end of the experiment, or by measuring the number of new infections per time (e.g. per day).

Before the development of marker vaccines, transmission could hardly be quantified, because it was not possible to determine whether contact animals were infected. The reason is that if animals do not show clinical signs or do not shed the virus, the infections has to be based on sero-conversion. Often, animals that are well vaccinated do not develop a clear sero-conversion after infection. The development of marker vaccines, however, allowed the detection of infected animals in vaccinated populations. Since then, many experiments have been carried out to determine the effectiveness of vaccines with respect to reduction of virus transmission in a vaccinated population. Below, the ones on list A diseases are reviewed.

CLASSICAL SWINE FEVER

The modified life vaccine (MLV) C strain is a very immunogenic vaccine that induces an immune response within few days after vaccination, and offers good clinical protection against CSF (e.g. De Smit *et al.*, 2001; Biront *et al.*, 1987). Many countries have applied the C strain in their attempts to eradicate the virus from their country. Nevertheless, whether or not the vaccine can prevent transmission after a single application was not clear until recently. However, De Smit *et al.* (2001) showed that the vaccine could induce a protective immunity and seemed to be able to prevent transmission within one week after application, since the contact pigs did not show any sign of CSF. Yet, the sero-response was inconclusive, which made it impossible to determine whether or not contact infections had actually occurred.

In 1993, a subunit vaccine was developed, based on the E2 glycoprotein of CSF (Hulst *et al.*, 1993). The accompanying discriminating test was based on the E^{rns} glycoprotein. Next, in transmission experiments, Bouma *et al.* (2000) showed that the vaccine was able to reduce the transmission of the very virulent CSF virus strain Brescia from 14 days after vaccination ($R < 1$). Moreover, all inoculated vaccinated pigs developed antibodies against E^{rns}, indicating that the test was able to discriminate between vaccinated and infected animals.

More transmission experiments have been done since. The EU has supported several experiments in different EU member states with the E2 vaccines of Bayer and Intervet (Uttenthal *et al.*, 2001, Depner *et al*, 2001). The vaccination-challenge intervals in these experiments were 7, 10, 14 and 21 days, using the Paderborn strain, that caused the huge 1997 epidemic, as challenge strain. The E^{rns} tests used were the Ceditest ELISA and Chekit ELISA, respectively. In an additional analysis, Klinkenberg *et al.* (2003) concluded that both E2 vaccines reduced transmission significantly from day 21 after vaccination. The Rs in the vaccinated populations were 0.04 and 0.4 respectively, whereas the R for the unvaccinated control group was 10. In addition, the sensitivity and specificity of the E^{rns} ELISAs were estimated in the same EU experiments (Floegel-Niesmann, 2001). The sensitivity of the Ceditest was estimated at 60-78%, whereas the sensitivity of the Checkit test was 73%-94%. The specificity of the first test was 99% whereas the Checkit test had a specificity of only 40% in that study. Moormann *et al.* (2000) estimated the specificity of the Ceditest also at 99%, based on negative field sera from the Dutch pig population in 1999.

Based upon all results, recently the European Commission has adopted a decision approving the discriminatory test to be used after vaccination against CSF (Commission Decision 2003/859/EC of 5 December 2003 amending Decision 2002/106/EC).

Klinkenberg *et al.* (2003) demonstrated the effectiveness of vaccination in groups of piglets with maternally derived antibodies. This study was essential, because piglets will be born from vaccinated sows during an epidemic and it is unclear whether their offspring should be vaccinated or not to control the epidemic. They showed that maternally immune piglets can be vaccinated effectively: the vaccine was found to be capable of sufficiently reducing virus transmission at 3 and 6 months after vaccination at two weeks of age.

Dewulf *et al.* (2001) studied the effectiveness of the E2 vaccine in gilts. Horizontal and vertical virus transmission was monitored in two inoculated, non-vaccinated and 10 vaccinated conventional gilts, housed in individual sow boxes. Within 10 days post-inoculation, all vaccinated gilts became infected. In three out of the eight vaccinated pregnant gilts vertical virus transmission occurred, resulting in infected offspring. It was concluded that double vaccination with an E2 sub-unit marker vaccine protected pregnant gilts from the clinical disease but could not prevent horizontal nor vertical spread of the CSF virus. However, since the two inoculated gilts themselves were not vaccinated, this conclusion cannot be justified from the results.

Recent developments in vaccinology resulted in genetically engineered chimeric marker vaccines and DNA vaccines. The chimeric vaccines have been tested in transmission experiments. The Flc11 vaccine, based on the glycoprotein E2 of CSFV and the E^{rns} of BVDV, the Flc9 vaccine, based on the E^{rns} glycoprotein of CSFV and the E2 of BVDV. Both chimeric viruses Flc9 and Flc11 provided good clinical protection against a challenge with virulent CSFV at 1 or 2 weeks after vaccination. The discriminating test, the same as for the E2 subunit vaccine, detected all inoculated pigs (De Smit *et al.*, 2001).

FOOT-AND-MOUTH DISEASE

The vaccines against FMD virus are conventional, killed vaccines, containing the structural proteins of the virus. In general they induce protective immunity on an individual level within days after vaccination (e.g. Doel, 2003). Before 1992, the vaccine had been successfully applied for years in several countries of the EU, although it has not been demonstrated clearly that the virus was eradicated due to the vaccination programs.

The effectiveness of a single emergency vaccination, that can prevent further spread of virus strains, has not been shown until recently. Alexandersen *et al.* (2003) did a transmission experiment, but did not quantify transmission nor determined the effectiveness of vaccination. Eblé *et al.* (2004) investigated whether and to which degree vaccination could reduce transmission of FMDV among pigs. They carried out transmission experiments with intervals between vaccination and challenge of 7 and 14 days. In the 7-days group, all contact animals became infected with a homologous FMD virus strain and in the 14 days group none of the animals contracted the infection. They concluded that protection against transmission of FMD virus by vaccination is probably attained between 7 and 14 days after vaccination.

Orsel *et al.* (2003) quantified the transmission of FMD virus among calves. The reproduction ratio for unvaccinated calves was estimated to be 3.3, and for the vaccinated groups 0.17. They concluded that transmission was significantly reduced in the vaccinated group. Experiments are currently being carried out to quantify the transmission among adult cattle.

A difficulty while interpreting the results of these experiments is that the discriminating ELISA, that is based on non-structural proteins (NS ELISA), appears not very sensitive. Moreover, it is not clear whether animals that become NS positive are infectious, when no virus is detected in either the saliva or other samples. On the other hand, when the NS ELISA is not positive, animals could be infected and become carriers. Carriers have been described mainly for cattle, but Mezencio *et al.* (1999) reported recently that swine might be persistently infected for at least 300 days. The question is whether these carriers are infectious to susceptible animals. Hedger and Condy performed an experiment with persistently infected buffalo and three susceptible cattle to determine whether the virus could spread from carriers to susceptible cattle. Only after transportation of one buffalo, the virus had spread to cattle (Hedger and Condy, 1985). This seems to be the only case where transmission from persistently infected animals to susceptible in contact animals was established under controlled conditions (Moonen and Schrijver, 2000).

The vaccine O1 Manisa was applied during the FMD epidemic in The Netherlands in 2001. To demonstrate the effectiveness of the vaccine, Orsel *et al.* (2002) took 5000 samples from cattle to be tested in the NS ELISA to demonstrate a possible sero-conversion. The blood samples were gathered in a high-risk area, where outbreaks still occurred at the time of vaccination, and in a 'control' area, where no outbreaks had occurred. All samples were from vaccinated cattle. They showed that the number of NS-positive samples was higher in the high-risk area than in the control area. However, the number of positive samples was only limited (29/5000 samples). This implies that conclusions about the effectiveness of vaccination or the characteristics of the NS ELISA under field conditions could not be drawn properly.

AVIAN INFLUENZA

Epidemics of highly pathogenic Avian Influenza (HPAI) have occurred in Italy in 1999, in The Netherlands in 2003 and in Asia in 2004. The epidemics had a devastating effect on the poultry industry. Millions of birds were killed and destroyed, and hundreds of farms were depopulated. In Europe, the authorities combated the epidemics of the HPAI by stamping out infected farms and pre-emptive culling. One of the reasons was that is was unclear whether vaccination was able to prevent further spread of the HPAI strains. There was a risk that the virus may spread unnoticed, when poultry became infected sub-clinically. However, vaccination was applied to control the subsequent outbreaks of low pathogenic AI strains in Italy.

During the epidemic in 2003 in The Netherlands, this question became urgent. Van der Goot *et al.* (in preparation) carried out experiments to investigate whether a single intra-muscular vaccination with killed AI vaccine could induce sufficient flock immunity. The experimental set up was a standard transmission experiment as described by De Jong and Kimman (1994). In the unvaccinated control group, all inoculated and all contact birds died of AI. In the vaccinated group, neither the inoculated animals nor the contact animals showed signs of AI. However, it cannot be concluded on the absence of clinical signs in vaccinated animals only whether or not virus transmission is reduced. More tests have to be done to give a solid interpretation of the experiments.

NEWCASTLE DISEASE

Vaccines against ND have been tested in all kinds of challenge experiments. The vaccine, based on lentogenic ND virus strains, is very effective in reducing clinical signs. Vaccines are also used routinely in the field, and vaccination is obligatory in the Netherlands. The effectiveness of an emergency vaccination, however, has not been determined until 1996. Since then, Koch *et al.* (in preparation) have performed various transmission experiments. The experiments were performed with conventional broilers. Vaccine was applied as a spray. The vaccine induced clinical protection, and appears to be able to reduce transmission in groups of broilers with high antibody titers. The difficulty with these experiments is that, quite often, no virus could be detected from inoculated vaccinated broilers. In such cases it was not possible to determine whether inoculated animals were also infectious.

Strategies to control livestock diseases

Since the implementation of the non-vaccination policy, epidemics of list A diseases are generally controlled by stamping out. Since mid-2003 emergency vaccination against FMD is allowed without the massive slaughter of vaccinated animals. In line with OIE requirements, the EU proposal describes the procedure how to recover the status "free of FMD without vaccination" within 6 months after the last outbreak or completion of vaccination, whichever occurs last. This revised and more flexible procedure is to be used by a country that used emergency vaccination in combination with eradication of infected herds and post-vaccination surveillance testing (EU press release, 2002),

As mentioned above, there is a lot of scientific evidence that shows that vaccines against list A diseases are effective in preventing further spread of the virus. Consequently, vaccines could be used as additional tool during a future outbreak. The question is when vaccines should be used, and when it is acceptable to control an outbreak without vaccination. Aspects that should be considered are the effectiveness of the vaccines, the characteristics of the diagnostic tests, the (in-) direct costs and benefit of vaccination compared to the non-vaccination policy and the economical consequences of vaccination.

CLASSICAL SWINE FEVER

In case of CSF, scenario studies have been undertaken by Klinkenberg *et al.* (2003) and Mangen *et al.* (2003). Klinkenberg *et al.* (2003) evaluated the effectiveness of emergency vaccination with two E2 sub-unit vaccines during epidemics of CSF. Results of animal experiments indicate that sufficient reduction of horizontal transmission was achieved, but that the vaccines reduced vertical transmission insufficiently. This may give problems with detection of herds with persistently infected piglets. Klinkenberg *et al.* (2003) suggested that this can be overcome by a control strategy in which sows are not vaccinated. This strategy would limit the size and duration of an epidemic effectively. Furthermore, potential screening problems with the specificity of the discriminatory E^{rns} ELISAs could be solved by declaring a herd infected only if the number of sero-positive samples is above a certain threshold (e.g. 5% prevalence) and by increasing sample size, provided that no other pesti viruses are prevalent. In contrast to Dewulf *et al.* (2001), Klinkenberg *et al.* (2003) saw an opportunity for the use of marker vaccines as additional tool in a future CSF outbreak.

An essential part in evaluating (alternative) control strategies is the cost-benefit analysis and ethical aspects. Mangen *et al.* (2001) simulated the development of an outbreak in combination with various control strategies. The adequate scenario was emergency vaccination as buffer in case of restricted destruction capacity, but this was only acceptable when trade with products of vaccinated animals was still possible. When these products could only be sold at the local market in The Netherlands, it would be a very expensive strategy. Culling of infected and contact farms would then still be the least expensive option. They further examined the importance of pig-population density in the area of an outbreak of classical swine fever (CSF) for the spread of the infection and the choice of control measures. A spatial, stochastic, dynamic epidemiological simulation model linked to a sector-level market-and-trade model for The Netherlands was used. The obligatory control strategy required by current EU legislation was predicted to be enough to eradicate an epidemic starting in an area with sparse pig population. By contrast, additional control measures would be necessary if the outbreak began in an area with high pig density. The economic consequences of using preventive slaughter rather than emergency vaccination as an additional control measure depended strongly on the reactions of trading partners. Reducing the number of animal movements significantly reduced the size and length of epidemics in areas with high pig density.

FOOT-AND-MOUTH DISEASE

Bates *et al.*, (2003a and b) and Keeling *et al.* (2003) performed scenario studies for FMD control. Using an individual farm-based model, they considered either national prophylactic vaccination campaigns in advance of an outbreak, or combinations of reactive vaccination and culling strategies during an epidemic. Consistent with standard epidemiological theory, mass prophylactic vaccination could reduce greatly the potential for a major epidemic, while the targeting of high-risk farms increases efficiency. Given sufficient resources and preparation, a combination of reactive vaccination and culling might control ongoing epidemics. They also found that, vaccination could reduce markedly the long tail that characterizes many FMD epidemics.

In those studies, the costs and economical benefit of the vaccination strategy were not taken into account. Using a simulation model for FMD, Tomassen *et al.* (2002) developed a decision-tree to support decision-making on control measures during the first days after declaration of an outbreak of FMD. Four strategies were modelled: a standard control program of stamping out, ring culling within a 1-km zone, ring vaccination in a 1-km zone of an infected herd and ring vaccination in a 3-km zone of an infected herd. An economic model converted outbreak and control effects of farming and processing operations into estimates of direct costs and export losses. They showed that ring vaccination was the economically optimal control strategy for densely populated livestock areas whereas ring culling was the economically optimal control strategy for sparsely populated livestock areas.

Since 2003, emergency vaccination is allowed. Trade between countries within the EU is now facilitated and allowed from 6 months after the last outbreak, and not 12 months after the last sero-positive animal is removed from the region (Anonymous, 2003).

AVIAN INFLUENZA AND NEWCASTLE DISEASE

Van Boven *et al.* (2003) carried out a simulation study to determine the epidemiological and economical effects of vaccination against AI. They concluded that, in case of a new AI outbreak in a poultry-dense area, vaccination should be applied in a 50 km zone to ensure that the outbreak would be limited within that region. They estimate that the total costs when vaccination is applied will be higher than when vaccination is not applied. In this analysis, it was assumed that the within herd transmission was sufficiently reduced (R<1). The evaluation highly depends on the EU policy whether or not products from vaccinated animals are accepted in other countries of the EU. If not accepted, vaccination seems to be a very costly measure; if accepted vaccination is to be preferred for economical reasons. Ethical considerations have not been made. Moreover, the epidemiological analysis is based on only limited data about the effectiveness of vaccination.

Only few economic evaluations are available on ND control programs, e.g. Ring *et al.* (1988) and Leslie (2002). The latter calculated the costs of controlling and eradicating the epidemic of Newcastle disease in Northern Ireland in 1973. The parameters of the 1973 epidemic had been adjusted to simulate the effect of the same epidemic in 1997, taking into account the relative changes of input and output prices, and the changes in the structure of the poultry industry. The costs and their distribution between producers, government and the industry, in 1973 and 1997 are compared, and the costs of an alternative vaccination strategy are compared with the eradication policy in both years.

Discussion

The devastating epidemics of CSF, FMD and AI, put serious questions towards stamping out infected herds as the primary way to control these diseases. Especially after the FMD epidemic in the UK and The Netherlands in 2001, the non-vaccination policy within the EU has been discussed. The culling and destruction of millions of, mainly healthy, animals, that are for most list A diseases of no harm for public health, is disputable, not only for the public but also for veterinarians.

List A diseases are defined as 'transmissible diseases that have the potential for very serious and rapid spread, irrespective of national borders, that are of serious socio-economic or public health consequence and that are of major importance in the international trade of animals and animal products'. It is remarkable that for the control of these diseases, hardly any use is made of one of the major breakthroughs in veterinary medicine, i.e. the development of vaccines. Vaccines can induce clinical protection and most likely induce sufficient herd immunity as shown above. An example of the control by means of vaccination is the FMD outbreak in The Netherlands in 2001. Moreover, vaccination was considered during the HPAI outbreak in 2003, but not applied. In Italy vaccination has been used to control the LPAI outbreaks in 2000 and later (Capua and Marangon, 2003).

Since mid-2003 emergency vaccination against FMD is allowed in the EU without the massive slaughter of vaccinated animals. This is an important step forward, and should be continued for other diseases. An important issue that remains is the discrimination between vaccinated and infected animals. At this moment, several types of marker vaccines against CSF virus have been developed with discriminating tests. For FMD virus, discriminating tests are available. The characteristics of the discriminating tests, however, have not yet been evaluated under field conditions. Moreover, the demands seem higher for tests against list A diseases than for tests against other diseases. The main reason for this is the risk of carriers for FMD or CSF.

Even if tests can detect animals that have been infected, the question is whether these animals are a risk for other animals. For FMD, an infection may result in carriers: persistently infected animals. Although it has not been shown that these animals are a risk for new outbreaks, it cannot be excluded as well. It remains questionable whether products from these infected animals are a risk. The prohibition of swill feeding can be an effective tool to minimize this risk. Moreover, testing cattle at the border with an antibody test is not very effective in preventing introduction of virus, since such a test runs at least one week behind. It is not clear for veterinary reasons why this is accepted world wide, in contrast to testing vaccinated animals. For pigs it seems even stranger. Evidence for FMD carriers in pigs is very thin; in contrast, pigs can clear the FMDV infection completely. Nevertheless, countries refuse products from sero-positive pigs, because they may be a risk for virus introduction.

Even when marker vaccines are not available, vaccination should be accepted as additional tool. Vaccination against ND is, in many EU countries, common to prevent introductions or secondary spread. The vaccines against ND are effective, conventional vaccines. Disadvantages are an increased risk for increased mortality of a small percentage of young broilers, and the risk of conjunctivitis in human. New introductions of ND, as well of AI, can be expected every moment because of the wild life reservoirs. In countries with a high density of poultry, it is very hazardous not to vaccinate. Although some EU countries do not vaccinate, it is not clear for veterinary reasons why they should have an economic benefit to third countries.

The discussion is mainly about the balance between economics and ethics. Until now, the policy was based mainly on the economical consequences of diseases for farmers. The effects on local economics in other branches, the social effects and protests from society, have not yet been considered. Since 2001, a discussion has started between politicians, consumers, farmers, scientists etc. The attitude to farm animals and the way outbreaks of list A diseases are controlled were the main topics of the discussions. As mentioned, it is necessary to have adequate scientific data about the effectiveness of vaccinations and the risk of carriers have to be available to decide what control strategy is optimal. This chapter has shown that there is a lot of information available to help making those decisions.

References

Alexandersen S, Quan M, Murphy C, Knight J, Zhang Z. (2003) Studies of quantitative parameters of virus excretion and transmission in pigs and cattle experimentally infected with foot-and-mouth disease virus. *J Comp Pathol* 129: 268-82.

Bates TW, Carpenter TE, Thurmond MC. (2003a) Benefit-cost analysis of vaccination and pre-emptive slaughter as a means of eradicating foot-and-mouth disease. *Am J Vet Res* 64: 805-12.

Bates TW, Thurmond MC, Carpenter TE. (2003b) Results of epidemic simulation modeling to evaluate strategies to control an outbreak of foot-and-mouth disease. *Am J Vet Res* 64: 205-10.

Biront P, Leunen J, Vandeputte J. (1987) Inhibition of virus replication in the tonsils of pigs previously vaccinated with a Chinese strain vaccine and challenged oronasally with a virulent strain of classical swine fever virus. *Vet Microbiol* 14: 105-13.

Bouma A, De Smit AJ, De Jong MC, De Kluijver EP, Moormann RJ. (2000) Determination of the onset of the herd-immunity induced by the E2 sub-unit vaccine against classical swine fever virus. *Vaccine* 18: 1374-81.

Capua I, Marangon S. (2003) Vaccination policy applied for the control of avian influenza in Italy. *Dev Biol* (Basel) 114: 213-9.

Dahle J, Liess B. (1992) A review on classical swine fever infections in pigs: epizootiology, clinical disease and pathology. *Comp Immunol Microbiol Infect Dis.* 15: 203-11.

Depner KR, Bouma A, Koenen F, Klinkenberg D, Lange E, De Smit H, Vanderhallen H. (2001) Classical swine fever (CSF) marker vaccine Trial II. Challenge study in pregnant sows. *Vet Microbiol.* 83: 107-120.

Doel TR. (2003) FMD vaccines. *Virus Res* 91: 81-99.

De Smit AJ, Bouma A, van Gennip HG, de Kluijver EP, Moormann RJ. (2001) Chimeric (marker) C-strain viruses induce clinical protection against virulent classical swine fever virus (CSFV) and reduce transmission of CSFV between vaccinated pigs. *Vaccine* 19 (11-12): 1467-76.

Di Trani L, Cordioli P, Falcone E, Lombardi G, Moreno A, Sala G, Tollis M. (2003) Standardization of an inactivated H17N1 avian influenza vaccine and efficacy against A/Chicken/Italy/13474/99 high-pathogenicity virus infection. *Avian Dis* 47 (Suppl): 1042-6.

Dewulf J, Laevens H, Koenen F, Mintiens K, de Kruif A. (2001) An E2 sub-unit marker vaccine does not prevent horizontal or vertical transmission of classical swine fever virus. *Vaccine* 20: 86-91.

De Smit AJ, Bouma A, de Kluijver EP, Terpstra C, Moormann RJ. (2001) Duration of the protection of an E2 subunit marker vaccine against classical swine fever after a single vaccination. *Vet Microbiol* 78: 307-17.

Eble PL, Bouma A, De Bruin MGM, Van Hemert-Kluitenberg F, Van Oirschot

JT, Dekker A. (2004) Vaccination of pigs two weeks before infection significantly reduces transmission of foot-and-mouth disease virus. *Vaccine* 22:1372-78.

Floegel-Niesmann G. (2001) Classical swine fever (CSF) marker vaccine. Trial III. Evaluation of discriminatory ELISAs. *Vet Microbiol* 83: 121-36.

Fritzemeier J, Teuffert J, Greiser-Wilke I, Staubach C, Schluter H, Moennig V. (2000) Epidemiology of classical swine fever in Germany in the 1990s. *Vet Microbiol.* 15 77(1-2): 29-41.

Hedger RS, Condy JB. (1985) Transmission of foot-and-mouth disease from African buffalo virus carriers to bovines. *Vet Rec.* 117: 205.

Keeling MJ, Woolhouse ME, May RM, Davies G, Grenfell BT. (2003) Modelling vaccination strategies against foot-and-mouth disease. *Nature* 421: 136-42.

Klinkenberg D. (2003) *Mathematical Epidemiology and the Control of Classical Swine Fever.* ID-Lelystad, WUR, The Netherlands, ISBN 9039333133.

Klinkenberg D, Everts-van der Wind A, Graat EA, de Jong MC. (2003) Quantification of the effect of control strategies on classical swine fever epidemics. *Math Biosci* 186: 145-73.

Klinkenberg D, Moormann RJ, de Smit AJ, Bouma A, de Jong MC. (2002) Influence of maternal antibodies on efficacy of a subunit vaccine: transmission of classical swine fever virus between pigs vaccinated at 2 weeks of age. *Vaccine* 20: 3005-13.

Leslie J. (2000) Newcastle disease: outbreak losses and control policy costs. *Vet Rec.* 146: 603-6.

Mangen MJ, Jalvingh AW, Nielen M, Mourits MC, Klinkenberg D, Dijkhuizen AA. (2001) Spatial and stochastic simulation to compare two emergency-vaccination strategies with a marker vaccine in the 1997/1998 Dutch Classical Swine Fever epidemic. *Prev Vet Med* 48:177-200.

Mangen MJ, Nielen M, Burrell AM. (2002) Simulated effect of pig-population density on epidemic size and choice of control strategy for classical swine fever epidemics in The Netherlands. *Prev Vet Med.* 56: 141-63.

Mezencio JM, Babcock GD, Kramer E, Brown F. (1999) Evidence for the persistence of foot-and-mouth disease virus in pigs. *Vet J.* 157(3): 213-7.

Moonen P, Schrijver R. (2000) Carriers of foot-and-mouth disease virus: a review. *Vet Q.* 22: 193-7.

Moormann RJ, Bouma A, Kramps JA, Terpstra C, De Smit HJ. (2000) Development of a classical swine fever subunit marker vaccine and companion diagnostic test. *Vet Microbiol.* 73: 209-19.

Orsel K, Bouma A., Dekker A., Frankena K., Elbers ARW, Bruschke CJM, Noordhuizen JTPM., De Jong MCM. (2002) *Analysis of the Effectiveness of Vaccination Against FMD and the Quality of the Discriminatory Test.* Report ID-Lelystad, September .

Orsel K, Dekker A, Bouma A, De Jong MCM. (2003) Vaccination against

FMD reduces virus transmission in groups of vaccinated calves as compared to non-vaccinated calves.*Proc. ISVEE,* Chile.

Ring C, Mayr A, Kandler J, Weinberg E. (1988) Economic valuation (cost-benefit analysis) of protective vaccination against Newcastle disease in poultry. *Zentralbl Veterinarmed B.* 35: 214-7.

Stone HD. (1987) Efficacy of avian influenza oil-emulsion vaccines in chickens of various ages. *Avian Dis.* 31: 483-90.

Tacken G.M.L. *et al. Economical Consequences of Vaccination Against Avian Influenza.* Report LEI, 2003.

Terpstra C. (1987) Epizootiology of swine fever. *Vet Q.* 9 Suppl 1: 50S-60S.

Terpstra C, Woortmeyer R, Barteling SJ. (1990) Development and properties of a cell culture produced vaccine for hog cholera based on the Chinese strain. *Dtsch Tierarztl Wochenschr* 97: 77-9.

Tollis M, Di Trani L. (2002) Recent developments in avian influenza research: epidemiology and immunoprophylaxis. *Vet J.* 164: 202-15.

Uttenthal A, Le Potier MF, Romero L, De Mia GM, Floegel-Niesmann G. (2001) Classical swine fever (CSF) marker vaccine. Trial I. Challenge studies in weaner pigs. *Vet Microbiol* 83: 85-106.

Van Boven M, Boender GJ, Elbers ARW, Nodelijk G, De Jong MCM, Dekker A, Koch G, Stegeman JA. (2003) Epidemiological consequences of vaccination against Avian Influenza. Report ID-Lelystad.

Van Oirschot JT. (2001) Present and future of veterinary viral vaccinology: a review. *Vet Q.* 23(3): 100-8.

19

LIVESTOCK - GENOMICS AND PRODUCTIVITY

CHRIS S. HALEY AND ALAN L. ARCHIBALD
Roslin Institute, Roslin, Midlothian, EH25 9PS, UK

Introduction

Since the dawn of agriculture about 8-10,000 years ago humans have been shaping the genetic make-up of animals raised for food. Over the last 50 years the scientific principles of animal breeding have been applied consistently in some livestock populations and the results have been remarkably successful. Without any knowledge of the underlying genes that control trait variation, selection on genetic merit has allowed improvement, for example, in growth rates and carcass lean composition in pigs that are estimated to be 0.5-1.5% of the mean per annum (Sellier and Rothschild, 1991). In poultry, Havenstein *et al.* (1994) have compared a broiler line from 1960 with one from 1990 on diets typical of the periods. They demonstrated that weight at 6 weeks of age had quadrupled over the thirty-year period and they estimated that this was approximately 85% due to genetic progress with the remaining 15% due to improved nutrition. Thus traditional breeding methods and their enhancements utilising advanced statistical techniques (initially the selection index, latterly best linear unbiased prediction or BLUP) have been very effective at generating genetic improvement, particularly for production and carcass quality traits.

Despite the success of animal breeding, there are a number of reasons why additional technologies are likely to be used to complement the traditional methods in future. Firstly, economic projections predict an increase in incomes worldwide and recent experience suggests that this will lead to an increased demand for livestock products. For example, Delgado *et al.* (1998) predict that this demand will double over the period from 1993 to 2020. Secondly, there are also increasing pressures to make agriculture more sustainable. Past genetic progress has already made a substantial contribution, as improved efficiency of animal growth leads to lower inputs being needed and fewer waste products being generated for a given level of output. For example, in pigs it has been estimated that progress over the last 40 years has led to a reduction of 50% in the amount of manure produced for every kilogram of lean meat produced. However, other traits that contribute to sustainability of production, such as disease resistance, welfare and product quality, are becoming increasingly important. Whilst it is clear that variation in such traits has a genetic component and that traditional breeding methods can contribute to their improvement, they are less amenable to these approaches than production traits. Hence there has been a search for alternative approaches to genetic improvement for such traits. Finally, the genetics revolution sweeping through biology is creating new understanding and opportunities not only in medicine but also in agriculture. Genomics is providing detailed information on some of the genes contributing to trait variation and is starting to deliver the complete DNA sequence of farm animal genomes and access to the information it contains. Ultimately it may be possible to identify many of the more important genes that control traits of economic and scientific interest. In parallel methods to manipulate reproductive processes, including cloning of animals and to create new genetic variation in the genome of farm animals, are being developed.

In this chapter, progress in livestock genomics and its potential impact on livestock production is considered. The prospects for direct genetic modification of farm animals and on how genomics might guide the choice of targets are also examined.

Genomics

The ultimate aim of farm animal genome research is to understand the genetic control of agriculturally, economically or biologically important traits. The direct application of such studies in genetic improvement is the use of marker-assisted or gene-directed selection. Such selection utilises naturally occurring genetic variants either to enhance the efficiency of what is currently possible or to allow selection for traits that are currently difficult to improve by traditional selection. There are also a number of secondary, but nonetheless potentially very important, applications of the results of genomics. For genetic improvement, this might include the identification of genes that might be targets of genetic modification or gene transfer. In addition, however, there are many other impacts on animal production that are not realised through genetic improvement, such as diagnostics and traceability and the identification of targets for drugs, as well as the potential myriad advantages that may flow from the improved understanding of the biology of farm livestock.

The genomes of the principle farm livestock (i.e. cattle, sheep, pigs, poultry) are similar in complexity to that of humans, comprising a sequence of between one billion (for chickens) and three billion (for the mammals) base pairs of DNA containing some 30,000 or so genes. The form and performance of an animal are determined by this blueprint defined in the genes, that it has inherited from its parents, plus environmental factors. The ultimate description of the structure of a genome is its complete DNA sequence. The sequence of the human genome was declared to be complete in April 2003. A draft sequence of the chicken has been published and a similar draft sequence of the cattle genome will be completed within a year at a cost of $40-$60 million per species, mainly from US sources. There are international efforts to raise the funding to sequence the pig genome.

However, even the full sequence of a genome does not reveal the function of the genes it contains. Many sorts of information can help unravel gene functions such as comparative studies (i.e. the function of similar genes in other species), *in vitro* and *in vivo* experimental studies (e.g. gene knockouts) and even computer-modelling-based predictions. Studies will continue for many years to assign functions to gene sequences and to build a picture of the complex net of interactions that lie between the sequence and the functioning of the whole organism. Nonetheless, even when this information is available it will not necessarily identify which of the genes contribute to variation within a population. Just because a gene plays a crucial role in the physiology of a particular trait, it does not mean that variation in that gene contributes to variation in the trait. This may be because mutations in the gene generally have too serious a consequence to be tolerated and are purged by selection, or perhaps because the advantageous mutations that have arisen have been fixed by selection, so none are segregating at present. Thus much of the current effort in livestock genomics is aimed at identifying the loci that do contribute to variation within and between breeds. This in itself will contribute to understanding of gene functions and interactions, but it has the more immediate consequence of allowing the application of marker assisted and gene-directed selection to utilise this natural genetic variation.

Functional genomics

Most agricultural traits (growth, fatness, meat quality, disease resistance and reproductive performance) are quantitative in nature with variation comprising a continuum between high and low-performing animals. Thus, the regions of the genome that control variation in such traits are termed quantitative trait loci (or QTL). There are two approaches to identify trait genes, either map-based approaches which look for the relationships between markers on the genetic map and phenotypic variation or physiological-based approaches.

Map-based approaches involve determining with ever greater precision the location of the trait gene. A scan of the genome for QTL is implemented by genotyping genetic markers which are spaced at intervals through the genome and which are polymorphic (variable) in the population being studied. The phenomenon of genetic linkage means that each marker can be used to follow the inheritance of a section of linked chromosome within a family or pedigree. The presence of a QTL in a particular region of the genome can be inferred if variation in a trait of interest cosegregates with genetic markers in that region. Using this approach, QTL loci with moderate effects on traits such as growth, carcass composition, milk production, litter size and host responses to infectious disease have been found, not only in experimental populations (e.g. see review by Bidanel and Rothschild, 2002) but also in commercial populations (Georges *et al.*, 1995). QTL can be mapped in this way to a region comprising perhaps 1% of the whole genome. Whilst this may seem a reasonable resolution, the positions of such loci are defined poorly in molecular terms, with the region containing perhaps 20-40 million base pairs of DNA and several hundred genes. Thus searching the physical map, even the completed DNA sequence, of the region of interest for the causative mutation is a daunting prospect.

The accuracy with which QTL are mapped can be improved by increasing the number of recombinations assessed, either by increasing the population size or analysing segregation over more generations. If a genetic marker and a (trait) gene are very close then there is a tendency for associations between specific marker and trait gene alleles to be maintained at a population level and not just within individual families. This phenomenon of linkage disequilbrium could be exploited to locate trait genes. With large numbers of markers it is possible to look for trait genes by testing for linkage disequilibrium between the markers and the trait genes in an association study. The most abundant class of genetic variation and hence potential markers are single nucleotide polymorphisms (SNPs), i.e. where two DNA sequences differ by a single base (letter). A resource of thousands or tens of thousands of SNPs is required for linkage disequilibrium studies – the necessary SNPs are being identified in chickens, cattle and pigs.

QTL can be defined as the marker interval that co-segregates with variation in the trait of interest. These flanking markers can be used in marker-assisted selection to introduce or retain the beneficial QTL allele. However, markers have to be very closely linked to the causative mutation in the trait gene if specific marker alleles are to remain associated with specific QTL alleles through several generations of selection and therefore be useful in practical breeding programmes. The ideal marker is the trait gene itself. Despite the inherent difficulties, there have been a few recent successes in which single base pair changes that are causative of major QTL effects have been identified (see below). The availability of new resources, such as the complete sequence of some species and high-density SNP maps, should make such successes more frequent.

Physiological-based approaches involve identifying candidate genes on physiological arguments rather than genetic arguments. The hypothesis is that variation in the physiological candidate gene contributes to variation in the trait of interest. Testing the hypothesis involves identifying variation in the candidate gene and looking for associations between this variation and differences in the trait.

Variation can be detected in the structure (DNA sequence) or expression of the candidate gene. Expression can be studied at the RNA or protein level. As it was often necessary to develop a specific assay for each gene of interest, until recently it was only practical to study a few candidate genes at a time. Recent technical developments have created the opportunities to study the expression of thousands of genes simultaneously. For example, transcript profiling using microarrays or 2-dimensional gel electrophoresis can be used to study expression at the RNA or protein level respectively. These technologies allow physiological approaches to be extended from considerations of one physiological candidate gene at a time to studies of all genes expressed in candidate tissues or cell types and thus approach the all-encompassing scope of genome scans.

GENOMICS - STATE OF THE ART

Many of the tools needed to dissect complex traits in livestock are now in place. The genome sequence of poultry and cattle is available, or will be soon, and that of pigs will hopefully follow shortly. Markers and maps are available and are being used, and other tools of functional genomics such as microarrays for expression analysis are being developed. The basic bioinformatic tools and statistical approaches are also available. The application of mapping tools over the last 10 years has seen a plethora of QTL and major genes mapped (see Bidanel and Rothschild (2002) for a recent summary for pigs). These QTL are found both in crosses between genetically divergent lines and within commercial populations. The latter result in particular underlines the potential for marker-assisted selection to be used to enhance genetic progress.

For some of the mapped major genes and QTL the underlying genetic lesion has been identified and some examples give an indication of the range of these results. Major genes associated with significant reductions in quality have been identified. For example, the halothane sensitivity allele that leads to a high risk of porcine stress syndrome (PSS) as well as pale, soft, exudative (PSE) meat was the first allele to be identified and then selected in breeding programmes using molecular genetic tools (Fujii *et al.*, 1991). More recently, the *RN* allele, which reduces yields of cured ham, has been identified and molecular tests developed for animals carrying the undesirable alleles (Milan *et al.*, 2000).

A number of genes with potential desirable effects on traits of importance to the meat industry have now been fully characterised and the underlying molecular basis for the effects determined. Several mutations in the gene encoding Growth Differentiation Factor 8 (GDF8) have been shown to be associated with double muscling, particularly in Belgium Blue cattle (e.g. Kambadur *et al.*, 1997). A single base pair change in the regulatory sequence controlling the post-natal expression of insulin-like growth factor 2 (*IGF2*) in pigs has been shown to cause a significant increase in lean muscle content (van Laere *et al.*, 2003). A single base pair change associated with the *callipyge* phenotype in sheep has been identified, but its mode of action is not yet resolved (Freking *et al.*, 2002).

Health is a central issue in livestock production with estimated costs due to disease of 17% of turnover in the developed world (£1.7 billion in the UK) and 35-50% in the developing world. Although there have been only a few attempts to use genomics to address the genetics of hosts responses to infectious diseases in farm animals, there have already been some notable successes. The gene encoding the receptor that allows *E. coli F18,* that cause oedema disease, to adhere to intestine has been identified (Meijerink *et al.*, 2000) and the gene which determines resistance / susceptibility to *E. coli F4* induced diarrhoea has been mapped to chromosome 13 (Edfors-Lilja *et al.*, 1995).

Where major genes influencing the trait of interest can be identified and a DNA test developed to distinguish between desirable and undesirable alleles or genotypes then selective animal breeding informed by knowledge of the DNA markers (marker assisted selection) offers an effective way to make more rapid progress. DNA markers for the *halothane*, *RN* and *E. coli F18* resistance genotypes are successful examples of this approach.

Gene transfer - technologies for direct modification of animal genomes

Selection is constrained by the genetic variation for the trait of interest and by species boundaries. The traditional means of acquiring new and desirable genetic variants is by cross-breeding, perhaps to exotic stock lines. However, this approach can only utilise variation that exists within a species and cross-breeding brings with it not only the desirable allele, but also the potentially undesirable aspects of the rest of the genome. For example, the genes for prolificacy presumed to be present in Chinese pig breeds, such as the Meishan, have been introduced to European pig stocks by cross-breeding. However, the resulting cross-bred individuals also inherit a proportion of the undesirable features of their exotic parents - such as the high fat content of Meishan meat. Transgenesis (or gene transfer), which brings together recombinant DNA and advances in reproductive technologies, offers the possibility of introducing single new genes or of modifying genes that already present without any unwanted genetic baggage.

Gene transfer can be achieved by direct injection of DNA, using sperm as a vector, by use of disabled viruses or by developing animals from cells modified during culture *in vitro*.

DNA (MICRO)INJECTION

The introduction of additional genetic material into fertilised mouse oocytes by microinjection and subsequent transmission of the modified genotype to subsequent generations is exemplified by the "supermouse" experiment (Palmiter *et al.*, 1982). In this, transgenic mice carrying additional copies of a growth hormone gene were shown to grow much larger than non-transgenic littermates, and this inspired much of the early work on transgenic livestock (Hammer *et al.*, 1985). The approach has been used to produce transgenic mice, rabbits, pigs, sheep, cattle and chickens although it suffers from a number of significant problems. First, less than 5% (often 1% for non-rodents) of the injected eggs develop into transgenic individuals. Second, the chromosomal site at which the transgene integrates and the number of tandem copies of the gene that are integrated cannot be pre-determined. As a result, the expression of the transgene often differs between independent lines, the transgene can cause insertional mutagenesis and only dominant phenotypes can be introduced. The technique is limited to adding new genetic material and cannot alter in a targeted way existing genetic material.

SPERM MEDIATED GENE TRANSFER

Lavitrano *et al.* (1989) reported that transgenic mice could be produced efficiently by *in vitro* fertilisation with sperm that had simply been incubated with DNA. Although sperm mediated gene transfer, as described by Lavitrano et al. (1989), could make gene transfer commonplace with its simplicity, the technology does not appear to have been transferred successfully to other laboratories.

VIRAL VECTORS

Gene transfer is not a domain populated solely by technically advanced humans; one class of viruses, retroviruses, have the capability to copy their genome into the chromosome(s) of the host that they infect. The retroviral genome can subsequently be excised from the host chromosome repackaged into an infectious particle and move on to infect another cell/host. As a result of errors in the integration and excision events DNA picked up from one host can be transferred to a subsequent host. Thus, modified and partially disabled retroviruses could form the basis of a gene transfer system.

Retroviral vectors have been created, which retain the ability to integrate copies of their genomes into host chromosomes, but which have lost the ability subsequently to produce further infective virus. Bosselman *et al.* (1989) described the integration of foreign DNA into the germline of chickens following injection of a replication-defective retroviral vector. Retroviral vectors have also been used for gene transfer into cattle, sheep and pig embryos (Kim *et al.*, 1993; Hettle *et al.*, 1989; Petters *et al.*, 1988). The disadvantage of using multicellular embryos is that the transgenic animals produced will be mosaics and the germline may be chimaeric. Transmission of the transgenes, therefore, may vary from the inefficient to the non-existent. Other disadvantages of the use of retroviral vectors are their limited capacity for foreign DNA sequences and the adverse effects of some retroviral sequences on expression of the transgene. Public acceptability may also prove a significant barrier to the use of retroviral vectors for gene transfer in livestock.

A recent development in this area that has potential for gene transfer in farm animals concerns lentiviruses. Lentiviruses are a class of non-oncogenic retroviruses that are being used to develop replication defective viral vectors for gene therapy (Thomas *et al.*, 2003). Lentiviral vectors have been demonstrated to be very efficient for gene transfer in mice (Lois *et al.*, 2002) and very recently have been shown to be similarly effective in pigs (Whitelaw, 2004; Whitelaw *et al.*, 2004). The early studies suggest that these approaches may be 50 times more efficient at generating individuals expressing a transgene than direct pronuclear injection of DNA (Table 19.1). Despite the potential drawbacks (e.g. addition of DNA only, limited size of DNA added, concern over use of viral vectors) this approach looks very promising at least as a research tool.

Table 19.1 THE RELATIVE EFFICIENCY OF LENTIVIRAL TRANSGENESIS.

	Lentiviral[1]	*Pronuclear injection*[2]	*Increase in efficiency*
Eggs injected	120	400	-
Piglets born	40	40	x 3.3
Transgenic piglets	37	4	x 30.8
Expressing transgene	35	2	x 58.3

[1]Data from Whitelaw *et al.*, 2004.
[2]Literature average, C. B. A. Whitelaw, personal communication.

CELLULAR VECTORS – EMBRYO STEM CELLS, NUCLEAR TRANSFER, GENE TARGETING

The problems with gene transfer by microinjection are two-fold: the uncertain nature of the genetic change effected and the low frequency of success. Thus, systems that allow the nature of rare successful events to be evaluated *in vitro*, for example in cultured cells, would be attractive. Methods for introducing additional genetic material into animal cells grown in culture had been established several years prior to the generation of transgenic mice. Whilst generating a genetically modified animal cell in culture may be relatively straightforward, creating an entire animal from the resulting cell(s) is not. Although liver, kidney or white blood cells grown in culture may be genetically identical (same DNA sequence) as the fertilised zygote that developed into the animal from which the cells were isolated, they have become specialised or terminally differentiated. As a result of epigenetic changes that have occurred during the development, such differentiated cells have lost their potential to develop into other cell types. Cells that have the potential to develop into every cell type found in an adult are termed totipotent.

In order to pursue a cellular based approach to gene transfer it is necessary either to carry out the genetic modification in cells that are totipotent or to reprogramme the nucleus of the modified cell in order to restore its totipotency. These options have been successfully effected using embryo stem cells and nuclear transfer respectively. Whilst embryo stem (ES) cell approaches were developed in mice, which are widely recognised as the experimental model mammal, the breakthrough in nuclear transfer came from an unexpected source – the pioneering work on sheep by Ian Wilmut and colleagues at Roslin Institute.

The key breakthrough in developing a cellular route for gene transfer in farm animals was the cloning of sheep by nuclear transfer where the donor nuclei were from cells cultured *in vitro* (Campbell *et al.*, 1996; Wilmut *et al.*, 1997). Although the production of 'Morag' and 'Megan' established the principle, it was 'Dolly' the sheep who was the product of nuclear transfer from a cultured somatic cell (from an adult mammary gland) into an enucleated oocyte (Wilmut *et al.*, 1997) that captured the headlines. Nuclear transfer cloning technology has also been extended to cattle, pigs and horses (Polejaeva *et al.*, 2000; De Sousa *et al.*, 2002; Galli *et al.*, 2003). The donor nuclei in these experiments were not genetically modified prior to transfer.

The next step in the development of a cell-based method for gene transfer in farm animals involved the creation of a sheep called 'Polly' by nuclear transfer where the donor nucleus was from cell that had been genetically modified by the addition of a human factor IX gene (Schnieke *et al.*, 1997). Finally, the production of sheep with targeted genetic modifications, including deletions of specific genes, has been achieved using nuclear transfer (e.g McCreath *et al.*, 2000). More recently, cloned pigs have been produced by nuclear transfer using nuclei from cells, in which targeted genetic changes have been made (e.g. Dai *et al.*, 2002).

Cellular routes to gene transfer in farmed mammals have addressed the requirement for precision: directed genetic changes can be confirmed *in vitro* prior to creating a genetically modified animal. However, the nuclear transfer elements of the method are profoundly inefficient.

TYPES OF GENETIC MODIFICATION

All the gene transfer methods described above can be used to introduce additional genetic material into the genomes and germlines of animals. There are limits on the size of the additional DNA fragments that can be introduced using viral vectors. Fragments of 100,000 to 500,000 bp or more, however, can be introduced using direct microinjection. If direct modification of the genomes of farm animals is contemplated, then adding new or extra copies of genes represents only one class of change that might be considered beneficial. A range of modifications to endogenous genes might be desirable including – knockouts, changing in coding or regulatory sequences, increasing or decreasing the level of expression. Targeted changes to an endogenous gene including changes in its sequences rely upon very rare homologous recombination events and thus can only realistically be achieved by the cellular route. Gene addition, which is deliverable by all the systems described, however, can be used to effect reductions in the expression of endogenous genes, with much recent interest focusing on RNA interference as a way of achieving this.

RNA interference is a phenomenon in which doublestranded RNA (dsRNA) silences genes that share sequence homology with the dsRNA. Short interfering RNAs (siRNA) are key intermediate in gene silencing by RNA interference. It has been demonstrated recently that siRNAs can be used to silence genes in animals cells (McManus and Sharp, 2002). Gene silencing has been demonstrated in transgenic mice carrying transgenes that are transcribed to yield siRNAs (Tiscornia *et al.*, 2003) or short-hairpin RNAs (shRNAs) (Rubinson *et al.*, 2003).

Thus, the technology exists to modify directly the genomes of farm animals. However, the procedures are inefficient and costly, or lack precision, or both. The technologies also raise both welfare and ethical concerns. Thus, the traits to which this technology might be applied need to be chosen with care.

CHOICE OF TRAITS AND TARGET GENES; DOES KNOWLEDGE OF THE GENOME HELP?

Mapping studies have demonstrated that major genes and QTL of moderate effect can be detected, mapped and in some cases the underlying DNA alteration has been identified. However, genome research in the target species is only one of several ways in which potential target genes could be identified. For example, the recent discovery of *DGAT1* as the gene underlying a QTL for milk fat in cattle (Grisart *et al.* 2002) and the earlier research which identified *GDF8* as the gene responsible for double muscling (Grobet et al., 1997; Kambadur et al., 1997; McPherron and Lee, 1997) were greatly accelerated by reports of the phenotypes of transgenic mice in which the *Gdf8* and *Dgat1* genes respectively had been knocked out (Smith *et al.* 2000; McPherron *et al.*, 1997). Thus, careful analysis of the extensive literature on transgenic mice, including notes of phenotypes that are incidental to the focus of the experiments reported, may point to genes that could be exploited by gene transfer in farm animals.

Gene transfer technology need not be considered to be competing with marker assisted selection on variation already present within the breeding population. Rather gene transfer could be used to pursue options not available through selective breeding. Gene transfer technology is not constrained by species boundaries, nor by the need to use transgenes that occur in nature. For example, skatole which is a primary cause of boar taint is produced by bacteria present in the hind gut rather than by the pig itself. Genetically modifying the pig genome to introduce a pathway to breakdown skatole in the gut might be an effective approach rather than trying to identify the pig genotype best able to clear skatole post-absorption. Similarly, creating transgenic animals that express siRNA molecules that would interfere with and effectively control specific viral pathogens might be an effective means of increasing disease resistance.

Prospects for genetic improvement

The primary impact of genome studies will be to enhance the efficiency and effectiveness of genetic improvement programmes. Amongst other things, genomic tools will allow improved traceability, quality assurance, pedigree control and maintenance of genetic diversity, but the most important aspect is likely to be the use of marker assisted and gene-assisted (or gene-directed) selection. This allows selection on the phenotype to be augmented by selection directly for naturally occurring genetic variation as determined from the DNA sequence. The advantages of this include the ability to collect and test DNA at any stage or from either sex. So animals can be tested before they have displayed a trait (e.g. meat quality) or for a trait they may never express (e.g. disease resistance) and males can be tested to determine their genotype for genes normally only expressed in females.

The genome selection tools all aim to choose animals on the basis of particular functional variants that they carry. However, they can be characterised in one of three broad categories:

Selection on markers showing linkage within families. This would be the basic material arising from a QTL mapping study, that is information on markers relatively loosely linked to a QTL such that any associations between markers and the desirable QTL variant only hold up within individual families. If a new family is included in the study, the presence and exact form of any association would need to be verified within that family. Marker assisted selection using such information can be more effective than that based on phenotypic selection only, however it is the least effective of the three types and requires the greatest level of genotyping (and hence cost) once it is up and running. In addition, it is difficult to protect by patent information on linked markers. Although most information coming out of QTL mapping programmes is of this sort, the problems mean that relatively little use of this type of selection has been made in breeding programmes.

Selection on markers showing population wide associations. Where genetic markers are identified that are very close to the site of a causative mutation, then there may be strong linkage disequilibrium between the two genes. This means that a particular marker allele (or perhaps a haplotype comprising alleles at several markers) may be associated with the desired functional allele at the level of the population, or even possibly across several or even all populations. Thus selection on the marker allele or haplotype may be a surrogate for selection directly for the desired functional allele. This may be as efficient as selection directly on the functional allele itself. The same association cannot be guaranteed to exist in all populations, so if a new population is studied, the presence of the association would need to be verified in that population.

Candidate gene studies may detect associations of this type. This is because if the candidate gene does indeed contain a polymorphism with a direct effect on a trait, any other polymorphism within the same gene will be closely linked and hence have a good chance of being in linkage disequilibrium. The estragen receptor (ER) polymorphism associated with litter size in pigs may represent a marker closely linked to the causative polymorphism rather than actually being the causative polymorphism. Indeed the ER polymorphism does appear to show effects that are consistent within populations but which differ between populations (Rothschild *et al.*, 1996). More recently, the availability of large number of SNP markers has made it possible to contemplate scanning the genome for population wide associations with anonymous markers (i.e. not markers that are associated with particular candidate genes). Such a scan would require thousands or perhaps tens of thousands of markers for there to be a good chance that any causative polymorphism is close enough to a marker or marker haplotype to be in strong linkage disequilibrium with it.

Selection directly on the causative polymorphism. This is the most straightforward and most effective form of selection; however, it carries the substantial caveat that the causative polymorphism has been identified. Such selection is being performed for a number of loci in pigs (e.g. *RYR/PSS, RN, IGF2*), cattle (*DGAT*) and sheep (*calipyge, PRP* variants for scrapie resistance). Such polymorphisms can be easily transferred between populations (i.e. they do not need to be verified within each population), are generally simple tests and can be protected by patent, making them very attractive for commercialisation.

All three categories of genome-based selection can be shown to give additional genetic progress compared to schemes that ignore such information and all three have been applied in livestock breeding (Dekkers, 2004). The additional progress achievable depends on a number of factors, with greater additional progress for traits that are difficult to record or not normally recorded at all and for traits recorded only in a single sex or late in life (Meuwissen and Goddard, 1996). Dekkers (2004) reviews much of the information on genome-based selection in livestock, both in the theoretical background and the practical applications and is cautiously optimistic for the future.

Implementing genetic modification. The basic technology is now available to undertake genetic modification in farm animals, in the form of additions of new genes and knock-outs or potentially modifications of existing genes. The technologies are largely experimental at present, limited to one gene at a time and most are costly and inefficient, so there is a long way to go before they could be routinely applied. For animal production, particularly for the European market, there will be strong ethical and public perception issues to be addressed before they could be used.

There are some stumbling blocks for the application of genetic modification in farm animals. Firstly, it is not clear that science has yet sufficient understanding of farm animal physiology to identify genes that are clear targets for modification, design the appropriate modification and have a good idea of the consequences of the modification. For example, although a specific 11 bp deletion in the *GDF8* gene has been shown to be the cause of the double muscling phenotype observed in Belgian Blue cattle (McPherron and Lee, 1997), yet the identical mutation is found in other breeds of cattle, including the South Devon, Dairy Shorthorns and Highland cattle, without the associated double muscling (Wiener *et al.*, 2002; John Williams, personal communication). It may be concluded from these observations that 11 bp deletion in *GDF8* is not the sole determinant of the resulting phenotype; the genetic background also has an impact.

However, if appropriate target genes can be identified and the consequences of manipulating the gene determined, there remains the issue of how best to integrate a gene into an existing breeding programme without jeopardising genetic progress generated by other means. The latter issue arises because of the time taken to produce and test transgenic animals. Thus even if animals, which are subject to genetic modification, are taken from the elite breeding population, in the several years that are likely to elapse whilst transgenic animals are produced, bred to homozygosity and tested (not only for the target trait but for others that might possibly have been affected), a lag of several years will have been introduced. This lag represents the normal genetic progress in the elite population over the period of testing. The lag can subsequently be reduced by backcrossing transgenic animals into the elite nucleus, but it is unlikely to be possible to reduce it much below the equivalent of two years of genetic progress (Gama *et al.*, 1992). As two years of genetic progress can amount to a change in several percent of the mean in a well-run breeding scheme, the value of an introduced transgene has to be at least this amount to make its use worthwhile.

The considerations of genetic lag, together with the initial cost of generating transgenic animals and potential associated impact on other factors, such as public perception means that only changes that provide substantial benefit are likely to be worthwhile. These could be changes that provide a quantum leap in production or quality parameters, or perhaps an alteration that opens up a completely new opportunity (for example, allowing exploitation of a new environment). In the former case it is possible to speculate about changes such as ablation of the porcine equivalent of the bovine double-muscling gene (myostatin or *GDF8*, McPherron and Lee, 1997). However, unforeseen consequences of other attempts at genetic modification, such as pigs carrying additional copies of growth hormone (Pursel *et al.*, 1989) should be borne in mind. Genetic modification to introduce disease resistance can be argued to benefit all of the animal, the producer and the consumer, and in some cases may open up an area where particularly virulent disease otherwise makes raising livestock impossible. However desirable such options might seem, we are still a long way from knowing what genes to change and what the consequences might be of making any changes.

Conclusions

It is considered that it is likely to be some time before genetic modification technologies see widespread use in farm animal / meat production. The obstacles still remaining include making the basic technologies work efficiently, cost, eliminating animal welfare concerns, identification of target genes for genetic modification and public perception. The latter may be a big hurdle to climb, especially in Europe. Genetic modification to eliminate livestock susceptibility to zoonotic disease (hence improving animal health and welfare and reducing risks to human health) might be acceptable to much of the public. However identifying appropriate target genes and desired changes to them and carrying out the desired changes with minimal deleterious effect remains beyond current capabilities.

The chapter has focused of genomics and gene transfer in terrestrial livestock, principally cattle, sheep, pigs and poultry. Whilst looking to the future of livestock production the potential for aquaculture should be mentioned. There are currently over 200 species being farmed and production from aquaculture has grown at around 10% per annum in recent years compared to around 3% per annum growth for traditional terrestrial livestock. Many of the species used in aquaculture are less than 10 generations from their wild ancestors, but rapid genetic progress from conventional selection is often possible. Furthermore, genomic and transgenic technologies are being rapidly developed in some of these species, with the possibility of shortcutting the 10000 years of domestication experienced by cattle, sheep and pigs.

In conclusion, future genetic progress will be driven by an increasing demand for animal protein from an ever more affluent world population and pressure to make animal production even more sustainable. New approaches for genetic modification have recently been developed and will certainly see experimental application. Whether these technologies see use in production will depend upon the identification of worthwhile gene targets for modification, as well as developing consumer attitudes. Genomic technologies for enhancing genetic progress through marker-assisted and gene-directed selection are already in use. These approaches only utilise naturally occurring variation and the prospects for their more widespread use in terrestrial and aquatic livestock are good. Thus overall the new genomic technologies are set to make a major contribution to future productivity of both terrestrial and aquatic livestock. The contributions of the new technologies are likely to be integrated with more traditional breeding tools in genetic improvement programmes that include focus on traits such as quality and disease resistance and not solely on productivity.

Acknowledgements

We are grateful to BBSRC, DEFRA and the EC for financial support.

References

Bidanel, J. P. and Rothschild, M. F. (2002), Current status of quantitative trait locus mapping in pigs. *Pig News and Information* **23**:39-54.

Bosselman, R. A., Hsu, R. Y., Boggs, T., Hu, S., Bruszewski, J., Ou, S., Kozar, L., Martin, F., Green, C., Jacobsen, F., Nicolson, M., Schultz, J. A., Semon, K. M., Rishell, W., and Stewart, R. G. (1989). Germline Transmission of Exogenous Genes in the Chicken. *Science* **243:** 533-535.

Campbell, K. H. S., McWhir, J., Ritchie, W. A., and Wilmut, I. (1996). Sheep cloned by nuclear transfer from a cultured cell line. *Nature* **380:** 64-66.

Dai, Y. F., Vaught, T. D., Boone, J., Chen, S. H., Phelps, C. J., Ball, S., Monahan, J. A., Jobst, P. M., McCreath, K. J., Lamborn, A. E., Cowell-Lucero, J. L., Wells, K. D., Colman, A., Polejaeva, I. A., and Ayares, D. L. (2002). Targeted disruption of the alpha 1,3-galactosyltransferase gene in cloned pigs. *Nature Biotechnology* **20:** 251-255.

De Sousa, P. A., Dobrinsky, J. R., Zhu, J., Archibald, A. L., Ainslie, A., Bosma, W., Bowering, J., Bracken, J., Ferrier, P. M., Fletcher, J., Gasparrini, B., Harkness, L., Johnston, P., Ritchie, M., Ritchie, W. A., Travers, A., Albertini, D., Dinnyes, A., King, T. J., and Wilmut, I. (2002). Somatic cell nuclear transfer in the pig: Control of pronuclear formation and integration with improved methods for activation and maintenance of pregnancy. *Biology of Reproduction* **66:** 642-650.

Dekkers, J. (2004) Commercial application of marker- and gene-assisted selection in livestock: Strategies and lessons. *J. Anim. Sci.* **82**(E. Suppl.): E313-E328.

Delgado, C., Rosegrant, M., Steinfeld, H., Ehui, S. and Courbois, C. (1999) *Livestock to 2020: The Next Food Revolution.* International Food Policy Research Institute, Washington.

Edfors-Lilja, I., Gustafsson, U., Duvaliflah, Y., Ellergren, H., Johansson, M., Juneja, R. K., Marklund, L., and Andersson, L. (1995). The Porcine Intestinal Receptor for Escherichia-Coli K88-Alpha- B, K88-Alpha-C - Regional Localization on Chromosome-13 and Influence of Igg Response to the K88 Antigen. *Animal Genetics* **26:** 237-242.

Freking, B. A., Murphy, S. K., Wylie, A. A., Rhodes, S. J., Keele, J. W., Leymaster, K. A., Jirtle, R. L., and Smith, T. P. L. (2002). Identification of the single base change causing the callipyge muscle hypertrophy phenotype, the only known example of polar overdominance in mammals. *Genome Research* **12:** 1496-1506.

Galli C, Lagutina I, Crotti G, Colleoni S, Turini P, Ponderato N, Duchi R, and Lazzari G (2003). Pregnancy: a cloned horse born to its dam twin. *Nature* **424:** 635.

Gama, L. T., Smith, C., and Gibson, J. P. (1992). Transgene Effects, Introgression Strategies and Testing Schemes in Pigs. *Animal Production* **54:** 427-440.

Georges, M., Nielsen, D., Mackinnon, M., Mishra, A., Okimoto, R., Pasquino, A.T., Sargeant, L.S., Sorensen, A., Steele, M.R., Zao, X., Womack, J.E. and Hoeschle, I. 1995. Mapping quantitative trait loci controlling milk production in dairy cattle by exploiting progeny testing. *Genetics* **139**: 907.

Grisart, B., Coppieters, W., Farnir, F., Karim, L., Ford, C., Berzi, P., Cambisano, N., Mni, M., Reid, S., Simon, P., Spelman, R., Georges, M., and Snell, R. (2002). Positional candidate cloning of a QTL in dairy cattle: Identification of a missense mutation in the bovine DGAT1 gene with major effect on milk yield and composition. *Genome Research* **12:** 222-231.

Grobet, L., Martin, L. J. R., Poncelet, D., Pirottin, D., Brouwers, B., Riquet, J., Schoeberlein, A., Dunner, S., Menissier, F., Massabanda, J., Fries, R., Hanset, R., and Georges, M. (1997). A deletion in the bovine myostatin gene causes the double- muscled phenotype in cattle. *Nature Genetics* **17:** 71-74.

Hammer, R. E., Pursel, V. G., Rexroad, C. E., Wall, R. J., Bolt, D. J., Ebert, K. M., Palmiter, R. D., and Brinster, R. L. (1985). Production of Transgenic Rabbits, Sheep and Pigs by Microinjection. *Nature* **315:** 680-683.

Havenstein, G.B., Ferket, P.R., Scheideler, S.E. and Rives, D.V. 1994. Carcass composition and yield of 1991 vs 1957 broilers when fed typical 1957 and 1991 broiler diets. *Poultry Science* **73**: 1795.

Hettle, S. J. H., Harvey, M. J. A., Cameron, E. R., Johnston, C. S. and Onions, D. E. (1989). Generation of transgenic sheep by sub-zonal injection of feline leukaemia virus. *Journal of Cell Biochemistry Suppl.* 13B, 180.

Kambadur, R., Sharma, M., Smith, T. P. L., and Bass, J. J. (1997). Mutations in myostatin (GDF8) in double-muscled Belgian blue and Piedmontese cattle. *Genome Research* **7:** 910-916.

Kim, T. A., Leibfriedrutledge, M. L., and First, N. L. (1993). Gene-Transfer in Bovine Blastocysts Using Replication-Defective Retroviral Vectors Packaged with Gibbon Ape Leukemia-Virus Envelopes. *Molecular Reproduction and Development* **35:** 105-113.

Lavitrano, M., Camaioni, A., Fazio, V. M., Dolci, S., Farace, M. G., and Spadafora, C. (1989). Sperm Cells As Vectors for Introducing Foreign Dna Into Eggs - Genetic-Transformation of Mice. *Cell* **57:** 717-723.

Lois, C., Hong, E. J., Pease, S., Brown, E. J., and Baltimore, D. (2002). Germline transmission and tissue-specific expression of transgenes delivered by lentiviral vectors. *Science* **295:** 868-872.

McCreath, K. J., Howcroft, J., Campbell, K. H. S., Colman, A., Schnieke, A. E., and Kind, A. J. (2000). Production of gene-targeted sheep by nuclear transfer from cultured somatic cells. *Nature* **405:** 1066-1069.

McManus, M. T. and Sharp, P. A. (2002). Gene silencing in mammals by small interfering RNAs. *Nature Reviews Genetics* **3:** 737-747.

McPherron, A. C. and Lee, S. J. (1997). Double muscling in cattle due to mutations in the myostatin gene. *Proceedings of the National Academy of Sciences of the United States of America* **94:** 12457-12461.

McPherron, A. C., Lawler, A. M., and Lee, S. J. (1997). Regulation of skeletal muscle mass in mice by a new TGF-beta superfamily member. *Nature* **387:** 83-90.

Meijerink, E., Neuenschwander, S., Fries, R., Dinter, A., Bertschinger, H. U., Stranzinger, G., and Vogeli, P. (2000). A DNA polymorphism influencing alpha(1,2)fucosyltransferase activity of the pig FUT1 enzyme determines susceptibility of small intestinal epithelium to Escherichia coli F18 adhesion. *Immunogenetics* **52:** 129-136.

Meuwissen, T. H. E. and Goddard, M. E. (1996) The use of marker haplotypes in animal breeding schemes. *Genet. Select. Evol.* **28**: 161-176.

Palmiter, R. D., Brinster, R. L., Hammer, R. E., Trumbauer, M. E., Rosenfeld, M. G., Birnberg, N. C., and Evans, R. M. (1982). Dramatic Growth of Mice That Develop from Eggs Micro-Injected with Metallothioneine-Growth Hormone Fusion Genes. *Nature* **300:** 611-615.

Petters, R. M., Johnson, B. H. and Shuman, R. M. (1988) Gene transfer to swine embryos using an avian retrovirus, *Genome* **30** (Suppl. 1), 448. (abstract 35.23.5).

Polejaeva, I. A., Chen, S. H., Vaught, T. D., Page, R. L., Mullins, J., Ball, S., Dai, Y. F., Boone, J., Walker, S., Ayares, D. L., Colman, A., and Campbell, K. H. S. (2000). Cloned pigs produced by nuclear transfer from adult somatic cells. *Nature* **407:** 86-90.

Pursel, V. G., Pinkert, C. A., Miller, K. F., Bolt, D. J., Campbell, R. G., Palmiter, R. D., Brinster, R. L., and Hammer, R. E. (1989). Genetic-Engineering of Livestock. *Science* **244:** 1281-1288.

Rothschild, M. F., Jacobson, C., Vaske, D., Tuggle, C., Wang, L., Short, T.,

Eckardt, G., Sasaki, S., Vincent, A., McLaren, D., Southwood, O., van der Steen, H., Mileham, A. and Plastow, G. 1996. The estrogen receptor locus is associated with a major gene influencing litter size in pigs. *Proceedings of the National Academy of Science* **93**: 201-205.

Rubinson, D. A., Dillon, C. P., Kwiatkowski, A. V., Sievers, C., Yang, L. L., Kopinja, J., Zhang, M. D., McManus, M. T., Gertler, F. B., Scott, M. L., and Van Parijs, L. (2003). A lentivirus-based system to functionally silence genes in primary mammalian cells, stem cells and transgenic mice by RNA interference. *Nature Genetics* **33:** 401-406.

Schnieke, A. E., Kind, A. J., Ritchie, W. A., Mycock, K., Scott, A. R., Ritchie, M., Wilmut, I., Colman, A., and Campbell, K. H. S. (1997). Human factor IX transgenic sheep produced by transfer of nuclei from transfected fetal fibroblasts. *Science* **278:** 2130-2133.

Sellier, P. and Rothschild, M. F. (1991) Breed identification and development, In Maijala, K. (ed.) *Genetic Resources of Pig, Sheep and Goat. World Animal Science*, **12**. Elsevier, Amsterdam, pp 125-143.

Smith, S. J., Cases, S., Jensen, D. R., Chen, H. C., Sande, E., Tow, B., Sanan, D. A., Raber, J., Eckel, R. H., and Farese, R. V. (2000). Obesity resistance and multiple mechanisms of triglyceride synthesis in mice lacking Dgat. *Nature Genetics* **25:** 87-90.

Thomas, C. E., Ehrhardt, A., and Kay, M. A. (2003). Progress and problems with the use of viral vectors for gene therapy. *Nature Reviews Genetics* **4:** 346-358.

Tiscornia, G., Singer, O., Ikawa, M., and Verma, I. M. (2003). A general method for gene knockdown in mice by using lentiviral vectors expressing small interfering RNA. *Proceedings of the National Academy of Sciences of the United States of America* **100:** 1844-1848.

Van Laere AS, Nguyen M, Braunschweig M, Nezer, C., Collette C, Moreau, L., Archibald, A. L., Haley CS, Buys N, Tally M, Andersson G, Georges, M., and Andersson, L. (2003). Positional identification of a regulatory mutation in IGF2 causing a major QTL effect on muscle growth in the pig. *Nature.*

Whitelaw, C. B. A. (2004) Transgenic livestock made easy. *Trends in Biotechnology* **22**: pp 157-158.

Whitelaw, C. B. A., Radcliffe, P. A., Ritchie, W. A., Carlisle, A., Ellard, F. M., Pena, R. N., Rowe, J., Clark, A. J., King, T. J. and Mitrophanous, K. A. (2004) Efficient generation of transgenic pigs using equine infectious anaemia virus (EIAV) derived vector. *FEBS Letters* **571**: 233-236.

Wiener, P., Smith, J. A., Lewis, A. M., Woolliams, J. A. and Williams, J. L. (2002) Muscle-related traits in cattle: The role of the myostatin gene in the South Devon breed. *Genetics Selection Evolution* **34**: 221-232.

Wilmut, I., Schnieke, A. E., McWhir, J., Kind, A. J., and Campbell, K. H. S. (1997). Viable offspring derived from fetal and adult mammalian cells. *Nature* **385:** 810-813.

20

YIELD TRENDS IN UK DAIRY AND BEEF CATTLE

P.C. GARNSWORTHY[1] AND P.C. THOMAS[2]
[1]Division of Agricultural and Environmental Sciences, School of Biosciences, University of Nottingham, Sutton Bonington Campus, Loughborough, Leics., LE12 5RD, UK; [2]Artilus Ltd, Edinburgh, EH14 5AJ

Introduction

In common with other types of agriculture in the UK, milk and beef production have experienced decades of continuous increase in productivity, characterised by increased yields per animal and increases in the efficiency of production. Underlying these changes has been the impact of science and technology on the genetic quality of UK livestock, on feeding, husbandry and resource management systems, and on the efficiency of land, labour and feed utilization.

Many of the improvements have reflected a continuous and incremental pattern of technology adoption, but in a few cases there have been technology watersheds where, over a brief few years, large parts of the industry have moved to introduce a 'new approach'. Examples of this would include: the replacement of hay by silage as winter forage; the shift from tethered housing to loose housing of stock; the adoption of 'controlled grazing' systems for grassland management; the use of genetic indices in breeding selection; embryo transfer for accelerated breed improvement; and the introduction of automated technologies in feed management, milking and the recording of animal performance.

As a general trend, dairy and beef production enterprises have each become progressively more specialised. In the mid 1990s, this pattern was reinforced as a result of the BSE regulations in the UK, which banned over-thirty-month (OTM) cattle from entering the food chain and so removed 'dairy beef' from the market. It is proposed that the OTM ban will be lifted in 2005 and 'dairy beef' will again become available. This, together with the planned 'decoupling' of Common Agricultural Policy (CAP) payments from beef and dairy production, could prompt some renewed consideration of 'dairy-beef' systems. However, there is a strong market demand for 'quality beef' from dedicated beef breeds or 'beef-crosses' from the dairy herd, and the present pattern of specialist dairy production and specialist beef production is likely to continue.

Moreover, 'decoupling' of CAP will in future intensify the drive for increased yield and efficiency and this in itself will reinforce the trend for specialisation.

This chapter will outline the historic trends in yield for UK dairy and beef production before going on to consider physiological aspects of yield regulation, together with technological developments and other issues that could create a significant departure from the historic trends by the end of this decade. An outline approach is adopted, intended to highlight main features and issues rather than providing a comprehensive summary of the literature. However, where appropriate, references are provided to give access to the original information or highlight key studies.

Yield trends for dairy cattle

According to national statistics, average milk yield in the UK increased from 3975 l/cow/annum in 1973 to 6531 l/cow/annum in 2002 (Figure 20.1). During that period, the total number of cows in the national herd remained fairly static until the introduction of milk quotas in 1984 and then decreased steadily until the present day (Figure 20.2). However, there has been a substantial restructuring of the dairy industry. Average herd size in the UK increased from 25 in 1965 to 74 in 2000. The distribution of herd sizes also changed; in 1970 only 3% of herds had over 100 cows and 80% had fewer than 50 cows; in 1999 45% of herds had more than 100 cows and 83% of herds had more than 50 cows (Figure 20.3). Total milk supply increased between 1973 and 1984, then decreased in response to quotas until 1989, but has remained at approximately 14.5 Gl for the past 15 years (Figure 20.4).

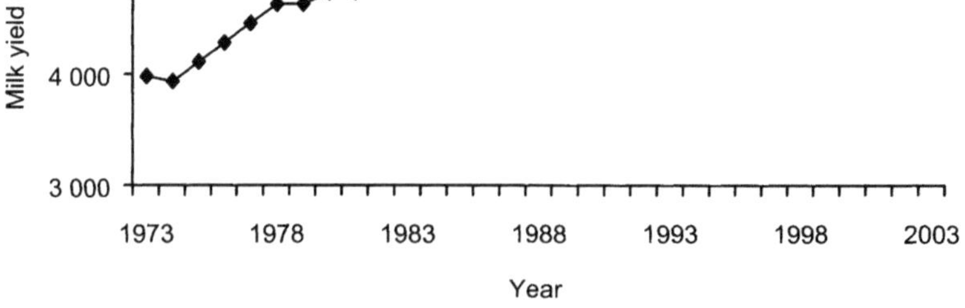

Figure 20.1 Annual average milk yield per cow in the UK from 1973-2002 (DEFRA, 2003a).

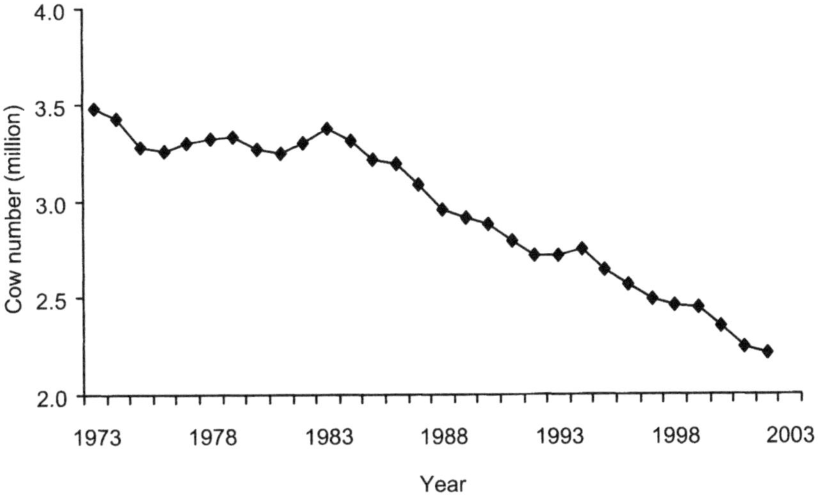

Figure 20.2 Annual number of dairy cows in the UK from 1973-2002 (DEFRA, 2003a).

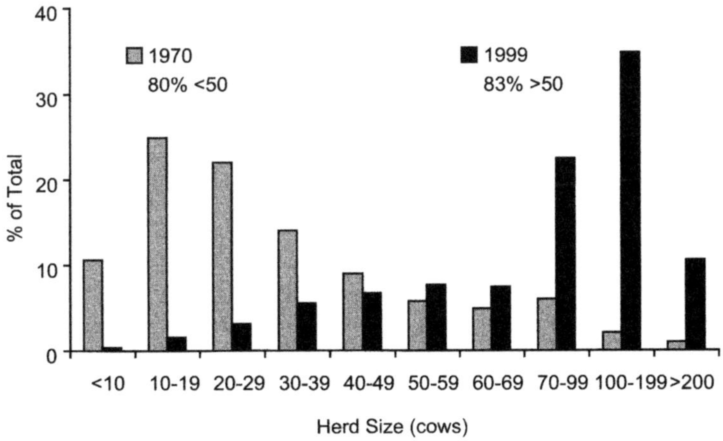

Figure 20.3 Distribution of UK dairy herds by size in 1970 and 1999 (MMB, 1970; 1999).

Pryce *et al.* (2004) estimated that the improvement in milk yield over the past 25 years has been 3-4% per year, up to 50% of which may be directly related to increased genetic merit. Predicted transmitting ability (PTA) for milk yield has increased in both bulls and cows by approximately 500 litres since 1990 (Figure 20.5), which is equivalent to 1000 litres of milk production. This has mainly been achieved by breed substitution,where Holstein genetics from North America have been used to replace the traditional UK Friesian cow, followed by within-breed selection whereby further increases in breeding value have been made within the Holstein breed.

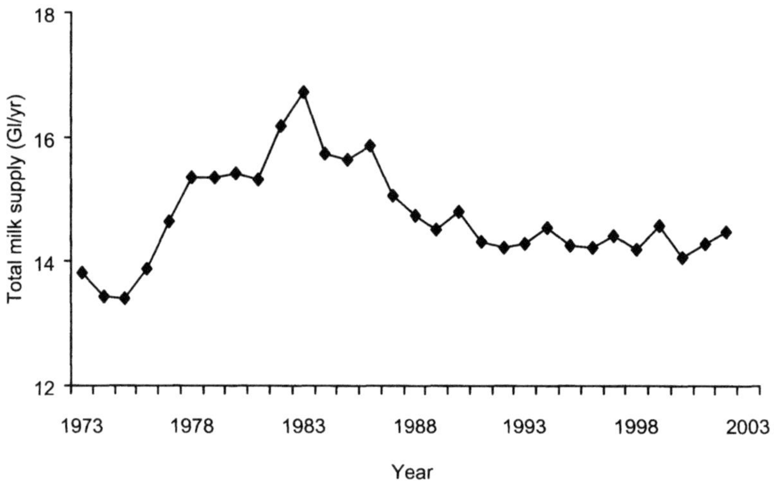

Figure 20.4 Annual total milk supply in the UK from 1973-2002 (DEFRA, 2003a).

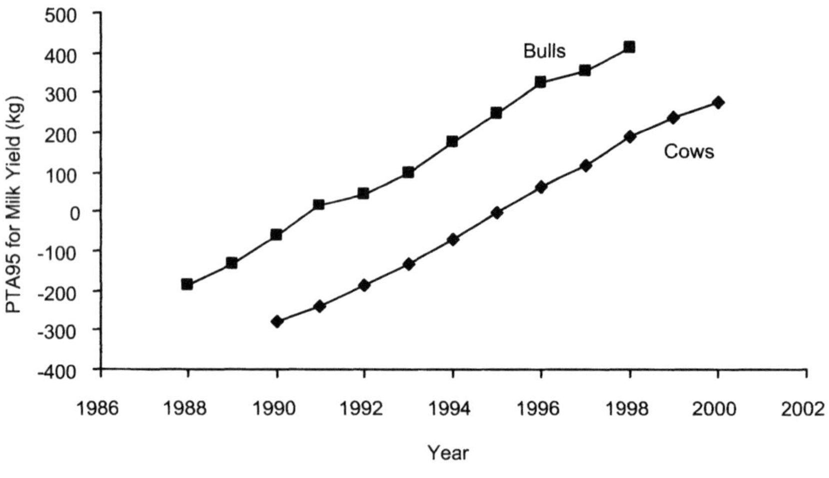

Figure 20.5 Change in predicted transmitting ability (PTA95) for milk yield in UK Holstein bulls and cows born between 1988 and 2000 (ADC, 2003).

The annual increases in PTA appear to be linear (Figure 20.5) and the coefficients of variation in PTA are 79 % in cows and 56 % in bulls (ADC, 2003), indicating that there is still scope for selection. Genetic and phenotypic trends for fat and protein yield were similar to the trend for milk yield in cows born between 1970 and 1994 (Lindberg *et al.*, 1998). Genetic progress in UK Holsteins is comparable with progress made in Canada, USA, Holland, France and Italy, ranging from 1.3 to 2.1 % per year for protein yield over the past five years (ADC, 2003). A comparison of the trends for lactation yields with those from PTAs for Holstein Friesians confirmed that approximately 65% of the improvement was due to breeding (Lindberg *et al.*, 1998). It appears, therefore, that genetic potential will not be a major limiting factor for future increases in milk yield in the UK.

Performance of top herds

Although the national average milk yield is 6531 l/cow/annum, there are many dairy herds in the UK with average yields in excess of 10,000 l/cow/annum.

There is nothing new about high-yielding herds. In 1954, Robert Boutflour described the elite "Steadings" herd at the Royal Agricultural College, Cirencester, UK as having an annual average milk yield of 1938 gallons (8810 litres), with one cow producing 3108 gallons (14 129 litres) in a single lactation (Boutflour, 1967). However, the national average milk yield in 1954 was only 600 gallons (2728 litres) per cow, so the superiority of the Steadings herd was much greater than that of elite herds today.

The highest yielding herd recorded by National Milk Records (NMR) for the year ending September 2001 produced an average of 13 472 kg milk per cow (NMR, 2002). The top 10% (for profitability) of herds recorded by Kingshay Dairy Costings, had an average annual milk yield of 8398 litres in 2001/2002 (Kingshay, 2003). The Genus Multiple Ovulation and Embryo Transfer (MOET) herd had an average annual milk yield of 9500 litres in 2002, with 85% of the herd being first-lactation heifers (Reaseheath, 2003).

In the USA, analysis of 11 000 dairy herds in the Northeast, mid-West, mid-South and South regions revealed an average milk yield of 9127 l/cow/annum, whilst the top 10% of herds (for milk yield) averaged 10 840 l/cow/annum in 2001 (Smith *et al.*, 2002). A preliminary study of 968 herds in the same regions found an average milk yield of 7762 l/cow/annum and the top 10% of herds averaged 9462 l/cow/annum in 1997 (Smith, 1998). This suggests that average herd yields in the USA increased by approximately 3.5% per annum over the past five years, similar to the rate of increase in the UK. It also suggests that milk yield has not reached a plateau in the top 10% of US herds, which increased at 3% per annum.

Performance of research herds

A long-term project at the University of Edinburgh was established in 1973 to examine the effect of bull selection on milk production (Langhill, 1999). Two genetic lines were established; one using bulls of average genetic merit for fat plus protein production and the other using semen from the top 4 or 5 bulls available. In 1988, an additional factor was added to the project when high and low forage feeding systems were examined in each line. Average milk yields for the four groups of cows in 1998 were High-Forage Control 6302 kg, High-Forage Select 7874 kg, Low-Forage Control 7538 kg, and Low-Forage Select 9536 kg. Corresponding yields of fat plus protein were 475, 602, 549 and 714 kg/annum respectively. Twenty-five years of selection therefore produced differentials of 25% in milk yield and 29% in fat plus protein yield. The differentials were marginally higher for the low-forage groups, reflecting the greater response to concentrates in cows of high genetic merit.

Yield trends for beef cattle

For many years the UK has had a unique system of beef supply, with almost equal numbers of slaughtered cattle originating in the dairy herd and specialist beef herds. Dairy cows not required for breeding dairy replacements have been mated to a beef bull. Female offspring are used as crossbred replacements in suckler herds and male offspring are reared for beef. Until 1996, 25% of UK beef came from cull cows; this source of beef was removed when a 30-month age restriction was imposed on animals entering the human food chain, in response to the BSE crisis. In 1996, 49% of home-produced beef came from dairy-bred calves and 43% from beef-bred calves; the remaining 8% was from cull cows slaughtered before the age restriction was imposed (MLC, 1997). Because of the links between the dairy herd and beef production, changes in the former have implications for the latter. The poor beef conformation of Holstein cattle reduced market demand for purebred calves. A calf processing aid scheme operated between 1996 and 1999, which removed 0.3-0.7 million Holstein bull calves from the market each year (DEFRA, 2003a). Calf slaughter has continued after the scheme was withdrawn. In 2000, therefore, only 34% of home-produced beef came from dairy-bred calves and 66% from beef-bred calves (MLC, 2000).

The number of cattle reared for beef production peaked in 1985 and then declined at a rate of 61 000 head per year (Figure 20.6). The number of cull cows and bulls used for beef declined by 19 000 head per year until 1995; since then, the number of animals slaughtered under the 30-month scheme has remained static, except for 2001 when many were lost in the Foot and Mouth Disease epidemic. Total supply of beef was linked to market demand from 1985 to 1995, so that self-sufficiency varied from 86 to 110% (Figure 20.7), in a period when market demand was declining. Since the removal of cull cows from the beef supply, self-sufficiency has varied between 72 and 81%.

Carcass weight per animal has increased in a linear fashion for the past 30 years (Figure 20.8). Due to the use of Holsteins, which have a greater mature weight than Friesians, carcass weight in cull cows increased by 1.7 kg/year. The increase in carcass weight of young cattle was greater at 2.3 kg/year; reflecting changes in terminal sire breeds as well as maternal influences.

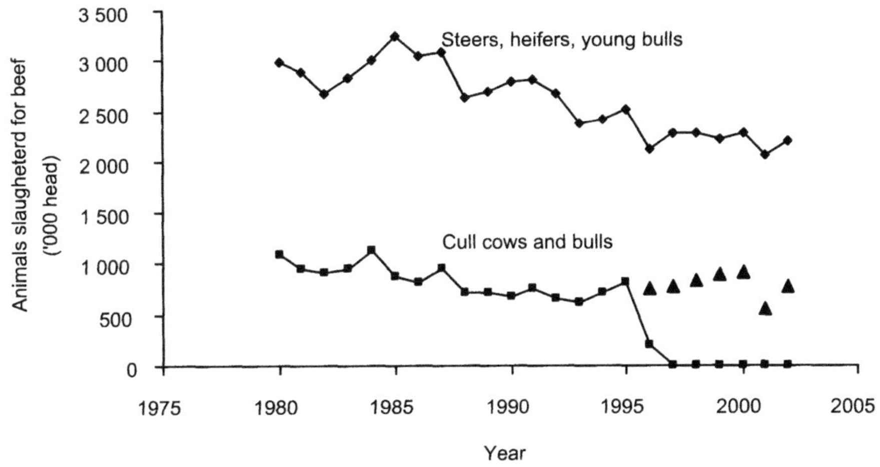

Figure 20.6 Number of cattle slaughtered for beef production in the UK 1980-2002 (DEFRA, 2003a).

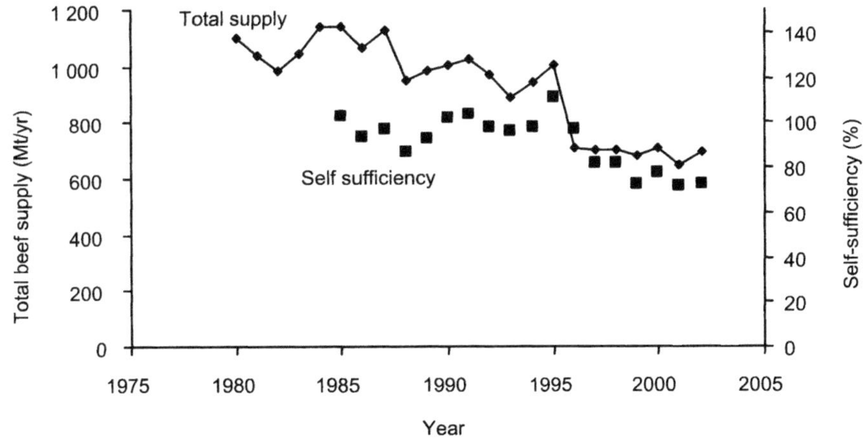

Figure 20.7 Total supply and self-sufficiency of beef in the UK 1980-2002 (DEFRA, 2003a).

Genetic progress in the dairy industry has had a detrimental effect on beef production from dairy-bred calves and the suckler herd. Increased use of Holstein genetics has led to poorer conformation in the resultant beef carcass (Figure 20.9). There was some recovery of conformation between 1992 and 1995, probably as a result of an increasing proportion of young bulls (Figure 20.10) and the popularity of the Belgian Blue breed for artificial insemination in dairy herds (Figure 20.11). These trends have reversed since 1996 and conformation has returned to 1992 levels.

Genetic progress for beef quality has been made in the beef terminal sire breeds over the past 20 years, mainly as a consequence of the Meat & Livestock Commission's 'Beefbreeder' programme. This uses statistical software (Best

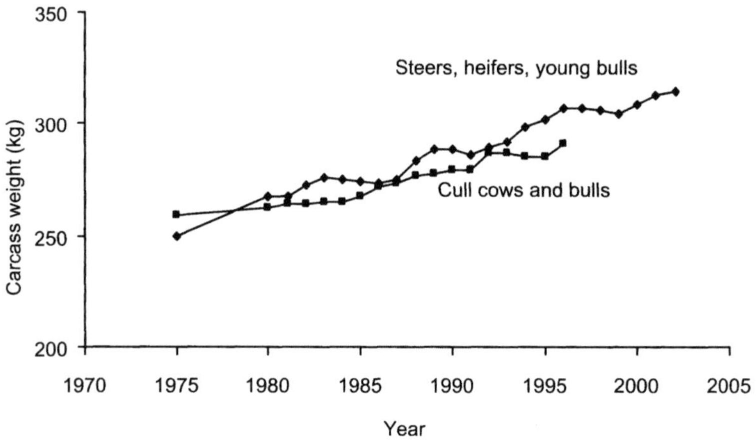

Figure 20.8 Carcass weight of beef cattle in the UK 1975-2002 (DEFRA, 2003a; Kempster *et al.*, 1986).

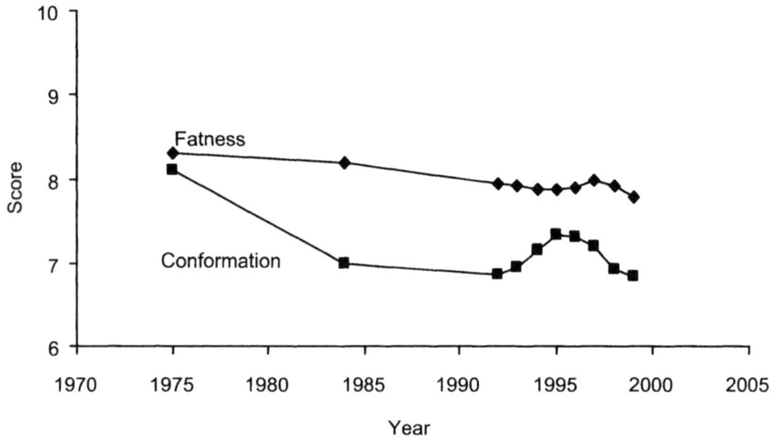

Figure 20.9 Changes in fatness and conformation of beef cattle in the UK 1975-2002 (MLC, 2000; Kempster *et al.*, 1986). (Fatness and conformation converted from MLC classification system to 15-point scales using the relationships of Kempster *et al.* (1986). MLC grade of R4L is equivalent to 8.0 conformation score and 8.25 fatness score.)

Linear Unbiased Prediction, BLUP) to attribute Estimated Breeding Values (EBV) for commercially important traits to individual bulls, enabling producers to select for traits that complement their current herd profile. The baseline used for calculating EBV is the breed average in 1980, and values are calculated separately for each breed. Progress within breeds is variable (Table 20.1), but is steady within major breeds, such as the Charolais (Figure 20.12).

Summary of yield trends per head

The milk yield information outlined above indicates that there has been an

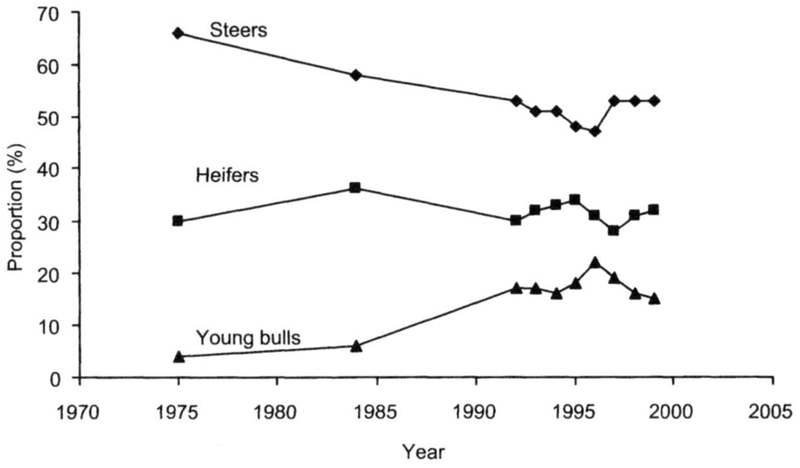

Figure 20.10 Steers, heifers and young bulls as proportions of cattle slaughtered for beef in the UK 1975-1999 (MLC, 2000; Kempster *et al.*, 1986).

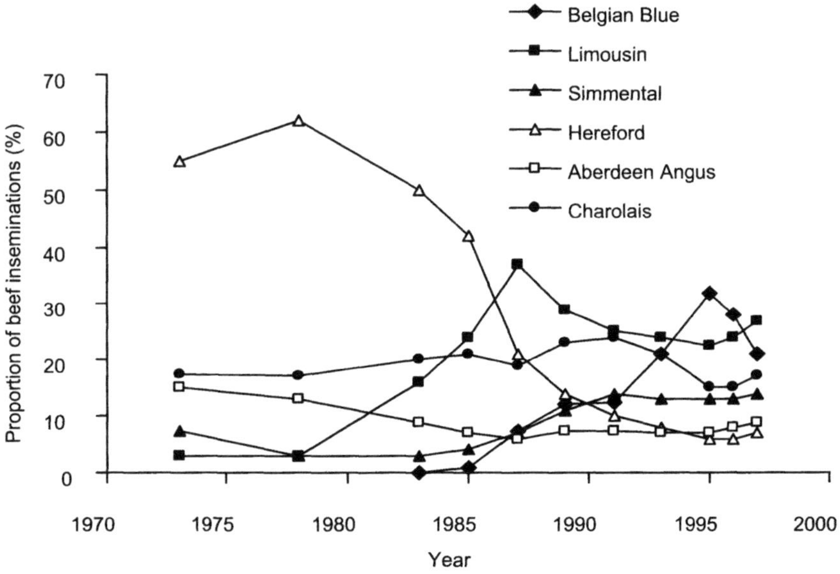

Figure 20.11 Trends in beef breeds used for artificial insemination by Genus (MLC, 1997).

underlying 30-year linear trend in the UK milk production averaging approximately 88 l/cow/year. Thus, based on the UK average yield figure of 6531 l/cow/year for 2002, there can be an expectation that by the end of the present decade average UK milk yields will be 7236 l/cow/year.

The figures for beef production are more complex to interpret, since the weight at which beef cattle are slaughtered is influenced by their body

Table 20.1 ESTIMATED BREEDING VALUES FOR GROWTH AND CARCASS TRAITS IN DIFFERENT BREEDS OF BEEF CATTLE. EACH VALUE IS AVERAGE WITHIN-BREED PROGRESS FROM 1980 TO 2003 (SIGNET, 2003)

Breed	*200-day growth (kg)*	*400-day growth (kg)*	*Muscling score*	*Muscle depth (mm)*	*Fat depth (mm)*
Aberdeen Angus	16	26	0.2	0.1	0.0
Belgian Blue	6	9	0.2	0.9	0.0
Charolais	16	26	0.3	1.4	0.0
Hereford	18	30	0.2	0.1	0.0
Limousin	11	19	0.3	1.3	0.0
Simmental	19	31	0.4	0.2	0.2
South Devon	12	22	0.4	1.3	0.1

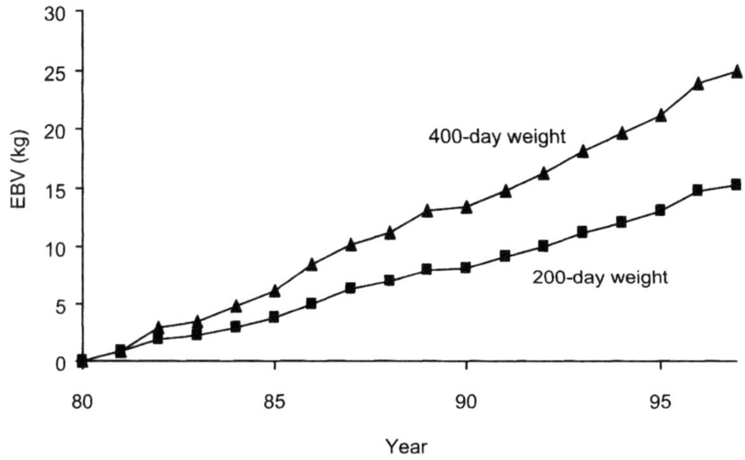

Figure 20.12 Estimated breeding value (EBV) for growth traits in Charolais cattle from 1980 to 1997 (MLC, 1997).

composition and, in particular, the relative proportions of lean tissue and fat. Thus the trend data for carcass weight reflect the slaughter of cattle at a particular compositional point in their development and reflects days to slaughter, as well as body-weight factors. However, with that qualification, the data show that there has been a 30-year linear trend for increased carcass weight per animal, averaging approximately 2.3 kg/year for steers, heifers and young bulls. Thus, based on the 2002 average UK slaughter weight of 315 kg/head, the corresponding figure for this main sector of the beef market will be approximately 333 kg/head by the end of the decade.

However, these estimates for milk and beef production are based on two underlying assumptions. Firstly, that there will be no quantum shift in dairy or beef cattle production fundamentally altering the long-term trend line. Secondly, that there will be a continual incremental improvement in the genetic capabilities of beef and dairy cattle, and that there will be the continued innovations in animal science and technology that will allow the historic trends in yield improvement to be maintained. These are matters that will be considered subsequently.

Physiological aspects of growth and milk yield

The rates of yield per animal, for milk production and beef production (growth) outlined in the previous sections, are influenced by a range of factors. In the context of the national herd, these include: the genetic capacity for growth and lactation; feed intake; the nutrient / energy content of diets; and the partition of nutrient / energy use between the various metabolic processes. In nutrient / energy partition, there is a first level of partition between body maintenance and net tissue or milk synthesis, which is determined largely by the level of feeding and nutrient / energy intake, and a second level of partition between the different types of net synthesis that may occur. In growing beef cattle this second partition includes the differential synthesis of body protein and body fat, whilst in lactating dairy cows it encompasses not only body protein and body fat but also the partition of nutrients / energy towards lactation and the differential synthesis of individual milk constituents.

Feed intake and nutrient / energy use in the tissues are under the influence of neural and endocrine regulations which respond to physiological state, e.g. stage of growth, lactation or pregnancy, and to the amount and composition of the mixture of metabolites released from the diet as products of digestion (see Thomas and Chamberlain, 1984). As a consequence, a complex array of regulatory mechanisms and control triggers influences tissue growth and milk yield. Notwithstanding genetic potential, yields may be restricted directly by the amount or balance of nutrients / energy derived from the diet, or indirectly by endocrine effects that alter the partition of nutrient / energy use between one synthetic process and another. In these ways environmental and management factors – nutrition, housing, health, metabolic stress etc. – have substantial impacts on the yields achieved under given conditions. Average statistics for milk yield or growth rates for cattle in different countries, or within countries for different systems of production, can therefore vary widely depending on the genetic potential of cattle, and the management and feeding systems that are in use.

Growth: overview

The physiology of growth is relevant both to the production of beef cattle and to the physical development of dairy cattle pre- and post-puberty. In colloquial terms growth may be interpreted as an 'increase in size and body weight as an animal matures'. However, it should be noted that 'increase in body size and weight' is associated with the accretion of skeletal tissue, muscle tissue and fat (adipose tissue), and the last two are particular relevant in the context of this chapter.

Even in young animals, growth involves a concurrent accretion of adipose tissue as well as muscle protein. However, since muscle protein synthesis is maximal in the early phase of development and adipose tissue deposition is dominant in mature animals which have reached their genetic limit for muscle tissue deposition, body protein synthesis tends to be associated with 'growth', whilst adipose synthesis is associated with 'fattening'. This is a useful distinction although, because of culling and replacement policies, UK dairy cows do not normally attain their potential mature size, so that both dairying and beef production employ relatively young cattle.

Muscle growth and adipose tissue deposition

The processes through which muscle cells are formed, from mesenchymal precursors through myoblasts and myotubes, to mature myofibrils, have long been recognised (Lindsay, 1983; Brameld, Buttery, Dawson and Harper, 1998), as has the biochemistry of the continuous synthesis and degradation of protein, the net balance of which determines protein accretion in the muscle tissues. Similarly, there is substantial understanding of the biochemistry of adipose tissue synthesis (Vernon, 1981; Vernon, Barber and Travers, 1999) and of the sequential pattern of development of adipose tissues in perirenal and omental fat, intermuscular and subcutaneous fat and intramuscular fat as animals move towards maturity (see Searle, Graham and O'Callaghan, 1972; Lindsay, 1983).

Annison and Bryden (1998, 1999) provided an excellent perspective of the relationships between diet, products of digestion, and the related utilization of metabolites for protein and fat synthesis in the ruminant. Fermentation in the rumen plays a crucial role in determining the composition of the mixture of short-chain fatty acids, amino acids, long-chain fatty acids and glucose that are absorbed from the digestive tract and become available for metabolism in the peripheral tissues. Notable features of tissue metabolism include: the importance of a balanced amino acid supply and of 'energy', predominantly as acetate, in muscle protein synthesis; the location of fatty acid and triglyceride synthesis in the adipose tissue, as distinct from the liver; and the major contribution of *de novo* synthesis of triglyceride fatty acids from blood plasma acetate as well as the uptake of preformed fatty acids from plasma triglycerides. Also distinctive is the restricted metabolism of glucose in ruminant muscle and adipose tissue and the fact that glucose supply mainly reflects liver metabolism of propionate, resulting from fermentation of dietary carbohydrates in the rumen.

Lactation: overview

During lactation the synthesised constituents of milk are formed in the alveolar cells of the mammary glands from metabolites that are extracted from blood plasma. These constituents are then transferred to the alveolar lumen together with water and some minor milk constituents, which are derived directly from blood plasma. The synthesis and secretion processes take place continuously and secreted milk is stored in the alveolar lumen, ducts and cisterns of the mammary glands. The act of milking provides a physical stimulus to the udder, which prompts release of oxytocin from the posterior pituitary gland, and this causes the contraction of mammary myoepithelial cells, resulting in 'milk ejection'. This allows milk to be removed from the udder, except for a small amount of 'residual milk' which is retained. Conventional machine milking is fast and effective. Average milk flow rates from the udder during milking are typically 2-3 l/min and peak rates are 5.5-5.7 l/min (Mein, 1998), so that removal of milk from the udder at each milking is normally accomplished in 5-10 minutes, although higher yielding cows generally take longer to milk.

THE LACTATION CYCLE

Under commercial conditions cows are separated from their calves shortly after calving and are established on a regular pattern of machine milking within two-three days. Milk yield rises quickly to reach a maximum level some 6-8 weeks later, and then progressively declines as lactation advances, until a point is reached where milking is discontinued and the cow 'dried off'. Based on mathematical analysis and modelling of the shape of the 'lactation curve' (see Wood 1969) it has been calculated that annual level of milk yield is maximised when the period from one calving to the next is 12-13 months, so that yield is close to peak levels for a high proportion of the time that the cow is lactating. Thus for more than 30-years a '12 month calving interval' has been a management objective for most UK dairy farms, and the consequence - that cows must be successfully rebred at or close to their peak of lactation - has been accepted. Typically, this has led to management systems with a nominal *ca* 305 d lactation period; a non-lactating 'dry period' of *ca* 42-60 days before the next calving; and a planned rebreeding within *ca* 60-75 days of calving.

Milk composition

In cows, as in most mammalian species, the largest component of milk is water, the other main constituents being milk fat, milk protein, lactose and the milk 'salts' – in aggregate estimated as milk 'ash content'. Milk composition is influenced by a range of nutritional and non-nutritional factors, the latter including breed, strain and individuality; stage of lactation; age; effects of udder diseases; stage of milking, milking interval, and completeness of milking; and extremes of environmental temperature (Thomas and Rook, 1983). However, over the lactation-period, water typically accounts for 859-878 g/kg whole milk, whilst concentrations of milk fat, protein, lactose and ash are typically 35-50 g/kg, 30-37 g/kg, 45-46 g/kg and 7.4-7.7 g/kg, respectively. Since milk has a similar osmotic pressure to blood, the volume or weight of whole milk secreted is determined largely by the secretion of milk water, which in turn depends on the secretion of lactose, the main osmotic constituent of milk.

Measurements of 'yield of total milk solids' or yields of 'milk fat', 'milk protein' and 'lactose', either alone or in combination, are widely used to describe the important traits and trends in milk production. However, 'milk yield' remains a general indicator of productivity, since whilst milk yield and the contents of fat and protein may be inversely correlated (see predicted transmitting ability, Figure 20.13, for example) an increase in milk yield is broadly reflected in increases in the yields of individual milk constituents.

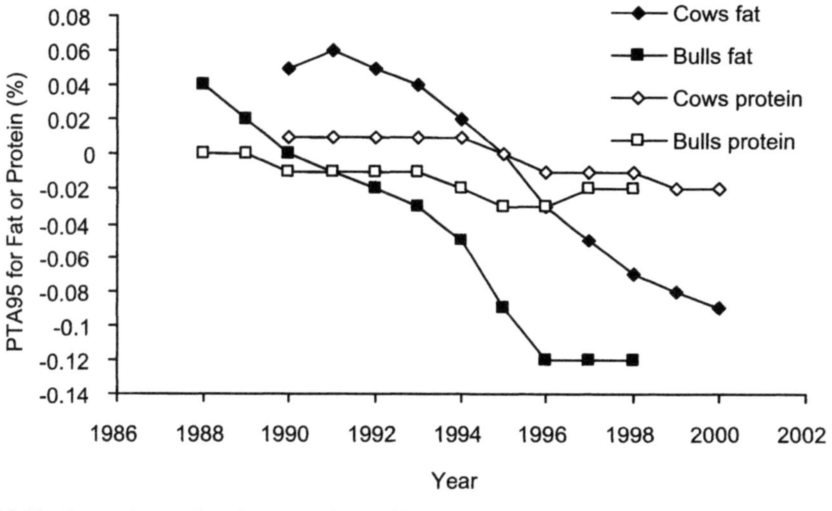

Figure 20.13 Change in predicted transmitting ability (PTA95) for milk fat and protein content in UK Holstein bulls and cows born between 1988 and 2000 (ADC, 2003).

Milk synthesis and secretion

The physiology and biochemistry of milk secretion have been extensively researched over the past four decades, and there is a substantial, although not complete, understanding of the key blood plasma metabolites that are used in the synthetic processes, the main pathways of synthesis, and the mechanisms by which the products are secreted (see Rook and Thomas, 1983; MacRae, Bequette and Crompton, 2000; Murphy, 2000).

Briefly, the main milk proteins (caseins, alpha-lactalbumin and beta-lactoglobulin) are wholly synthesised in the mammary gland from blood plasma amino acids, some of which are partially metabolised or interconverted in the process (see Bequette, Hanigan and Lapierre, 2003). Protein synthesis takes place at the endoplasmic reticulum of the secretory cells, and as the polypeptide chains are elongated during synthesis they are transported through the reticular lumen to the Golgi region where they are assembled within Golgi vesicles. Lactose is wholly synthesised from blood plasma glucose, through a pathway in which the key enzyme, lactose synthetase, is formed within the Golgi vesicles through an association of membrane bound galactosyl transferase and the milk protein alpha-lactalbumin. Milk proteins and lactose are transported in the Golgi vesicles to the apical surface of the cell, the ongoing synthesis of lactose continuously drawing in water. At the cell surface the vesicular contents are expelled into the alveollar lumen, so that the milk proteins and lactose are secreted together.

Milk fat synthesis and secretion involve a complementary process (see Murphy, 2000). Triglyceride formation depends on three main precursor components: preformed long-chain fatty acids taken up from the blood plasma lipoproteins; short- and medium-chain fatty acids synthesised *de novo* from blood plasma acetate and beta-hydroxybutyrate; and glycerol derived either from the breakdown of plasma triglycerides or mammary metabolism of plasma glucose. As they are formed at the endoplasmic reticulum of the secretory cells, triglyceride droplets coated with a phospholipid layer are released into the cell cytoplasm. These droplets migrate towards the apical surface of the cell progressively 'growing' through the accumulation of triglyceride. Secretion is thought to involve a combination of two processes. In one mechanism the triglyceride globules are envisaged to bud off through the apical membrane so that they are secreted surrounded by a film of membrane coating. In the second, the triglyceride droplets are thought to attract a coating of Golgi vesicles, which eventually merge to create an intracytoplasmic vacuole, which moves to the cell surface and expels its contents.

Physiological regulation of meat and milk yield

The yield potential of an animal for beef and/or dairy production is ultimately limited by the number and metabolic activity of the cells in the muscles, adipose tissues and mammary glands, which reflect not only the genetic characteristics but also hormonal and metabolic influences on cell development and activity, both in the short- and long-term. This raises some still-to-be-addressed questions about the ways that contemporary rates of yield can be influenced by earlier physiological experience, including the animal's experience *in utero* and at the very early stages of development (Robinson, 1990; Brameld *et al.,* 1998). However, there is clear evidence that yield can be related to the amount and activity of metabolically active mammary tissue (see Davis, 1997) and myofibres (see Brameld *et al.*, 1998), and that muscle growth and milk production can be promoted by increased levels of cellular growth factors, metabolic and steroid hormones, and nutrients / energy (Brameld *et al.*, 1998; Vernon, 1998). These physiological control points therefore offer possibilities for technologies designed to promote yield.

Technological considerations

NUTRITIONAL MANIPULATION OF METABOLISM

Since the development of metabolisable protein systems (Agricultural Research Council, 1980) it has been acknowledged that feeding systems for ruminants should seek to address responses to changes in nutrient / energy supply (see Biotechnology and Biological Sciences Research Council, 1998). Whilst that ultimate goal is still some way off, recently developed feeding systems (Chalupa and Sniffen, 1994; National Research Council, 2001; Offer *et al.*, 2002) have recognised the objective and have provided an improved basis for ration design and for the prediction of protein and amino acid supply, in particular.

In parallel with this, there has been continued research to understand the basis of milk production responses to changes in the level and type of dietary protein. From a detailed evaluation of experiments undertaken in their own laboratory and elsewhere, Chamberlain and Yeo (2003) have made significant progress in developing a nutritional rationale for the effects. They concluded that, depending on the circumstances, one of several amino acids can be limiting for milk synthesis. Moreover they highlighted that, in addition to the amino acid composition of undegraded dietary protein reaching the small intestine, variations in the composition of rumen microbial protein are probably important in determining which amino acid(s) are most limiting in supply. However, an understanding of the metabolic mechanisms for mammary gland responses to amino acid supply remains elusive, and current thinking is that regulation could involve a combination of at least six regulatory steps – in amino acid transport, initiation of protein synthesis, t-RNA charging, regulation of mammary blood flow and mammary catabolism of amino acids (see MacRae, Bequette and Crompton, 2000).

As a result of this and other gaps in understanding of regulation of metabolism, there may be halting progress in optimising nutrition for milk production in dairy cows. Nonetheless, the understanding of the relationships between specific nutrient / energy supply and milk secretion that exists (see Thomas and Chamberlain, 1984; Annison and Bryden, 1999) will continue to allow semi-empirical development of new nutritional approaches, so that incremental advances in feeding practice will probably be maintained.

Developments in biotechnology

As new developments in biology and biotechnology have emerged there has been widespread recognition of their potential directly to enhance ruminant meat or milk production. The 'first wave' technologies, such as steroids, beta-adrenergic agonists and bovine somatotrophin to reduce carcass fatness and increase muscle growth in beef cattle (see Hanrahan *et al.,* 1986; Buttery and Dawson 1990; Bell *et al.*, 1998) and of bovine somatrotrophin to promote milk yield in dairy cows (Bauman, 1992), have not been developed in the EU for technical and market reasons. However, notwithstanding the recognised scepticism of European consumers about the application of 'biotechnology' in the primary production of food, new technological approaches are in prospect. These include: the development of antibodies capable of destroying adipocytes, thereby reducing fatness and increasing lean meat production (Flint, 1966); the use of anti-body mediation intervention to potentiate natural growth hormone activity (see Pell and Flint, 1997); direct insertion of genes into the mammary gland through transfection (Schanbacher and Amstutz, 1997); and modification of mammary function through transgenesis (Vilotte *et al.*, 1997). There is potential to manipulate milk synthesis using transgenic constructs that would downregulate or upregulate specific enzyme steps in mammary metabolism, and Knight, Muir and Sorensen (2000) have discussed this approach with particular reference to lactose synthesis.

However, whilst these new scientific advances appear to offer great potential, it would be very optimistic to envisage their practical application in meat or milk production within the next decade.

Management

Many technological constraints to increasing the yield of cattle products have been addressed and/or overcome in recent years. For example reproductive technologies, such as artificial insemination, embryo transfer and ultrasound scanning, have become common practice. In addition sexed semen for cattle is now commercially available, which could enable dairy herds virtually to eliminate the purebred Holstein bull calf. Sexed semen has an accuracy of about 90%, so the number of dairy cows needed to breed dairy replacements would be reduced by almost 50%. This would increase the selection intensity in dairy herds and increase the rate of genetic gain. The numbers of beef cross calves would double, leading to a potential increase in total beef supply of 14%. If sexed semen were also used to increase the numbers of beef cross male calves, total beef production would be increased by 22% (MLC, 2000).

In dairying, milking machines continue to be improved and refined by manufacturers and there is a slow but steady appearance of automatic milking systems (AMS), colloquially referred to as 'robotic milking', in commercial practice. It is too early to determine the full impact of this development, since whilst it offers very significant benefits in terms of labour use there are still some uncertainties about the ultimate impact of putting the cow 'in charge' of her frequency of milking. There is evidence that milking serves to remove from the udder a small protein which acts locally in the gland as a 'feed back inhibitor of lactation (FIL)', exerting its effects after milk synthesis but before milk secretion (Rennison *et al.*, 1993; Wilde *et al.*, 1995). As a consequence of that, as well as other factors, more frequent milking has been shown to increase milk yield in both the short- and longer-term (see Knight, Peaker and Wilde, 1998). Early evidence is that cows on AMS may choose to be milked up to six times per day, but whether this will significantly increase milk production as AMS becomes more widespread has yet to be established.

Notwithstanding the general progress that has been made, there are a number of areas in which conventional management strategies for dairy cows have created problems as levels of animal productivity have increased. For example, fertility in the dairy cow appears to be negatively correlated with milk yield, and it is estimated that conception rate to first service has declined by 1% per annum over the past 25 years (Figure 20.14).

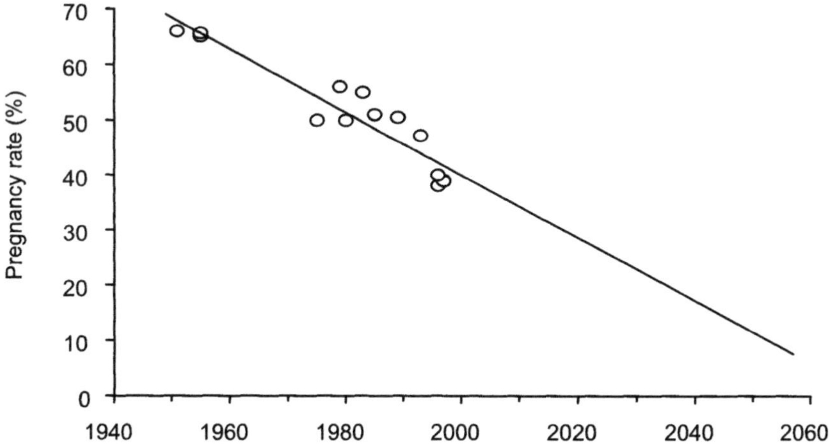

Figure 20.14 Decline in fertility of UK dairy cows (Mann, 2002).

Esslemont and Kossaibati (2002) analysed data from 52 herds recorded between 1987/88 and 1997/98 under the DAISY system. Over this 10-year period, milk yield increased from 6300 litres per cow to 7200 litres per cow. Over the same period pregnancy rate decreased from 50% to 43%; heat detection rate decreased from 51% to 48%; calving to conception interval increased from 96 days to 110 days; number of services per conception increased from 1.96 to 2.36; calving interval increased from 379 days to 390 days; and total culling rate increased from 21% to 27%. The net result of these changes was that the economic cost of poor fertility increased by 94%. Fertility has, therefore, become a major concern for dairy farmers and continued reductions in reproductive performance could offset potential yield gains from genetic improvement. As milk yields increase over the first 3-4 lactations, premature culling for failure to conceive will reduce the average milk yield of a herd.

As a result of this problem, a DEFRA LINK programme is underway in the UK to address the nutritional management of cows in early lactation - there is evidence that particular types of diets may improve hormonal balance and conception rates (see Webb *et al.*, 1999). More broadly, both in research and in commercial practice, renewed attention is being given to management strategies based on increasing lactation persistency (see McFadden, 1997), extending lactation period, reducing the lifetime number of pregnancies and potentially increasing cow longevity. As yet, there is little information from controlled long-term experiments using these alternative strategies. However, in one study Osterman (2003) has measured energy corrected milk (ECM) production in 4 groups of cows with either a 12-month or 18-month calving interval and milked either twice-daily or thrice-daily. The results suggest that more frequent milking with an extended calving interval may offer a realistic alternative approach, with no loss of milk yield (Table 20.2).

Table 20.2 ENERGY CORRECTED MILK (ECM) PRODUCTION PER DAY BETWEEN CALVING INTERVALS FOR DAIRY COWS WITH 12- OR 18-MONTH CALVING INTERVALS AND MILKED TWO- OR THREE TIMES PER DAY (DATA ARE FROM OSTERMAN, 2003).

	Group			
	12-2	*12-3*	*18-2*	*18-3*
No cows	24	29	16	21
Lactation length (weeks)	41.7[a]	45.8[b]	63.8[c]	64.4[c]
Dry period (weeks)	9.3[a]	9.6[a]	14.0[b]	11.2[a]
ECM kg/day	22.7	23.4	21.3[a]	24.2[b]
Feed intake (ME MJ/day)	227[a]	238[b]	212[c]	223b[c]
Feed efficiency (ME MJ/kg ECM)	5.9[a]	5.7[b]	5.6[b]	5.5[b]

Notes:
1. Different superscript letters on the same line indicate significant (P<0.05) differences.
2. Time between calving interval is lactation period plus dry period.

Health and welfare

Despite the potential for genetic progress, future development of yield of milk could be constrained by health and welfare factors. Mastitis and lameness continue to pose significant welfare problems and, in conjunction with fertility problems mentioned earlier, are the principal reasons for premature culling from the herd.

Whitaker (2002) compared the incidence of mastitis in over 300 herds in 1998 and 2000; somatic cell counts increased from 88 700 to 140 000 cells/ml and the incidence of clinical mastitis increased by 36.6%, often in conjunction with an increase in herd size. However in another study Esslemont and Kossabati (2002) reported the incidence of disease from 1988/89 to 1998/99 in herds recorded under the DAISY system. They found no trend in the incidence of mastitis, calf mortality, retained placenta or vulval discharge. Moreover the incidence of several health problems had decreased: milk fever from 7% to 4%; lameness per 100 cows from 29 cases to 22 cases; assisted calving from 9-11% to 7%. This might be taken to suggest disease incidence had been unaffected or reduced as average milk yields had increased. However, since the incidence of many diseases increases with age, premature culling for fertility could be masking other effects on disease incidence.

Esslemont and Kossabati (2002) found culling rate had been increased from 20% to 28% and equal proportions of cows were culled for poor fertility as for other reasons (Figure 20.15). Premature culling itself is an important efficiency factor, since reduced longevity means a higher replacement rate to maintain herd size; more female replacements are required, reducing the availability of feed for lactating cows, as well as the number of cattle available for beef production.

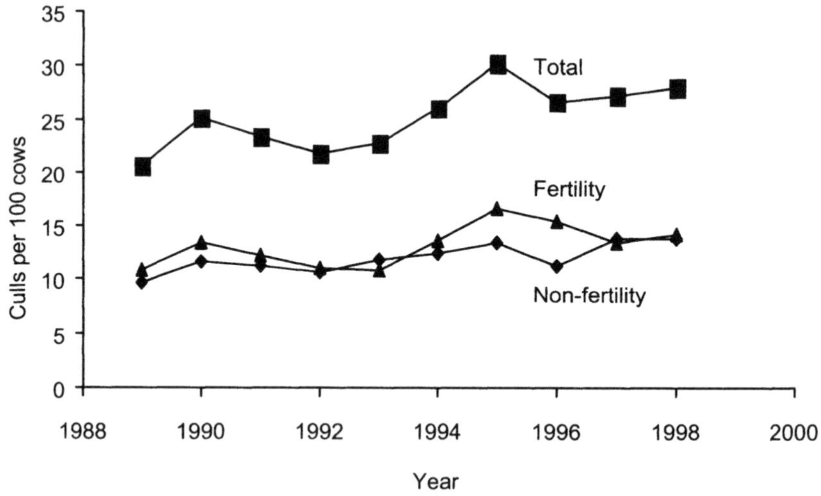

Figure 20.15 Culling for fertility and non-fertility in herds recorded under the DAISY system over a ten-year period (Esslemont and Kossabati, 2002).

Whilst potential causes of health and welfare problems in dairy cows are varied, there is a growing view that metabolic stress linked with high yields could act as a contributory factor (Neilsen, 1999). In practice, metabolic stress can be created by a variety of environmental, nutritional and management factors. However, an underlying feature appears to be the impairment of immune function with potential implications for disease resistance. Stress causes changes in the release of cytokinins and hormones such as growth hormone, insulin, glucagon and cortisol (see Annison and Bryden, 1999), and potentially shifts the partition of nutrient / energy use away from growth and milk synthesis. Stress can also reduce feed intake.

Sinclair *et al* (1999) have shown that nutrition is important in modulating immune responses in dairy cows, and this emphasises the need for holistic strategies of feeding and management to deal with stress problems. The implementation of national strategies for animal health and welfare improvement (DEFRA, 2003b; SEERAD, 2003) should have a significant beneficial impact on beef and dairy herds over the next decade. However, whilst resultant increases in beef and dairy yields can be anticipated, their scale is difficult to estimate at this stage,

Land use and environment resources

The UK has a unique climate and land distribution which favours grass production in the West and arable crops in the East. Milk production, suckler-cow herds and beef fattening have therefore been concentrated in the grass-growing areas to maximise the use of this low-cost source of nutrients / energy, either as grazed grass or winter forage. Some meat animals are transported across the country to be finished intensively on cereals, but the location of national livestock resources essentially reflects a West to East divide.

As the average dairy cow spends up to 8 hours per day grazing, nutrient / energy intake is limited by herbage availability and sward composition. In spring, lush grass can support milk yields of 30 l/d, but this reduces as grass matures so late season grazing can only support 15 l/d. Nutrient / energy resources thus impose a limit on milk production of 5000-6000 l/cow/year for production systems that are predominantly grass reliant.

Traditionally, in the UK, this issue has been addressed through feeding cows cereal and protein concentrate feeds, imported from the East of the country or from overseas. However, with changes in environmental legislation, there is an increasing prospect of regulatory controls on farm 'total nutrient budgets', and these may limit the allowable use of concentrate feeds on livestock enterprises. As for cropping, the advent of new maize varieties has enabled forage maize to be grown throughout the Southern half of the UK (Phipps, 1994) and the area grown has almost quadrupled since 1990 (Table 20.3). Maize silage has a higher energy concentration than grass silage, allowing greater nutrient / energy intakes necessary for high milk yields. As new varieties allow expansion of the areas in which maize may be grown, the use of maize silage in dairy and beef systems will increase, with associated changes in the appearance of the UK landscape.

Table 20.3 TOTAL AREA (HA) OF FORAGE MAIZE GROWN IN THE UK

1990	*1995*	*2000*	*2001*	*2002*
33,265	100,432	97,624	119,557	111,333

Source: DEFRA June Annual Census

Internationally, there is environmental concern about the contribution of cattle to global methane and ammonia emissions. This has received less consideration in the UK, because non-agricultural industries provide the bulk of greenhouse gas emissions (as compared with New Zealand, for example). However, it is reasonable to assume that all industrial sectors will be required ultimately to contribute to reducing the threat of global warming, and that UK livestock farming will need to develop emission mitigation strategies.

In this context, milk yield has a major effect on methane and ammonia emissions. For example, Garnsworthy (2004) has calculated that a dairy herd producing one million litres of milk using cows of average yield (6 000 litres/year) emits 37 t methane/year and 7.9 t ammonia/year, whilst a herd producing one million litres of milk using high-yielding cows (9 000 litres/year) emits only 19 t methane/year and 5.2 t ammonia/year. Moreover, these calculations are based on current levels of fertility (50% oestrous detection rate, 38% conception rate) and if fertility could be restored to ideal levels (70% oestrous detection rate, 60% conception rate), emissions could be further reduced by up to 25%. Thus, in this case, strategies for environmental protection and for increased yield and efficiency of dairy production go hand in hand.

Conclusions

On the basis of the assessments outlined above it can be concluded that, subject to any major market changes which may develop as a result of CAP reform, milk yield and beef production will continue to follow their historic trends. Thus estimates are that milk yield will increase to 7236 l/cow/year and beef carcass weight to 333 kg/head by the end of the decade. These predictions can be considered to be fairly sound since production over the next decade will be based on the genetics available today.

Over the longer term, if current yield trends continue, average milk yield in the UK will increase to 9000 l/cow/year in 2030 and 10 760 l/cow/year in 2050, and beef carcass weight will increase to 380 kg/head in 2030 and 426 kg/head in 2050. The predictions are obviously speculative, but those for milk yield are certainly achievable. There are many dairy herds in the UK already producing herd averages equal to the 2030 prediction, and the top 10 % of herds in the USA are already producing the 2050 prediction. The predictions for beef carcass weight might not be achieved if market requirements dictate that optimum carcass weight cannot exceed 400 kg. However, an average carcass weight greater than 300 kg might have been considered unachievable in 1975.

Whilst there will be continued progress in the development and introduction of new technologies and some significant changes in livestock feeding and management systems, the pattern will be one of evolution. It is not considered that the more futuristic applications of transgenic technology will be applied in practice within a 10-year time scale.

In the UK context, there are very significant economic, social and environmental benefits from increased yields and efficiencies in dairy and beef production, although this may appear counterintuitive in the light of much of current public and policy debate about the future for 'low-intensity' systems of farming. However, whilst the socio-political climate may have changed, the underlying need for technological development to achieve higher yields and improved efficiencies in livestock production remains.

Finally, it is worth noting that, in both beef and dairy production, size matters. As indicated by Bergen and Merkel (1991), the most successful strategy to increase lean beef production has been the breeding and selection of large, late maturing cattle. Similarly, Linzell (1972) pointed out that in lactating animals milk yield increases with the weight of active mammary tissue and that, across species, this results in the general 'scaling rule' that milk yield increases with body weight$^{0.79}$. The implication is that, at a constant level of genetic merit for milk production, large cows will have higher yields than small cows.

The Ayrshire, Friesian and Friesian-cross cows that traditionally formed the dominant part of the UK dairy herd (typically 450-550 kg body weight) were smaller than their modern Holstein and Holstein-Friesian counterparts (typically 600-725 kg body weight) and cattle sizes generally continue to creep upwards. However, even large modern cows are unlikely to reflect the maximum size that could be achieved through breeding and selection. There are historical accounts of the breeding of 'oversized black and white oxen', reaching heights of over 3 metres and weights of 2,136 kg (Avery, 1900-1906). Such animals are clearly not comparable with modern-day cows, but they suggest that current perceptions of cow size may be conditioned more by the size of animals with which we have become familiar, than by what might theoretically be achieved.

References

ADC (2003) *Animal Data Centre Statistics*. MDC Evaluations Ltd. http://www.animaldata.co.uk accessed Sept 2003

Agricultural Research Council (1980) *The Nutrient Requirement of Ruminant Livestock*, Technical Review by an ARC Working Party, CAB, Farnham

Royal.

Annison, E.F. and Bryden, W.L. (1998) Perspectives on ruminant nutrition and metabolism. I. Metabolism in the rumen. *Nutrition Research Reviews*, **11**, 173-198.

Annison, E.F. and Bryden, W.L. (1999) Perspectives on ruminant nutrition and metabolism. II Metabolism in ruminant tissues. *Nutrition Research Reviews*, **12**, 147-169.

Avery, J.D. (1900-1906) Mack the Giant Ox Banner. Memorial Hall Museum, Pocumtuck Valley Association, Deerfield, Massachusetts, USA.

Bauman, D.E. (1992) Bovine somatotrophin: review of an emerging animal technology. *Journal of Dairy Science*, **75**, 3432-3451.

Bell, A.W., Bauman, D.E., Beermann, D.H. and Harrell,R.J. (1998) Nutrition, development and efficacy of growth modifiers in livestock species. *Journal of Nutrition*, **128**, 360S-363S.

Bequette, B.J., Hanigan, M.D. and Lapierre, H. (2003) Mammary uptake and metabolism of amino acids by lactation ruminants. In *Amino Acids in Animal Nutrition*, pp. 347-365. Edited J.P.F. D'Mello, CAB International, Wallingford.

Bergen, W.G. and Merkel, R.A. (1991) Body composition of animals treated with partitioning agents: implications for human health. *Journal of the Federation of American Societies for Experimental Biology*, **5**, 2951-2957.

Biotechnology and Biological Sciences Research Council (1998) *Responses in the Yield of Milk Constituents to the Inatke of Nutrients by Dairy Cows.* Ed G. Alderman, pp1-96. CAB, Wallingford.

Boutflour, R. (1967) *The High Yielding Dairy Cow.* Crosby Lockwood, London.

Brameld, J.M., Buttery, P.J., Dawson, J.M. and Harper, J.M.M. (1998) Nutritional and hormonal control of skeletal-muscle cell growth and differentiation. *Proceedings of the Nutrition Society*, **57**, 207-217.

Buttery, P.J. and Dawson, J.M. (1990) Growth promotion in farm animals. *Proceedings of the Nutrition Society*, **49**, 459-466.

Chalupa, W. and Sniffen, D.J. (1994) Carbohydrate, protein and amino acid nutrition of dairy cows. In *Recent Advances in Animal Nutrition -1994*, pp 265-275. Edited P.C. Garnsworthy and D.J.A. Cole. Nottingham University press, Nottingham.

Chamberlain, D.G. and Yeo, J.M. (2003) Effects of amino acids on milk production. In: *Amino Acids in Animal Nutrition.* pp 367-387. Edited by J.P.F. D'Mello, CABI Publishing, Wallingford.

DEFRA (2003a) http://www.defra.gov.uk accessed Sept 2003

DEFRA (2003b) *Animal Health and Welfare Strategy Implementation Plan for England: A Work in Progress*, pp 3-64. Department of Environment and Rural Affairs, London.

Esslemont, D. and Kossabati, M. (2002) The Costs of Poor Fertility and Disease in UK Dairy Herds. *Daisy Research Report No. 5*, Intervet UK Limited,

Milton Keynes.
Flint, D.J. (1996) Immunological manipulation of adipose tissue. *Biochemical Society Transactions*, **24**, 418-422.
Garnsworthy, P.C. (2004) The environmental impact of fertility in dairy cows: a modelling approach to predict methane and ammonia emissions. *Animal Feed Science and Technology*, **112**, 211-223.
Hanrahan, J.P., Quirke, J.F., Bomann, W., Allen, P., McEwan, J.C., Fitzsimons, J.M., Kotzian, J. and Roche, J.F. (1986) Beta-agonists and their effects on growth and carcase quality. In *Recent Advances in Animal Nutrition -1986*, pp 125-138. Edited W. Haresign and D.J.A. Cole. Butterworths, London.
Kempster, A.J., Cook, G.L. and Grantley-Smith, M. (1986) National estimates of the body composition of British cattle and pigs with special reference to trends in fatness. A review. *Meat Science*, 17, 107-138.
Kinsghay (2003) *Kingshay Dairy Costings*. http://www.countrywidefarmers.co.uk/farming/kingshaycostings.htm accessed Sept 2003.
Knight, C.H., Peaker, M. and Wilde, C.J. (1998) Local control of mammary development and function. *Reviews of Reproduction*, **3**, 104-112.
Knight, C.H., Sorensen, A. and Muir, D.D. (2000) Non-nutritional (novel) techniques for manipulation of milk composition. In *Milk Composition*, pp223-238. Edited R. E. Agnew, K.W. Agnew and A.M. Fearon. Occasional Publication No 25, British Society of Animal Science, Edinburgh.
Langhill (1999) *Annual Report, Langhill Farm*. University of Edinburgh.
Lindberg, C.M., Swanson, G.J.T. and Mrode, R.A. (1998) Genetic and phenotypic trends in production traits in the United Kingdom (UK) dairy herd. *Proceedings of the British Society of Animal Science 1998*, 191.
Linzell, J.L. (1972) Milk yield, energy loss in milk and mammary gland weight in different species. *Dairy Science Abstracts*, **34**, 351-360.
MacRae, J.C., Bequette, B.J. and Crompton, L.A.(2000) Synthesis of milk protein and opportunities for nutritional manipulation. In *Milk Composition*, pp 179-199. Edited R. E. Agnew, K.W. Agnew and A.M. Fearon. Occasional Publication No 25, British Society of Animal Science, Edinburgh.
Mann G.E. (2002) Reproduction-mating management. In: *Encyclopaedia of DairySciences*, pp1770 - 1777. Eds J.W. Fuquay and P.F. Fox. Academic Press, San Diego, USA.
Mein, G. A. (1998) *Design of milk harvesting systems for cows producing one hundred pounds of milk daily*. British Mastitis Conference, Stoneleigh, UK.
MLC (1997) *Beef Yearbook 1997*. Meat and Livestock Commission, Milton Keynes.
MLC (2000) *Beef Yearbook 2000*. Meat and Livestock Commission, Milton Keynes.

MMB (1970) *United Kingdom Dairy Facts and Figures*. Federation of UK Milk Marketing Boards, Thames Ditton.

MMB (1999) *United Kingdom Dairy Facts and Figures*. Federation of UK Milk Marketing Boards, Thames Ditton.

Murphy, J. J. (2000) Synthesis of milk fat and opportunities for nutritional manipulation. In *Milk Composition*, pp 201-222. Edited R. E. Agnew, K.W. Agnew and A.M. Fearon. Occasional Publication No 25, British Society of Animal Science, Edinburgh.

National Research Council (2001) *Nutrient Requirements of Dairy Cattle*, Seventh Revised Edition, pp 381. National Academic Press, Washington DC.

Neilsen, B.L. (1999) Perceived welfare issues in dairy cattle, with special emphasis on metabolic stress. In *Metabolic Stress in Dairy Cows*, pp 1-7. Edited J.D. Oldham, G. Simm, A.F.Groen, B.L. Neilsen, J.E. Pryce and T.J. Lawrence. Occasional Publication 24, British Society of Animal Science, Edinburgh.

NMR (2002) Annual Production Report, National Milk Records.

Offer, N.W., Agnew, R.E., Cottrill, B.R., Givens, D.I., Keady, T.W.J., Mayne, C.S., Rymer, C., Yan, T., France, J., Beever, D.E. and Thomas, C. (2002) Feed into milk – an applied feeding model coupled with a new system of feed characterisation. In *Recent Advances in Animal Nutrition -2002*, pp 167- 194. Edited P.C. Garnsworthy and J. Wiseman. Nottingham University Press, Nottingham.

Osterman, S. (2003) *Extended Calving Interval and Increased Milking Frequency in Dairy Cows*. Doctoral thesis, Swedish University of Agricultural Science, Uppsala.

Pell, J.M. and Flint, D.J. (1997) Immunomodulation of lactation. In *Milk Composition, Production and Biotechnology*, pp 307-317. Edited R.A.S. Welch, D.J.W. Burns, A.I. Popay and C.G. Prosser. CAB International, Hamilton, New Zealand.

Phipps, R.H. (1994) Complementary forages for milk production. In . In *Recent Advances in Animal Nutrition -1994*, pp 215- 230. Edited P.C. Garnsworthy and D.J.A. Cole. Nottingham University Press, Nottingham.

Pryce, J.E., Royal, M.D., Garnsworthy, P.C. and Mao I.L. (2004) Fertility in the high producing dairy cow. *Livestock Production Science*, **86**, 125-135.

Reaseheath (2003) *Farm Guide* http://www.reaseheath.ac.uk/AgSector/fmguide/dairy.html accessed Sept 2003.

Rennison, M.E., Kerr, M., Addey,C.V.P., Handel, S.E., Turner, M.D., Wilde, C.J. and Burgoyne, R.D. (1993) Inhibition of constitutive protein secretion from lactating mouse mammary epithelial cells by FIL (feedback inhibitor of lactation), a secreted milk protein. *Journal of Cell Science*, **106**, 641-648.

Robinson, J.J. (1990) Nutrition in the reproduction of farm animals. *Nutrition*

Research Reviews, **3**, 253-276.
Schanbacher, F.L. and Amstutz, M.D. (1997) Direct transfection of the mammary gland: opportunities for modification of mammary function and production, composition and qualities of milk. In *Milk Composition, Production and Biotechnology*, pp 243-264. Edited R.A.S. Welch, D.J.W. Burns, A.I. Popay and C.G. Prosser. CAB International, Hamilton, New Zealand.
Searle, T.W., Graham, N.McC and O'Callaghan, M. (1972) Growth in sheep. 1, Chemical composition of the body. *Journal of Agricultural Science*, **79**, 371-382.
SEERAD (2003) *Animal Health and Welfare in Scotland: Implementing the Animal Health and Welfare Strategy*, pp. 3-36. Scottish Executive Environment and Rural Affairs Department, Edinburgh.
Signet (2003) *Breed Benchmarks*. Signet Breeding Services, Milton Keynes.
Sinclair, M.C., Nielsen, B.L. Oldham, J.D. and Reid, H.W. (1999) Consequences for immune function of metabolic adaptations to load. In *Metabolic Stress in Dairy Cows*, pp. 113-118. Edited J.D. Oldham, G. Simm, A.F.Groen, B.L. Neilsen, J.E. Pryce and T.J. Lawrence. Occasional Publication 24, British Society of Animal Science, Edinburgh.
Smith, J.W. (1998) Using benchmark values in herd management analysis. 1998 *Annual Report*, pp 112-117, University of Georgia.
Smith, J.W., Chapa, A.M., Ely, L.O. and Gilson, W.D. (2002) Dairy Production and Management Benchmarks. *Bulletin 1193*, University of Georgia.
Thomas, P.C. and Chamberlain, D.G. (1984) Manipulation of milk composition to meet market needs. *In Recent Advances in Animal Nutrition – 1984*, pp 219-243. Ed W. Haresign and D.J.A. Cole,. Butterworths, London.
Thomas, P.C. and Rook, J.F.A. (1983) Milk Production. In: *Nutritional Physiology of Farm Animals*, pp 558-622. Eds J.A.F. Rook and P.C. Thomas. Longman, London and New York.
Vernon, R.G. (1981) Lipid metabolism in the adipose tissue of ruminants. In *Lipid Metabolism in Ruminant Animals*, pp 279-362. Ed W.W. Christie. Pergamon Press, Oxford.
Vernon, R.G. (1998) Homeorhesis. In *Hannah Research Institute Yearbook*, pp 64-73.Edited E.Taylor. Hannah Research Institute, Ayr.
Vernon, R.J., Barber, M.C. and Travers, M.T. (1999) Present and future studies on lipogenesis in animals and human subjects. *Proceedings of the Nutrition Society*, **58**, 541-549.
Vilotte, J., Solier, S., Persuy, M., Lepourry, L., Legrain, S., Printz, C., Stinnakre, G., L'Huiller, P. and Mercier J. (1997) Application of transgenesis to modifying milk protein composition. In *Milk Composition Production and Biotechnology*, pp 231-242. Edited R.A.S. Welch, D.J.W. Burns, A.I. Popay and C.G. Prosser. CAB International, Hamilton, New Zealand.
Webb, R., Garnsworthy, P.C., Gong, J.G., Robinson, R.S. and Wathes, D.C.

(1999) Consequences for reproductive function of metabolic adaptation to load. In *Metabolic Stress in Dairy Cows*, pp 99-112. Edited J.D. Oldham, G. Simm, A.F.Groen, B.L. Neilsen, J.E. Pryce and T.J. Lawrence. Occasional Publication 24, British Society of Animal Science, Edinburgh.

Whitaker, D.A. (2002) Clinical mastitis in British dairy herds. *Veterinary Record*, **151**, 248.

Wilde, C.J., Addey, C.V.P., Boddy, L.M. and Peaker, M. Autocrine regulation of milk secretion by a protein in milk. *Biochemical Journal*, **305**, 51-58.

21

YIELD OF SHEEP: PHYSIOLOGICAL AND TECHNOLOGICAL LIMITATIONS

R. WEBB[1], L. STUBBINGS[2], K. GREGSON[1] AND J.J. ROBINSON[3]
[1]Division of Agricultural and Environmental Sciences, School of Biosciences, University of Nottingham, Sutton Bonington Campus, Loughborough, Leics., LE12 5RD, UK; [2]LSSC Ltd, 3 Fullers Close, Aldwincle, Kettering NN14 3UU; [3]Scottish Agricultural College, Craibstone Estate, Bucksburn, Aberdeen AB21 9YA, UK

Background

To predict future yields it is necessary to review previous trends in sheep production, particularly to investigate the major factors that have impacted on the sheep industry. This chapter will review changes in the industry and provide an assessment of physiological and technological yield.

Those involved in analysing agricultural systems in the 1960s and 1970s invariably drew attention to the low efficiency of ruminants in general and sheep in particular as producers of human food (Blaxter, 1968; Wilson, 1968). They also went on to identify that, in the case of sheep, the low average fecundity of the ewe and the obligatory production of adult meat and wool, for which there was little demand, were central to the inefficiency of sheep in converting their food to high quality edible protein. There was also the problem of the fatty nature of lamb carcasses.

In the intervening years there have been significant advances in the knowledge of the physiology of reproduction (Gordon, 1997), lactation (Cant *et al.* 1999) and lean tissue growth (Grizard *et al.* 1999). Nutrient and energy requirements have been defined (ARC, 1980; AFRC, 1993) and there have been significant improvements in the reproductive technologies as reviewed previously in these proceedings (Sinclair and Webb, 2005). It is therefore appropriate to look briefly at the main drivers of change in the industry during the last 25 years, consider in detail the current factors that limit production and set out the physiological and technological opportunities for future improvements in productivity.

From the early 1970s onwards

In 1972 the Meat and Livestock Commission (MLC) published the first ever results of a survey to establish the breed and system structure of the UK sheep industry. It provided a comprehensive description of the stratified nature (hill, upland and lowland sectors) of the industry, the interdependence of the sectors and the breeds involved. Of the 12.1 million breeding ewes in 1971 (Figure 21.1), 44%, 17.5% and 38.5% were in the hill upland and lowland sectors respectively. Despite the subsequent expansion of the breeding flock (Figure 21.1) there was little change in the proportions in each land-type category by 1980 (Figure 21.2). By 2000 the national breeding flock was over 20 million (Figure 21.1), with the slightly lower percentage in the hill sector balanced in equal proportions by the increases in the upland and lowland sectors (Figure 21.2).

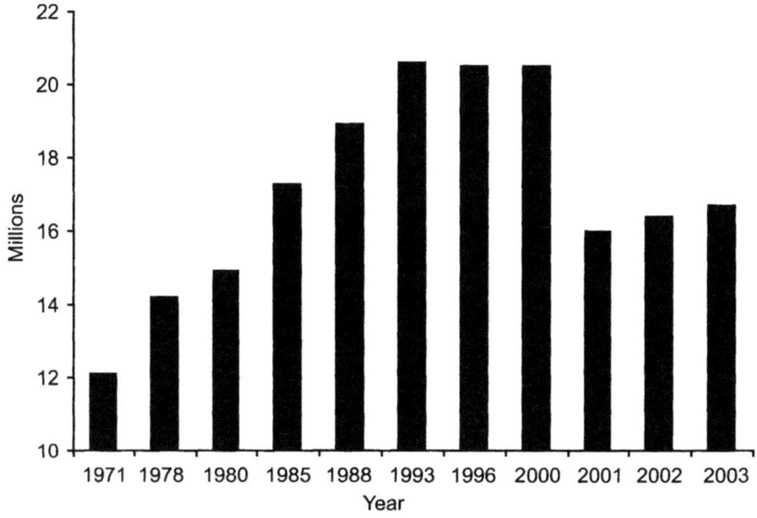

Figure 21.1 Breeding ewe numbers in the UK

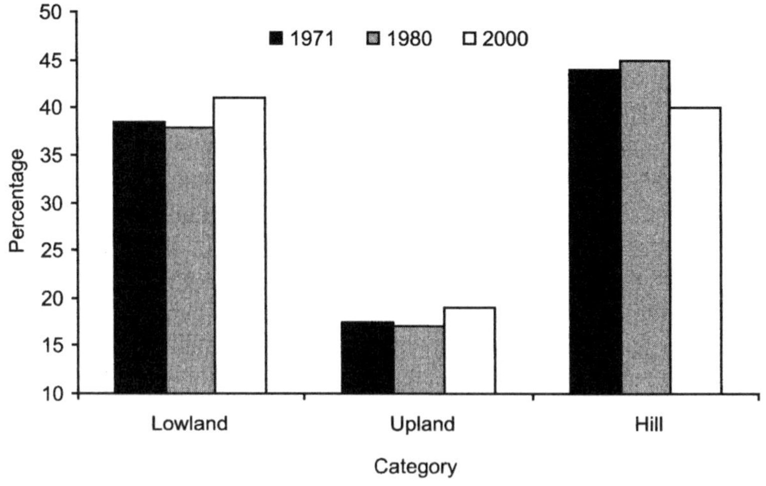

Figure 21.2 Structure of the national flock over the last three decades

The stratified system, which is largely unique to the UK, has been the subject of much discussion and, in recent years, criticism mainly due to the intrinsic need to transport sheep, and potentially disease, around the country. However, this structure has developed to optimise land use and sustain fragile rural communities in the hills and uplands. Indeed in 1970 the UK sheep industry, while still not vibrant, had one major advantage over the more intensive livestock sectors, for example pigs, which was its much lower cost base. While the intensive high input sectors struggled, the sheep industry remained relatively insulated from input cost increases. However, there were some clear trends emerging as farmers tried to reduce their cost base. For example, flock sizes were increasing and the number of holdings with sheep contracted sharply. In 1970, 51% of sheep in England and Wales were in flocks of 500 or more; in 1980, this had risen to 57% and there was a further sharp rise in the number of flocks with over 2,000 ewes.

The UK entered the EU in 1973 but, unlike many other sectors, there was no EU scheme for sheep meat until October 1980. Prior to 1973, the UK had a subsidy system called 'deficiency payments', which turned into 'guarantee payments' after entry into the Common Market. However, due to relatively buoyant prices in the latter part of the decade, monies were only paid out under the latter scheme for a few weeks of the year.

1980: INTRODUCTION OF THE EU SHEEPMEAT REGIME (SMR)

The SMR was based on a set price (guide price) for lambs sold during each week of the year, with any deficit between the market price being paid to the farmer, providing lambs met basic grading standards. In addition, each of the six regions of the EU agreed a reference price for sheepmeat and at the end of the year any deficit between this and actual income was paid as a single payment on the breeding ewe. The net effect of the SMR was to give the sheep industry a guaranteed level of income, which was also predictable throughout the ensuing season.

Profitability of the sheep sector increased correspondingly and, throughout the decade, sheep numbers increased, particularly in the lowland. Figure 21.1 shows how the national flock expanded rapidly during this period. In parallel, home consumption of sheepmeat was falling, from just under 10 kg/capita in 1970 to 7 kg/capita by 1992. However, the opening up of EU trade, in particular the French market, meant that the UK began to export significant amounts of sheepmeat, at peak ~ 30% of production (Figure 21.3). At the same time, imports from New Zealand fell sharply, from a peak of 330,000 tonnes to a low of 90,000 tonnes in 1990. The confidence that the SMR brought to the sheep industry encouraged lowland farmers back into sheep production and it was this that changed the balance of ewes in the hill, upland and lowland categories over the last 20 years (Figure 21.4).

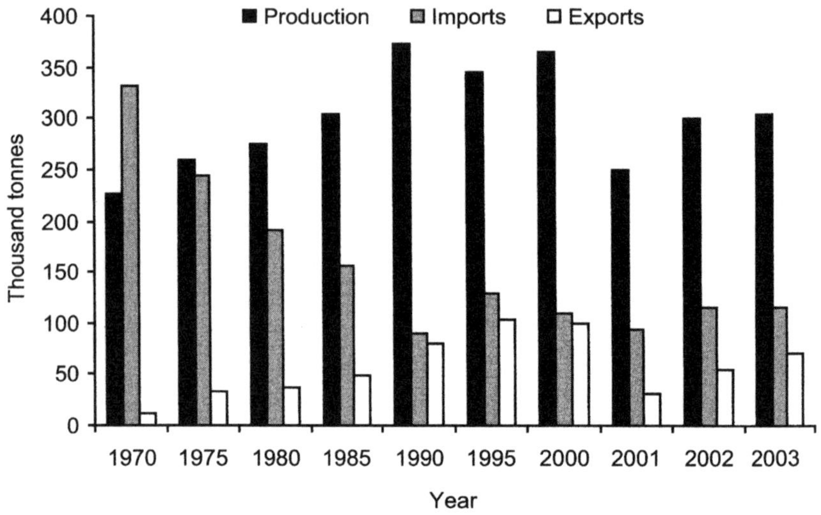

Figure 21.3 UK sheep meat supplies over the last 30 years, including the proportion exported and imported.

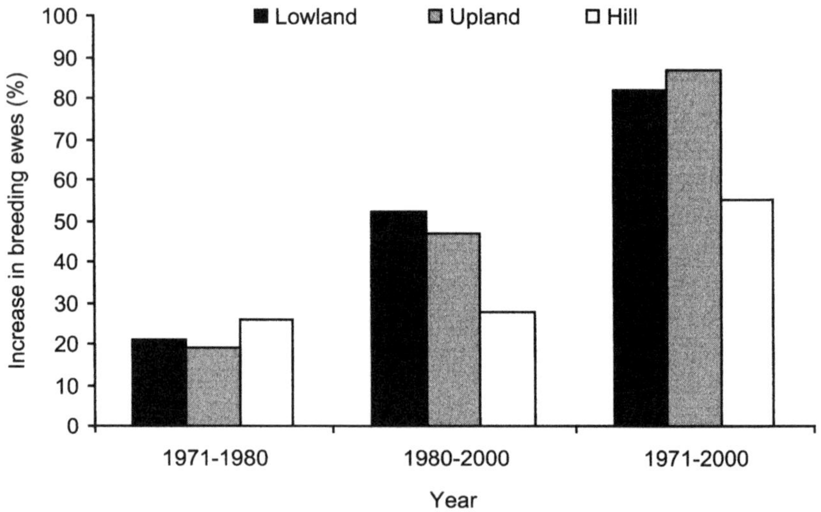

Figure 21.4 Percentage changes in the number of ewes in the stratified UK sheep production system.

One of the more contentious aspects of the export trade was, and remains, that of live sheep exports. Historically there had been a trade in live animals, mainly to France. This amounted to around 250,000 animals per annum in the 1970s. However, this increased at a rapid pace and by 1990 had reached 600,000 per annum, peaking at 1.4 million in 1992 and falling back to 780,000 in 2000, the year before the foot and mouth disease (FMD) outbreak.

The fortunes of the sectors are of course inextricably linked through the stratified structure of the industry. In the early years of the SMR, up to the introduction of quotas in 1992, most of the subsidy paid was on the weight of finished lamb sold, so the lowland farmer strove for output in terms of the weight and number of lambs produced. Hill and upland farmers benefited because they could supply crossbred breeding ewes, mainly of the Mule type (Bluefaced Leicester x Swaledale or Scottish Blackface) which satisfied the lowland farmer (Figure 21.5). They could also sell on males in store condition for finishing in the lowlands where there were resources available in the autumn and winter. Hence, while much of the EU monies went directly to the lowland sector, the whole industry benefited and breeding ewe numbers in the hill sector also expanded to fill demand.

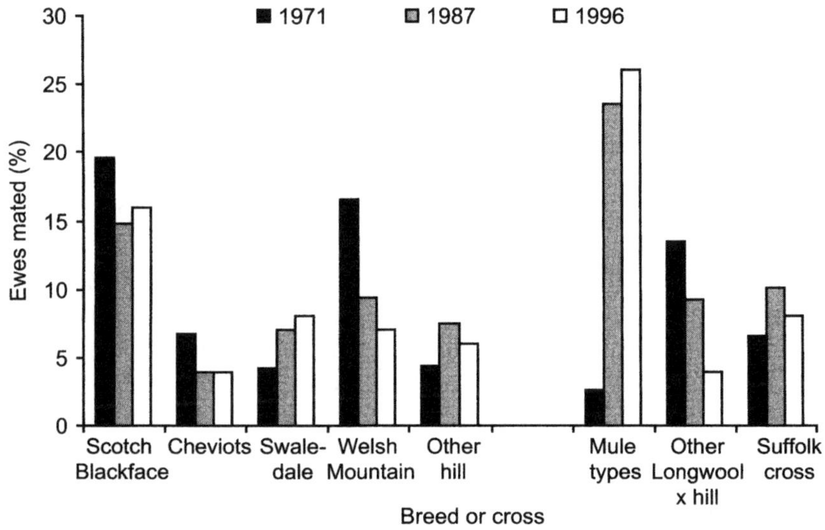

Figure 21.5 Changes during the last three decades in different breeds of sheep mated within the UK.

The balance of the SMR payment in the period 1980-1991 was paid as a Ewe Premium and was the deficit between the reference price and that actually achieved in the EU region (initially there were 6 regions). The scale of these payments is shown in Table 21.1 and an indication of the value of the variable premium to a lowland producer selling 1.5 lambs per breeding ewe at an average carcass weight of 17.5 kg is also included.

The hill sector was not totally dependent on the SMR for support during this period. Hill Livestock Compensatory Allowances (HLCA) were paid per head of stock, at varying rates according to the severity of the 'disadvantage' and whether the flock was of a 'hardy' breed. The link beween premium and the number of sheep kept was therefore already quite strong in the hills and uplands at the start of the 1990's when the shift to all payments, on a headage basis, was heralded as the cause of over stocking in some of these areas.

Table 21.1 EWE PREMIUM PAYMENTS (POUNDS STERLING)

Year	*Ewe premium*	*Average variable premium value per ewe*	*Total estimated payment per lowland ewe*
1982	2.73	14.70	17.43
1984	4.68	13.20	17.88
1986	5.12	10.72	15.84
1988	6.11	15.33	21.44
1990	9.88	12.70	22.58

As the UK industry flourished in the 1980s, the French industry continued to decline, while demand for sheepmeat in France remained strong, giving the UK an excellent opportunity to export and fill the gap, particularly as NZ lamb is not favoured in France.

1992 – 2000: Recent trends

The cost of the SMR, while small in the context of the CAP budget, had risen dramatically from 33 million ECU in 1980 to 1.4 billion ECU in 1991. Initially the UK took 98% of this money, but by 1988 this reduced to 27%. Moves to limit further expansion were applied in the form of stabilisers. From 1988, every 1% increase in breeding flock numbers reduced the basic price by 1% and in 1990, full rate Ewe Premium was limited to the first 500 ewes (lowland) and 1000 ewes (hill). By 1993, the variable premium also had to be phased out as all EU intra-community trade barriers had to be removed and in 1992 a system of quotas was introduced, based on sheep numbers in 1990. This locked national flock size into the 1990 figures (20.3 million) and prevented further expansion of the number of ewes receiving premium payments. Quotas can be traded, but only within ring fences to prevent any move either in or out of the hills and uplands (Less Favoured Areas – LFAs). Indeed, the national flock had stabilised at around this figure throughout the 1990s, until the FMD outbreak of 2001. A further change was that ewes no longer needed to be productive to receive subsidy, so unmated ewe lambs became eligible, which had a small negative effect on overall production of sheepmeat.

The positive benefits to sheep farmers of the SMR started to change when quotas were imposed in 1992, along with the Arable Area Payments Scheme (AAPS) and Integrated Administration and Control System (IACS) for the arable sector and, of course, extensification payments for beef. These changed the incentives for the enterprises and signalled an end to any significant integration of sheep on arable farms, with consequences for biodiversity and the environment. It also introduced an element of competition between sheep and beef cattle for grazing and set an artificial baseline for the rental value of land in line with extensification payments.

These changes in support mechanisms put sheep and arable into direct competition for most of the ploughable land in the lowlands. This meant that very quickly, sheep were forced on to permanent grass with all the associated problems of gastrointestinal parasitism and anthelmintic resistance, and an enterprise forced to utilise areas of grazing spread across and between farms. The sheep flock has struggled to maintain outputs in recent years due to this change in policy and the lack of integration with arable farming; even the use of catch crops and by-products from the arable sector has diminished.

This shift to permanent grass was not too serious for the lowland sheep flock in terms of numbers while cereal prices were high, although there were increasing management problems associated with moving between grazings and associated health issues. Surprisingly, however, a buoyant arable sector helped to keep sheep on many arable farms through the 1990s because overall farm profitability could support the reduced returns from the flock. However, as cereal prices have declined, arable units have been put under severe pressure to reduce costs and this has resulted in a reduction in lowland sheep production.

More recently, exposure of the cereal trade to world prices has forced cereal farmers to reduce their unit costs through farm amalgamation, more powerful machinery and lower manpower. The sheep flock has been a direct casualty of the reduction in on-farm labour. This started even before the sterling/euro imbalance made exports harder and the FMD crisis reduced sheep numbers further (Figure 21.6). Historically many of the most profitable sheep enterprises were mid-sized, on smaller farms run by family labour. Many of these farms have not been able to surmount the challenges of the arable sector and have disappeared into larger units, often under contract. On larger farms, the sheep are usually isolated in terms of both the land and labour that they use. The flock is either big enough to justify the specialist labour, or it is not there at all. While there has been a trend towards bigger flocks, with one man looking after greater numbers, this has limited application in the sheep sector with pressures on animal welfare and traceability.

BREEDS

The UK has over 70 breeds of sheep, both indigenous and imported, and over 170 crossbred types. However only a few breeds have more than 100,000 ewes and production is, in reality, influenced by a small number of breeds. Indeed there is only a relatively small gene pool used to sire the slaughter generation. Figure 21.5 illustrates the main breeds and crosses that make up the UK flock and shows how these have changed in the last three decades. The most notable feature is the increase from 1971 in the Mule type ewe which is a product of the stratified industry. With regard to terminal sires, the Suffolk is still dominant, but losing ground to the Texel and Charollais imported during the 1980s. It has been claimed that the stratification of the British sheep industry has started to disintegrate in recent years. However, the dominance of the Mule type of ewe refutes this suggestion. Maintaining a range of breeds may prove to be important in the future in order to exploit the benefits of genetic diversity and hybrid vigour.

PRODUCT TRENDS

Since 1977 there has been a smoothing out of the regional differences in the average carcass weights at slaughter with those in England and Wales increasing from 17.5kg and 15.5kg to 18.8kg and 16.5 kg respectively and those in Scotland decreasing from 19.9kg to 18.4 kg. At the same time the proportion of overfat carcasses (fat classes 4 and 5) has fallen from 14% to 8%.

WOOL

The value of wool has nearly halved since 1980. Most of the UK wool clip is coarse and hairy, and hence used in the carpet industry and other specialised sectors such as mattresses. In 1980, the UK was only 37% self sufficient in wool, by 1993 this had increased to 75% due to the increase in the size of the breeding flock and the contraction in manufacturing. Following the Tianamen Square massacre, exports to China of both wool and skins were terminated in the 1980s with disastrous effects on wool and skin prices on the home market. Currently the UK exports ~ 50 million square feet of dressed sheepskin leather and 10,000 tonnes of semi-finished leather annually. However, this has been jeopardised due to skin parasite damage following the removal of compulsory dipping for sheep scab in 1992. Lower skin prices are a particular problem to the slaughter trade, which traditionally relies on this source of income.

FMD

The FMD outbreak of 2001 had devastating short-term effects on the UK sheep industry. Breeding ewe numbers were reduced by 18% as a direct result of the slaughter policy for disease eradication. By 2002/3 there had been a small recovery in sheep numbers, but forecasts are that there will be only a modest increase in the future. In the longer-term, however, the impact is likely to be minimal and confined to logistical issues such as animal identification, traceability and transport limitations rather than any direct effects on production and trade in sheep meat. In conclusion, trends in the number of sheep slaughtered over the past 35 years exhibited a significant increase followed by a downward trend due to a variety of factors (see Figure 21.6).

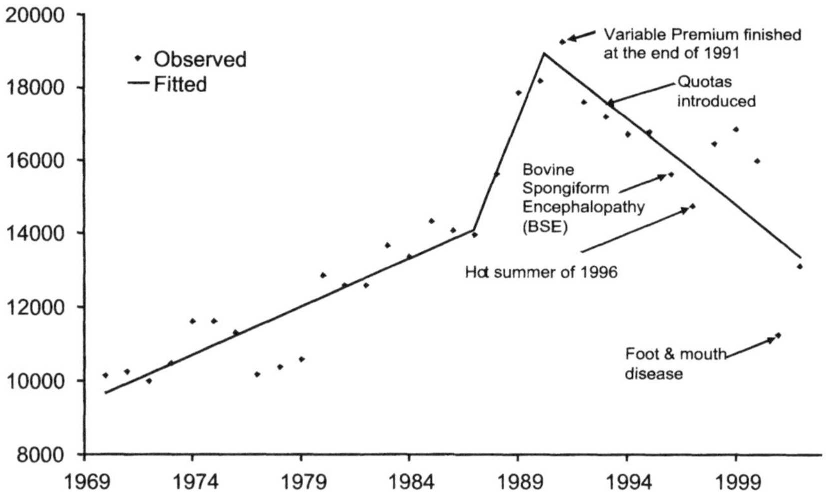

Figure 21.6 Trends in the numbers (thousands) of clean sheep slaughtered in the UK over the past 35 years. Annotations indicate possible causes of variation.

The future

The challenges that now face the sheep industry are those related to the CAP reform and the effects of decoupling and globalisation; also food safety, particularly in relation to the spongiform encephalopathies.

DE-COUPLING

The UK Government has indicated that it intends to fully de-couple support payments from 2005. This is likely to have the following effects on the sheep industry:

- The lowland sector will no longer have to compete against support for other sectors, namely the beef extensification subsidies and AAPs.
- Marginal arable land is likely to come out of cereal production and into grass, giving sheep an opportunity to use additional grazing land.
- The lowland beef herd will fall dramatically, possibly by 50%, but lowland farmers are unlikely to go back into sheep production. In many cases businesses have been downsized and have neither the labour, equipment, infrastructure nor probably the determination to return to sheep production.
- The profitability of sheep production may increase marginally as farm-gate prices improve and the UK export market should present opportunities, particularly to France.

- Overall, the lowland breeding flock is expected to remain fairly stable in numbers, albeit probably managed in fewer, larger, flocks.
- In the LFAs the beef herd is expected to contract further. Only cross-compliance or agri-environmental payments to reverse the degradation of some moorland areas will keep cattle numbers up in these areas.
- Sheep numbers will fall, although this will not necessarily be across the board and there must be some concern about the potential for vast areas to be under grazed unless cross-compliance requires minimum stocking rates.
- Estimates range from 25 – 50% or more reduction in LFA sheep numbers. This has implications for the supply of breeding females to the lowlands.
- It is forecast that many LFA farmers in particular will become 'part-time' farmers, similar to many in Continental Europe. There are implications for the productivity and welfare of the sheep in these areas.
- It is unlikely there will be any significant direct effects of globalisation on the UK sheep sector. Australia and New Zealand are unlikely to export more sheepmeat to the UK since alternative markets have been identified and their own production has fallen in recent years, rather than increased.

The key question is whether the UK sheep industry will revert to a mirror image of the situation described 30 years ago, since this would impact on the possible uptake of technological and/or physiological developments. In the past, it is the lowland sector that has bore the brunt of the reduction in numbers. In future it looks as though it will be the LFAs. It is the longer-term implications of this and the structure of production systems in the lowland, coupled with physiological and technological developments, that will determine future output, level of profitability and production efficiency within the sheep sector.

Biological ceilings to production

Physiological characteristics such as age of puberty, litter size and growth rates can all impact on biological production. For example, to our knowledge, the youngest recorded age for a ewe to lamb following a natural-occurring oestrus is 7½ months (Robinson, 1982). This is ~ 4½ months less than in farming practice. Litter sizes as high as seven have been recorded in ewes with a major gene for prolificacy (Bindon *et al.* 1996) and five has been taken as the biological ceiling (Wilson, 1968). However four is a more realistic upper limit when lamb survival issues are taken into account. A proportion of ewes can also lamb naturally twice per year (40% in the Finnish Landrace flock recorded by Walton and Robertson, 1974). Combined with a litter size of four this gives a biological ceiling for production of eight lambs per ewe per year, or five fold the level currently achieved by the top third of lowland flocks (MLC, 2002). Furthermore, lamb growth rates for terminal sire breeds such as the Suffolk are now in excess of 600 g/day with every indication that these can be increased further by genetic selection (see later discussion).

Production efficiency

The theoretical improvements in production efficiency that could be obtained from increasing lamb output through a combination of increases in litter size and breeding frequency are illustrated in Figure 21.7. In this exercise it is assumed that lamb numbers in excess of triplets are reared artificially. Appropriate adjustments (Large, 1970) have therefore been made to food inputs to allow for this assumption. As well as illustrating that there is an almost 2.5 fold difference in efficiency of metabolisable energy (ME) utilization for meat production between the 'biological ceiling' (8 lambs per ewe per year) and the levels of production currently achieved by the lowland sector of the industry, Figure 21.7 provides a clear estimate of the relative importance of increases in litter size and breeding frequency on the efficiency of utilization of ME. In this regard a 'three-times-in-two-years' breeding frequency combined with a litter size of two is a realistic goal and would improve current production efficiency on the best lowland farms by over 30%. Such a system has been tested using the best available technology at the time (artificial daylength control) for increasing breeding frequency (Robinson, 1978). Annual production was 3.5 lambs per ewe and there were no adverse effects on ewe health. It was subsequently shown that the same level of production could be achieved without resorting to daylength control, but gonadotrophin hormones were required to stimulate ovulation during seasonal anoestrus (Gordon, 1997).

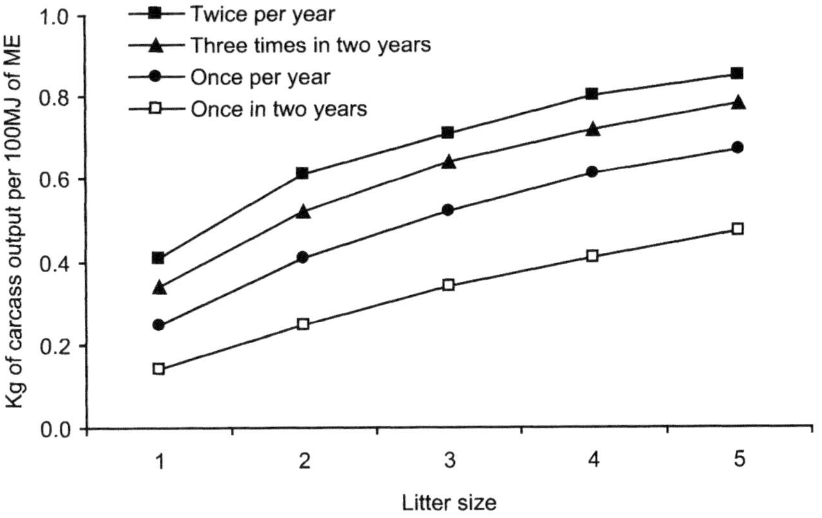

Figure 21.7 Effect of litter size and lambing frequency on efficiency of use of metabolizable energy (ME) for the production of sheep meat. Efficiency = kg carcase output per 100 MJ of ME.

Integration of the system into a UK lowland mixed farming enterprise (grassland and arable) provided, for the first time, experimental as opposed to theoretical data, for a comparison of the requirements for metabolizable energy from specific foods (grazed herbage, preserved forage, concentrates), for carcass production from a 'three-times in two years' as opposed to a 'once-a-year' breeding programme (Figure 21.8). It therefore provided the necessary information for optimising, at a specific farm level, the efficiency of utilization of food resources in relation to the degree of intensity of the production system, and the availability of food by-products from other farm enterprises (Tempest, 1983).

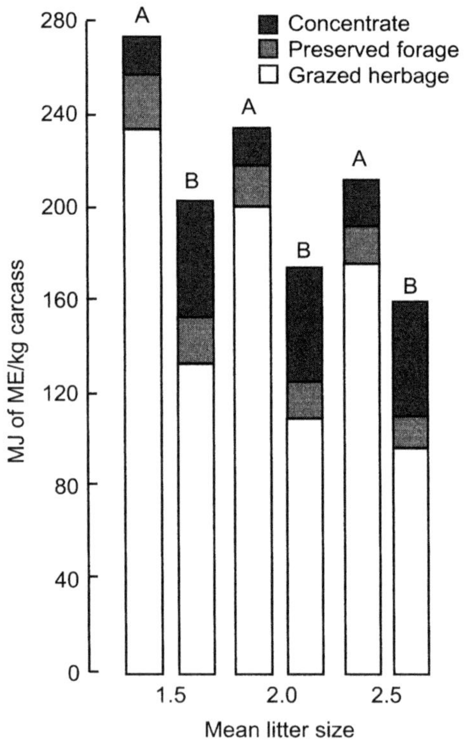

Figure 21.8 The effect of mean litter size and breeding frequency (A = breeding once per year; B = breeding three times in two years) on the total food energy (MJ of ME) and of the individual components of the diet required to produce 1 kg of lamb carcass (From Robinson 1978).

Another component of the early theoretical calculations, on the efficiency of sheep meat production that translated into farming practice was the effect of ewe size. For example, the theoretical calculations showed that the small Welsh Mountain ewe crossed with a large terminal sire (Suffolk) is almost 40% more efficient in the production of meat than the much larger Scottish Half bred ewe when each produces a single lamb. For twins the improvement in efficiency is ~ 24%. Thus, in terms of food utilization, the efficiency of a small ewe producing twins to a large sire is higher than for a large ewe producing triplets. This principle has been exploited at regional level by Vipond *et al.* (1987), using crossbred ewes derived from the small Shetland breed. Also, it can be readily accommodated within the existing stratified nature of the UK industry where small hill ewes such as the Scottish Blackface, Swaledale and Welsh Mountain are crossed with the large long-wool sires for the production of medium-sized crossbreds for use in intensive grassland production systems. With careful selection of ewe and sire sizes and of breed types for ease of lambing, the improvement in production efficiency from the use of smaller ewes has beneficial environmental effects. These effects may become even more relevant with a move towards more efficient nutrient and energy budgeting within animal production enterprises.

Impact of production system analyses on research and development

The preceding system analyses have provided the major stimulus for research studies into improving the yield potential of sheep systems. These studies have embraced digestive physiology, nutrient and energy utilization, and appetite regulation. They have also probed the underlying physiology and endocrinology controlling seasonal breeding, with a view to developing practical methods for its elimination. They have focussed on breeding, feeding and pharmacological intervention strategies for improving ovulation rate and embryo survival as well as reducing neonatal mortality, enhancing growth and reducing the fat content of carcasses. Naturally prolific breeds have been identified, so too have major genes for ovulation rate within breeds. There have been advances in artificial insemination, semen cryopreservation, semen sexing, embryo transfer, gene mapping and cloning (see Sinclair and Webb, 2005; Chapter 4 of these proceedings). Research in all these areas provides new opportunities for improving production, given the market incentive to do so. The following sections, therefore discuss the physiological and technological opportunities that could influence the yield of sheep meat. Attention is also drawn to problems requiring further research effort.

Physiological limitations and opportunities

DIETARY

In most sheep systems the overriding factor limiting production is nutrient and energy supply which in turn is a function of the diet and its intake relative to the animal's maintenance needs. The factors controlling intake are food availability, food quality, rumen fill and a series of signals to the brain that reflect the animal's post-absorptive state, and the size and rate of change of its lipid reserves (Forbes, 1996; Langhans, 1999). When the physical nature of forages constrains their intake, and therefore their nutritive value, chopping and grinding them are well established processes for improving their intake and production potential. Others such as sodium hydroxide and ammonia treatment, which increase the availability of the structural carbohydrate fraction, improves their digestibility and intake, but inadequate structural carbohydrate degradation is still regarded as a major limiting factor to production (Satter *et al.* 1999). Attempts to combat the problem are now the subject of gene manipulation at the level of the enzymes involved in the production of lignin (o-methyl transferase and cinnamoyl alcohol dehydrogenase) and its precursors (cinnamoyl CoA reductase and 4-coumarate: CoA ligase) (reviewed by Robinson and McEvoy, 1999).

Another approach to overcoming dietary induced limitations to production is the breeding of forage varieties containing elevated water-soluble carbohydrate concentrations. In the study of Lee *et al.* (2001) a conventionally-bred perennial ryegrass with this attribute increased lamb growth rate by 12% and production per unit land area by 23%. Supplementation of forage diets with readily fermented carbohydrate in the form of cereal grains is widely used to increase production, but the effectiveness of this approach is limited by its adverse effect on the utilization of the energy-yielding fibre component of the diet. This arises from the reduction in rumen pH that accompanies concentrate supplementation inhibiting the growth of the rumen bacteria that digest cellulose. What is required is an acid-tolerant cellulolytic bacteria; thus leading to the concept of using gene transfer to introduce cellulolytic function into the robust, acid-tolerant bacteria such as *Prevotella ruminicola* (see Satter *et al.* 1999). As was stated, an associated environmental advantage of a low rumen pH is the inhibition of methanogensis.

Despite the high levels of crude protein (6.25 x nitrogen) in many temperate grasslands, animal production from these systems is often limited by inadequate amino acid supply. Thus responses to high quality rumen undegradable protein (RUP) during late pregnancy have been obtained for lamb birthweight, wool production, colostrum production and maternal tissue nitrogen balance in triplet bearing ewes (Barry and Manley, 1985), for milk production in twin suckling ewes (Dove *et al.* 1985 and Penning and Treacher, 1988) and also for growth rate in lambs following weaning (Poppi *et al.* 1988). Despite the high protein content of the basal forage diets such responses cannot be achieved by carbohydrate supplementation alone (Table 21.2). This clearly demonstrates the relative inefficiency of nitrogen utilization from fresh forage and the associated adverse impact on the environment.

The recent demonstration of the important role that additional undegraded dietary protein (UDP), during late pregnancy and early lactation, can have in suppressing gastrointestinal worm burdens (Donaldson *et al.* 2001) emphasises further the need to develop procedures for optimising amino acid supply to the animal within the context of nitrogen budgeting that minimises environmental pollution.

Table 21.2 PRODUCTION RESPONSES TO FEED SUPPLEMENTS WHICH INCREASE THE POST-RUMINAL SUPPLY OF AMINO ACIDS IN SHEEP OF DIFFERENT PHYSIOLOGICAL STATES AND RECEIVING HIGH PROTEIN FORAGE DIETS TO APPETITE

Physiological State	*Nature of Supplement*	*Production Response*	*Reference*
Late Pregnancy (Triplet–bearing)	Abomasal infusion of casein	Increases in: Lamb birthweight, colostrum production and wool production. Maternal Tissue nitrogen balance moved from –5.5g/d to +3.6 g/d	Barry and Manley (1985)
Lactation (Twin suckling)	Molassed sugar beet pulp (600 g/d)	No significant effect on milk yield	Dove *et al.* (1985)
	Molassed sugar beet pulp and formaldehyde-treated soya bean in a 1:1 mixture (600 g/d)	Daily milk yield increased from 2.1kg to 2.9 kg	Dove *et al.* (1985)
	Barley + Maize starch	No significant effect on milk yield	Penning and Treacher (1981)
	10 MJ of ME/d from a 1:1 mixture of barley and soya bean meal	Daily milk yield increased from 2.6g to 3.2 kg	Penning and Treacher (1981)
Lamb growth	3.25 g fish meal DM/kg lamb weight/d	Daily growth rate increased from 201g to 266 g on tall fescue and from 331g to 388 g on white clover	Poppi *et al.* (1988)

DISEASE

Gastro-intestinal parasitism is now regarded as the number one sheep health problem worldwide, reducing lamb growth rate by more than 25%. As well as the nutritional dimension to its expression there is a genetic component (Gray, 1997; Bishop and Stear, 2001). However, in the absence of a genetic marker, selection for resistance is, so far, limited to nematode faecal egg output (h^2 = 0.2 to 0.4 with large variability) or associated host responses such as plasma IgA activity, fructosamine concentrations (Stear *et al.* 2001) or blood eosinophil counts (Stear *et al.* 2002).

Genetic selection programmes for gastrointestinal parasitic resistance, as a consequence of anthelmintic resistance in the New Zealand and Australian sheep industries, have demonstrated what can be achieved by genetic selection in their worm faecal egg count (FEC) and Nemesis commercial breeding programmes. Some producers are now operating without anthelmintics which gives them the added advantage of marketing lambs as 'organic'.

Meantime, in the UK it has recently (Walling *et al.*, 2004a) been shown that there is increased heterozygosity and allele variants in the major histocompatability gene complex (MHC) on chromosome 20 in the Texel breed, compared with the Suffolk. Since MHC is associated with the ability of the immune system to respond to parasite infection, this may account for the greater natural resistance of Texels than Suffolks to gastrointestinal infections. Heterozygosity in the MHC is also being used by Lincoln University in New Zealand as a genetic marker in selection for resistance to footrot (Hickford, 2004).

Understanding the genetic basis of scrapie susceptibility, in which vulnerability to the disease is due to variability at PrP gene codons 136, 154 and 171, has lead to the UK National Scrapie Plan directed at the elimination of susceptible genotypes from the UK sheep population. This has raised concerns, at industry level, that there may be associated beneficial traits and that these could be lost in a mass cull of susceptible genotypes. This is one of the reasons why the recently established 'Sheep Trust', which followed on from the Heritage Gene Bank set up during the 2001 FMD outbreak to cryopreserve gametes from endangered breeds, is a central campaigner in the fight to cryopreserve gametes from scrapie susceptible genotypes (Bowles *et al.* 2004). It also underlines the requirement to have a balanced selection process where the full implications of selection on a single trait have been modelled.

In view of public health concerns, high health status initiatives at national level, in the case of the Maedi Visna accreditation scheme and, on a more regional basis, the Highlands and Islands of Scotland enzootic abortion free flock scheme, are likely to be given much greater prominence in future sheep production systems.

GROWTH, CARCASS COMPOSITION AND QUALITY

The introduction of sire referencing schemes for the improvement of lamb growth rate and carcass lean based on ultrasonic scanning and computerised tomography is providing the industry with new opportunities to meet consumer needs (Simm, 1998). Index data for 7 breeds are given in Figure 21.9. Genetic trends in the Suffolk sire reference scheme are also shown in Figure 21.10 for a range of characteristics included in the scheme index. There is no indication, after ten years of selection, that there is a slowing down in the rate of progress, with extrapolation of a best fit mathematical relationship to the data predicting steady improvements in the Suffolk (Figure 21.11), Texel and Meatlinc beyond 2025 and even larger improvements in the Charollais presumably reflecting a lower selection pressure prior to entering the scheme. Indeed for the Meatlinc, a composite terminal sire breed of relatively recent origin (Fell, 1987) and now fixed as a pure breed, recent improvements in the index are particularly dramatic. An indication of how the index translates into increases in 8 week weights (a measure of ewe milk yield and lamb growth potential), scan weights at 20 weeks of age and the corresponding improvements at 20 weeks in the depth of the *longissimus dorsi* muscle in the Suffolk sire reference scheme is given in Figure 21.10.

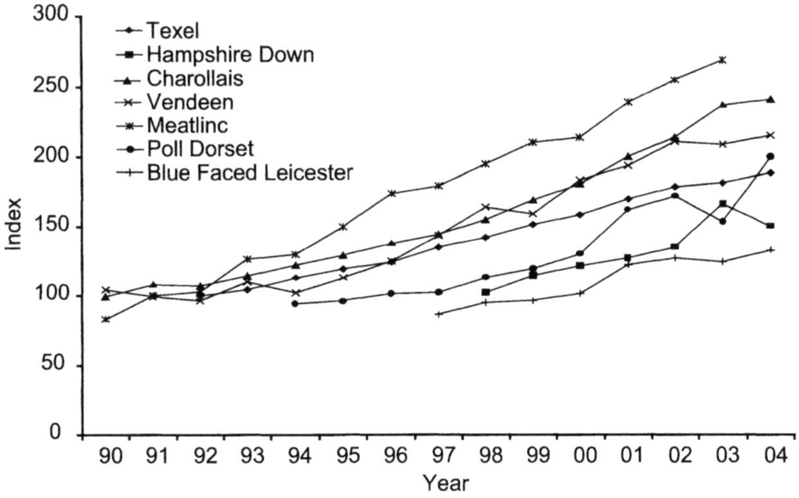

Figure 21.9 The influence of Sire Referencing Schemes on improving index score in different breeds of sheep (MLC's Signet Breeding Services).

In view of the recent evidence that there appears to be quantitative trait loci which influence muscle depth (on chromosomes 1 and 18) and fat depth (on chromosomes 2 and 3) in sheep involved in these commercial breeding programmes (Walling *et al.*, 2004b), there is every likelihood that the use of

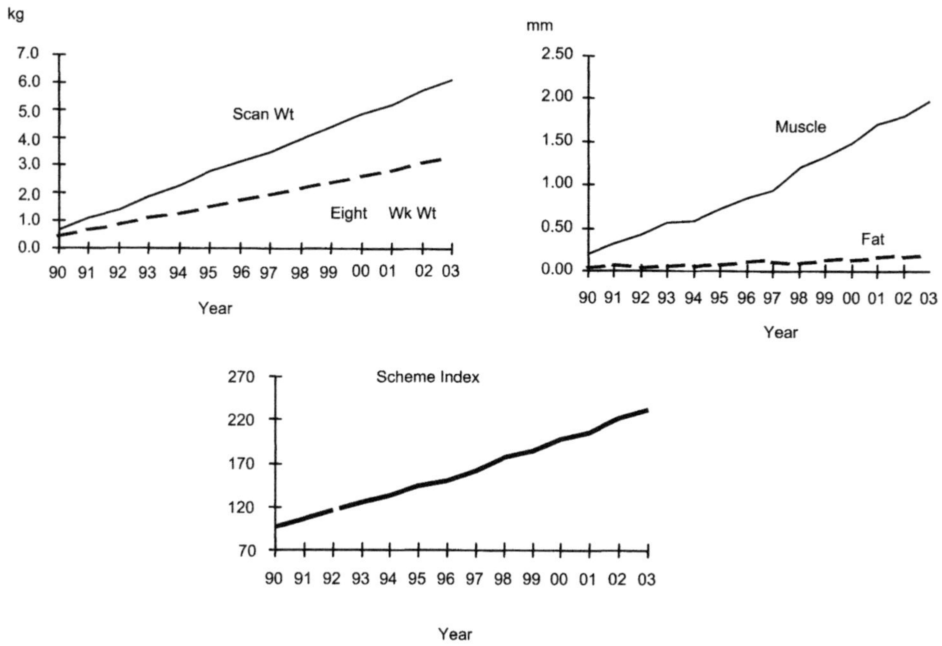

Figure 21.10 Genetic trends in the Suffolk sire reference scheme. Lamb liveweights are recorded at 8 and 20 weeks of age and ultrasonic scanning for muscle and back fat depths at 20 weeks.

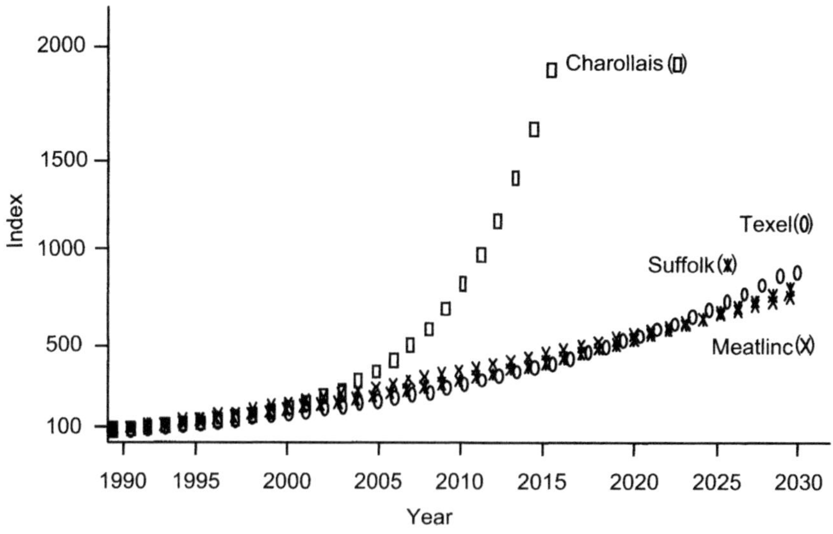

Figure 21.11 Predicted Genetic Trends in the MLC's Signet Breeding Services SRS Selection Index.

these QTLs in marker-assisted selection, and possibly candidate genes in the future, will increase the rate of improvement of carcass traits. Furthermore, concerns that by placing the selection emphasis on leaner carcasses, there may be corresponding adverse effects on ewe reproductive rate, particularly in hill ewes which are known to exhibit a strong positive relationship between body condition score and ovulation rate (Doney *et al.*, 1982) have so far been unwarranted [J. Conington, 2004, personal communication].

In terms of the eating quality of meat, genetic markers for meat tenderness have been identified (Meadus, 1998) implying that there may also be opportunities to develop DNA probes for use in genetic selection programmes, aimed at improving both the production and eating quality of lean meat. Relationships between muscle fibre type distribution and both slow and fast myosin heavy chain content on calpastatin are being investigated in order to gain a more fundamental understanding of the factors controlling tenderness of sheep meat (Sazali *et al.*, 2005). Current research into the effects of diet on the fatty acid content of meat is also demonstrating new possibilities for bringing it more in line with medical recommendations for improved human health (Cooper *et al.*, 2004).

REPRODUCTION

This is an area where research has moved on from the use of hormone manipulation to enhance litter size (Boland and Crosby, 1993) and stimulate out-of-season breeding (Gordon, 1997). The emphasis now is on the more consumer-acceptable marker assisted selection approach. Table 21.3 provides an overview of the recent advances in the identification of major genes for improvements in ovulation rate and aseasonality of breeding (see also Sinclair and Webb, 2005; Chapter 4 these proceedings).

These advances provide a range of production-system ceilings for output based on the natural production potential of the animal. However, within these options, in many sheep production systems the desirable goal is twin lambs of uniform birthweight. This minimises the adverse effects on neonatal survival of extremes in birthweight and inequality between siblings in competition for the fixed pool of maternal resources. Evidence of genetic variation in the distribution of ovulation rate (Hanrahan, 2003) indicates that there is scope, through selection, for minimising the incidence of unwanted triplets and higher multiples in prolific breeds. Through a better understanding for each genotype, of the relationship between ovulation rate and energy/nutrient supply, it should be possible to select genotypes that optimise the use of feed resources at individual farm level. Where there is unlimited food supply the economic benefits of increases in litter size occur over a much wider range than where food supply is limiting; indeed in the latter situation relatively small increases in litter size can become counter productive economically, with the cost of additional inputs exceeding the extra returns (Conington *et al.*, 2004).

Table 21.3 SINGLE GENES CONTROLLING REPRODUCTIVE TRAITS IN EWES, THE BREEDS THAT EXPRESS THEM AND THE MAGNITUDE OF THEIR EFFECTS

Trait Ovulation rate		*Number of copies of gene*	*Average Increase in:-*		*Trait Ability to breed out of season*	
Breed	*Gene*		*Ovulation Rate*	*Litter Size*	*Breed*	*Gene*
Inverdale (Fec X^I)	BMP15	1	1.1	0.7	Merinos D'Arles	Mel_{1A}
		2	Sterile			
Booroola (Fec^B)	BMPR-IB	1	1.3	0.9	Small Tail	
		2	3.1	1.4	Han Sheep	Mel_{1A}
+Cambridge	BMP15, GDF9	N/D	4.2	2.7		
+Belclare	BMP15, GDF9	N/D	2.6	2.1		
Icelandic Thoka	N/D	1	1.2	0.7		

N/D:- not determined. +Values for ovulation rate and litter size are absolute.
Note: BMP - bone morphogenetic proteins; BMPR - BMP receptor; GDF - growth differentiation factor; Mel_{1A} - melatonin receptor.
Data sources: - Hanrahan and Owen (1985), Fahmy and Davis (1996), Hanrahan (1989), Owen (1996), Russel *et al.* (1997), Galloway *et al.* (2000), Pelletier *et al.* (2000), Walling *et al.* (2002), Souza *et al.* (2003), McNatty *et al.* (2003), Mulsant *et al.* (2003), Chu *et al.* (2003) and Hanrahan *et al.* (2004)

Although restrictions in lamb production caused by seasonal breeding can be overcome by pharmacological intervention most notably in the form of melatonin (reviewed by Haresign, 1990) which transmits information on daylength, out-of-season breeding occurs in a small number of individuals of most breeds, implying that selection for aseasonal breeding may be possible. In this regard a genetic marker would be particularly valuable for the identification of the trait; thus the relevance of the recent investigation by Chu *et al.* (2003,) which adds to the earlier observation of Pelletier *et al.* (2000), showing an association between alleles for the melatonin receptor 1a (Mel_{1A}) gene and reproductive seasonality is of particular interest. Working with the Small Tail Han sheep of China (see Figure 21.12), which is noted for its aseasonality in ovulatory activity, Chu *et al.* (2003) have found a polymorphism of the Mel_{1A} receptor gene which is associated with their aseasonal breeding activity (see Table 21.3). Rate of response could also be increased by selecting in young animals, but the appropriate marker would need to be well characterised to ensure that the trait of interest is altered, rather than associated traits.

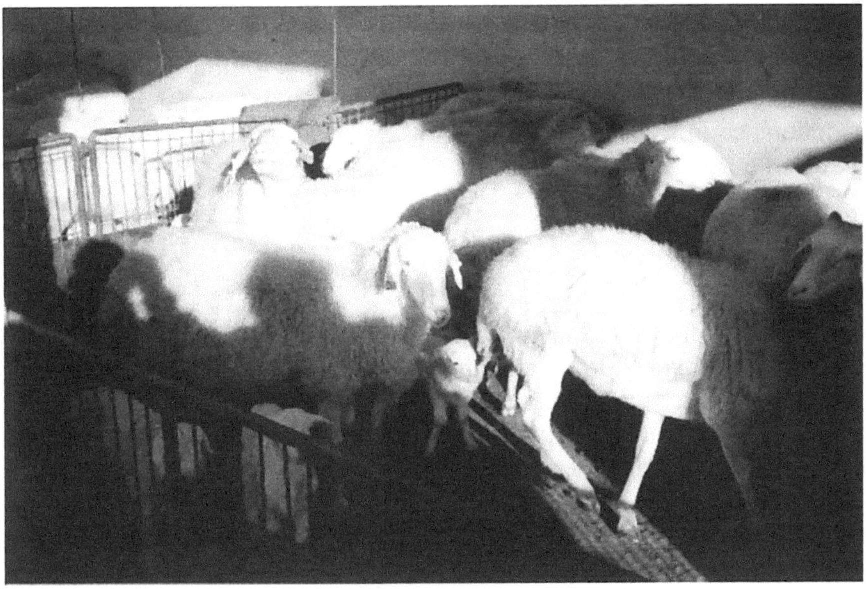

Figure 21.12 Han Sheep in China, with lambs at foot in October 2004 (courtesy of K. A. Linklater).

LAMB SURVIVAL

The rapid decrease in lamb survival as litter size increases above two (Hanrahan, 2003) is a major impediment to improving output. There are a number of reasons for this decrease and understanding the underlying process can avoid excessive neonatal losses. Firstly the curvilinear nature of the relationships between litter size and ovulation rate means that there are intermediate optima for ovulation rate, and these vary between breed, largely as a result of breed differences in the distribution of ovulation rates (Davis *et al.*, 1998). Exceeding the optimum ovulation for a particular breed leads to excessive embryo loss during implantation; the result is below average birthweights of those that survive and an associated greater risk of neonatal death from environmental factors such as hypothermia. As litter size increases there is also the problem of greater competition between litter mates for the fixed pool of maternal resources. In this regard more information is required on factors influencing within-litter variation in birthweight, since lambs with below average birth weight are less able to compete in the acquisition of colostrum, with a subsequent reduction in passive immunity to infection.

Despite these preceding limitations to increasing output, there are examples of the successful incorporation of single genes for ovulation rate into production systems. For example Russel *et al.* (1997) introduced the Icelandic Thoka gene for fecundity (see Table 21.3) into Cheviot ewes and increased the mean number of lambs born per ewe mated from 1.2 to 1.8 with only a marginal increase in lamb mortality (10% as opposed to 8% for non-carriers of the gene) at 6 weeks of age. A similar programme to introduce the Inverdale gene into the Texel for the production of lowground crossbred ewes of comparable prolificacy to the current Mule, but with improved carcass attributes, is now underway in Scotland. The advantage of incorporation of genes such as the Thoka into sheep would result in reduced variation in litter size with a preponderance in twin births. This would be of a major advantage to producers, for example in the management of nutrition.

LACTATION

Despite the wide variation in milk yields, both between and within breeds (Treacher and Caja, 2002), and the importance of a high potential for milk production to meet the increased demand for milk created by improvements in ewe prolificacy, remarkably little effort has been devoted to selecting for improved milk production. Undoubtedly, there are many situations where nutrient supply is inadequate to meet the genetic potential for milk production, as discussed. However, when corrected for the much higher total solids content of ewe's milk compared with cow's (see Garnsworthy *et al.*, 2005; Chapter 20 these proceedings), many prolific ewe breeds fall short of today's highest yielding dairy cows in terms of peak energy and protein production per unit metabolic size. They are also relatively poor regarding the persistency of their lactations. A major limitation therefore to production has been the failure to select, within prolific breeds, for increased milk production. Where single genes for prolificacy are being incorporated into ewe breeds of low or moderate prolificacy, greater emphasis should be placed on the selection of 'gene-recipients' that are at the upper end of the milk yield potential for the breed.

A further factor limiting the translation of major increases in prolificacy into significant improvements in productivity is the limited number of milk dispensary points (2 teats). Although teat number is highly heritable, and flocks of 4-teated ewes are relatively easy to establish, the small proportion of total milk that is usually produced by the front two teats limits their usefulness. However, Davies (1999) points out that there is considerable variation in the amount of milk produced by the front teats indicating that there may be selection opportunities for improving fore udder development and therefore fore teat milk yields. In this regard, up to 30% of total milk production from the front two teats of a 4-teated ewe has been recorded (D.A.R. Davies, 2004, personal communication).

Technological developments

This aspect has already been discussed (see Sinclair and Webb 2005, Chapter 4 these proceedings), hence it will only be reviewed in the context of sheep production.

SEMEN SEXING

In view of the important role of the hill sheep sector in the production of female breeding stock for lowland systems and the likely contraction of this sector in the future (as discussed), the use of female-producing spermatozoa could improve the efficiency in the production of females. Although current technology can guarantee high levels of purity (85 – 92%) for both female and male enriched spermatozoa fractions segregated by flow cytometry (O'Brien *et al.*, 2003), there are technical difficulties in the application of the technology. For example, the necessity for a low-dose inseminate makes intrauterine insemination by laparoscope a necessity. This is a veterinary procedure with associated welfare concerns and prohibitive costs. Even with intrauterine insemination, conception rates, albeit much improved from the first reported use of sexed semen in sheep (Cran *et al.*, 1997) and now involving post-sorting cryopreservation (Hollinshead *et al.*, 2002) as opposed to fresh semen originally, are still too low (~ 25% and 15% for female and male-enriched fractions respectively) for commercial application. Further advances in both the sorting and insemination techniques are required to make this fragile technology commercially viable. However, development of a significantly robust technology should ensure future widespread commercial application in hill flocks.

CLONING

In terms of biotechnology, sheep are at the forefront of cloning technology (Wells, 2003) and gene transfer both by the pronuclear injection of numerous copies of a gene and the nuclear transfer of cultured cells already known to express the desired gene (Thomson *et al.*, 2003). This has put them centre stage in the creation of new products in their milk for combating human diseases such as emphysema and cystic fibrosis. There are three main methods for producing genetically identical mammals:

- Individual separation of embryonic blastomeres up to the four cell-stage
- Embryo bisection at the morula or blastocyst stages
- Nuclear transfer (or nuclear cloning)

The first two cloning methods rely on the inherent cellular totipotency of very early embryonic cells. This therefore limits the number of viable embryos and offspring that can be obtained. The animals produced from founder embryos are true clones, in that they share identical mitochondrial and genomic DNA. Although embryo splitting has been applied to commercial multiple ovulation and embryo transfer (MOET) programmes, aimed at multiplying valuable genotypes (Wells *et al.*, 1990), it is the nuclear transfer (NT) methodology that has greatest potential application for animal agriculture, since many offspring can be produced from each donor embryo or animal. The major arguments made by commercial breeding companies in favour of cloning are its potential in the production of disease resistant animals and more consistency in meat quality. It will also depend on major improvement in the efficiency of this process, which is still low. In addition, progress in the speed and extent of application of this technology and the assisted reproductive technologies in general will, quite properly, also be controlled by their animal welfare implications (FAWC, 2004) and public acceptability.

ANIMAL IDENTIFICATION

The development of a practical, affordable electronic identification (EID) system should be a major immediate objective for the UK sheep industry. The benefits are numerous (Dziuk, 2003). In terms of animal health and biosecurity the FMD outbreak in 2001 made both industry and Government fully aware of the need for better traceability of stock. The current AMLS (Animal Movement Licensing Scheme) works on the basis of batch movements of sheep, but is paper-based and therefore has a significant time-lag between movement and registration on the system. An electronic data capture and handling system would vastly improve the traceability of stock with advantages for producers, including trade links outside the UK, and consumers.

Consumer health and confidence, particularly in relation to spongiform encephalopathies, are essential to the sheep industry and one way of winning this confidence is an assurance that sheep meat can be traced to its farm of origin. Hence, an ability to identify sheep, but with their individual details (health status, genetic make-up, ancestry, progeny, performance etc) would be of huge value to breed improvement programmes. The supply trade for the sheep sector would also gain valuable information of commercial value from a reliable EID system that also recorded for example, health product inputs. Furthermore, improved prediction of market demand trends would be extremely important in marketing terms. Current databases still have severe limitations (see Groeneveld, 2004, for discussion), although undoubted future advances in electronic technologies should help to overcome this.

The ability to record individual carcass data and process them into a commercial value index would enable both the farmer and retailer to gain maximum benefits. The farmer could hone selection skills and breed improvement programmes could be enhanced with much more accurate and trustworthy feedback. The abattoir / retailer would be able to make commercial decisions on the most economically viable sheep which would aid in the development of new products, for example cuts from heavier carcasses if this was shown to be cost effective.

Current identification systems are still very much in the development phase (Rossing, 1999) and include radio frequency (RF) boluses and tags together with the use of retinal imaging linked to an analogue tag. There are four major technical issues to be resolved:

- How to make sure the RF id stays on/in the animal (bolus + tag)
- The need to make the system(s) affordable
- Compatibility between systems – currently they are not interchangeable
- Handling the digital data – transferring from farm to database and architecture of the latter

Finally, a major missing link in transforming the sheep industry into one that is market focussed, and can prove to its consumers that it is devising management and breeding systems to meet their demands, is the lack of an objective carcass grading system. In this regard a digital imaging system and/or one that takes certain pre-determined measurements from the carcass is required.

Conclusions

Despite their ability to provide all the essential ingredients for sustaining communities in arid and hostile environments, sheep are relatively inefficient in utilizing high quality temperate grassland for the production of human food. Currently, their output is limited by their low average fecundity, seasonality of breeding and high perinatal mortality. Small improvements in average fecundity invariably result in a proportion of very large litters, with the associated counter-productive effects on neonatal viability. Sub-optimal milk yields and a limited number of dispensary points (2 teats) impair lamb growth rates and, as a result of suckling competition, predispose the ewe to udder infections. Unwanted fat, a significant amount of which is intramuscular, is deposited in the carcass relatively early in the growth phase.

At the level of the digestive system production is restricted by appetite (particularly during late pregnancy), inadequate structural carbohydrate degradation in the rumen (accentuated when high starch containing concentrate supplements are given) and the inability of microbial protein to meet the nutrient needs of late pregnancy, early lactation and lamb growth. These can result in adverse effects on foetal and neonatal growth and development, and an impaired immune system. In combination with increasing anthelmintic resistance, production is further impeded by gastrointestinal parasitism.

At the level of the reproduction system there are also a number of physiological and technological limitations. For example, due to the impenetrable nature of the sheep cervix the industry has not been able to exploit fully, for commercial production, the benefits of artificial insemination using frozen semen from superior sires. In addition, the costs of laparoscopic insemination restrict its use to pedigree flocks. The same is true for sexed semen, yet because of the stratified nature of the UK sheep industry, hill sheep flocks producing lowland breeding females are an obvious target for its application. Welfare concerns about laparoscopic insemination also limit its application and therefore restrict the effectiveness of MOET, *in vitro* fertilisation (IVF) and cloning technologies. Indeed the last two of these are also hampered in their application by aberrant gene expression (Young *et al.* 2001), which can result in a range of foetal and neonatal lamb phenotypes including the Large Offspring Syndrome (see Sinclair and Webb, 2005).

Despite these limiting factors there is reason for optimism. Advances in the identification of genetic markers and/or candidate genes for improved ovulation rate (with a reduced ovulation rate range), for aseasonality of breeding, for improving meat quality and growth, and for disease resistance already show promise. Importantly the availability of the genome sequence of sheep will enhance these changes. Although breed diversity within the UK sheep industry has been regarded by some as a hindrance, in the new era of breeding and genome technologies, diversity can be a strength. Provided the underlying physiology is properly researched and understood, to ensure the correct genes can be identified, selection of desirable traits in juvenile animals or indeed in embryos prior to their transfer to recipients should enhance the rate of genetic selection. This should ensure a reduction in the influence of negatively correlated responses.

Furthermore, ryegrass species with enhanced carbohydrate content have been produced recently, leading to improvements in lamb growth rate and increases in lamb output per unit land area. Better understanding of ruminant fatty acid metabolism shows potential for bringing the fatty acid content of sheep meat more in line with current health recommendations. Sire referencing schemes have also illustrated the potential of genetic selection for improved growth rate and carcass composition, with considerable untapped improvements still to be achieved.

Technological advances in electronic technology for animal identification, semen freezing and sexing should improve the application of genetic improvement and ensure product authenticity. In terms of biotechnology, sheep have been one of the model species for cloning and gene transfer technologies. Moreover, the ability to make a rapid and precise assessment of carcase composition would have another major impact on the industry.

It is anticipated that the enlargement of the EU will have very little impact on the sheep sector, adding less than 2 million ewes. With no history of consuming large quantities of sheep meat, the new member states are not seen as providing any immediate new opportunities for exports above the current export trade which is mainly from Hungary to Italy (Ashworth, 2004). For the UK, where returns from sheep production as a percentage of total agricultural production are second only to Greece, the emphasis will be on improving profitability through improved production efficiency; thus the need to exploit the range of physiological opportunities and technical developments, as discussed, assumes even greater significance.

Over the next few decades there are major challenges for the sheep industry. However the genetic diversity of the species, coupled with ongoing technological and physiological advances should ensure improvements in efficiency, quality and sustainability. It is this background, together with changing trends in public attitude, consumer demand and industry investment that will influence future production systems and the rate of change in the UK sheep industry.

References

Agricultural Research Council (1980). *The Nutrient Requirements of Ruminant Livestock.* Commonwealth Agricultural Bureaux, Slough, UK.

Agricultural and Food Research Council (1993). *Energy and Protein Requirements of Ruminants.* An advisory manual prepared by the AFRC Technical Committee on Responses to Nutrients. CAB International, Wallingford, UK.

Ashworth, S. (2004). *The importance of sheep in Scotland.* The Scottish Farmer No 5823; Vol 1112 p. 20.

Barry, T.N. and Manley, T.R. (1985). Glucose and protein metabolism during late pregnancy in triplet-bearing ewes given fresh forages ad lib. 1. Voluntary intake and birthweight. *British Journal of Nutrition* **54**: 521-533.

Bindon, B.M., Piper, L.R. and Hillard, M.A. (1996). Reproductive Physiology and Endocrinology of Prolific Sheep. In: *Prolific Sheep* 453-469. [Ed. M. H. Fahmy] CAB International, Wallingford, UK.

Bishop, S.C. and Stear, M.J. (2001). Inheritance of faecal egg counts during early lactation in Scottish Blackface ewes facing mixed natural nematode infections. *Animal Science* **73**: 389-395.

Blaxter, K.L. (1968). Relative efficiencies of farm animals in using crops and byproducts in production of foods. In: *Proceedings Second World Conference on Animal Production*, Maryland, pp.31-40. American Dairy Science Association, Urbana, Illinois, USA.

Boland, M.P and Crosby, T.F. (1993). Fecundin: An immunological approach to enhance fertility in sheep. *Animal Reproduction Science* **33**: 143-158.

Bowles, D. Gilmartin, P., Holt, W., Leese, H., Mylne, J. Picton, H. Robinson, J., and Simm, G. (2004). Evolution of Heritage Genebank into The Sheep Trust: conservation of native traditional sheep breeds that are commercially farmed, environmentally adapted and contribute to the economy of rural communities. In: *Farm Animal Genetic Resources* pp. 45-55 [Eds. G. Simm, B. Villanueva, K.D. Sinclair and S. Townsend]. BSAS Publication 30, Nottingham University Press.

Cant, J.P., Qiao, F. and Toerien, C.A. (1999). Regulation of mammary metabolism. In: *Protein Metabolism and Nutrition*, pp. 203-219. [Eds. G.E. Lobley, A. White and J.C. MacRae], Wageningen Pers.

Chu, M.X., Ji, C.L. and Chen, G.H. (2003). Association between PCR-RFLP of melatonin receptor 1a gene and high prolificacy in Small Tail Han sheep. *Asian-Australasia Journal of Animal Sciences* **16**: 1701-1704.

Conington, J., Bishop, S.C., Waterhouse, A. and Simm, G. (2004). A bioeconomic approach to derive economic values for pasture based sheep genetic improvement programs. *Journal of Animal Science* **82**: 1290-1304.

Cooper, S.L., Sinclair, L.A., Wilkinson, R.G., Hallett, K.G., Enser, M. and Wood, J.D. (2004). Manipulation of the n-3 polyunsaturated fatty acid content of muscle and adipose tissue in lambs. *Journal of Animal Science* **82**: 1461-1470.

Cran, D.G., McKelvey, W.A.C., King, M.E., Dolman, D.F. McEvoy, T.G., Broadbent, P.J. and Robinson, J.J. (1997). Production of lambs by low dose intrauterine insemination with flow cytometrically sorted and unsorted semen. *Theriogenology* **47**: 267.

Davies, D.A.R. (1999). Performance of ewes selected to have four functional teats. *Proceedings of the British Society of Animal Science*, p.117.

Davis, G.H., Morris, C.A. and Dodds, K.G. (1998). Genetic studies of prolificacy in New Zealand sheep. *Animal Science* **67**: 28-297.

Donaldson, J., van Houtert, M.F.J. and Sykes, A.R. (2001). The effect of dietary fish-meal supplementation on parasite burdens of periparturient sheep. *Animal Science* **72**: 149-158.

Doney, J.M., Gunn, R.G. and Horák, F. (1982). Reproduction. In: *Sheep and Goat Production*, pp. 57-80. [Ed. I.E. Coop], Elsevier Scientific Publishing Company, Amsterdam.

Dove, H., Milne, J.A., Lamb, C.S., McCormack, H.A. and Spence, A.M. (1985). The effect of supplementation on non-ammonia nitrogen flows at the abomasium of lactating grazing ewes. *Proceedings of the Nutrition Society* **44**: 63A.

Dzuik, P. (2003). Positive, accurate animal identification. *Animal Reproduction Science* **73**: 319-323.

Fahmy, M.H. and Davis, G.H. (1996). Breeds with newly discovered genes for prolificacy. In: *Prolific Sheep*, pp. 174-177. [Ed. M.H. Fahmy]. CAB International.

FAWC (2004). *Farm Animal Welfare Council Report on the Welfare Implications of Animal Breeding and Breeding Technologies in Commercial Agriculture*. Farm Animal Welfare Council, London.

Fell, H.R. (1987). Formation of new breeds. In: *New Techniques in Sheep Production* pp. 147-156. [Eds. I. Fayez, M. Marai and J.B. Owen], Butterworths, London.

Forbes, J.M. (1996). Integration of regulatory signals controlling forage intake

in ruminants. *Journal of Animal Science* **74**: 3029-3035.
Galloway, S.M., NcNatty, K.P., Cambridge, L.M., Laitinen, M.P.E., Juengel, J.L., Jokiranta, T.W., McLaren, R.J., Luiro, K., Doods, K/G., Montgomery, G,W., Beattie, A.E., Davis, G.H. and Ritvos, O. (2000). Mutations in an oocyte-derived growth factor gene (*BMP15*) cause increased ovulation rate and infertility in a dosage-sensitive manner. *Nature Genetics* **25**: 279-283.
Garnsworthy, P.C. and Thomas, P.C. (2005). Yield trends in UK dairy and beef cattle. In: *Yields of Farmed Species*, pp. 435-462. [Eds. R. Sylvester-Bradley and J. Wiseman], Nottingham University Press.
Gordon, I. (1997). *Controlled Reproduction in Sheep and Goats.* CAB International, Wallingford.
Gray, G.D. (1997). The use of genetically resistant sheep to control nematode parasitism. *Veterinary Parasitology* **72**: 345 –366.
Grizard, J., Picard, B., Dardevet, D., Balage, M. and Rochon, C. (1999). Regulation of muscle growth and development. In: *Protein Metabolism and Nutrition*, pp. 177-201. [Eds. G.E. Lobley, A. White and J.C. MacRae], Wageningen Pers.
Groeneveld, E. (2004). An adaptable platform independent information system in animal production: framework and generic database structure. *Livestock Production Science* **87**: 1-12.
Hanrahan, J.P. (1989). Reproductive performance of Belclare sheep – a new composite breed. *Proceedings of the Annual Meeting EAAP*. Dublin, Ireland, S6.5, pp.1-4.
Hanrahan, J.P. (2003). Aspects of reproductive performance in small ruminants-opportunities and challenges. *Reproduction* **61**: 15-26.
Hanrahan, J.P. and Owen, J.B. (1985). Variation and repeatability of ovulation rate in Cambridge ewes. *Animal Production* **40**: 529.
Hanrahan, J.P., Gregan, S.M., Mulsant, P., Mullen, M., Davis, G.H., Powell, R. and Galloway, S.M. (2004). Mutations in the genes for oocyte-derived growth factors GDF9 and BMP15 are associated with both increased ovulation rate and sterility in Cambridge and Belclare sheep (*Ovis aries*). *Biology of Reproduction Supplement* **70**: 900-909.
Haresign, W. (1990). Controlled reproduction in sheep. In: *New Developments in Sheep Induction.* Occasional Publication No. 14 [Eds C.F.R. Slade and T.L.J Lawrence] pp23-37. British Society of Animal Production.
Hickford, J. (2004). *Gene Markers for Breeding Sheep That are Tolerant to Footrot.* http://www.maf.govt.nz/about-projects/pastoral-farming/01042attachment.htm.
Hollinshead, F.K., O'Brien, J.K., Maxwell, W.M.C. and Evans, G. (2002). Production of lambs of predetermined sex after the insemination of ewes with low numbers of frozen-thawed sorted x- or y-chromosome-bearing spermatozoa. *Reproduction, Fertility and Development* **14**: 503-508.
Langhans, W. (1999). Appetite regulation. In: *Protein Metabolism and*

Nutrition [Eds. G. E. Lobley, A. White and J.C. MacRae] pp. 225-251. Wageningen Pers.

Large, R.V. (1970). The biological efficiency of meat production in sheep. *Animal Production* **12**: 393-401.

Lee, M.R.F., Jones, E.L., Moorby, J.M., Humphreys, M.O., Theodorou, M.K., MacRae, J.C. and Scollan, N.D. (2001). Production responses from lambs grazed on Lolium perenne selected for an elevated water-soluble carbohydrate concentration. *Animal Research* **50**: 441-449.

McNatty, K.P., Juengel, J.L., Wilson, T., Galloway, S.M., Davis, G.H., Hudson, N.L., Moeller, C.C., Cranfield, M., Reader, K.L., Laitinen, M.P.E., Groome, N.P., Sawyer, H.R. and Ritvos. O. (2003). Oocyte-derived growth factors and ovulation rate in sheep *Reproduction Supplement* **61**:339-351.

Meadus, W.J. (1998). Molecular techniques used in the search for genetic determinants to improve meat quality. *Canadian Journal of Animal Science*, **79**: 483-492.

Meat and Livestock Commission (1972). *Sheep Improvement*. Scientific Study Group Report.

Meat and Livestock Commission (1983). *Feeding the Ewe*. Sheep Improvement Services. 78p.

Meat and Livestock Commission (2002). *Sheep Year Book*.

Mulsant, P., Lecerf, F., Fabre, S., Bodin, L., Thimonier, J., Monget, P., Lanneluc, I., Monniaux, D., Teyssier, J. and Elsen, J.M.. (2003). Prolificacy genes in sheep: the French genetic program. *Reproduction Supplement* **61**:353-359.

O'Brien, J.K., Hollinshead, F.K., Evans, K.M., Evan, G. and Maxwell, W.M.C. (2003). Flow cytometric sorting of frozen-thawed spermatozoa in sheep and non-human primates. *Reproduction, Fertility and Development* **15**: 367-375.

Owen, J.B. (1996). The Cambridge Breed. In: *Prolific Sheep*. pp. 161-173. [Ed. M.H. Fahmy]. CAB International.

Pelletier, J., Boden, L., Hanoiq, E., Lampaux, B., Teyssier, J., Thimonier, J. and Chemineau, P. (2000). Association between expression of reproductive seasonality and alleles of the gene for Mel_{1A} receptor in the ewe. *Biology of Reproduction* **62**: 1096-1101.

Penning, P.D. and Treacher, T.T. (1988). Responses of lactating ewes, offered fresh herbage indoors and when grazing, to supplements containing differing protein concentrations. *Animal Production* **46**: 403-415.

Poppi, D.P., Cruickshank, G.J. and Sykes, A.R. (1988). Fish meal and amino acid supplementation of early-weaned lambs grazing roa tall fescu (*Festuca arundinacea*) or huia white clover (*Trifoluim repens*). *Animal Production* **46**: 491.

Robinson, J.J. (1978). Intensive systems. In: *The Management and Diseases of Sheep*. pp431-446. Commonwealth Agricultural Bureaux, Slough, UK.

Robinson, J.J. (1982). The nutrition and management of sheep for improved productivity. *Journal of the Royal Agricultural Society of England* **143**: 112-131.

Robinson, J.J. (1983). Nutrition of the Pregnant Ewe. In: *Sheep Production.* pp.111 – 131. Ed. W Haresign. Butterworths, London.

Robinson, J.J. and McEvoy, T.G. (1999). Opportunities and constraints into the next millenium. In: *Protein, Metabolism and Nutrition*, pp. 265-282. [Eds. G.E. Lobley, A. White and J.C. MacRae], Wageningen Pers.

Rossing, W. (1999). Animal identification: Introduction and history. *Computers and Electronics in Agriculture* **24**: 1-4.

Russel, A.J.F., Alexieva, S.A. and Elston, D.A. (1997). The effect of the introduction of the Thoka gene for fecundity on lamb production from Cheviot ewes. *Animal Science* **64**: 503-507.

Satter, L.D., Jung, H.G., van Vuuren, A.M. and Engels, F.M. (1999). Challenges in the Nutrition of high-producing ruminants. In: *Nutritional Ecology of Herbivores* [Eds H-J.G. Jung and G.C. Fahey, Jr] pp. 609-646. American Society of Animal Science, Savoy, Illinois, USA.

Sazili, A.Q., Parr, T., Sensky, P.L., Jones, S.W., Bardsley, R.G. and Buttery, P.J. (2005). The relationship between slow and fast myosin heavy chain content, calpastatin and meat tenderness in different ovine skeletal muscles. *Meat Science* **69**: 17-25.

Simm, G. (1998). *Genetic Improvement of Cattle and Sheep.* Farming Press. Ipswich.

Sinclair, K.D. and Webb, R. (2005). Reproductive rate in farm animals: strategies to overcome biologial constraints through the use of advances reproductive technologies. In: *Yields of Farmed Species*, pp. 51-88. [Eds. R. Sylvester-Bradley and J. Wiseman], Nottingham University Press.

Souza, C.J.H., Campbell, B.K., McNeilly, A.S. and Baird, D.T. (2003). What the known phenotypes of the Booroola mutation can teach us about its mechanism of action? *Reproduction Supplement* **61**:361-370.

Stear, M.J., Eckersall, P.D., Graham, P.A., McKellar, Q.A., Mitchell, S. and Bishop, S.C., (2001). Fructosamine concentration and resistance to natural, predominantly Teladorsagia circumcincta infection. *Parasitology* **123**: 211-218.

Stear, M.J., Henderson, N.G., Kerr, A., McKellar, Q. A., Mitchell, S., Seeley, C. and Bishop, S.C. (2002). Eosinophilia as a marker of resistance to Teladorsagia circumcincta in Scottish Blackface lambs. *Parasitology* **124**: 553-560.

Tempest, W.M. (1983). Management of the frequent lambing flock. In: *Sheep Production.* pp. 467-481. Eds. W. Haresign, Butterworths; London.

Thomson, A.J., Marques, M.M. and McWhir, J. (2003). Gene targeting in livestock. *Reproduction Supplement* **61**: 495-508.

Treacher, T.T. and Caja, G. (2002). Nutrition during lactation. In: *Sheep Nutrition* pp. 213-236. [Eds. M. Freer and H. Dove]. CABI International,

Wallingford, UK in association with CSIRO Publishing, Collingwood, Australia.

Vipond, J.E., Clark, C.F.S., Peck, D.N. and King, M.E (1987). The effect of ewe size on efficiency of lowland lamb production and the consequences of fulfilling ideal demand for ewe replacements in North Eastern Scotland from crosses derived from small Shetland ewes through a stratified breeding system. *Research and Development in Agriculture* **4** (**1**); 13-19.

Walling, G.A., Bishop, S.C., Pong-Wong, R., Gittus, G., Russel, A.J.F. and Rhind, S.M. (2002). Detection of a major gene for litter size in Thoka Cheviot sheep using Bayesian segregation analyses. *Animal Science* **75**: 339-347.

Walling, G.A., Wilson, A.D., Mcteir, B.L. and Bishop, S.C. (2004a). Increased heterozygosity and allele variants are seen in Texel compared to Suffolk sheep. *Hereditary* **92**: 102-109.

Walling, G.A., Visscher, P.M., Wilson, A.D., McTeir, B.L., Simm, G. and Bishop, S.C. (2004b). Mapping of quantitative trait loci for growth and carcass traits in commercial sheep populations. *Journal of Animal Science* **82**: 2234-2245.

Walton, P. and Robertson H.A. (1974. Reproductive performance of Finnish Landrace ewes mated twice yearly. *Canadian Journal of Animal Science* **54**: 35-40.

Wells, D.N. (2003). Cloning in livestock agriculture. *Reproduction Supplement* **61**: 131-150.

Wells D.N., Thompson J.G.E., Tervit H.R., James R.W. and Udy G.B. (1990). Experiences in the application of embryo bisection in sheep MOET programmes *The Proceedings of the New Zealand Society of Animal Production* **50**: 431-435.

Wilson P.N. (1968). Biological ceilings and economic efficiencies for the production of animal protein, AD 2000. *Chemistry and Industry* 899-902.

Young, L.E., Fernandes, K., McEvoy, T.G., Butterwith, S.C., Gutierrez, C.G., Carolan, C., Broadbent, P.J., Robinson, J.J., Wilmut, I. And Sinclair, K.D. (2001). Epigenetic change in IGF2R is associated with fetal overgrowth after sheep embryo culture. *Nature Genetics* **27**: 153-154.

22

LIVESTOCK YIELDS NOW, AND TO COME: CASE STUDY PIGS

JULIAN WISEMAN[1], MIKE VARLEY[2], ANDREW KNOWLES[3], REX WALTERS[4]

[1]*Division of Agricultural and Environmental Sciences, School of Biosciences, University of Nottingham, Sutton Bonington Campus, Loughborough, Leics., LE12 5RD, UK;* [2]*Provimi Ltd, Dalton, SCA Mill, Thirsk, North Yorkshire, YO7 3HE, UK;* [3]*British Pig Executive, PO Box 44, Winterhill House, Snowdon Drive, Milton Keynes, MK6 1AX, UK;* [4]*Livestock Genetics Ltd, Rye Cottage, Peasemore, Newbury, Berks, RG20 7JN, UK*

Introduction

The nature of the UK pig industry sector has changed considerably over the last 4 decades, a scenario which has been mirrored in many other developed and, more recently, developing countries. Of all the higher mammals kept for production, the pig is unique in that it is an omnivore; it can thrive on a wide range of dietary raw materials including those of both high energy and nutrient concentration (where the efficiency of conversion to live weight is very good) as well as those, at the other extreme, characterised as being highly fibrous as often found in extensive production systems (where biological efficiency may not be as high but it is important to appreciate that raw materials in these situations are often waste products not consumed by humans). Reproductive performance is excellent as measured by, for example, the number of viable offspring produced per adult breeding female per annum; numbers can exceed 25 under well-managed systems, although the UK national average is currently - spring 2004 - around 18.5. This is due in all probability to viruses (which have compromised reproductive efficiency through poor fertility), general herd health and herd parity structure which has been disrupted by the restrictions on animal movements associated with the Foot and Mouth epidemic of 2001. Finishing animals rapidly produce a carcass in well-under 6 months which is both lean and, increasingly, utilised in a wide range of both fresh and cured / processed / added value products.

It should not be forgotten that pig production in the UK has developed in the almost complete absence of any form of subsidy payments which are such a central feature of the beef and sheep sectors.

Table 22.1 illustrates, using a simple comparative calculation, the relative efficiencies of the various farmed species in terms of their ability to turn a hectare of intensive cereals into meat products. It can be seen that the pig appears, according to this calculation, to be 184% more efficient than beef steers under feedlot conditions and 221% more efficient than intensive fat lamb production

Table 22.1 RELATIVE EFFICIENCIES OF FARMED LIVESTOCK

Meat produced from 1 hectare of land growing 10 tonnes of cereals per annum

	Food conversion ratio (kg feed / kg live weight gain)	*Weight gain kg*	*Killing out% (carcass weight / live weight*	*Meat produced from 1 hectare of land (tonnes)*
Poultry	1.8	5.5	85	4.7
Pigs	2.8	3.57	75	2.67
Beef	5.3	1.88	50	0.94
Sheep	6.0	1.66	50	0.83

Before considering biological aspects of pig production and how these have changed, it is worth outlining the structural developments which have taken place in recent times.

The data in figure 22.1 demonstrate the dramatic increases in herd size since 1962 when over 80% of all UK sows (within the MLC PigPlan system which does tend to record the bigger herds) were within herds of less than 50 breeding animals; data since 1999 confirm that the trends are continuing and there is an increasing proportion of herds with over 1000 sows; even enterprises of 10,000 sows are found. Pig production has thus become the domain of the specialist unit, often linked to an arable business and less as a small enterprise within a mixed livestock / arable farming system.

Interestingly pig production is concentrated currently (and as expected as grain is a component of pig diets) predominantly in the grain-growing regions of the UK (Humberside, East Anglia and Aberdeenshire), whereas traditionally it was to be found associated with dairy areas (pigs being fed on co-products) or even large urban areas where animals were fed on wastes from, for example, breweries and starch manufacturers (e.g. Wiseman, 2001).

Accompanying the increase in herd size has been a reduction in the number of holdings containing breeding pigs from 25,000 in 1980 to around 6,000 in 2002 (Figure 22.2). There are more subtle aspects of the British Pig Industry which are of interest. Ownership could be finishing pigs by a third party; this leads to less management control by the owner with the third party being interested primarily only in the provision of buildings and labour to secure a rental and income. There has been considerable integration of producers and processors which has improved efficiency and shortened the supply chain. Multi-site production is becoming more common (as in other pig-producing parts of the World) in an attempt to improve disease control although there are

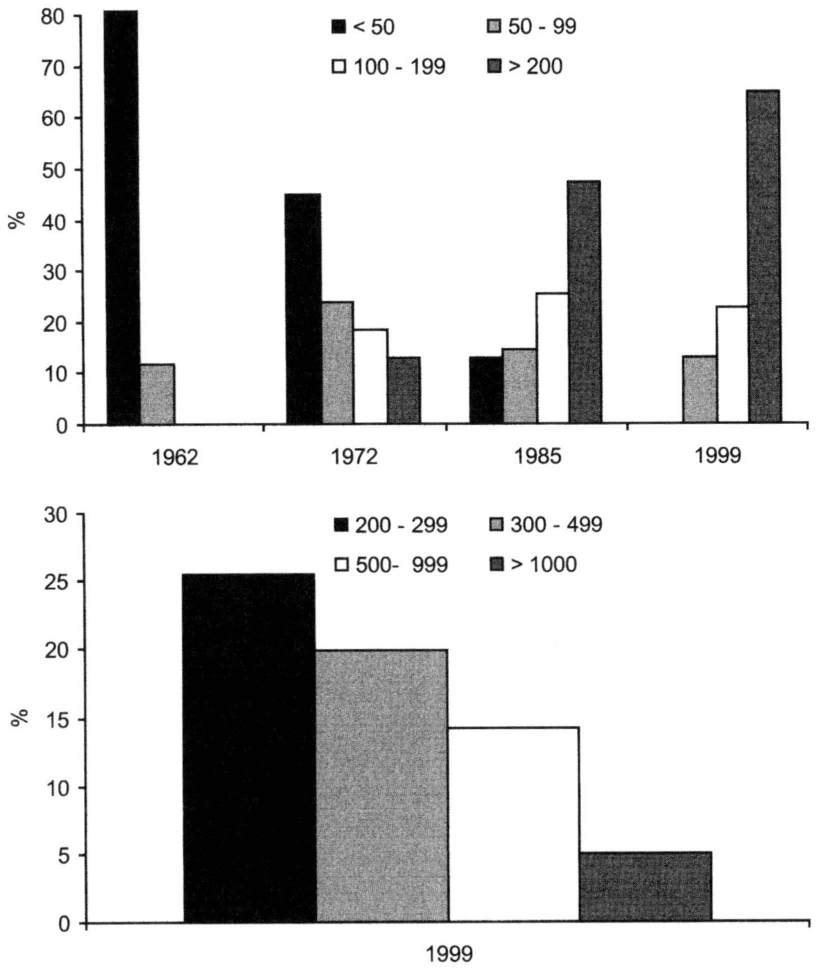

Figure 22.1 Changes in UK sow herd size (% of farms) over the last 40 years (MLC 2000 PigPlan; the last year data were available))

accompanying problems including increased need for transport which may have welfare implications. Around 33% of breeding sows are now housed outdoors, a system in the UK rarely found elsewhere. Because of lower entry costs, however, this has in part led to greater volatility in the breeding herd as moving in and out of pig breeding is consequently easier and associated with reduced financial risk.

It is difficult to estimate future developments in numbers of holdings; certainly as the pig industry becomes increasingly more specialised, there will probably be further reductions associated also with the recent national decline in sow numbers (Figure 22.3) which, however, has now (spring 2004) stabilised.

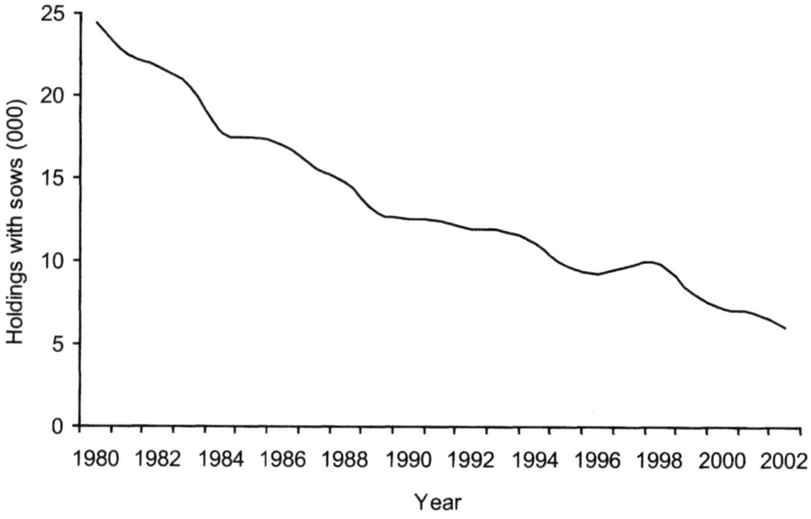

Figure 22.2 Numbers of holdings in the UK with breeding sows

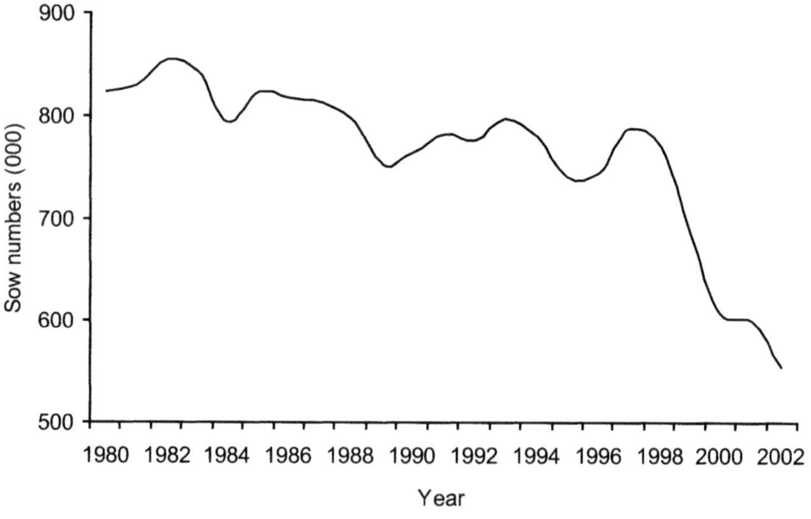

Figure 22.3 Numbers of breeding sows in the UK

A key factor likely to influence the future size and structure of the industry is that of environmental legislation. Controls on nitrates and phosphate in particular have led many European Union pig industries effectively to place a production quota on the size of their domestic pig industries. Within the United Kingdom the introduction of Integrated Pollution Prevention and Control (IPPC) regulations has made it extremely unattractive to establish large (650+ sow) units or expand existing units.

Developments in the late 20th century

There have been a number of key developments accompanying the structural changes described above. Before considering individual aspects it should be noted that data presented are average within the MLC recording scheme; this frequently disguises considerable individual variation which is a serious problem in the pig industry.

BREEDING

In the 1960s, breeding was almost completely dominated by pedigree pure breds; the notion of cross-breeding was an anathema to many, even the most progressive, members of the industry. There was much inter-breed rivalry (a characteristic which remains within some sectors of the industry) based often on quite unquantifiable traits ('good mother', 'alert', 'easily bred', etc.) and appearance (which could be altered positively in time for the show ring by astute and experienced owners). This is not to say that pedigree breeding did not (and still does through the maintenance of Great Grand Parent - GGP - nucleus herds) have a positive impact on the industry. Indeed, conscientious pig farmers who kept accurate records of key performance traits (and, crucially but uncommonly, carcass characteristics from the abattoir) were able to select improved sire and dam lines generation by generation.

The dominant breed of the time was the Large White (often referred to as the Yorkshire in many other parts of the World) which had been the backbone of the industry for decades. A more recent addition was the Landrace. The good reputation of the Danish pig industry was widespread and considered to be a consequence of the breed of animal used (Landrace; ironically descended in part from animals imported from the UK at the end of the 19th century); the earliest imports into the UK of Landrace pigs were in the early 1950s where they changed hands for several thousand pounds each, with eager purchasers seemingly unaware that the name 'Landrace' is a generic term for pigs from the Low Countries of Europe and Scandinavia. Those entering the UK then were in fact from Sweden.

However, during the late 1960s there was increasing interest in breeding animals of superior performance irrespective of the actual name of the breed. The abolishment in 1972 of boar licensing in the UK (a system which had reinforced the notion of pedigree as a key component of quality) allowed the development of hybrid lines where biological performance was the ultimate selection criterion. Many of the global pig breeding companies emerged in the UK at that time and the continued worldwide excellence of UK pig genetics owes much to the foresight of those involved at the time supported subsequently by the GB Meat and Livestock Commission Pig Improvement Scheme. Cross-breeding also carries with it the advantage of hybrid vigour, particularly in reproductive performance.

Pig breeding has now become a highly sophisticated business with a number of tiers (classically these are arranged as the 'pyramid' of nucleus, multiplier and production units). The adoption of Best Linear Unbiased Prediction (BLUP) has revolutionised breeding methodologies in the selection of improved genotypes, particularly for traits of low heritability.

Although the Large White and Landrace remain the backbone of the UK pig breeding sector, there have been more recent imports. Many are used as terminal sires in meat production (Hampshire and Duroc from USA, Pietrain from Belgium) whereas others are renowned for their high reproductive rate (e.g. Meishan which, however, has very poor carcass characteristics)

NUTRITION

The UK of the 1950s did not have the luxury of vast areas of land growing cereals and plant protein crops destined for the livestock sector. Accordingly, developments in nutrition were also a key feature of an industry unable to rely on, for example, maize and soya of North America (although the latter raw material has assumed increasing importance with the withdrawal of meat and bone meal from pig diets). Whilst peas and field beans are employed, rapeseed meal is perhaps the major 'home-grown' plant protein feedstuff employed widely in pig diets although predominantly in the finishing phase. Accurate characterisation of the nutritive value of dietary raw materials was accompanied by detailed examination of energy and nutrient requirements such that these could be met in an accurate and cost-effective manner. As an example, the concept of lysine as an essential amino acid emerged in the 1960s and quality pig diets were sold on the basis of 'containing 0.6% lysine'. A further development (perfectly acceptable at the time but currently much more controversial) was the label on pig food 'contains antibiotics'; the Swan report of 1968 led to the banning in animal diets of many antibiotics with potential human therapeutic use, and all non-prescription feed grade antibiotics will be phased out of EU pig diets in 2006 (the current proposal – a number have already been phased out and the remaining 4 will be phased out on January 1st 2006).

Despite the advances in selection of improved genotypes, the problem of excessively fat carcasses remained, a situation which was compounded by widespread castration of entire males. It was thus common to restrict feed intake towards the end of the finishing (which unfortunately had the negative effect of increasing days to slaughter), a practice which is not easy to manage in group-housed pens. Selection has indirectly resulted in lower voluntary feed intake capacity so that the problem has effectively been solved such that around 90% of all UK carcasses are within the top 2 grades. This has been helped by the exclusion of castration, for animals entering pig meat assurance schemes, on welfare grounds. It has been argued by some, however, that this does restrict the weight to which entire males can grow in order to avoid boar taint – a problem not found anywhere else where castration is still the norm – although the problem is no more severe at higher carcass weights. Boar carcases exceeding 85kg must, however, be taint tested under EU legislation.

MANAGEMENT

A further major development in the pig industry has been a reduction in the age at weaning mainly in the 1960s and 1970s from 8 weeks to around 25 days. This has had a dramatic effect on sow performance.

Table 22.2 EFFECT OF INCREASING LACTATION LENGTH (THE ONLY VARIABLE FACTOR INCLUDED IN THE MODEL) FROM 25 TO 56 DAYS ON PIGLET OUTPUT FROM SOWS

Gestation (d)		115	
Lactation (d)	25		56
Weaning to re-breeding (d)		7	
Cycles per annum	2.48		2.05
Pigs reared per litter		10	
Pigs reared per sow per annum	24.8		20.5

Thus, all other things being equal, the effect of this reduction in lactation length alone is an increase of 4.3 piglets reared per sow per year. This is a highly significant difference in the context of pig unit output; under current conditions where the UK pig industry is struggling to survive, a return to later weaning systems would have serious consequences. Increases in weaning age are being actively considered from the welfare perspective; the European Union has already decided to impose a minimum weaning age of 28 days and there are suggestions that this should be longer.

Consequences of changes in the late 20th century on output

There are many pig recording schemes available and, for consistency, it was decided that the use of only one would allow better comparisons as to how pig performance changed in the later 20th century. Data below are all taken for the GB Meat and Livestock Commission and are average; it should not be forgotten that one of the major problems arising from data base analyses is the considerable variability which exists both between units and within.

REPRODUCTIVE PERFORMANCE

The conventional means of assessing breeding herd output are based on numbers of cycles per annum (litters per sow per year), live animals born and their subsequent mortality up to weaning. There have been considerable advances in management of the pre-pubertal gilt; the 'boar effect' has been used to great advantage in both stimulating puberty and also synchronising it in a group of animals.

Figure 22.4 shows that litters per sow per year have increased from the baseline values of 1.8 in 1970 to well over 2 in the late 1990s.

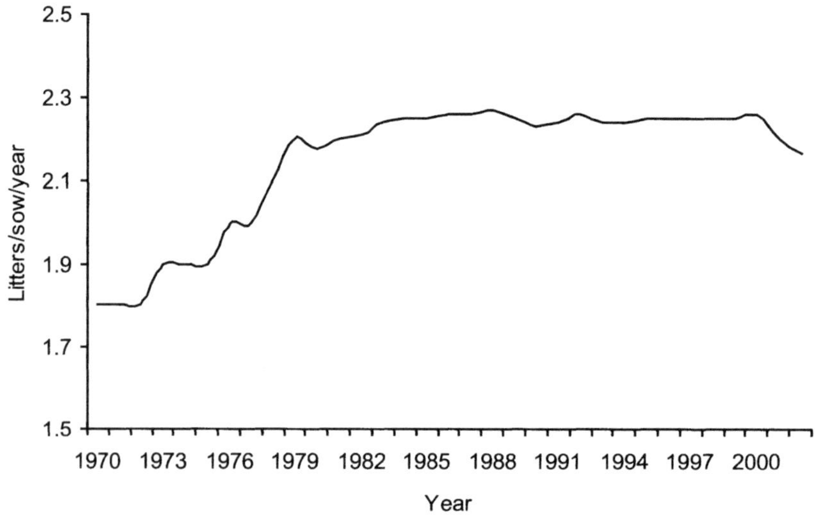

Figure 22.4 Litters per sow per year

Whilst this increase in productivity has been of major importance, it is probable that it is not due to any biological improvement *per se* but to the indirect effect of a reduction in weaning age (see Table 22.1 above) and also increased producer awareness of mating management and heat detection procedures. Interestingly, there has only been a reduction of 4 days in average weaning age since 1979 (29 to 25); the major reductions were before then (and the response in figure 4 confirms this effect on cycle number).

With regard to the use of on farm artificial insemination (AI), over recent years there have been few major breakthroughs in either fresh or freezing-thawing technology that have represented significant progress in reproductive technology. What there has been however in the industry is a steady uptake of existing technology coupled with developments in the application of AI techniques in practice. Improved equipment such as the use of flat pack semen delivery systems and disposal catheter systems has meant improving performance on farms and a more repeatable and better level of conception and fertility. The UK industry has moved to a position where it is likely that more than 50 % of all inseminations are via AI. The hope is that this progress will continue and, at some future point, semen sexing will become a commercial reality and semen freezing systems will be available to bring with it more flexibility and opportunities for disseminating good genes more quickly.

There has been a steady increase in piglets born alive (Figure 2.5) amounting to on average 0.026 per annum (equivalent to an extra 0.84 of pig over the period 1970-2002 or 1.86 piglets per sow per annum assuming 2.2 cycles). Although the heritability of litter size is low, genetic improvements have accrued largely from the adoption of Best Linear Unbiased Prediction (BLUP) approaches, in which the UK is a world-leader. Currently, Specialist Breeding Companies are also becoming increasingly interested in specific genes which may boost future output. At the same time, non-genetic improvements in performance are attributable essentially to management improvement.

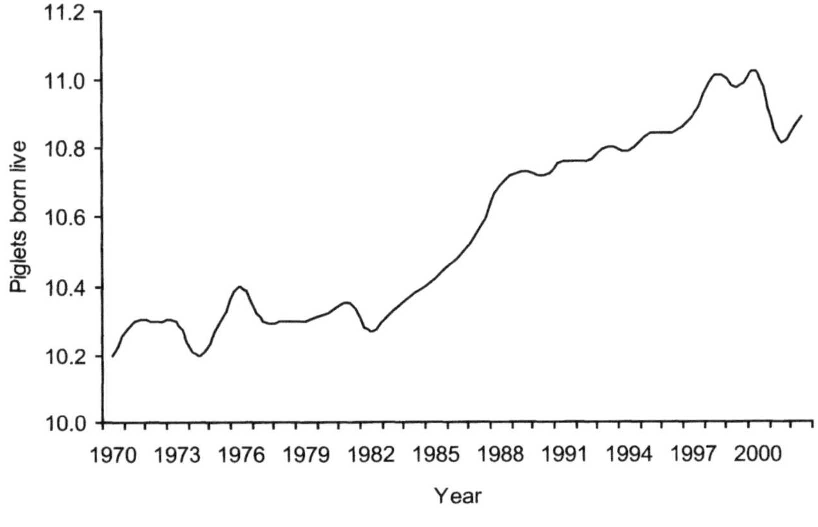

Figure 22.5 Piglets born alive per litter

It is currently felt by some that litter size is at an optimum. There is an inverse relationship between prolificacy and birth weights (the phenotypic interaction between litter size and birth weights does however require investigation) and further increases in litter size must be placed alongside smaller piglets with greater within-litter variation (there has not been much success in selection for reducing this variability) and higher subsequent mortality. Smaller pigs at weaning will also take longer to reach slaughter weight (although subject to some debate, it is generally held that pigs do not exhibit compensatory growth; Whittemore, Green and Knap, 2000). Pluske *et al* (1995) concluded that a 0.9 kg advantage one week after weaning becomes a 12 kg difference in weight at slaughter (which is equivalent to around 15 days difference to slaughter at the same liveweight).

Increased litter size / survivability is being investigated through a number of routes including improved implantation / embryo survival and improved uterine capacity. The importance of oestrogen receptor (location 1p25 - p24) is also being studied. It is felt that foetal development / piglet maturity in late gestation is more important to increased farrowing and post-natal survival than differences in early neonatal behaviour. Monitoring muscle glycogen reserves and cortisol levels may be a useful subject for study. Certainly breeding companies are selecting for increased litter size (although the heritability is very low at 0.05, there is high genetic variance) which has to be accompanied by selection for increased feed intake.

Piglet pre-weaning mortality has also reduced (Figure 22.6) amounting to 4.37% absolute between 1970 and 2002, although there has been little change since the mid 1980s.

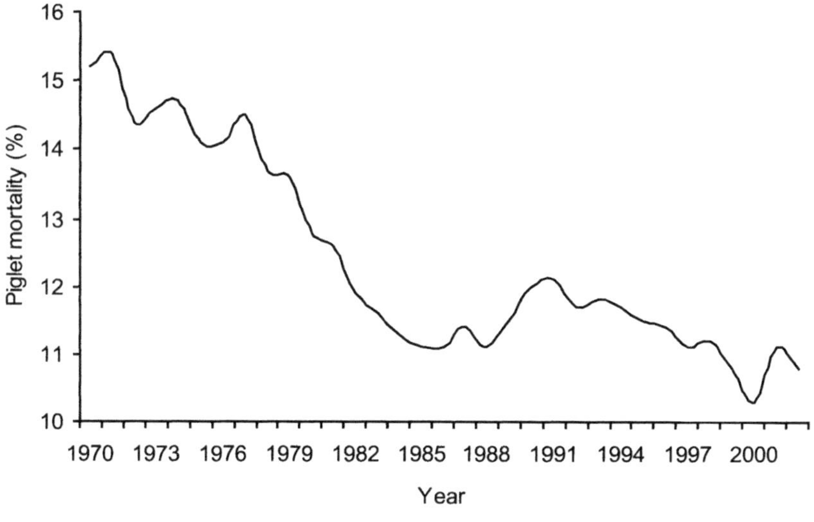

Figure 22.6 Pre-weaning piglet mortality

The net effects of all these changes has been an increase of 5.6 piglets reared (15.5 to 21.1) from 1970 to 2002 (Figure 22.7).

Data from the GB Meat and Livestock Commission have, since 1987, included two additional categories being performance of the top1/3 of herds and that of the top 10%. Figure 22.7 indicates the improvement of these two compared with the overall mean; litters per sow per year were improved (although weaning ages are the same which suggests better sow management) as were piglets born alive indicating better service management as mortality did not differ between categories.

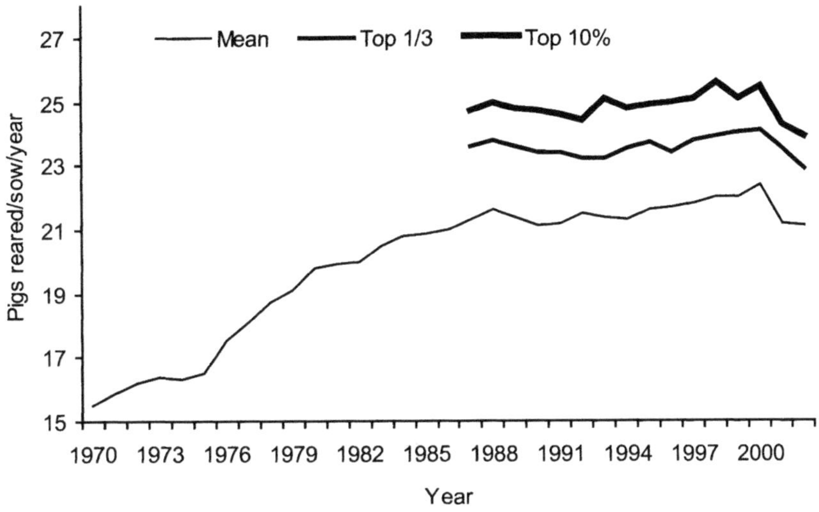

Figure 22.7 Pigs reared per sow per year

The improvements to reproductive performance outlined above, however, need to be placed alongside increasing problems of lifetime performance. A well-managed sow ought to provide 6 litters which, if each reared 10 piglets, would result in lifetime output of 60 pigs. A recent estimate (Gill, 2000) suggests that the UK average is only 52.8, while the US average is only 36.8 pigs per sow lifetime. One of the major reasons for culling prematurely is failure to return to oestrus following weaning. This is particularly evident in the gilt where up to 50% of animals culled after parity one is because of reproductive failure (Gill, 2000). Changes to genotype have resulted in superior lean tissue mass with lower feed intake capacities (to be placed alongside higher milk yield and maintenance costs, leading to increased energy requirements, and reduced body fat reserves). Whilst this is of evident benefit to the meat generation, it is probably not so for breeding animals. As a result many breeders now use separate selection objectives in dam-lines and sire-lines within breed. Another cause of premature culling is lameness (around 14% of gilts culled following their first parity) suggesting that there is inadequate bone growth during rearing (a nutritional issue), although pig bones are inherently weaker than those of other large mammals.

Gilts during their first pregnancy are still growing and nutritional regimes often fail to appreciate this. Animals have not reached a critical lean tissue mass assumed to be essential for good reproductive performance and are devoid of fat reserves which are important to support the rigours of lactation; short term flushing (increased dietary provision around puberty) is ineffective and it seems the only solution is to re-design gilt diets. However, it is interesting to note that the recent MLC Gilt Management Trial Report (2004) suggests that gilts, after weaning their first litter, are comparatively resilient so that body fat may act as a 'buffer' when feeding is inadequate and is not an issue when feeding is consistently adequate; this does however presume that fat reserves are adequate which may not be the case. There is evidence (see figures 22.4, 22.5, 22.6 and 22.7) that reproductive performance currently is decreasing. This could be attributable to the recent infertility syndrome but also to other health issues which have disrupted the UK pig industry including Foot and Mouth Disease movement restrictions.

GROWING / FINISHING PIG PERFORMANCE

Over the past 10 to 15 years there have been very significant changes in pig performance due to selection. The result has been an increase in maximum protein deposition, a shift in the distribution of energy towards more protein and less lipid deposition, an increase in maintenance requirements and a decrease in voluntary feed intake (Knap, 2000). Using MLC national UK data collected between 1984 and 1999, Walters (2000) estimated that actual annual percentage changes in key production traits were as presented in table 22.3. The current section will develop this theme further.

Table 22.3 ANNUAL PERCENTAGE CHANGES IN KEY PRODUCTION TRAITS OVER THE GROWING / FINISHING PHASE

Growth rate	+0.41
Lean growth	+0.79
Feed conversion	-0.69
Backfat	-1.66
Feed intake	-0.32

The two major estimates of performance of the growing / finishing pig are daily live weight gain (DLWG) and food conversion ratio (FCR); both have shown improvements as shown in Figures 22.8 and 22.9 respectively for DLWG and FCR

DLWG data for the top 1/3 and top 10% of producers (Figure 22.8) are between 25 and 40g better respectively than mean figures. Interestingly, both categories experienced the same collapse as the mean for 1992 (mean went from 628 to 587, top 1/3 from 659 to 591, top 10% from 676 to 591 g DLWG); recovery for top 1/3 and top 10% was more rapid than the mean but all 3 categories have shown a down-turn. The recent increase may partly be a reflection of the move to higher carcass weights (see Figure 22.9) and immune challenge through PMWS (see below).

The subdivisions introduced in 1987 into the top 1/3 and top 10% of herds revealed interestingly that the former gave FCRs (Figure 22.10) of 2.46 and 2.37 respectively for that year (compared with 2.70 for the average); thereafter

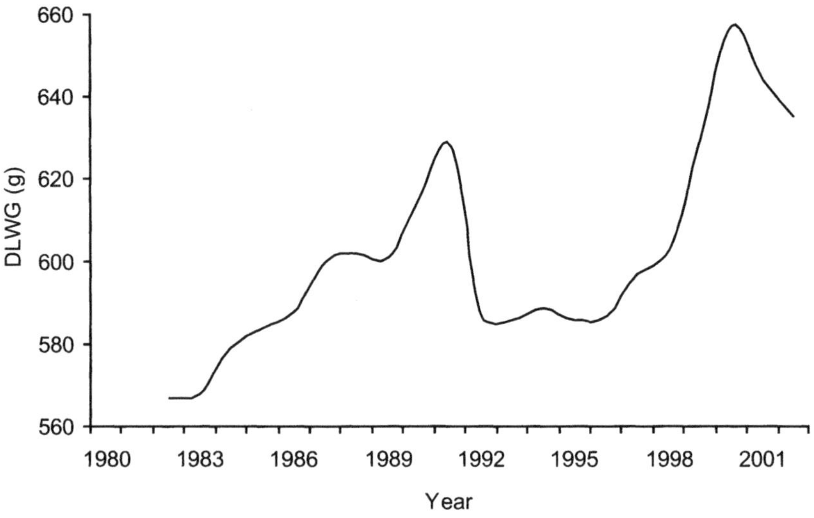

Figure 22.8 Changes in daily live weight gain (DLWG, g) since 1982

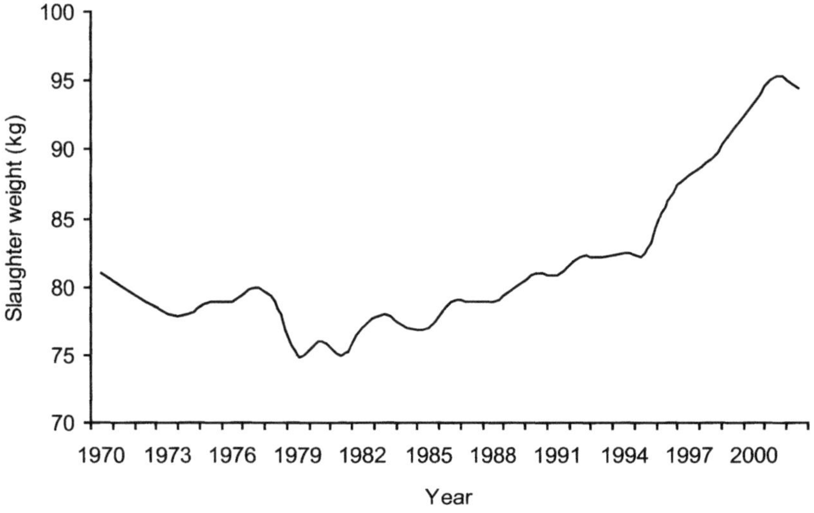

Figure 22.9 Slaughter weights (kg) since 1970

there has been little improvement and the data for 2001 were 2.70 (average), 2.65 (top 1/3) and 2.73 (top 10%). All 3 categories have shown a deterioration since 1998 for reasons which are difficult to establish. There are those who believe that diseases such as post-weaning multisystemic wasting syndrome (PMWS) have had an important negative impact on performance. In addition other diseases have struck the UK herd in recent years including PDNS (Porcine Dermatits Nephropathy Syndrome) and Classical Swine Fever.

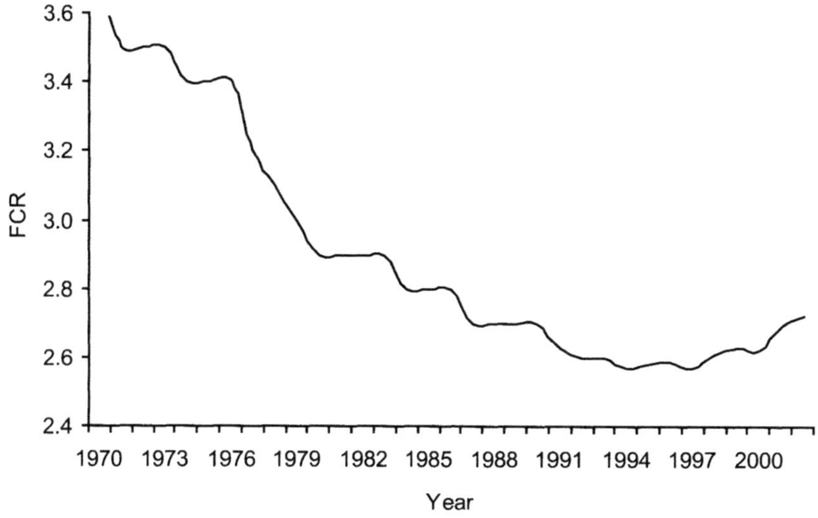

Figure 22.10 Changes in food conversion ratio (FCR) since 1970

It is not easy to identify causative factors for the improvements recorded up until the late 1990s ; they are in all likelihood a combination of improved genotype, better nutrition and management inputs. Data for feed intake (Figure 22.11) confirm that intake has declined; a commonly-held view is that this is an indirect effect of selection for better FCR and lowered backfat. However data for top 1/3 and top 10% herds (supposedly those with greater interest in obtaining animals of better genetic potential) indicate little improvement with constant figures of around 4.1 and 3.8 kg/d since 1987 (admittedly lower than the mean which fell from 4.5 to 4 from 1987-2001).

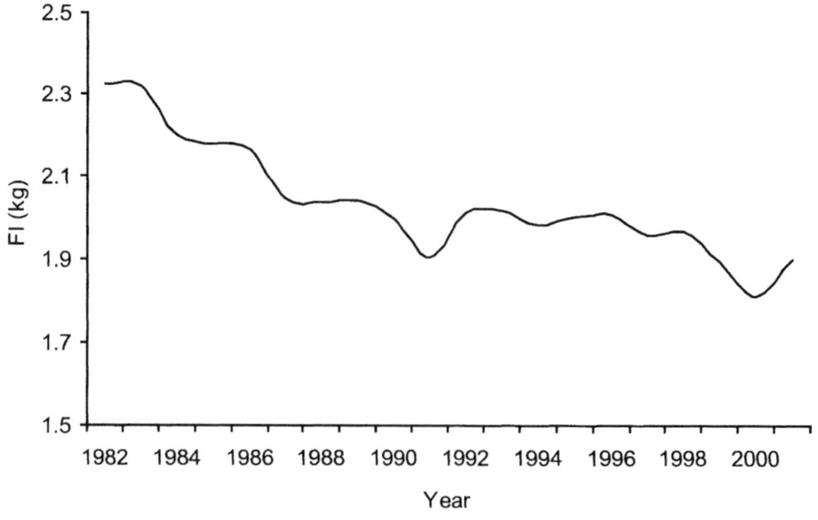

Figure 22.11 Changes in feed intake (FI, kg/d) since 1982

Feed intake is a major determinant of growth rate. Thus, when feed intake increases the usual result is an increase in growth. Growth models typically indicate general rules for the effect of increasing intake by 0.1kg:

- Feed conversion ratio improves by 0.05
- DLWG increases by 35 g
- Backfat (P_2) increases by 0.45 mm
- Killing-out increases by 0.25%

However, this is a simplified analysis as the response is usually not linear between traits. Figure 22.12 presents the curvilinear relationship between feed intake and feed conversion ratio, although it should be noted that there is a large range of feed allowances over which there is very little change in feed conversion ratio.

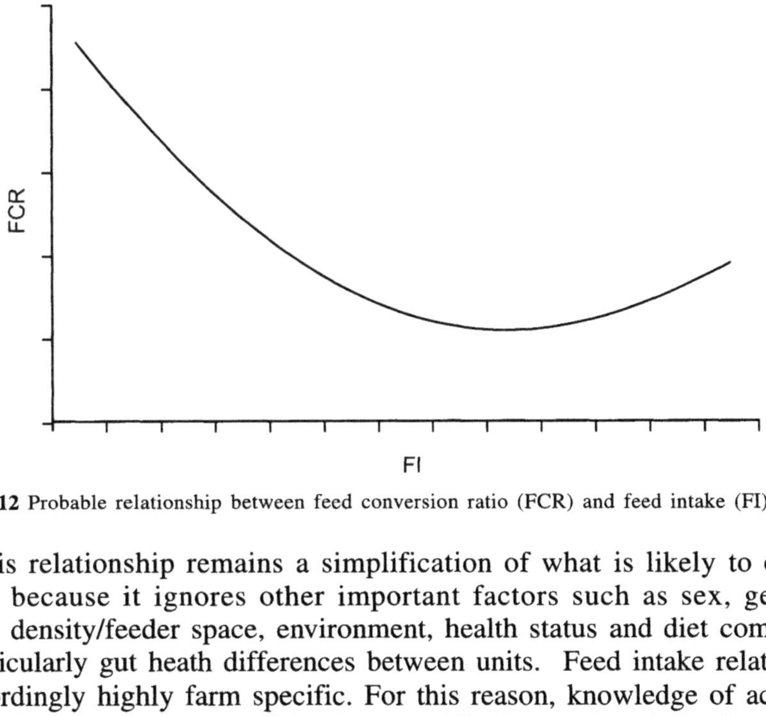

Figure 22.12 Probable relationship between feed conversion ratio (FCR) and feed intake (FI)

Even this relationship remains a simplification of what is likely to occur in practice because it ignores other important factors such as sex, genotype, stocking density/feeder space, environment, health status and diet composition and particularly gut heath differences between units. Feed intake relationships are accordingly highly farm specific. For this reason, knowledge of actual on-farm feed intake is a prerequisite for the effective application of nutritional standards in any particular production circumstance (De Lange, Marty, Birkett, Morel and Szkotnicki, 2001).

What is worrying about the general performance in the growing / finishing herd is that the genetic potential of the modern genotype is never realised in commercial situations. Thus, for example, the University of Nottingham pig trial facility regularly generates DLWG data from 25-110kg live weight of over 950g in animals individually housed. Data are close to genetic potential.

Stress is associated with mixing litters post-weaning and poor performance which results will influence subsequent growth rates. However, despite the introduction of high-health breeding pyramids, split-site systems and all-in / all-out strategies, health remains far from optimal in most commercial operations. When pigs are subjected to disease burden, the immune system is challenged and there is an associated loss of appetite and reduced performance (Table 22.4) with the problems being more severe in animals with greater lean potential, i.e. genetically superior pigs (Table 22.5) .

Table 22.4 INFLUENCE OF HEALTH STATUS ON PIG PERFORMANCE IN PIGS OVER A WEIGHT RANGE 5 TO 113kg (WILLIAMS, STAHLY AND ZIMMERMAN, 1997)

	Health Status	
	High	*Low*
Daily feed intake (kg/day)	2.22	2.00
Daily gain (g/d)	830	720
Feed conversion ratio	2.53	2.92
Backfat (mm)	25	31
Estimated muscle (%)	57.0	52.3

Table 22.5 PERFORMANCE AS INFLUENCED BY HEALTH STATUS AND LEAN ACCRETION POTENTIAL (WILLIAMS, STAHLY AND ZIMMERMAN, 1997)

	Lean potential	*Health status*	
		High	*Low*
Feed Intake (kg/day)	Low	2.69	2.58
	High	2.54	2.22
Daily Gain (g/d)	Low	680	599
	High	826	625
Feed Conversion Ratio	Low	3.98	4.28
	High	3.02	3.60

Another important health component has been the concentration of production in large units and the emergence of 'new' diseases. For example, porcine respiratory syndrome (PRRS) has been a major problem in many countries with increasing evidence of 'waves' of infection, extensive mutation (particularly in the American strain) and variation in virulence levels between strains. Of even greater current concern is the rapid spread of PMWS and PDNS worldwide. Unfortunately, despite typical mortality levels of up to 20%, relatively little is known about the real cause of the disease. Although the ubiquitous virus Porcine Circovirus-2 is implicated, various 'trigger' factors as yet poorly understood are also involved.

Under commercial conditions, the ultimate assessment of performance is based more on live weight gain per unit of floor space per annum. There is obviously a trade-off between individual pig performance and that of the pen / barn, and the optimum output of the later is at a point where the former is not. In conclusion, the biological potential of the pig for growth even on the best units is not being realised under commercial conditions, due probably to stocking densities (although there are official limits imposed) increasing competition between animals and encouraging spread of disease.

A final analysis which combines both the growing/finishing and breeding herds is the lean pigmeat production per sow per year. Although the UK has one of the best grower / finisher FCRs in Europe (average in 2002 of 2.72, matched only by The Netherlands) and one of the highest carcass lean meat percentages, it has a low output of lean meat per sow per year. The reasons for this are due largely to the light live weights at slaughter some 10-20kg lighter than many other countries. Welfare conditions in the UK are among some of the best in the World and stock have high potential lean tissue growth rates. However, the UK is constrained somewhat by having only entire males rather than castrates; whilst this will give better FCRs, there is the perception by some that heavier carcasses carry with them a greater risk of boar taint, although as indicated earlier this is probably not the case.

CREEP FEEDING

Much work over the years has focussed on feeding both the pre and post-weaned piglet but understanding of the principles involved are still far from complete. The young piglet below 20 kg liveweight is far removed from a growing / finishing pig in the higher weight range not only in terms of nutrient / energy requirements but also in the way in which the immune system and the gut microflora interact with these requirements. When attempting to extrapolate backwards from the clear principles of feeding a grower / finisher diet, this is fraught with huge errors and variations.

In commercial practice working strategies have been devised based on high nutrient / energy concentration feeds used in detailed phase feeding programmes. The creep feeds used before weaning have to be both palatable and contain high quality protein sources such as skim milk powder, but these feeds also must provide a priming to the gut wall and serve to initiate enzyme systems to make the transition to post-weaning life as smooth as possible. Following weaning the piglet requires continuity in ingredient exposure but also progressive and rapid changes in energy / nutrient balance and specification to maintain both feed intake and efficient growth. It is also very easy in feeding young post-weaned piglets to incur either severe growth checks due to anorexia or perhaps worse, acute diarrhoea. There is hence much work to be done in this area.

Yield projections

REPRODUCTIVE OUTPUT

The potential for increasing annual sow output will require both improved management and higher feed intake capacity of the sow. Eissen (2000) showed that the modern gilt could rear up to 11 piglets, but that larger litter sizes resulted in high maternal weight / backfat loss and poorer litter growth due to inadequate feed intake As a result, Eissen, Kanis and Kemp (2000) argued that sow appetite in lactation should ideally be included in all dam-line breeding programmes.

Initial data from a major co-ordinated trial in the Netherlands (Appeldoorn, 1999) reported that gilts with high feed intakes on performance test go on to have high feed intakes in subsequent lactation. This helps to explain why animals bred for low intakes have a poor record for longevity. Karsten, Rohe, Schulze, Looft and Kalm (2000) estimated the genetic correlations between performance test traits measured in boars and the reproductive traits in their offspring. The correlations between feed intake and reproductive traits ranged between 0.12 to 0.27, suggesting that appetite is a limiting factor on sow performance and that the conflict between production and reproduction increases with reduced feed intake. Cameron, Kerr, Garth, Fenty and Peacock (2002) showed (in a comparison of different selection lines) that selection strategies which result in reduced feed intake during lactation must be avoided as, otherwise, lipid mobilisation would be required to support lactation resulting in reduced reproductive performance.

GROWING / FINISHING PIGS

Many experiments (eg Campbell and Dunkin, 1983) have shown that young pigs fail to maximise protein deposition because of low appetite (which will reduce energy intake, a key driver of lean tissue growth rate). Tullis, Henderson and Whittemore (1980) observed faster growing pigs and better carcass quality with lower variation in the time taken to reach 25 kg. More recent data (eg Lawlor, Lynch, Caffrey and Doherty, 2002) have shown in 'modern genotypes' that pigs with the fastest early growth potential continue to have a growth advantage through to slaughter. Weatherup, Beattie and Walker (1998) reported that a 3kg spread at 11 weeks (35 kg) resulted in a spread of 15 kg at slaughter. Thus nutritional inputs in the young pig should be as high as possible to ensure that they grow as close to their genetic potential as possible.

The consequences of selecting for various traits have been studied by many. Initially it is important to appreciate that design of programmes themselves may influence responses obtained. Thus Whittemore, Kyriazakis, Emmans and Tolkamp (2001) showed that optimum feed intake is that which allows genetic potential to be achieved. If this potential is not achieved, it is because intake is probably being constrained. Accordingly testing must provide a regime where feed is supplied such that provision of energy and nutrients is in excess of requirements necessary to achieve maximum lean tissue growth rate in the "best" pigs, otherwise differences in ranking will not be identified.

The key question for breeders is to identify pathways to genetic progress where feed intake is not reduced. Various selection experiments (e.g. Cameron and Curran, 1995) have confirmed that the optimum is probably reached when selection is for lean growth on a semi-restricted diet a little below *ad libitum* intakes through the growth period.

In the past, geneticists have had little involvement with veterinarians, nutritionists and production specialists. The result has often been poor targeting of priorities and the failure to use integrated models of production systems. In particular, there has been a failure to match nutrition to genetic potential across the range of sexes of different genotypes at different growth stages which can be achieved by growth modelling. A key feature of growth modelling allows the forecasting of future requirements with continuing genetic progress. As an example, Walters (2001) projected the performance of pigs growing from 40 - 110kg at 950gm per day with an annual progress of 15g per day (Table 22.6), typically claimed by Breeding Companies for nucleus herd boars. These data were modelled over a fifteen year period using equations developed by Dr. W. Close.

Table 22.6 PROJECTED PIG PERFORMANCE (DESCRIBED AS GENETIC POTENTIAL)

Year	*DLWG (g)*	*Daily lean growth (g)*	*Fat P2 (mm)*	*FCR**	*Feed intake* (kg)*
2005	950	431	11.9	2.50	2.37
2010	1025	471	11.3	2.38	2.44
2015	1100	513	10.7	2.28	2.51
2020	1175	554	10.1	2.18	2.57

The model predicts that animals will be leaner at a given weight and less mature at that weight. In combination with the increased growth the result will be an increase in mature size and a resulting change in the nutrient requirements:

Table 22.7 PROJECTED ENERGY AND NUTRIENT REQUIREMENTS OF GROWING FINISHING PIGS

Year	*Energy MJ DE/day*	*Lysine g/day*
2005	31.2	26.2
2010	32.1	28.1
2015	32.9	30.1
2020	33.7	32.1

·Assumes 5% wastage; 13.89 MJDE/kg diet

The model (Table 22.7) predicts an increase in daily energy and lysine to support the genetic potential for lean growth. The result is an on-going requirement for increased daily feed intake. As there is currently a considerable shortfall between genetic potential and commercial performance in this trait there is growing awareness that this is an area requiring considerable emphasis for the immediate future. Even if the model is projected to commercial rather than nucleus performance, the required increase in feed intake to fuel future biological requirements is highly significant. Table 22.8 shows commercial projections every five years for fifty years from the current MLC grower/finisher average (MLC, 2003). The annual increase in growth rate used was 3.5 g/d which is based on the 0.41% reported by Walters (2000):

Table 22.8 PROJECTED IMPROVEMENTS IN PIG PERFORMANCE FOR THE NEXT 50 YEARS

	Growth (g/d)	*Backfat (mm)*	*FCR**	*Intake (kg/d)**
2005	635	10.7	2.72	1.73
2010	652	10.6	2.67	1.75
2015	670	10.4	2.63	1.76
2020	687	10.3	2.59	1.78
2025	705	10.1	2.55	1.80
2030	722	10.0	2.51	1.82
2035	740	9.8	2.47	1.83
2040	757	9.7	2.43	1.85
2045	775	9.5	2.40	1.86
2050	792	9.4	2.37	1.88
2055	810	9.2	2.34	1.89

* Assumes 5% wastage; 13.89 MJDE/kg diet

It should be noted that the 'commercial' model again predicts that higher feed intakes will be required in order to maximise future lean growth.

One possible route to future overall improvements will be to exploit developments in gene technology. As genes, gene markers and Quantitative Trait Loci (QTL) are identified it may be possible to introduce cost-effective selection for specific trait combinations (as considered in by Haley and Archibald, 2004; Chapter 19 of these proceedings).

Technical limitations to yield

There are many factors that could be proposed as major limiting technical / regulatory factors to the achievement of maximum performance yields. Stocking density and group size are known to have direct impact on feed intake, growth rates and probably food conversion ratios. There are of course optimum values for stocking density and group size in terms of maximising performance, but these are curtailed currently by more detailed regulation. The UK Department of Environment, Food and Rural Affairs (DEFRA) have introduced new legislation on stocking densities for different weight ranges and this means that in some cases producers have to operate at lower levels than would otherwise optimise performance and economics.

Since the beginning of 2004, new EU regulations limit weaning age to a minimum of 28 days unless weaned piglets are moved into separate accommodation to the sow. This limitation could potentially reduce the ceiling on sow productivity and piglets per sow per year will accordingly be less than if the currently accepted norm of 21 –25 days is continued. In reality this regulation has been introduced on animal welfare grounds; however in commercial practice many farms had moved in this direction anyway because of PMWS (immune status is higher with later weaning).

With regard to legislation, the other items that need mentioning are environmental issues and the IPPC (Integrated Pollution Prevention Control) legislation that is imminently being introduced and which will in effect be a new environmental tax. Pig producers will become increasingly aware of the overall impact of their farming practices on the environment because, if they get it wrong, it will have a rapid bearing on their economic viability. The effect of this on biological yields and performance is that producers may well opt out or operate at reduced levels of performance to maximise their own net margins.

Environmental legislation such as IPPC will have a direct impact on the structure of the industry as noted earlier in limiting the size of new or replacement units.

Limitations imposed on nitrate and phosphate content in farm waste such as pig slurry may also result in biological yield potential not being fulfilled unless genetics can be altered to more effectively metabolise these compounds thereby reducing their content in effluent discharge.

Similar new regulations on food safety in pig production, such as the new ZAP (Zoonosis Action Plan) scheme for controlling salmonella organisms in the food chain, may also impact on biological yields. Farms are rated as 1, 2 or 3 with 3 being the worst. Farms scoring repeatedly in category 3 may quickly be unable to sell their pigs under national assurance schemes. Producers will hence place more emphasis on hygiene and health status generally rather than planning for maximum lean meat produced per unit area of building system.

In practice however it could be speculated that, with a greater focus on health and hygiene, pig performance might well improve accordingly.

New legislation is always seen by producers initially as a limit to their own yield performance, whereas very often (but not always) new legislation brings new opportunities. The new sow housing legislation introduced in 1999 for example in the UK, which banned the use of individual sow stalls, was initially seen as adding cost to the systems to the tune of 20% extra housing costs. Four years on from this position, this hardly gets a mention in pig discussion groups and most producers are reasonably happy with the alternative group housing systems now they have acquired the skills to manage them. However, there is considerable anger that the increase in production costs for UK producers and subsequent erosion in terms of competitiveness with other EU producers is not recognised by other elements of the supply chain or valued ultimately by the consumer.

This high welfare production system operated in the UK ought to bring new opportunities in premium pricing for UK products against competition from overseas including Denmark and the Netherlands. Sadly this has not happened and the supermarkets can legally still supply pork and bacon products to British consumers that have been produced in systems that are not legal in the UK. Clearly this situation needs resolution in the future.

For the future commercial producers will need to maximise productivity and profitability by making use of all the technologies available. The current situation, where the results from the service house through to delivery at the abattoir are sub-optimal, must be overcome. As scientific knowledge of the biology of the pig expands it will be increasingly important to ensure that the lessons from research are rapidly implemented in user-friendly cost-effective terms. It could even be argued that the failure to use an integrated approach, the poor take-up of available technology on-farm and lack of exploitation of the benefits of research findings are all important in understanding why practical performance of pigs falls below that of the biological potential.

References

Appeldoorn, E. (1999) *Relation Between Feed* ad libitum *Feed Intake of Gilts During Rearing and Feed Intake Capacity of Lactating Sows.* Pig Breeders' Round Table, Wye, UK.

Campbell, R.G. and Dunkin, A.C. (1983) The effects of birth weight and level of feeding in early life on growth and development of muscle and adipose tissue in the young pig. *Anim. Prod.* **36**, 415-423.

Cameron, N. D. and Curran, M. K. (1995) Genotype with feeding regime interaction in pigs divergently selected for components of efficient lean growth rate. *Anim. Prod.* **61**: 123-132.

Cameron, N. D., Kerr, J. C., Garth, G. B., Fenty, R. and Peacock, A. (2002) Genetic and nutritional effects on lactational performance of gilts selected for components of efficient lean growth. *Anim. Sci.* **74**, 25-38.

De Lange, C. F. M., Marty, B. J., Birkett, S., Morel P. and Szkotnicki, B. (2001) Application of pig growth models in commercial pork production. *Can J. Anim. Sci.* **81**, 1-8.

Eissen, J. J. (2000) *Breeding for Feed Intake Capacity in Pigs.* Thesis,

Wageningen University, The Netherlands.

Eissen, J. J., Kanis, E. and Kemp, B. (2000) Sow factors affecting voluntary feed intake during lactation. *Liv. Prod. Sci.* **64**, 147-165.

Gill, B.P (2000). Nutritional influences on lifetime performance of the sow. In *Recent Advacnes in Animal Nutrition 2000*, ed. P.C. Garnsworthy and J. Wiseman, Nottingham University Press, Nottingham.

Karsten, S., Rohe, R., Schulze, V., Looft, H. and Kalm, E. (2000) Genetic association between individual feed intake during performance test and reproduction traits in pigs. *Archiv. fur Tierzucht* **43** (5), 451-461.

Knap, P (2000) *Variation in Maintenance Requirements of Growing Pigs in Relation to Body Composition.* Thesis, Wageningen University, The Netherlands.

Lawlor, P. G., Lynch, P. B., Caffrey, P. J. and O'Doherty, J. V. (2002) Effect of pre- and post-weaning management on subsequent pig performance to slaughter and carcass quality. *Anim.Sci.* **75**, 245-256.

Meat and Livestock Commission (2003) *Pig Yearbook 2003.*

Pluske, J.R., Williams, I.H. and Aherne, F.X. 1995., Nutrition of the neonatal pig. In. *The Neonatal Pig: Development and Survival.* pp187-235. ed M.A.Varley. CABI Publishing Oxford.

Tullis, J. B., Henderson, R. and Whittemore, C. T. (1980) Growth and body composition of young entire male pigs fed diets of differing ingredient composition and nutrient quality. *J. Sci. Food Agric.* **31**, 573-577.

Walters, J. R. (2000) *UK Observations on Lost Genetic Potential and Future Possibilities for Improved Sow Performance.* Pig Genetics Workshop, Armidale, Australia.

Walters, J. R. (2001) *Commercial Needs and the Delivery of Genetic Improvement – Success or Failure?* Pig Genetics Workshop, Armidale, Australia.

Weatherup, R. N., Beattie, V. E. and Walker, N. (1998) The effect of cereals or by-product based finishing diets on growth performance and fatty acid profile of carcase fat. *Irish J. Ag. and Food Res.* **37**, 191-200.

Whittemore, C., Green, D.M. and Knap, P. (2001) Technical review of the energy and protein requirements of growing pigs: food intake. *Anim. Sci.* **73**, 3-17.

Whittemore, E.C., Kyriazakis, I., Emmans, G.C. and Tolkamp, B.J. (2001) Tests of two theories of food intake using growing pigs: 1. The effect of ambient temperature on the intake of foods of different bulk content. *Anim. Sci.* **72**: 351-360.

Williams, N. H., Stahly, T. S. and Zimmerman, D. R. (1997) Effect of chronic immune system activation on the rate, efficiency, and composition of growth and lysine needs of pigs fed from 6 to 27 kg. *J. Anim. Sci.* **79**, 2463-2471.

Wiseman, J. (2001) *The Pig - A British History*. Duckworths, London.

23

LIMITS TO THE PERFORMANCE OF POULTRY

A WALKER[1], F SHORT[1] AND MG MACLEOD[2]
[1]*ADAS Gleadthorpe, Meden Vale, Mansfield, Nottingham, NG20 9PF, UK*
[2]*Roslin Institute, Roslin, Midlothian, EH25 9PS, UK*

Introduction

The biological performance of any farm species, including the chicken, is finite. Yet performances have continued to increase over long periods of time, despite repeated claims that they must be nearing their limits. This chapter attempts to provide some information on the factors affecting the limits, with a view to anticipating performances up to 2010 and then 2050. The basis of the forecasts is a review of the progress of the past 50 years. This is a particularly appropriate time span in the poultry sector because the industry in its modern form developed after the ending of feed rationing 50 years ago in 1953/54, following the post World War II austerity years. Conditions were different enough before and after that date to justify separately named historic periods. Thus a Defra-funded study group, reported by Gordon and Charles (2002), defined the period before that date as the *traditional* period of the history of the British poultry industry, and after that date as the *conventional* period. Charles (2002a) reviewed the history of the poultry sector in the context of the history of food, agriculture and the countryside.

Forecasting and projection techniques

2050 is a convenient date to choose for the final analysis, because it is the chosen forecasting date of a recently published major review of the trends in many issues (Cole, 1999), and considered the likely trends in the use and supply of many commodities, including some which affect the potential prospects for poultry performance.

Cole (1999) pointed out some important principles of forecasting. When extrapolating from past trends there are often many possible extrapolations. It is necessary to examine the factors conceivably affecting the shapes of the trend lines, and to analyse and surmise which are likely to be the most influential. It is often possible to make several forecasts based upon different assumptions, and the assumptions can be overtly declared so that the reader may make informed choices between forecasts. Sometimes detailed models are used. Over a projection as long as 47 years (2003 to 2050) there could be many factors at work, some of which may not yet have begun to exert their influence. For example Cole (1999) used nine different projections for the world human population by 2050. The highest estimate was nearly twice the lowest, at 15.0 billion compared with only 8.3 billion, based on a 1997 estimate of 5.84 billion.

Market background

Many of the biological forecasts offered below depend upon the maintenance of the *status quo* in several aspects of the global market. It must be remembered however that the UK poultry industry is a very small part of the expanding world industry. Data from the website of the Food and Agriculture Organisation (FAO, 2003) and from their publication of 2001, suggest that UK hen egg production of 630,000 t/year in 2002 was only about 1.2% of world production. By contrast China accounted for 38.5% and USA for 9.6%. Furthermore production in China was expanding at an estimated 6.6% per year.

Similarly, the same information sources suggested that UK poultry meat production of 1,554,000 t/year was about 2.2% of world production in 2002, compared with 24.0% in USA, 18.7% in China and 9.5% in Brazil.

Thus in that year UK was the world's 7th largest importer of eggs in shell, though it also ranked 12th as an exporter. It was the 6th largest importer of poultry meat and the 9th largest exporter. Clearly the UK industry is part of global trading and cannot be considered in isolation.

The scale of production in the world makes the global poultry and pig industries major consumers of feed ingredients, in particular grains and pulses. As production expands in major producing countries, such as China and the USA, there will be growing demand on world ingredient markets. FAO (2001) figures for 2000 suggest, for example, that the annual increase in pigmeat production in China is more than twice the UK total production. It is worth remembering that consumer demand for poultry products in China may expand considerably as even small rises in standard of living are multiplied by 1220.2 million consumers (compared with 58.1 million in UK); (population figures from Economist, 1997). Consumption of poultry products tends to increase as developing economies become better fed. For example data for India from Panda and Mohapatra (1993) indicated egg consumption increasing from 6 eggs per person in 1961 to 19 in 1983 and 30 by 1991.

Thus it cannot necessarily be assumed that the pattern of supplies of feed ingredients will necessarily remain as it is now for the duration of the periods forecasted. The mix of materials available at prices economic to the poultry industry and the feed trade may change, so that current assumptions about economic feed formulation may change. However the biological principles will not change: only the economic interpretation of response curves and the choice of ingredients. It is tempting to speculate for example that, by 2050 or even by 2020, there may be more dependence on synthetic amino acids and perhaps on novel protein sources not currently appearing in the matrices used by formulators. If this happens there may be particular ingredient supply, and therefore cost, difficulties for the organic sector of the poultry industry.

It is hardly surprising that the breeding of hybrid poultry is by international companies who must, presumably, be interested primarily in the needs of world markets, though no doubt they will be able to offer different breeds for different markets and purposes, just as they have in the past.

Another key commodity which is now global is information, including poultry science and technology. Scientific journals, scientific and technical books, feed and poultry trade magazines, consultancy, company information and the internet all operate globally, and frequently in English. This may have the effect of speeding up the adoption of the biological concepts reviewed below. This has been taken into account in the estimation of speculative yields at future dates.

Unfortunately there is also a tendency for diseases to be global, so this may need to be countered by the global operations of the pharmaceutical and vaccine industries if the potential of the birds is to be met.

Biological principles

GROWTH OF CHICKENS

By definition performance at any given time cannot exceed the genetic potential of the species and the breed at that time. Limits to the genetic potential are dealt with later. For the present suffice it to recognise that the growth of broilers follows a sigmoid pattern when body weight is plotted against age. The principles of sigmoid growth have been reviewed by Lawrence and Fowler (1999) and much of the application of the principles to the growth of chickens was developed by Emmans (1977) and by Emmans and Fisher (1986). One of the parameters of the growth curve is the breed and line specific maximum body weight attained under non-limiting conditions. Limiting factors depress growth rates so that the progress of individual birds towards the maximum is retarded.

EGG PRODUCTION

Egg production is limited by the maximum rate of ovulation per unit time and by the maximum fractional rate of maturation of the ova into fully formed eggs. Both of these are likely to be genetically controlled. These are dealt with elsewhere. Limiting factors depress egg production below the maximum imposed genetically.

FLOCK VERSUS INDIVIDUAL MAXIMA

Commentators are often impressed by the very high levels of egg production quoted in anecdotes about flocks of traditional breeds during the traditional period (as defined above), and before the advent of the hybrids bred by the international breeding companies. This is because in the traditional period of the British poultry industry outstanding birds were occasionally recorded as individuals, whereas it is now usual to describe performance on a flock (i.e. population) basis. It would be highly impractical to record any other way, even in relatively small flocks. It is important to note that it is the potential for and limits to the performance of flocks not individuals which is important in poultry.

Flock average performances in the traditional period were very low by modern standards, even though the flocks sometimes contained outstanding individuals. For example a flock egg production of 180 eggs per bird per year was regarded as good in the 1940s (Robinson, 1948), even though the flocks were generally small by modern standards, whereas modern flocks of hybrids may regularly produce 305 or more eggs per hen housed in large flocks.

Historic performance and improvements in performance

By quoting data from four sources Charles (2002) summarised the improvements in eggs per hen per year in UK from 1898 to 1999. One of the sources was ADAS, and the others were Hawk (1910), Thompson (1952) and Whittle (1995). The data show an improvement from approximately 150 eggs per bird in 1898 to just over 300 in 1999, i.e. 101 years later. The improvement has continued to be steady in recent years from 1970 to 1999, with no evidence of a slowing of the rate of increase in eggs per bird per year.

Charles (2002) tabulated the improvements in commercial broiler performance in UK from 1958 to 1996 by quoting data from ADAS and the Museum of British Poultry, for which information had already been reviewed. In 1958 a liveweight (as hatched, i.e. averages of males and females) of 1.9 kg was normally reached at 70 days of age. By 1996 a weight of 2.2 kg was typically reached at 42 days.

Variation in performance

HISTORIC RECORDS OF VARIATION

Genetic progress has been in flock performance rather than in that of outstanding individuals. As long ago as 1949 Romanoff and Romanoff noted that 300 eggs per hen per year was possible. The record for an individual bird was 361 eggs in 364 days, held by a Black Orpington in 1929/30, although this yield did include three soft shelled eggs. Flock means of the time were more like 180 eggs per bird for White Leghorns and 150 for Rhode Island Reds. The distribution around the means was skewed to the right.

VARIATION BETWEEN FLOCKS

ADAS (1985) and Charles (1991) reported the results of an analysis of the performance of a large number of commercial caged laying flocks in UK, as recorded by the ADAS costings scheme in the 1980s. For flocks recorded in 1984 the flock potential was estimated by taking the best performance of small flocks in non-limiting treatments at ADAS Gleadthorpe, and the commercial potential by considering the top 4% of 221 flocks on the ADAS costing scheme. By these criteria flock potential for egg production (eggs/hen housed in 365 days) was estimated at 300, on a feed intake of 130 g/d per bird, while commercial potential was >280 eggs/hen housed. Mean commercial performance of the 221 flocks was 273 eggs/hen housed on a feed intake of 117 g/d per bird.

There was considerable variation between the commercial flocks. The variation was examined by classifying the flocks by production classes with class band widths of 20 eggs per bird. The analysis showed that while 6 flocks produced less than 220 eggs per bird per year 19 produced more than 280. 136 flocks fell within the 260 to 280 eggs per bird class. Yet it was thought that most of the flocks were of similar hybrid stock.

INDIVIDUAL VARIATION

The examples quoted by Romanoff and Romanoff (1949) illustrate that there were individual birds of the traditional breeds capable of quite remarkable biological performance. It was recognised by Fisher, Morris and Jennings (1973) that there is substantial individual variation within flocks of modern hybrids. In both the traditional breeds and the hybrids flock averages included the performances of individuals of a range of performances.

Physiological literature which accounts for the yield trends

Although some of the improved performance of broiler chickens over the past several decades has been due to improved nutrition and management, most has resulted from genetic selection. Initially, most of the increase in growth rate of the broiler chicken was produced by selection for body weight. Until relatively recently, most of the effect was due to increased food intake. The accompanying improvement in conversion efficiency would largely have been a consequence of "dilution of maintenance". This occurs when maintenance requirements for energy and nutrients remain relatively static while intakes increase. A greater proportion of the energy and nutrients consumed are, therefore, available for growth and this leads to a decreased food:gain ratio (increased gross efficiency). However, since the mid-1980s, there has been targeted selection for conversion efficiency. This was partly because it was shown that selection simply on rate of weight gain caused an unwelcome increase in body fat content (Leenstra and Pym, 1995). Selection on efficiency of weight gain will indirectly select against fat deposition, since proteinaceous tissue not only has a lower energy content per unit weight but is also almost always characterised by high water content. Growth can be described crudely as the deposition of bone, muscle and fat. However, this produces an extremely limited description of the processes of growth and development. Selection for growth rate and commercially desirable body conformation has led to a relatively large increase in muscle mass without a parallel increase in the mass of other tissues. This has sometimes been summarised as an imbalance between demand (muscle) and supply (heart, lungs *etc.*) tissues and organs.

FOOD INTAKE

The increased food intake of broilers may be associated with altered activity of neuropeptide Y (NPY), which is known to stimulate feeding (Kuenzel and Fraley, 1995). The difference seems to lie at the receptor level; NPY receptor density was 2-3 times greater in the brains of growing broilers than in those of layers (Merckaert and Vandesande, 1996). Boswell *et al.* (1999) tested the hypothesis that genetic selection for rapid growth rate is associated with increased hypothalamic NPY mRNA by comparing NPY gene expression in a commercial broiler line with a control line in which selection for growth rate had been relaxed since 1976. No significant differences were detected between the two lines, suggesting that selection for increased growth rate was not associated with altered gene expression of NPY in the hypothalamus. However, selection for rapid growth may be linked to other changes in the signalling pathway (Merckaert and Vandesande, 1996).

DIGESTION AND ABSORPTION

Selection for rapid growth has been associated with increases in the absolute weight and length of small intestine but reductions in relative weight and length. There has been a decrease in the relative mass of the mucosa, especially in the upper part of the small intestine (Mitchell and Smith, 1991). However, villus area and amino acid accumulation rates have increased (Mitchell and Smith, 1990). There appears, therefore, to have been an increase in digestive and absorptive efficiency in birds selected for improved growth rate and feed conversion efficiency.

HEAT PRODUCTION AND LOSS

Increased food intake and growth rate would both be expected to increase heat production. The change in the body shape of broilers to a more spherical conformation would also be expected to make it more difficult to dissipate heat, since surface:volume ratio has been reduced. However, selection for improved food conversion efficiency has probably been associated with reduced locomotor activity, which will have the effect of reducing heat production or at least offsetting the rate of increase. The time broilers spend sitting and resting increases from 50% in week 1 to 75% in week 6, while the time spent running or walking decreased from 9% to 1% (McLean *et al.*, 2002). Layers at the same age spend only 20% of their time sitting (Savory and Mann, 1997).

BALANCE BETWEEN PROTEIN SYNTHESIS AND BREAKDOWN

The rate of protein accretion is a function of the rates of protein synthesis and protein breakdown. An increase in protein accretion rate accompanied by improved energetic efficiency and food conversion would be more likely to result from a reduced ratio of breakdown to synthesis. Tomas *et al.* (1988) used N^t-methylhistidine excretion as an indicator of the myofibrillar breakdown associated with muscle protein turnover and found a significantly lower rate of breakdown in birds selected for food conversion efficiency and an intermediate rate of breakdown in birds selected for rate of weight gain. A similar pattern was demonstrated by Saunderson and Leslie (1988) in a comparison of layer and broiler lines. Tesseraud *et al.* (2000) compared fast and low-growing lines obtained by divergent selection within a population of meat-type chickens and demonstrated more clearly that the difference in growth rate was related to rate of protein degradation rather than protein synthesis.

MUSCLE FIBRE GROWTH

Muscle fibre diameters are greater in broilers than in layers throughout the growth period. This is true both in the glycolytic breast muscle and in the more oxidative leg muscle. In broilers, there was a divergence in the size of breast muscle and thigh muscle fibres, which may have resulted from selection for increased breast muscle yield (Cooke *et al.*, 2003). Increases in growth rate of broilers and turkeys may have been accompanied by detrimental effects on skeletal muscle, including an increased incidence of spontaneous and stress-induced myopathy (Mitchell, 1999).

MUSCLE DAMAGE

Plasma creatine kinase activity, an indicator of skeletal muscle damage, was correlated with body weight and muscle yield in 37 different lines of domestic fowl (12 broiler, 12 layer and 13 traditional). There were significant increases in plasma creatine kinase activity with age but these were removed by regression on body weight. However, differences between broilers and other lines could not be removed by adjusting for body weight. The authors suggested that changes have occurred in muscle function and membrane integrity of broiler lines which may be due to genetic changes in muscle fibre dimensions (Sandercock *et al.*, 2002). Mitchell *et al.* (2001) found that increased muscle damage in turkeys was likely to be found when muscle fibre diameter exceeded 65 μm.

ASCITES

Mortality in broilers associated with fluid accumulation in the body cavity (ascites or "water belly") is the result of high blood pressure in the pulmonary circulation. The symptoms include cardiac (heart) hypertrophy, particularly in the right ventricle, and generalised oedema (fluid accumulation), seen particularly in hydropericardium and the accumulation of ascitic fluid in the abdominal cavity. The root cause has often been considered to be an imbalance between demand and supply of oxygen. The rapid synthesis of muscle protein is an energy demanding and hence an oxygen demanding process. There have also been alterations in the relative masses of muscle protein (meat) and the support systems (respiratory, cardiovascular, *etc.*). The ratio of glycolytic to oxidative muscle cells has also increased with selection for breast muscle yield. The tendency for the ascites syndrome to develop is exacerbated by environmental factors such as low oxygen tension (high altitude or poor ventilation), low temperature, and high nutrient density. The latter two factors produce an elevated metabolic rate. There have been a number of reviews on the topic, including Decuypere *et al.* (2000) and Julian (2000). More recently, the complex aetiology of ascites has been highlighted by the demonstration that the fast-growing birds most susceptible to ascites actually have a lower basal metabolic rate per unit of weight than ascites-resistant slower-growing birds. Ascites appeared to be associated with relative hypothyroidism and proportionally smaller lungs. The latter, in turn, may have been the cause of low arterial pO_2 and high arterial pCO_2 (Malan *et al.*, 2003).

SKELETAL DISORDERS

The problem of leg weakness in broilers has been reduced in recent years by a combination of genetic selection, altered feeding regimes (*e.g.* meal feeding) and longer dark periods in the lighting cycle. Some of the effects may have been due to increased locomotor activity. The most frequent problem has been tibial dyschondroplasia. This is caused by a defect in tibial growth plate chondrocyte differentiation, which leads to the formation of a mass of unvascularised cartilage at the joint (Whitehead, 1998). It often results in abnormal bone formation leading to leg weakness. Osteoporosis is an important factor in the welfare problem of bone fractures in laying hens. Bishop *et al.* (2000) showed that bone strength in end-of-lay hens was moderately to strongly inherited and responded well to selection. There was a lower incidence of fractures in hens selected for bone strength. The authors suggested that application of a selection approach would greatly improve the welfare of laying hens and also improve public perception of the egg industry.

Technology which might overcome constraints

GENOME MAPPING AND "TRAIT-GENES"

Bulfield (1994) pointed out that "the chicken was an early model for genetic analysis (Hutt, 1949), an early subject for quantitative genetic techniques (Lerner, 1950) and the possessor of the second gene ever to be cloned using molecular techniques (ovalbumin in 1976; see Bulfield, 1990)". In more recent years, there has been rapid progress towards producing a comprehensive genetic map of the chicken genome. Research on livestock (including poultry) genetics has benefited from the human genome project (Lander and Weinberg, 2000), with detailed genetic marker maps and mapping of trait-genes (Andersson, 2001). There is now a map containing more than 2500 loci and over 700 genes have now been mapped (Burt, 2002). Public databases (*e.g.* Arkdb-CHICK, Roslin) have been established to provide access to chicken genome data. Genetic marker maps have been used in a number of resource populations for mapping quantitative trait loci for traits such as growth, follicle number, muscling, fatness, feeding behaviour, meat quality, disease resistance and length of the intestine. The next step is to identify the genes at these loci. The first step in mapping the genes that control a trait is to provide a robust definition of the trait and then to identify population resources showing a difference in this trait. If the trait can not be measured reliably then mapping QTLs will be difficult (Burt *et al.*, 2002). The generation of large numbers of genotypes from the large pedigrees (500-2000 animals) required for QTL mapping has necessitated developments in database management (Law and Archibald, 2000).

GENOME MAPPING AND GENE DISCOVERY

Some genes have a large enough effect to be individually recognisable, such as the sex-linked *dwarf* gene in the chicken. However, traits such as growth rate or conversion efficincy are under the control of a number of genes (Burt *et al*, 2002). The genes that control these quantitative traits are located at quantitative trait loci (QTL). QTL mapping is the first stage in the process of identifying the trait-genes at these loci. QTL can be located in the genome through associations between performance and the inheritance of genetic markers in a suitable pedigree (Mackay, 2001). This process requires a map of genetic markers evenly spaced throughout the genome.

GENETICALLY DEFINED LIMITS TO SELECTION

Genetic improvement of growth rate has continued in an almost linear fashion for many generations. However, there is clearly a theoretical limit. Robertson (1960) describes how this is defined by population size, initial gene frequency and selection coefficient. Limits may have to be imposed on selection for growth rate but the future for genetic improvement of poultry remains optimistic. There is scope for the improvement of traits related to welfare, reproductive fitness and environmental impact.

BREEDING STRATEGIES

The technologies which are most relevant to genetic selection are information technology and genomics. Information and communication technology speed up collection, processing and analysis of data. More complex genetic models, more accurate data and data from a wider range of environmental and geographical conditions should lead to improvements in breeding value estimation and genetic management of breeding populations. Albers and van Sambeek (2002) suggested that this would increase the rate of genetic progress by 10-20%. Layer breeding companies now use BLUP (best linear unbiased prediction) to estimate breeding values for a large number of candidate breeders using data from ten or more generations.

CANDIDATE GENE APPROACH

When several genes interact in a complex manner, the consequences of variation at one locus are difficult to predict. This can make the "candidate gene" approach very laborious, unless a key step can be identified in the metabolic pathway under study (Tixier-Boichard, 2002).

WELFARE AND GENETICS: A CAVEAT

There is an ethical dimension, in that selection for tolerance of poor welfare standards should not be used as an excuse for failing to improve welfare directly.

Possible limits to the performance of laying hens imposed by environmental factors and substrate supply

GENERAL PRINCIPLES AND REPRODUCTIVE PHYSIOLOGY

It is evident from the examples quoted above that there have often been individual birds capable of performance very much higher than the flock averages of their time, or even modern flock averages. This suggests that constraints to performance such as the ability of the species to digest, absorb and utilise nutrients, and the ability to transfer them to the oviduct, are unlikely to be limiting.

Thus this could be taken to indicate that performance improvement trends could be extrapolated, at least up to performances equivalent to one egg per day (i.e. 365 eggs per bird per year). Presumably selection for improved egg production must effectively be increasing the skewness to the right in the frequency distribution of individual egg production classes observed by Romanoff and Romanoff (1949). Individual performances above 365 eggs per bird per year seem unlikely, since that would involve more than one ovulation per day, and Etches (1995) described ovipositions as, by definition, on consecutive days. He went on to describe the hormonal control of ovipositions by luteinising hormone and progesterone. Earlier Gilbert (1971) had reviewed a considerable volume of literature on poultry reproductive physiology.

An extrapolation of the data quoted by Charles (2002) suggests that, in the absence of limiting factors, this performance would be reached by about 2020 to 2030, followed by little further improvement. However, despite the ability of outstanding individuals to reach very high rates of lay (eggs/bird per unit time), it should be remembered that the eggs of modern hybrids are considerably larger (g/egg) than they used to be. Thus the amounts of energy and nutrients to be absorbed and utilised are also correspondingly larger. Therefore there could be limits to performance improvement imposed by the supply of substrate to the oviduct, and these constraints must be considered.

Clearly high performance requires an adequate supply of essential amino acids, vitamins and minerals. In circumstances where no marketing or regulatory constraints are placed on the use of synthetic amino acids it seems unlikely that their supply need be limiting, except by cost. It is more likely that the supply of labile calcium may be limiting, and the balance between calcium and phosphorus, and this will be examined further below. Meanwhile it is important to realise that cost constraints are real, and increasing the supply of amino acids may become uneconomic in some egg and ingredient markets, before biological constraints are reached. Dillon (1977) discussed the concepts and methods of analysis of agricultural response in general and Fisher *et al.* (1973) and Curnow (1986) described the analysis of the response of layers to amino acids. There is a diminishing response to increasing dose rates, fairly well described by inverse polynomial functions, and therefore the production model of Charles (1984) used these functions. However Morris (1989) cautioned that the inverse polynomial function may indicate responses continuing to too high a level, thereby slightly over-estimating the economic optimum amino acid intake.

Since it is unlikely that eggs will reach very high prices cost benefit constraints cannot be ignored in forecasting the use of inputs. The price of eggs is market driven: it is not necessarily related to the cost of production. Thus even in the complete absence of brand or regulatory constraints economic constraints could become primary limiting factors. Nevertheless the following husbandry and biological factors are worth noting.

SPECIFIC LIMITING FACTORS

Charles (1991) observed that the spread of performance was difficult to account for since the same genetic material, equipment, housing and nutrition were available to all participating producers. He therefore offered a list of factors potentially relevant at the time, including:

Inadequate feed (unlikely)
Wrong temperature (possible, see below)
Wrong lighting pattern (possible, see below)
Poor ventilation (possible)
Poor cage or system design, incorrect stocking rate (unlikely)
Poor trough or drinker design
Poor vaccination programme

Light

Photoperiod exerts powerful influences on ovulation and practical lighting patterns have been developed and recommended since the late 1950s (reviewed by Morris, 1968a; Morris, 1994; Charles, Lewis and Tucker, 2002). They had been well publicised, and were believed to be commonly applied, by the time of the costings examination of ADAS (1985) and Charles (1991). Yet evidence was found of events such as inadvertent mismatches between rearing and laying patterns, due to factors such as poor communications at the time of pullets changing hands, and light proofing inadequacies. For some hatch dates the mismatches could conceivably have accounted for depressions in performance during lay.

Light intensity also affects egg output, and the performance of indoor birds could be depressed if the intensity in all or parts of the housing falls below about 10 lux (Morris, 1968a). In a later review Charles, Lewis and Tucker (2002), quoting Lewis and Morris (1999), concluded that the economic optimum may be close to 5 lux, but 10 lux was still recommended in the interest of working conditions and perceived bird welfare. It is important to note that even free range birds may experience lower than optimal light intensity if they spend much time in poorly lit housing. Thus the design of light for range housing is important. At some times of the year, for some hatch dates, the reliance on natural light from windows may be insufficient.

Light colour (wavelength) and light source (e.g. tungsten filament or fluorescent) may have some influence on egg output. However Lewis and Morris (1998) concluded from a review that there was no evidence of consistent detrimental effects of flourescent sources. Red light is more stimulatory to sexual reproduction than blue or green, whereas the growth of broilers and turkeys was found to be inferior in red (Lewis and Morris, 2000).

Equipment and furniture

The relevant and important aspects of feeder and drinker include the need to ensure that all individual birds may receive genuinely *ad libitum* intakes of both feed and water despite social and competitive effects.

Temperature and ventilation

Egg production is maximised at environmental temperatures of about 19-22°C, depending upon feed formulation and breed, according to several authors reviewed by Charles (2002b). These temperatures have, by convention, normally been recorded in the gangways between battery cages. Optimum temperatures at all times of the day and night and all the year round can only be provided in indoor systems. Therefore if the proportion of the national flock in these systems continues to decline there may be a thermal effect on performance resulting in a depression in egg production of the order of, perhaps, about 5%. However this effect is not inevitable, since the temperature response is at least partly to the 24 hour mean (reviewed by Marsden, 1981), and it may therefore be possible to alleviate cold free range conditions by warm housing at night.

The control of poultry house temperature was found in the field studies of ADAS (1985), Sutcliffe, King and Charles (1987) and Charles (1991) to be less accurate and less precise than was physically possible. Yet the published effects of temperature on performance are considerable (e.g. Payne, 1966;

Emmans and Charles, 1977; Marsden and Morris, 1987; Charles, 2002b). Good control of temperature would have been desirable, and it will be important in meeting future bird performance potential. It was also observed in the field studies, and it has often been observed during consultancy work, that it can be difficult to provide uniformity of environment within houses, so that even if the average temperature is correct some birds are too warm and others are too cold. Therefore the design and installation of ventilation systems and equipment will continue to be important.

Even in indoor systems temperature control is sometimes imperfect, due to practical difficulties such as interference with the proscribed ventilation rate by external wind forces, and inadequate standards of insulation (Charles, Clark and Tucker, 2002). Sutcliffe, King and Charles (1987) monitored practical poultry house environmental control and highlighted some of the practical constraints, many of which probably still apply, particularly in older houses. A continuing difficulty is that in practice it is difficult to obtain measurements of the real ventilation rate of a housing system under working conditions, after interference by natural wind forces.

Despite the practical emphasis on preventing extraneous excess air supply in cold windy weather it is important to note that the birds have a minimum ventilation rate requirement for the provision of oxygen and the removal of carbon dioxide, dust, excess moisture, pollutant gases such as ammonia, and airborne organisms. Charles, Clark and Tucker (2002) reviewed the literature from several decades of work on requirements and the reasons for them, and tabulated some example requirements against liveweight. In summary layers require about (1.6 to 2.0) x 10^{-4} m^3/s per $kg^{0.75}$. This is, for example, 0.32 to 0.40 m^3/s per 1000 layers of 2.5 kg liveweight.

When designing housing, even housing for free range birds, it is important to ensure that the system can also supply the maximum ventilation rate requirement. This is defined as the amount of air needed to prevent metabolic heat causing the house temperature to rise too far above outside temperature. For UK conditions the rate is usually calculated to prevent equilibrium temperature rise exceeding 3°C. In practice, due to the thermal capacity of the masses of the building, the contents, the floor and the birds, inside temperature takes several hours to rise to its equilibrium level. Therefore maximum ventilation rates based on a 3°C rise usually provide acceptably cool conditions in most warm weather in Britain. The maximum requirement tabulated by Charles *et al.*, (2002) was 1.55 x 10^{-3} m^3/s per $kg^{0.75}$, or 3.1 m^3/s per 1000 layers of 2.5 kg liveweight.

The ventilation system must be capable of supplying the maximum ventilation rate, whether it is driven by electric fans or by natural forces. It must also be provided with some means of thermostatically reducing the air supply down towards the minimum when house temperature falls below the temperature required by the birds. Some design principles for fan systems were summarised by Charles et al. (2002). Systems for ventilation by natural forces are often based on calculations by Bruce (1978). Natural forces include wind and convection. It is necessary to design naturally ventilated systems to provide enough air under windless conditions, but to be aware during operating them that wind often enhances ventilation rate at times of cold weather when the minimum is required. The performance of layers in naturally ventilated free range housing could be depressed by the failure to apply these principles.

Calcium supply

Simons (1986) calculated the calcium (Ca) turnover of an individual hen with an egg mass output of 75 g on an egg laying day. He considered the calcium content of shell material and assumed a utilisation efficiency of 0.5. Maintenance requirement was taken as 0.10 g/bird day. On a day when 75 g of egg material was produced the calcium requirement was estimated at 5.592 g/bird, compared with 4.512 on a day when 60 g of egg was produced. Sturkie (1976) calculated that a hen laying 250 eggs per year secreted 20 times the Ca content of its whole body. During the first 5 hours of shell formation calcium carbonate deposition was at a rate of 300 mg/hour. While hens appear to keep this up for a few days at a time, or even over fairly long clutches, it seems probable that Ca metabolism may be a constraint on long term continuing egg output performance. The risk of skeletal Ca depletion seems high, and weak bones might be the first manifestation of such a constraint on performance. Whitehead (1994) noted the risk that bone fragility might be induced by the loss of structural bone mass during lay.

Amino acid supply

In organic systems prohibited from using synthetic amino acids, a limit to performance may be intake of methionine plus cystine and of lysine, unless fish meal continues to be permitted. The principles of response to amino acids were rationalised and quantified on the basis of partition between maintenance and production by Fisher *et al.* (1973) and by McDonald and Morris (1985). Typical requirements have been tabulated by McDonald, Edwards, Greenhalgh and Morgan (2002). Protein and amino acid sources have been reviewed by Gordon and Charles (2002). The realisation of genetic potential may depend upon the availability on the market of suitably priced organic non-GM soya, lupins, sunflower, naked oats and low glucosinolate rapeseed as methionine sources; and soya and rapeseed as lysine sources. Unfortunately antinutritional factors tend to limit acceptable inclusion levels in practical feeds, therefore Gordon and Charles (2002) reviewed and tabulated suggested maximum inclusion rates for various vegetable protein sources.

Possible limits to the performance of table chickens imposed by environmental factors and substrate supply

SPECIFIC FACTORS

Charles (1991) identified nine limiting factors potentially affecting the performance of broiler flocks:

Inadequate feed (unlikely under well managed conditions)
Wrong temperature (see below)
Incorrect lighting
Poor ventilation (see below)
Incorrect vaccination programme
Poor litter (see below)
Incorrect stocking rates
Poor feeder or drinker design, or inadequate provision of feeders and drinkers (see below)
Poor chick quality (see below)

Amino acid supply

The points described above on amino acid sources for organic layers apply equally to organic table chickens, except that the references are not all the same. Fisher *et al.* (1973) and McDonald and Morris (1985) are replaced by Boorman and Burgess (1986), who applied the concepts of Fisher *et al.* (1973) to growing birds.

Removal of antibiotic growth promoters

Most antibiotic growth promoters were removed from European Union poultry diets in 1999. This has presented new challenges in disease control and nutrition. Using ingredients with low digestibility may encourage excessive bacterial growth. Control of coccidiosis becomes critical because of its relationship with necrotic enteritis. Bedford (2000) reviewed nutritional and other strategies for minimising the consequences of removing antibiotic growth promoters.

Temperature and ventilation

Growth rate after the brooding stage is probably fastest in the environmental temperature range 18 to 22°C, according to references reviewed by Charles (2002b), and the optimum temperature has tended to fall as genetic potential growth rates have increased over the years. MacLeod and Dabutha (1997) demonstrated the relevance of a version of Payne's hypothesis in growing poultry. Japanese quail offered a choice maintained growth rate over a wide range of temperatures (from 20 to 35°C) by selecting a diet mix which maintained protein intake. However the authors made the important point that the literature indicates that while broilers sometimes do this, their response is inconsistent (they quoted Cowan and Michie, 1978; Sinurat and Balnave, 1986; Mastika and Cumming, 1987).

In the UK climate it is theoretically feasible to maintain such temperatures in controlled environment broiler housing, though some fossil fuel input may be necessary in severe winter weather. However practical temperature control does not always match theoretical possibility due to factors such as wind interference with ventilation rates, and the deterioration of insulation materials in older buildings (ADAS, 1985; Sutcliffe *et al.*, 1987; Charles, 1991). Therefore the design of ventilation systems and equipment will continue to be important, including the provision of uniformity within buildings so that a high proportion of the birds may experience the intended environmental conditions. The principles of ventilation and environmental control were discussed by Charles, Clark and Tucker (2002).

Growing chickens of any given liveweight have a minimum and a maximum ventilation requirement. The principles and the methods of calculation are as described for layers above. The minimum tabulated by Charles *et al.* (2002) was given as (1.9 to 2.0) x 10^{-4} m^3/s per $kg^{0.75}$. This is equivalent to 0.34 to 0.36 m^3/s per 1000 birds of 2.2 kg liveweight. The maximum was quoted as 1.6 x 10^{-3} m^3/s per $kg^{0.75}$, or, for example 3.4 m^3/s per 1000 birds of 2.7 kg. Note that when installing housing it is important to anticipate the heaviest weight of birds ever likely to occupy the building.

Light

It was traditionally thought that broiler chickens needed long daylengths of 23 or even 23.5 hours in order to have the opportunity for adequate feed intake (e.g. Classen, 1992). However Gordon (1994) showed that 16 hours is probably sufficient. The potential benefits of the shorter daylength included sleep, improved immuno-responsiveness, the establishment of activity rhythms, and possible improvements in bone metabolism and leg health.

Litter condition

Poor litter can be a potential limiting factor, due to miscellaneous constraints ranging from foot damage to ammonia in the air. Tucker and Walker (1992) reviewed factors affecting litter condition, including work at ADAS Gleadthorpe. The influences of nutritional factors such as excessive dietary salt, dietary fat quality and quantity, and dietary protein quality and quantity; and environmental factors such as condensation on surfaces, poor ventilation, poor drinker design leading to water spillage and ground moisture were described. The nutritional factors were capable of influencing droppings moisture content and stickiness.

For example excessive protein, or protein of poor amino acid balance, led to higher rates of nitrogen excretion. Ammonia in the house air can limit performance in its own right, partly by direct effects and partly by increasing the susceptibility of the birds to certain respiratory disorders (reviewed by Charles, 1993).

Chick quality and parent flock management

Broiler growth performance is likely to be limited by chick quality, and therefore by some aspects of the management of the parent flock. Pearson (1982) reviewed several relevant factors and pointed out that yolk lipid is the main energy source for the developing embryo. Dietary fats in the breeder feeds did not affect total yolk lipid but did affect its fatty acid composition. Linoleic and arachidonic acids in the feed were essential to the production of viable chicks. Wilson (1997) noted in a review that dietary palmitic acid increased yolk oleic acid and spared linoleic acid for embryonic growth. The dependence of the embryo on parental nutrient supply means that the vitamin and mineral content of breeder diets are important (Leeson and Summers, 1997). Reviews of vitamin and mineral relationships have been provided by Pearson (1982), by Whitehead, Pearson and Herron (1985) and by Wilson (1997). The role of carrier proteins in transferring nutrients into the egg was discussed by Muniyappa and Adiga, (1979) and Pearson (1982). The literature on nutrition and hatchability and chick quality is vast, so it is fortunate that there have been reviews such as those of Pearson (1982) and more recently of Wilson (1997). The latter contained 132 references.

Factors potentially limiting the realisation of the genetic potential of both layers and table chickens

FEED INTAKE AND INGREDIENT AVAILABILITY

Improvements in both egg output and growth rate depend upon nutrient intake. The rates of egg production or of growth can only increase as the intake of essential nutrients increases. This process may become constrained simply by the ability of the bird to eat enough. Emmans and Fisher (1989) realised this and noted that feed intake increases when the birds are kept colder. The concept that the effects of keeping birds too warm include a performance depression which reflects nutrient intake is sometimes called Payne's hypothesis, after Payne (1966). A further complication is that feed intake falls as dietary energy concentration increases, as quantified for layers by Morris (1968b) and for broilers by Fisher and Wilson (1974). Thus it may be increasingly necessary to formulate low energy high protein feeds to birds at lower temperatures than currently recommended. (Temperature requirements have been recently reviewed by Charles, 2002b). This type of feed formulation may not always be feasible or economic, depending on ingredient availability and cost.

The crucial importance of feed intake as genetic progress is made is illustrated by the findings of MacLeod, McNeill, Knox and Bernard (1998). They compared two 1998 strains of broiler with a relaxed selection line with the growth characteristics of 40 years earlier. Lysine requirement as a proportion of the feed was found to be similar, and this was attributed to feed intake having kept pace with improvements in growth rate. This is fortunate, since tables of the nutrient content of feed ingredients are usually expressed as proportions by weight (g/kg)(e.g. Leeson and Summers, 1997; McDonald, Edwards, Greenhalgh and Morgan (2002).

The addition of enzymes to feeds is effectively a method of making more nutrients available in the same quantity of feed, so it is a technology which may be increasingly relevant in future years as the ability of the birds to consume enough feed becomes limiting. Enzyme addition is also a way of permitting otherwise poorly digested ingredients, such as barley, to become useful sources of nutrients and energy. Enzyme technology has a vast literature of its own. Reviews include that of Hotten (1993), and Bedford and Morgan (1996). Modes of action include the reduction of viscosity, including in wheat based feeds, and the cleaving of bonds within complex molecules.

AVAILABILITY OF LAND AND FEED INTAKE ON RANGE

If consumer pressures should eventually demand that all production takes place on free range then the availability of suitable land could become a constraint.

More importantly, and more likely to be a constraint, is the effect of the consumption of herbage material on nutrient intake. It is probably true that herbage can make a contribution to the nutrition of the birds, but if performance levels were to increase by very much then it could be difficult to achieve adequate nutrient intakes if bulky herbage at low dry matter contents were to be eaten. The capacity of the digestive tract is finite. As described above there is scope for increasing feed intake by the use of low energy feeds fed to birds kept at low temperatures outdoors, but there must be limits to this imposed by the limits to bulk intake. In the warmer months it would sometimes be difficult to keep the temperatures low enough in the range accommodation, even at night. The temperature within housing, including range housing, depends upon the ventilation rate, the stocking density and thence the amount of metabolic heat, and the structural heat losses. The physics and the quantification of these influences were recently reviewed and described in detail by Charles, Clark and Tucker (2002).

DISEASE AND PARASITE CONTROL

The forecasts below assume that there are no new and unexpected clinical or sub-clinical diseases or parasites. Continuing veterinary care and the continuing availability of vaccines and anti-parasitic agents such as coccidiostats will be important. It will also be important that the development of new vaccines and controls keeps up with the advent of variants of the existing pathogens and parasites or the arrival of new ones. The latter may include the acquisition of virulence by organisms which are currently harmless, or the arrival from abroad of infectious agents not currently in UK.

MISCELLANEOUS LIMITING FACTORS

It is conceivable that shortages of skilled labour and management could become constraints to performance.

The forecasts also assume no new and unforeseen regulatory, planning or consumer driven constraints on factors capable of influencing performance.

Speculative yield forecasts

Based on the likely genetic improvements and on the limiting effects of the factors discussed above the following are some speculations about future yields on a flock (population) basis. These are not the yields of outstanding individuals.

Layers: By 2010 yields of about 320 eggs per bird per year might be expected, at an average egg weight of 65 g/egg. The ultimate yield may be about 360 eggs per bird per year, and this may be attained by about the years 2020 to 2030.

Table chickens: By 2010 a liveweight (as hatched) of 2.2 kg may be reached by 39 days of age, and the ultimate yield may be the same weight reached at 34 days by 2025. This assumes, of course, that the desirable sale weight remains about 2.2 kg.

Acknowledgement

The authors gratefully acknowledge the contribution of Dr David Charles of DC R&D Ltd, Loughborough, UK.

References

ADAS (1985) *Getting the Basics Right*. Ministry of Agriculture, Fisheries and Food;

Albers, G.A.A. and van Sambeek, F.M.J.P. (2002) Breeding strategies for layers in view of new technologies. *11th European Poultry Conference*, September 6-10 2002, Bremen, Germany.

Andersson, L. (2001) Genetic dissection of phenotypic diversity in farm animals. *Nature Reviews Genetics* **2:** 130-138.

Bedford, M.R. and Morgan, A.J. (1996) The use of enzymes in poultry diets. *World's Poultry Science Journal* **52:** 61-68;

Bedford, M. (2000) Removal of antibiotic growth promoters from poultry diets: implications and strategies to minimise subsequent problems. *World's Poultry Science Journal* **56:** 347-365.

Boswell, T., Dunn, I.C. and Corr, S.A. (1999) Hypothalamic neuropeptide Y mRNA is increased after feed restriction in growing broilers. *Poultry Science* **78:** 1203-1207.

Bishop, S.C., Fleming, R.H., McCormack, H.A., Flock, D.K. and Whitehead, C.C. (2000). Inheritance of bone characteristics affecting osteoporosis in laying hens. *British Poultry Science* **41:** 33-40.

Bruce, J.M. (1978) Natural convection through openings and its application to cattle building ventilation. *Journal of Agricultural Engineering Research* **23:** 151-161;

Bulfield, G. (1990) Molecular Genetics. In: R.D. Crawford (ed.) Poultry *Breeding and Genetics.* Elsevier. pp 543 – 584.

Bulfield, G. (1994) *Biotechnology and the Poultry Industry.* Temperton Fellowship, Report No. 3. Harper Adams Agricultural College. 65 pp.

Burt, D.W., Hocking, P.M. and Haley, C. (2002) Chicken genomics and its application in mapping simple and complex traits. *11th European Poultry Conference*, September 6-10 2002, Bremen, Germany.

Boorman, K.N. and Burgess, A.D. (1986) Responses to amino acids. In: *Nutrient Requirements of Poultry and Nutritional Research.* Edit. Fisher, C. and Boorman, K.N. (1986), Butterworths, London, 99-123;

Charles, D.R. (1984) A model of egg production. *British Poultry Science* **25:** 309-322;

Charles, D.R. (1991) Getting the basis right. In: *National Poultry Housing Seminars, Australia.* Edit. Embury, I., NSW Agriculture and Fisheries, 3-8;

Charles, D.R. (1993) Air quality problems in poultry production: ammonia, temperature and performance. In: *Recent Advances in Reducing Environmental Pollution While Improving Performance.* Proceedings of the Alltech European Lecture Tour, 11-20;

Charles, D.R. (2002a) *Food, Farming and the Countryside: Past, Present and Future.* Nottingham University Press, Nottingham.

Charles, D.R. (2002b) Responses to the thermal environment. In: *Poultry Environment Problems: A Guide to Solutions.* Eds. Charles, D.R and Walker, A.W., Nottingham University Press, Nottingham. pp 1-16;

Charles, D.R., Clark, J.A. and Tucker, S.A. (2002) Ventilation rate requirements and principles of air movement. In*: Poultry Environment Problems: A Guide to Solutions.* Eds. Charles, D.R. and Walker, A.W., Nottingham University Press, Nottingham. pp. 17-46;

Charles, D.R., Lewis, P.D. and Tucker, S.A. (2002) Lighting programmes for laying hens. In: *Poultry Environment Problems: A Guide to Solutions.* Eds. Charles, D.R. and Walker, A.W., Nottingham University Press, Nottingham, pp. 69-81;

Cole, J. (1999) *Global 2050. A Basis for Speculation.* Nottingham University Press, Nottingham.

Cooke, V.E., Gilpin, S., Mahon, M., Sandercock, D.A. and Mitchell, M.A. (2003) A comparison of skeletal muscle fibre growth in broiler and layer chickens. *British Poultry Science* 44, Supplement 1, S33-S34.

Cowan, P.J. and Michie, W. (1978) Environmental temperature and choice feeding of the broiler. *British Journal of Nutrition* **40:** 311-315;

Classen, H.L. (1992) Management factors in leg disorders. In: *Bone Biology and Skeletal Disorders in Poultry.* Eds. Whitehead, C.C., Carfax, Abingdon, 195-211;

Curnow, R.N. (1986) The statistical approach to nutrient requirements. In: *Nutrient Requirements of Poultry and Nutritional Research.* Eds. Fisher, C. and Boorman, K.N. (1986), Butterworths, London, 79-90

Decuypere, E., Buyse, J. and Buys, N. (2000) Ascites in broiler chickens: exogenous and endogenous structural and functional causal factors. *World's Poultry Science Journal* **56:** 367-377.

Dillon, J.L. (1977) *The Analysis of Response in Crop and Livestock Production.* Pergamon, Oxford

Economist (1997) *Pocket World in Figures.* Profile Books, London

Emmans, G.C. (1977) A model of the growth and feed intake of *ad libitum* fed animals, particularly poultry. In: *Computers in Animal Production*, British Society of Animal Production, Occasional Publication No.5, 103-110

Emmans, G.C. and Charles, D.R. (1977) Climatic environment and poultry feeding in Practice. In: *Nutrition and the Climatic Environment.* Eds. Haresign, W., Swan, H. and Lewis, D., Butterworths, London, 31-49

Emmans, G.C. and Fisher, C. (1986) Problems in nutritional theory. In: *Nutrient Requirements of Poultry and Nutritional Research.* Eds. Fisher, C. and Boorman, K.N. (1986), Butterworths, London, 9-39

Etches, R.J. (1995) Physiology of reproduction: the female. In: *Poultry Production.* Eds. Hunton, P., World Animal Science Sub-series C9, Elsevier, Oxford, 221-241

Fisher, C., Morris, T.R. and Jennings, R.C. (1973) A model for the description and prediction of response of laying hens to amino acid intake. *British Poultry Science* **14:** 469-484

Fisher, C. and Wilson, B.J. (1974) Response to dietary energy concentration by growing chickens. In: *Energy Requirements of Poultry.* Eds. Morris,

T.R. and Freeman, B.M., Poultry Science Symposium No. 9, British Poultry Science Ltd., 151-184

Food and Agriculture Organisation (2001) *Production 2000*. Rome

Gilbert, A.B. (1971) The female reproductive effort. In: *Physiology and Biochemistry of the Domestic Fowl*. Eds. Bell, D.J. and Freeman, B.M., Academic Press, London, Volume 3, 1153-1162

Gordon, S.H. (1994) Effects of daylength and increasing daylength on broiler welfare and performance. *World's Poultry Science Journal* **50:** 269-282

Gordon, S.H. and Charles, D.R. (2002) *Niche and Organic Chicken and Egg Products: Their Technology and Scientific Principles*. Nottingham University Press, Nottingham.

Hawk, W. (1910) *Poultry Keeping for Profit*. Cornwall County Council

Hotten, P. (1993) Enzymes as feed additives for poultry, pigs and ruminants: current practice and future developments. In: *Recent Advances in Reducing Environmental Pollution While Improving Performance*. Proceedings of the Alltech 7th European Lecture Tour, 21-34

Hutt, F.B. (1949) *Genetics of the Fowl*. McGraw-Hill.

Julian, R.J. (2000) Physiological, management and environmental triggers of the ascites syndrome, a review. *Avian Pathology* **29:** 519-527.

Kuenzel, W.J. and Fraley, G.S. (1995) Neuropeptide Y: its role in the neural regulation of reproductive function and food intake in avian and mammalian species. *Poultry and Avian Biology Reviews* **6:** 185-209.

Lander, E.S. and Weinberg, R.A. (2000) Genomics: journey to the center of biology. *Science* **287:** 1777-1782.

Law, A.S. and Archibald, A.L. (2000) Farm animal genome databases. *Briefings in Bioinformatics* **1:** 151-160.

Lawrence, T.L.J. and Fowler, V.R. (1997) *Growth of Farm Animals*. CAB International, Wallingford

Leenstra, F.R. and Pym, R.A.E. (1995) OECD Workshop on Growth and quality in broiler production (Celle, Germany, June 1994). *Archiv für Geflügelkunde* (Special Issue), 3-6.

Leeson, S. and Summers, J.D. (1997) *Commercial Poultry Nutrition*. University Books, Guelph

Lerner, I.M. (1950) *Population Genetics and Animal Improvement as Illustrated by the Inheritance of Egg Production*. Cambridge University Press.

Lewis, P.D. and Morris, T.R. (1998) Responses of domestic fowl to various light sources. *World's Poultry Science Journal* **54:** 7-25

Lewis, P.D. and Morris, T.R. (1999) Light intensity and performance of domestic pullets. *World's Poultry Science Journal* **55:** 241-250

Lewis, P.D. and Morris, T.R. (2000) Poultry and coloured light. *World's Poultry Science Journal* **56:** 189-207

Marsden, A. (1981) *The Effect of Environmental Temperature on Energy Intake and Egg Production in the Fowl*. PhD thesis, University of Reading

Marsden, A. and Morris, T.R. (1987) Quantitative review of the effects of environmental temperature on food intake, egg output and energy balance in laying pullets. *British Poultry Science* **28:** 693-704

MacKay, T.F.C. (2001) The genetic architecture of quantitative traits. *Annual Review of Genetics* **35:** 303-339.

Malan, D.D., Scheele, C.W., Buyse, J., Kwakernaak, C., Siebrits, F.K., van der Klis, J.D. and Decuypere, E. (2003) Metabolic rate and its relationship with ascites in chicken genotypes. *British Poultry Science* **44:** 309-315.

Mastika, M. and Cumming, R.B. (1987) Effect of previous experience and environmental variations on the performance and pattern of feed intake of choice-fed and complete-fed broilers. In: *Recent Advances in Animal Nutrition in Australia.* Eds. Farrell, D.J., Armidale, 260-282

McDonald, M.W. and Morris, T.R. (1985) Quantitative review of optimum amino acid intakes for young laying pullets. *British Poultry Science* **26:** 253-264

McDonald, P., Edwards, R.A., Greehalgh, J.F.D. and Morgan, C.A. (2002) *Animal Nutrition.* Pearson Education, Harlow

McLean, J.A., Savory, C.J. and Sparks, N.H.C. (2002) Welfare of male and female broiler chickens in relation to stocking density, as indicated by performance, health and behaviour. *Animal Welfare* **11:** 55-73.

MacLead, M.G. and Dabutha, L.A. (1997) Diet selection by Japanese quail (Coturnix coturnix japonica) in relation to ambient temperature and metabolic rate. *British Poultry Science* **38**: 586-589.

MacLeod, M.G., McNeill, L., Knox, A.I. and Bernard, K. (1998) Comparison of bodyweight responses to dietary lysine concentration in broilers of two commercial lines and a "relaxed-selection" line. *World's Poultry Science Association (UK Branch) Proceedings of Spring Meeting*, Scarborough, 41-42

Merckaert, J. and Vandesande, F. (1996) Autoradiographic localization of receptors for neuropeptide Y (NPY) in the brain of broiler and leghorn chickens (Gallus domesticus). *Journal of Chemical Neuroanatomy* **12:** 123-134.

Mitchell, M.A. (1999) Muscle abnormalities – pathophysiological mechanisms. In: *Poultry Meat Science* Ed. R.I. Richardson and G.C. Mead. CABI Publishing, Oxon, England. pp 65-98.

Mitchell, M.A. and Smith, M.W. (1990) Jejunal alanine uptake and structural adaptation in response to genetic selection for growth rate in the domestic fowl (*Gallus domesticus*) *in vitro. Journal of Physiology, London* **424:** 7P.

Mitchell, M.A. and Smith, M.W. (1991) The effects of genetic selection for increased growth rate on mucosal and muscle weights in the different regions of the small intestine of the domestic fowl (*Gallus domesticus*). *Comparative Biochemistry and Physiology* **99A:** 251-258.

Mitchell, M.A., Mills, L.J., Mahon, M. and Gilpin, S. (2001) Idiopathic myopathy in commercial turkeys: A relationship with muscle fiber diameter? *Poultry Science* 80, Supplement 1, 237.

Morris, T.R. (1968a) Light requirements of the fowl. In: *Environmental Control in Poultry Production*. Eds. Carter, T.C., Oliver and Boyd, Edinburgh, 15-39

Morris, T.R. (1968b) The effect of dietary energy level on the voluntary calorie intake of laying birds. *British Poultry Science* **9:** 285-295

Morris, T.R. (1989) The interpretation of response data from animal feeding trials. In: *Recent Developments in Poultry Nutrition*. Eds. Cole, D.J.A. and Haresign, W., Butterworths, London, 1-11

Morris, T.R. (1994) Lighting for layers – what we know and what we need to know. *World's Poultry Science Journal* **50:** 283-287

Muniyappa, K. and Adiga, P.R. (1979) Isolation and characterisation of thiamin-binding protein from chicken egg white. *Biochemical Journal* **177:** 887-894

Panda, B. and Mohapatra, S.C. (1993) Poultry development strategies in India. *World's Poultry Science Journal* **49:** 265-273

Payne, C.G. (1966) Environmental temperature and egg production. In: *The Physiology of the Domestic Fowl*. Eds. Horton-Smith, C. and Amoroso, E.C., Oliver and Boyd, Edinburgh, 235-241

Pearson, R.A. (1982) Influence of nutritional factors on hatchability. In: *Recent Advances in Animal Nutrition*. Eds. Haresign, W., Butterworths, London, 141-156

Robertson, A. (1960) A theory of limits in artificial selection. *Proceedings of the Royal Society of London B* **153:** 234-239.

Robinson, L. (1948) *Modern Poultry Husbandry*. Crosby Lockwood, London

Romanoff, A.L. and Romanoff, A.J. (1949) *The Avian Egg*. Wiley, New York

Sandercock, D.A., Hunter, R.R., Mitchell, M.A. and Hocking, P.M. (2000) Genetic selection for production traits in chickens: correlations with muscle damage. *Twenty-first World's Poultry Congress,* Montreal, Canada, August 20th-24th, 2000.

Saunderson, C.L. and Leslie, S. (1988) Muscle growth and protein degradation during early development in chicks of fast and slow growing strains. *Comparative Biochemistry and Physiology* **89A:** 333-337.

Savory, C.J. and Mann, J.S. (1997) Behavioural development in groups of pen-housed pullets in relation to genetic strain, age and food form. *British Poultry Science* **38:** 38-47.

Simons, P.C.M. (1986) Major minerals in poultry nutrition. In: *Nutrient Requirements of Poultry and Nutritional Research*. Eds. Fisher, C. and Boorman, K.N. (1986), Butterworths, London, 141-154

Sinurat, A.P. and Balnave, D. (1986) Free-choice feeding of broilers at high temperatures. *British Poultry Science* **27:** 577-584

Sturkie, P.D. (1976) *Avian Physiology*. Springer-Verlag, New York, 302-330

Sutcliffe, N.A., King, A.W.M. and Charles, D.R. (1987) Monitoring poultry house environment. In: *Computer Applications in Agricultural*

Environments. Eds. Clark, J.A., Gregson, K. and Saffell, R.A., Butterworths, London, 207-218

Tesseraud, S., Chagneau, A.M. and Grizard, J. (2000) Muscle protein turnover during early development in chickens divergently selected for growth rate. *Poultry Science* **79:** 1465-1471.

Thompson, A. (1943) *Feeding for Eggs.* Faber and Faber, London

Tixier-Boichard, M. (2002) From phenotype to genotype: major genes in chickens. *World's Poultry Science Journal* **58:** 65-75.

Tomas, F.M., Jones, L.M. and Pym, R.A. (1988) Rates of muscle protein breakdown in chickens selected for increased growth rate, food consumption or efficiency of food utilisation as assessed by N^t-methylhistidine excretion. *British Poultry Science* **29:** 359-370.

Tucker, S.A. and Walker, A.W. (1992) Hock burn in broilers. In: *Recent Advances in Animal Nutrition.* Eds. Garnsworthy, P.C., Haresign, W. and Cole, D.J.A., Butterworth Heineman, Oxford, 33-50

Whitehead, C.C. (1994) Nutritional factors and bone structure in laying hens. *Proceedings 9th European Poultry Conference.* World's Poultry Science Association, Glasgow, 129-136

Whitehead, C.C. (1998) A review of nutritional and metabolic factors involved in dyschondroplasia in poultry. *Journal of Applied Animal Research* **13:** 1-16.

Whitehead, C.C., Pearson, R.A. and Herron, K.M. (1985) Biotin requirements of broiler breeders fed diets of different protein content and effect of insufficient biotin on the viability of progeny. *British Poultry Science* **26:** 73-82

Whittle, T.E. (1995) Application of science and market forces in agriculture. *Auchincruive College Association Journal* **73:** 9-12

Wilson, H.R. (1997) Effects of maternal nutrition on hatchability. *Poultry Science* **76:** 134-143

NUTRITIONAL VALUE AND YIELD OF FORAGES FOR LIVESTOCK: MEANS OR LIMITATIONS TO INCREASING ANIMAL PRODUCTION?

B. R. COTTRILL, M. J. GOODING AND D. I. GIVENS
School of Agriculture, Policy and Development, Department of Agriculture, The University of Reading, Earley Gate, PO Box 237, Reading, RG6 6AH, UK

Introduction

Grassland occupies over 11 million ha, or 0.68 of the total agricultural land in Great Britain (SEERAD, 2004), and produces in excess of 80 million tonnes of dry matter (DM) annually. Approximately half of the grassland area is 'rough grazing' where the scope for improving productivity is limited by soil type, topography or location. For the remainder, permanent or temporary leys occupy about 0.10 and 0.41 of the grassland area, but probably account for > 0.7 of the total dry matter yield. In addition to grassland, some 120,000 ha are sown to forage maize, producing 1.4 million tonnes of DM annually. Forages therefore represent major feed reserves for the 8.7 million cattle and 33.4 million sheep in Great Britain. This chapter reviews developments in breeding and the use of grass and other forages as feeds for ruminants.

The UK Government's Strategy for Sustainable Farming and Food, published in December 2002, identified healthy eating as one of its key principles with a major objective to produce safe, healthy products. This Strategy aims to secure a profitable and internationally competitive future for food and farming whilst contributing to the environment and improving nutrition and public health. To date, the productivity of forages has usually been measured in terms of DM yield. In view of the significant contribution of ruminant livestock products to the amount and form of fat consumed by the human population (DEFRA, 2003), it is suggested that additional measures of 'productivity' that relate specifically to the fatty acid content of animal products should be incorporated into plant breeding programmes.

Increasing nutrient / energy supply from forages: plant breeding and agronomy

GRASSES

Improvements in grassland technology and forage grasses have been reviewed by Aldrich (1987), Frame *et al.* (1995), Camlin (1997), and Wilkins and

Humphreys (2003). Progress in breeding has lagged behind other major crops in the UK. Wilkins and Humphreys (2003) considered that grass breeding is still at an early stage, with most of the genetic variation within and between species yet to be utilized. Grasses were the last of the major crops to be included in recommended list programmes (Aldrich, 1987). This lag in development reflects a number of challenges facing the breeder and assessor:

- Sales of forage grasses are much lower than for other major crops and in 2000 only 0.01 of seed sown to UK farmers was grass seed (Burgon *et al.*, 1997). The crop is not resown every year.
- Yield of spaced plants does not correlate with yield in plots and index selection on spaced plants does not usually increase yield.
- Yield is complicated by heading date, thus the requirement is either to harvest frequently or only to compare families with similar heading dates.
- Yield performance does depend upon harvesting strategy (conservation v grazing), e.g. there is some genetic independence between dry matter yield of vegetative growth and reproductive growth.
- Yield needs to be considered over the whole season and over several seasons. Yield in the short term is often at the expense of persistence and therefore yield in the long term (Camlin, 1997).
- There has been scepticism as to whether yields in cut plots accurately reflect the performance of grazing animals, but it is not practical to include actual grazing in breeding or recommendation programmes. Aldrich and Elliot (1974) did demonstrate a high correlation between the yield of grass in rotationally grazed plots and a 9-cutting regime, but animal performance depends on intake, digestibility and utilizable / metabolizable energy. There are negative associations between digestibility and both forage yield and seed yield.
- It is not easy to increase harvest index and yields may also have approached their theoretical maximum (Leafe, 1988).

The issue of maximum yield potential from grass swards in NW Europe has been addressed several times. Alberda (1977) used the model of deWit (1965) to calculate a maximum production of 45 t dry matter (DM)/ha/yr. A harvest index of 0.60 would suggest a yield of 27 t/ha. Leafe (1978) calculated the harvestable dry matter (H) to be the net photosynthesis during the day (P_{nc}) minus root respiration and respiration of the shoot at night (R_{rs}), death and decay of root and shoot material (D_{rs}), and dry matter in parts not harvestable (W). In a system with a high degree of harvesting efficiency R_{rs}, D_{rs} and W were estimated to be 0.45, 0.15 and 0.13 of P_{nc} respectively. Therefore a well watered, manured and managed sward yielding (H) 12 t/ha reflected values of P_{nc}, R_{rs}, D_{rs} and W of 45, 20, 7 and 6 t/ha respectively. Leafe (1978) then calculated that a sward under continually sunny (300 W m^{-2}, 400-700 nm), watered, and protected conditions with adequate fertilizer from April-October in the UK would yield 45 t DM/ha/yr (P_{nc}=166; CO_2 assimilation 9g m^2/h). Substituting actual irradiance in Cambridgeshire reduced predicted yields to 30.5 t/ha. Leafe (1978) then considered that potential annual yield would be further reduced to 25 t/ha to allow for reduced light interception just after cutting, and then a further reduction to 19.5 t/ha because assimilation efficiency was progressively lower in leaves of successive regrowths compared to newly expanded leaves in the first period of reproductive growth. This last value is not dissimilar to the yields obtained in well-watered and fertilized experimental plots of the 1960s to 1980s (Cooper, 1968; Alberda, 1968; Alberda, 1971; Leafe, 1988). Leafe (1988) reported maximum yields of 15 t DM/ha/yr from 'best agricultural practice'. This is comparable to 15.8 t DM/ha from the 1st harvest year of Aberystwyth S 24 early perennial ryegrass, cut to simulate conservation practice, as described in the 1982/83 recommended list of varieties of the National Institute of Agricultural Botany (NIAB, 1982). Under this management, however, yields of S 24 declined to 12.3 t DM/ha by the second year of harvest. In the current recommended list (NIAB, 2003) the yield of the highest yielding early perennial ryegrass cultivar under conservation management, averaged over the first and third harvest year, was 14.7 t DM/ha/yr. This is an average figure from different sowings at eight sites, with plots cut four or five times and a total nitrogen fertilizer application of up to 350 kg N/ha.

Despite the challenges, over the past fifty years there have been a number of regulatory and evaluation developments that have encouraged the improvement of forage species by formal plant breeding. Frame *et al.*, (1995) mentioned the commercialisation of many varieties that developed from continental breeding programmes that were initiated after the Second World War; the introduction of the 1964 Plant Varieties and Seed Act; the publication of the first NIAB 'Descriptive List' of grasses in 1960 and its first Recommended List in 1968; and the statutory performance, and distinctness, uniformity and stability (DUS) testing that followed the UK accession to the EEC in 1973.

Evaluation and assessment of cultivar performance has been principally with respect to yields of DM. Yields of digestible material, as well as palatability and other intake characteristics, are also of great importance. Selection programmes in the first half of the twentieth century were for plants with a high proportion of leaf because nutritive value was equated with protein content (Lazenby, 1988), and there were later attempts to select for quality (Hughes, 1971). As mentioned previously, digestibility is strongly confounded with crop maturity and therefore yield and heading date. Nonetheless, Wilkins and Humphreys (2003) observed from data presented in NIAB (2001) that content of digestible organic matter (DOMD) of perennial ryegrass cultivars during late summer appeared to be increasing by at least 10 g/kg per decade. Analysis of similar data in the most recent recommended list (NIAB 2003), however, presents a more complicated picture (Figure 24.1). No improvement in early heading cultivars is discernable. There is significant improvement in the aftermath DOMD of intermediate heading cultivars (about 14 g/kg per decade) but a decline (about 5 g/kg per decade) in the DOMD of late heading cultivars. Lazenby (1988) was not optimistic about the future improvement in the quality of ryegrasses. The following description, therefore, concentrates on yield improvements.

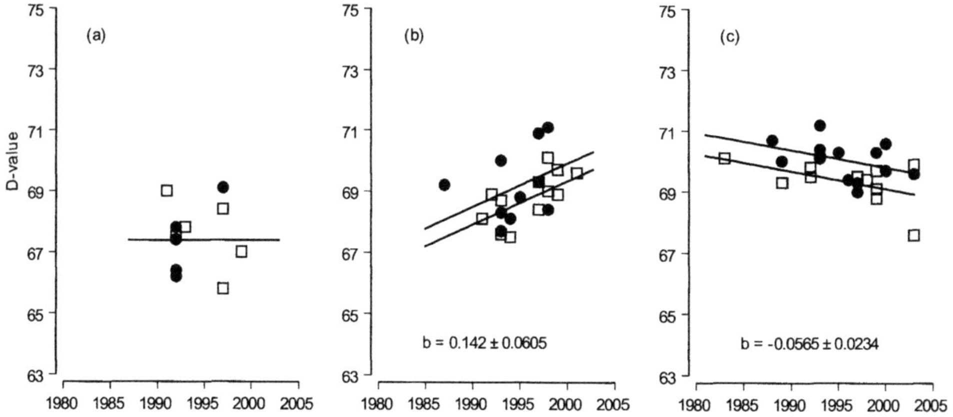

Figure 24.1 The relationship between date of first listing of (a) early, (b) intermediate, and (c) late heading cultivars of perennial ryegrass cultivars and aftermath digestibility. From data presented in NIAB (2003). b = regression coefficient ± standard error.

Ryegrasses (**Lolium** *spp.*)

Over the last fifty years, ryegrasses have become an increasingly dominant component of the sown grasses, reaching 0.80 of the weight sold to farmers in the mid 1990s. This trend reflects the combination of high yield and digestibility, particularly under intensive, high nitrogen availability systems of management (Frame, 1991).

The high yields of leafy, dense cuts, early in the season have been exploited and encouraged by technological developments in silage making. The ryegrasses are also larger seeded and thus afford more rapid and reliable establishment compared to some of the species they have tended to replace such as Timothy and Cocksfoot. The ryegrasses are normally diploid, but their comparatively low chromosome number ($2n = 14$) lends them to improvement by inducing polyploidy. Tetraploidy is relatively easy to induce, for example, by treating germinating seeds with colchicine. During the 1960s tetraploid cultivars bred in The Netherlands became available in the UK and by 1969/70, half of the perennial ryegrass cultivars on the NIAB recommended list were tetraploids. The tetraploids have larger, but fewer tillers and a more open growth habit, which can be associated with reduced persistence (Figure 2). However, tetraploids have higher sugar and water soluble carbohydrate contents, are more palatable and digestible, and often result in higher animal intakes.

*Perennial ryegrass (*Lolium perenne*)*

Perennial ryegrass accounts for most grass seed sales in the UK. Traditionally it has been used as an indicator of productive grassland, being high yielding on fertile land with pH 6-7, with intensive grazing management. Yields and persistency have been improved in varieties (Wright, 1981; Camlin, 1997). Aldrich (1987) presented data for the NIAB recommended list trials for the mean yield of varieties recommended for the first time in the years 1970-87. When expressed as a proportion of the yield of S.23 the mean annual rate of improvement was 0.006 per year (or 0.06 per decade). Interestingly, the relationship suggested that the yield would have been predicted to be the same as S.23 in 1967, i.e. that there had been no or little improvement in cultivar performance over S.23 since its release in 1935 until after 1967. Figure 24.2 is a response to the request of Camlin (1997) that similar analyses are done over a longer time frame to the present day. This is derived from a database that includes all data in published NIAB recommended lists and shows a genetic contribution to yield of between 0.003 and 0.004 per year (or 0.03 to 0.04 per decade). This rate of improvement is less than other estimates (Wilkins and Humphreys, 2003), although may yet be a slight over-estimate if yields of 'long-lived' cultivars decline in absolute terms over time due, for instance, to increased disease susceptibility. The deviation from the earlier work by Aldrich (1987) is partly because the former work calculated yields as a percentage of a low yielding cultivar. If yields were calculated as a proportion of the highest yielding cultivar the annual rate of increase declines from 0.0061 to 0.0055. The results in Figure 24.2 largely avoid this because the REML analysis uses information from all cultivars in all years and the control variety changes over time. Figure 24.2 demonstrates that the rates of improvement depend on heading date class and ploidy level.

Early heading cultivars produce reproductive tillers early in the spring and have been seen traditionally as suited for conservation. This class was typified initially by 'Irish Commercial' (Wright, 1981), a local strain that had poor persistence. Persistence (measured as ground cover in the second and third years of use) was greatly improved (1.7) by the introduction of Aberystwyth S. 24 in the 1930s. In more modern times, persistence in this class has continued to improve (Figure 24.2) for both diploids and tetraploids, as have yields under 'grazing' and conservation management regimes.

Late-heading cultivars have traditionally been leafy, wear tolerant and persistent, typified by the local strain, Kent Indigenous (Wright, 1981), and used extensively in long-term grazing pastures. Modest improvements in yield and persistence were achieved by the release of S 23, and in winter hardiness by Perma, both in the 1930s (Camlin, 1997). Further improvements are apparent in modern times, particularly under simulated grazing (Figure 24.2). Yields under conservation management have increased at a slower rate for both diploids and tetraploids. Persistence appears to have reached a plateau

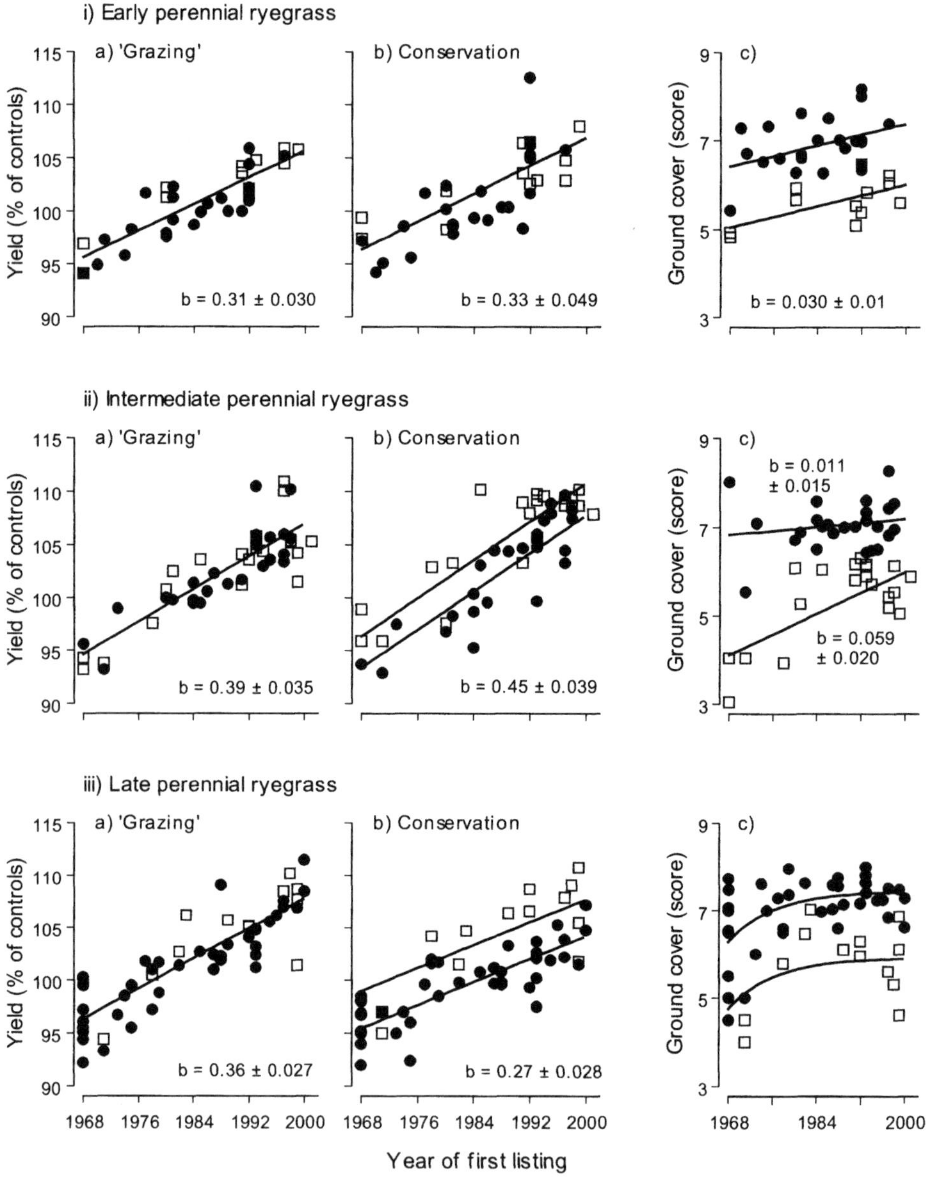

Figure 24.2. The relationship between date of first listing of diploid (l) and tetraploid (?) perennial ryegrass cultivars of different heading date classes and: their dry matter yields under a) simulated grazing (8-9 cuts at 30-40 mm) and b) for conservation (4-5 cuts at 60 mm); and c) their retention of ground cover. Points are predicted means following a residual maximum likelihood analysis of cultivar performance presented in successive lists of recommended varieties (NIAB 1982-2004). *b* = regression coefficient ± standard error.

for both ploidy groups, with no improvement in ground cover evident for the last twenty years.

Intermediate heading cultivars have been seen as a way of combining the high yielding conservation cuts of the early cultivars with the persistence and grazing quality of the late cultivars. Camlin (1997) made special mention of Talbot, released in 1973, as being the first truly 'flexible' intermediate. Yield increases over the last 35 years have been fastest for this group for both grazing and conservation management. There has been no improvement in the persistence of the diploid cultivars, but there has been significant increase in the ground cover of the tetraploids (Figure 24.2).

*Italian ryegrass (*Lolium multiforum*) and Hybrid ryegrass (*Lolium perenne x multiflorum*)*

Italian ryegrass cultivars are grown for short-term leys. Autumn sowings in Southern England can be used in the following spring (Sheldrick, 2000) and be productive for two years. The species can provide early grazing followed by high yielding conservation cuts. The yields and persistence of diploid cultivars have averaged above those of tetraploids (Figure 24.3) but yield potential has been increasing in both groups by about 0.04 per decade. The highest yielding cultivar in the current recommended list (NIAB 2003), under a 7-cut regime receiving up to 400 kg N/ha, achieved an average yield over seven sites of just more than 17 t DM/ha. Winter hardiness in the second winter is less than that of perennial ryegrass although persistence has been improving steadily (Figure 24.3).

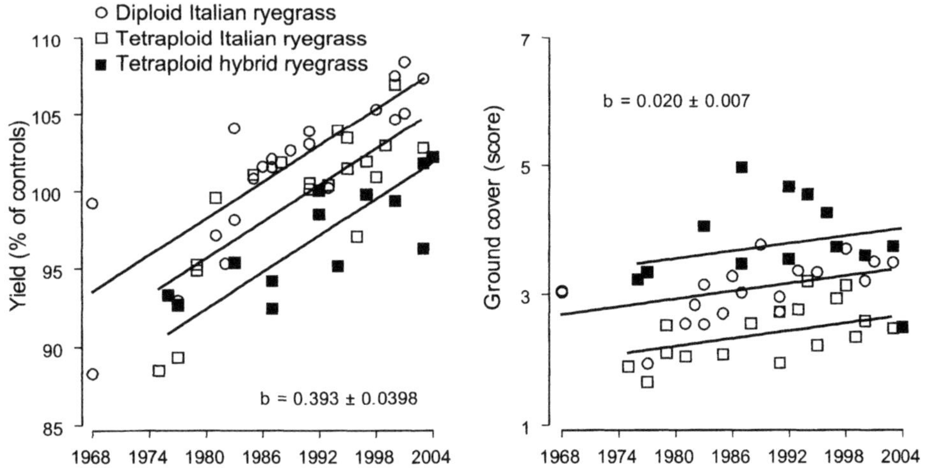

Figure 24.3. The relationship between date of first listing of diploid and tetraploid Italian ryegrass cultivars, and of tetraploid hybrid ryegrass cultivars, and dry matter yields and retention of ground cover. Points are predicted means following a residual maximum likelihood analysis of cultivar performance presented in successive lists of recommended varieties (NIAB 1982-2004). *b* = regression coefficient ± standard error.

Breeders have attempted to combine the higher yields of Italian ryegrass with the greater persistency of perennial ryegrass. Hybrid varieties with poor winter hardiness from New Zealand were available in the UK in the 1950s but significant improvements were achieved by cultivars introduced from Aberystwyth in the 1970s (Camlin, 1997). Since then, the rate of improvement in yield and persistency of hybrid ryegrasses does not appear to have been significantly greater than that achieved in the Italian ryegrasses (Figure 24.3). Camlin (1997) pointed out, however, that hybrids are usually only tested over the short durations relevant for Italian ryegrasses and, therefore, improvements in persistence may not be expressed fully.

*Timothy (*Phleum pratense *L.)*

Timothy is tolerant of cooler and wetter conditions than perennial ryegrass (Wilkins and Humphreys, 2003) but is small seeded and slow to establish (Hopkins, 1998). Later heading and palatability is particularly valued in its role in hay systems. Cultivar yields in recommended list trials have increased at about 0.025 per decade over the last thirty years (Figure 24.4); the most recent list (NIAB, 2003) suggest highest yields of 10.5 and 13.3 t DM/ha under 'grazing' and conservation management respectively.

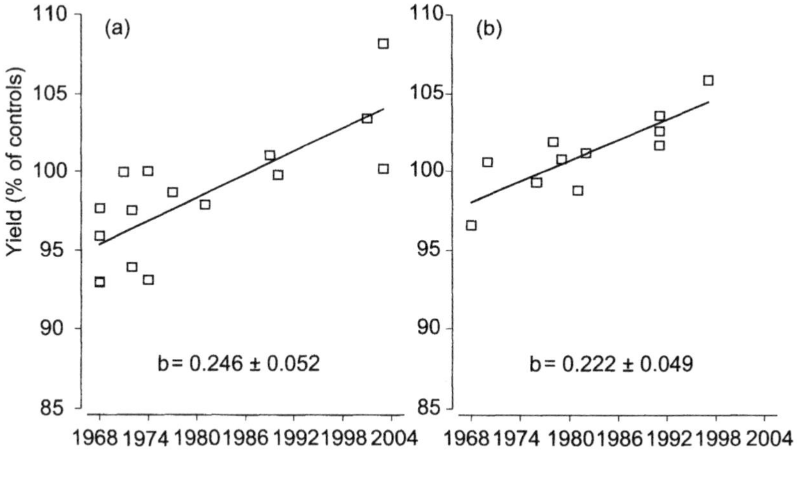

Figure 24.4 The relationship between date of first listing of (a) Timothy and (b) cocksfoot cultivars under simulated grazing management. Points are predicted means following a residual maximum likelihood analysis of cultivar performance presented in successive lists of recommended varieties (NIAB 1982-2004). *b* = regression coefficient ± standard error.

*Cocksfoot (*Dactylis glomerata*)*

Cocksfoot is more tolerant of drier and warmer conditions than perennial ryegrass and, despite being slower to establish, less digestible, less palatable and less tolerant of treading than perennial ryegrass, is still used on sandy and drought-prone soils in the UK (Edmund, 1964; Hopkins, 1998; Wilkins and Humphreys, 2003). Despite the much lower UK sales of Cocksfoot, breeding efforts continue and yields have increased at a modest rate of about 0.02 per decade (Figure 24.4). Highest dry matter yields of 12 and 13.5 t DM/ha have been achieved by varieties in recommended list trials for 'grazing' and conservation (NIAB, 2003)

The basis for yield improvements in forage grasses

Although the rate of yield improvement in forage grasses is lower than that seen in cereals, given the challenges outlined at the start of this chapter, the yield improvements that have been achieved are noteworthy. Grain yields in cereals have been achieved mostly by increasing the proportion of biomass that is partitioned to the grain, rather than by increasing biomass. Yield increases in grasses are increases in harvested biomass. There has been little detailed work attempting to explain the bases of the yield improvements in grass. As described above, there have been marked early improvements in winter hardiness in some groups and there does appear potential for further improvement (Hofgaard *et al.*, 2003). There has also been some improvement in disease resistance (Charles, 1973; Wilkins and Humphreys, 2003). An increase in persistence is also evident in most groups. This contributes directly to yield for species harvested over a number of years. Varieties and species differ markedly in their rates of decline in tiller numbers and ground cover under cutting regimes. Regrowth after the harvesting of reproductive tillers is dependant on the mobilization of carbohydrate and nitrogen reserves and genotypes differ significantly in the rates of regrowth.

Future yield improvements in forage grasses

Wilkins and Humphreys (2003) considered that the rates of grass improvement over the last fifty years should be maintained for at least another thirty years, i.e. the result of another 10 generations of hybridisation and recurrent selection from within elite lines. This is supported by observations that selection is still at a comparatively early stage and that much genetic variation still exists within current forage cultivars. If the existing rate of improvement is taken to be 0.0033 per year (Figure 24.2) a number of projections can be made. Perennial ryegrass yields under conditions of high nitrogen availability and the conservation management of the recommended list testing regime increases from about 15 t DM/ha to about 16.5 t DM/ha in 2030 and, if continued, 17.5 t DM/ha by 2050. If starting with 15 t DM/ha as 'best agricultural practice' in the 1980s, the predicted equivalent in 2050 would approach 19 t DM/ha. More realistically, if starting with 12 t DM/ha as an example of a good, well watered, fertilized and managed sward in the 1980s (Lazenby, 1988), the equivalent in

2050 is 15 t DM/ha. Assuming a utilization of 0.75 and a metabolizable energy (ME) of 10.5 MJ/kg DM is maintained, the utilisable ME (UME) output (GJ/ha) of 94.5 in the 1980s becomes 118 in 2050. Table 24.1 gives yields and target UME for optimum N rates for different site classes (based on soil-type and rainfall) together with the equivalent prediction for 2050.

Table 24.1 YIELDS OF GRASS AND UTILISABLE METABOLISABLE ENERGY (UME) FROM SWARDS RECEIVING ECONOMICALLY OPTIMAL LEVELS OF NITROGEN FERTILIZER

Site class[a]	*Situation c. 1990*				*Prediction for 2050*		
	N application[a] *kg N/ha*	*Dry matter yield*[a] *(t/ha)*		*Target UME*[b] *(GJ/ha)*	*Dry matter yield (t/ha)*		*Target UME (GJ/ha)*
		Conserved	*Grazed*		*Conserved*	*Grazed*	
1	430	15.4	14.3	126	18.7	17.4	153
2	380	14.4	14.4	115	17.5	17.5	139
3	320	13.4	13.4	105	16.3	16.3	127
4	300	12.6	12.6	93	15.3	15.3	113
5	255	11.7	11.7	83	14.2	14.2	101

[a]Baker *et al.* (1991)
[b]Fisher and Mayne (1991)

This discussion assumes that grass is grown and used to its potential in highly fertilized and intensively stocked situations. As already mentioned, the nitrogen applied to the recommended list trials exceeds 300 kg N/ha. In practice, the average amount of nitrogen applied as fertilizer to managed grassland (as opposed to rough grazing) falls well below this value. Nitrogen fertilizer rates did increase from very low average levels (less than 10 kg N/ha) in the 1940s to reach 130 kg N/ha by the late 1980s (Figure 24.5; Frame *et al.*, 1995). The reasons for this increase given by Goodlass *et al.* (2003) included a high cost/benefit ratio; a reliable response; and need in intensive, high stocking rate, high output systems. Nonetheless, average nitrogen fertilizer levels never approached recommendations for high stocking densities (MAFF, 2000). For example, even where soil nitrogen supply is high and grass growing conditions only moderate, recommended nitrogen applications can still reach 300 kg N/ha for a dairy grazing system (MAFF, 2000). Since the early 1990s, nitrogen fertilizer applications to grassland have declined, partly associated with a reduction in numbers of livestock and possibly also extended grazing practices on dairy farms reducing the demand for silage (Goodlass *et al.*, 2003). The recent expansion of the area categorised as a Nitrate Vulnerable Zone (NVZ), and associated restrictions on N-use may also have reduced the amount of fertilizer applied (DEFRA 2002), as would the revival in the use of clover-rich swards. Some of the shortfall in N is made up by returning slurry and farm yard manure from housed animals, or as dung and urine of grazed animal. Nonetheless, Wilkins and Humphreys (2003) concluded that nitrogen levels are usually well below the optimum for growth. On low output farms, say with grass yields of 6 t DM/ha, 0.50 utilization and ME of 8 MJ/kg DM, the UME output of 24 in the 1980s (Lazenby, 1988) rises to 30 by 2050.

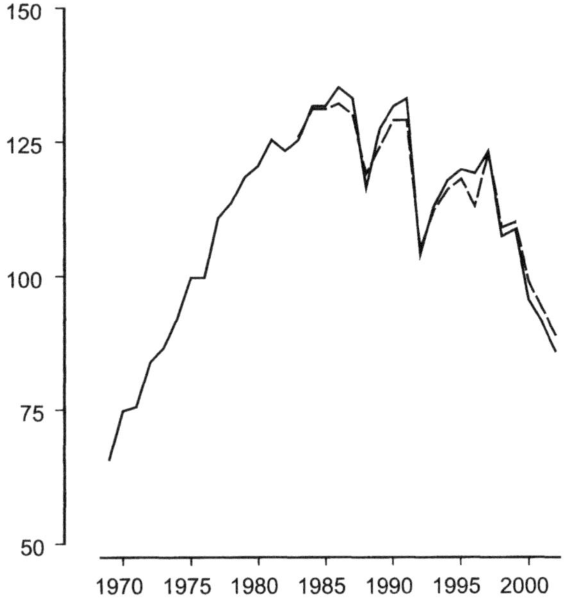

Figure 24.5 Application of nitrogen fertilizer on grassland (excluding rough grazing) in England and Wales (solid line) and Great Britain (dashed line). Data from Goodlass *et al.* (2003).

FORAGE LEGUMES

Forage legumes have an important role to play in many livestock enterprises (Sheldrick *et al.*, 1987; Beever and Thorp, 1996; Sheldrick, 2000). They are;

- the mainstay of lower input systems because of symbiotic relationships with rhizobia for nitrogen fixation;
- preferentially grazed and consumed compared with grass, so that voluntary intakes can be 0.20 to 0.30 higher;
- often lower in total fibre content and digested more rapidly than grass;
- associated with increased microbial population in the rumen compared to grass and the microbes work more efficiently;
- often higher in protein concentration than grass;
- relatively more productive in mid-season than grass such that there is less variation between spring and summer;

*White clover (*Trifolium repens*)*

White clover is nearly always sown in mixtures with grass. When sown with grass, the persistence and yield of clover under grazing is strongly related to length and density of stolons and leaf area (Caradus and Chapman, 1996). In realistic grazing scenarios in the UK, yields of clover in mixtures with grass are linearly related to stolon length (20 to 100 m/m^2) in the spring. Stolon length may not be limiting at lengths longer than 100 m/m^2, e.g. in cut swards after mild winters when yield may become more closely related to leaf production per unit of stolon (Rhodes and Ortega, 1997). Important breeding targets and achievements have been reviewed by Rhodes and Ortega (1997). In summary, stolon survival over winter is probably the most important single factor determining subsequent yield potential of white clover. One cultivar, first listed in 1994 (AberHerald), was claimed to be amongst the first to combine cold hardiness with leaf expansion in cool conditions, contributing to 0.50 more production than the old standard medium-leaved cultivar, Grasslands Huia (first listed in 1969). Average yield reductions caused by pests and diseases are difficult to quantify, but unreliability of clover contents have been ascribed locally to clover rot (*Sclerotinia trifoliorum*), stem eelworm (*Ditylenchus dipsaci*), and *Sitona* weevil. Genetic variability is present for these biotic factors and further yield improvement may derive from including resistant material in breeding programmes. Yields may also be improved by identifying and exploiting variation that contributes to better compatibility with grass, and improved nitrogen fixation.

Varieties have been classified according to leaf size, which is loosely related to ability to survive under different types of management. Small leaved cultivars have thin stolons that branch profusely, have short petioles and survive under close sheep grazing. The largest leaved varieties have thicker, but less branched stolons and longer petioles. The larger leaved varieties can persist in higher swards, for example in silage or hay systems, or grazing by cows, and are better able to compete with grass when nitrogen is applied. There is, therefore, a clear negative relationship between leaf size and persistency under grazing, and a positive relationship between leaf size and persistency under cutting (Figure 24.6). Some selection for overcoming these relationships may be possible, i.e. by breeding for large leaf size but also for stolon quantity, short internode length and or vigorous rooting at nodes (Rhodes and Ortega, 1997). The fact that 0.80 of the variation in persistency of current white clovers under hard defoliation can be explained on the basis of leaf size alone, suggests that variation in stolon characteristics are either much less important, are closely linked to leaf characteristics (e.g. Caradus *et al.*, 2000) or have yet to be fully exploited in breeding programmes. There is some evidence that important stolon characteristics have some genetic independence from leaf characteristics (Collins *et al.*, 1998), suggesting that progress is possible.

Maximum yield potential from grass / white clover swards is considerably

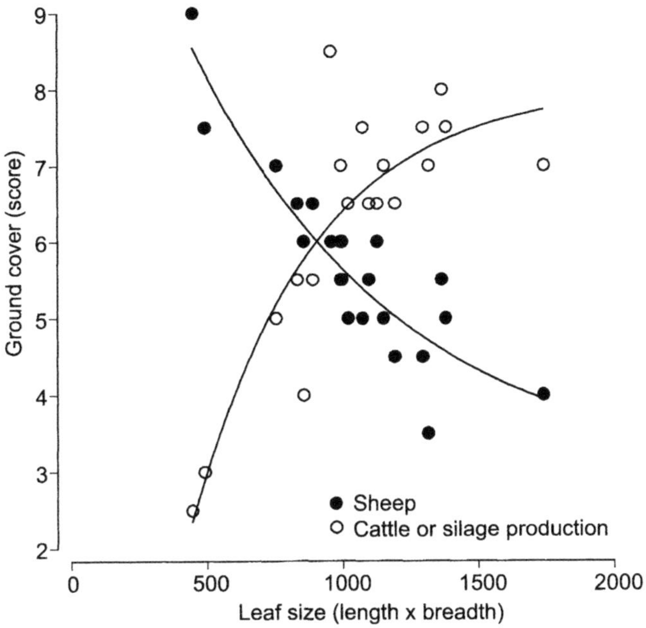

Figure 24.6 Relationship between leaf size and autumn ground cover of white clover cultivars under hard defoliation (e.g. by sheep) or light defoliation (e.g. by cattle or silage production). Plotted from data in NIAB (2003).

less than that from pure ryegrass swards supplied with nitrogen. Annual dry matter production in grass / white clover swards equals that of pure grass swards supplied with 200 kg N/ha when the clover contributes 0.30-0.50 of the DM (Morrison, 1981; Sheldrick *et al.*, 1987). However, grass yields were 1.3 times higher when pure grass was given 400 kg N/ha (Sheldrick *et al.*, 1987). Yields of white clover / grass swards can be improved in the short term by applying fertilizer nitrogen but this compromises the competitive position of the clover, and, therefore long term yield. Calculations that have placed theoretical potential yields of pure grass swards at 27-30 t DM/ha estimate potential production from grass / clover swards to be 18.5 – 22.5 t DM/ha (Frame and Newbould, 1984). Yields of perennial ryegrass, associated with high nitrogen and water availability, of 12-15 t DM/ha compare with estimates of only 5-10 t DM/ha for grass / white clover swards. Differences between the systems are not always this great and Baker *et al.* (1991) reported an average yield of 8.3 t DM/ha for grass / white clover with 0 kg N/ha compared with 11.7 t DM/ha for grass with 340 kg N/ha.

Rhodes and Ortega (1997) consider that classical breeding methods will continue to increase yield potential, which they calculate to have been increased by 0.50 between 1969 and 1994, mostly through improvements in survival and growth in cool conditions. There is also potential of using interspecific hybrids of *Trifolium* to improve persistence, drought tolerance (Marshall *et al.*,

2001), water soluble carbohydrate content (Marshall *et al.*, 2004), and seed yield (Marshall *et al.*, 2002). Further improvements appear possible through marker assisted selection, while Webb (1996) identified potential for genetic transformation. Although precise values for improvement rates are difficult to determine, it would appear that these are probably faster than perennial ryegrass and it does not seem unreasonable to suppose that they will remain quick enough to maintain at least their current competitiveness within ryegrass swards over the next few decades.

FORAGE MAIZE

It has long been recognised that the inclusion of forage maize in grass-silage based diets can increase food intake, milk yield and milk protein content (Pain & Phipps, 1975; Phipps *et al.*, 1988; Phipps *et al.*, 1995). The area of forage maize in the UK has increased dramatically over the last 30 years; from about 30 000 ha in the mid 1970s to 130 000 ha in 2001, since when it has declined slightly to about 120 000 ha (Figure 24.7). Initial increases in the growth and utilisation of forage maize undoubtedly reflected a growing awareness of the potential contribution of the crop to ruminants. The rapid expansion during the 1990s was due to a combination of factors, including the introduction of new varieties which were better adapted to UK conditions, warmer weather and the introduction in 1993 of an arable payment scheme under the Common Agricultural Policy. In future, the area under production might be expected to increase if predicted increases in temperatures are realised. However, elevated temperatures will favour higher maize yields only where there is sufficient soil moisture. Simulation studies have predicted that higher temperatures accompanied by 0.10 reduction in precipitation would result in a reduction in DM yields in the south and South-east of England (Cooper and McGechan, 1996; Davies *et al*, 1996). Since the value of the crop is insufficient to justify irrigation, expansion in the production of maize may be concentrated in the western and northern regions of the UK, which receive higher rainfall but are currently too cool for maize production.

The increasing popularity of maize occurred alongside the development of earlier maturing varieties, which attain higher dry matter contents at harvest. Phipps (2002) reported the increase in dry matter content (g/kg) at harvest of selected cultivars from about 250 in 1973, to 275 in 1982 and 375 in 1993. Over this range, silage intake by lactating dairy cows is positively related to DM content, and Phipps (2002) regarded it necessary for crops to reach 300 – 350 g/kg dry matter at harvest if the high energy and intake potential of maize as a complementary forage was to be adequately exploited.

Since the mid-1980s it appears that the DM yield of maize cultivars has increased at over 0.08 per decade (Figure 24.8) and, during this time, the rate of increase appears independent of maturity class. On average, selecting more

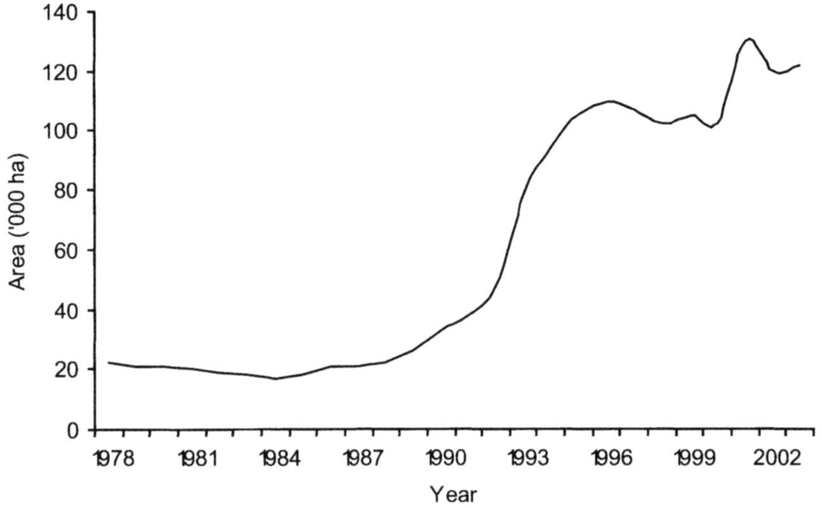

Figure 24.7 The area sown to forage maize in England and Wales, 1978-2003.

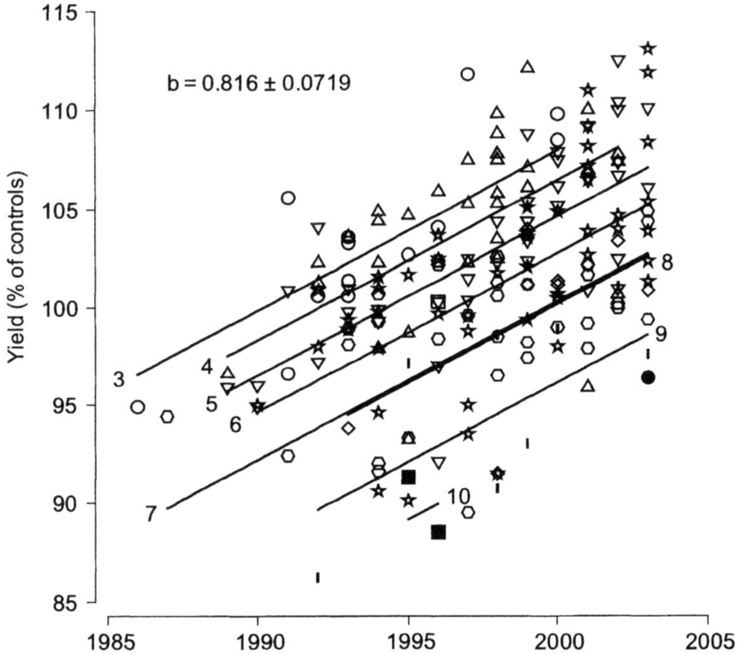

Figure 24.8 The relationship between date of first listing of forage maize cultivars and yield of dry matter. Points are predicted means following a residual maximum likelihood analysis of cultivar performance presented in successive lists of recommended varieties (NIAB 1991*b*-2002*b*, NIAB 2002, NIAB 2003*b*). Different symbols denote different maturity classes to which regression lines have been fitted and labelled with the maturity class (high numerals denote earlier maturation). *b* = regression coefficient ± standard error.

rapidly maturing varieties does reduce yield potential (Figure 24.8) but, as intimated above, the maturity class has a major influence on feeding quality. Not only is greater DM content and intake achieved with earlier maturing crops, but that DM has higher ME and starch contents (Figure 24.9). In maize, starch is the primary energy source and starch content is a good predictor of the digestible substrate supply (Sutton *et al.*, 2000). Starch content increases with maturity as water soluble carbohydrates are translocated from the stover and deposited as starch in the grain.

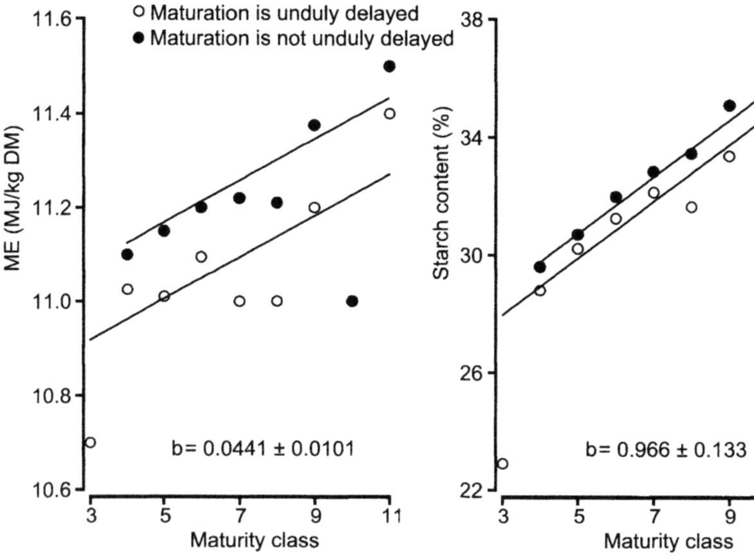

Figure 24.9 The relationship between feeding quality and maturity class (higher values represent earlier maturation) of maize. Points are means of all cultivars within a particular maturity class (NIAB, 2003) grown at sites where maturation is or is not unduly delayed by climate and sowing date. Regression lines are weighted according to the inverse of the variances around each point and are, therefore, less influenced by means for extreme maturity classes (3, 10 & 11) which are only represented by a very few cultivars.

There is no evidence in Figure24. 8 that the yield potential of forage maize cultivars is reaching a limit. The extent to which this can continue is unclear. As benefits of forage maize rely on it maturing sufficiently rapidly to reach dry matter contents of at least 300 g/kg with good starch deposition, and adequate maturation is difficult to achieve in many areas of the UK, there appears little benefit in extending the growing season in autumn. This appears to have been demonstrated by comparing cultivars with and without 'stay green' characteristics (Wilkinson & Hill, 2003). There is also competition between stover and grain for assimilates, particularly in cool wet conditions where only compact, low-stover hybrids may produce adequate quantities of starch (Frei, 2000). Adaptation of maize to northern Europe has involved faster early growth, earlier flowering and improved cold tolerance. Frei (2000) considered that indirect selection for lower base temperatures has and will continue to improve these characteristics. There would seem to be benefit in selecting for short lag phase between anthesis and starch accumulation (Frei, 2000). Furthermore as starch accumulates the digestibility of the stover declines. In areas where starch accumulation is adequate, therefore, there is potential to improve digestibility further by exploiting variation in, for example, cell wall composition (Fontaine *et al.*, 2003). Indeed there is concern that the breeding efforts over the last 20 years, which has been directed towards producing earlier maturing varieties for the UK, has given rise to cell walls of increased lignification and lower nutritional value (Givens and Deaville, 2001). Developments from the detailed knowledge obtained from the higher digestibility brown midrib maize mutants may be possible.

Increasing nutrient and energy supply from forages: forage conservation

In the UK, herbage growth is restricted to the milder spring and summer months, with relatively less or no growth during the winter. As a result, livestock farmers have exploited the surplus forage produced during the growing season by conserving it, traditionally as hay or silage, and these conserved forages now represent a major source of nutrients and energy during the period of little or no herbage growth. Developments in forage conservation have therefore had a major impact on the production of livestock products (milk, meat) over the last 50 years. Until the middle of the last century, haymaking was the predominant method of forage conservation and development in silage making started slowly, with only some 0.35m t of silage being ensiled in 1947 (Brassley, 1996). Thereafter silage production has increased at the expense of the hay crop. By 1954 grass silage production had exceeded 2 mt, and by 1980 silage was as popular as hay (in terms of conserved DM) with ~ 46 mt being conserved. Estimates of recent hay and silage production in the UK are given in Table 24.2 and suggest that the amount of grass conserved as hay in 2002 was substantially higher than in 2001, marking a change in a long period of decline. The reason for this increase is not clear, nor is it possible from these data to know whether this represents a longer-term change in the pattern of hay production.

Table 24.2 ESTIMATED QUANTITIES OF CONSERVED FORAGE IN THE UK ('000 TONNES FRESH WEIGHT)

Year	*2000*	*2001*	*2002*	*% change 2002/2001*
Hay (including barn-dried hay and artificially dried fodder)	2 956	2 948	3 242	+ 10.0
Silage	40 873	40 123	38 779	- 3.3
Other arable crops cut and harvested for silage	4 120	4 847	5 207	+ 7.4

The long term increase in the popularity of silage as a means of conserving excess forage has been the result of developments in a number of critical areas, in particular in handling and storage of silage and the development of additives. Solutions to many of the problems associated with making high quality silage arose from the considerable amounts of public and commercially financed research undertaken during the past 50 years (Wilkins, 1996). The development of big-bale silage in the mid 1970s overcame many of the problems associated with making high quality grass silage on small farms, and in particular the costs associated with silo construction and effluent control. Increases in the cost of concentrate feeds relative to forages in the 1970s and 1980s provided further incentive to improve conservation and utilisation of silage. Some of the key issues affecting the development of silage making in the UK and factors likely to influence future trends are discussed below.

WILTING

Grass silage made during the first half of the last century was frequently characterised by low dry matter contents, poor fermentation, high ensiling losses and low voluntary intake. Increasing the DM content of the grass before ensiling was shown in many studies to lead to an improvement in fermentation quality and silage intake. Mean increases in silage intake of 9% in growing cattle and 4% in dairy cows as a result of increasing DM content from about 220 to 350 g/kg were reported in the Eurowilt studies (Zimmer and Wilkins, 1984). In addition, wilting provided benefits in terms if reducing the volume of grass to ensile, with associated reductions in ensiling costs and reducing silage effluent production. However, animal production (liveweight gain, milk yield) on wilted silage was often lower than on unwilted material, despite the higher intakes. While the reasons for this have not been conclusively determined, they are likely to have been as a result of changes in plant composition during the wilting period. For the livestock farmer, the potential reduction in costs and environmental impact associated with feeding higher dry matter silage have to be set against lower levels of production. Nevertheless, climatic changes arising as a result of global warming might result in an increase in the degree of field drying.

Chopping

Evidence that consolidation of grass in a silage clamp was an important factor in achieving a rapid and stable fermentation led to the development of multi-cut forage harvesters. Double-chop silage was common on many farms by the mid 1970s and the development of machinery capable of chopping grass to short and defined lengths followed. In addition to improvements in fermentation quality, research identified benefits in terms of improved intake. In their assessment of silage produced in N Ireland, Porter and Agnew (2002) suggested that the intake of precision-chop silage was some 7.5% higher than that of single or double cut silage.

Additives

Reliance on dry weather to achieve wilted silage, and the benefits associated with it, led researchers to develop ways of improving the fermentation quality and animal production using unwilted (high moisture content) grass by using chemical additives that were intended to increase the speed of fermentation. The first additives to be widely used in Europe were based on mixtures of sulphuric and hydrochloric acids, and for many years thereafter the use of organic acids or their salts were widely used. These had the effect of augmenting the acids produced as a result of natural fermentation of the grass sugars present in the grass. Significant improvements in silage additive applicators were made in the 1960s. As a result, a number of chemicals that had previously only been shown to be effective in small-scale trials were introduced. The most widely used of these was formic acid, used either alone or in mixtures. Other products marketed as silage additives included mixtures of organic acids and formalin, other sterilising agents and sugars. The use of these additives increased substantially and they were undoubtedly a key factor in the increase in silage making that took place in the 1970s.

During the last twenty years or so, a plethora of biological additives have been developed. Bacterial inoculants were first imported into the UK in the early 1980s but they were relatively ineffective, probably due to inappropriate application rates or poor product quality. Today they represent the largest number of products currently on the UK register of silage additives, with the main active ingredient being *Lactobacillus plantarum*. Products tend to be differentiated by the strain of the bacteria, although all aim to ensure that at least 1 million viable (colony forming) bacteria are applied to each gram of fresh forage ensiled.

The benefits of using additives, particularly under difficult ensiling conditions, have been well documented. These include improved fermentation, reduced in-silo losses and improved nutrient utilisation and animal production under a wide range of conditions. Such has been the success of silage additives in reducing ensiling losses and improving production, that by the mid 1990s there were over 150 different products on the market.

Wilkins (1996) reviewed some 109 papers presented at Silage Research Conferences between 1970 and 1993 concerned with chemical or biological additives and their effects on preservation or feeding value. A summary of the effects of bacterial inoculants compared with either no additive or with formic acid is given in Table 24.3. It was concluded that inoculants produced silages with intake and liveweight gains intermediate between silages without an additive and those made with formic acid. In the review, milk quality did not appear to be affected by additive type, but fat and protein yields tended to be slightly higher with formic acid than inoculant-treated silages.

Table 24.3 THE RELATIVE CHANGE IN INTAKE AND ANIMAL PRODUCTION RESULTING FROM THE USE OF INOCULANTS COMPARED WITH EITHER NO ADDITIVE OR FORMIC ACID (ADAPTED FROM WILKINS, 1996)

Contrast:	*Enzyme treated as % of no additive**	*Enzyme treated as % of formic acid treated**
Silage DM intake	1.06 (19)	1.10 (16)
Liveweight gain (cattle) (g/day)	1.07 (9)	1.13 (7)
Milk yield (cows) (kg/day)	1.01 (7)	1.00 (8)
Fat (g/kg)	1.01 (7)	1.08 (8)
Protein (g/kg)	1.01 (7)	1.02 (8)
Fat + Protein (kg/day)	1.02 (7)	1.05 (8)

* Number of comparisons in parentheses.

These results are broadly in line with the survey of silages made at the Agricultural Research Institute of Northern Ireland (Mayne and Steen, 1993). In other studies, however, the use of formic acid has resulted in higher ($P<0.001$) D-value and intakes than an inoculant or no additive (Keady and Mayne, 1996; Sheils *et al.*, 1996).

It would appear that crop type, stage of maturity and climatic conditions all influence the efficacy of additives, with effects differing between different additive type. On the basis of analyses of 10,000 first cut silages produced in N Ireland over a 6 year period, Porter and Agnew (2002) showed that intake value and metabolisable energy (ME) concentrations were significantly ($P<0.001$) higher in the silages produced from early-cut grass and treated with inoculants than with other types of additive. These effects were not observed in silages produced after mid-May.

However, all of these comparisons were based on silages made using a range of inocula, application rates and composition of the crops ensiled, and the results therefore need to be treated with caution. Owen (2002) compared the results of 26 studies in which the DOMD in grass silages prepared either untreated or treated with a strain of *L plantarum* (MTD/1) and fed to beef or dairy cattle was determined. The mean values for the untreated and MTD/1 treated silages were 713 g/kg and 740 g/kg, respectively ($P<0.001$). These data clearly illustrate the benefits of silage additives, and it is not unreasonable to assume that developments of new strains of micro-organisms will continue, improving the efficacy of the additives in which they are used and increasing the range of crops and conditions in which they are effective.

Aerobic instability in silage during silage feeding is a common problem, particularly but not exclusively associated with whole-plant maize and small grain cereal silages. When the silo is opened, oxygen has unrestricted access to the exposed feeding face. Aerobic micro-organisms, e.g. the yeasts, *Hansenula* and *Candida* spp., which are the most prevalent, and other fungi present in the silage can utilise soluble components including lactic acid. This microbial activity leads to an increase in the temperature and pH of the silage. If allowed to continue, the deterioration and spoilage can result in considerable losses of DM and nutritive value. In addition to improving in-silo fermentation, a number of acid and inoculant-based additives have been shown to provide robust control of aerobic spoilage when silage is removed from the clamp. New technologies are being developed for the production of additives that will both improve fermentation and reduce aerobic spoilage.

Changes have recently been introduced in EU legislation,[1] which mean that all silage additives will need to undergo an approval process by October 2011. This will apply both to products currently in use and any new products prior to being marketed[2]. The approval process, which will include a requirement to demonstrate safety and efficacy, will have significant cost implications for silage additive manufacturers and is likely to markedly reduce the number of new additives coming on to the market, as well as removing from sale some of those currently being used.

While grass silage has been a major feed for ruminant livestock, there have been significant developments in the conservation and utilisation of alternative forages.

WHOLE CROP CEREAL SILAGE

The early harvesting of whole-crop cereals offers a practical way of providing an economical source of home produced forage for feeding to ruminant livestock. They enable farmers outside the geographical limits for forage maize production to produce alternative conserved forage to grass silage. Yields of over 10 t DM/ha can be achieved, and since the crops are harvested at DM contents of between 300 and 600 g/kg, no wilting is required. The inclusion of an additional dietary component, such as whole crop cereal silage constitutes an option to increase energy intake (Phipps *et al.,* 1995). Whole-crop may be either ensiled or preserved with an alkali. A number of technological factors are likely to be influential in the use of whole-crop cereals as feeds for ruminant livestock. These will include developments in additives to both improve the ensiling process and the stability of the silage once it has been exposed to air, the cost of production relative to costs of other forages, and changes in support under the Common Agricultural Policy.

---

[1] Regulation (EC) No. 1831/2003 of the European Parliament and of the Council.

[2] All current silage additives will need to have been approved for use as such by 2011.

FORAGE LEGUMES

An EU-funded study[3] on the use of forage legumes has recently concluded that forage legumes can play an increasing role in developing more sustainable farming practices. Because they capture nitrogen from the air, they reduce the need for inorganic fertilisers. In addition, using these crops as silage can also reduce the need for high-protein concentrate feeds on the farm. The results of this study have confirmed that silage made from legumes, or from legume-grass mixtures, outperforms silage made from grass alone, and that using silage made from forage legumes can significantly increase the profitability of dairy systems.

Nutritional limitations associated with achievement of increased forage yields

It is vital that systems of animal production based on forages allow rumen function to be optimised so that microbial metabolism is not compromised and the efficiency of transfer of ingested nutrients to the animal and the animal product is maximised. Optimisation of rumen function is also critical if voluntary intake of forages is to be maximised. Grass breeding over the last 30-40 years has been extensive but has focused primarily on yield and responsiveness to fertiliser applications. Whilst there has been some nutritional considerations in breeding strategies this has largely been restricted to digestibility and its relationship with maturity. As discussed by Beever and Reynolds (1994), there has been little attempt to manipulate the key nutritional entities in grass to optimise rumen function and increase the efficiency of nutrient utilisation.

VOLUNTARY INTAKE

One of the greatest concerns is the effect of increasing maturity of grasses that can seriously reduce voluntary feed intake. This is normally considered to be a function of increased lignification, which reduces the availability of the major carbohydrate fractions in grasses. Since a key factor influencing voluntary intake is the rate of particle breakdown in the rumen, increased lignification will increase rumen residence time (Jung and Allen, 1995). In this respect it is interesting to compare grasses and legumes. Legumes are generally consumed in greater quantities than grasses even when overall digestibility is similar. Wilson (1993) proposed that this was due to the fact that legumes are

---

[3] Legumes for silage in low input systems of animal production (LEGSIL)

dicotyledenous with a cell wall structure that allows rapid physical breakdown in the rumen.

Plant breeding objectives for the next 10 years at least should therefore concentrate on the proposal made by Wilson (1993) that 'in the next decade a major aim in biology should be to establish the ability to limit or direct cell developmental processes'. In particular it would seem necessary to understand what controls are present which switch on thickening and lignification of the secondary cell wall.

EFFICIENCY OF PROTEIN UTILISATION

A number of studies with perennial ryegrass and white clover fed to cattle have shown that losses of dietary nitrogen (N) from the rumen are considerable (Beever and Reynolds, 1994). The work of Beever *et al.* (1986) with ryegrass and clover indicated that the main reason for the low efficiency of capture of dietary N was an imbalance in the supply of degraded N relative to ATP available, resulting in substantial amounts of ammonia being absorbed from the rumen. As a result of this the protein value of fresh forages and silages is generally discounted with additional protein sources being fed to the animal. Although the use of additional protein often elicits a production response it inevitably leads to a greater loss of N to the environment.

In fresh grass some 0.75 to 0.90 of the total N is present as protein, with a high proportion of the protein N being derived from enzyme proteins in the chloroplasts with one enzyme, ribulose 1, 5 carboxylase (Rubisco) being particularly abundant. The fact that this protein is very rapidly degraded in the rumen goes some way to explaining the low efficiency of capture of dietary N. Flint and Chesson (1999) have argued that the lack of progress in this area is due to research and breeding being focussed mainly on seed proteins rather than leaf proteins. A major effort in the next 10 years on leaf proteins seems to be vital. An alternative approach may be to introduce into grass compounds such as tannins from other plants that will reduce their rate of hydrolysis in the rumen.

The situation with grass silages is of even more concern. After harvesting, rapid proteolysis begins due to the action of the plant proteases. After ensiling either directly or after wilting, proteolysis continues and within 24 h the protein content of the total N can have fallen from 0.8 to 0.6. The effect of this is to make the N compounds in silage even more rapidly available in the rumen than in fresh grass (see Hvelplund and Weisbjerg, 2000). An associated problem with silages is that the ensiling process effectively converts most of readily available carbohydrate to lactic and volatile fatty acids that yield little ATP in the rumen for microbial synthesis. This, coupled with the changes in the N fractions, results in a very low efficiency of microbial protein synthesis. Table 24.4 shows the mean values for diets based on various types of silage compiled by Givens and Rulquin (2004).

Table 24.4 SUMMARY OF MEAN EFFICIENCY OF MICROBIAL PROTEIN SYNTHESIS (EMPS) VALUES FOR DIETS BASED ON VARIOUS SILAGE TYPES (FROM GIVENS AND RULQUIN, 2004)

	Diets based on:			
	Maize silage	*Grass silage*	*Cereal silage*	*Legume silage*
Mean EMPS				
(g microbial N/kg OMADR)	48.4	30.1	35.9	19.5
Standard deviation	19.71	7.20	4.21	3.44
No. of observations	86	17	9	6

These data support the idea that maize silage based diets support greater EMPS than grass silage probably due to ATP supply being higher as a result of the starch present in maize silage. The very low values in legume silages are particularly concerning.

There is little doubt that one of the major problems of grazed and conserved grasses and legumes is the poor efficiency of protein utilisation and in silage particularly, the low efficiency of microbial protein synthesis. For the future there appears to be little to be gained from increasing protein concentrations in forages, but efforts should be concentrated on increasing carbohydrate availability in the rumen. Failure to achieve this for grasses and legumes will mean that increased productivity by ruminants will need to increasingly depend on maize or similar forages.

Nutritional limitations associated with description of nutritive value

Much research over the last 50 years has been directed towards improvement in knowledge on the nutritional requirements of animals and the nutritive value of forages and other feeds. Most emphasis has been placed on energy and protein, but as reported by AFRC (1998), all current systems are based on the idea of estimating how much dietary energy and protein is needed to meet a predetermined animal production output. In the future, and particularly if animal productivity is to be optimised, it will be necessary to predict the output of animal product and its composition from diets themselves optimised nutritionally and economically. This will require complex models that integrate knowledge on the nature of forages and how they present nutrients and energy to the animal at both the digestive tract and tissue levels. This will need to be coupled with a detailed model of the metabolism of the animal.

The framework for a nutrient -based approach for forages has been set out by López *et al.* (2000) but much remains to be done. In relation to energy supply the main drawback of most current systems is the fact that ascribed energy values say little about the range of commodities that the energy value represents. As an example, one problem that current energy systems have failed to explain is the poorer performance of animals on autumn compared with spring grazed grass. Some detailed work has indicated that, per unit of energy consumed, spring grass led to higher intestinal absorption of α-amino N than autumn grass (MacRae *et al.*, 1985). More recently van Vuuren et al. (1992) showed that this was due to enhanced microbial protein synthesis with spring grass.

Future forage breeding programmes will need to embrace these nutrient and energy supply concepts if the potential of the substantial capacity for dry matter yield is to be realised in terms of useful animal products whilst minimising environmental impact.

Towards a new definition of yield

To date, yield of animal product (milk, meat) has been the main criterion used to judge output. There is however growing concern with diet-health relationships which reflects the enormous and mounting economic and social costs of diseases such as cancer, Type 2 diabetes, cardio- and cerebro-vascular diseases. It is generally accepted that their incidence is closely related to diet, and in particular the fatty acid composition of ingested fat. Increasing prevalence of diet-related diseases is predicted to result in unsustainable health care costs, with greater restriction on surgical and drug treatments and driving a move towards a preventative health care model. More than half of all saturated fatty acids in the diet arise from animal-derived products produced by the rural economy (DEFRA, 2003), and indeed the contribution may be even greater if indirect sources such as biscuits, cakes etc. are taken into account.

There is now evidence (Chilliard *et al.*, 2000) that use of traditional fresh herbage-based diets for ruminants is likely to lead to an improved fatty acid profile in milk and meat, in particular a reduction in saturates and increases in a-linolenic acid (C18:3 n-3) and conjugated linoleic acid (CLA). Fresh grass contains a high proportion (0.5 – 0.75) of its total fatty acids as a-linolenic acid (C18:3 n-3) (Harfoot and Hazlewood, 1988) but factors such as stage of maturity and plant genotype affect this. In addition the now very common practice of conservation of grass by cutting, wilting and ensiling substantially reduces a-linolenic acid content (Dewhurst and King, 1998). In the future forage breeding must take these factors into account and develop plants with an enhanced lipid composition but also with factors which reduce the saturation of these fatty acids in the rumen. It will be necessary to consider yield not simply as volume of milk or mass of meat but in terms of individual fatty acids and other nutrients. Failure to consider animal products as foods and to change their composition to match that needed for disease prevention will result in a continued decline in consumption of milk and ruminant meat products, which would have an immense negative impact on rural land use, landscape and economy.

References

AFRC (1998) *Response in the Yield of Milk Constituents to the Intake of Nutrients by Dairy Cows.* AFRC Technical Committee on Responses to Nutrients, Report No 11, CABI, Wallingford, 96pp.

Alberda, Th. (1968) Dry matter production and light interception of crop surfaces. IV. Maximum herbage production as compared with predicted values. *Netherlands Journal of Agricultural Science* **16**, 142-153.

Alberda, Th. (1971) Potential production of grassland. In *Potential Crop Production*, pp 159-171. Edited by P.F. Wareing and J.P. Cooper. Heinemann, London.

Alberda, Th. (1977) Possibilities of dry matter production from forage plants under different climatic conditions. *Proceedings of the 13th International Grassland Congress.* pp 61-69. Leipzig, Germany.

Aldrich, D. T. A. and Elliott, C. S. (1974) A comparison of the effects of grazing and of cutting on the relative herbage yields of six varieties of perennial ryegrass (*Lolium perenne*). *Proceedings of the 12th International Grassland Congress*, pp 11-17. State Committee on Science and Technique of the USSR Council of Ministers, Moscow.

Aldrich, D.T.A. (1987) Developments and procedures in the assessment of grass varieties at NIAB 1950-87. *Journal of the National Institute of Agricultural Botany* **17**, 313-327.

Baker, R., Doyle, C. and Lidgate, H. (1991) Grass production. In *Milk from Grass*, pp 1-26. Edited by C. Thomas, A. Reeve and G.E.J. Fisher. The British Grassland Society, Reading.

Beever, D. E. and Reynolds, C. K. (1994) Forage quality, feeding value and animal performance. In *Grassland and Society*, pp 48-60. Edited by L. t'Mannetje and J Frame. Wageningen Pers.

Beever, D. E., Dhanoa, M .S., Losada, H. R., Evans, R. T., Cammell, S. B. and France, J. (1986) The effect of forage species and stage of harvest on the processes of digestion occurring in the rumen of cattle. *British Journal of Nutrition* **56**, 439-454.

Beever, D.E. and Thorp, C. (1996) Advances in the understanding of factors influencing the nutritive value of legumes. In *Legumes in Sustainable Farming Systems,*). pp 194-207. Edited by D. Younie. British Grassland Society, Reading.

Brassley, P. (1996) Silage in Britain 1880-1990: the delayed adoption of an innovation. *Agricultural History Review,* 44, 61-82.

Burgon, A., Bondesen, O.B., Verburgt, W.H., Hall, A.G., Bark, N.S., Robinson, M. and Timm, G. (1997) The forage seed trade. In *Forage Seed Production. 1. Temperate Species,* pp 271-286. Edited by D.T. Fairey and J.G. Hampton. CAB International, Wallingford, UK.

Camlin, M.S. (1997) Plant breeding – achievements and progress. In *Seeds of Progress. BGS Occasional Symposium No. 31*, pp 2-14. Edited by. J.R.

Weddell. British Grassland Society, Reading.

Caradus, J.R. and Chapman, D.F. (1996) Selection for and heritability of stolon characteristics in two cultivars of white clover. *Crop Science* **36**, 900-904.

Caradus, J.R., Woodfield, D.R., Dunn, A. and Cousins, G. (2000) Response to selection for node appearance rate of white clover. *New Zealand Journal of Agricultural Research* **43**, 157-162.

Charles, A.H. (1973) A comparison of ryegrass populations from intensively managed permanent pastures and leys. *Journal of Agricultural Science, Cambridge* **81**, 99-106.

Chilliard, Y., Ferlay, A., Mansbridge, R.M. and Doreau, M. (2000) Ruminant milk plasticity: nutritional control of saturated, polyunsaturated, *trans* and conjugated fatty acids. *Ann. Zootech.* **49**,181-206.

Collins, R.P., Abberton, M.T., Michaelson-Yeates, T.P.T., Marshall, A.H. and Rhodes, I. (1998) Effects of divergent selection on correlations between morphological traits in white clover (Trifolium repens L.). *Euphytica* **101**, 301-305.

Cooper, G. and McGechan, M.B. (1996) Implications of an altered climate for forage conservation. *Agricultural and Forest Meteorology*, 79, 253–269.

Cooper, J.P. (1968) Energy and nutrient conversion in a simulated sward. *Report of the Welsh Plant Breeding Station, 1967*, pp 10-11.

Davies, A., Shao, J., Brignall, P., Bardgett, R.D., Parry, M.L. and Pollock, C.J. (1996) Specification of climatic sensitivity of forage maize to climate change. *Grass and Forage Science*, 51, 306–317.

de Wit, C.T. (1965) *Photosynthesis of Leaf Canopies*. Agricultural Research Reports 663, IBS, Wageningen.

DEFRA (2002) *Guidelines for Farmers in NVZs – England. Revised.* London: Department for Environment and Rural Affairs. 34 pp.

DEFRA (2003). *National Diet and Nutrition Survey*. http://www.defra.gov.uk

Dewhurst, R.J. and King P.J. (1998). Effects of extended wilting, shading and chemical additives on the fatty acids in laboratory grass silages. *Grass and Forage Science,* **53**. 219-224.

Eagles, C., Pollock, C., Thomas, H. and Wilson, D. (1984) Breeding for better winter hardiness and drought resistance. In *Grassland Research Today*, pp 6-8. Edited by J. Hardcastle.

Fisher, G. and Mayne, S. (1991) The integration of conservation with grazing. In *Milk from Grass*, pp 73-88. Edited by C. Thomas, A. Reeve and G.E.J. Fisher). The British Grassland Society, Reading.

Flint, H. J. and Chesson, A. (1999) The impact of gene technology used in animal feedstuffs. In *Proceedings from the 25th International Dairy Congress 1998 Vol III Future Milk Farming*. International Dairy Books, Aarhus, Denmark.

Fontaine, A.S., Bout, S., Barriere, Y. and Vermerris, W. (2003). Variation in cell wall composition among forage maize (Zea mays L.) inbred lines and its impact on digestibility: Analysis of neutral detergent fiber composition by pyrolysis-gas chromatography-mass spectrometry. *Journal of Agricultural and Food Chemistry* **51,** 8080-8087.

Frame, J. (1991) Herbage production and quality of a range of secondary grass species at five rates of fertilizer nitrogen application. *Grass and Forage Science* **46**, 139-151.

Frame, J. and Newbould, P. (1984) Herbage production from grass/white clover swards. In *Forage Legumes. BGS Occasional Symposium No. 16*, pp 15-35. Edited by D.J. Thompson. British Grassland Society, Reading.

Frame, J., Baker, R.D. and Henderson, A.R. (1995) Advances in grassland technology over the past fifty years. In *Grassland into the 21st Century. BGS Occasional Symposium No. 29*, pp 31-65. Edited by G.E. Pollot. British Grassland Society, Reading.

Frei, O.M. (2000). Changes in yield physiology of corn as a result of breeding in northern Europe. *Maydica* **45**, 173-183.

Givens, D. I. and Deaville, D. I. (2001) Comparison of major carbohydrate fractions and cell wall digestibility in silages made from older and newer maize genotypes grown in the UK. *Animal Feed Science and Technology*, **89**, 69-82.

Givens, D. I. and Rulquin, H. (2004) Utilisation by ruminants of nitrogen compounds in silage-based diets. *Animal Feed Science and Technology*, **114**, 1-18.

Goodlass, G., Wilshin, S. and Allin, R. (2003) *The British Survey of Fertilizer Practice. Fertiliser Use on Farm Crops for Crop Year 2002*. Department for Environment, Food and Rural Affairs, London. 105 pp.

Harfoot, C. G. and Hazlewood, G. P. (1988) Lipid metabolism in the rumen. In *The Rumen Microbial Ecosystem*, pp 285–322. Edited by P. N. Hobson. Elsevier Applied Science, New York, NY.

Hofgaard, I.S., Vollsnes, A.V., Marum, P., Larsen, A., Tronsmo, A.M. (2003) Variation in resistance to different winter stress factors within a full-sib family of perennial ryegrass. *Euphytica* **134**, 61-75.

Hopkins, A. (2000). Herbage production. In *Grass: Its Production and Utilisation*, pp. 90-110. Edited by A. Hopkins. Blackwell Science, Oxford.

Hughes, R. (1971) Grassland Agronomy. *Report of the Welsh Plant Breeding Station 1970.* pp 20-24. Aberystwyth: University College of Wales

Hvleplund, T., and Weisbjerg, M. R., (2000) *In situ* techniques for the estimation of protein degradability. In *Forage Evaluation in Ruminant Nutrition,* pp 233-258. Edited by D. I. Givens, E. Owen, R. F. E. Axford and H. M. Omed. CABI Publishing, Wallingford.

Jung, H. J. and Allen, M. S. (1995) Characteristics of plant cell walls affecting intake and digestibility of forages by ruminants. *Journal of Animal Science*, **73**, 2774-2790.

Keady, T.W.J. and Mayne, C.S. (1996). The effects of nitrogen fertiliser, harvesting dates and additive treatment on fermentation, in-silo losses and predicted feeding value of grass silage. In *Proceedings of the XIth International Silage Conference, Aberystwyth, Wales*, pp52-53. Edited by D.H.I. Jones, R. Jones, R. Dewhurst, R. Merry and P.M. Haigh).

Lazenby, A. (1988) The grass crop in perspective: selection, plant performance and animal production. In *The Grass Crop* pp 311-360. Edited by M.B. Jones and A. Lazenby. Chapman and Hall, London.

Leafe, E.L. (1978) Physiological, environmental and management factors of importance to maximum yield of the grass crop. In *Maximizing Yields of Crops,* Edited by J.K.R. Gasser and B. Wilkinson. ARC Symposium Proceedings, HMSO, London.

Leafe, E.L. (1988) Introduction – the history of improved grasslands. In *The Grass Crop*, pp 1-23. Edited by M.B. Jones and A. Lazenby. Chapman and Hall, London.

López, S., Dijkstra, J. and France, J. (2000) Prediction of energy supply in ruminants with emphasis on forages. In *Forage Evaluation in Ruminant Nutrition,* pp 63-94. Edited by D. I. Givens, E. Owen, R. F. E. Axford and H. M. Omed. CABI Publishing, Wallingford.

MacRae, J. C., Smith, J. S., Dewey, P. J. S., Brewer, A. C., Brown, D. S. and Walker, A. (1985) The efficiency of utilisation of metabolisable energy and apparent absorption of amino acids in sheep given spring and autumn-harvested dried grass. *British Journal of Nutrition,* **54**, 197-209.

MAFF (2000) *Fertilizer Recommendations for Agricultural and Horticultural Crops (RB209). 7th edn.* Ministry of Agriculture, Fisheries and Food, London.

Marshall, A.H., Rascle, C., Abberton, M.T., Michaelson-Yeates, T.P.T. and Rhodes, I. (2001). Introgression as a route to improved drought tolerance in white clover (*Trifolium repens* L.). *Journal of Agronomy and Crop Science* **187**, 11-18.

Marshall, A.H., Williams, T.A., Abberton, M.T., Michaelson-Yeates, T.P.T., Olyott, P. and Powell, H.G. (2004) Forage quality of white clover (Trifolium repens L.) x Caucasian clover (T-ambiguum M. Bieb.) hybrids and their grass companion when grown over three harvest years. *Grass and Forage Science* **59**, 91-99.

Marshall, A.H., Williams, T.A., Powell, H.G., Abberton, M.T. and Michaelson-Yeates, T.P.T. (2002) Forage yield and persistency of Trifolium repens x Trifolium nigrescens hybrids when sown with a perennial ryegrass companion. *Grass and Forage Science* **57**, 232-238.

Mayne, C.S. and Steen, R.W.J. (1993). A review of animal production responses to formic acid and inoculant treatment of grass silage in trials at the Agricultural Research Institute of Northern Ireland. *Proceedings of the Xth International Conference on Silage Research, Dublin,* pp 178-179.

Morrison, J. (1981) The potential of legumes for forage production. In *Legumes and Fertilizers in Grassland Systems.* British Grassland Society Winter Meeting 1981. 1.1-1.10.

NIAB (1982-1985) *Recommended Varieties of Grasses. Farmers Leaflet No. 16.* Cambridge: National Institute of Agricultural Botany.

NIAB (1986-1993) *Recommended Varieties of Grasses and Herbage Legumes. Farmers Leaflet No. 4.* National Institute of Agricultural Botany, Cambridge.

NIAB (1991*b*-1994*b*) *Recommended Varieties of Forage Maize. Farmers Leaflet No. 7.* National Institute of Agricultural Botany, Cambridge.

NIAB (1994-1996) *NIAB Recommended List of Grasses and Herbage Legumes.* National Institute of Agricultural Botany, Cambridge.

NIAB (1995*b*-1996*b*) *Recommended List of Forage Maize.* National Institute of Agricultural Botany, Cambridge.

NIAB (1997-2001) *Grasses and Herbage Legumes Variety Leaflet.* National Institute of Agricultural Botany, Cambridge.

NIAB (1999*b*-2002*b*) *Forage Maize Variety Leaflet.* National Institute of Agricultural Botany, Cambridge.

NIAB (2002) *Pocket Guide to Livestock Crops 2003.* National Institute of Agricultural Botany, Cambridge.

NIAB (2003) *Pocket Guide to Grasses and Herbage Legumes 2004.* National Institute of Agricultural Botany, Cambridge.

NIAB (2003*b*) *Pocket Guide to Maize Varieties 2004.* National Institute of Agricultural Botany, Cambridge.

Owen, T. R (2002) The effects of a combination of a silage inoculant and a chemical preservative on the fermentation and aerobic stability of whole-crop cereal and maize silage. In *Proceedings of the XIIIth International Silage Conference, Auchincruive, Scotland*, pp 196-197. Edited by L. M. Gechie and C Thomas.

Pain, B.F. & Phipps, R.H. (1975). The energy to grow maize. *New Scientist* **66**, 394-396.

Phipps, R.H. (2002). Complementary forages for milk production. In *Recent Developments in Ruminant Nutrition 4.* Edited by J. Wiseman & P.C. Garnsworthy. pp. 121-138. Nottingham University Press, Nottingham.

Phipps, R.H., Sutton, J.D. & Jones, B.A. (1995). Forage mixtures for dairy cows: the effect on dry matter intake and milk production of incorporating either fermented or urea treated whole-crop wheat, brewers' grain, fodder beet or maize silage into diets based on grass silage. *Animal Science* **61**, 491-496.

Phipps, R.H., Weller, R.F., Elliot, D. & Sutton, J.D. (1988). The effect of level and type of concentrate and type of conserved forage on dry matter intake and milk production of lactating dairy cows. *Journal of Agricultural Science, Cambridge* **111**, 179-186.

Porter, M.G. and Agnew, R.E. (2002). The effect of silage additive and harvesting method on the quality of first cut silages in Ireland. In *Proceedings of the XIIIth International Silage Conference, Auchincruive, Scotland*, pp 228-229. Edited by L. M. Gechie and C Thomas.

Rhodes, I. and Ortega, F. (1997) Forage legumes. In *Seeds of Progress. BGS Occasional Symposium No. 31* pp 15-27. Edited by. J.R. Weddell). British Grassland Society, Reading.

SEERAD (2004) *Agriculture Facts and Figures* Scottish Executive Environment and Rural Affairs Department, http://www.scotland.gov.uk

Sheils, P., O'Kiely, P.O., Moloney, A.P. and Caffrey, P.J. (1996) The effects of a bacterial inoculant or formic acid on the fermentation or nutritive value of grass silage and the interactions between silage and quantity of supplementary concentrates. In *Proceedings of the XIth International Silage Conference, Aberystwyth, Wales*, pp56-57. Edited by D.H.I. Jones, R. Jones, R. Dewhurst, R. Merry and P.M. Haigh.

Sheldrick, R.D. (2000) Sward establishment and renovation. In *Grass: Its Production and Utilization. Third Edition* pp 13-30. Edited by A. Hopkins. British Grassland Society, Reading.

Sheldrick, R.D., Thomson, D. and Newman, G. (1987) *Legumes for Milk and Meat.* Marlow: Chalcombe. 101 pp.

Steen, R.W.J., Gordon, F.J., Dawson, L.E.R., Park, R.S., Mayne, C.S., Agnes, R.E., Kilpatrick, D.J. and Porter, M.G. (1998) Factors affecting the intake of grass silage by cattle and prediction of silage intake. *Animal Science*, **66**,115-127.

Sutton, J.D., Cammell, S.B., Phipps, R.H., Beever, D.E. & Humphries, D.J. (2000). The effect of crop maturity on the nutritional value of maize silage for lactating dairy cows. 2. Ruminal and post ruminal digestion. *Animal Science* **71**, 391-400.

van Vuuren, A. M., Krol-Kramer, F., van der Lee, R. A. and Corbijin, H. (1992) Protein digestion and intestinal amino acids in dairy cows fed fresh *Lolium perenne* with different nitrogen contents. *Journal of Dairy Science*, **75**, 2215-2225.

Webb, K.J. (1996) Opportunities for biotechnology in forage legume breeding. In *Legumes in Sustainable Farming Systems. BGS Occasional Symposium No. 30* pp 77-85. Edited by D. Younie. British Grassland Society, Reading.

Wilkins, P.W. and Humphreys, M.O. (2003) Progress in breeding perennial forage grasses for temperate agriculture. *Journal of Agricultural Science* **140**, 129-150.

Wilkins, R.J. (1996). Silage production and utilisation – Perspectives from the Silage Research Conferences. In *Proceedings of the XIth International Silage Conference, Aberystwyth, Wales*, pp 1-10. Edited by D.H.I. Jones, R. Jones, R. Dewhurst, R. Merry and P.M. Haigh.

Wilkinson, J.M. & Hill, J. (2003). Effect on yield and dry-matter distribution of the stay-green characteristic in cultivars of forage maize in England. *Grass and Forage Science* **58**, 258-264.

Wilson, J. R. (1993). Organisation of forage plant tissues. In *Forage Cell Wall Structure and Digestibility*, American Society of Agronomy, Madison, pp 1-32.

Wright, C.E. (1981). Introductory remarks. In *Plant Physiology and Herbage Production. Proceedings of British Grassland Society Occasional Symposium No 13*, pp. 1-3. Edited by C.E. Wright. British Grassland Society, Reading.

Zimmer, E. and Wilkins, R.J. (1978). Efficiency of silage systems: a comparison between unwilted and wilted silages. *Landbauforschong Völkenrode,* 69-84.

25

POSSIBLE FUTURES FOR AGRICULTURE IN NORTHERN EUROPE DURING A PERIOD OF POLICY REFORM

SEAN RICKARD*, JOE MORRIS**, ERIC AUDSLEY+
** Cranfield School of Management, Cranfield University, Cranfield, Beds, MK43 0AL, UK; **Institute of Water and Development, Cranfield University, Silsoe, Bedford, MK45 4DT, UK; + Silsoe Research Institute, Wrest Park, Silsoe, Bedford, MK45 4DT, UK*

Introduction

In fundamentally switching support from production based subsidies to decoupled single farm payments (SFPs), the Medium–Term Review (MTR)–reform might correctly be viewed as the start of a new chapter in the life of the Common Agricultural Policy (CAP). The reform was concerned with combinable crops, beef, sheep and dairying, but the European Commission intends that the underlying philosophy of decoupled support will in the near future be applied to other sectors. Indeed, attempts are underway to reform other regimes, notably sugar, olive oil and tobacco. Although many of the details concerning implementation of the MTR–reform are still to be announced by member states, it is clear that if the 'new' CAP develops as intended, market forces will to a large extent, if not completely, determine farm–gate prices. Farm support will be provided wholly or largely by a SFP and these payments will to varying degrees be subject to conditionality regarding the protection of the countryside.

The precise details of the MTR–reform are not of concern to this chapter. Interest is in the thrust of the reform and the longer–term implications for the pattern and levels of production, associated farming practices, productivity and sustainability. These elements are themselves the product of a host of endogenous and exogenous influences including the openness of EU agriculture to foreign of competition, farming's terms–of–trade, the real value of support payments, the extent of regulation and the scope of legislation. Many uncertainties hang over these variables, all of which will be influenced by developments outside agriculture. For example, there is uncertainty over the future costs of the CAP in an enlarged EU and the intention, at some point to 'buy–out' on–farm milk quotas. Increasing demands are being made on the Community's budget, casting doubt as to whether the current real level of funding for the CAP will continue. The willingness of member states to provide the necessary funds is but one of the many uncertainties that will significantly influence the future development of the EU agricultural industry.

This chapter describes a methodology that has been employed to rationalise the many uncertainties facing agriculture so as to explore future possible outcomes and also identify opportunities and threats that may arise in the future. The methodology is that prescribed by UK Foresight Futures 2020 (DTI, 2002) and consists of a systematic framework for developing exploratory scenarios. This approach to envisioning the future has been developed over a number of years by a team of researchers at SPRU – Science and Technology Policy research –University of Sussex in consultation with stakeholders from business, government and academia (Berkhout and Hertin, 2002). Exploratory scenarios are a tool for dealing with uncertainty involving a time horizon beyond normal commercial considerations. They are a response to the recognition that the future cannot be extrapolated solely from historical data and past relationships. Rather than attempting to predict the future, they are tools for mapping a range of possible emergent and only partially knowable outcomes given the large and diverse set of factors capable of influencing the future.

Berkhout and Hertin (2002) described future exploratory scenarios as a set of partially viewable alternatives that define an outcome 'possibility space'. This concept of a future possibility space populated by the outcomes of alternative exploratory scenarios lies at the heart of the Foresight Futures' framework. It provides a systematic framework in which to develop plausible, alternative assumptions about the future; a process of drawing out, challenging and refining, knowledge of the driving factors and their influence on the future. Although the Foresight Futures exploratory scenario framework incorporates trends, they serve only as a starting point and a key outcome of the process is to reveal potential discontinuities and reversals. The Foresight Futures framework is particularly suitable for visualising the alternative futures for agriculture where past relationships are now subject to changes in not only economic conditions, but also social attitudes and concerns about the environment. Underpinning the exploratory scenario framework set out in this paper are the Foresight Futures' four key assumptions:

- The future is not only a continuation of past relationships and dynamics but also can be shaped by human choice and activities;
- The future cannot be foreseen but exploration of the future can inform the decisions of the present;
- There is not one but a range of possible futures that can be mapped within a 'possibility space';
- Development of scenarios involves both rational analysis and subjective judgment necessitating interactive and participative methods.

Against the background of the MTR–reform of the CAP, Defra sponsored an Agricultural Futures and Implications for the Environment Project under the auspices of Cranfield University's Institute of Water and Environment in collaboration with the Silsoe and Macaulay Research Institutes. The objective was twofold: firstly, given the forces for change bearing down on EU agriculture to set out the possible futures for farming in England and Wales and the consequences for the natural environment; and secondly, to inform policy by inclusion of commentary as to what might be done to exploit potential environmental benefits and mitigate environmental risks.

This chapter reports the results of the first stage of the project utilising the Foresight Futures framework but, where judged appropriate adjusted to take account of agriculture's particular circumstances. Research conducted by SPRU suggests that sectoral futures scenarios tend to fall into a limited set that can be represented within a four quadrant, two–dimensional futures possibility space. Although the global futures literature identifies five main dimensions of change (Berkhout *et al* 1998) – demography and settlement patterns; the composition and rate of economic growth; the rate and direction of technological change; the nature of governance; and social and political values – the Foresight Futures framework focuses primarily on the last two when applying exploratory scenarios to an industrial sector. This is judged sufficient when the sectors under study are located in developed countries. In developed nations human demography is relatively well characterised in current population models and hence there would be little to gain from scenario elaboration. According to the Foresight Futures framework economic growth can be regarded as the outcome of a set of institutional factors (fiscal and monetary policies, trade liberalisation, regulation and legislation) rather than an autonomous driver of change. Similarly, perhaps more controversially, the rate and direction of technological change is regarded as being shaped by market, regulatory, political and cultural factors and is not regarded as an exogenous factor of change. This corresponds with theories in evolutionary economics that view the rate and direction of innovation as the product of economic and institutional contexts within which innovation occurs (Nelson and Winter, 1982). Thus, the Agricultural Futures project's output is determined by the two key exogenous dimensions of change, as set out in Figure 25.1.

The two dimensions that define the matrix set out in Figure 1 are: social and political values; and the nature of governance. Although there are obvious connections between values – taken here as contemporary tastes, beliefs and norms – and levels of governance – defined here as the way in which authority and control is exercised, whether local, national or global – the Foresight Futures framework treats the two dimensions as independent. This reflects the view that although values play a role in legitimating structures of power they also have an independent role in disrupting and changing those structures. Similarly structures of power can be seen as creating the essential conditions within which values, ideas and knowledge are generated.

In Figure 25.1 the horizontal dimension captures alternative possible developments in core values. At the 'individualism/consumerism' end of the continuum values are dominated by the right to private consumption and personal freedom. The rights of the individual and the needs of the present have precedence over those of the collective and the needs of future generations. Incomes and resources are distributed overwhelmingly by free and competitive markets with the economic and social functions of governance limited to the protection and enforcement of property rights and minimum social safety nets. At the other end of the continuum 'community/conservation' represents values that are determined by an overwhelming concern for the common good. The individual is seen as part of a collective, with rights and responsibilities to the community. High value is attached to equity and participation and there is a general concern to ensure a sustainable future. Civil society is strong and highly valued and resources are allocated via heavily regulated markets.

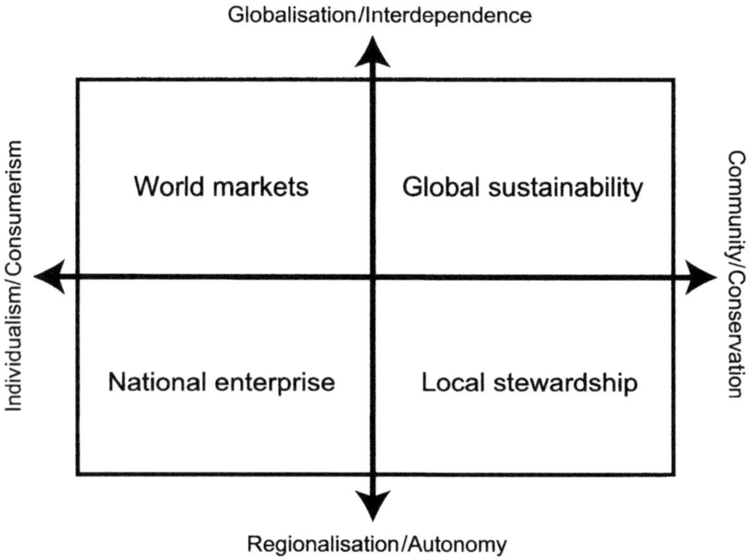

Figure 25.1 Foresight futures possibility space.

The vertical governance dimension reflects alternative structures of political and economic power: in essence who has the power of decision–making. At the 'globalisation/interdependence' end of the continuum the power to govern is distributed. National governments divest sovereignty upwards to international bodies for overseeing and regulating issues that are deemed to have global dimensions eg, trade, the environment. The outcome at this end of the continuum is fully aligned with the process characterised as globalisation. At the other end of the vertical governance continuum, 'regionalisation/ autonomy' reflects the retention of economic and political power at the national and sub–national level defined here as the regional level. Sovereignty is retained over all key areas of policy, national governments or regional authorities have autonomy in decision making and the process of globalisation is reversed.

The UK, as a member of the European Union (EU) sits somewhere in the middle of the vertical continuum reflecting the fact that it has transferred large areas of sovereignty eg, agricultural policy to the authority and regulation by the EU.

These two dimensions determine the futures possibility space for agriculture. The space is divided into quadrants reflecting four distinct exploratory scenarios named by the Foresight Futures as: world markets; global sustainability; national enterprise; and local community. The world market scenario reflects a libertarian outcome in which people aspire to personal freedoms and material wealth to the exclusion of wider social goals. Integrated global markets are viewed as the best way to deliver the internationally co–ordinated policy sets. Under the national enterprise scenario people also aspire to personal freedoms and material wealth, but within a nationally co–ordinated cultural identity. International co–operation is limited to traditional domains such as defence. In contrast, within the global sustainability quadrant people seek high levels of welfare within communities that share values and a concern to protect the environment that are in common throughout the world. The outcome of these shared values is an active public policy set as part of integrated global governance systems in areas such as economic development and environmental protection. Finally, the local stewardship scenario reflects people aspiring to sustainable levels of welfare within regional and networked communities. Markets are subject to social regulation to ensure a more equal distribution of opportunities and a high quality local environment. Under this scenario international co–operation is limited to coalitions with other regions, some of which are characterised by conflict and tensions.

Maximising the learning benefits of scenario planning requires close attention to the process. The key challenge is to engage stakeholders in a process that accommodates a diversity of viewpoints and technical expertise. Such a process will necessarily be interactive, combining creative, participative workshops and additional analysis by individuals or small groups to synthesise and elaborate scenarios. Accordingly, as part of the first phase of the Agricultural Futures and Implications for the Environment Project, a one day workshop attended by representatives of various stakeholder interests in the farming, food and related environmental sectors was held at Cranfield University at Silsoe. The stakeholder workshop was asked to:

- Identify the main factors and the attributes that could possibly shape the future structure and production practices of British farming with particular reference to the implications for the environment;
- Map these factors over a thirty year horizon using the Foresight Futures analytical framework set out in Figure 1;
- Identify in general terms key environmental risks and possible benefits associated with the future scenarios.

The workshop participants were selected by researchers to provide balanced representation from key stakeholder groups with interests in farming and environmental futures. They included representatives from commercial farming, land management, the food industry, environmental organisations, Defra, academics, researchers, regulators and lobbyists. The target number of participants was between 20 and 25. In the event a total of 27 people participated in the workshop; comprising 21 external representatives and 6 members from the project research team (see Appendix for a list of attendees).

The plan of the chapter is as follows. The next section, Section II, outlines the CAP's MTR–reform including its rural development and agri–environment aspects, as presented to the workshop participants in order to provide a factual base upon which to develop alternative agricultural futures. It identifies the key drivers for agriculture and the CAP, and provides an assessment of the links between drivers and broad trends. Section III sets out four alternative agricultural futures, generated by participants at the Cranfield workshop using the Foresight Futures exploratory scenarios framework. The intention is to identify the key differences between the four alternative futures arising from different assumptions regarding developments in social values and governance structures. Section IV reports the details of the four exploratory agricultural scenarios that emerged from the workshop. This section outlines the implications of the alternative futures at the aggregate industry or sector level. It describes the policy environment, the sources of pressure on farm incomes, the resulting industry structure, the outcome for consumer prices and choice, together with the broad implications for sustainable farming practices and the rural economy as revealed by the workshop's deliberations. A subsequent chapter in the proceedings, *Agricultural Productivity and Futures: Drivers and Consequences of Technology Change in Farming in England and Wales* builds on these aggregate future outcomes to analyse the implications at the farm–level for production techniques, productivity, conservation and sustainability.

The New CAP

The 2003 MTR–reform of the CAP turned out to be more radical than envisaged at the outset of the review process. It amounts to a full–blown reform involving for all practical purposes the removal of price support instruments for combinable crops and beef – save as a safety net in periods of very low prices – and the replacement of targeted payments with a system of 'decoupled' SFPs. The reform also confirmed the widening of decoupling to the dairy sector, though it extended the life of the quota regime until 2012. Significantly it gave member states considerable discretion in the implementation of the SFP and it included a mechanism to reduce SFPs in nominal terms if budgetary pressures build up in future years. In the preamble to the Commission's explanatory memorandum on the reform (Commission, 2003) the following objectives were set:

- To enhance the competitiveness of EU agriculture by setting intervention as a real safety net measure, allowing EU producers to respond to market signals while protecting them from extreme price fluctuations;
- To promote a more market orientated, sustainable agriculture by completing the shift from product to producer support with the introduction of a decoupled single farm payment based on historical references and subject to compliance with environmental, food safety and animal welfare requirements;
- To provide a better balance of support and strengthen rural development by transferring funds from the first to the second pillar of the CAP via the introduction of an EU–wide system of modulation and expanding the scope of currently available instruments for rural development to promote food quality, meet high standards and foster animal welfare.

In an article published to coincide with the publication of the Commission's MTR, the Farm Commissioner, Franz Fischler explained that the proposed reforms would help the EU agricultural industry become internationally competitive (Fischler, 2002). He also explained that the proposed replacement of support prices and targeted payments which SFPs would reduce the ecological costs associated with modern farming. In his opinion the combination of the switch to decoupled SFPs and associated rural development measures would specifically aid smaller farm business and hence slow, if not arrest, the decline in the numbers of farm businesses in the EU.

The concern to make EU agriculture more internationally competitive reflects, arguably the major source of pressure for reform; namely, the World Trade Organisation (WTO). The importance of including agriculture within the multilateral Uruguay GATT Round cannot be overstated. Multilateral trade negotiations are now a key driving force and without the pressure from this source it is unlikely that the CAP would have advanced so far, if at all, towards the adoption of pure decoupling (Rieger, 2000). In ratifying the Uruguay GATT Agreement the Community embarked on a process whereby its ability to support farming, and associated environmental outcomes, would in future, be subject to multilateral negotiations within the WTO. The EU as a signatory to the Uruguay Agreement is therefore committed to a reform process involving further reduction in tariffs, export subsidies and production related support payments so as to advance the cause of free trade (Josling and Tangermann, 1999).

A second external force for CAP reform is its cost. In December 1995, just months after the Uruguay Agreement came into effect the EU Heads of State, concerned at the growing public expenditure burden of the CAP, invited the European Commission to prepare a thorough analysis of the financial implications of the CAP post enlargement of the EU. Two years later the Commission reported with the publication *Agenda 2000* (Commission, 1997) which in turn led to a further reform of the CAP in 1999. But the growing costs of the CAP had not been adequately addressed and EU leaders had little choice but to call for the Commission to . . . *submit a report to the Council in 2002 on the development of agricultural expenditure, accompanied if necessary by appropriate proposals and requests to the Council to take the necessary steps in line with the objective of the (1999) reform* (Agrafocus, 1999). Thus, what has become the MTR–reform is very largely the product of the twin forces of the WTO membership and the budgetary costs of the CAP.

A third force for reform was growing concern regarding the actual and potential negative environmental impacts of modern farming. According to Fischler (2002) the MTR–reform is widely supported by populations across the EU because it will encourage less intensive farming. Whether or not the MTR–reform does deliver a less intensive agricultural industry than would otherwise have been the case and whether populations will continue to believe that the cost of the CAP represents value for money must for the moment remain a matter of conjecture. At a more practical level, as Tangermann and Buckwell (1999) emphasise, the methodological difficulties in determining the effects of existing, let alone future, farm support policies are immense. They point out that the future will be shaped by many factors not directly influenced by farm support policies. This observation has direct relevance for the Foresight Futures framework's exploratory scenarios discussed in detail below. It reminds us that agricultural change will not take place in a vacuum and that changes in other sectors of the economy, in technology and in economic policy will all impact on agriculture's profitability, its use and conservation of natural resources, as well as its contribution to the rural economy.

The heart of the 2003 MTR–reform was the introduction of a completely, or pure, decoupled SFP. The European Commission initially intended that the SFP would be based on historical payments (Commission, *op cit*). However, EU Farm Ministers negotiated a Council Compromise so that the final agreement included the option for member states to calculate the SFP on the basis of either: an individual farm's historical support payments; as a regional flat rate payment; or as a mixture of the two, the so–called hybrid option. Moreover, member states also negotiated the option to link a proportion of the SFP to production. At the time of writing, member states are coming forward with their variants of the SFP. For example, England and Ireland have opted for full decoupling, whereas most other member states have retained an element of coupling, particularly in the beef sector. The important point, for the purpose of this paper, is that the MTR–reform has overseen a significant further shift towards denationalisation ie, effective national and regional control over the level and conditions attached to SFPs.

Understandably, Fischler (2002), in presenting the MTR–reform emphasised its positive aspects and in particular the benefits he believes will follow for a sustainable farming industry. However, sustainability has developed into a multi–dimensional concept – captured in the Community's description of multi–functionality – which embraces economic, ecological, environmental and social concepts. These dimensions of sustainability are to a large extent complementary, but they also involve trade–offs. In practice, there is no generally accepted definition of sustainability; the concept's definition depending on where the emphasis is placed (Pezzey, 1995). The absence of a clear definition means that in assessing potential future outcomes for agriculture, care must be taken when defining a particular outcome as sustainable. The definition that has achieved the widest acceptance is that put forward by the World Commission on Environment and Development, now more generally referred to as the Bruntland Commission; namely, . . . *sustainable development is development that meets the needs of the present without compromising the ability of future generations to meet their own needs* (WCED, 1987).

The Bruntland Commission's definition has important implications for exploratory scenarios of agricultural futures. Its vision of sustainability views global productivity growth as essential in order to deal with poverty and inequality, but in future it must be based on sustainable production techniques. Contrary to the Bruntland Commission's recognition of the need for continued productivity growth there is a widespread view when considering the sustainability of modern farming that the pursuit of ever higher levels of productivity are in conflict with sustainability. Many environmentalists view modern, highly productive farming techniques as being the cause of an array of ecological, environmental and social problems (see for example, Magdoff *et al* 2000). As noted above Fischler (2002) has argued that the introduction of a SFP would reduce the intensity of production and this is presumably the basis for his claim that the ecological costs associated with modern farming will be reduced by the MTR–reform.

It is a fact that within many developed nations contemporary farming practices have resulted in a range of adverse ecological effects associated with soil erosion, run off and diffuse pollution to water and air, calling for intervention to minimise or prevent these effects (eg, Environmental Agency 2002, Defra 2003). However, this is not in itself grounds for either expecting or encouraging less intensive production techniques and certainly not if less intensive production is interpreted as *turning back the clock* to a more labour intensive technology. What is certain is that a return to a more labour intensive technology would increase the relative price of food and if such a move also resulted in a slowing or a cessation of productivity growth the relative price of food would rise further over the longer–term (Brouwer, 2000).

In agriculture levels of productivity are commonly divided into two categories, 'intensive' or 'extensive'. Intensity relates to the volume of physical, biological and human resources, including units of livestock, that are combined with a unit of land for the purpose of production. Viewed in this way intensity is often equated with land productivity measured for example in terms of yields or cops per hectare. But an index of land productivity is not a measure of overall productivity nor is it a measure of sustainability once environmental and social factors are included. It does not make intuitive sense to single out just one measure of productivity as an indicator of sustainability, the more so as sustainability is a function of environmental stress rather than the level of output *per se*.

In measuring productivity of more relevance is an index of total factor productivity (*TFP*) which measures the ratio of output to all inputs used in its production. When projected over a period of years *TFP* will, or should, reflect the influence of technological advance and improvements in the quality of inputs. Unfortunately, *TFP* as conventionally measured does not take account of outputs such as damage to or enhancement of natural capital and hence the services provided by the ecosystem.

In principle, exploratory scenarios for agriculture should attempt to project total social factor productivity (*TSFP*) incorporating both market and non–market inputs and outputs. However, while a *TSFP* index is an intuitively appealing concept, measuring and valuing non–market inputs and outputs is a formidable exercise. Any attempt to value a non–market input or output must rest on a range of (frequently arbitrary) assumptions (Byerlee and Murgai, 2000). However, when environmental impacts are included, estimates of productivity are likely to be lower as a result (Barnes, 2002). In a similar vein, Pretty and Howes (1993) assemble convincing evidence from Britain to show that farming practices which integrate natural processes within farm production processes, thereby reducing external inputs and potentially enhancing environmental outcomes, can achieve similar levels of *TFP* and income to that of conventional farming and by implication, significantly higher *TSFP*.

Although the MTR–reform is designed to allow market forces greater influence over the pattern and volumes of agricultural output it also places reliance on regulations and conditionality to encourage codes of good agricultural practice ie, sustainable practices. Fore more than 20 years agriculture has been constrained by a succession of Directives designed to limit damage to the environment or pollution. In principle the reform is aligned with the EU's Environmental Liability Directive which after 15 years of discussion was finally agreed in April 2004 (Council, 2004). The Directive is based on the 'polluter pays principle' which implies the possibility in future of greater emphasis on direct action to limit the excessive use of inputs and farming practices judged injurious to the natural environment. By making the SFP conditional on fulfilling minimum standards regarding the environment, food safety, animal health and welfare standards the reform opens up the possibility of a much greater degree of direct, farm–level regulation and intervention than heretofore. However, by allowing regions of the Community to determine the conditions and sanctions (if any) standards will vary across the EU.

As regards rural development, the MTR–reform has embedded the concept of modulation; whereby, the SFP will be reduced by 5 percent by 2007 and the saved funds redistributed for rural development policies. Rural development has emerged as a formal component of the CAP – under its second pillar – though as currently proposed the purposes to which these new rural development funds will be put are limited. In the UK, rural development policy can be defined in terms of the government's statement: . . . *Our vision is of rural areas evolving in ways which enhance the landscape and biodiversity. It is of a forward looking and competitive farming industry delivery good stewardship of the environment as well as producing our food. It is of a rural economy based on information technology as well as on traditional skills. In short, not a theme park, but a living working countryside for real people. We want a countryside which can shape its own future, with its voice heard by government at all levels* (Defra, 2000). An interpretation of this vision statement that aligns with the objectives of Pillar–2, is political support for a long–term trend of reducing direct support for agriculture and increased funding for rural enterprises in general.

Applying the Foresight Futures framework

The MTR–reform could be viewed as restricting exploratory scenarios of EU agricultural futures to an industry in which its output prices are set by market forces and SFPs, that are subject to regulations and conditionality, are provided for income support. Such a view would however, be wrong. It is perhaps very difficult to conceive of a CAP reverting to open–ended price support given the fact that the EU now produces, or is capable of producing in excess of its demands for combinable crops, beef, sheep and milk. It is however, very likely that the post–MTR reform CAP will continue to develop and this gives rise to alternative futures that would lie within the Foresight Futures framework, possibility space set out in Figure 25.1 above.

The genesis of MTR–reform can be traced back to the 1992 MacSharry reform of the CAP. Although there had been some policy adjustments to the CAP during the 1980s – including the introduction of on–farm milk quotas in 1984 – it would generally be true to say that prior to the 1992 reform, open–ended price support and its attendant intervention buying, minimum import prices and export subsidies (ie, restitutions) defined the CAP. Building on Figure 25.1 we would place the pre–1992 – or 'old' – CAP as being represented by point A in Figure 25.2. By 1992 the CAP had developed from a set of national (prior to membership of the Community) agricultural policies to a Community wide policy focussed on the production of private outputs. The industry was protected from global competition by high tariff barriers and internal prices were held above their market clearing levels by the use of intervention purchases and other output restricting policies.

In most respects the recent MTR–reform can be represented by the shift from A to B in Figure 25.2. This represents a weighting of two distinct movements: a northerly shift towards globalisation/interdependence reflecting the prospect of a further opening of EU agricultural markets to global competition; and a westerly shift towards community/conservation by the encouragement of an increased output of environmental goods and also rural development under Pillar–2. The CAP's weighted north–westerly shift in the future possibilities space represents the belief that community concerns to protect and enhance natural capital outweigh the greater freedom of market forces to determine the private actions of farmers. In both cases the move is only partial. We await the conclusion of the Doha Development Round, but even a successful outcome is likely to involve lower levels of, rather than the removal of, tariff walls and export subsidies. As noted above, the introduction of 'cross–compliance' and modulation is not only subject to regional interpretation, but also it is not yet clear how efficiently the policy will be policed and the extent of sanctions.

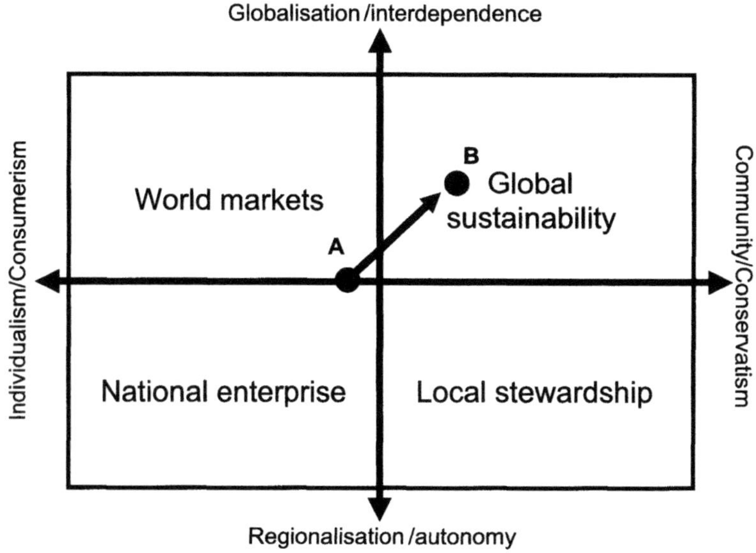

Figure 25.2 Futures Framework for EU Agriculture.

Point B also implies that the globalisation elements of the reform far outweigh the local community elements. That is, the right of regions to chose an historical, flat–rate or hybrid basis for the SFP and also to retain an element coupled to production – eg, the retention of suckler cow payments – can both be interpreted as a south–easterly shift. Allowing the authorities at a more local level to choose payment systems that better align with local priorities such as maintaining the maximum number of farms or avoiding land abandonment demonstrates the possibility that in the future this could become a more dominant feature of the CAP. Starting at point B – which represents the post MTR–reform within the Foresight Futures framework possible space – we can represent the four possible agricultural futures exploratory scenarios discussed by the workshop participants with the dotted lines shown in Figure 25.3.

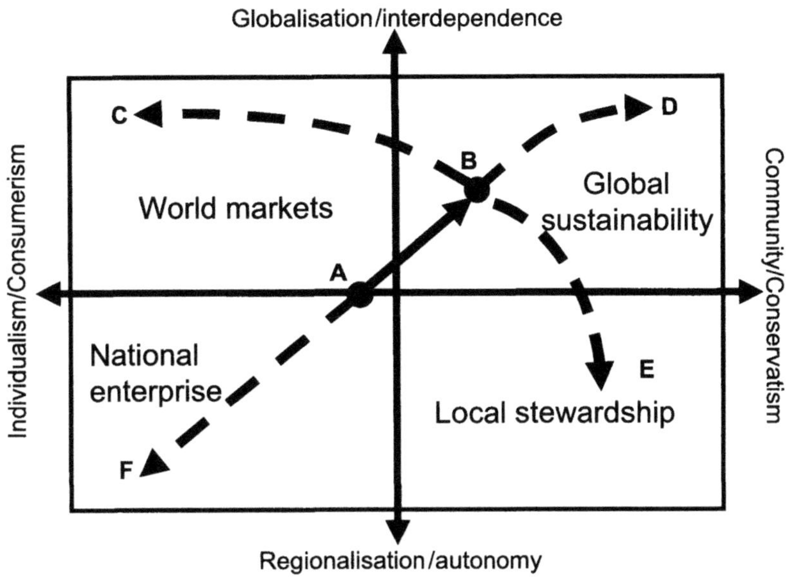

Figure 25.3 Four possible futures.

A north–easterly move towards C implies further economic and trade liberalisation. In the extreme this would involve the complete absence of any trade distorting instruments eg, tariffs, quotas and export subsidies, as well as the abandonment of any direct income payments. This 'hard' interpretation gives primacy to markets and private interests. The delivery of a sustainable environment would depend on the 'marketisation' of the environment ie, the pricing of environmental goods to reflect their value and the taxing or marketing of pollution permits to limit environmentally harmful inputs eg, nitrogenous fertilisers. Point C indicates that the marketisation of the environment is only partial as it would be subject to private values whose utilitarianism outweighs a concern for future generations. The outcome for the structure of farming, farming practices and indeed other non–land using rural industries would be determined on the basis of market forces – including free access to EU markets by the other world producers – implying an absence of a Community concern to influence change. This scenario would see farm–gate prices being set by world markets, implying generally low prices, and therefore reduced farm incomes per hectare.

A 'softer' interpretation of this scenario would be represented by a point to the east of C. Such a possible outcome would involve a free market in agricultural commodities, but the retention of some form of SFP for income support and the delivery of environmental goods and services. This 'softer' approach is not explored as a separate scenario and might more usefully be interpreted as lying somewhere between the possible future outcomes C and D in the global sustainability quadrant.

At point D, community responsibility towards a sustainable future outweighs the benefits of free markets. Thus, point D would be consistent with a WTO multilateral trading system whose rules and regulations have been adjusted so that internationally co–ordinated protection and enhancement of the environment takes precedence over free and unfettered trade. Under this scenario legislation and regulation would be set and enforced at the EU level to meet global level commitments under the auspices of the WTO or possibly a Global Environmental Organisation (GEO). This vision of the future would involve slower economic growth, but *TSFP* would be higher. This would be reflected in agriculture where the growth of output would be constrained by regulations and strict conditionality, limiting the freedom to adopt high yielding production practices. Some academics (see for example Marsh, 2001) have argued that a future possibility such as that represented by point D would be consistent with maintaining the multifunctional nature of the 'new' CAP. Farm–gate prices would be lower, but farm incomes would be supported according to the level of public funding for the SFP which in turn would deliver the multi–functional benefits desired by the EU. In effect all support would be provided under Pillar–2 funding. The assumption made by Marsh (2001) is that multi–functionality can be set at the level of the EU. It further presupposes that a sufficient level of funding would be available to offset the fall in farm incomes arising from lower commodity prices.

A third possible future outcome is represented by point E. Under this scenario if the EU continues to exist it does so as a loose association between European regions and the CAP has been largely, or completely, re–nationalised. Levels of support and any conditions attached to support, would be determined at a local level in order to meet the local community's priorities. This scenarios raises a number of issues. The first relates to the shielding of agriculture from world markets. This implies the weakening or the demise of the WTO, enabling the re–introduction of tariff barriers to protect regions from cheaper imports from third countries and other EU regions. A second issue concerns the method or predominant method of support in a particular region. A 'hard' version of this possible future could involve the re–introduction of price support at varying levels in different regions. Limiting the free movement of agricultural commodities within the Community – though presumably under this possible future restrictions would apply to a wide range of products – would also necessitate quotas to limit over–supply in surplus regions and bilateral trade agreements with deficit regions to balance aggregate supplies with demand. This 'hard' local price scenario also presupposes a willingness on the part of consumers in localities to pay different prices for agricultural products and to limit their demands for a wider range of foods grown in other parts of the world, if not the Community.

A 'softer' local stewardship scenario would envisage support only in the form of direct payments, but this still raises issues as to how viable such a system would be if accompanied by free trade within the Community. Different regions would require different per capita levels of funding according to the level of farm–gate prices, the conditions attached to the payments and the opportunities for diversification. In a loose association of EU regions it is doubtful if this could be achieved by a redistribution of CAP funds and this implies a local funding with adverse consequences for poorer regions. Apart from these funding and distributional issues a focus on local food production also implies the encouragement of mixed farming and the siting of food processors within regions. This has implications for not only agricultural production costs but also the processing and distribution costs of food. The solution to these issues would rest on an overwhelming concern to protect and enhance the multi–functionality of farming at a local level. Such a future outcome would envisage considerable autonomy at the regional level in both the level and the conditionality attached to SFP.

As illustrated, point E implies a local concern to support the environment and in this case in particular, characteristics associated with the region. But a point to the west of E might be suggestive of a greater community emphasis on local employment, smaller farms and rural businesses in general. Responsibility at a local level for rural development policies need not imply a focus on food production. Scenarios in this area of the possibility space, would be consistent with agriculture *per se* losing its statues as *primus inter pares* amongst rural businesses. At a local level non–food producing rural businesses might be judged to offer a greater prospect of employment growth as well as alternative revenue generating land uses.

Finally, there is the possible future outcome represented by point F. In the context of the CAP this would represent a reversion to something akin to UK agricultural policy prior to membership of the European Community. Support mechanisms and levels would be set at a national level. A 'hard' variant of this scenario would see the EU revoking its membership of the WTO – or the demise of the WTO – and the process of EU integration would be reversed to the status of a trading bloc which, depending on the relationship agreed, might even involve the reintroduction of border protection at national levels for some or all agricultural produce. The restriction of global trade means that economic development grows more slowly under this scenario. The outcome of this scenario in a developed nation such as the UK would in part depend on how it came about. If it was a reaction to a breakdown in the global trading system, the policy objects at the national level might be to limit the damage to trade by seeking multiple bilateral trade agreements. Alternatively, it could be the result of national concerns within the EU concern to support farm incomes. If this social concern was the driver then behind border protection, price support could be reintroduced but it would be unlikely to be open–ended as this would increase the costs of storing or subsidising the export of surpluses. On farm quotas would sit within this scenario with the effect that consumers – presumably willingly – would be prepared to pay the higher food prices that would follow.

A 'soft' variation of a national enterprise future scenario would involve little, if any, price–support but greater variation between countries in the level and conditions attached to SFPs. With national farming industries becoming the focus of policy, the level of SFPs would vary according to a member state's ability and willingness to pay. Consumers' values are such that despite an emphasis on private values they would be prepared to meet the costs of preserving a national farm structure through their taxes. However, given the dominance of private values, people would presumably not be so demanding as to require the production of environmental goods and the enhancement of natural capital so as to achieve sustainability. The policy would also need continual increases in funding. The history of the CAP shows that levels of support are largely absorbed in land values and the prices of farm inputs with the effect that incomes fall in real terms and are ultimately unable to prevent a drift from the land. Unilateral withdrawal from the WTO would almost certainly provoke retaliation. This rather suggests that such a scenario might conceivably be more of a reaction to what the rest of the world is doing. If other nations adopt a more protectionist stance then the future implied by Point F becomes a more serious possibility. But a more protectionist future would also imply slower economic growth, higher unemployment and therefore the likelihood of less funding for agriculture.

It is worth observing however, that Figure 25.3 captures rather neatly the opportunity costs – indeed the conflicts – of achieving any one of the four scenarios. Free trade appears to threaten local control over the levels and objectives of support and conservation threatens reduced rates of economic growth and the private choices of individuals. Another intriguing possibility is that the CAP could simultaneously follow the scenarios represented by points C, D and E. Following categorisation of three principle farm types (Marsh, 2001), scenario C might represent the future for commercial farms, scenario D the future for farms who focus on the production of public goods and E the future for part–time farms who earn their living in other sectors of the economy.

Workshop results

As noted above, the key to a successful possible futures scenario workshop is to fully engage stakeholders in the process. This section reports the outcome of the Cranfield agricultural futures scenario workshop as regards the implications of the four alternative exploratory scenarios outlined in the previous section at the sector or industry level. These sector level outcomes are the responses to the forces set up by different weightings of the two dimensions that define the Foresight Futures framework, possibility space. In turn, these responses will give rise to farm level and supply chain developments. As noted previously, a sister paper reports the agricultural futures scenarios workshop outcomes for farmer practices, productivity and yield.

WORLD MARKETS FOR AGRICULTURE

The workshop viewed this scenario as a high risk outcome for farmers as competition from low cost producers located in other parts of the world pushed farm–gate prices to low levels seriously reducing farming's terms–of–trade. The accompanying phasing–out of SFPs implies that this agricultural future would represent the extreme of economic pressures which in turn would provoke and incentivise actions from farmers to ease their financial difficulties. The workshop participants thought that the economic pressures would result in a rapid polarisation of farming into commercial and 'lifestyle' or 'hobby farms'. Commercial farms would seek to earn a living from the production of agricultural products by economies of scale, high levels of productivity, super efficiency and the capture of some of the downstream value added. Commercial farms would utilise technology and machines to minimise on hired labour and they would be managed by highly skilled individuals. Lifestyle or hobby farmers would earn the bulk of their incomes from some other activity. Lifestyle/hobby farms would be very small producing as a group a very small share of the total agricultural output, and despite accounting for the majority of farms would farm only a small proportion of the total agricultural area. Lower yielding land in remoter areas would be abandoned by farmers even though its value and rents would decline markedly.

The implications of free trade in foodstuffs would spread beyond farming. The whole food chain would find itself facing heightened competition and new sources of supply for raw materials. A key issue for the workshop participants was to what extent this might result in the development of new trading relationships within food chains to improve efficiencies and also to counter the intense competition. In addition to increasing scale, a complementary offset to falling net–income per hectare is for farmers to seek to capture some of the value created downstream. One way this might be helped is by the adoption of GM technology which the workshop considered would become widespread. In the future this technology may facilitate the production of highly specific agricultural products eg, to counter a particular medical problem, and the wide adoption of GM foods would give rise to the creation of a large number of higher value niche markets. Under this scenario organically produced food would remain a niche market serving only the purpose of extending choice to consumers. Another, or complementary approach would be a new emphasis on the formation of farmer controlled businesses (FCB). The advantage of well managed FCBs is that they increase the influence and financial power of farmers within the food chain. A possible strategy for coping with unfettered free trade would be for farmers to utilise the power of FCBs to enter into joint ventures and alliances with processors or even to invest in their own processing, distribution and retail activities.

The main beneficiaries of this exploratory scenario, according to the workshop, would be consumers. Consumers' affluence was judged to continue rising at its historical rate and this – together with the utilitarianism implied by this scenario – would significantly increase the demands for new food and taste experiences. Not only would the satisfaction of these demands be one of benefits arising from competition, but also competition would deliver improved quality, wider choice and low prices. The dominance of individualism and consumerism would be reflected in high demands for convenience foods and eating out. In such a world, brands would take on the credence role of safety and assurance as well as protection of the environment though in the latter case workshop participants thought standards would be minimal. The workshop also considered the possible outcome under this scenario for diets. The *laissez–faire* approach that underpins this scenario implies minimal government intervention to regulate diets or control advertising. There would be trending towards global diets with an emphasis on standardised, high specification produce with strong convenience elements. Consumer acceptance of GM technology would create opportunities for new functionally engineered foods and this could provide an offset to obesity and diet–related illnesses.

The outcome for the protection and enhancement of the environment would, the participants judged, suffer under this scenario. As noted above, the domination of market forces would result in the 'marketisation' of the environment. For farms located in scenic locations this would offer scope to generate a new income stream. Taxes and permits on environmentally harmful inputs would have the effect of more than offsetting the lower prices implied by global competition for these inputs and consequently they would incentivise precision farming and impart a premium for more productive land. As this scenario would generally involve lower land prices and rents there would also be opportunities for extensive, large scale farming systems eg, beef. Marginal land would be taken out of farming, especially in remote areas and some, perhaps all of it, taken into other commercial activities eg, recreational pursuits. Although there would be no annual income payments for farms, public support for the provision of specific public goods eg, the creation of wildlife habitats, is likely though participants considered that the prevailing self–interest of consumerism would result in this provision being minimal. This vision of the future would see only a limited role for legislation and regulation to protect and enhance natural capital. As regards rural development, this again would be subject largely to market forces. Rising affluence would see a continued rise in rural households – including lifestyle/hobby farms – but the location and growth of rural diversification opportunities and hence employment would be subject to market forces. These possible outcomes are summarised in Figure 25.4.

GLOBAL SUSTAINABILITY AND AGRICULTURE

Workshop participants essentially viewed this possible future as a continuation

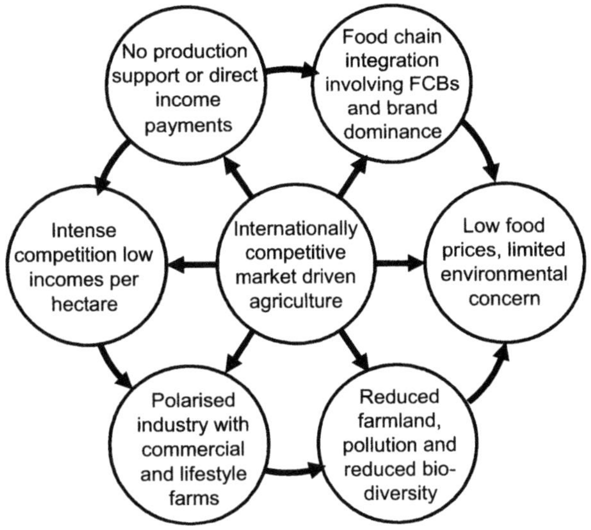

Figure 25.4 World markets for agriculture.

of the two trends prominent in the MTR–reform; namely, full engagement by the EU with the WTO to achieve integrated global governance of agricultural trade to ensure environmental production; and the use of an EU co–ordinated SFP to deliver environmental and social objectives. This outcome is set in the context of a WTO, possibly in partnership with a GEO that has elevated sustainability to a condition for free trade. Hence, not only the products traded but also their production processes would be required to meet protocols regarding environmental protection, animal welfare and ethical standards. Such a world would be characterised by a plethora of non–tariff barriers (NTB) preventing trade in products failing to meet globally set minimum standards. These global standards would be reinforced at the EU level with a comprehensive suite of conditions and regulations underpinning SFPs in support of sustainable farming practices. Participants considered that regulations would spread to placing limits on the use of chemical inputs and the conversion of permanent grasses and woodlands to arable crops. A concern to maximise carbon sequestration would result in controls on tillage and other farming practices. In short a vision of a highly regulated EU agricultural industry engaged in free trade, but balanced to achieve social and sustainability objectives.

The implications of this vision for farming would be continued competitive pressures arising from overseas competition, but these forces would be tempered by NTBs and the need for exporters to meet high environmental and animal welfare standards. The ascendancy of sustainability would increase the growth of productivity measured by a *TSFP* index, but this would be accompanied by a slower rate of growth in global production of foodstuffs. Although population growth is not explicitly included in the Foresight Futures framework it is relevant here to observe that a combination of a slow down in the growth of agricultural output and a global population heading towards 9 billion by 2050 is likely to tighten world commodity prices. Within the EU, this scenario implicitly assumes that budgetary constraints have been reduced and sufficient funds are forthcoming to encourage farmers to adopt a high commitment to conservation at some cost to agricultural productivity and efficiency as traditionally measured. The greater weight attached by society to a *TSFP* index would justify higher food prices, though farmers would find that high standards of conservation become the minimum standards for the commercial sale of foodstuffs.

The trend towards the consolidation of production on larger scale farmers and the growth of lifestyle–hobby farms would continue, but at a slower pace than envisaged under the World Markets scenario as the incomes of smaller scale full–time farms would be under less pressure. This follows from the continued existence, indeed the possible enhancement of the SFP, together with higher product prices, but larger scale farms would continue to benefit from economies of scale and higher farm incomes would be reflected in land prices and other costs. Indeed, this scenario could see increased interest in large scale as the option of very high levels of output per hectare would be rendered uneconomical by SFP conditionality, regulation and legislation. The effect would be an increased demand for the services of land and hence even greater upward prices on land prices. Participants believed that the heightened importance of sustainability would offer more economically viable opportunities for diversification and multi–functional activities, though the affluence driven demand for lifestyle farming would also be as strong under this scenario.

The emphasis on a competitive but sustainable farming industry would be reflected both upstream and downstream in the food chain. As noted above, under this scenario commercial consideration would ensure positive reactions to consumers' concerns about production processes. Supply chain assurance and traceability demands would encourage closer supply chain relationships discussed under World Markets scenario. Arguably here the price and income pressures on farmers to co–operate would not be as great, but marketing and competitive pressures to achieve and demonstrate the highest standards of care and protection would be highly influential. Thus, FCBs are also likely to play an important role under this scenario in part to capture downstream value, but also here to provide the food chain and consumers with the assurance and traceability associated with sustainability.

As trade and competition would be subject to global regulation and productivity growth (as conventionally measured) and output growth would be slower, consumer prices would be higher than they would be if the future was one of unfettered trade. This however, may not be viewed as a lowering in potential living standards. Participants judged that consumers would attach high value to the assurance that their food prices incorporated high environmental and social standards. They also believed that associated with these values would be a greater concern for healthy balanced diets, though there is no reason to support any diminution in consumer demands for new food and taste experiences. Participants did not consider that under this future scenario health–conscious consumers would eat less meat. Acceptance of GM foods would be dependent upon a demonstration of dietary or environmental benefits – including non–food crops – and again, this would offer new opportunities for value creating market segments and niche products. Participants thought organic production might also benefit from official encouragement under environmental and/or diversification options though it is unclear how commercially attractive organic output would be in an industry where sustainable farm practices are the norm.

As this exploratory scenario is predicated on the delivery of a more sustainable agricultural industry it would deliver greater protection to natural capital, improved biodiversity and reduced pollution. Land abandonment would be unlikely as firstly low input–low output systems – particularly on marginal lands – would be integral to the conditionality attached to SFPs and secondly alternative sustainable uses for land retired from agricultural production would attract public funding. Integrated farming systems promoted by payments for adhering to codes of good farming practice and compliance would result in lower use of agro–chemicals, thought this development would encourage technological improvements in precision farming. Under this scenario the area of farm land would decline only in line with demands for homes and other regulated industrial activity. Within the constraints commercial farms would need to strive for high levels of efficiency and competitiveness. Low yielding cereals land would return to grassland where extensive livestock systems would absorb the additional supply. And farms generally would devote more resources to the protection and enhancement of their natural capital. As noted above, GMOs would be accepted only if an environmental or dietary benefit is demonstrated, but this might amount to an important proportion of production in 30 years time. A summary of the key outcomes of this exploratory scenario, as expressed by the workshop participants are set out in Figure 25.5.

LOCAL STEWARDSHIP AND AGRICULTURE

It was agree by participants that this possible future would involve the demise of the CAP and a reversion to national, and for larger countries regional support regimes focussed on local needs and priorities. This scenario would also involve the demise of the WTO or its replacement by a GEO. Alternatively, this scenario would be consistent with the withdrawal of the EU from the WTO. Within the EU, such a policy could be the result of the development of a loose association of an 'EU of the Regions' in which considerable weight was attached to protecting and enhancing social, culture and environmental diversity. The concept of a single market and the emphasis on economic growth would no longer be the priority and economic policy in general would sacrifice economic growth for a greater emphasis on equity, social inclusion and democratic values. Participants believed that the encouragement of these values would result in a parallel emphasis on environmental protection and the prudent use of natural resources.

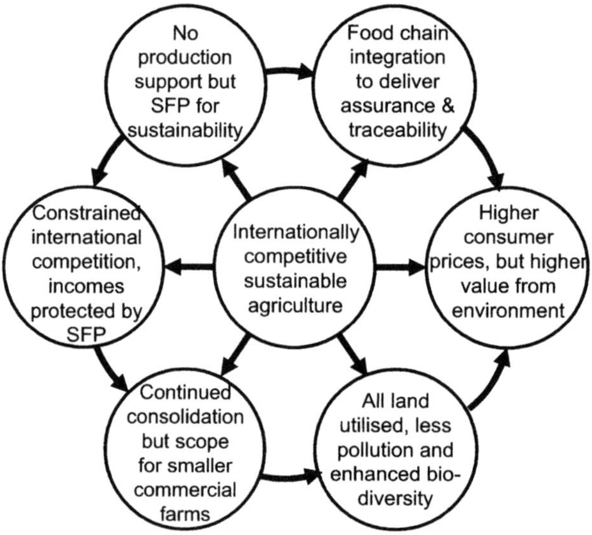

Figure 25.5 Global sustainability and agriculture.

The very essence of this possible future would be local procurement and self–sufficiency though in practice as regards agriculture this would be constrained by climate, geography and cost. Changed consumer values and import controls would result in a much reduced volume of imports of exotic and out–of–season fruit and vegetables: in keeping with a general desire to minimise 'food–miles'. However, the change in values envisaged under this scenario would also have implications for intra–community trade. This scenario implies increasing self–sufficiency in deficit regions of the EU and enforced cut backs for surplus producing specialised regions of the EU. The demise of the WTO and the abandonment of an EU single market would, in principle, provide national or regional authorities with the scope to re–introduce price–support. In deficit regions this scenario would see farm–gate prices rising in the initial stages but eventually these regions would also suffer a declining terms–of–trade once self sufficiency was achieved. Alternatively, and perhaps more likely, regional authorities would seek to achieve their agricultural objectives by setting the levels of and conditions attached to SFPs. Self sufficient and more limited trade opportunities imply the imposition of quantitative restrictions on farms in some regions and the encouragement of mixed farms in all regions. The outcome would be higher cost production and behind high levels of protection, less emphasis on improving efficiency and seeking new commercial opportunities.

Participants were clear that the outcome of such a scenario for farming would influence the farm structure and pattern of output. If a small nation or region was primarily concerned to protect the structure of its farming industry then, subject to the cost being acceptable to local taxpayers and/or local food consumers, a local community scenario could conceivably arrest the decline in farm numbers in a specific region, thereby protecting farm structure and farm based employment. Participants considered this likely outcome would be reinforced as a focus on a regional economy would also offer farmers the greatest likelihood of supplementary off–farm income sources. However a predominately farming region would have to rely largely on consumers in its region paying the higher prices set for their output and/or their taxpayers providing the funds for farm payments. It is a further consideration is that a regional focus would not be confined to agriculture and other business would also demand help from the authorities thereby reducing the extent to which agriculture would be treated as a special case.

Viewed at the level of the loose association of EU regions, agricultural output and *TFP* would be lower under this scenario. In part this is because the benefits of specialisation would be curtailed and technological and management developments would be constrained by a greater reliance on regional resources and a reduced incentive to investigate new possibilities. In contrast a local emphasis attaching greater weight to environmental and local community objectives would result in improved *TSFP* growth. It would not be possible for all regions to achieve self–sufficiency; climate, geography and urbanisation would consign some regions to enter into bilateral trade agreements with surplus regions. But the desire for regional self–sufficiency and efforts to minimise the transporting of food suggests a tension in some regions if restrictions to protect and enhance the environment lowers output. In deficit regions farmers could be encouraged, via official support to increase output, particularly if it was judged beneficial to a local food supply chain. Participants saw a future in which 'local' brands dominated local food markets and the re–emergence of regional processors and supply chains. As local supply chains would be dependent on local producer –processor–retailer connectivity, the scenario also sees an enhanced role for farmer co–operation – horizontally in FCBs and vertically with their customers. Participants believed that direct sales would increase and in general sustainability would improve resulting from lower levels of output per hectare and regional controls on pollution.

Consumers would pay higher prices for food and they would necessarily suffer a reduced choice. The major multiples would lose market share to regional outlets, including farmers markets. In response to the changing values of consumers, sales of processed and convenience foods would decline as sales of fresh produce increased. Participants thought that this would be acceptable to consumers who attach high value to the protection of a regional farming industry and a preference for local foods, particularly as this would be viewed as providing assurance on safety, nutrition, as well as care and protection for the environment. Participants saw the opportunity costs of these benefits for consumers as being lower product specification, reduced variety and non–seasonal foods. They thought that a generally greater emphasis on healthy eating would benefit producers of fruit and vegetables and GM food would find few markets.

Subject to the caveats entered above, the workshop saw the commitment to regional production as co–existing with a commitment to sustainable production; that is, a strong conservation and community ethic. The overall reduction in productivity would result in the agricultural area increasing in deficit regions though the use of land for non–farming purpose could increase in regions capable of producing excess supplies or high local demands for non–farming activities eg, tourism. An emphasis on local production would involve a reversal of the trend towards specialisation and the re–emergence of mixed farming in some regions viewed by the participants as being, on balance, a more sustainable form of agriculture. The participants saw a commitment to the community as being synonymous with greater care for natural capital, an enhanced biodiversity and reduce pollution. The implication of this scenario, as expressed by the workshop's participants are summarised in Figure 25.6.

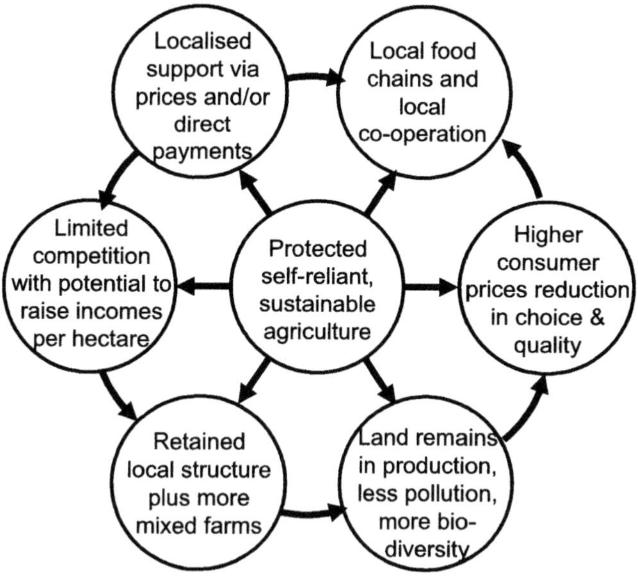

Figure 25.6 Local stewardship and agriculture.

NATIONAL ENTERPRISE AND AGRICULTURE

This scenario also envisions a possible future involving the demise of the WTO and a reversion, behind tariff protection to a re–nationalised form of CAP, whereby national governments, operating within a broad framework, determined the nature of agricultural support, a situation that could involve the re–introduction of support prices, alongside a range of targeted direct payments and production control instruments. Participants believed that the objective of this policy would be self–sufficiency in food at the level of the European Community, but where possible – and subject to a minimal risk of multilateral retaliation – surpluses would exported with the aid of subsidies. Food imports would be restricted to non–indigenous produce or to meet seasonal shortages. The consensus of the workshop was that this scenario would involve limited export opportunities and therefore the CAP would have to resort to output and input quotas where the level of production and its potential rate of growth was in excess of demand. The need for food production quotas would be reduced by increased scope for agro–industrial and bio–energy crops to meet national needs, possibly enhanced by GM technology.

Participants took the view that the effect of a reversion to a pre–CAP support framework would see the bulk of support provided via direct payments. The effect of tariff protection and supply control policies would allow farm–gate prices, and returns in general, to be higher under this scenario than envisaged under the world markets scenario. However, reduced exposure to international competition is likely to lead to higher prices than under the World Markets scenario, such that pressure on farm incomes would persist resulting in continued structural change, albeit at a slower rate. Although production would be steadily concentrated on larger sale farms, medium sized family farms would retain a high degree of financial viability due to high levels of support. In terms of the structure of farming, the polarisation discussed above would be evident, but to a more limited extent. Land that was freed due to quota restrictions would be absorbed by part–time lifestyle farmers. The workshop took the view that under this scenario input intensive farming would be the norm though there would be a need for regulation and legislation to control pollution and environmental damage. Participants also doubted if such an inward looking policy would encourage sufficiently high levels of Research and Development and farm level investment to maintain the rate of growth of *TFP*.

This scenario also has implications for farming's relationship with the food chain. Continued protection for farmers would limit the incentive to enter into either horizontal or vertical collaborative business ventures. However, participants considered that food safety considerations would result in the widespread adoption of traceability and quality assurance schemes. They also took the view that diets would not change markedly, indigenous foods would dominate with limited global influence. Trends towards processed foods and convenience would continue including passive acceptance of GM foods. The consensus was that the outcome would be high quality, reliable suppliers of traditional foods at relatively high but affordable prices. And, in terms of the environment, animal welfare and rural society consumers would expect minimum standards to be secured, but there would be less concern to bring about marked improvements.

As regards sustainability, this scenario sees the emphasis as remaining with commercial production utilising technology and scale to lower unit costs. Yields would continue to rise reducing the area of land needed to meet essentially national demands for food. Although farming in relatively disadvantaged areas would draw benefit from generally higher levels of support, given the emphasis on self–sufficiency, there would be little concern if areas of more remote, less productive farm land were abandoned. Care and protection of natural capital and ecosystem services will be set largely by levels of regulation and legislation, together with modestly funded schemes. The countryside, or more correctly the land area, would remain overwhelmingly agricultural, though less so on the uplands and marginal land. The rural economy would retain an agricultural focus even though farming will account for a diminishing share of rural economic activity and employment. The Workshop discussion of this scenario is summarised in Figure 25.7.

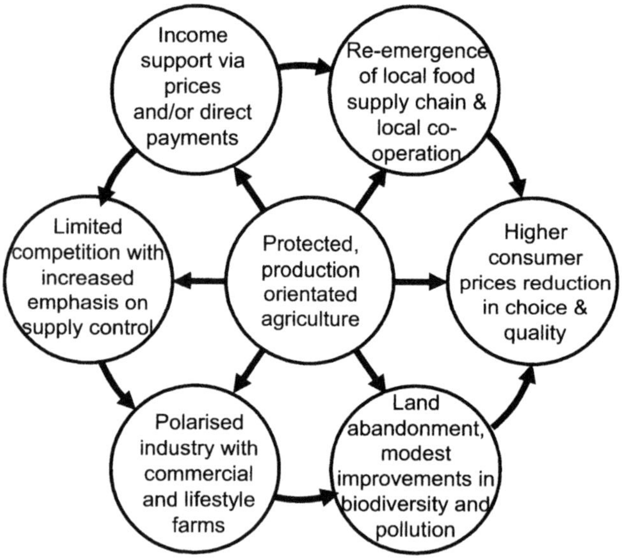

Figure 25.7 National enterprise and agriculture.

Concluding observations

Although the four exploratory scenarios outlined above are driven by different social values and systems of governance, they point very clearly to a fundamental trade–off. Economic liberalisation is likely to deliver higher levels of productivity, as conventionally measured, increased levels of output, wider choice and lower prices. However, if a *TFP* index was replaced by a *TSFP* index economic liberalisation would probably be accompanied by a declining index; a situation that at some point will adversely influence the quality of life possibly provoking a change in values and governance. In contrast, a greater emphasis on sustainability is likely to deliver the mirror image of economic liberalisation and would only be possible if people viewed a protected and enhanced environment as sufficient compensation for a lower rate of growth of material wealth, perhaps even some reversal.

As regards farming, the key issues raised by the foregoing are the definition of efficiency and productivity, and the mechanisms for delivering the highest possible levels of both. The role of markets and unfettered trade is to drive efficiency improvements, new ways of delivering higher productivity and also new products that meet the consumers' demands for value, quality and choice. In principle, markets and competition can delivery sustainability, but only if accurate market values are attached to environmental outcomes and those who contribute positive and negative outcomes are rewarded or penalised respectively. It would be true to say that generally the workshop participants believed the emphasis for the delivery of sustainability would remain with regulation and legislation.

Apart from the World Markets scenario, workshop participants foresaw an agricultural future with greater emphasis on the environment and rural development. In the case of the Global Sustainability and Local Stewardship scenarios agriculture would reflect the positive commitment and strategic interventions to protect and enhance natural resources and the services of the eco–system. Despite resort to regulation and legislation, such policies would require funding at the farm level and the reduced emphasis on productivity as conventionally measured would result in food prices being higher than would otherwise be the case. As regards rural development, a key issue is the scope and need for farm level diversification. Under the National Enterprise scenario, participants considered there would be less need for diversification and under the World Markets scenario diversification would depend on profitable market opportunities. The other two scenarios, in the view of participants, would ensure that farming provided its prominent multi–functional role within the rural economy even though its contribution as a proportion of economic activity and employment is unlikely to grow.

Acknowledgement

The assistance of participants of the workshop on Agricultural Futures held at Silsoe, September 25th 2003 is gratefully acknowledged (see Appendix), along with the support provided by Defra.

References

Barnes, A.P. (2002) Publicly-funded UK agricultural R&D and 'social' total factor productivity. *Agricultural Economics*, **27:** 65-74

Berhout, F, Eames, M and Skea, J (1998) *Environmental Futures Scoping Study*, SPRU, Brighton

Berhout and Hertin (2002) *Foresight Futures Scenarios: Developing and Applying a Participative Strategic Planning Tool*, SPRU, Brighton

Brouwer, F (2002) *Effects of Agricultural Policies and Practices on the Environment*: Review of the Empirical Work in OECD Countries, OECD, Directorate for Food, Agriculture and Fisheries

Byerlee, D and Murgai, R (2001) *Sense and Sustainability Revisited: the Limits of Total Factor Productivity Measures of Sustainable Agricultural Systems*, Agricultural Economics **26:** 227–236

Commission (1997) *Agenda 2000: For a Stronger and Wider Europe*, Com, Brussels

Commission (2003) *Explanatory Memorandum: A Long–Term Policy Perspective for Sustainable Agriculture*, COM (2003) 23 final

Council (2004) *Council Directive 2004/35/EC* of 21 April

Defra (2000) *Our Countryside: The Future*. Department for Environment, Food and Rural Affairs

Defra. (2003) *Diffuse Water Pollution from Agriculture*: The Government's Strategic Review: Stakeholder Discussion Paper. Department for Environment, Food and Rural Affairs

DTI (2002) *UK Foresight Futures 2020: Revised Scenarios and Guidance*, DET, London

Environment Agency (2002) *Agriculture and Natural Resources: Benefits, Costs and Potential Solutions.* Environment Agency , Bristol

Fischler, F (2002) Agricultural Policy for the Future: A Synthesis of Competency Concerns, *Euro Choices*, Vol 1, 2

Magdoff, F, Foster, J and Buttel, F editors (2000) Hungry for Profit: The Agribusiness Threat to Farmers, Food and the Environment, *Monthly Review Press*, New York

Marsh, J (2001) *Agriculture in the UK – its Role and Challenges*, Food Chain and Crops Industry Panel

Nelson, R and Winter, S (1982) *An Evolutionary Theory of Economic Change*, Belknap Press/Harvard University, Press, Cambridge MA/London

Pezzy, J (1997), Sustainability Constraints Versus 'Optimality' Versus Intertemporal Concerns and Axioms Versus Data, *Land Economics* **73(4):** 448

Pretty, J.N. and House, J. (1993). *Sustainable Agriculture in Britain: Recent Achievements and the New Policy Challenge.* IIED, London.

Reiger, E (2000) *The Common Agricultural Policy*, in H Wallace and W Wallace eds *Policy Making in the European Union,* 4th ed, Oxford

Tangermann, S and Buckwell, A (1999), *The Purpose and Methodology of Evaluation in Regard to EU Agricultural Expenditure*, working paper BUDG–102 EW Directorate–General for Research, European Parliament, Luxembourg

WCED (1987), *Our Common Future,* World Commission on Environment and Development, Oxford University Press

APPENDIX

Workshop, Cranfield University, Silsoe, Bedfordshire
25 September 2003

List of Participants

Name	*Organisation*
Mr Philip Andrews	Defra
Mr Eric Audsley	Silsoe Research Institute
Dr Alison Bailey	University of Reading
Mr Brian Bibby	Defra
Dr Paul Burgess	Cranfield University
Ms Saffron Clackson	Defra
Mr Andy Dalziel	The Macaulay Institute
Mr Andrew Dart	Defra
Alastair Dickie	Home-Grown Cereals Authority
Dr Pinder Gill	Meat & Livestock Commission
Professor Jim Harris	Cranfield University
Dr Ian Holman	Cranfield University
Mr Robert Kynaston	Farmer
Mr Henry Leveson-Gower	Environment Agency
Professor Sir John Marsh	University of Reading
Mr Archie Montgomery	Farmer
Professor Joe Morris	Cranfield University
Mr Charles Neame	Cranfield University
Ms Tanya Olmeda-Hodge	Country Land & Business Association
Ms Kerry Pearn	Silsoe Research Institute
Mr Sean Rickard	Cranfield University
Mike Roper	Defra
Dr Mike Segal	Defra
Ms Kirsty Shaw	Countryside Agency
Ms Carmen Suarez	National Farmers Union
Dr Alison Vipond	Defra
Dr Keith Weatherhead	Cranfield University
Professor Paul Webster	Imperial College at Wye, London University
Professor Julian Wiseman	University of Nottingham
Dr Iain Wright	The Macaulay Institute

26

FUTURE DIRECTIONS FOR PRODUCTIVITY IN UK AGRICULTURE

JOE MORRIS*, ERIC AUDSLEY**, KERRY PEARN**, SEAN RICKARD+
** Institute of Water and Development, Cranfield University, Silsoe, Bedford, MK45 4DT, UK; **Silsoe Research Institute, Wresj Park, Silsoe, Bedford, MK45 4DT, UK; + Cranfield School of Management, Cranfield University, Cranfield, Beds, MK43 0AL, UK*

Introduction

Agriculture in the UK and Europe as a whole is experiencing unprecedented change as the policy drivers which have hitherto shaped the characteristics of the farming sector are realigned. Dependency on high levels of taxpayer support, concerns about the negative environmental impacts of intensive farming, the growing dominance of agri-business and the demise of the family farm have questioned the long term sustainability of modern farming systems. Further side swipes in the form of BSE and Foot and Mouth Disease have, it is said, damaged consumer confidence in the ability of the farming sector to deliver safe food.

In this context, the relevance of conventional measures of success to assess the productivity and performance of UK farming has been challenged. Although increasingly intensive, specialised, mechanised farming has been the dominant trend for the last 50 years or so, a mix of external and internal drivers and pressures could produce very different futures for farmers, with consequences for farming systems, technology and productivity.

Following an overview of the characteristics of agricultural systems, this paper explores definitions of agricultural productivity, and reports on recent trends in key indicators. Reference is made to the need to include environmental outcomes in the assessment of productivity. The paper goes on to identify the drivers and pressures which determine the 'state' of agriculture. These aspects are further developed through the application of scenario analysis to determine possible futures and implications for productivity. Some preliminary results are generated from modelling selected farming systems which provide insights into conventional and broader environmental measures of productivity. The paper draws on work conducted under a Defra funded project: Agricultural Futures and Implications for the Environment Project.[1]

[1] Agricultural Futures and Implications for the Environment: Defra Project IS 0209. Cranfield University at Silsoe, Silsoe Research Institute and Macaulay Research Institute

Characteristics and context of farming systems

Figure 26.1 describes agricultural systems in terms of the processes by which inputs are transformed into outputs, where the latter comprise traded and priced commodities such as crop and livestock commodities, and untraded, unpriced environmental outcomes such as landscape (positive) and agrochemical pollution of surface waters (negative). Agricultural systems are set in a broad social, economic and political framework that, along with the endowment of natural resources and climate, define the outputs required and the resources available (Dalton, 1975).

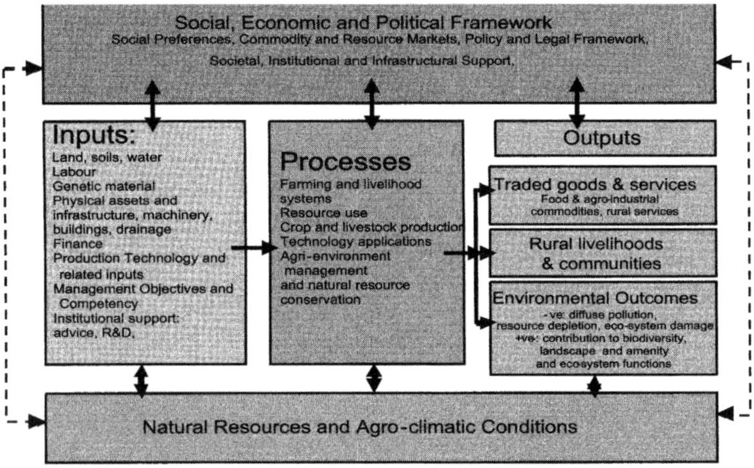

Figure 26.1 Components of agricultural systems.

Productivity in agriculture implies some physical or value-based measure of the relationship between crop and livestock outputs, and inputs such as seeds, agro-chemicals, labour, machinery and land. The greater the ratio between outputs and inputs, the greater is the perceived productivity. Increased productivity implies greater efficiency, whether this results in greater outputs for the same inputs, or the same outputs for reduced inputs.

Agricultural productivity is usually associated with the development and application of production technologies which contain the indigenous or scientific knowledge-based products, processes, tools and equipment by which inputs are converted into outputs. These technologies, combined with the inherent properties of soils and climate in a given geographical location, determine the physical possibilities and limits of production systems: the technical frontier. Improvements in productivity may be achieved by improved use of existing technologies; that is by working within the existing technical frontier, for example by exploiting economies of scale. Alternatively, technology change can extend the technical frontier by enhancing output for all previous levels of input, such as for example when a new disease resistance crop variety increases yields without a change in other inputs. (In much of the economics literature, the terms technology change and technical change are used to imply the same meaning; i.e. a change in the set of available technologies).

Although productivity has tended to focus on 'tradeable' ouputs, greater attention has been paid more recently to accounting for environmental outcomes, as well the contribution of agriculture to sustainable livelihoods and communities (Defra, 2002a, b; Pretty, 1998). These aspects are captured under the mantle of sustainable agriculture as discussed later.

There is a hierarchy of exploitation of potential crop productivity as shown in Figure 26.2 using the example of take-up of potential crop yield. The maximum technological potential is set by the bio-physical limits of the crop type such as those defined by photosynthesis. The existing technological boundary of productivity is evident in the trials of 'researchers'. There is often a small but recognisable gap between the yields of research trials and those achieved by the best farm practitioners which can be attributable to a 'research attention' bias. The best farmers represent the technical frontier, fully exploiting available technologies to maximise productivity. They can be regarded as the most technically 'efficient', against which other 'less efficient' farms can be compared.

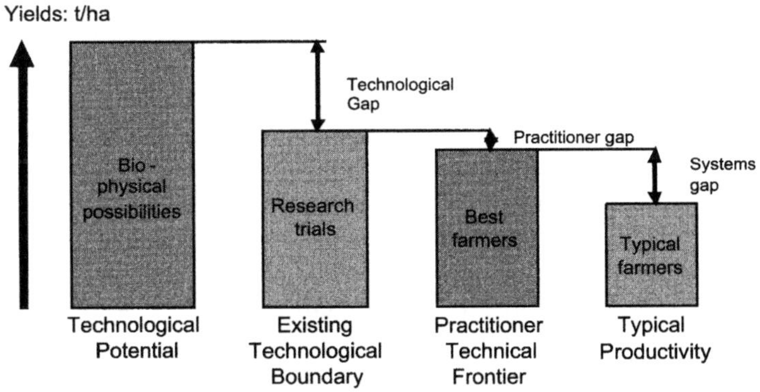

Figure 26.2 Exploitation of technological potential.

For a variety of reasons farmers may not be able, or may choose not to take up the potential benefits of well established and new technologies: there is often a considerable 'systems gap' between the productivity of the most efficient farm (or group of farms) and that of the mean or typical farm. This gap may due to a range of economic, social and environmental factors associated for example with farmer knowledge, motivation, resources, and perceptions of the potential advantage and suitability of the technology. For technologies to be adopted, they must be perceived to be relevant by, and therefore suited to the needs of the adopter community. It is often the case that when farmers are under pressure, the systems gap widens, as illustrated later. Although Figure 26.2 refers to yield in t/ha, the principle can be extended to incorporate a broader definition of productivity which compares the ratio of all outputs and all inputs used by a given farm unit.

Measuring agricultural productivity

There are various measures of productivity according to the purpose of the assessment.

- Physical scientists tend to focus on maximising total output per unit of fixed input (e.g. t wheat/ha) or average output per unit of variable input (kg wheat /kg N). The latter measure is used to denote the most technically efficient use of a variable input. At a farm level, the concept of Total Factor Productivity provides a measure of the relationship between the total volume of outputs and the total volume of inputs and how this can change over time as a result of the process of technical change.
- Environmental scientists tend to focus on preventing or minimising the negative environmental outcomes of farming systems (kg/ha N emitted, loss of birds/ha) in order to protect the integrity and diversity of ecological systems, with varying degrees of emphasis on levels of physical output. There is now increased awareness of the synergy between environmental stability and physical productivity in the conventional sense.
- Economists measure productivity in terms of 'value added' whereby prices are attached to outputs and inputs in order to derive measures of benefits and costs. Here the purpose is to maximise value-added per unit of limiting resource (e.g. £/ha). This is achieved by using variable inputs such as fertiliser only as long as the benefit derived (e.g. £ wheat output/kg N) is greater than or equal to the cost involved (£/kg N). By adopting this principle for all variable inputs, it is possible in theory to maximise value added per unit of limiting resource (e.g. £/ha or £/labour hour). At farm level, productivity can be expressed as profit, best regarded as a return on management and investment.
- Sociologists also focus on values and preferences, but they go beyond monetary valuation to include measures of the quality of life and the capacity of individuals and groups to sustain rewarding and secure livelihoods. Some aspects of technology change have increased rather than lessened the vulnerability of farming communities.

These views of productivity may not converge to give a single optimum. Technical possibilities are sometimes not taken up because they are not worth it, given relative prices of inputs and outputs. For example, irrigation can maximise wheat yields but in most UK circumstances the value of extra output will not recover the extra costs involved (Morris et al., 2004). There may be social resistance to improved productivity because technologies are not compatible with motivation or values, or may increase exposure to perceived risk (Rogers, 1995; Olhmer, Olson and Brehmer, 1998). Technology change, however, is commonly induced by the economic incentives provided by high output prices and, in the case of input-saving technologies, by high input prices (Grilliches, 1957; Binswanger & Ruttan, 1978). Similarly the development or transfer of new technologies for agriculture is also driven by the potential economic benefits to researchers and technologists – which of course depends on research products being fit for purpose and taken up by farmers. Once technologies which increase productivity are adopted by the majority of producers, it may be difficult for others to resist adoption without disadvantage. Much depends on the way in which a technology delivers potential benefit. For example, biological technologies which augment productivity per unit of land (i.e. t/ha, £/ha) tend to be relatively scale-neutral compared to technologies such as mechanisation which augment returns per unit of labour (t/hour, £/hour) and which tend to favour larger scale operators. For this reason, mechanisation technologies that can reduce total average costs per unit of output may disadvantage small farms compared to larger ones, potentially forcing them out of business as the industry scales-up and they are rendered relatively inefficient. By comparison, bio-technologies which enhance the stock of genetic material are less scale dependent in their application, although they may require greater amounts of complementary inputs such as fertilisers and greater management competency.

Not only may technical and economic optima differ, but attempts to fully exploit the potential physical productivity of crop and livestock systems may prove counterproductive. The price inelasticity of demand for raw food products means that markets may be unable to absorb increased output without big reductions in prices and therefore revenues to producers (Ritson, 1977). Perversely, producers have increased output in the face of falling prices in order to maintain incomes, further exacerbating the problem. Furthermore, as household incomes rise, a smaller relative share of income is spent on raw food, and a concomitant smaller share of national income goes to farming. Unless people move out of the sector, farm incomes lag behind national levels.

Improved physical productivity can be a threat to the farming industry in the absence of market potential, whether domestic or international. Measures to deal with this have included: restrictions on production, intervention storage, segmented industrial markets, protection from cheaper imports or provision of export subsidies, helping farmers to capture a greater share of final consumer food prices through processing, quality assurance or targeted marketing, withdrawal of land from production, retirement schemes, alternative land use such as forestry, protein or energy cropping, and payments for environmental services. It is an irony that increased productivity and efficiency in UK farming have not prevented a decline in real farm incomes in spite of high levels of support and mass exodus from the sector to the point where farming now employs less than 1.5% of the UK work force and contributes less than 1% to GDP (Defra/SEERAD/DARD/NAWAD, 2003).

Trends in farm productivity

Conventional measures of productivity in terms of crop and livestock yields have shown considerable gains in the last 50 years or so, as reported in other papers presented in this conference, associated with changes in the technologies available to and adopted by farmers.

As referred to earlier, the most commonly applied indicator of productivity is Total Factor Productivity (TFP) which measures the volume of outputs from agriculture per unit of all inputs (expressed as indices). Figure 26.3 (from Defra 2003a) shows that TFP increased by 45% since 1973. Up to the mid 1980s, TFP increased sharply as a result of increased volume of output with the volume of inputs remaining stable, although decreased labour inputs were substituted by other types of input. TFP stagnated from 1985 to 2000, but has increased more recently mainly because inputs have fallen more sharply than outputs.

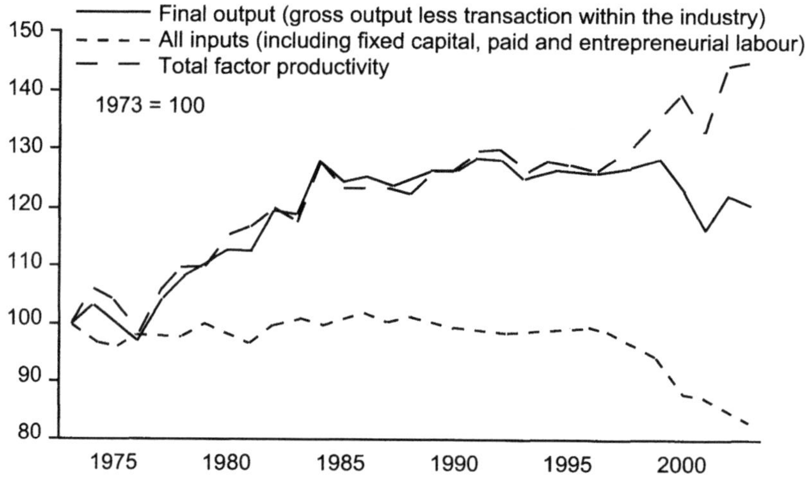

Figure 26.3 Total Factor Productivity in UK agriculture (source Defra 2003).

A recent review of the productivity of UK agriculture commissioned by Defra (Thirtle and Holding, 2003) used farm business survey data to derive estimates of TFP for selected crop enterprises and for types of farms. Value based estimates were derived for the ratio between output and labour, land and all inputs respectively over the period 1970 to 1997 for sugar beet, potatoes, oilseed rape (post 1974), winter wheat and spring barley. Figure 26.4 shows the derived estimates of TFP relative to the 1970 baseline. The estimates allow for changing yields, output prices, input levels and prices, and thereby value added, revealing considerable differences in the growth of productivity amongst crops during the period. Although yields per ha increased by 150%-200% during the period for most crops, output per worker increased by between 200% and 400% due mainly to increased mechanisation, except for potatoes which remained relatively labour intensive. Oilseed rape (from a low base) and sugar beet showed the highest productivity gains, with potatoes the least. As might be expected, there was a noticeable shift in land use towards higher productivity crops, into wheat, oilseed rape and sugar beet, and out of potatoes and barley: 23% of the growth in productivity was associated with this shift, while 77% was due to improved productivity of existing cropping. The authors refer to beneficial effects of crop research and advice which promoted increased productivity in sugar beet, raising the technological boundary and plugging the systems gap respectively, as shown in Figure 26.2.

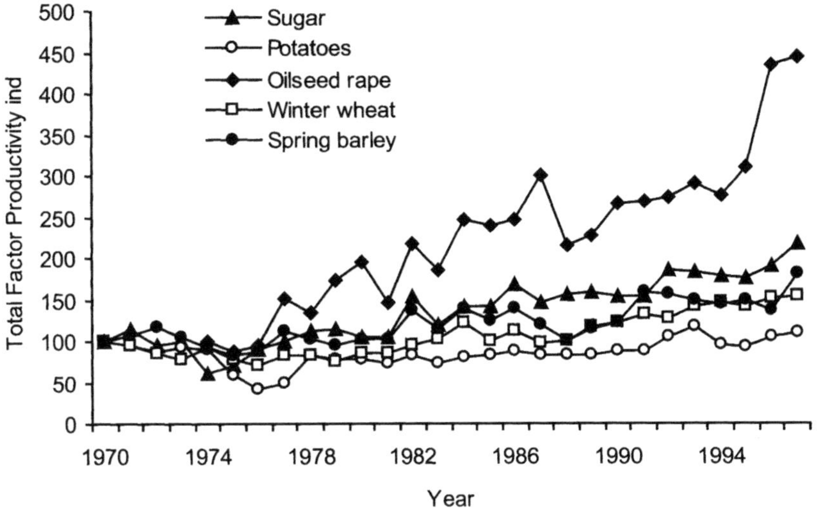

Figure 26.4 Trends in Total Factor Productivity for selected enterprises in Eastern England, 1970-1997. Source: Thirtle and Holding, 2003

The same study (Thirtle and Holding, 2003) also measured efficiency at farm level for cereal, dairy, pig, sheep and poultry farms in terms of the value of outputs and inputs during the period 1982 to 1997 (Figure 26.5). The most efficient farms of each type (those at the practitioner technical frontier in Figure 26.2) increased their productivity considerably during the period as a result of technical change; by about 2.5% per year for pig and poultry farms, 1.5% for dairy, 0.7% for sheep and 5% for cereal farms. Farms were benchmarked against the most efficient farm of the same type, whereby a farm with an efficiency of 0.8 achieves 80% of the outputs that the most efficient farm would achieve with the same level of inputs. Overall, average efficiency declined in the period (Figure 26.5) as the majority of farms fell further behind the most efficient, suggesting a failure to exploit potential productivity gains (a widening systems gap in Figure 26.2). The most efficient farms were generally larger, and the least efficient were associated with older farmers, indebtedness and a lower degree of specialisation. The results require careful interpretation because they are affected by changes in relative prices of commodities and exchange rates over time (although adjustments were made for this) and policy interventions in later years that encouraged extensification, including set-aside of arable land.

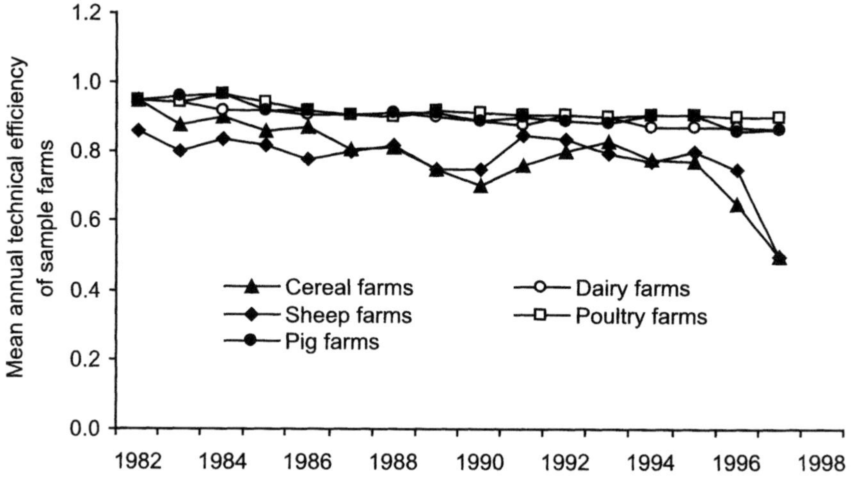

Figure 26.5 Comparative assessment of efficiency by farm type in the UK, 1982-1997. Source: Thirtle and Holding, 2003.

Take-up of potential productivity is strongly associated with farmer management characteristics. A study of wheat farmers in eastern England, using data for the period 1992-1997 (Wilson, Hadley and Asby, 2001), found a mean efficiency of 87%, and that there was a positive correlation between efficiency and management characteristics such as information seeking, experience and farm size. A managerial component was also apparent amongst the more efficient Dutch sugar beet growers (Koeijer, Wossink, Struik and Renkema, 2002).

International comparisons of productivity

Comparisons of TFP in agriculture have been made between countries to assess relative efficiency and competitiveness (Gopinath, Arnade, Shane and Roe, 1997; Ball, Bureau, Butaultand, and Nehring, 2001). In Europe, northern member states demonstrated significantly higher TFP than southern states in 1973, but by 1993 the gap had narrowed as the latter gained from 'the diffusion of technical knowledge and proceeded to grow most rapidly' (Ball et al., 2001). In 1973, the UK started off in the high TFP growth group as shown in Figure 26.6, but has lagged behind since the mid 1980s, with growth in TFP less than 50% of that achieved in France for example. TFP is an important determinant of a country's international comparative advantage, and in the event of further liberalisation of agricultural trade, a key determinant of whether a country imports or exports. The competitive advantage of USA agriculture has rested on its ability to sustain and increase growth in TFP (Gopinath *et al.*, 1997).

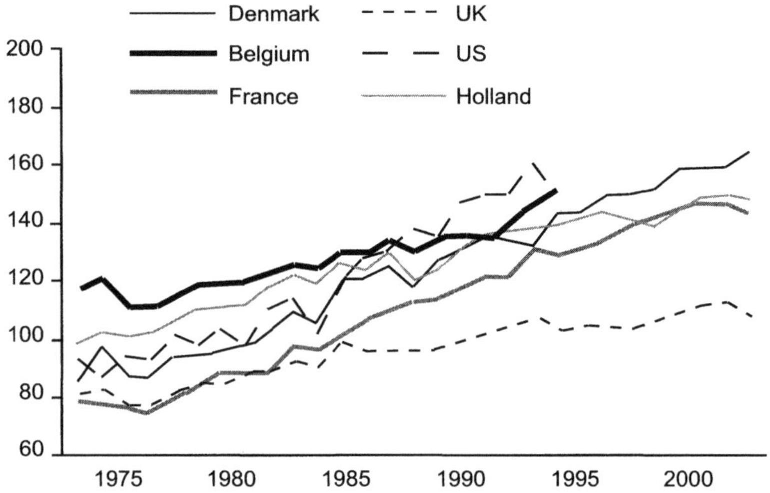

Figure 26.6 International comparisons of agricultural TFP, 1975-2001. Source: DEFRA, 2002a.

The environmental dimension of productivity

As alluded to earlier, there is increasing concern that modern farming technologies have external impacts which are often unrecognised or unaccounted for (Environment Agency, 2002). For example, energy intensive technologies deplete finite fossil fuel reserves which can increase the cost for subsequent users. Agrochemical use may result in pollution of water with loss of value and increased costs for third parties without compensation. Indeed, agriculture is now regarded as the major source of uncontrolled discharges of pollutants to water (HCEFRAC, 2003; Defra, 2003). These costs, as well as the environmental benefits from farming, have not been included in the conventional estimates of productivity.

Measures of the economic productivity of agriculture should take this broader view. For example, the estimated external costs to the environment of agriculture in the UK have been estimated at over £1.2 bn per year, equivalent to about £200/ha (for the UK: Pretty, Brett, Gee, Hine, Mason, Morison, Raven, Rayment, van der Bijl, 2000; Hartridge and Pearce, 2001; and for England and Wales: Environment Agency, 2000). This is offset to a degree by positive environmental contributions equivalent to £0.3 bn per year according to Hartridge and Pearce (2001). For these reasons, sets of sustainability indicators have been developed (MAFF, 2000; CEC, 2000; Defra, 2002a; Pannell and Glenn, 2000) and there is a move to account for the natural resource and environmental impacts of farming in the national accounts for agriculture, following guidance by UNSEEA (2003).

In an attempt to explore a more inclusive definition of productivity, Barnes (2002) revisited UK TFP indices to include environmental impacts associated with agrochemical pollution, finding that this reduced estimated annual TFP growth rates by about 20%. This also resulted in reduced rates of return to public expenditure on agricultural research and development. Given the potential for some agricultural inputs to cause environmental damage, it seems reasonable that enhancing technical efficiency can also reduce the environmental burden, especially as inefficiency is associated with waste and pollution. Wilson et al. (2001) found that the most efficient wheat farmers gave greatest weight to 'maintaining the environment', concluding that some synergy was possible. This point was also made by de Koeijer et al. (2000) who suggest that there is convergence of economic and environmental efficiency.

The preceding examples refer to mainly conventional farming practices. The adoption of alternative farming practices, however, may enhance rather than merely protect environmental quality. Those which integrate natural processes into agricultural production processes by reducing dependence on external inputs have been shown to equal or better the financial performance of conventional methods, even though yields may be reduced by 10 to 20% (Pretty and Howes, 1993, Pretty, 1998). Focussing on horticultural producers, Rigby, Woodhouse, Young, and Burton (2001) developed a composite indicator of sustainable agricultural practice to reflect resource use and environmental impacts. They concluded that, although there was much variation amongst organic growers, they appeared to achieve a higher sustainability rating than conventional growers. Given the increasing importance attached to sustainability, it seems reasonable that measures of productivity should explicitly include social and environmental aspects. That is, TFP should be extended to become TSFP, where S refers to a measure of sustainability.

Key drivers and influencing factors

From the foregoing, it is important to consider productivity of farming systems and agriculture as a whole in a broad, dynamic context. Figure 26.7 applies the Drivers-Pressure-State-Impact-Response framework to provide an understanding of the factors which drive change and generate economic, social and environmental pressures. These pressures have consequences for the state of farming systems and related processes which can be measured in terms of performance indicators, such as farm productivity, food supply and environmental quality. Variations in these state or process indicators result in impacts on people and the environment.

Anthropogenic concerns arise when agricultural systems do not deliver the required outputs 'efficiently' and there are unacceptable impacts on people and society. This encourages responses by government and others to intervene accordingly, as shown in Figure 26.7. Interventions may be taken to modify drivers such as agri-environment schemes or price support for organic production. Attempts may be made to relieve pressures by promoting codes of good agricultural practice, placing limits on nitrogen use and providing economic incentives for cleaner technologies. Actions may be taken to protect the state or condition of agriculture by designating protected area status or biodiversity targets, or providing income support for vulnerable farming communities. Responses may also include measures to mitigate undesirable impacts through, for example, river and groundwater recharge during drought periods and compensatory or insurance payments for farmers experiencing loss due to disease or adverse climatic conditions. Figure 26.7 reinforces the argument that measures of agricultural productivity must take a broad perspective, and demonstrates that the factors that influence productivity are diverse and subject to change. It also helps to frame responses to direct agricultural productivity in accordance with societal preferences.

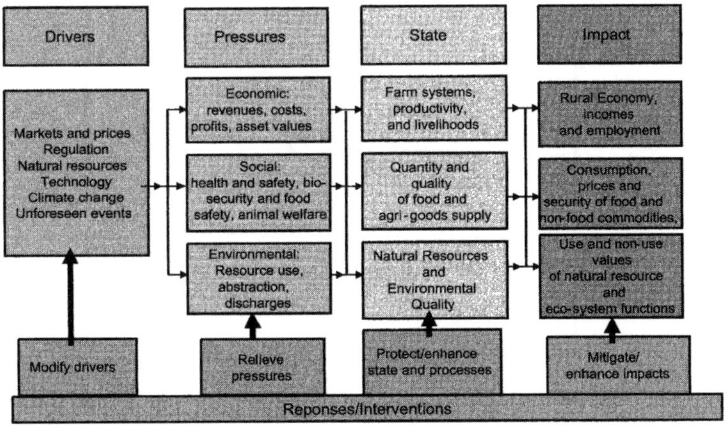

Figure 26.7 DPSIR framework applied to farming systems.

Generic scenarios

The DPSIR framework can help to understand and explain the considerable changes that have occurred in UK agriculture in the last 50 years. It can also help to identify and interpret possible futures, especially regarding the drivers and consequences of change, and the implications for policy. The current focus here is with mapping possible futures for UK agriculture and the implications for resource use and the environment.

The construction of future agricultural and related environmental scenarios draws on the methodology developed for the UK Foresight programme (Berhout and Hertin, 2002; DTI, 1999, 2002) which considers long term futures and possible implications for UK industry and society. Scenarios are not intended to predict the future, but rather to help think about how it might turn out. Scenario analysis can help to map out the features and consequences of possible futures and how decisions made in the interim might help shape or better cope with possible outcomes.

Scenarios are generated as a consequence of modelling drivers of economic and social change, new trends and innovation, and of unexpected events. They are usually made up of a qualitative story-line and a set of quantitative indicators. The sister paper (Rickard, Morris and Audsley, 2004) describes in more detail how the Foresight Futures framework was used to construct four possible scenarios, distinguished in terms of social values and governance. Figure 26.8 summarises their key features. Unforeseen events, such as international conflict or major technological advances or failures, can also shape possible futures.

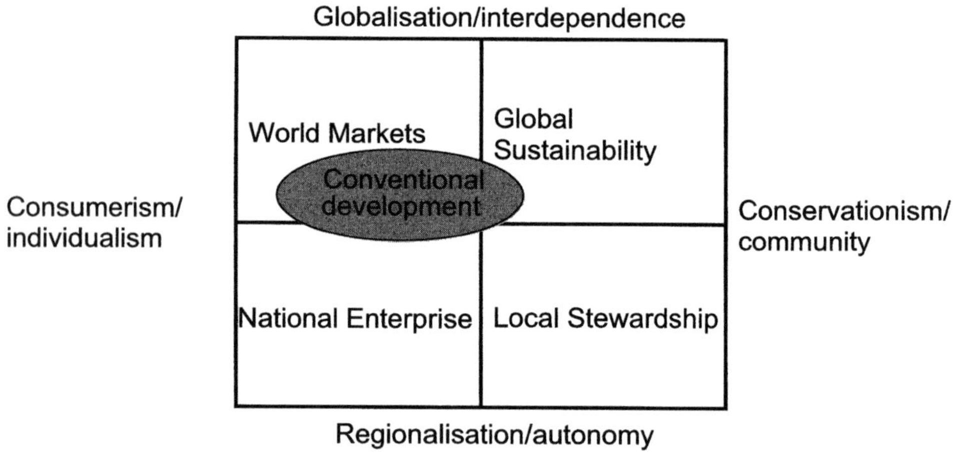

Figure 26.8 Possible futures, based on Foresight (DTI, 2002).

Notes to Figure 26.8

World Markets: high economic growth, an emphasis on private consumption and a highly developed and integrated world trading system. Average incomes increase, but with marked disparities in distribution. Regulation focuses on supporting trade.

Global Sustainability (also referred to as Global Responsibility): pronounced social and ecological values, evident in global institutions and trading systems. There is collective action to address social and environmental issues. Growth is slower but more equitably distributed compared to the World Markets scenario.

National Enterprise: emphasis on private consumption but with decisions made at national and regional level to reflect local priorities and interests. Although market values dominate, this is within national/ regional boundaries.

Local Stewardship: strong local or regional governments which emphasise social values, encouraging self-reliance, self sufficiency and conservation of natural resources and the environment.

Agricultural scenarios

These scenarios vary in terms of the key drivers and pressures (as indicated in Figure 26.6) which shape the agricultural sector. Not only do they vary in terms of the macro-economic conditions which define the context of agriculture (notably growth in economic output and incomes, income distribution, commodity and resource prices including energy prices and wage levels, currency exchange rates, population growth and rates of urbanisation) but also in terms of the detail of agricultural policy, consumer demand and prices, environmental policy, farmer motivation and agricultural systems and technologies. Agricultural scenarios, which mimic the generic scenarios given in Figure 26.8 are summarised in Table 26.1.

A mixture of past evidence and expert opinion obtained through a participatory workshop undertaken as part of the Defra Agricultural Futures project (Rickard *et al.*, 2005) concluded that the main factors affecting agriculture under each scenario are the prices of farm inputs and outputs, the extent of environmental regulation, and the development and adoption of technologies that potentially enhance the performance of farming in accordance with market or regulatory needs, such as genetic enhancement, crop fertilisation and protection, animal nutrition, and mechanisation. These technologies are mainly induced by economic incentives or regulation, although some such as transport, communication and refrigeration largely transfer from applications in other sectors.

Factors influencing exploitation of yield potential of crops

In order to assess the likely role of productivity in farming futures, a matrix of indicators considered to influence farm yields was drawn up and presented to a panel of crop and livestock specialists in the Defra sponsored Yield Limitation Project (Table 26.2). Specialists were asked to confirm the validity of the relative weights given to the indicators for each scenario and then to estimate likely achievement of yields expressed as a percentage of the current yield levels, bearing in mind the genetic potential of crops. A future 'business as usual' scenario was also included to represent extrapolation of recent trends.

The perceived strength and direction of factors influencing yield uptake are shown in Table 26.2. Commodity farm gate prices, as determined by the interaction of market demand and supply and interventions by government, are perceived to be strongly positively correlated with farm crop yields. In some cases, however, markets may offer greater reward for products which are differentiated in terms of quality, placing less emphasis on high yields. The application of production inputs is clearly associated with yields. These include a varied array of physical and knowledge-based inputs and processes, such as crop and livestock genetics, crop nutrition and crop protection,

Table 26.1 FUTURE SCENARIOS FOR UK AGRICULTURE

	World Markets	Global Sustainability	National Enterprise	Local Stewardship
Agricultural and rural policy	Abandonment of CAP. WTO led free trade in agricultural commodities. Limited interventions for social or environmental purposes. Increased global trade in agricultural commodities. Rural diversification opportunities based on market potential.	Reformed CAP. WTO promoted liberalisation. Decoupled agric. support. Promotion of sustainable agriculture, including agri-environment and animal welfare regimes. Global rules seek ethical rural development. Multi-functional agriculture produces public goods.	Protectionist agricultural policies, involving input and commodity subsidies, deficiency payments and marketing/intervention regimes. Limited environmental and social concerns. Rural economy is based primarily on agriculture and food. Farming is the main agent of development.	Support regimes in accordance with local needs and priorities reflecting self reliance, social and environmental objectives as defined at local level. Development defined in terms of conservation and community: a living/working countryside.
Food markets and prices	Market led, consumer driven, but with increased domination of major food retailers. International procurement and market integration. Producer and consumer food prices fall for global products, with premia for niche products.	Food supply chain accepts responsibility for promoting and responding to consumer concerns about safe, healthy and ethical foods. Consumer food prices rise due to quality assurance and compliance costs, providing incentives to producers.	Supply driven food chain. Food industry, especially producers and processors define product offering and criteria for food quality. Government supported supply side interventions maintain high producer prices, but cheap consumer food prices.	Greater connectivity between consumer and producer. Local area produce and market. Local 'brands' emphasise environmental and social attributes. Farmers join co-operative production and marketing schemes to add value and raise prices.
Environmental policy	Limited restrictions on chemical use, other than market imposed. Limited interest in soil and water conservation unless affecting production. Environmental risk managed through economic instruments.	Comprehensive, integrated approach to prevention/minimisation of diffuse pollution from agriculture. Policy mix includes regulation, voluntary measures and economic instruments reflecting a commitment to 'stewardship' and biodiversity.	Input intensive farming, limited controls on agro-chemicals and farming practices on environmental grounds. Regulation for controlling high risks which prejudice commercial interests.	Generally lower environmental risk but fragmented and selective regulation and control. Sustainable soil and water management embedded in farming culture, with policies, including regulation, to promote and support.

Table 26.1 CONTD.

	World Markets	Global Sustainability	National Enterprise	Local Stewardship
Farmer attitudes/ motivation	Polarisation into commercial and lifestyle farmers: 'real' and 'hobby' farmers. Biodiversity in farmed areas to suit commercial farming, or a commercial activity in itself.	Production oriented farmers tempered by increasing interest in conservation. Conservationists find expression in agri-environment schemes.	Commercially driven production focus, emphasis on output and production. Environmental motivations mainly commercially based and remedial.	Welfare maximising custodians, embracing commitment to sustainable livelihoods. Strong conservation and community ethic. Varied income sources, on and off-farm.
Agricultural production and farming systems	Competition leads to moderate to highly intensive, high technology, commercially driven large scale production by specialists, industrialised and global in scope, emphasis on efficiency through reduced unit costs for bulk commodity crops, with focused high quality production to gain price advantage where possible. Marginal land abandoned. GMOs widely promoted and adopted. Differentiated organic produce are an important niche market. Intensive feedlot livestock systems, but some extensive grazing on abandoned cropland.	Moderate increases in agricultural productivity and production. Agri-environment contributes to global services. Diversification/multi-functionality important. Strong 'compliance' requirements. Large scale farms, but with policy to retain family farms. Areas taken out of production used to support nature conservation. Selected adoption of GMOs, driven by environmental benefits. Limits on stocking rates, extensification incentives, strong welfare controls. High quality assurance. Some differentiated organic produce.	Broad based, relatively high input:high output farming to provide self sufficiency. Vegetables and agro-industrial raw materials are growth sectors. Mixed arable and livestock farming systems, intensive lowland dairy and cattle, with beef and sheep maintained in disadvantaged areas. Moderate trend towards large farms but family farms remain viable. Patchy adoption of GMOs, given limited economic incentives and little concern about side effects. Limited by investment. Organics limited.	Decreased productivity but total agricultural area increases. Commitment to sustainable rural livelihoods reflecting community priorities. Mix of intensive and extensive and greatly diversified systems. Retention of small scale, family based farming units. Low input systems an important part of sustainable farming. Widespread adoption of Integrated Farming Systems. GMOs rejected. Relatively extensive livestock systems, part of mixed farming systems. Emphasis on environment and welfare. Undifferentiated organic produce widespread.

mechanisation, irrigation and general levels of farm husbandry. Although generally working in a positive direction, they have a differential effect on yield and quality. For example, irrigation of wheat and sugar beet delivers mainly yield benefits, whereas irrigation of potatoes is predominantly for quality assurance. Other positive drivers include area based payments linked to particular crop types (although single farm payment regimes attempt to break this link), farm size in that larger farms exhibit greater take-up of technological possibilities, and genetic modification which may be able to enhance yield potential or overcome constraints.

Table 16.2 RELATIVE VALUES OF FACTORS INFLUENCING EXPLOITATION OF YIELD POTENTIAL BY FUTURE SCENARIO.

Characteristics	*Yield assoc.*	*Now*	*Future scenarios* Business as usual (BAU)	World markets (WM)	Global sustain-ability (GS)	National Enter-prise (PE)	Local Steward-ship (LS)
Commodity prices	+++	L-M	L	L	M	H	H
Area based payments	+	H	H	0	L	0	0
Input prices	—	M	M	L	H	M	M-H
Input levels	+++	M	M	M-H	M-L	H	L
Farm size	+	M-H	M-H	H	M-H	M	L
Uncertainty	—	L-M	L-M	H	M	L	L
Organics*	—	L	L-M	L*	M	L	H*
GMOs	+	0	L	H	L-M	L-M	0
Environ. regulation	—	L-M	L-M	L	H	L	H
Climate change signal	+/-	L	M	H	L	M-H	L-M

Notes to Table 2 : Relationship with yields + positive correlation, - negative correlation, weighted by strength of association. Relative value of parameter amongst scenarios: 0 = not applicable or zero, L= low, M = medium, H = high. *Organics: under WM organics are low as a % of total crop production but there is an important market in differentiated organic products compared to business as usual, driven by concern about food quality and facilitated by high incomes. Under LS, food production using organic methods is a common feature accounting for a relatively high % of total cropping.

Factors perceived to be negatively correlated with yield include input prices which have the effect of reducing profits, and environmental regulation as it constrains the levels of input use. The adoption of organics is similarly perceived to be associated with lower yields, partly as a result of reduced use of artificial inputs. Business uncertainty, particularly linked with unpredictable variation in prices of outputs and inputs and outcomes of farm business decisions is also likely to influence take up of yield benefits negatively. For the most part farmers are risk averse, and business risk tends to reduce willingness to invest in new technology. Although climate change may affect yields of crops in a particular location negatively (due to water stress) or positively (due to increased temperatures or radiation), this is likely to be offset by changes in the geographical distribution of crops to exploit potential advantage or minimise loss.

Table 26.2 also describes the present situation and possible future scenarios in terms of relative (along the row) values of these selected indicators, including a Business as Usual (BAU) case which extrapolates recent trends. For example, farmgate output prices are perceived to be relatively high under the protected National Enterprise (NE) and Local Stewardship (LS) scenarios, and relatively low under the Business As Usual case and World Markets as agriculture is exposed to international competition. Global Sustainability (GS) demonstrates moderate farmgate prices, partly reflecting increased supply costs associated with greater regulation.

Each column in Table 26.2 describes each scenario as a package of indicators. For example, World Markets are characterised relative to other scenarios by low output prices (with no price or income support) which are partly offset by relatively low input prices, giving an incentive for medium to high input levels, unconstrained by environmental regulation. (The panel reduced the original estimates of input levels under the WM scenario due to relatively low commodity prices). Farm size is relatively high, promoting technically efficient farms, with widespread use of GM technology. There are clear drivers to achieve high levels of efficiency in terms of TFP, with an emphasis on cost saving for bulk commodities, and generally moderate to high incentives to exploit yield potential. Farm businesses adopt technologies which reduce exposure to business risk and allow a degree of flexibility in the face of changing market conditions. In this high economic growth, high income scenario, farmers are rewarded for differentiated products which address niche market needs, placing emphasis on quality as much as quantity.

Global Sustainability exposes farmers to international competition, but world market prices are moderate, reflecting universal commitment to 'fair trade'. Input prices are high, reflecting 'green taxes'. There is considerable environmental regulation, and this results in low to medium aggregate input levels. Although there is a reduction of some inputs such as agrochemicals, the precision of application increases, as does farmer competency in sustainable agriculture. Organics become more widespread and GMOs are adopted where there are clear environmental benefits. Productivity under this scenario takes a broad view to include environmental effects. Yields in the conventional sense are likely to be moderate. The conventional concept of TFP is paramount.

By comparison, National Enterprise is characterised by high output prices associated with protection from international competition in order to protect farm incomes and encourage self sufficiency in agricultural products. Inputs are subsidised and environmental regulation is low. Input levels are high in response to economic incentives to maximise output. Business uncertainty is low. Under this regime producers face relatively secure market prospects, being encouraged to exploit fully the technical potential of production systems. Yields are likely to be high and capable of sustaining medium sized farms.

Local Stewardship demonstrates a strong commitment to sustainable agriculture and rural communities through a mix of market protection, income support and regulation. Farmgate prices are high, as are input prices especially of artificial inputs. The use of agrochemicals is relatively low, but within a small farm system, this is to some extent offset by intensive labour and management inputs. Organic production is widespread. There is less emphasis on maximising yields where this compromises the quality of 'natural foods' or results in unacceptable environmental burden. Thus productivity is defined in the context of sustainable farming, environmental protection and rural livelihoods, that is, TSFP.

Estimates of realisable yield by future scenario

Table 26.2 was used to derive estimates of the relative magnitude of crop yields on farms under each of the scenarios for the years 2012, 2025 and 2050 drawing on the panel of crop and livestock experts specialists engaged in the Defra project on Yield Limitations. Future yields were expressed as an index of the current (2000-2003) average farm yields. Panel members were asked to consider the main factors affecting yields under each scenario and the main uncertainties affecting estimation. The derived estimates are given in Table 26.3 for crops. Two messages emerge: there is considerable difference amongst scenarios in terms of perceived exploitation of yield potential for a given crop type, and there are differences in yield take-up amongst crops for any one scenario.

With respect to wheat, the physical applications of inputs to land together with managerial effort are perceived to be the key determinants of yield and these are driven by output price and environmental regulation (Sylvester-Bradley, pers. comm.) The greatest potential take-up occurs under the BAU and NE, with farm yields at 160% of current levels, equivalent to over 12t/ha in response to high prices and relatively low regulation. Regulation of inputs under GS and LS result in limited take-up of potential yield. There is a move to extensive systems under LS, with yields at 75% (6t/ha) of present levels. For the international scenarios of WM and GS there is uncertainty of price associated with variations in economic growth and demand (from countries such as China and India, including demand for wheat as an animal feed) and supply (from exporting countries such as Russia assuming agricultural rehabilitation).

Table 26.3 INDICES OF ESTIMATED CROP YIELDS BY FUTURE SCENARIO (CURRENT YIELDS = 100).

Crop yields by year	*Business as usual*	*World markets*	*Global sustainability*	*National enterprise*	*Local stewardship*
Wheat (8t/ha)*					
2012	120	115	100	120	95
2025	140	130	105	140	90
2050	160	150	115	160	75
Oilseed Rape (3.2t/ha)					
2012	100	108	100	117	90
2025	100	115	90	130	75
2050	100	140	90	180	75
Potatoes (42t/ha)					
2012	100	100	100	100	95
2025	100	105	90	110	80
2050	100	110	80	120	80
Beans (4.5t/ha)					
2012	100	100	108	120	110
2025	95	95	115	142	120
2050	95	95	140	160	130
Peas (3.6t/ha)					
2012	100	100	108	120	105
2025	110	110	115	133	115
2050	110	115	125	150	120

*current average farm yields shown in brackets

With respect to oilseed rape, current average farm yields at 3.2 t/ha are about 70% of potential yields of about 4.5 t/ha following recommended practices (Berry, pers. comm.). The take-up of genetic potential (the highest technological boundary in Figure 26.2), perceived to be about 8 t/ha, amongst scenarios mainly reflects input levels and management intensity in response to market incentives and regulation. Even though there has been an increase in yields of 0.5 t/ha per decade in the last 20 years, under BAU, there is little incentive to exploit further genetic potential and yields stagnate. Under WM, lower input prices and reduced regulation encourages take-up of about 50% of yield potential. Under GS, restrictions on inputs will constrain yields, and there may be a switch to spring sown crops, with possible decline in average yields. The incentives for production are greatest under NE, encouraging a continued improvement of 0.5t/ha, giving a farm yield of about 5.5 t/ha in 2050. Under LS, a mix of integrated farming methods (yields at 80-90% of conventional) and organic methods (yields at 50-60% of conventional) result in average yields stabilising at about 75% of existing levels.

Potato yields show considerable variation amongst scenarios, largely reflecting the use of inputs and managerial effort, but also the relative incentives given for quality as opposed to quantity (Sparkes, pers. comm.). Take-up of genetic yield potential is relatively modest under all scenarios, the greatest increase being about 20% under PE where prices are high for all qualities of produce and regulation is limited. The WM scenario sees a modest improvement associated with GM technology and high input levels, including irrigation. However, willingness to pay for quality assurance under this and the GS scenario reduce the relative importance of high yields. Yields are lower under GS and LS stewardship mainly due to high environmental regulation which reduce scope for chemical control of cyst nematode infestation. Under GS, GM technology is used to reduce pesticides as a means of retaining quality without unduly compromising yields. Although under LS there is less emphasis on supermarket defined quality, reduced chemical applications and rejection of GM technology result in lower yields.

Peas and beans make an important contribution to crop rotations and diversity, with peas being best suited to lighter and beans to heavier soils. Whereas beans are regarded as relatively low input:low output crops, peas are more dependent on pesticide use. Potential yields of beans and peas (at the technological boundary in Figure 26.2) are about 6 t/ha and 8 t/ha respectively (Weightman, pers.comm.). Current research trials and best farm yields are about 65% of these yield levels. Under the BAU scenario, the recent trend in stagnating yields of beans continues, associated with soil damage and compaction, use of contractors, and limited use of seed dressing. Peas show some improvement towards French yield levels of about 4.5 t/ha which have benefited from targeted breeding programmes. Under WM, similar price regimes to BAU offer similar incentives and responses in beans, but lower herbicide prices and limited regulation favour increased yields in peas. Under GS, higher output prices, nitrogen fixing potential and relatively low input levels increase the comparative advantage of beans and encourage improved practices and some improvement in genetic material. In peas, herbicide costs and related environmental risk restrain yield enhancement and uptake, although this is partly offset by GM technology. Under NE, a combination of strong price drivers and moderate input prices encourage relatively high input levels and yield improvement in beans and peas towards their genetic potential. Under LS, bean and pea yields have the potential to increase due to strong relative prices, their important contribution to crop rotations and diversity, and their relatively low use of inputs, although in the case of peas regulation of agrochemicals restricts yield uptake.

Modelling farming systems and productivity

It is apparent from the foregoing that different futures are likely to give rise to different outcomes in terms of land use, farming practice, productivity, farm incomes and environmental impacts. Quantitative estimates of key parameters are being used in a whole farm systems modelling framework (Annetts and Audsley, 2002) to explore the implications of possible futures for agriculture in England and Wales. To date, the focus has been on modelling conventional measures of farm outputs, inputs and productivity. It is proposed to include environmental outputs, such as potential emissions of pollutants to air and water, as well as the potential contribution to environmental goods such as biodiversity, landscape and amenity. A preliminary example analysis of the effect of these scenarios applied to lowland farms is presented to illustrate the approach.

METHOD OF ANALYSIS

The list of main input parameters used in modelling is given in Table 26.4. These cover costs of inputs such as labour and chemicals as well as changes in machinery input to farms associated with increased mechanisation (e.g. machine size and automation), farm sizes (hectares per man) and increased efficiency such as improved use of irrigation water through better scheduling and application. The scenarios also define prices and subsidies for the main crops. Allowance is also made for regional differences in soil and climatic conditions. In this way, it is possible to predict what would happen on typical farms under these scenarios. Assuming that farmers aim to maximise their profits within the restrictions imposed upon them – both physical (soil and climate) and legal (restrictions on inputs, quotas) – an optimising farm model can determine activities that do this under each scenario. Applying these factors to the farms, the analysis determines the proportion of the modelled crops grown on each farm for each scenario. Combining these typical farms for each scenario, it is possible to derive estimates of aggregate land use and agricultural production, and assess how these vary from the existing and likely 'business as usual' scenario.

For the purpose of the preliminary analysis, England was divided into four types of land: low and high rainfall and light and heavy soil. A full analysis will use far more divisions. Two thirds of the soils are light and half have low rainfall. Each type of land is treated as a separate farm. These farms are analysed using a whole farm management model (Annetts and Audsley, 2002) to determine what a group of profit maximising farmers would choose to do. A degree of observed variability in yields and prices amongst farms is included, allowing for the fact that performance and choices within farm types vary, resulting in different outcomes even on the same land type. The results are then summed to give an overall cropping pattern for the whole farming sector. The modelled initial distribution of cropping, with current prices and yields, is very similar to the observed current distribution of cropping, with perhaps a slight under-representation of barley. 70% of the high rainfall heavy soil farms are currently grass, mainly for dairying, but with the possibility of conversion to cropping if financially attractive.

Table 26.4 INDICES OF MODEL INPUT PARAMETERS FOR 2025 SCENARIOS (CURRENT = 100)

	BAU	World Markets	Global Sustainability	Enterprise	Local Stewardship
Costs of fertiliser (€)	100	90	122	116	120
Costs of machinery (€)	100	90	125	105	135
Cost of labour (€/person)	105	115	120	100	95
Area subsidy	0	0	80	0	0
Cereal prices (€/t)	70	85	95	133	165
Cereal area subsidy (€/ha)	95	0	0	0	0
Sugar beet price (€/t)	75	75	90	95	105
Oilseeds price (€)	95	85	90	140	170
Oilseed area subsidy (€/ha)	95	0	80	0	0
Roots and tubers price (€/t)	100	88	100	110	125
Milk prices (€/l)	95	90	90	105	100
Crop yield changes due to technological advances	105	115	105	110	90
Reduction in labour required due to technology, % of h/ha	80	60	80	90	90
Wheat yields	140	130	105	140	90
Barley yields	122	117	103	122	94
OSR yields	100	115	90	130	75
Bean yields	95	95	115	142	120
Pea yields	110	110	115	133	115
Potato yields	100	100	100	100	100
Grass yields	110	115	110	120	110

In some cases, the initial results of scenario simulation can appear very unrealistic, such as when all land is allocated to grass. Given the likely market conditions for agricultural outputs, this would indicate that the scenario had been incorrectly constructed, rather than that this is a possible future. In order to determine feasibility, typical demands for agricultural commodities have been derived from other studies (adjusted for differences between scenarios), namely from SRES (Special Report on Emissions Scenarios) (Nakicenovic and Swart, 2000), allowing for population and income growth under each scenario (Table 5). The production derived from simulation is compared then with these predicted demands. Where the simulated production of any commodity varies more than +/-50% from the demand estimates for the scenario, the scenario values are deemed impossible and the appropriate value is adjusted – usually a commodity price. It is also noted that agricultural profitability needs to lie within a reasonable range for farming to be sustainable.

It was found that the initial set of prices used in modelling produced inconsistent results for some scenarios. For example, the initial price estimates resulted in overproduction of milk in the LS and GS scenarios, and of sugar beet in the BAU scenario, and the general underproduction of some other crops. The results generated by these initial parameter values were clearly infeasible in so much as major a mismatch of demand and supply would result in an adjustment of market prices or some intervention by a regulatory authority. The price values were therefore adjusted to give levels of production which matched (within the limits previously referred to) the market demand for each scenario. The final values for input parameters are those shown in Table 4. The estimates of production for each scenario are shown in Table 6 (which can be compared to the market demand estimates in Table 5). In this preliminary analysis the model does not include the option of land abandonment or biofuels, which would allow greater flexibility in land use.

Table 26.5 PERCENTAGE INCREASES IN DEMAND FOR FARM PRODUCTS FROM THE SRES SCENARIOS 2000 TO 2025.

	WM	*GS*	*NE*	*LS*
Cereals	20	12	16	5
Roots	13	10	12	4
Oils	25	14	18	6
Animal products	11	10	12	4
Biofuels (000t)	170	95	197	246

Thus the method of analysis attempts to ensure that a set of scenario values are self-consistent. It confirms that it is the 'within scenario' (rather than between scenario) relative prices and, more critically, gross margins of crop and livestock activities that influence farmer choice of land use and farm enterprise options.

RESULTS

The results show that farm productivity in the conventional sense is most strongly influenced by the relative prices of outputs and technological changes associated with yield exploitation. Most scenarios lead to changes in the current patterns of land uses. Most scenarios (with the exception of LS) exceed the levels of demand for food products, thus allowing scope for alternative land uses such as biofuels, and also reductions in the area of productive land due to entry into environmental schemes under some scenarios. The latter aspects were not considered in this preliminary analysis.

The NE and LS scenarios are relatively profitable for farmers (Figure 26.7) but generate increased winter arable cropping with possible consequences for runoff, erosion risk and diffuse pollution (although in the case of LS measures would be taken to prevent this). Given the range of possible futures for farming, broader definitions of productivity, as discussed earlier, are needed to incorporate economic, social, and environmental objectives, especially for scenarios which emphasise the multi-functionality of farming. Even under predominantly market-driven scenarios which emphasise agricultural production, there remains a need to take an holistic view of productivity in order to inform policy measures to address unacceptable risk to society and the environment.

Figure 26.9 shows the changes in cropping in the different scenarios resulting from the analysis. For WM, the analysis (Figure 26.10) shows that the balance between crops is fairly well maintained. There is a tendency, however, for dairy farming to move to light land using forage maize and arable cropping to move to heavy land, giving a more uniform cropping over all land types (except for roots). Total production (Table 26.6) of cereals has increased by 45% and roots by 85%, but oil seeds have reduced by 23%. To be more consistent with the predicted increases in demand, the prices of cereals and roots need further adjustment downward, whereas the prices of oil seeds may need to rise. Of course under this regime, market requirements are matched through trade, but in the case of bulky or perishable crops, such as roots, this may be less feasible due to high transport and storage costs. Furthermore, relative price changes, either up or down, may have a secondary influence on farmer uptake of yield potential.

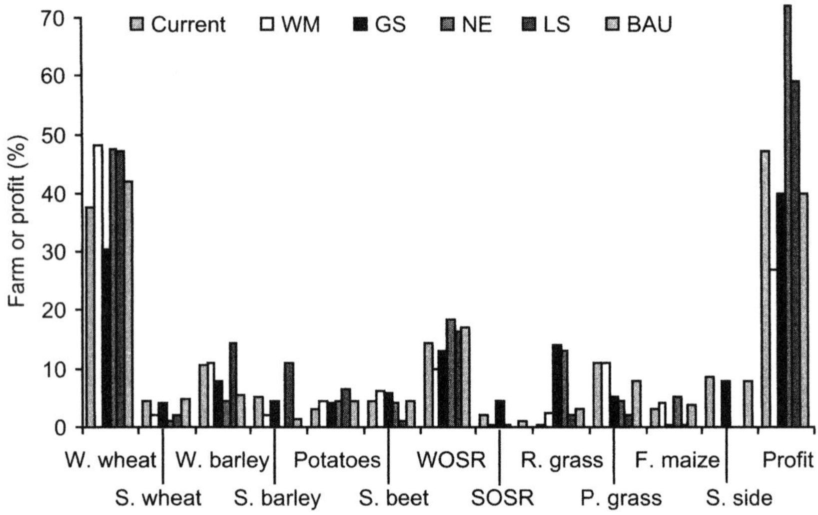

Figure 26.9 Comparison of overall cropping and farm profit in current and future scenarios.

Table 26.6 PERCENTAGE INCREASES IN TOTAL PRODUCTION WHEN ALL LAND IS CROPPED IN EACH SCENARIO, COMPARED WITH CURRENT SITUATION.

	BAU	*WM*	*GS*	*NE*	*LS*
Cereals	31	45	-15	35	20
Roots	43	85	32	27	10
Oils	22	-23	-2	62	-20
Animal products	20	49	81	131	-80

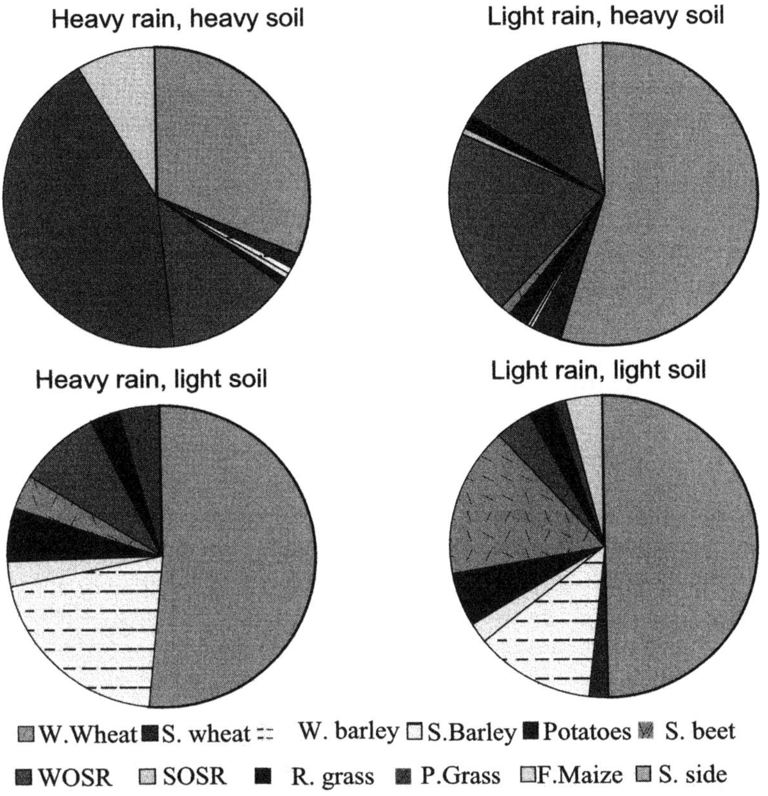

Figure 26.10 Expected cropping in WM scenario for the four land types.

In NE there is a clear polarisation of arable cropping into winter wheat and winter oilseed rape, with the consequent environmental effects of this type of cropping. Barley is almost eliminated. As with WM, there is the same tendency to uniformity of cropping over all land types. Profits are substantially increased as a consequence of relatively higher farm gate prices for all crops. Increased production is also substantially in excess of the increase in demand, suggesting, as discussed in the accompanying paper (Rickard et al., 2004) that production quotas would be needed under this scenario. Alternatively, much land would be used for biofuels or set-aside.

In GS, the area of cereals is reduced and substituted by animals and roots, consistent with a return to mixed farming systems. Animals are fed on grass leys rather than forage maize or permanent grass. Farm profit is similar to current levels.

In LS the reductions in yields in the scenario result in sugar beet being almost eliminated and grass production also reduced and restricted to areas of heavy soil and rainfall. 75% of the cropping is cereals, which suggests corresponding environmental issues and reduced diversity. This appears inconsistent with the commitment to sustainable farming and, most likely, actions would be needed to address this. Total crop production is a little in excess of demand, so that prices may change to increase the area of grass. There is a possibility that there may not be enough land to meet the total demand for agricultural products.

In BAU, the results are similar to the current scenario, but there is a trend to wheat rather than barley and, as with WM, a trend to dairying on all land types using more forage maize.

PRELIMINARY FINDINGS AND PROPOSED WORK

This preliminary analysis has shown that whole farm analysis can help to determine the likely impact of alternative future scenarios on landuse, production, productivity and farm incomes. It can also help to determine the likely environmental impacts associated with these future outcomes, whether associated with unregulated intensive farming or farming which is required to conform to high standards of environmental protection. The analysis also shows that estimates of agricultural production under each scenario must be linked to market demands and likely prevailing prices. An iterative approach was adopted here to balance supply with likely demand. The scenario approach can help to devise strategies for research and technology development that enhance the ability to adapt to and cope with a range of possible futures as well as the inevitable but unknown challenges that actual futures will present.

It is proposed to develop the scenario approach by considering more land types and more crop and livestock alternatives, including options for land abandonment, biofuels, and environmental schemes, within major upland and lowland farm systems. The model will iteratively adjust the prices in order to attain a suitable level of production, productivity and farm profit. The prices of inputs will be used to determine the economic optimum level of inputs and hence the yield level for each crop. The likely environmental effects associated with variations in land use, agrochemicals, cultivation practices and other farming practices will be identified.

Summary and conclusions

This paper has explored the productivity of farming systems, making reference to the conventional measures used to assess performance, as well as more recent attempts to include aspects of natural resource use and environmental effects. The limits of productivity are determined by the technologies available to farmers. It is these which define the processes used within farming systems to convert inputs to outputs. Productivity could be enhanced through changes in available technologies as a result of, for example, research and development into new crop varieties and production methods. The extent to which potential productivity is exploited in practice depends on the factors, whether linked to regulation or market forces, which drive research and innovation, and influence the adoption and diffusion of technologies by end users. The evidence in the UK over the last 50 years is that farm productivity in the conventional sense has continued to increase as a result of technological change, albeit at a slower rate in recent years. There is, however, evidence of some overall decline in the efficiency with which existing inputs are used by farmers, partly because of failure to adopt new technologies which are potentially more productive.

Attention was drawn to the need to take a wider definition of productivity, to include the impacts of farming on natural resources and the environment, as well as farming livelihoods. In future, technologies which protect and enhance the latter are likely to be viewed more favourably than those which do not. Emerging evidence suggests that there is some convergence between farming efficiency and environmental performance, given that pollution often arises from wasteful practices. Changing priorities in the countryside, whereby farmers deliver environmental goods as well as food or industrial products, will increase the need for a more comprehensive definition of productivity.

Productivity was explored under possible future scenarios, drawing out the differences in market prices and regulatory systems that are likely to provide incentives for the development and application of farming technologies. Using expert opinion, it was concluded that different possible futures for agriculture in the UK could result in very different take-up of yield potential on major crops by farmers. A modelling approach showed that land use, farming systems and indicators of productivity from a production and environmental viewpoint are likely to vary according to the incentives offered to land managers under the different scenarios. Much depends, however, on the markets for farm produce and the relative importance placed on environmental and social objectives as these provide the incentives to develop and adopt technologies in accordance with needs and priorities. The scenario approach can help to devise strategies for research and technology development that enhance the ability to adapt to and cope with a range of possible futures, as well as some of the inevitable but unknown challenges that actual futures will present.

ACKNOWLEDGEMENT

The assistance of participants of the workshop on Agricultural Futures held at Silsoe, September 25th 2003 is acknowledged (see Rickard et al., 2004), together with that of the members of the Defra Yield Limitations Project, IS 0210. Section 10 of the paper drew on personal communications with Roger Sylvester-Bradley, R. (Cereals, ADAS), Debbie Sparkes, (Potatoes, University of

Nottingham), Richard Weightman, (Legumes, ADAS) and Peter Berry, (Oil Seeds, ADAS). Defra funding for the work reported in this paper is gratefully acknowledged.

References

Annetts, J.E. and Audsley, E. (2002). Multiple objective linear programming for environmental farm planning. *Journal of the Operational Research Society* **53**, 933-943.

Ball, V.E., Bureau, J., Butault, J. and Nehring, R. (2001). Levels of farm sector productivity: an international comparison. *Journal of Productivity Analysis* **15**, 5-29.

Barnes, A.P. (2002). Publicly-funded UK agricultural R&D and 'social' total factor productivity. *Agricultural Economics* **27**, 65-74.

Berhout, F. and Hertin, J. (2002). Foresight futures scenarios: developing and applying a participative strategic planning tool. *Greener Management International* **37**, 37-52. Greenleaf Publishing, Sheffield.

Binswanger, H.P. and Ruttan, V.W. (1978). *Induced Innovation: Technology, Institutions and Development.* Johns Hopkins University Press, Baltimore.

CEC (2000). *Indicators for the Integration of Environmental Concerns into the Common Agricultural Policy.* Communication from the Commission to the Council and the European Parliament. COM(2000) 20 final, Commission of the European Communities, Brussels.

Dalton, G.E. (Ed.) (1975). *Study of Agricultural Systems.* Applied Science Publishers.

Defra (2002a). *Farming and Food's Contribution to Sustainable Development. Economic and Statistical Analysis.* Department for Environment, Food and Rural Affairs, Defra Publications, London.

Defra (2002b). *The Strategy for Sustainable Farming and Food. Facing the Future.* Department for Environment, Food and Rural Affairs, Defra Publications, London.

Defra (2003). *Diffuse Water Pollution from Agriculture*: The Government's Strategic Review: Stakeholder Discussion Paper. Department for Environment, Food and Rural Affairs, London.

Defra/SEERAD/DARD/NAWAD (2003). *Agriculture in the United Kingdom, 2001.* The Stationery Office, London.

DTI (1999). *Environmental Futures.* PB 4475. http://www.foresight.gov.uk/

DTI (2002). *Foresight Futures 2020 Revised Scenarios and Guidance.* The Stationery Office, London.

English Nature (2002). *Policy Mechanisms for the Control of Diffuse Agricultural Pollution, with Particular Reference to Grant Aid.* English Nature Research Report No. 455. English Nature, Peterborough.

Environment Agency (2002). *Agriculture and Natural Resources: Benefits, Costs and Potential Solutions.* Environment Agency, Bristol.
Gopinath, M., Arnade, C., Shane, M. and Roe, T. (1997). Agricultural competitiveness; the case of the United States and major EU countries. *Agricultural Economics* **16**, 99-109.
Grilliches, Z. (1957). Hybrid corn: an exploration of the economics of technology change. *Econometrica* **25**, 501-522.
Hartridge, O. and Pearce, D. (2001). *Is UK Agriculture Sustainable? Environmentally Adjusted Economic Accounts for UK Agriculture.* CSERGE-Economics, University College, London.
HCEFRAC (2003). *The Water Framework Directive, House of Commons, Environment, Food and Rural Affairs Committee*, Fourth Report of Session 2002-2003, March 2003. The Stationery Office, London.
Koeijer, T.J. de, Wossink, G.A.A., Renkema, J.A. and Struik, P.C. (2002). Measuring agricultural sustainability in terms of efficiency: the case of Dutch sugar beet growers. *Journal of Environmental Management* **66**, 9-17.
MAFF (now Defra) (2000). *Towards Sustainable Agriculture: A Pilot Set of Indicators.* London: MAFF Publications, available on www.defra.gov.uk/farm/sustain/pilot.htm
Morris, J., Weatherhead, E. K., Knox, J., Vasilieou, K., de Vries, T., Freeman, D., Leiva, F., Twite C. (2004). *The Sustainability of European Irrigation under Policy Change. Country Report: England and Wales.* EU 5th Framework Project 'WADI'. The Sustainability of European Irrigation under Policy Change, Cranfield University.
Nakicenovic, N. and Swart, R. (Eds.) (2000). *Special Report on Emissions Scenarios. A Special report of Working Group III of the Intergovernmental Panel on Climate Change.* Intergovernmental Panel on Climate Change, Cambridge University Press, Cambridge. pp. 599.
Olhmer, B., Olson, K., and Brehmer, B. (1998). Understanding farmers' decision making processes and improving managerial assistance. *Agricultural Economics* **18**, 273-290.
Pannel, D.J. and Glen, N.A. (2000). A framework for the economic evaluation and selection of sustainability indicators in agriculture. *Ecological Economics* **33**, 135-149.
Pretty, J. (1998). *The Living Land.* Earthscan, London.
Pretty, J.N., Brett, D., Gee, D., Hine, R.E., Mason, C.F., Morison, J.I.L., Raven, H., Rayment, M.D. and van der Bijl, G. (2000). An assessment of the total external costs of UK agriculture. *Agricultural Systems* **65**, 113-136.
Pretty, J.N. and House, J. (1993). *Sustainable Agriculture in Britain: Recent Achievements and the New Policy Challenge.* IIED, London.
Rickard, S., Morris, J. and Audsley, E. (2005). Possible futures for agriculture in Northern Europe during a period of policy reform. In: *Yields of Farmed Species.* Edited by R. Sylvester-Bradley and J. Wiseman. Nottingham

University Press, Nottingham. pp 577-606.

Rigby, D., Woodhouse P., Young, T. and Burton, M. (2001). Constructing a farm level indicator of sustainable agricultural practice. *Ecological Economics* **39**, 463-478.

Ritson, R. (1977). *Agricultural Economics: Principles and Policy.* Crosby Lockwood Staples, London.

Rogers, E. M. (1995). *Diffusion of Innovations.* 4th Edition. Free Press, New York.

Thirtle, C., and Holding, J. (2003). *Productivity in UK Agriculture: Causes and Constraints.* Report to Department for Environment, Food and Rural Affairs. Imperial College at Wye, University of London.

UNSEEA (2003). *Systems for Economic and Environmental Accounting (SEEA)* http://unstats.un.org/unsd/envAccounting/seea.htm

Wilson, P., Hadley, D. and Asby, C. (2001). The influence of management characteristics on the technical efficiency of wheat farmers in eastern England. *Agricultural Economics* **24**, 329-338.

List of Delegates

Aguilar Perez, Mr C	University of Nottingham, Sutton Bonington Campus, Loughborough, Leics LE12 5RD, UK
Alderson, Dr P	University of Nottingham, Sutton Bonington Campus, Loughborough, Leics LE12 5RD, UK
Allen, Mr E J	Cambridge University Farm, Huntingdon Road, Cambridge CB3 OLH, UK
Allison, Dr M	Cambridge University Farm, Huntingdon Road, Cambridge CB3 OLH, UK
Aubry, Mr J	University of Nottingham, Sutton Bonington Campus, Loughborough, Leics LE12 5RD, UK
Audsley, Mr E	Silsoe Research Institute, Wrest Park, Silsoe, Beds UK,
Baenziger, Prof P S	University of Nebraska, 330 Keim Hall, Lincoln NE 68583-0915, USA
Bailey, Mr S W	RSPB, The Lodge, Sandy, Beds SG19 2DL, UK
Barraclough, Dr P B	Rothamsted Research, West Common, Harpenden, Herts , UK
Berry, Dr P	ADAS High Mowthorpe, Duggleby, Malton, N Yorks YO17 8BP, UK
Bingham, Dr I	Crop and Soil Research Group, SAC, Craibstone Estate, Bucksburn AB21 9YA, UK
Broadley, Dr M R	University of Nottingham, Sutton Bonington Campus, Loughborough, Leics LE12 5RD, UK
Brook, Dr R	University of Wales Bangor, School of Agric. & Forest Sciences, Bangor, Gwynedd LL57 2UW,
Byrne, Mr J	Kildalton College, Crop Variety Testing Division, Dept of Agriculture & Food, Piltown Co Kilkenny, Ireland

Byrne, Mr R	Matthews Agri Services Ltd, Callow, Co Louth , Ireland
Chambers, Ms P	Farmacy plc, 42 Manor Street, Ruskinston, Sleaford, Lincs NG34 9EP, UK
Clark, Mr D A	Dexcel Ltd, Private Bag 3221, Hamilton New Zealand,
Clough, Dr J	Syngenta, Jealott's Hill Int Research Centre, Bracknell, Berks RG42 6EY, UK
Collier, R	University of Arizona, Department of Animal Sciences, Tucson A2 85745,
Cooper, Dr M	RSPCA, Farm Animals Dept, Wilberforce Way, Southwater, West Sussex UK,
Cottrill, Dr B	University of Reading, School of Agriculture, Policy & Develop., Reading RG6 6AR, UK
Davies, Mr M H	ADAS Rosemaund, Preston Wynne, Hereford HR1 3PG, UK
Duran-Melendez, Mr L	University of Nottingham, Sutton Bonington Campus, Loughborough, Leics LE12 5RD, UK
Foulkes, Dr J	University of Nottingham, Sutton Bonington Campus, Loughborough, Leics LE12 5RD,
Garnsworthy, Dr P C	University of Nottingham, Sutton Bonington Campus, Loughborough, Leics LE12 5RD, UK
Garwes, Dr D	DEFRA Livestock Science Unit, 703 Cromwell House, Dean Stanley St, London SW1P 3JH, UK
Gill, Sir B	Home Farm, Hawkhills, Easingwold, York YO61 3EG, UK
Gooding, Dr M	University of Reading, School of Agriculture, Policy & Develop., Reading RG6 6AR, UK
Goodliffe, Dr P	DEFRA, NRRA Science Unit, Rm 705, Cromwell House, Dean Stanley St, London SW1P 3JH, UK
Gregson, Dr K	University of Nottingham, Sutton Bonington Campus, Loughborough, Leics LE12 5RD, UK
Haley, Prof C	Roslin Institute, Roslin, Midlothian EH25 9PS, UK
Harris, Mr B	Kildalton College, Crop Variety Testing Division, Dept of Agriculture & Food Backweston, Leixlip, Co Kildare, Ireland
Holdsworth, Prof M	University of Nottingham, Sutton Bonington Campus, Loughborough, Leics LE12 5RD,
Johnson, Dr B	English Nature, Roughmoor, Taunton TA1 5AA, UK

Kettlewell, Dr P	Harper Adams University College, Crop & Environment Research Centre, Newport, Shropshire TF10 8NB, UK
Kindred, Mr D	University of Reading, Dept of Agriculture, Earley Gate, Reading RGG 6AR, UK
Larkham, Mrs A	Macaulay Institute, Craigiebuckler, Aberdeen AB15 8QH, UK
Lee, Miss E	Univesity of Nottingham, Sutton Bonington Campus, Loughborough, Leics LE12 5RD, UK
Legg, Prof B	NIAB, Huntingdon Rd, Cambridge CB3 OLE, UK
Long, Prof S	University of Illinois, 190 Plant & Animal Biotech Lab, 1201 W Gregory Drive, Urbana IL 61801-3838, USA
Maguire, Mr M	Dept of Agriculture & Food, Crop Variety Testing Division, c/o TEAGASC, Moorepark, Fermoy Co Cork, Ireland
Mayes, Dr S	University of Nottingham, Sutton Bonington Campus, Loughborough, Leics LE12 5RD, UK
McKay, Mr A	PGG Seeds, P O Box 194, Ashburton New Zealand,
McKay, Mrs T	PGG Seeds, P O Box 194, Ashburton New Zealand,
Miflin, Dr B	Lawes Trust Senior Fellow, Rothamsted Research, Haprenden, Herts AL5 2JQ, UK
Morris, Prof J	Institute of Water and Environment, Cranfield University, Silsoe, Bedford MK45 4DT, UK
Murphy-Bokern, Dr D	DEFRA, Cromwell House, Rm 607, Dean Stanley St, Westminster London SW1P 3JH, UK
Mwale, Mr S	University of Nottingham, Sutton Bonington Campus, Loughboorugh, Leics LE12 5RD, UK
Naylor, Prof R E L	Trelareg Consultants, Glenrock, Finzean, Banchory, Scotland AB31 6NE, UK
Nugent, Mr L	Farmhill Agri Services Ltd, Rathangan, Co Kildare , Ireland
Onions, Miss V	University of Nottingham, Sutton Bonington Campus, Loughborough, Leics LE12 5RD, UK
Papadopoulos, Mr J	University of Nottingham, Sutton Bonington Campus, Loughborough, Leics LE12 5RD, UK
Parry, Dr M	Rothamsted Research, Harpenden, Herts AL5 2JQ, UK
Paveley, Dr N	ADAS High Mowthorpe, Duggleby, Malton, N Yorkshire
Rannou, Mr R	University of Nottingham, Sutton Bonington Campus, Loughborough, Leics LE12 5RD, UK

Reynolds, Prof M	CIMMYT, AD-6-641, 06600, Mexico D.F. MEXICO
Rickard, Mr S	Cranfield School of Management, Cranfield University, Cranfield, Beds MK43 OAL, UK
Roberts, Prof J	University of Nottingham, Sutton Bonington Campus, Loughborough, Leics LE12 5RD, UK
Robinson, Prof J	Animal Breeding & Development, SAC, Ferguson Building, Bucksburn, Aberdeen AB21 9YA, UK
Robinson, Mr S	Nottingham University Press, Manor Farm, Church Lane, Thrumpton, Nottingham NG11 OAX, UK
Russell, Mrs H	University of Nottingham, Sutton Bonington Campus, Loughborough, Leics LE12 5RD, UK
Salter, Mrs J	Agricultural Industries Confederation, East of England Showground, Peterborough PE2 6XE, UK
Shewry, Prof P	Rothamsted Research, Harpenden, Herts AL5 2JQ, UK
Simm, Prof G	SAC Sustainable Livestock Systems, SAC Central Office, Kings Building, West Mains Rd Edinburgh, EH9 3JG, UK
Sinclair, Dr K	University of Nottingham, Sutton Bonington Campus, Loughborough, Leics LE12 5RD, UK
Smil, Dr V	University of Manitoba, Manitoba, R3T 2N2 Canada,
Sparkes, Dr D L	University of Nottingham, Sutton Bonington Campus, Loughborough, Leics LE12 5RD, UK
Spink, Mr J	ADAS Rosemaund, Preston Wynne, Hereford HR1 3PC,
Stegeman, Prof A	Utrecht University, Faculty of Veterinary Medcine, Yarlelan 7, 3584 CL, Utrecht The Netherlands,
Sylvester-Bradley, Prof R	ADAS, Centre for Sust. Crop Management, Boxworth, Cambridge CB3 8NN, UK
Tomkins, Mr C H	C H Tomkins Ltd, Moorfield Lodge, Orlingbury, Kettering, Northants NN14 1JF, UK
Tucker, Prof G	University of Nottingham, Sutton Bonington Campus, Loughborough, Leics LE12 5RD, UK
Turley, Mr D B	Central Science Laboratory, Sand Hutton, York YO41 1LZ, UK
Vereijken, Dr P H	Plant Research International, P O Box 16, 6700 AA Wageningen Netherlands,
Walker, Mr A	ADAS Gleadthorpe, Meden Vale, Mansfield, Notts NG20 9PF, UK

Walters, Dr R	Livestock Genetics Ltd, Rye Cottage, Peasemore, Newbury, Berks RG20 7JN, UK
Webb, Prof R	University of Nottingham, Suton Bonington Campus, Loughborough, Leics LE12 5RD, UK
Weightman, Dr R	ADAS Arthur Rickwood, Nepal, Ely, Cambs CB6 2BA, UK
White, Mr G	University of Nottingham, Sutton Bonington Campus, Loughborough, Leics LE12 5RD, UK
Wiseman, Prof J	University of Nottingham, Sutton Bonington Campus, Loughborough, Leics LE12 5RD, UK

INDEX

A

B

C

F

G

H

I

L

Q

R

S

ND - #0096 - 110926 - C6 - 234/156/31 - PB - 9781789183221 - Matt Lamination